Thomas W. Wieting

1994

~~71^00~~
44^95
geom
disc

MATHEMATICAL SURVEYS AND MONOGRAPHS SERIES LIST

Volume

39 **Geometric analysis on symmetric spaces,** Sigurdur Helgason

38 **Analysis of and on uniformly rectifiable sets,** Guy David and Stephen Semmes

37 **Structural properties of polylogarithms,** Leonard Lewin, Editor

36 **The theory of subnormal operators,** John B. Conway

35 **Algebraic geometry for scientists and engineers,** Shreeram S. Abhyankar

34 **Inverse source problems,** Victor Isakov

33 **Spectral theory and analytic geometry over non-Archimedean fields,** Vladimir G. Berkovich

32 **An introduction to CR structures,** Howard Jacobowitz

31 **Representation theory and harmonic analysis on semisimple Lie groups,** Paul J. Sally, Jr. and David A. Vogan, Jr., Editors

30 **The Markoff and Lagrange spectra,** Thomas W. Cusick and Mary E. Flahive

29 **Amenability,** Alan L. T. Paterson

28 **Direct and inverse scattering on the lines,** Richard Beals, Percy Deift, and Carlos Tomei

27 **Basic hypergeometric series and applications,** Nathan J. Fine

26 **Operator theory and arithmetic in** H^∞, Hari Bercovici

25 **Asymptotic behavior of dissipative systems,** Jack K. Hale

24 **Noetherian rings and their applications,** Lance W. Small, Editor

23 **Introduction to various aspects of degree theory in Banach spaces,** E. H. Rothe

22 **Noncommutative harmonic analysis,** Michael E. Taylor

21 **The Bieberbach conjecture: Proceedings of the symposium on the occasion of the proof,** Albert Baernstein, David Drasin, Peter Duren, and Albert Marden, Editors

20 **Partially ordered abelian groups with interpolation,** Kenneth R. Goodearl

19 **Contributions to the theory of transcendental numbers,** Gregory V. Chudnovsky

18 **Essentials of Brownian motion and diffusion,** Frank B. Knight

17 **Approximation by polynomials with integral coefficients,** Le Baron O. Ferguson

16 **Symplectic groups,** O. Timothy O'Meara

15 **Vector measures,** J. Diestel and J. J. Uhl, Jr.

14 **Geometric asymptotics,** V. Guillemin and S. Sternberg

13 **Topics in operator theory,** A. Brown, R. G. Douglas, C. Pearcy, D. Sarason, A. L. Shields; C. Pearcy, Editor

12 **Uniform spaces,** J. R. Isbell

11 **Fixed points and topological degree in nonlinear analysis,** J. Cronin

10 **An introduction to the analytic theory of numbers,** R. Ayoub

9 **Linear approximation,** Arthur Sard

MATHEMATICAL
Surveys and Monographs

Volume 39

Geometric Analysis on Symmetric Spaces

Sigurdur Helgason
Massachusetts Institute of Technology

American Mathematical Society
Providence, Rhode Island

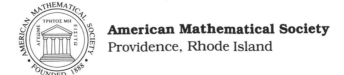

1991 *Mathematics Subject Classification*. Primary 43A85, 53C35, 22E46, 22E30, 43A90, 44A12, 32M15; Secondary 53C65, 58G35, 31A20, 43A35, 35L05.

Library of Congress Cataloging-in-Publication Data

Helgason, Sigurdur, 1927–
Geometric analysis on symmetric spaces/Sigurdur Helgason
p. cm. — (Mathematical surveys and monographs, ISSN 0076-5376; v. 39)
Includes bibliographical references (p. –) and index.
ISBN 0-8218-1538-5 (acid-free)
1. Symmetric spaces. I. Title. II. Series: Mathematical surveys and monographs; no. 39.
QA649.H43 1994 93-48556
516.3′62— dc20 CIP

♾ The paper used in this book is acid-free and falls within the guidelines
established to ensure permanence and durability.
♻ Printed on recycled paper.
This volume was printed directly from copy prepared by the author using AMS-TEX,
the American Mathematical Society's TEX macro system.

10 9 8 7 6 5 4 3 2 1 99 98 97 96 95 94

To my Danish mathematical friends

past and present

CONTENTS

PREFACE

Among Riemannian manifolds the symmetric spaces in the sense of É. Cartan form an abundant supply of elegant examples whose structure is particularly enhanced by the rich theory of semisimple Lie groups. The simplest examples, the classical 2-sphere $\boldsymbol{S}^2$ and the hyperbolic plane $\boldsymbol{H}^2$, play familiar roles in many fields in mathematics.

On these spaces, global analysis, particularly integration theory and partial differential operators, arises in a canonical fashion by the requirement of geometric invariance. On $\boldsymbol{R}^n$ these two subjects are related by the Fourier transform. Also harmonic analysis on *compact* symmetric spaces is well developed through the Peter-Weyl theory for compact groups and Cartan's refinement thereof. For the *noncompact* symmetric spaces, however, we are presented with a multitude of new and natural problems.

The present monograph is devoted to geometric analysis on noncompact Riemannian symmetric spaces X. (The Euclidean case and the compact case are also briefly investigated in Chapter III, §§7–9, and Chapter IV, §5, but from an unconventional point of view). A central object of study is the algebra $\boldsymbol{D}(X)$ of invariant differential operators on the space. A simultaneous diagonalization of these operators is provided by a certain Fourier transform $f \to f^{\sim}$ on X which is the subject of Chapter III. Just as is the case with $\boldsymbol{R}^n$ the symmetric space X turns out to be self-dual under the mentioned Fourier transform; thus range questions like the intrinsic characterization of $(C_c^\infty(X))^{\sim}$ in analogy with the classical Paley-Wiener theorem in $\boldsymbol{R}^n$ become natural and their answers useful.

Chapters II and IV are devoted to the theory of the Radon transform on X, particularly inversion formulas and range questions. The space Ξ of horocycles in X offers many analogies to the space X itself and this gives rise to the study of conical functions and conical distributions on Ξ which are the analogs of the spherical functions on X. They have interesting connections with the representation theory of the isometry group G of X, discussed in Chapter II, §4, and in Chapter VI, §3, where the conical distributions furnish intertwining operators for the spherical principal series. In Corollary 3.9, Ch. VI, these intertwining operators are explicitly related to the above-mentioned Fourier transform on X.

While the Fourier transform theory in Chapter III gives rise to an explicit simultaneous diagonalization of the algebra $\boldsymbol{D}(X)$, the Radon transform theory in Chapter II is considered within the framework of a general integral transform theory for *double fibrations* in the sense of Chapter I, §3. This viewpoint is extremely general: *two dual integral transforms arise whenever we are given two subgroups of a given group G.* In the introduction to Chapter I we stress this point by indicating five such examples arising in this fashion from the single group $G = \boldsymbol{SU}(1,1)$ of the conformal maps of the unit disk, namely the *X-ray transform, the horocycle transform, the Poisson integral, the Pompeiu problem, theta series, and cusp forms.* When range results are considered, this viewpoint of the Poisson integral as a Radon transform offers a very interesting analogy with the X-ray transform in $\boldsymbol{R}^3$ (Chapter I, §3, No. 5).

With the tools developed in Chapters I–IV we study in Chapter V some natural problems for the invariant differential operators on X, solvability questions, the structure of the joint eigenfunctions, with emphasis on the harmonic functions, as well as the solutions to the invariant wave equation on X. In Chapter VI we consider in some detail the representations of G which naturally arise from the joint eigenspaces of the operators in the algebra $\boldsymbol{D}(X)$ and the algebra $\boldsymbol{D}(\Xi)$.

The length of this book is a result of my wish to make the exposition easily accessible to readers with some modest background in semisimple Lie group theory. In particular, familiarity with representation theory is not needed. Occasionally, results and proofs rely on material from my earlier books, "***Differential Geometry, Lie groups, and Symmetric Spaces***" and "***Groups and Geometric Analysis***". In the text these books are denoted by [DS] and [GGA].

Some of the material in this book has been the subject of courses at MIT over a number of years and feedback from participants has been most beneficial. I am particularly indebted to Men-chang Hu, who in his MIT thesis from 1973 determined the conical distributions for X of rank one. His work is outlined in Chapter II, §6, No. 5–6, following his thesis and in greater detail than in his article Hu [1975]. I am also deeply grateful to Adam Korányi for his advice and generous help with the material in Chapter V, §§3–4, as explained in the notes to that chapter. Similarly, I am grateful to Henrik Schlichtkrull for beneficial discussions and for his suggestions of Proposition 8.6 in Chapter III and Corollary 5.11 in Chapter V, indicated in the text. I have also profited in various ways from expert suggestions from my colleague David Vogan. I am grateful to the National Science Foundation for support during the writing of this book.

Many people have read at least parts of the manuscript and have furnished me with helpful comments and corrections; of these I mention Fulton Gonzalez, Jeremy Orloff, An Yang, Werner Hoffman, Andreas Juhl, François Rouvière, Sönke Seifert, and particularly Frank Richter. I thank them all. Finally, I thank Judy Romvos for her expert and conscientious TEX-setting of the manuscript.

A good deal of the material in this monograph has been treated in earlier papers of mine. While subsequent consolidation has usually led to a rewriting of the proofs, texts of theorems as well as occasional proofs have been preserved with minimal change. I thank Academic Press for permission to quote from the following journal publications of mine, listed in the bibliography: [1970a], [1976], [1980a], [1992b], [1992d], as well as the book [1962a].

CHAPTER I

A DUALITY IN INTEGRAL GEOMETRY

The *Radon transform* in Euclidean space $\boldsymbol{R}^n$ associates to a function f on $\boldsymbol{R}^n$ a function $\widehat{f}$ on $\boldsymbol{P}^n$, the set of (nonoriented) hyperplanes in $\boldsymbol{R}^n$ by the formula

$$\widehat{f}(\xi) = \int_{\xi} f(x)dm(x) \qquad \xi \in \boldsymbol{P}^n, \tag{1}$$

dm being the Euclidean measure on the hyperplane ξ. Along with the mapping $f \longrightarrow \widehat{f}$ we consider the *dual transform* $\varphi \longrightarrow \overset{\vee}{\varphi}$ which to a function φ on $\boldsymbol{P}^n$ associates the function $\overset{\vee}{\varphi}$ on $\boldsymbol{R}^n$ given by

$$\overset{\vee}{\varphi}(x) = \int_{\xi \ni x} \varphi(\xi)d\mu(\xi), \tag{2}$$

the average of φ over the set of hyperplanes ξ passing through x. In §2 we present a detailed study of these transforms $f \longrightarrow \widehat{f}, \varphi \longrightarrow \overset{\vee}{\varphi}$ with emphasis on their inversion formulas and range properties. For the proofs not given here we refer to [GGA], Ch. I.

The spaces $\boldsymbol{R}^n$ and $\boldsymbol{P}^n$ are both homogeneous spaces of the same group $\boldsymbol{M}(n)$, the group of isometries of $\boldsymbol{R}^n$,

$$\boldsymbol{R}^n = \boldsymbol{M}(n)/\boldsymbol{O}(n), \quad \boldsymbol{P}^n = \boldsymbol{M}(n)/\boldsymbol{M}(n-1) \times \boldsymbol{Z}_2 \tag{3}$$

where $\boldsymbol{O}(n)$ is the orthogonal group fixing the origin $0 \in \boldsymbol{R}^n$ and $\boldsymbol{M}(n-1) \times \boldsymbol{Z}_2$ is the subgroup of $\boldsymbol{M}(n)$ leaving a certain hyperplane $\xi_o \ni 0$ stable ($\boldsymbol{Z}_2$ consists of the identity and the reflection in ξ_o).

Motivated by this example we consider two homogeneous spaces

$$X = G/K, \qquad \Xi = G/H \tag{4}$$

of the *same* group G. Putting $L = K \cap H$ we define (in (5) (6), §3) the integral transforms $f \longrightarrow \widehat{f}$, $\varphi \longrightarrow \overset{\vee}{\varphi}$ by

$$\widehat{f}(\gamma H) = \int_{H/L} f(\gamma h K)dh_L, \tag{5}$$

$$\overset{\vee}{\varphi}(gK) = \int_{K/L} \varphi(gkH)dk_L. \tag{6}$$

General features of these transforms are developed in §3, No. 1.

The transforms (1) and (2) being inverted in Theorems 2.1 and 2.3, the first problem would be that of inverting (5) and (6). We now illustrate this by some examples for the group $G = \boldsymbol{SU}(1,1)$, given by

$$G = \left\{ \begin{pmatrix} a & b \\ \bar{b} & \bar{a} \end{pmatrix} : |a|^2 - |b|^2 = 1 \right\},$$

which acts on the unit disk $|z| < 1$ by

$$g : z \longrightarrow \frac{az+b}{\bar{b}z+\bar{a}}.$$

Consider now the subgroups of G given by

$$K = \left\{ \begin{pmatrix} e^{i\theta} & 0 \\ 0 & e^{-i\theta} \end{pmatrix} : \theta \in \boldsymbol{R} \right\}, \qquad A = \left\{ \begin{pmatrix} \operatorname{ch} t & \operatorname{sh} t \\ \operatorname{sh} t & \operatorname{ch} t \end{pmatrix} : \quad t \in \boldsymbol{R} \right\},$$

$$M = \left\{ \begin{pmatrix} \epsilon & 0 \\ 0 & \epsilon \end{pmatrix} : \epsilon^2 = 1 \right\}, \qquad M' = \left\{ \begin{pmatrix} e^{i\theta} & 0 \\ 0 & e^{-i\theta} \end{pmatrix} : \theta = 0, \pm\frac{\pi}{2}, \pi \right\},$$

$$N = \left\{ \begin{pmatrix} 1+ix & -ix \\ ix & 1-ix \end{pmatrix} : x \in \boldsymbol{R} \right\}, \qquad \Gamma = C\boldsymbol{SL}(2,\boldsymbol{Z})C^{-1},$$

where C is the map $w \longrightarrow (1+iw)/(w+i)$ of the upper half-plane onto the unit disk.

Example I.

$$X = G/K, \qquad \Xi = G/M'A.$$

Here X is the unit disk (with the non-Euclidean metric) and Ξ is the space of geodesics in X. The transform (5) is the *X-ray transform,* inverted in Theorem 3.12.

Example II.

$$X = G/K, \qquad \Xi = G/MN.$$

Here Ξ is the space of horocycles in the non-Euclidean disk X (cf. Chapter II,§1, No. 1). The transform (5) is the *horocycle transform;* it is inverted in Theorem 3.13, Chapter II.

Example III.

$$X = G/MAN, \qquad \Xi = G/K.$$

Since MAN is the subgroup of G fixing the point $(1, 0)$, X is identified with the boundary of the unit disk Ξ. Here the transform (5) is just the classical *Poisson integral* (see this chapter, §3, No. 5) and is inverted by Schwarz' classical theorem,

$$f(b) = \lim_{z \to b} \widehat{f}(z) \qquad (f \text{ continuous}).$$

Example IV.

$$X = G/K, \qquad \Xi = G/H,$$

where H is the subgroup of G fixing a point in X of non-Euclidean distance r from the origin. Then Ξ can be identified with the set of circles in X of radius r. The problem of inverting (5) then becomes the so-called *Pompeiu problem* (see Exercise 2, Ch. V).

Example V.

$$X = G/N, \qquad \Xi = G/\Gamma.$$

Putting $\Gamma_\infty = N \cap \Gamma$ the transforms (5) and (6) become

$$\widehat{f}(g\Gamma) = \sum_{\Gamma/\Gamma_\infty} f(g\gamma N), \qquad \check{\varphi}(gN) = \int_{N/\Gamma_\infty} (\varphi(gn\Gamma) dn_{\Gamma_\infty}.$$

The series for $\widehat{f}$ are called *theta series* and the kernel of the map $\varphi \longrightarrow \check{\varphi}$ consists of the so-called *cusp forms.* One has the orthogonal decomposition

$$L^2(G/\Gamma) = L^2_c(G/\Gamma) \oplus L^2_d(G/\Gamma),$$

where the first summand is the closure of the range $C_c(X)^\wedge$ and the second summand consists of the cusp forms.

The variety of these examples shows that, while the definitions (4)–(6) lead to some general results (cf. §3, No. 1), the individual detailed results can hardly be captured by a single general theory. However, the general set-up in §3, No. 1 for the transforms (5) and (6) is useful as a framework for examples. With that viewpoint an individual result for a single example automatically suggests a host of new problems.

The examples above (except for V) will be discussed in this book for general semisimple G; the example where X is the set of p-planes in $\boldsymbol{R}^n$, Ξ the set of q-planes in $\boldsymbol{R}^n$, that is,

$$X = \boldsymbol{M}(n)/\boldsymbol{O}(n-p) \times \boldsymbol{M}(p), \qquad \Xi = \boldsymbol{M}(n)/\boldsymbol{O}(n-q) \times \boldsymbol{M}(q),$$

is studied in further detail in the present chapter, §3 No. 2.

The Radon transform for constant curvature spaces relative to totally geodesic submanifolds is studied in §3, No. 3. Here the generalized incidence via group theory enters in a crucial fashion. In §3, No. 4–5 we report on recent results for the d-plane transform in $\boldsymbol{R}^n$ and discuss its group-theoretic analogies with the Poisson integral. Other examples are developed in Ch. II and Ch. IV.

§1. GENERALITIES.

1. Notation and Preliminaries.

As usual, $\boldsymbol{R}$ and $\boldsymbol{C}$ denote the fields of real and complex numbers, respectively, and $\boldsymbol{Z}$ the ring of integers. Let $\boldsymbol{R}^+$ denote the set $\{t \in \boldsymbol{R} : t \geq 0\}$ of nonnegative real numbers and put $\boldsymbol{Z}^+ = \boldsymbol{R}^+ \cap \boldsymbol{Z}$. On $\boldsymbol{R}^n$ we let (x,y) denote the inner product

$$(x,y) = x_1 y_1 + \cdots + x_n y_n$$

for $x = (x_1, \ldots, x_n), y = (y_1, \ldots, y_n)$ in $\boldsymbol{R}^n$ and we put $|x| = (x,x)^{\frac{1}{2}}$, $\partial_i = \partial/\partial x_i$. Also, if $\alpha = (\alpha_1, \ldots, \alpha_n)$ is an n-tuple of integers $\alpha_i \geq 0$ we put

$$D^\alpha = \partial_1^{\alpha_1} \ldots \partial_n^{\alpha_n}, \qquad x^\alpha = x_1^{\alpha_1} \ldots x_n^{\alpha_n}, \qquad |\alpha| = \alpha_1 + \cdots + \alpha_n.$$

If X is a topological space (always assumed Hausdorff) and $A \subset X$ then $\overline{A}$ (or $C\ell(A)$) denotes the closure of A in X and $\mathring{A}$ the interior of A. Also $C(X)$ (resp. $C_c(X)$) denotes the space of complex-valued continuous functions (resp. of compact support) on X. If X has metric d then $B_R(x)$ denotes the open ball $\{y \in X : d(x,y) < R\}$ and $S_R(x)$ denotes the sphere $\{z \in X : d(x,z) = R\}$. Let X be a locally compact Hausdorff space. The *measures* on X are the linear forms on $C_c(X)$ which for each compact set $K \subset X$ are continuous on the space $C_K(X) = \{f \in C(X) :$ support $(f) \subset K\}$ in the uniform topology. The space $C_o(X)$ of continuous functions vanishing at ∞ on X is the uniform closure of $C_c(X)$ in $C(X)$.

If M is a manifold and $p \in M$, the tangent space to M at p will be denoted by M_p. We shall usually assume that M has a countable basis for the open sets. Then the space $C^\infty(M) = \mathcal{E}(M)$ of complex-valued C^∞ functions is a Fréchet space in the topology given by the requirement: a sequence (f_n) in $\mathcal{E}(M)$ converges to 0 if and only if for each differential operator D on M the sequence (Df_n) converges to 0 uniformly on each compact subset of M. Let us indicate a family of seminorms giving this topology of $\mathcal{E}(M)$. Let $K \subset M$ be a compact subset and $m \in \boldsymbol{Z}^+$. Then there is a finite set of differential operators $D_1, \ldots, D_r$ on M of order $\leq m$ such that each differential operator on M of order $\leq m$ is of the form $\sum_1^r f_i D_i$ in a neighborhood of K, the f_i being suitable functions in $\mathcal{E}(M)$. This is readily seen by covering K with a finite family of coordinate neighborhoods. Now let (K_s) be an increasing sequence of compact sets with union M. Then by the remarks above we can construct a family (L_j) of differential operators on M such that the seminorms

$$\mathcal{V}_{r,s}(f) = \sup_{1\leq j\leq r}\left(\sup_{p\in K_s} |L_j f(p)|\right)$$

define the topology of $\mathcal{E}(M)$. Let $\mathcal{D}_K(M)$ denote the topological subspace of $\mathcal{E}(M)$ consisting of the functions with support contained in K. Then the space

$$\mathcal{D}(M) = C_c^\infty(M) = C^\infty(M) \cap C_c(M)$$

is given the inductive limit topology $\lim ind_{i\to\infty} \mathcal{D}_{K_i}(M)$. With this topology of $\mathcal{D}(M)$ the dual space $\mathcal{D}'(M)$ of $\mathcal{D}(M)$ is identified with the space of distributions on M; also the dual space $\mathcal{E}'(M)$ of $\mathcal{E}(M)$ is naturally identified with the subspace of $\mathcal{D}'(M)$ consisting of the distributions of compact support. We often write supp(T) for support (T).

Let φ be a diffeomorphism of M onto a manifold N. If $f \in \mathcal{D}(N)$, $g \in \mathcal{E}(N)$, $T \in \mathcal{D}'(M)$ and D a differential operator on M we put

$$g^{\varphi^{-1}} = g \circ \varphi, \qquad T^\varphi(f) = T(f^{\varphi^{-1}}), \qquad D^\varphi(g) = (Dg^{\varphi^{-1}})^\varphi.$$

Then $g^{\varphi^{-1}} \in \mathcal{E}(M)$, $T^\varphi \in \mathcal{D}'(N)$ and D^φ is a differential operator on N, the *image* of D under φ. Similar notions apply to operators on other function spaces.

The Fourier transform on $\boldsymbol{R}^n$ is defined for an integrable function f on $\boldsymbol{R}^n$ by

$$\widetilde{f}(u) = \int_{\boldsymbol{R}^n} f(x) e^{-i(x,u)} dx, \qquad u \in \boldsymbol{R}^n,$$

dx being the Lebesgue measure. The map $f \longrightarrow \widetilde{f}$ gives a bijection of the space $\mathcal{S}(\boldsymbol{R}^n)$ of rapidly decreasing functions on $\boldsymbol{R}^n$ onto itself. With the topology given by the seminorms

$$\|f\|_{m,P} = \sup_x |x|^m |P(\partial_1, \ldots, \partial_n) f(x)|$$

where $m \in \boldsymbol{Z}^+$ and P a polynomial, $\boldsymbol{S}(\boldsymbol{R}^n)$ is a Fréchet space. Since the injection $\mathcal{D}(\boldsymbol{R}^n) \longrightarrow \mathcal{S}(\boldsymbol{R}^n)$ is continuous with a dense image the members of the dual space $\mathcal{S}'(\boldsymbol{R}^n)$ are distributions, the so-called *tempered distributions.* Since the map $f \longrightarrow \widetilde{f}$ is a homeomorphism of $\mathcal{S}(\boldsymbol{R}^n)$ onto itself and satisfies

$$\int_{\boldsymbol{R}^n} f(x)\widetilde{g}(x)dx = \int_{\boldsymbol{R}^n} \widetilde{f}(u)g(u)du$$

the Fourier transform extends to a bijection of $\mathcal{S}'(\boldsymbol{R}^n)$ onto itself by the formula

$$\widetilde{T}(f) = T(\widetilde{f}) \qquad f \in \mathcal{S}(\boldsymbol{R}^n),\ T \in \mathcal{S}'(\boldsymbol{R}^n)$$

and then

$$(\widetilde{T})^{\sim}(x) = (2\pi)^n T(-x).$$

Let G be a group, H a subgroup, G/H the set of left cosets gH, $g \in G$. If $g_o \in G$ the left translation $g \longrightarrow g_o g$ and the right translation $g \longrightarrow g g_o$ will be denoted by $L(g_o)$ and $R(g_o)$, respectively. The bijection $gH \longrightarrow g_o gH$ of G/H onto itself is denoted by $\tau(g_o)$. The adjoint representation of a Lie group G on its Lie algebra $\mathfrak{g}$ is denoted Ad_G (or just Ad) and the adjoint representation of $\mathfrak{g}$ on itself is denoted $ad_{\mathfrak{g}}$ (or just ad). We have $ad(X)(Y) = [X, Y]$ and

$$\exp Ad(g)X = g \exp X g^{-1}, \qquad Ad(\exp X) = e^{adX}.$$

2. Principal Problems.

Let f be a function on $\boldsymbol{R}^n$, integrable over each hyperplane and $\widehat{f}(w,p)$ the integral

$$\widehat{f}(w,p) = \int_{(x,w)=p} f(x)dm(x) \tag{1}$$

over the hyperplane $(x, w) = p$, w being a unit vector, $(\ ,\ )$ the inner product and $p \in \boldsymbol{R}$. On the space $\mathcal{S}(\boldsymbol{R}^n)$ we have the *Radon inversion formula* (cf. [GGA], Ch. I, §2)

$$c_o f(x) = (-L_x)^{(n-1)/2}\Big\{ \int_{\boldsymbol{S}^{n-1}} \widehat{f}(w,(w,x))dw \Big\}, \qquad f \in \mathcal{S}(\boldsymbol{R}^n), \tag{2}$$

where dw is the surface element on the unit sphere $\boldsymbol{S}^{n-1}$ and c_o the constant

$$c_o = (2\pi)^n/\pi.$$

In (2) L_x denotes the usual Laplacian

$$L = \frac{\partial^2}{\partial x_1^2} + \cdots + \frac{\partial^2}{\partial x_n^2}$$

operating on the variable $x \in \boldsymbol{R}^n$. If n is even, $(-L)^{(n-1)/2}$, and more generally $(-L)^p$ is defined in terms of the Riesz potential ([GGA] p. 135)

$$(3) \qquad (I^\gamma f)(x) = \frac{1}{H_n(\gamma)} \int_{\boldsymbol{R}^n} f(y)|x-y|^{\gamma-n} dy, \qquad \gamma \in C_n,$$

where

$$(4) \qquad H_n(\gamma) = 2^\gamma \pi^{\frac{n}{2}} \frac{\Gamma(\frac{1}{2}\gamma)}{\Gamma(\frac{1}{2}(n-\gamma))}, \qquad C_n = \{\gamma \in \boldsymbol{C} : \gamma - n \notin 2\boldsymbol{Z}^+\}.$$

By definition

$$(-L)^p f = I^{-2p}(f), \qquad f \in \mathcal{S}(\boldsymbol{R}^n)$$

for $-2p \in C_n$ and then we have the desired Fourier transform relation

$$(5) \qquad ((-L)^p f)^\sim = |u|^{2p} \widetilde{f} \qquad (-2p \in C_n)$$

valid in the sense of tempered distributions. In fact, $\alpha \longrightarrow r^\alpha$ is a meromorphic mapping from $\boldsymbol{C}$ to $\mathcal{S}'(\boldsymbol{R}^n)$ with simple poles at $\alpha = -n-2h$ ($h \in \boldsymbol{Z}^+$) and the residue

$$\mathrm{Res}_{\alpha=-n-2h}\, r^\alpha = \lim_{\alpha \longrightarrow -n-2h} (\alpha + n + 2h) r^\alpha$$

is tempered. Moreover, the Fourier transform is given by

$$(r^\alpha)^\sim = 2^{n+\alpha} \pi^{n/2} \frac{\Gamma(\frac{1}{2}(n+\alpha))}{\Gamma(-\frac{1}{2}\alpha)} r^{-\alpha-n} \qquad \alpha, -\alpha-n \notin 2\boldsymbol{Z}^+$$

([GGA], p. 134) and by analytic continuation this holds also for $\alpha \in 2\boldsymbol{Z}^+$. Since the relation $(f * S)^\sim = \widetilde{f}\widetilde{S}$ holds for $f \in \mathcal{S}(\boldsymbol{R}^n)$, $S \in \mathcal{S}'(\boldsymbol{R}^n)$, $*$ denoting convolution and since by (3)

$$I^{-2p} f = \frac{1}{H_n(-2p)} f * r^{-2p-n}, \qquad -2p \in C_n,$$

we obtain

$$\big((-L)^p f\big)^\sim(u) = |u|^{2p} \widetilde{f}$$

for $-(-2p-n)-n \notin 2\boldsymbol{Z}^+$ and $-2p \in C_n$. The relation being obvious for $p \in \boldsymbol{Z}^+$ (5) is proved.

Let $\mathcal{H}$ denote the Hilbert transform

$$(\mathcal{H}F)(t) = \frac{i}{\pi} \int_{-\infty}^{\infty} \frac{F(p)}{t-p} dp, \qquad F \in \mathcal{S}(\boldsymbol{R}). \tag{6}$$

Considering $\mathcal{H}F$ as a tempered distribution we have

$$(\mathcal{H}F)^{\sim}(s) = \operatorname{sgn} s \, \widetilde{F}(s) \tag{7}$$

where

$$\operatorname{sgn} s = \begin{cases} 1 & s \geq 0 \\ -1 & s < 0. \end{cases}$$

By Schwartz [1966], p. 248, $\mathcal{S}(\boldsymbol{R}) * \mathcal{S}'(\boldsymbol{R}) \subset \mathcal{E}(\boldsymbol{R})$ (where $*$ denotes convolution) so $\mathcal{H}F$ is a smooth function. We define the operator Λ by

$$(\Lambda\varphi)(w,p) = \begin{cases} \frac{d^{n-1}}{dp^{n-1}} \varphi(w,p) & n \text{ odd}, \\ \mathcal{H}_p \frac{d^{n-1}}{dp^{n-1}} \varphi(w,p) & n \text{ even}. \end{cases} \tag{8}$$

Here $\varphi(w,p)$ is a C^∞ function on $\boldsymbol{S}^{n-1} \times \boldsymbol{R}$ satisfying the symmetry condition $\varphi(-w,-p) = \varphi(w,p)$ and rapidly decreasing in p (uniformly for $w \in \boldsymbol{S}^{n-1}$). Note that $(\Lambda\varphi)(-w,-p) = (\Lambda\varphi)(w,p)$.

The inversion formula (2) can then also be written

$$c_0 f(x) = (-i)^{n-1} \int_{\boldsymbol{S}^{n-1}} \{(\Lambda\widehat{f})(w,p)\}_{p=(w,x)} dw \tag{9}$$

(cf. for example [GGA], Ch. I, §2).

Formulas (2) (9) suggest the following general questions the first of which was already considered by Radon [1917]. Let X be a manifold in which certain submanifolds ξ are singled out. Let Ξ denote the family of these submanifolds ξ. Given a function f on X its *Radon transform* $\widehat{f}$ is the function on Ξ defined by

$$\widehat{f}(\xi) = \int_{\xi} f(x) dm(x), \qquad \xi \in \Xi, \tag{10}$$

dm being some preassigned measure on ξ.

The basic problems in this framework are the following:

A. Inversion Problem. *Determine f on the basis of the integrals $\widehat{f}$.*

B. Range Problem. *Determine the range and the kernel of the mapping $f \longrightarrow \widehat{f}$ as f runs through some function space on X* (say $C_c(X)$ or $C_c^\infty(X)$).

In the example above $X = \boldsymbol{R}^n$ and Ξ is the set of hyperplanes in $\boldsymbol{R}^n$. In the next section we shall consider X and Ξ on equal footing and deal with Problems **A** and **B** for both of them. We shall use results and tools developed in [GGA], Ch. I, §2; we also refer to this book for proofs and documentation of Theorems 2.1, 2.2 and Theorem 2.4 stated in the next section where we carry the discussion of Problems **A** and **B** quite a bit further.

A restricted form of Problem B is the following:

C. Support Problem. *Show that the conditions*

a) *f "small" at ∞,* b) *$\widehat{f}$ of compact support,*

imply

c) *f of compact support.*

In the case when X is a Riemannian manifold one would state this problem more precisely:

Let $B \subset X$ be a ball, f a function on X small at ∞. Assuming

b′) $\widehat{f}(\xi) = 0$ for $\xi \cap B = \emptyset$,

prove that

c′) *$f(x) = 0$ for $x \notin B$.*

§2. THE RADON TRANSFORM FOR POINTS AND HYPERPLANES.

1. The Principal Results.

As a special case of our formalism above let $X = \boldsymbol{R}^n$ and let $\Xi = \boldsymbol{P}^n$, the set of all hyperplanes in $\boldsymbol{R}^n$. If we write a hyperplane ξ in the form $(x, w) = p$ ($|w| = 1, p \in \boldsymbol{R}$) then $\boldsymbol{P}^n$ acquires a natural manifold structure and the mapping $(w, p) \longrightarrow \xi$ is a double covering map of $\boldsymbol{S}^{n-1} \times \boldsymbol{R}$ onto $\boldsymbol{P}^n$.

By analogy with $\mathcal{S}(\boldsymbol{R}^n)$ we define $\mathcal{S}(\boldsymbol{S}^{n-1} \times \boldsymbol{R})$ as the space of C^∞ functions φ on $\boldsymbol{S}^{n-1} \times \boldsymbol{R}$ which for any integers $k, l \geq 0$ and any differential operator D on $\boldsymbol{S}^{n-1}$ satisfy

$$(1)\qquad \sup_{w\in\boldsymbol{S}^{n-1},p\in\boldsymbol{R}}\left|(1+|p|^k)\frac{d^\ell}{dp^\ell}(D\varphi)(w,p)\right|<\infty.$$

The space $\mathcal{S}(\boldsymbol{P}^n)$ is then defined as the set of $\varphi\in\mathcal{S}(\boldsymbol{S}^{n-1}\times\boldsymbol{R})$ satisfying $\varphi(-w,-p)=\varphi(w,p)$. We denote by $\square$ the differential operator

$$\square:\ \varphi(w,p)\longrightarrow\frac{d^2}{dp^2}\varphi(w,p)$$

on $\mathcal{E}(\boldsymbol{P}^n)$ and in analogy with (3)–(5) §1 the fractional power $\square^k$ is defined by

$$(2)\quad (-\square)^k\varphi(w,p)=\frac{1}{H_1(-2k)}\int_{\boldsymbol{R}}\varphi(w,p)|p-q|^{-2k-1}dq\qquad(-2k-1\notin 2\boldsymbol{Z}^+)$$

and then we have by (5)§1

$$(3)\qquad (-\square^k\varphi)^{\sim}(w,s)=|s|^{2k}\widetilde{\varphi}(w,s)\qquad(-2k-1\notin 2\boldsymbol{Z}^+).$$

Let f be a function on $\boldsymbol{R}^n$ integrable on each hyperplane. Its *Radon transform* is defined as the function $\widehat{f}$ on $\boldsymbol{P}^n$ given by

$$(4)\qquad \widehat{f}(\xi)=\int_\xi f(x)dm(x),\qquad\qquad \xi\in\boldsymbol{P}^n,$$

dm being the Euclidean measure on ξ. The *dual transform* associates to a continuous function φ on Ξ the function $\check{\varphi}$ on $\boldsymbol{R}^n$ given by

$$(5)\qquad \check{\varphi}(x)=\int_{x\in\xi}\varphi(\xi)d\mu(\xi),$$

where $d\mu$ is the rotation-about-x invariant measure on the set of hyperplanes through x such that the total measure is 1.

Let us write ξ for the hyperplane $(x,w)=p$. Then (4) and (5) can also be written

$$(4a)\qquad \widehat{f}(w,p)=\int_{(x,w)=p}f(x)\,dm(x)$$

and

$$\check{\varphi}(x) = \Omega_n^{-1} \int\limits_{S^{n-1}} \varphi(w, (x,w))\, dw, \tag{5a}$$

where $\Omega_n = 2\pi^{n/2}\Gamma(n/2)^{-1}$ is the area of S^{n-1}. The connection of $\widehat{f}$ with the Fourier transform

$$\widetilde{f}(u) = \int\limits_{R^n} f(x) e^{-i(x,u)} dx$$

is given by

$$\widetilde{f}(sw) = \int\limits_{R} \widehat{f}(w,p) e^{-ips} dp. \tag{6}$$

Consider now the convolution

$$f(x) = (f_1 * f_2)(x) = \int\limits_{R^n} f_1(x-y) f_2(y)\, dy$$

whose Radon transform is given by

$$\widehat{f} = \widehat{f}_1 * \widehat{f}_2, \text{ i.e. } \widehat{f}(w,p) = \int\limits_{R} \widehat{f}_1(w, p-q) \widehat{f}_2(w,q)\, dq. \tag{7}$$

This is clear by letting x and y above run over the hyperplanes $(x,w) = p$ and $(y,w) = q$, respectively. Formula (7) is for example valid if f_1 and f_2 are in $L^1(R^n) \cap C(R^n)$; later we shall see other instances.

The transforms $f \longrightarrow \widehat{f}$ and $\varphi \longrightarrow \check{\varphi}$ intertwine L and $\square$ as follows:

$$(Lf)^\wedge = \square \widehat{f}, \qquad (\square\varphi)^\vee = L\check{\varphi}. \tag{8}$$

Formulas (2) (9) in §1 can now be written in simplified notation.

Theorem 2.1. *The transform $f \longrightarrow \widehat{f}$ is inverted by each of the formulas,*

$$Cf = (-L)^{(n-1/2}((\widehat{f})^\vee), \qquad f \in \mathcal{S}(R^n), \tag{9}$$

$$Cf = \left((-\square)^{(n-1)/2} \widehat{f}\right)^\vee, \qquad f \in \mathcal{S}(R^n), \tag{10}$$

where

$$C = (4\pi)^{(n-1)/2}\frac{\Gamma(n/2)}{\Gamma(1/2)}. \tag{11}$$

Remark. This theorem supersedes Theorems 2.13, 2.15, Ch. I in [GGA] where the constant is less explicitly given.

We can now ask for analogous inversion formulas for the dual transform $\varphi \longrightarrow \check{\varphi}$. There is now a basic difference in that $\check{\varphi}$ is defined for all continuous functions φ but on the other hand $(\check{\varphi})^\wedge$ does not necessarily exist.

Let $\mathcal{S}^*(\boldsymbol{R}^n)$ denote the space of functions $f \in \mathcal{S}(\boldsymbol{R}^n)$ which are orthogonal to all polynomials i.e.

$$\int_{\boldsymbol{R}^n} f(x)P(x)dx = 0 \quad \text{for all polynomials } P.$$

Note that under the Fourier transform the space $\mathcal{S}^*(\boldsymbol{R}^n)$ corresponds to the subspace $\mathcal{S}_o(\boldsymbol{R}^n)$ of functions in $\mathcal{S}(\boldsymbol{R}^n)$ all of whose derivatives vanish at 0. [Recall in this connection (Schwartz [1966], Theorem VII, II) that the functions in $\mathcal{S}(\boldsymbol{R}^n)$ can be viewed as the functions on the one point compactification $\boldsymbol{S}^n = \boldsymbol{R}^n \cup \{\infty\}$ that vanish at ∞ together with all their derivatives; thus the space $\mathcal{S}_o(\boldsymbol{R}^n)$ is given by such a vanishing condition at *two* points of $\boldsymbol{S}^n$].

Similarly, we define $\mathcal{S}^*(\boldsymbol{P}^n)$ as the space of functions $\varphi \in \mathcal{S}(\boldsymbol{P}^n)$ satisfying

$$\int_{\boldsymbol{R}} \varphi(w,p)Q(p)dp = 0 \quad \text{for all polynomials } Q.$$

On these spaces we have the desired analog of (9)(cf. [GGA], Ch. I, §2).

Theorem 2.2. *The transform $f \longrightarrow \widehat{f}, \varphi \longrightarrow \check{\varphi}$ are bijections of $\mathcal{S}^*(\boldsymbol{R}^n)$ onto $\mathcal{S}^*(\boldsymbol{P}^n)$ and of $\mathcal{S}^*(\boldsymbol{P}^n)$ onto $\mathcal{S}^*(\boldsymbol{R}^n)$, respectively. Moreover,*

$$C\varphi = (-\square)^{(n-1)/2}((\check{\varphi})^\wedge), \qquad \varphi \in \mathcal{S}^*(\boldsymbol{P}^n),$$

where the constant C is given by (11).

As indicated earlier the formula in Theorem 2.2 does not hold for all $\varphi \in \mathcal{S}(\boldsymbol{P}^n)$ because $(\check{\varphi})^\wedge$ may not be defined. However, the following modified version of the inversion formula always holds.

Theorem 2.3. *The dual transform $\varphi \longrightarrow \check{\varphi}$ is inverted by the formula*

$$i^{n-1}C\varphi = ((\Lambda\varphi)^\vee)^\wedge, \qquad \varphi \in \mathcal{S}(\boldsymbol{P}^n),$$

with the operator Λ *defined in* §1, (8).

The proof, using Fourier transforms, will be given later. First we state the principal results about the range and kernel questions.

In order to describe the images $\mathcal{S}(\boldsymbol{R}^n)^\wedge$ and $\mathcal{D}(\boldsymbol{R}^n)^\wedge$ we consider the spaces

$$\mathcal{S}_H(\boldsymbol{P}^n) = \left\{ \varphi \in \mathcal{S}(\boldsymbol{P}^n) : \begin{array}{l} \text{For each } k \in \boldsymbol{Z}^+, \int \varphi(w,p)p^k dp \text{ is} \\ \text{a homogeneous } k^{th} \text{ degree poly-} \\ \text{nomial in } w_1, \ldots, w_n \end{array} \right\}$$

$$\mathcal{D}_H(\boldsymbol{P}^n) = \mathcal{D}(\boldsymbol{P}^n) \cap \mathcal{S}_H(\boldsymbol{P}^n).$$

Then we have (cf. [GGA]. Ch. I, §2),

Theorem 2.4. *The Radon transform $f \longrightarrow \widehat{f}$ is a bijection of*

(i) $\mathcal{S}(\boldsymbol{R}^n)$ *onto* $\mathcal{S}_H(\boldsymbol{P}^n)$

(ii) $\mathcal{D}(\boldsymbol{R}^n)$ *onto* $\mathcal{D}_H(\boldsymbol{P}^n)$.

Moreover, $f \in \mathcal{S}(\boldsymbol{R}^n)$ has support in the ball $|x| \leq A$ if and only if $\widehat{f}(\xi) = 0$ for $d(0,\xi) \geq A$, d denoting distance.

Remark. It is pointed out in [GGA], Ch. I (Remark 2.9) that the condition $|x|^k f(x) \longrightarrow 0$ for *each* $k > 0$ is needed in Theorem 2.4. For the inversion formula (9) a much weaker condition is sufficient; consequently Theorem 2.4 could not follow from the inversion formula.

On the other hand, if n is odd the following support result is immediate from the inversion formula (9):

If $f \in \mathcal{S}(\boldsymbol{R}^n)$, $\widehat{f}(\xi) = 0$ for $d(0,\xi) < A$ then $f(x) = 0$ for $|x| < A$.

For the dual transform the description is of course completely different. To state it let E_k denote the eigenspace of the Laplacian $L_{\boldsymbol{S}} = L_{\boldsymbol{S}^{n-1}}$ on $\boldsymbol{S}^{n-1}$ for eigenvalue $-k(k+n-2)$, i.e.

$$E_k = \{F \in \mathcal{E}(\boldsymbol{S}^{n-1}) : \; L_{\boldsymbol{S}}F = -k(k+n-2)F\}.$$

It is well known that each eigenspace of $L_{\boldsymbol{S}}$ is of this form for some $k \in \boldsymbol{Z}^+$. Let $E_k \otimes p^\ell$ denote the space of functions $\varphi(w,p)$ of the form $\varphi(w,p) = \psi(w)p^\ell$ as $\psi \in E_k$.

Note that since ψ is the restriction to $\boldsymbol{S}^{n-1}$ of a homogeneous k^{th} degree polynomial the condition $\varphi \in \mathcal{E}(\boldsymbol{P}^n)$ implies that $k - \ell$ is even.

Theorem 2.5.

(i) *Let $\mathcal{N}$ denote the kernel of the dual transform $\varphi \longrightarrow \overset{\vee}{\varphi}$ on $\mathcal{E}(\boldsymbol{P}^n)$. Then $\mathcal{N}$ is the closed subspace of $\mathcal{E}(\boldsymbol{P}^n)$ generated by the spaces*

$$E_k \otimes p^\ell \qquad k, \ell \in \boldsymbol{Z}^+, \quad k - \ell > 0 \quad \text{and even.}$$

(ii) *The dual map $\varphi \longrightarrow \overset{\vee}{\varphi}$ is injective on $\mathcal{S}(\boldsymbol{P}^n)$.*

(iii) *The dual transform $\varphi \longrightarrow \overset{\vee}{\varphi}$ maps $\mathcal{E}(\boldsymbol{P}^n)$ onto $\mathcal{E}(\boldsymbol{R}^n)$.*

The Radon transform and its dual can be extended to the distribution spaces $\mathcal{E}'(\boldsymbol{R}^n)$ and $\mathcal{D}'(\boldsymbol{P}^n)$, respectively. Denoting by $d\xi$ the invariant measure $d\xi = \Omega_n^{-1}\, dw\, dp$ on $\boldsymbol{S}^{n-1} \times \boldsymbol{R}$ (Ω_n = area of $\boldsymbol{S}^{n-1}$) we have the formula

$$\int_{\boldsymbol{P}^n} \widehat{f}(\xi)\varphi(\xi)\,d\xi = \int_{\boldsymbol{R}^n} f(x)\overset{\vee}{\varphi}(x)\,dx \tag{12}$$

for $f \in C_c(\boldsymbol{R}^n), \varphi \in C(\boldsymbol{P}^n)$. More generally, (12) holds in the following two situations (cf. [GGA], Ch. I, Lemma 2.18):

a) $f \in L^1(\boldsymbol{R}^n)$ vanishing outside a compact set; $\varphi \in C(\boldsymbol{P}^n)$;

b) $f \in C_c(\boldsymbol{R}^n)$; φ Borel measurable and locally integrable.

In both cases (12) is obtained by using the Fubini theorem on the function $(x, w) \longrightarrow f(x)\varphi(w, (w, x))$ on $\boldsymbol{R}^n \times \boldsymbol{S}^{n-1}$.

Formula (10) leads to the definition of the Radon transform $S \longrightarrow \widehat{S}$ on $\mathcal{E}'(\boldsymbol{R}^n)$ and of the dual transform $\Sigma \longrightarrow \overset{\vee}{\Sigma}$ on $\mathcal{D}'(\boldsymbol{P}^n)$; in fact we put

$$\widehat{S}(\varphi) = S(\overset{\vee}{\varphi}) \qquad \text{for } \varphi \in \mathcal{E}(\boldsymbol{P}^n);$$

$$\overset{\vee}{\Sigma}(f) = \Sigma(\widehat{f}) \qquad \text{for } f \in \mathcal{D}(\boldsymbol{R}^n).$$

For this extension we shall prove the following result.

Theorem 2.6. *The range $\mathcal{E}'(\boldsymbol{R}^n)^\wedge$ of $\mathcal{E}'(\boldsymbol{R}^n)$ under the Radon transform can be described by the two following equivalent conditions:*

(i) $\mathcal{E}'(\boldsymbol{R}^n)^\wedge = \{\Sigma \in \mathcal{E}'(\boldsymbol{P}^n) : \quad \Sigma(\mathcal{N}) = 0\}$, *where $\mathcal{N}$ is the kernel of the dual transform described in Theorem 2.5.*

(ii) *$\mathcal{E}'(\boldsymbol{R}^n)^\wedge$ consists of the distributions $\Sigma \in \mathcal{E}'(\boldsymbol{P}^n)$ with the following property:*

For each $k \in \boldsymbol{Z}^+$ the distribution

$$\psi \in \mathcal{E}(\boldsymbol{S}^{n-1}) \longrightarrow \int\limits_{\boldsymbol{S}^{n-1}\times\boldsymbol{R}} \tfrac{1}{2}(\psi + (-1)^k\psi')(w)p^k\, d\Sigma(w,p)$$

is a homogeneous polynomial in $w_1, \ldots, w_n$ of degree k. Here $\psi'(w) = \psi(-w)$.

(iii) *Furthermore, we have the inversion formula*

$$i^{n-1}CS = (\Lambda\widehat{S})^\vee, \qquad\qquad S \in \mathcal{E}'(\boldsymbol{R}^n)$$

with C as in (11) and Λ as in §1 (8).

Because of the remarks at the end of §1 and since (iii) is proved in [GGA], I, §2 we shall just give proofs of Theorems 2.3, 2.5 and 2.6 (i) (ii). Although they may seem to have been stated above in their natural order we shall see that the proof of Theorem 2.5 (iii) depends on Theorem 2.6.

2. The Kernel of the Dual Transform

We shall now prove Theorem 2.5 (i) describing the kernel $\mathcal{N}$. To begin with we decompose the functions $f \in \mathcal{E}(\boldsymbol{R}^n)$, $\varphi \in \mathcal{E}(\boldsymbol{P}^n)$ by means of spherical harmonics. For each $k \in \boldsymbol{Z}^+$ consider the Laplacian eigenspace $E_k \subset L^2(\boldsymbol{S}^{n-1})$ and take the inner product $\langle\ ,\ \rangle$ on $L^2(\boldsymbol{S}^{n-1})$ in terms of the normalized measure $d'w = \Omega_n^{-1}dw$ on $\boldsymbol{S}^{n-1}$. We recall that the natural representation $\delta = \delta^k$ of $K = \boldsymbol{O}(n)$ on E_k is irreducible. Let $o = (0, \ldots, 0, 1) = \boldsymbol{e}_n$ be the north pole and $M = \boldsymbol{O}(n-1)$ the subgroup of K fixing o. Let $\widehat{K}$ denote the set of equivalence classes of unitary irreducible representations of K and let $\widehat{K}_M$ be the subset of $\delta \in \widehat{K}$ such that the group $\delta(M)$ has a nonzero fixed vector. We recall that the representations $\delta = \delta^k$, $k \in \boldsymbol{Z}^+$, are all inequivalent and (up to equivalence) exhaust the set $\widehat{K}_M$. Let $S_{k1}, \ldots, S_{kd(k)}$ be an orthonormal basis of E_k and χ_δ the character of δ.

Proposition 2.7. *Let $f \in \mathcal{E}(\boldsymbol{R}^n)$. Then we have the series expansion*

$$f(x) = \sum_{k=0}^{\infty} d(k) \int_K \chi_{\delta^k}(u) f(u \cdot x)du = \sum_{k=0}^{\infty} f_k(x) \tag{13}$$

converging in the topology of $\mathcal{E}(\boldsymbol{R}^n)$. Furthermore, if $x = rw$ $(|w| = 1,\ r \in \boldsymbol{R})$,

$$f_k(x) = \sum_{m=1}^{d(k)} f_{km}(r) S_{km}(w), \tag{14}$$

where $r^{-k} f_{km}(r)$ *is* C^∞ *and even.*

Proof. Let $f \in \mathcal{E}(\boldsymbol{R}^n)$ and fix $x \in \boldsymbol{R}^n$. The function $F(u) = f(u \cdot x)$ on $K = \boldsymbol{O}(n)$ has a Peter-Weyl expansion

$$F(u) = \sum_{\delta \in \widehat{K}} d(\delta)(\overline{\chi}_\delta * F)(u)$$

where $d(\delta)$ and χ_δ, respectively, denote the degree and character of δ, and $*$ denotes convolution on K. Putting $u = e$ we get the expansion

$$f(x) = \sum_{\delta \in \widehat{K}} d(\delta) \int_K \chi_\delta(k) f(k \cdot x) dk \tag{15}$$

dk being the normalized Haar measure. Now $x = |x|\ell \cdot \boldsymbol{e}_n$ where $\ell \in \boldsymbol{O}(n)$. Let $m \in M$ and dm the normalized Haar measure. Then

$$\int_K \chi_\delta(k) f(k\ell \cdot |x|\boldsymbol{e}_n) dk = \int_K \chi_\delta(k\ell^{-1}) f(k \cdot |x|\boldsymbol{e}_n) dk$$

$$= \int_K \chi_\delta(\ell^{-1}k) f(km \cdot |x|\boldsymbol{e}_n) dk = \int_K \left(\int_M \chi_\delta(\ell^{-1}km) dm \right) f(k \cdot |x|\boldsymbol{e}_n) dk.$$

However,

$$\int_M \chi_\delta(\ell^{-1}km) dm = Tr\left(\delta(\ell^{-1}k) \int_M \delta(m) dm \right) = Tr(\delta(\ell^{-1}k)P)$$

where P is the orthogonal projection of the representation space of δ onto the space of fixed points of $\delta(M)$. This shows that the integral in (15) vanishes unless $\delta \in \widehat{K}_M$ and in this case δ is equivalent to δ^k for a unique $k \in \boldsymbol{Z}^+$. Thus (15) can be written

$$f(x) = \sum_{\delta \in \widehat{K}_M} d(\delta) \int_K \chi_\delta(u) f(u \cdot x) du = \sum_{k \geq 0} f_k(x) \tag{16}$$

and the function f_k is $\boldsymbol{O}(n)$-finite of type $\delta = \delta^k$ (see e. g. [GGA], p. 395 for the general definition of a K-finite vector of type δ). By Harish-Chandra's vector-valued version of the Peter-Weyl theorem the series (16) converges in the topology of $\mathcal{E}(\boldsymbol{R}^n)$ (cf. [GGA], Ch. V, Cor. 3.4).

Next we note that if $\ell \in K$,

$$S_{km}(\ell^{-1}w) = \sum_{s=1}^{d(k)} \delta^k(\ell)_{sm} S_{ks}(w),$$

so by the orthogonality relations

$$\int_K \chi_\delta(\ell^{-1}) S_{km}(\ell^{-1}w)\,d\ell = \frac{1}{d(\delta)} S_{km}(w) \tag{17}$$

if δ is equivalent to δ^k, otherwise the integral is 0. For a given $r \in \boldsymbol{R}$ we can expand the function $w \longrightarrow f(rw)$ in spherical harmonics

$$f(rw) = \sum_{k=0}^{\infty} \sum_{1}^{d(k)} f_{km}(r) S_{km}(w) \tag{18}$$

and then (16)-(18) show that

$$f_k(rw) = \sum_{1}^{d(k)} f_{km}(r) S_{rm}(w). \tag{19}$$

Then

$$f_{km}(r) = \int_{\boldsymbol{S}^{n-1}} f(rw)\overline{S_{km}}(w)\,d'w. \tag{20}$$

Expanding $f(rw)$ in a Taylor series around 0 (with remainder) and remembering that $S_{km}(w)$ is orthogonal to the restriction to $\boldsymbol{S}^{n-1}$ of polynomials of degree $< k$ we see from (20) that $f_{km}(rw)/r^k$ is a smooth and even function. This proves the proposition.

Proposition 2.8. *Let $\varphi \in \mathcal{E}(\boldsymbol{P}^n)$. Then we have the series expansion*

$$\varphi(\xi) = \sum_{k=0}^{\infty} d(k) \int_K \chi_{\delta^k}(u)\varphi(u\cdot\xi)\,du = \sum_{k=0}^{\infty} \varphi_k(\xi) \tag{21}$$

converging in the topology of $\mathcal{E}(\boldsymbol{P}^n)$. Furthermore, if ξ is the hyperplane $(x, w) = p$,

$$\varphi_k(\xi) = \varphi_k(w,p) = \sum_{m=1}^{d(k)} \varphi_{km}(p) S_{km}(w), \tag{22}$$

and $\varphi_{km}(-p) = (-1)^k \varphi_{km}(p)$.

The proof is the same as that of Prop. 2.7. The difference between the statements about f_{km} and φ_{km} comes from the fact that in contrast to the mapping $(w,p) \longrightarrow \xi$ of $\boldsymbol{S}^{n-1} \times \boldsymbol{R}$ onto $\boldsymbol{P}^n$ the polar coordinate representation $(r,w) \longrightarrow rw$ on $\boldsymbol{R}^n$ is singular at 0.

Turning now to the kernel $\mathcal{N}$ we have by (12)

$$\mathcal{N} = \left\{ \varphi \in \mathcal{E}(\boldsymbol{P}^n) : \int_{\boldsymbol{P}^n} \widehat{f}(\xi)\varphi(\xi)d\xi = 0 \quad \text{for } f \in \mathcal{D}(\boldsymbol{R}^n) \right\}. \tag{23}$$

With the bilinear form

$$\langle \psi, \varphi \rangle = \int_{\boldsymbol{P}^n} \psi(\xi)\varphi(\xi)d\xi$$

on $\mathcal{D}(\boldsymbol{P}^n) \times \mathcal{E}(\boldsymbol{P}^n)$ we can state (23) in the form

$$\mathcal{N} = \left(\mathcal{D}(\boldsymbol{R}^n)^\wedge\right)^\perp, \tag{24}$$

where $\perp$ denotes annihilator. Now if $f \in \mathcal{D}(\boldsymbol{R}^n)$ then

$$\int_{\boldsymbol{R}} \widehat{f}(w,p)p^\ell dp \tag{25}$$

can be written as a homogeneous ℓ^{th} degree polynomial in $w_1, \ldots, w_n$. The space P_ℓ of homogeneous polynomial functions $p(x_1, \ldots, x_n)$ of degree ℓ on $\boldsymbol{R}^n$ has the decomposition

$$P_\ell = H_\ell + |x|^2 H_{\ell-2} + \cdots + x^{2m} H_{\ell-2m}, \quad m = [\tfrac{1}{2}\ell], \tag{26}$$

where $H_\ell \subset P_\ell$ is the subspace of harmonic polynomials in P_ℓ. It follows that (25) is a linear combination of members of $E_\ell, E_{\ell-2}, \ldots$ and therefore is orthogonal to E_k for $k > \ell$. Consequently (24) shows that

$$E_k \otimes p^\ell \in \mathcal{N} \qquad \text{if} \qquad k - \ell > 0 \text{ and even.} \tag{27}$$

The evenness condition is required for $E_k \otimes p^\ell$ to represent functions φ satisfying $\varphi(-w,-p) = \varphi(w,p)$ which is needed for them to be considered as functions on $\boldsymbol{P}^n$. Next we prove that if $\varphi \in \mathcal{D}(\boldsymbol{P}^n)$ then

$$\varphi \in \mathcal{D}(\boldsymbol{R}^n)^\wedge \Longleftrightarrow \langle \varphi, E_k \otimes p^\ell \rangle = 0 \text{ for } k - \ell > 0, \text{ even.} \tag{28}$$

Part $\Longrightarrow$ is already proved. For the converse we consider the decomposition (22) and deduce

$$\int_{\boldsymbol{R}} \varphi_{km}(p)p^{\ell}dp = 0 \quad \text{for} \quad k-\ell>0, \text{ even} .$$

Since $\varphi_{km}(-p) = (-1)^k\varphi_{km}(p)$ this integral vanishes for $k-\ell$ odd. Hence (22) implies that

$$\int_{\boldsymbol{R}} \varphi(w,p)p^{\ell}dp \tag{29}$$

is a linear combination of $S_{km}(w)$ with $\ell - k \geq 0$ and even. Multiplying by suitable powers of $w_1^2+\cdots+w_n^2$ we see that expression (29) is a homogeneous polynomial of the ℓth degree so by Theorem 2.4, $\varphi \in \mathcal{D}(\boldsymbol{R}^n)^\wedge$ as desired.

Remark. From (28) we note that the subspace $\mathcal{D}(\boldsymbol{R}^n)^\wedge \subset \mathcal{D}(\boldsymbol{P}^n)$ is closed.

To prove Theorem 2.5 suppose $\psi \in \mathcal{N}$. As we saw,

$$\psi_k(\xi) = d(\delta)\int_K \chi_\delta(u)\psi(u\cdot\xi)du \tag{30}$$

and since the dual transform $\varphi \longrightarrow \overset{\vee}{\varphi}$ commutes with the K-action,

$$\overset{\vee}{\psi}_k(x) = d(\delta)\int_K \chi_\delta(u)\overset{\vee}{\psi}(u\cdot x)du = 0. \tag{31}$$

Consider now the spaces

$$\mathcal{D}(\boldsymbol{R}^n)_k = \{f_k \ : \ f \in \mathcal{D}(\boldsymbol{R}^n)\}, \tag{32}$$

$$\mathcal{D}(\boldsymbol{P}^n)_k = \{\varphi_k \ : \ \varphi \in \mathcal{D}(\boldsymbol{P}^n)\}. \tag{33}$$

By (28) we have for each k and each $\varphi \in \mathcal{D}(\boldsymbol{P}^n)_k$:

$$\varphi \in \big(\mathcal{D}(\boldsymbol{R}^n)_k\big)^\wedge \Longleftrightarrow \langle \varphi, E_k \otimes p^\ell\rangle = 0 \ \text{ for } \ k-\ell>0, \text{ even} .$$

For a fixed k the spaces $E_k \otimes p^\ell$ ($\ell < k, k-\ell$ even) span a finite-dimensional space $\widetilde{E}_k$ whose annihilator $(\widetilde{E}_k)^\perp$ in $\mathcal{D}(\boldsymbol{P}^n)_k$ is $\big(\mathcal{D}(\boldsymbol{R}^n)_k\big)^\wedge$. However, the equation $\overset{\vee}{\psi}_k = 0$ implies that ψ_k is orthogonal to this annihilator $\big(\mathcal{D}(\boldsymbol{R}^n)_k\big)^\wedge$, in other words, ψ_k belongs to the double annihilator $\big((\widetilde{E}_k)^\perp\big)^\perp$ which by the finite-dimensionality equals $\widetilde{E}_k$. Since (22) converges in the topology of $\mathcal{E}(\boldsymbol{P}^n)$ we conclude that ψ belongs to the closed subspace generated by the spaces $E_k \otimes p^\ell$ ($k-\ell > 0$, even). This proves Theorem 2.5(i).

For Theorem 2.5(ii) suppose $\varphi \in \mathcal{S}(\boldsymbol{P}^n), \overset{\vee}{\varphi} = 0$. Then by (31) $(\overset{\vee}{\varphi})_k = 0$ for each $k \geq 0$. As we saw this means that for each k

$$\varphi_k(w,p) = \sum_{1 \leq m \leq d(k)} S_{km}(w) Q_m(p)$$

where $Q_m(p)$ is a polynomial. This contradicts the rapid decrease of φ and φ_k in the variable p.

Part (iii) will be proved at the end of subsection 5.

3. The Radon Transform and its Dual on the K-types.

We shall now put the Radon transform $f \longrightarrow \widehat{f}$ and it dual $\varphi \longrightarrow \overset{\vee}{\varphi}$ into a more explicit form when specialized to the spaces $\mathcal{D}(\boldsymbol{R}^n)_k$ and $\mathcal{D}(\boldsymbol{P}^n)_k$ in (32) and (33).

With $M = \boldsymbol{O}(n-1)$ as the isotropy subgroup of $K = \boldsymbol{O}(n)$ at the north pole $o = \boldsymbol{e}_n$ let φ_k be the unique M-invariant function in E_k satisfying $\varphi_k(o) = 1$. Then φ_k is the restriction to $\boldsymbol{S}^{n-1}$ of a unique M-invariant harmonic polynomial on $\boldsymbol{R}^n$ (see e.g. [GGA] p. 19) homogeneous of degree k. Since this polynomial is a polynomial in $x_1^2 + \cdots + x_{n-1}^2$ and x_n we see that for $s \in \boldsymbol{S}^{n-1}, \varphi_k(s)$ is a kth degree polynomial in s_n, the socalled *Legendre polynomial* $P_k(n, s_n)$ of degree k in n dimensions. It is a constant multiple of the ultraspherical polynomial $P_k^{(n-2)/2}(s_n)$. For $n \geq 3$ this has the integral representation

$$P_k(n,\ \cos\theta) = \frac{\Omega_{n-2}}{\Omega_{n-1}} \int_0^\pi \big(\cos\theta + i\ \sin\theta\ \cos\varphi\big)^k\ \sin^{n-3}\varphi\ d\varphi.$$

(cf. [GGA], p. 23). For the case $n = 2$ see Exercise **A3**. We note also that since $|\varphi_k(s)| \leq 1$ ([GGA], p. 20) we have

$$|P_k(n,t)| \leq 1 \qquad \text{for } |t| \leq 1. \tag{34}$$

Let $Y_k \in E_k$ be arbitrary. We integrate Y_k over the $(n-2)$-dimensional sphere

$$\big\{s' \in \boldsymbol{S}^{n-1} :\ (s')_n = s_n\big\} = M \cdot s \qquad |s_n| \neq 1.$$

The result is a constant multiple of $\varphi_k(s)$ and the constant is evaluated by letting $s \longrightarrow o$. If $d\omega_{n-2}$ is the volume element on $M \cdot s$ we have

$$(35)\qquad \int\limits_{(s,e_n)=s_n} Y_k(s')dw_{n-2}(s') = \Omega_{n-1}\int\limits_M Y_k(m\cdot s)dm = \Omega_{n-1}Y_k(o)\varphi_k(s).$$

We have now the following result of Funk-Hecke.

Proposition 2.9. *Suppose φ is a continuous function on the interval $-1 \le t \le 1$. Then for each $Y_k \in E_k$, $\xi \in S^{n-1}$,*

$$(36)\qquad \int\limits_{S^{n-1}} \varphi((\xi,\eta))Y_k(\eta)dw(\eta) = c_{k,n}Y_k(\xi),$$

where

$$c_{k,n} = \Omega_{n-1}\int_{-1}^{1} \varphi(t)P_k(n,t)(1-t^2)^{(n-3)/2}dt.$$

Proof. Since the scalar product and the space E_k are rotation-invariant we can take $\xi = 0$. Then we evaluate the integral (36) by decomposing S^{n-1} into spheres with centers on the e_n-axis. Since the intersection of S^{n-1} with the plane $s_n = t$ has radius $(1-t^2)^{\frac{1}{2}}$ and dimension $n-2$ formula (36) follows very easily from (35) (cf. [GGA], Lemma 2.30 p. 127).

Corollary 2.10. *Let $Y_k \in E_k$ and let $\varphi \in \mathcal{S}(\boldsymbol{R})$ satisfy $\varphi(-p) \equiv (-1)^k\varphi(p)$. Let $Y_k\otimes\varphi$ denote the function on $\boldsymbol{P}^n$ defined by $(Y_k\otimes\varphi)(w,p) = Y_k(w)\varphi(p)$. Then*

$$(37)\qquad (Y_k\otimes\varphi)^\vee(rw) = \frac{\Omega_{n-1}}{\Omega_n}Y_k(w)\int_{-1}^{1} P_k(n,t)\varphi(rt)(1-t^2)^{(n-3)/2}dt.$$

In fact we have

$$(Y_k\otimes\varphi)^\vee(rw) = \frac{1}{\Omega_n}\int\limits_{S^{n-1}} \varphi(r(w,\eta))Y_k(\eta)d\eta$$

so (37) follows immediately.

Next we put the Radon transform on $\mathcal{D}(\boldsymbol{R}^n)_k$ in a similarly explicit form, keeping in mind the decomposition in Prop. 2.7.

Proposition 2.11. *Let $f \in \mathcal{S}(\boldsymbol{R}^n)$ have the form*

$$f(rw) = \big(F \otimes Y_k\big)(rw) = F(r)Y_k(w), \qquad r \in \boldsymbol{R}, |w| = 1$$

where the function $F(r)r^{-k}$ is C^∞ and even. Then

$$\big(F \otimes Y_k\big)^\wedge(w,p) = \Omega_{n-1}Y_k(w)\int_{|p|}^{\infty} P_k\left(n, \frac{p}{r}\right) F(r)r^{n-2}\left(1-\frac{p^2}{r^2}\right)^{(n-3)/2} dr.$$

Proof. We shall derive this from Prop. 2.9. From the definition of $\widehat{f}$ we have for $\psi \in \mathcal{S}(\boldsymbol{R})$

$$\int_{\boldsymbol{R}} \widehat{f}(w,p)\psi(p)dp = \int_{\boldsymbol{R}^n} \psi((x,w))f(x)dx$$

$$= \int_0^\infty \int_{\boldsymbol{S}^{n-1}} \psi(r(\eta,w))F(r)Y_k(\eta)r^{n-1}dr\, d\eta,$$

which by Prop. 2.9 equals

$$\Omega_{n-1}Y_k(w)\int_0^\infty F(r)r^{n-1}\left(\int_{-1}^{1} P_k(n,t)\psi(rt)(1-t^2)^{(n-3)/2}dt\right) dr.$$

We interchange the integrations and then split the integral over $(-1,1)$ into two integrals over $(0,1)$. Replacing r by the variable $p = rt$ the double integral above becomes (after ignoring $\Omega_{n-1}Y_k(w)$ and using $P_k(n,-t) = (-1)^k P_k(n,t)$)

$$\int_0^1 P_k(n,t)(1-t^2)^{(n-3)/2}\int_0^\infty F(p/t)p^{n-1}t^{-n}\psi(p)dp\, dt$$

$$+(-1)^k\int_0^1 P_k(n,t)(1-t^2)^{(n-3)/2}\int_0^\infty F(p/t)p^{n-1}t^{-n}\psi(-p)dp\, dt.$$

In the last integral put $q = -p$ and use $F(-q) = (-1)^k F(q)$. Next interchange $\int_0^1 dt$ and $\int_0^\infty dp$ and put $t = p/s$. Then interchange $\int_0^1 dt$ and $\int_{-\infty}^0 dq$ and put $t = |q|/s$. We find both for n odd and n even that the integrals add up to

$$\int_{-\infty}^{\infty} \psi(p)dp \int_{|p|}^{\infty} P_k\left(n, \frac{p}{s}\right) F(s) \left(1 - \frac{p^2}{s^2}\right)^{(n-3)/2} s^{n-2} ds$$

and this proves the proposition.

4. Inversion of the Dual Transform.

To begin with we prove a general result about the Fourier transform $f \longrightarrow \widetilde{f}$ on $\boldsymbol{R}^n$.

Proposition 2.12. *Let F bc a function satisfying*

(i) $F \in L^1(\boldsymbol{R}^n) \cap C^\infty(\boldsymbol{R}^n - \{0\})$;

(ii) *For each n-tuple $\nu = (\nu_1, \ldots, \nu_n)$*

$$|x|^{|\nu|} D^\nu F(x) \text{ is bounded on } \boldsymbol{R}^n.$$

Then

$$\widetilde{F}(u) = O(|u|^{-n}). \tag{38}$$

Proof. Assuming (i) and (ii) we must show the function

$$\widetilde{F}(u) = \int_{\boldsymbol{R}^n} F(x) e^{-i(x,u)} dx$$

satisfies

$$\left|\widetilde{F}(u)\right| \leq C|u|^{-n}, \qquad |u| > 1,$$

where C is a constant. Writing $s = |u|, u' = u/s$ we have

$$\widetilde{F}(u) = \int F(x) e^{-i(sx,u')} dx = s^{-n} \int F(x/s) e^{-i(x,u')} dx = s^{-n} \widetilde{F}_s(u') \tag{39}$$

if $F_s(x) = F(x/s)$ so we must show that the restriction $\widetilde{F}_s|S_1(0)$ has a bound, independent of s. Let $\varphi \in \mathcal{D}(\boldsymbol{R}^n)$ such that $0 \leq \varphi \leq 1$, $\varphi \equiv 1$ on $|x| \leq 1$, $\quad \varphi \equiv 0$ for $|x| \geq 2$. Then we decompose

$$F_s = F' + F'', \qquad \text{where} \qquad F' = F_s\varphi,\ F'' = F_s(1 - \varphi). \tag{40}$$

Then if $\| \ \|_\infty$ denotes the uniform norm and $\| \ \|_1$ the L^1-norm we have

$$\|\widetilde{F}'\|_\infty \leq \|F'\|_1 \leq \int\limits_{|x|\leq 2} |F(x/s)|dx \leq \sup_x |F(x)|c_1,$$

where c_1 is independent of s. Thus by (ii) for $\nu = 0$ the Fourier transform $\widetilde{F}'$ has a bound which is uniform in s.

Next we express $D^\nu F''$ by means of Leibniz' product rule. We obtain

$$\begin{aligned}(D^\nu F'')(x) &= (1-\varphi(x))\frac{\partial^{|\nu|}}{\partial x_1^{\nu_1}\dots\partial x_n^{\nu_n}}(F(\tfrac{x}{s})) + B(x)\\ &= (1-\varphi(x))(D^\nu F)(x/s)\cdot s^{-|\nu|} + B(x)\\ &= \big[(|x|/s)^{|\nu|}(D^\nu F)(x/s)\big](1-\varphi(x))|x|^{-|\nu|} + B(x).\end{aligned}$$

Here $B(x)$ is a linear combination of terms

$$\big[(|x|/s)^{|\mu|}(D^\mu F)(x/s)\big]|x|^{-|\mu|}\{D^w(1-\varphi(x))\},$$

with $w \neq 0$. The expressions inside the brackets are bounded (by (ii)) and the expression inside the braces has support in $1 \leq |x| \leq 2$. Consequently,

$$\big|(D^\nu F'')(x)\big| \leq c_2|x|^{-|\nu|}$$

where c_2 is independent of s. Taking p sufficiently large we obtain a bound on $\|L^pF''\|_1$ (and hence on $|v|^{2p}\widetilde{F}''(v)$) which is uniform in s. Thus by (40) we have the desired bound on $\widetilde{F}_s\big|S_1(0)$ and the proposition is proved.

Theorem 2.13. *Let $\varphi \in \mathcal{S}(\boldsymbol{P}^n)$ and define F on $\boldsymbol{R}^n - (0)$ by*

$$F(sw) = \int\limits_{\boldsymbol{R}} \varphi(w,p)e^{-ips}dp. \tag{41}$$

Then the Fourier transform $f(x) = (2\pi)^{-n}\widetilde{F}(-x)$ is continuous, $f(x) = 0(|x|^{-n})$, and $\widehat{f} = \varphi$.

Proof. Since $\varphi(-w,-p) = \varphi(w,p)$ equation (41) gives a valid definition of F. For the derivatives of F we have by [GGA], I, §2 (11) the following expression

$$|u|^q\frac{\partial^q F}{\partial u_{i_1}\dots\partial u_{i_q}} = \sum_{1\leq i+j\leq q, 1\leq k_1,\dots k_i\leq n-1} a_{j,k_1\dots k_i}(w)s^j\frac{\partial^{i+j}F(sw)}{\partial w_{k_1}\dots\partial w_{k_i}\partial s^j}$$

on a part of the sphere $|w| = 1$ where w_n is bounded away from 0 and $w_1, \dots w_{n-1}$ serve as coordinates. Here $u_i = sw_i$. Since

$$-s^j \frac{\partial^j F(sw)}{\partial s^j} = \int_{\boldsymbol{R}} \frac{\partial^j}{\partial p^j}(\varphi(\omega, p)p^j)e^{-isp}\,dp$$

and since $\varphi \in \mathcal{S}(\boldsymbol{P}^n)$ it follows that F satisfies condition (ii) of Prop. 2.12. Condition (i) is obviously satisfied too so the function $f(x) = (2\pi)^{-n}\widetilde{F}(-x)$ is continuous and $f(x) = O(|x|^{-n})$.

Consider now any function G in the subspace $\mathcal{S}_o(\boldsymbol{R}^n)$ (see No. 1) and the function $g \in \mathcal{S}^*(\boldsymbol{R}^n)$ such that $G = \widetilde{g}$. Then

$$G(sw) = \int_{\boldsymbol{R}} \widehat{g}(\omega, p)e^{-isp}\,dp \tag{42}$$

so by (41)

$$F(sw)G(sw) = \int_{\boldsymbol{R}} \left(\int_{\boldsymbol{R}} \varphi(w, p-q)\widehat{g}(w, q)\,dq \right) e^{-isp}\,dp. \tag{43}$$

Next we claim that

$$(FG)^{\sim} = (2\pi)^{-n}\widetilde{F} * \widetilde{G}. \tag{44}$$

If F were in $\mathcal{S}(\boldsymbol{R}^n)$ this would be trivial by the inversion formula. Approximation of $\widetilde{F}$ in the L^2-norm by a sequence of functions in $\mathcal{S}(\boldsymbol{R}^n)$ gives (44) very easily.

Now since F satisfies (ii) in Proposition 2.12 and since all derivatives $(D^\nu G)(0)$ vanish we see from considering the Taylor series with remainder for G that all derivatives of $(FG)(u)$ tend to 0 as $u \longrightarrow 0$. Using (ii) in Prop. 2.12 for large argument we conclude that $FG \in \mathcal{S}_o(\boldsymbol{R})$ so by (44), $\widetilde{F} * \widetilde{G} \in \mathcal{S}^*(\boldsymbol{R}^n)$.

Using the inversion formula on (44) we get

$$(\widetilde{F} * \widetilde{G})^{\sim}(u) = (2\pi)^{2n}(FG)(-u) \tag{45}$$

so

$$(2\pi)^{2n}F(sw)G(sw) = \int_{\boldsymbol{R}} (\widetilde{F} * \widetilde{G})^{\wedge}(w, p)e^{isp}\,dp. \tag{46}$$

Hence by (43)

$$(\widetilde{F} * \widetilde{G})^{\wedge}(w,p) = (2\pi)^{2n} \int_{\boldsymbol{R}} \varphi(w, -p-q)\widehat{g}(w,q)dq. \tag{47}$$

We would now like to have in analogy with (7)

$$(\widetilde{F} * \widetilde{G})^{\wedge}(w,p) = \int_{\boldsymbol{R}} (\widetilde{F})^{\wedge}(w,p-q)(\widetilde{G})^{\wedge}(w,q)dq \tag{48}$$

and the proof of (7) is indeed still valid since $\widetilde{F} * \widetilde{G}$, $\widetilde{G} \in \mathcal{S}(\boldsymbol{R}^n)$ and $\widetilde{F}(x) = 0(|x|^{-n})$. Now $\widetilde{G}(x) = (2\pi)^n g(-x)$ and

$$(\widetilde{G})^{\wedge}(w,p) = (2\pi)^n \widehat{g}(w,-p), \qquad (\widetilde{F})^{\wedge}(w,p) = (2\pi)^n \widehat{f}(w,-p).$$

Hence from (47)–(48)

$$\int_{\boldsymbol{R}} \left(\widehat{f}(w,p-q) - \varphi(w,p-q)\right) \widehat{g}(w,q)dq = 0 \tag{49}$$

for all $p \in \boldsymbol{R}, w \in \boldsymbol{S}^{n-1}, g \in \mathcal{S}^*(\boldsymbol{R}^n)$. Now the map $g \longrightarrow \widehat{g}$ maps $\mathcal{S}^*(\boldsymbol{R}^n)$ onto the set $\mathcal{S}^*(\boldsymbol{P}^n)$ of functions $\psi \in \mathcal{S}(\boldsymbol{P}^n)$ satisfying $\int \psi(w,p)p^k dk = 0$ for all $k \in \boldsymbol{Z}^+$. (Theorem 2.2). In particular, using this for functions ψ independent of w we can say that $\mathcal{S}^*(\boldsymbol{P}^n)$ contains the space $\mathcal{S}_e^*(\boldsymbol{R})$ of even functions in $\mathcal{S}^*(\boldsymbol{R})$. Since $\mathcal{S}_o(\boldsymbol{R})$ is dense in $L^2(\boldsymbol{R})$ we see, taking Fourier transforms, that $\mathcal{S}^*(\boldsymbol{R})$ is dense in $L^2(\boldsymbol{R})$ so $\mathcal{S}_e^*(\boldsymbol{R})$ is dense in the space of even functions in $L^2(\boldsymbol{R})$. Also since f is continuous and $f(x) = 0(|x|^{-n})$ we can state that $p \longrightarrow \widehat{f}(w,p)$ is continuous and $0(p^{-1})$ (hence in $L^2(\boldsymbol{R})$). In view of these facts, (49) implies

$$\widehat{f}(w,p-q) - \varphi(w,p-q) + \widehat{f}(w,p+q) - \varphi(w,p+q) = 0$$

for all p, q and w. Hence $\widehat{f} = \varphi$ and the theorem is proved.

Corollary 2.14. *Let $\varphi \in \mathcal{S}(\boldsymbol{P}^n)$. Then the function f in Theorem* 2.13 *can be written*

$$i^{n-1}Cf = (\Lambda\varphi)^{\vee} \tag{50}$$

with C as in (11), and consequently

$$i^{n-1}C\varphi = ((\Lambda\varphi)^{\vee})^{\wedge}, \qquad \varphi \in \mathcal{S}(\boldsymbol{P}^n).$$

For n odd this formula can be written

$$C\varphi = \left((-L)^{\frac{n-1}{2}} \overset{\vee}{\varphi}\right)^{\wedge}.$$

In fact, expressing $f(x) = (2\pi)^{-n}\widetilde{F}(-x)$ in polar coordinates we obtain

$$f(x) = (2\pi)^{-n} \int\limits_{S^{n-1}} dw \int\limits_0^\infty \left(\int\limits_{-\infty}^{\infty} \varphi(w,p)e^{-isp}dp \right) e^{is(x,w)} s^{n-1} ds.$$

We write this formula as

$$f(x) = (2\pi)^{-n} \int\limits_{S^{n-1}} F(x,w)dw$$

$$= (2\pi)^{-n} \int\limits_{S^{n-1}} \tfrac{1}{2}\big(F(x,w) + F(x,-w)\big)dw,$$

so

$$f(x) = \tfrac{1}{2}(2\pi)^{-n} \int\limits_{S^{n-1}} dw \int\limits_{R} |s|^{n-1} e^{is(x,w)} ds \int\limits_{R} \varphi(w,p)e^{-isp}dp$$

which is also

$$f(x) = \tfrac{1}{2}(-i)^{n-1}(2\pi)^{-n} \int\limits_{S^{n-1}} dw \int\limits_{R} \operatorname{sgn} s \; e^{is(x,w)}ds \int\limits_{R} \varphi^{(n-1)}(w,p)e^{-isp}dp.$$

Now we have for each $F \in \mathcal{S}(\boldsymbol{R})$ the identity

$$(*) \qquad \int\limits_{R} \operatorname{sgn} s \; e^{ist} \left(\int\limits_{R} F(p)e^{-isp}dp \right) ds = 2\pi(\mathcal{H}F)(t),$$

where $\mathcal{H}$ is the Hilbert transform. In fact, if we apply both sides to $\widetilde{\psi}$ with $\psi \in \mathcal{S}(\boldsymbol{R})$, the left hand side is by (7) §1,

$$\int\limits_{R} \left(\int\limits_{R} \operatorname{sgn} s \; e^{ist}\widetilde{F}(s)ds \right) \widetilde{\psi}(t)dt$$

$$= \int\limits_{R} \operatorname{sgn} s \; \widetilde{F}(s) \; 2\pi \; \psi(s)ds = 2\pi\widetilde{\mathcal{H}F}(\psi).$$

Since the right hand side of (*) yields the same result the identity is proved. Using (*) for $F(p) = \varphi^{(n-1)}(w,p)$ we obtain

$$f(x) = \pi(-i)^{n-1}(2\pi)^{-n} \int_{S^{n-1}} (\Lambda\varphi)(w,(w,x))dw$$

which proves (50).

5. The Range Characterization for Distributions and Consequences.

Next we prove Theorem 2.6 characterizing the range $\mathcal{E}'(\boldsymbol{R}^n)^\wedge$ of the space of compactly supported distributions on $\boldsymbol{R}^n$ under the Radon transform. By the definition

$$\widehat{S}(\varphi) = S(\check{\varphi}) \tag{51}$$

it is trivial that each $\Sigma \in \mathcal{E}'(\boldsymbol{R}^n)^\wedge$ vanishes identically on $\mathcal{N}$. Moreover, the distribution $\widehat{S}$ has compact support: if S has support inside the ball $B_R(0)$ then if φ in (51) satisfies $\varphi(w,p) = 0$ for $|p| \leq R$ we have $\check{\varphi}(x) = 0$ for $|x| \leq R$ so $\widehat{S}(\varphi) = S(\check{\varphi}) = 0$.

Conversely, suppose Σ is a compactly supported distribution on $\boldsymbol{P}^n$ vanishing identically on $\mathcal{N}$. We write the inversion formula (9) §1 in the form

$$cf = (\Lambda\widehat{f})^\vee. \tag{52}$$

Because of (12) this implies

$$c\int_{\boldsymbol{R}^n} f(x)g(x)dx = \int_{\boldsymbol{P}^n} (\Lambda\widehat{f})(\xi)\widehat{g}(\xi)d\xi, \quad f,g \in \mathcal{D}(\boldsymbol{R}^n). \tag{53}$$

This suggests defining a functional S on $\mathcal{D}(\boldsymbol{R}^n)$ by

$$cS(f) = \Sigma(\Lambda\widehat{f}), \tag{54}$$

with the hope of proving $\widehat{S} = \Sigma$. The definition of Λ (§1, (8)) shows quickly that S is a distribution on $\boldsymbol{R}^n$. Our first task is now to prove that it has compact support.

Let G denote the group $\boldsymbol{M}(n)$ of isometries of $\boldsymbol{R}^n$ and dg a Haar measure on G. Since G is the semidirect product of $\boldsymbol{R}^n$ and $K = \boldsymbol{O}(n)$ the measure dg is bi-invariant. For $g \in G$ we write as usual,

$$f^g(x) = f(g^{-1}\cdot x), \qquad \varphi^g(\xi) = \varphi(g^{-1}\cdot\xi)$$

$$S^g(f) = S(f^{g^{-1}}), \quad \Sigma^g(\varphi) = \Sigma(\varphi^{g^{-1}}).$$

The operator $\Lambda : \mathcal{D}(\boldsymbol{P}^n) \longrightarrow \mathcal{E}(\boldsymbol{P}^n)$ is *invariant* under each $g \in G$, i.e. $(\Lambda\varphi)^g = \Lambda(\varphi^g)$; this is obvious if g is a rotation around 0; if g is a translation T we have

$$(\varphi \circ T)(w,p) = \varphi(w, p + q(w,T)),$$

where $q(w,T)$ is independent of p. Since Λ commutes with translations in the p-variable, the stated invariance of Λ under T, and hence under all $g \in G$, follows. Hence (54) implies

$$cS^g(f) = \Sigma^g(\Lambda\widehat{f}), \qquad f \in \mathcal{D}(\boldsymbol{R}^n),$$

and if $F \in \mathcal{D}(G)$,

$$c\int_G F(g)S^{g^{-1}}(f)dg = \int_G F(g)\Sigma^{g^{-1}}(\Lambda\widehat{f})dg. \tag{55}$$

Consider the distribution σ on $\boldsymbol{P}^n$ given by

$$\sigma(\varphi) = \int_G F(g)\Sigma^{g^{-1}}(\varphi)dg = \int_G F(g)\left(\int_{\boldsymbol{P}^n} \varphi(g^{-1}\cdot\xi)\, d\Sigma(\xi)\right) dg.$$

Let H be the subgroup of G leaving invariant the hyperplane $\xi_o : x_n = 0$ in $\boldsymbol{R}^n$. We normalize the Haar measure dh on this unimodular group $H = \boldsymbol{Z}_2 \times \boldsymbol{M}(n-1)$ such that

$$\int_G F(g)dg = \int_{G/H}\left(\int_H F(gh)dh\right) dg_H, \tag{56}$$

where dg_H is the invariant measure $d\xi$ on $G/H = \boldsymbol{P}^n$. The mapping $F \longrightarrow \dot{F}$ given by

$$\dot{F}(gH) = \int_H F(gh)dh$$

is a continuous surjective map of $\mathcal{D}(G)$ onto $\mathcal{D}(G/H)$; the dual map $\Psi \longrightarrow \widetilde{\Psi}$ given by

$$\widetilde{\Psi}(F) = \Psi(\dot{F}), \qquad F \in \mathcal{D}(G),$$

is therefore an injective map of $\mathcal{D}'(\boldsymbol{P}^n)$ into $\mathcal{D}'(G)$. With $\varphi \in \mathcal{D}(\boldsymbol{P}^n)$ select $\Phi \in \mathcal{D}(G)$ such that $\dot{\Phi} = \varphi$. Then $(\Phi^{L(g)})^\bullet = \varphi^g$ so

$$\int_{\boldsymbol{P}^n} \varphi(g^{-1}\cdot\xi)d\Sigma(\xi) = \Sigma(\varphi^g) = \widetilde{\Sigma}(\Phi^{L(g)}),$$

whence

$$\sigma(\varphi) = \int_G F(g)\widetilde{\Sigma}(\Phi^{L(g)})dg.$$

By the Fubini theorem for distributions (Schwartz [1966], Ch. IV, §3) this is

$$\sigma(\varphi) = \int_G \Big(\int_G F(g)\Phi(g^{-1}h)dg\Big)d\widetilde{\Sigma}(h) = (F' * \widetilde{\Sigma})(\Phi),$$

where $F'(g) = F(g^{-1})$ and $*$ denotes the convolution

$$(F * T)(g) = \int_G F(gh^{-1})dT(h),$$

for $F \in \mathcal{D}(G)$, $T \in \mathcal{D}'(G)$. Because of the unimodularity of $H, \widetilde{\Sigma}$ is right invariant under H and so is $F' * \widetilde{\Sigma}$. Consequently,

$$\sigma(\varphi) = \int_{G/H} (F' * \widetilde{\Sigma})(g)\varphi(gH)dg_H.$$

In other words, the distribution σ on $\boldsymbol{P}^n$ is identified with the smooth function $gH \longrightarrow (F' * \widetilde{\Sigma})(g)$ which belongs to $\mathcal{D}(\boldsymbol{P}^n)$ since the definition of σ shows immediately that it has compact support.

Since Σ vanishes on $\mathcal{N}$ and since $\mathcal{N}$ is G-invariant it follows that σ vanishes on $\mathcal{N}$. Because of (28) and Theorem 2.4, 2.5 (i) we conclude that $\sigma = \widehat{s}$ for some $s \in \mathcal{D}(\boldsymbol{R}^n)$. Consequently,

$$c\int_{\boldsymbol{R}^n} s(x)f(x)dx = \int_{\boldsymbol{R}^n} s(x)(\Lambda\widehat{f})^{\vee}(x) = \sigma(\Lambda\widehat{f})$$

so by (55)

$$\int_{\boldsymbol{R}^n} s(x)f(x)dx = \int_G F(g)S^{g^{-1}}(f)dg. \tag{57}$$

We denote by $\widetilde{S}$ the "lift" of S to G defined by

$$\widetilde{S}(F) = S(\dot{F}),$$

where

$$\dot{F}(gK) = \int_K F(gk)dk.$$

The proof above of the relation $\sigma(\varphi) = F' * \widetilde{\Sigma}(\Phi)$ can be used to reduce the right hand side of (57) to

$$\int_G (F' * \widetilde{S})(gK) f(gK)dg$$

so

$$s(gK) = (F' * \widetilde{S})(g). \tag{58}$$

This shows that $\mathcal{D}(G) * \widetilde{S} \subset \mathcal{D}(G)$. However, by combining the definition of σ with the last remark of Theorem 2.4 we can deduce a more precise statement: For each compact set $C \subset G$ there exists another compact set $C' \subset G$ such that $C_1 \subset C_2 \Longrightarrow C_1' \subset C_2'$ and

$$\mathcal{D}_C(G) * \widetilde{S} \subset \mathcal{D}_{C'}(G). \tag{59}$$

In particular, suppose C is a compact neighborhood of e and let (C_j) be a sequence of neighborhoods of e, all contained in C, converging to e. Let $f_j \in \mathcal{D}_{C_j}(G)$, $f_j \geq 0$, $\int f_j(g)dg = 1$. If $\varphi \in \mathcal{D}(G)$ then

$$(f_j * \widetilde{S})(\varphi) = \widetilde{S}(f_j' * \varphi) \longrightarrow \widetilde{S}(\varphi) \tag{60}$$

as $j \longrightarrow \infty$. If $\operatorname{supp}(\varphi) \cap C' = \emptyset$ then by (59) the left hand side of (60) is 0 so $\widetilde{S}(\varphi) = 0$. In particular, $\widetilde{S}$ has compact support and so does S.

The equation $\widehat{s} = \sigma$ implies $s(\check{\varphi}) = \sigma(\varphi)$. Since S has compact support, (57) remains valid for $f \in \mathcal{E}(\boldsymbol{R}^n)$. Thus we obtain from the above

$$\int_G F(g) S^{g^{-1}}(\check{\varphi})dg = (F' * \widetilde{\Sigma})(\Phi). \tag{61}$$

Letting again $F_j \longrightarrow \delta$ we deduce

$$S(\check{\varphi}) = \widetilde{\Sigma}(\Phi) = \Sigma(\varphi)$$

so $\widehat{S} = \Sigma$ as desired. This proves Theorem 2.6 (i).

It is now easy to prove Theorem 2.6 (ii). Because of Theorem 2.5 (i) we have for $T \in \mathcal{E}'(\boldsymbol{P}^n)$:

$$T \in \mathcal{E}'(\boldsymbol{R}^n)^\wedge \Longleftrightarrow T(E_k \otimes p^\ell) = 0 \text{ for } k - \ell > 0, \text{ even }. \tag{62}$$

For $\ell \geq 0$ let T^ℓ denote the distribution

$$\psi \in \mathcal{E}(\boldsymbol{S}^{n-1}) \longrightarrow T(\tfrac{1}{2}(\psi + (-1)^\ell \psi') \otimes p^\ell),$$

where $\psi'(w) = \psi(-w)$. Then (62) shows that

$$T \in \mathcal{E}'(\boldsymbol{R}^n)^\wedge \Longleftrightarrow T^\ell(E_k) = 0 \text{ for } k - \ell > 0, \text{ even }. \tag{63}$$

On the other hand, the definition of T^ℓ shows that $T^\ell(E_k) = 0$ if $k - \ell$ is odd. Thus (63) shows that $T \in \mathcal{E}'(\boldsymbol{R}^n)^\wedge$ if and only if for each $\ell > 0$, T^ℓ has its spherical harmonics expansion terminate at the ℓth degree terms and this is just the statement of Theorem 2.6 (ii).

It remains to prove Theorem 2.5 (iii), namely that $\mathcal{E}(\boldsymbol{P}^n)^\vee = \mathcal{E}(\boldsymbol{R}^n)$. If we invoke the general Theorem 2.16 below this becomes an immediate consequence of Theorem 2.6. In fact, we let α there play the role of the dual transform $\varphi \longrightarrow \overset{\vee}{\varphi}$ of $\mathcal{E}(\boldsymbol{P}^n)$ into $\mathcal{E}(\boldsymbol{R}^n)$. The transpose ${}^t\alpha$ is then the Radon transform $T \longrightarrow \widehat{T}$ of $\mathcal{E}'(\boldsymbol{R}^n)$ into $\mathcal{E}'(\boldsymbol{P}^n)$ which is injective (cf. [GGA], p. 120) and has closed range by Theorem 2.6 (i). Here "closed" refers to the strong topology (cf. Prop. 2.18 below).

6. Some Facts about Topological Vector Spaces.

It will be convenient now as well as later to collect here some well known facts from the theory of topological vector spaces.

Let F and G be two vector spaces over $\boldsymbol{C}$, $\langle\ ,\ \rangle$ a non-singular bilinear form on $F \times G$. Then $\sigma(F, G)$ denotes the weakest topology on F for which all the maps $x \longrightarrow \langle x, y\rangle$ from F to $\boldsymbol{C}$ are continuous, y running through G.

If E is a locally convex (Hausdorff) topological vector space and E' its dual the *evaluation mapping* $(x, \lambda) \longrightarrow \lambda(x)$ $(x \in E, \lambda \in E')$ is a nonsingular bilinear form on $E \times E'$. The topology $\sigma(E, E')$ is often called the *weak topology* of E and the topology $\sigma(E', E)$ the *weak* topology* of E'. A subset $B \subset E$ is said to be *bounded* if for each neighborhood N of the origin in E there exists an $\alpha > 0$ such that $\alpha B \subset N$. The *strong topology* on E' is by definition the topology of uniform convergence on bounded subsets of E.

Let E be a locally convex topological vector space with dual E'. A locally convex topology τ on E is said to be *compatible with this duality* if the dual $(E_\tau)'$ of E with the topology τ equals E'. The following elementary results then hold (cf. Bourbaki [1955] Ch. IV, §2).

Proposition 2.15.

(i) *The topology $\sigma(E, E')$ is compatible with the duality.*

(ii) *The closed subspaces of E are the same for all topologies compatible with the duality.*

If F and G are topological vector spaces and $\alpha : F \longrightarrow G$ a continuous linear mapping, the *transpose map* ${}^t\alpha : G' \longrightarrow F'$ is given by ${}^t\alpha(\lambda)(f) = \lambda(\alpha f)$ for $\lambda \in G'$, $f \in F$. For Fréchet spaces we have the following extension of Banach's open mapping theorem (see e.g. Trèves [1966], Ch. I).

Theorem 2.16. *Let E and F be two Fréchet spaces and $\alpha : E \longrightarrow F$ a continuous linear mapping. Then*

(i) If α is surjective it is an open mapping.

(ii) $\alpha(E)$ is closed in $F \Longleftrightarrow {}^t\alpha(F')$ is weak-closed in E'.*

Furthermore,

(iii) $\alpha(E)$ is dense in $F \Longleftrightarrow {}^t\alpha : F' \longrightarrow E'$ is injective.

To check that the image under a linear map is closed the following criterion is useful.

Theorem 2.17. *Let E be a Fréchet space. Then*

(i) *A convex set $A' \subset E'$ is weak* closed if for each subset $B' \subset E'$ which is weak* bounded and closed, the set $A' \cap B'$ is weak* closed.*

(ii) *A set $B' \subset E'$ is weak* bounded if and only if it is strongly bounded.*

For a proof see e.g. Bourbaki [1955], Ch. III. Part (i) is usually stated for B' as the polar

$$\{x' \in E' : \sup_{x \in U} |x'(x)| \leq 1\} \tag{64}$$

of a neighborhood U of 0 in E. Since such sets (64) are weak* bounded, Theorem 2.17 is a special case.

If M is a manifold with a countable base the space $\mathcal{E}(M)$ is a Fréchet space. Taking $\mathcal{E}'(M)$ with the strong topology its dual coincides with $\mathcal{E}(M)$, i.e. $\mathcal{E}(M)$ is semireflexive (Schwartz [1966], Ch. III), that is $\mathcal{E}(M)$ fills up the second dual space $(\mathcal{E}'(M))' = \mathcal{E}''(M)$.

Proposition 2.18.

(i) *The closed subspaces of $\mathcal{E}'(M)$ are the same for both the strong topology of $\mathcal{E}'(M)$ and the topology $\sigma(\mathcal{E}', \mathcal{E}) = \sigma(\mathcal{E}', \mathcal{E}'')$.*

(ii) *If $B' \subset \mathcal{E}'(M)$ is weak* bounded there exists a compact subset $C \subset M$ such that*

$$B' \subset \mathcal{E}'_C(M),$$

the set of $T \in \mathcal{E}'(M)$ *with* $\text{supp}(T) \subset C$.

Part (i) is clear from Prop. 2.15 because $\mathcal{E}'' = \mathcal{E}$ is the dual of $\mathcal{E}'(M)$ both for the strong topology on $\mathcal{E}'$ and for $\sigma(\mathcal{E}', \mathcal{E}'') = \sigma(\mathcal{E}', \mathcal{E})$. Because of Theorem 2.17, B' is strongly bounded so by Schwartz [1966] p. 90, $B' \subset \mathcal{E}'_C(M)$ for a suitable compact set C.

The space $\mathcal{D}'(M)$ is similarly given the strong topology. Again $\mathcal{D}'(M)$ is semireflexive (Schwartz, *loc. cit.*) so the closed subspaces are the same for the weak and strong topologies.

§3. HOMOGENEOUS SPACES IN DUALITY.

1. The Radon Transform for a Double Fibration.

In §1, No. 2 we have outlined the principal integral geometric problems suggested by the theory of the Radon transform in $\boldsymbol{R}^n$, treated in §2. A natural framework for such problems is provided by group theory. This is motivated by the fact that for the example $X = \boldsymbol{R}^n$, $\Xi = \boldsymbol{P}^n$, the isometry group $\boldsymbol{M}(n)$ acts transitively on both X and on Ξ. In fact,

$$\boldsymbol{R}^n = \boldsymbol{M}(n)/\boldsymbol{O}(n), \quad \boldsymbol{P}^n = \boldsymbol{M}(n)/\boldsymbol{M}(n-1) \times \boldsymbol{Z}_2, \tag{1}$$

where $\boldsymbol{O}(n)$ is the orthogonal group in $\boldsymbol{R}^n$ and the group $\boldsymbol{M}(n-1)\boldsymbol{Z}_2$ is the subgroup of $\boldsymbol{M}(n)$ leaving a certain hyperplane ξ_o through the origin invariant.

We note that a point $x = g\boldsymbol{O}(n)$ lies on a hyperplane $\xi = \gamma \cdot \boldsymbol{M}(n-1)\boldsymbol{Z}_2$ if and only if these cosets considered as subsets of $\boldsymbol{M}(n)$ are not disjoint. In fact $x \in \xi$ if and only if $g \cdot 0 \in \gamma h \cdot 0$ for some $h \in \boldsymbol{M}(n-1)\boldsymbol{Z}_2$ which in turn amounts to $gk = \gamma h$ for some $k \in \boldsymbol{O}(n)$ as stated.

Let G be a locally compact group, X and Ξ two left coset spaces

$$X = G/K, \qquad \Xi = G/H, \tag{2}$$

where K and H are two closed subgroups. We put $L = K \cap H$.

The remarks in the example above characterizing a point lying on a hyperplane suggest the following definition.

Definition. Two points $x \in X, \xi \in \Xi$ are said to be *incident* if as cosets in G they are not disjoint. We put

$$\check{x} = \{\xi \in \Xi \ : \ x \text{ and } \xi \text{ incident }\}$$
$$\widehat{\xi} = \{x \in X \ : \ x \text{ and } \xi \text{ incident }\}.$$

The spaces X and Ξ in (2) will be called *homogeneous spaces in duality.*

Proposition 3.1. *We have the identification*

$$G/L = \{(x, \xi) \in X \times \Xi : \ x \text{ and } \xi \text{ incident } \}$$

via the bijection $\tau : gL \longrightarrow (gK, gH)$.

In fact, the map is well-defined and injective and the surjectivity is clear, because if $gK \cap \gamma H \neq \emptyset$ then $gk = \gamma h$ for suitable k and h whence $(gK, \gamma H) = \tau(gkL)$.

Note that, writing $x_o = \{K\}$, $\xi_o = \{H\}$, we have

$$\check{x}_o = \{kH : \ k \in K\} = K/L; \quad \widehat{\xi}_o = \{hK : \ h \in H\} = H/L.$$

More generally, if $x = gK$, $\xi = \gamma H$, then

$$(3) \qquad \check{x} = \{gkH : \ k \in K\} = K^g/L^g; \quad \widehat{\xi} = \{\gamma hK : \ h \in H\} = H^\gamma/L^\gamma$$

the superscript denoting conjugation (cf. [GGA], Ch. I, §3).

We shall usually make the following assumptions (i) and (ii) about X and Ξ.

(i) The groups G, K, H, and L have bi-invariant Haar measures dg, dk, dh, and $d\ell$, respectively.

(ii) The subset $KH \subset G$ is closed.

Note that (ii) is automatic if K is compact.

The Haar measures in (i) give rise to invariant measures $dx = dg_K$, $d\xi = dg_H$, $dm = dh_L$ and $d\mu = dk_L$ on the homogeneous spaces G/K, G/H, H/L, and K/L, respectively.

Remark. Note that definitions (5) and (6) below only require the existence of the invariant measures dk_L and dh_L which (cf. [GGA], I, §1) amounts to

$$|\det Ad_H(\ell)| = |\det Ad_L(\ell)| = |\det Ad_K(\ell)| \qquad \ell \in L.$$

This is for example guaranteed if L is compact.

We now define the Radon transform $f \longrightarrow \widehat{f}$ and its dual $\varphi \longrightarrow \check{\varphi}$ for the double fibration

$$(4) \qquad \begin{array}{ccc} & G/L & \\ p \swarrow & & \searrow \pi \\ X = G/K & & \Xi = G/H, \end{array}$$

where $p(gL) = gK$, $\pi(\gamma L) = \gamma H$, by associating with each $f \in C_c(X)$ its integral over each $\widehat{\xi}$ and to each $\varphi \in C_c(\Xi)$ its integral over each $\check{x}$. More specifically, in view of (3) we define

$$\widehat{f}(\gamma H) = \int_{H/L} f(\gamma hK)dh_L = \int_{\widehat{\xi}_o} f(\gamma \cdot x)dm(x) \qquad f \in C_c(X), \tag{5}$$

$$\check{\varphi}(gK) = \int_{K/L} \varphi(gkH)dk_L = \int_{\check{x}_o} \varphi(g \cdot \xi)d\mu(\xi) \qquad \varphi \in C_c(\Xi). \tag{6}$$

For the convergence we note that for the natural mapping $\Pi : G \longrightarrow G/H$ we have

$$\Pi^{-1}(\Xi - \{\check{x}_0\}) = \{g \in G : gH \notin K \cdot H\} = G - KH.$$

In particular, $\Pi(G - KH) = \Xi - \{\check{x}_o\}$ so by (ii) and the fact that Π is an open mapping, $\check{x}_o$ and hence *each* $\check{x} \subset \Xi$ *is a closed subset.* Thus (6) is an integral of φ over a compact subset of Ξ, hence convergent. We also note that (3) can be written

$$\check{x} = \pi(p^{-1}(x)), \qquad \widehat{\xi} = p(\pi^{-1}(\xi)) \tag{7}$$

in terms of the natural maps π and p.

The maps $f \longrightarrow \widehat{f}$, $\varphi \longrightarrow \check{\varphi}$ are not only geometrically dual, they are also adjoint operations in the sense that

$$\int_X f(x)\check{\varphi}(x)\ dx = \int_\Xi \widehat{f}(\xi)\varphi(\xi)\ d\xi. \tag{8}$$

This can be seen by integrating $(f \circ p)(\varphi \circ \pi)$ over G/L according to the two fibrations in (4) (cf. [GGA], Ch. I, §3).

Lemma 3.2. *Consider the subgroups*

$$K_H = \{k \in K : kH \cup k^{-1}H \subset HK\}, \quad H_K = \{h \in H : hK \cup h^{-1}K \subset KH\}.$$

The following properties are equivalent:

(a) *Transversality:*

$$K \cap H = K_H = H_K.$$

(b) *The maps* $x \longrightarrow \check{x}$ $(x \in X)$, $\xi \longrightarrow \widehat{\xi}$ $(\xi \in \Xi)$ *are injective.*

(c) *The functions in* $C_c(X)^\wedge$ *separate points on* Ξ *and the functions in* $C_c(\Xi)^\vee$ *separate points on* X.

Proof. Assume (a) and suppose $x_1 = g_1K$, $x_2 = g_2K$ and $\check{x}_1 = \check{x}_2$. Then $g_1 \cdot \check{x}_o = g_2 \cdot \check{x}_o$ so putting $g = g_1^{-1}g_2$ we have $g \cdot \check{x}_o = \check{x}_o$. In particular, $g \cdot \xi_o \in \check{x}_o$ so $g \cdot \xi_o = k \cdot \xi_o$ for some $k \in K$. Thus $k^{-1}g = h \in H$ whence $h \cdot \check{x}_o = \check{x}_o$, that is, $hK \cdot \xi_o = K \cdot \xi_o$. This implies $hK \subset KH$ and similarly $h^{-1}K \subset KH$ so by (a), $h \in K$ which gives $x_1 = x_2$.

On the other hand, suppose the map $x \longrightarrow \check{x}$ is one-to-one and suppose $h \in H$ satisfies $h^{-1}K \cup hK \subset KH$. Then $hK \cdot \xi_o \subset K \cdot \xi_o$ so $h \cdot \check{x}_o \subset \check{x}_o$. Similarly $(h^{-1} \cdot x_o)^\vee \subset \check{x}_o$ so $(h \cdot x_o)^\vee = \check{x}_o$ whence $h \cdot x_o = x_o$ and $h \in K$. Together with the same reasoning for Ξ this verifies (a). Next we observe that the right hand side of (5) is the integral of f over $\gamma \cdot \widehat{\xi}_o = (\gamma \cdot \xi_o)^\wedge$. If (b) holds and if $\xi_1 \neq \xi_2$ then $\widehat{\xi}_1$ and $\widehat{\xi}_2$ are distinct closed subsets of X so for a suitable $f \in C_c(X)$, $\widehat{f}(\xi_1) \neq \widehat{f}(\xi_2)$ so (c) holds. If (b) fails we have $\xi_1 \neq \xi_2$ with $\widehat{\xi}_1 = \widehat{\xi}_2$ so $\widehat{f}(\xi_1) = \widehat{f}(\xi_2)$ for all $f \in C_c(X)$ so (c) fails. This proves the lemma.

Remark. This lemma shows that under the transversality condition (a) on the groups K and H the elements ξ can be viewed as closed subsets $\widehat{\xi}$ of X and the elements x as closed subsets $\check{x}$ of Ξ.

Lemma 3.3. *Assume the transversality* (a). *Then there exists a nonzero measure on each* $\check{x}$, *coinciding with* $d\mu = dk_L$ *on* $\check{x}_o = K/L$ *such that whenever* $g \cdot \check{x}_1 = \check{x}_2$, *the measures on* $\check{x}_1$ *and* $\check{x}_2$ *correspond under* g. *A similar statement holds for each* $\widehat{\xi}$.

Proof. If $\check{x} = g \cdot \check{x}_o$ we transfer $d\mu = dk_L$ over on $\check{x}$ by the map $\xi \longrightarrow g \cdot \xi$. If $g \cdot \check{x}_o = g_1 \cdot \check{x}_o$ then $(g \cdot x_o)^\vee = (g_1 \cdot x_o)^\vee$ so by Lemma 3.2, $g \cdot x_o = g_1 \cdot x_o$ so $g = g_1 k$ $(k \in K)$. Since $d\mu$ is K-invariant the lemma follows.

The measures hereby defined on each $\check{x}$ and each $\check{\xi}$ will be denoted by $d\mu$ and dm, respectively. Assuming property (a) and taking the Remark above into account, the transforms (5) and (6) can be written

$$\widehat{f}(\xi) = \int_{\widehat{\xi}} f(x)\,dm(x), \qquad f \in C_c(X) \tag{9}$$

$$\check{\varphi}(x) = \int_{\check{x}} \varphi(\xi)\, d\mu(\xi), \qquad \varphi \in C_c(\Xi). \tag{10}$$

This brings our abstract transform into the context of the general problems discussed at the end of §1.

However, the above abstract setup does not just bring increased generality. It brings a new and valuable viewpoint to some simple concrete problems. For example, in No. 3 we shall investigate the X-ray transform $f \longrightarrow \widehat{f}$ on the hyperbolic plane where

$$\widehat{f}(\xi) = \int_{\xi} f(x)\, dm(x), \qquad \xi \text{ geodesic}.$$

For the inversion it becomes useful not just to use the inclusion $x \in \xi$ as the incidence condition but more generally to choose the subgroups K and H such that x incident to ξ means $d(x, \xi) = p$, d denoting distance and $p \in \boldsymbol{R}^+$ being fixed.

Similarly in No. 2 where both X and Ξ are the space of lines in 3-space the naive incidence definition: "x is incident to ξ iff $x \cap \xi \neq \emptyset$" leads to serious convergence difficulties in the corresponding Radon transform (Helgason, [1965a], Theorem 8.2). These convergence difficulties are resolved by replacing this naive incidence by the group-theoretic incidence above which here becomes: "x is incident to ξ iff x and ξ intersect under a right angle". This leads to the fully satisfactory Theorem 3.10 of Gonzalez, proved later.

We now state two further group-theoretic interpretations of the incidences (cf. Exercise B1). The property

$$HK \cap KH = K \cup H \tag{d}$$

is equivalent to the property: for any $x_1 \neq x_2$ in X there is at most one $\xi \in \Xi$ incident to both. By symmetry, this is then equivalent to the property: given $\xi_1 \neq \xi_2$ in Ξ there is at most one $x \in X$ incident to both.

The property

$$KHK = G \tag{e}$$

is equivalent to the property: any two $x_1, x_2 \in X$ are incident to some $\xi \in \Xi$. Similar statement for $HKH = G$.

We assume now that G is a connected Lie group. Then the subgroups K, H and L are also Lie groups and the subsets $\check{x} \subset \Xi$, $\widehat{\xi} \subset X$ are submanifolds. The maps

$$f \in \mathcal{D}(X) \longrightarrow \widehat{f} \in \mathcal{E}(\Xi), \qquad \varphi \in \mathcal{D}(\Xi) \longrightarrow \check{\varphi} \in \mathcal{E}(X) \tag{11}$$

being continuous ([GGA], Ch I, §3) we can for $s \in \mathcal{E}'(X), \sigma \in \mathcal{E}'(\Xi)$ define $\widehat{s} \in \mathcal{D}'(\Xi), \overset{\vee}{\sigma} \in \mathcal{D}'(X)$ by

$$(12) \qquad \widehat{s}(\varphi) = s(\overset{\vee}{\varphi}), \qquad \overset{\vee}{\sigma}(f) = \sigma(\widehat{f}), \qquad f \in \mathcal{D}(X), \varphi \in \mathcal{D}(\Xi).$$

Because of (8) these definitions extend the definitions (5) and (6).

While the properties of Lemma 3.2 are desirable for example because of the canonical choice of measures on $\overset{\vee}{x}$ and $\widehat{\xi}$ (via Lemma 3.3) such a canonical choice may well be possible without properties (a)–(c), for example if X and Ξ are Riemannian manifolds.

For an example where property (a) fails let us look at the case $X = \boldsymbol{S}^3$, $\Xi =$ the space of totally geodesic two-spheres in X. The group $G = \boldsymbol{O}(4)$ acts transitively both on X and on Ξ. We choose K as the isotropy group for the north pole x_o and H the isotropy group of a $\xi_o \in \Xi$ passing through x_o. Then x is incident to ξ if $x \in \xi$. Clearly property (b) fails because if ξ passes through x it also passes through the antipodal point Ax. This is reflected in the inversion formula

$$(13) \qquad f + f \circ A = c(L_X - 1)((\widehat{f})^{\vee})$$

where L_X is the Laplacian on X and c is a constant ([GGA], Ch. I, §4).

Returning now to the general manifolds

$$X = G/K, \qquad \Xi = G/H$$

we shall now relate the Radon transform to certain differential operators on X and on Ξ.

Let M be a manifold and φ a diffeomorphism of M. If f is a function on M we put $f^{\varphi} = f \circ \varphi^{-1}$. If D is a differential operator on M we denote by D^{φ} the differential operator given by

$$D^{\varphi}(f) = (Df^{\varphi^{-1}})^{\varphi}.$$

The operator D is said to be *invariant under* φ if $D^{\varphi} = D$, in other words if $D(f \circ \varphi) = (Df) \circ \varphi$ for $f \in \mathcal{D}(M)$.

Suppose now D is a differential operator on the Lie group G and suppose D is invariant under all right translations $R(g) : h \longrightarrow hg$ of G. If F is a function on G right-invariant under K then the function DF is also right-invariant under K. If $\pi' : G \longrightarrow G/K$ denotes the natural map we have therefore a differential operator D' on G/K given by

$$(14) \qquad (D'f) \circ \pi' = D(f \circ \pi') \qquad f \in \mathcal{D}(G/K).$$

Similarly if $\pi'' : G \longrightarrow G/H$ is the natural map we have a differential operator D'' on G/H given by

$$(15) \qquad (D''\varphi) \circ \pi'' = D(\varphi \circ \pi''), \qquad \varphi \in \mathcal{D}(G/H).$$

The maps $D \longrightarrow D'$ and $D \longrightarrow D''$ are homomorphisms.

We can also consider the representation ν' of G on $\mathcal{E}(X)$ given by

$$\nu'(g)f = f^{\tau(g)}.$$

This induces a Lie algebra homomorphism of $\mathfrak{g}$ into the Lie algebra of vector fields on X and hence a homomorphism

$$\nu' : \boldsymbol{D}(G) \longrightarrow \boldsymbol{E}(X)$$

of the algebra $\boldsymbol{D}(G)$ of left invariant differential operators on G into the algebra $\boldsymbol{E}(X)$ of all differential operators on X. If $Y \in \mathfrak{g}$ and $\overline{Y}$ and $\widetilde{Y}$ denote, respectively, the right and left invariant vector fields on G such that $\overline{Y}_e = Y$, $\widetilde{Y}_e = Y$, then (cf. [DS], Ch. II, Lemma 3.5) the mapping $\widetilde{Y} \longrightarrow -\overline{Y}$ extends to an isomorphism τ of $\boldsymbol{D}(G)$ onto the algebra $\tau(D(G))$ of all right invariant differential operators on G. Since $-\overline{Y}' = \nu'(\widetilde{Y})$ for $Y \in \mathfrak{g}$ ([GGA], Ch. II, §5 (76)) we have

$$(16) \qquad \tau(E)' = \nu'(E) \qquad \text{for } E \in \boldsymbol{D}(G).$$

Similarly we have a homomorphism

$$\nu'' : \boldsymbol{D}(G) \longrightarrow \boldsymbol{E}(\Xi),$$

where $\boldsymbol{E}(\Xi)$ is the algebra of all differential operators on Ξ, and again

$$\tau(E)'' = \nu''(E), \qquad E \in \boldsymbol{D}(G).$$

Proposition 3.4. *Let D be a right invariant differential operator on G and let D' and D'' be defined by (14) and (15). Then for $f \in \mathcal{D}(X)$, $\varphi \in \mathcal{D}(\Xi)$*

$$(D'f)^\wedge = D''\widehat{f}, \qquad (D''\varphi)^\vee = D'\check{\varphi}.$$

Equivalently,

$$(17) \quad (\nu'(E)f)^\wedge = \nu''(E)\widehat{f}, \qquad (\nu''(E)\varphi)^\vee = \nu'(E)\check{\varphi}, \qquad E \in \boldsymbol{D}(G).$$

Proof. Since $(f^{\tau(g)})^\wedge = (\widehat{f})^{\tau(g)}, (\varphi^{\tau(g)})^\vee = (\varphi^{\tau(g)})^\vee$, relation (17) follows for $E = \widetilde{Y}$ by putting $g = \exp tY$ and differentiating with respect to t (cf. (11)). Since ν' and ν'' are homomorphisms, (17) follows in general.

An interesting special case of Prop. 3.4 occurs when D is a bi-invariant differential operator on G; in that case D' and D'' are invariant under the action of G on X and on Ξ (cf. Exercises **B6-B8**).

In this subsection we have been examining the Radon transform and its dual for the double fibration (4), G being any connected Lie group. We now make the additional assumption that K is compact and shall derive some consequences for analysis on X and on Ξ.

Lemma 3.5. *Let G be a connected Lie group and K compact. Then*

(i) $f \longrightarrow \widehat{f}$ *is a continuous mapping of $\mathcal{D}(X)$ into $\mathcal{D}(\Xi)$.*

(ii) $\varphi \longrightarrow \overset{\vee}{\varphi}$ *is a continuous mapping of $\mathcal{E}(\Xi)$ into $\mathcal{E}(X)$.*

Proof. (i) The assumptions on G and K imply that $X = G/K$ has a G-invariant Riemannian structure. In the corresponding metric d, X is complete (see e.g. [DS], p. 148) and this in turn implies that for each $R > 0$, the closed ball F^R with radius R, center x_o, is compact. Let

$$\Phi^R = \{\xi \in \Xi : d(x_o, \widehat{\xi}) \leq R\}.$$

Then Φ^R is closed; for this consider a sequence $\xi_n \subset \Phi^R$ converging to $\xi_o \in \Xi$. For each ξ_n select a point $g_n \cdot x_o \in \widehat{\xi_n}$ such that $d(x_o, g_n \cdot x_o) \leq R$. Since K is compact the elements $g_n \in G$ are contained in a fixed compact subset of G. Passing to a subsequence if necessary we may assume the sequence (g_n) convergent to a $g_o \in G$. But then

$$d(x_o, \widehat{\xi_o}) \leq d(x_o, g_o \cdot x_o) \leq R$$

so $\xi_o \in \Phi^R$ as claimed. Also, if $\xi \in \Phi^R$ then for a suitable $g \in G$, $d(x_o, g \cdot x_o) \leq R$ and $g \cdot x_o \in \widehat{\xi}$. However,

$$g \cdot x_o \in \widehat{\xi} \Longleftrightarrow \xi \in (g \cdot x_o)^\vee \Longleftrightarrow \xi \in g \cdot \overset{\vee}{x}_o$$

so

$$\Phi^R \subset (\pi')^{-1}(F^R) \cdot \overset{\vee}{x}_o.$$

Since the right hand side is compact (K, F^R and $\overset{\vee}{x}_o$ are all compact) and Φ^R closed, Φ^R is compact.

Suppose $f \in \mathcal{D}_{F^R}(X)$, that is f has support contained in F^R. If $\xi \notin \Phi^R$ then $d(x_o, x) > R$ for $x \in \widehat{\xi}$ so $\widehat{f}(\xi) = 0$. Thus $\mathrm{supp}(\widehat{f}) \subset \Phi^R$ so $f \longrightarrow \widehat{f}$ maps $\mathcal{D}(X)$ into $\mathcal{E}(\Xi) \cap C_c(\Xi) = \mathcal{D}(\Xi)$ and $\mathcal{D}_{F^R}(X)$ into $\mathcal{D}_{\Phi^R}(\Xi)$. This proves (i) because of the continuity of the maps (11).

For (ii) we use the formula (10) for $\check{\varphi}$. Assuming $\varphi_n \longrightarrow 0$ in $\mathcal{E}(\Xi)$ we must show that for each $R > 0$ and each differential operator D on X, $D\check{\varphi}_n \longrightarrow 0$ uniformly on F^R. If $gK \in F^R$ then

$$d(x_o, gk \cdot \widehat{\xi}_o) \le d(x_o, g \cdot x_o) \le R,$$

so gkH lies in the compact set Φ^R. Fix $\epsilon > 0$, a compact set C and a function $\varphi \in \mathcal{D}_C(\Xi)$ such that $\varphi \equiv 1$ on $\Phi^{R+\epsilon}$. Put $\psi_n = \varphi\varphi_n$. Then $\psi_n = \varphi_n$ on $\Phi^{R+\epsilon}$ so by (10), $\check{\varphi}_n = \check{\psi}_n$ on $F^{R+\epsilon}$ whence $D\check{\varphi}_n = D\check{\psi}_n$ on F^R. Since $\varphi_n \longrightarrow 0$ in $\mathcal{E}(\Xi)$ we have $\psi_n \longrightarrow 0$ in $\mathcal{D}_C(\Xi)$ so by the continuity in (11), $\check{\psi}_n \longrightarrow 0$ in $\mathcal{E}(X)$. Thus $D\check{\varphi}_n \longrightarrow 0$ uniformly on F^R, as desired.

Corollary 3.6. *With assumptions as in Lemma 3.5 we have*

$$\mathcal{E}'(X)^\wedge \subset \mathcal{E}'(\Xi), \qquad \mathcal{D}'(\Xi)^\vee \subset \mathcal{D}'(X).$$

More precisely,

(i) If $s \in \mathcal{E}'(X)$ then the mapping $\varphi \in \mathcal{E}(\Xi) \longrightarrow s(\check{\varphi})$ extends $\widehat{s}$ to an element of $\mathcal{E}'(\Xi)$;

(ii) The mapping $\sigma \in \mathcal{E}'(\Xi) \longrightarrow \check{\sigma} \in \mathcal{D}'(X)$ extends to a mapping $\sigma \in \mathcal{D}'(\Xi) \longrightarrow \check{\sigma} \in \mathcal{D}'(X)$ by

$$\check{\sigma}(f) = \sigma(\widehat{f}), \qquad f \in \mathcal{D}(X).$$

As we already saw in §2, the closedness of the range $\mathcal{E}'(\boldsymbol{R}^n)^\wedge$ implies by general functional analysis that $\mathcal{E}(\boldsymbol{P}^n)^\vee = \mathcal{E}(\boldsymbol{R}^n)$. In particular, in the plane any smooth point function $f(x)$ is a line function integral

$$f(x) = \int_{l \ni x} \varphi(\ell)\, d\mu(\ell)$$

for a smooth line function φ ((5) §2).

Using Corollary 3.6 and a result from §2, No. 6 we can generalize this formula.

Theorem 3.7. *Assume that K is compact and the range $\mathcal{E}'(X)^\wedge \subset \mathcal{E}'(\Xi)$ closed. Then*

$$\mathcal{E}(\Xi)^\vee = N^\perp,$$

where N is the kernel

$$N = \{s \in \mathcal{E}'(X) \; : \; \widehat{s} = 0\} \tag{18}$$

and

$$N^\perp = \{f \in \mathcal{E}(X) \; : \; s(f) = 0 \text{ for } s \in N\}. \tag{19}$$

Proof. Let $(N^\perp)^\perp$ denote the annihilator

$$\{t \in \mathcal{E}'(X) \; : \; t(g) = 0 \text{ for } g \in N^\perp\}$$

of $N^\perp$ in $\mathcal{E}'(X)$. By general theory this is the closure of N in the topology $\sigma(\mathcal{E}'(X), \mathcal{E}(X))$ so by Prop. 2.18 it equals N. The mapping $\alpha : \varphi \longrightarrow \overset{\vee}{\varphi}$ is a continuos mapping of $\mathcal{E}(\Xi)$ into $N^\perp$ and the transpose ${}^t\alpha$ maps $(N^\perp)'$ into $\mathcal{E}'(\Xi)$. By the Hahn-Banach theorem each member of $(N^\perp)'$ extends to an element of $\mathcal{E}'(X)$ and two such extensions agree if and only if they coincide on $N^\perp$, that is, by the above, differ by a member of N. Thus $(N^\perp)' = \mathcal{E}'(X)/N$ and the map ${}^t\alpha$ becomes the mapping

$${}^t\alpha : \; s + N \longrightarrow \widehat{s}.$$

Since this is an injective mapping the image $\alpha(\mathcal{E}(\Xi))$ is dense in $N^\perp$. On the other hand a continuous linear map $T : \; E \longrightarrow F$ of one Fréchet space into another has a closed range if and only if ${}^tT(F')$ is closed for the weak* topology $\sigma(E', E)$ (Theorem 2.16). Applying this for $T = \alpha$ and noting that the range $\mathcal{E}'(X)^\wedge$ of ${}^t\alpha$ is closed we find that $\alpha(\mathcal{E}(\Xi))$ is closed; this proves that $\mathcal{E}(\Xi)^\vee = N^\perp$.

2. The Radon Transform for Grassmannians.

As an example of homogeneous spaces in duality we shall now consider the space $X = \boldsymbol{G}(p, n)$ of $p-$planes in $\boldsymbol{R}^n$ and the space $\Xi = \boldsymbol{G}(q, n)$ of q planes in $\boldsymbol{R}^n$, p and q related by $p + q = n - 1$. The case $p = 0, q = n - 1$ was treated in §2. Both X and Ξ are acted on transitively by the group $G = \boldsymbol{M}(n)$ of isometries of $\boldsymbol{R}^n$. To each $g \in \boldsymbol{M}(n)$ we associate the translation $t(V)$ by the vector $V = g \cdot 0$ and the rotation $k \in \boldsymbol{O}(n)$ given by $g = t(V)k$. Then $kt(V)k^{-1} = t(k \cdot V)$ so

$$g_1 g_2 = t(V_1)k_1 t(V_2)k_2 = t(V_1 + k_1 \cdot V_2)k_1 k_2. \tag{20}$$

Thus $G = \boldsymbol{M}(n)$ is isomorphic to the $(n+1) \times (n+1)$ matrix group

$$(21) \qquad \left\{ \begin{pmatrix} k & V \\ 0 & 1 \end{pmatrix} : \ k \in \boldsymbol{O}(n), V \in \boldsymbol{R}^n \right\}$$

acting on $\boldsymbol{R}^n$ by

$$\begin{pmatrix} k & V \\ 0 & 1 \end{pmatrix} \cdot \begin{pmatrix} Y \\ 1 \end{pmatrix} = \begin{pmatrix} k \cdot Y + V \\ 1 \end{pmatrix}, \ Y \in \boldsymbol{R}^n.$$

The Lie algebra $\mathfrak{g}$ of G is given by the set of matrices

$$(22) \qquad \begin{pmatrix} T & Z \\ 0 & 0 \end{pmatrix} : \qquad T \in \mathfrak{o}(n), Z \in \boldsymbol{R}^n,$$

$\mathfrak{o}(n)$ being the Lie algebra of $\boldsymbol{O}(n)$. Since the adjoint representation $Ad_G = Ad$ of G on $\mathfrak{g}$ satisfies $Ad(g)X = gXg^{-1}$ a simple computation shows that

$$(23) \qquad Ad \begin{pmatrix} k & V \\ 0 & 1 \end{pmatrix} \begin{pmatrix} T & Z \\ 0 & 0 \end{pmatrix} = \begin{pmatrix} kTk^{-1} & k \cdot Z - kTk^{-1} \cdot V \\ 0 & 0 \end{pmatrix}$$

With the standard basis $\boldsymbol{e}_1, \ldots, \boldsymbol{e}_n$ of $\boldsymbol{R}^n$ let the origins in X and Ξ be

$$x_o = \sum_1^p \boldsymbol{R}\boldsymbol{e}_i, \qquad \xi_o = \sum_{p+2}^n \boldsymbol{R}\boldsymbol{e}_j.$$

The corresponding isotropy groups H_p and H_q are then the matrix groups

$$(24) \qquad H_p = \left\{ h_p = \begin{pmatrix} a & 0 & v \\ 0 & b & 0 \\ 0 & 0 & 1 \end{pmatrix} : a \in \boldsymbol{O}(p), b \in \boldsymbol{O}(n-p), v \in \boldsymbol{R}^p \right\}$$

$$(25) \qquad H_q = \left\{ h_q = \begin{pmatrix} \alpha & 0 & 0 \\ 0 & \beta & w \\ 0 & 0 & 1 \end{pmatrix} : \alpha \in \boldsymbol{O}(n-q), \beta \in \boldsymbol{O}(q), w \in \boldsymbol{R}^q \right\}$$

and

$$(26) \qquad X = G/H_p, \qquad \Xi = G/H_q.$$

The spaces are homogeneous spaces in duality in the sense of (i), (ii) at the beginning of subsection 1. In fact, each of the groups G, H_p and H_q is the semidirect product of a compact group and an abelian group and

$L = H_p \cap H_q$ is compact so all of them are unimodular. Also using (20) for $g_1 \in H_p$, $g_2 \in H_q$ it is straightforward to verify that H_pH_q is closed. Finally, we verify the transversality (a) in Lemma 3.2, in the stronger form that

$$h \in H_p, \qquad hH_q \subset H_qH_p \tag{27}$$

imply $h \in H_q$. For this note that

$$H_q \cdot x_o = \{x \in X : x \perp \xi_o, \quad x \cap \xi_o \neq \emptyset\}, \tag{28}$$

the set of all p-planes intersecting ξ_o under a right angle and

$$h \cdot H_q \cdot x_o = \{y \in X : y \perp h \cdot \xi_o, \qquad y \cap h \cdot \xi_o \neq \emptyset\}. \tag{29}$$

Because of (27), $h \cdot H_q \cdot x_o \subset H_q \cdot x_o$ which by (28)-(29) implies $\xi_o = h \cdot \xi_o$ whence $h \in H_q$. Thus conditions (i), (ii) and (a) are indeed satisfied. Also (28) shows in connection with (3) that

$$x \text{ is incident to } \xi \Longleftrightarrow x \perp \xi, x \cap \xi \neq \emptyset. \tag{30}$$

Note that

$$L = H_p \cap H_q = \boldsymbol{O}(p) \times \boldsymbol{Z}_2 \times \boldsymbol{O}(q). \tag{31}$$

We also put

(32)

$$K_p = H_p \cap \boldsymbol{O}(n) = \boldsymbol{O}(p) \times \boldsymbol{O}(n-p); \quad K_q = H_q \cap \boldsymbol{O}(n) = \boldsymbol{O}(n-q) \times \boldsymbol{O}(q).$$

Let $d\ell, dk_p$ and dk_q be normalized Haar measures on the groups L, K_p, and K_q, respectively. Then Haar measures on G, H_p and H_q are given by

$$dk\, dV, \qquad dk_p\, dv, \qquad dk_q\, dw, \tag{33}$$

respectively, in terms of the description (21), (24) and (25) (cf. [GGA], p. 222). The Radon transform and the dual are now given by

$$\widehat{f}(\xi) = \int\limits_{\substack{x \cap \xi \neq \emptyset \\ x \perp \xi}} f(x)dm(x) \quad , \qquad \check{\varphi}(x) = \int\limits_{\substack{\xi \cap x \neq \emptyset \\ \xi \perp x}} \varphi(\xi)d\mu(\xi), \tag{34}$$

so by (5) (6) and (33)

$$\widehat{f}(g \cdot \xi_o) = \int\limits_{K_q} \int\limits_{\xi_o} f(g \cdot (w + k_q \cdot x_o))dw\, dk_q, \tag{35}$$

$$\check{\varphi}(\gamma \cdot x_o) = \int_{K_p} \int_{x_o} \varphi(\gamma \cdot (v + k_p \cdot \xi_o)) dv\, dk_p. \tag{36}$$

The simplest nontrivial example of these transforms occurs for $p = q = 1$, $n = 3$. Then $X = \Xi$ is the space of lines in $\boldsymbol{R}^3$ and the transforms $f \longrightarrow \widehat{f}$, $\varphi \longrightarrow \check{\varphi}$ assign to a line function its integrals over the set of lines intersecting a given line under a right angle.

Let $\boldsymbol{G}_{p,n}$ denote the space of p-planes in $\boldsymbol{R}^n$ through the origin and let

$$\pi_p : \boldsymbol{G}(p,n) \longrightarrow \boldsymbol{G}_{p,n}$$

denote the mapping obtained by parallel translating a plane to one through 0. If $\sigma \in \boldsymbol{G}_{p,n}$, the fiber $F = \pi_p^{-1}(\sigma)$ is naturally identified with the Euclidean space $\sigma^{\perp}$. Let L_F denote the corresponding Laplacian. If $f \in \mathcal{E}(\boldsymbol{G}(p,n))$ let $f|F$ denote the restriction to F. Consider the linear operator $\square_p$ on $\mathcal{E}(\boldsymbol{G}(p,n))$ given by

$$(\square_p f)|F = L_F(f|F). \tag{37}$$

It is clear that $\square_p$ is a differential operator; it is also invariant under the action of G on $X = \boldsymbol{G}(p,n)$. To see this we first remark that under an isometry between two Riemannian manifolds, the Laplace-Beltrami operators correspond ([GGA], p. 246). Thus if $g \in G$, and we write $\square$ instead of $\square_p$, we have

$$\begin{aligned}(\square^g f)|F &= (\square f^{g^{-1}})^g|F = \left((\square f^{g^{-1}})|g^{-1}F\right)^g \\ &= \left(L_{g^{-1}F}(f^{g^{-1}}|F)\right)^g = L_F(f|F) = (\square f)|F.\end{aligned}$$

With E_{ij} denoting the matrix with 1 where the i-th row and the j-th column intersect, 0 elsewhere, consider the vectors $V_k = E_{k,n+1}$ in the Lie algebra $\mathfrak{g}$. Let

$$L = \sum_1^n V_k^2 \tag{38}$$

as a member of the symmetric algebra $S(\mathfrak{g})$. Then (23) shows immediately that L is $Ad(G)$-invariant. Hence if $\widetilde{V}_k$ is the extension of V_k to a left-invariant vector field on G, the differential operator

$$\widetilde{L} = \sum_1^n \widetilde{V}_k^2, \tag{39}$$

is a bi-invariant differential operator on G (cf. [GGA], p. 283, even though G has two components).

Lemma 3.8. *Let $\widetilde{L}'$ be the differential operator on G/H_p induced by $\widetilde{L}$ (cf. (14)). Then*

$$\widetilde{L}' = \square_p.$$

Proof. If $F \in \mathcal{E}(G)$ then

$$(\widetilde{L}F)(e) = \left\{\left(\frac{\partial^2}{\partial t_1^2} + \cdots + \frac{\partial^2}{\partial t_n^2}\right) F(exp(t_1V_1 + \cdots + t_nV_n))\right\}_{(t)=0}.$$

Suppose F has the form $F(g) = f(g \cdot x_o)$, $f \in \mathcal{E}(X)$. Since the vectors $V_1, \ldots, V_p$ belong to the Lie algebra of H_p we obtain by (14)
(40)

$$(\widetilde{L}'f)(x_o) = \left\{\left(\frac{\partial^2}{\partial t_{p+1}^2} + \cdots + \frac{\partial^2}{\partial t_n^2}\right) f(exp(t_{p+1}V_{p+1} + \cdots + t_nV_n) \cdot x_o)\right\}_{(t)=0}$$

On the other hand

$$\{exp(t_{p+1}V_{p+1} + \cdots + t_nV_n) \cdot x_o : (t_{p+1}, \ldots, t_n) \in \boldsymbol{R}^{n-p}\}$$

is just the fiber F_o over x_o so the right hand side of (40) is just

$$\big(L_{F_o}(f|F_o)\big)(x_o).$$

Consequently, the operators $\widetilde{L}'$ and $\square_p$ coincide at the point x_o in X. Since both are invariant under the action of G on X the lemma follows.

Combining this lemma with Prop. 3.4 we deduce the following intertwining result.

Proposition 3.9. *For $f \in \mathcal{D}(X)$, $\varphi \in \mathcal{D}(\Xi)$*

$$(\square_p f)^\wedge = \square_q \widehat{f}, \qquad (\square_q \varphi)^\vee = \square_p \overset{\vee}{\varphi}.$$

Theorem 3.10. (Gonzalez). *Let $p, q \in \boldsymbol{Z}^+$ such that $p + q + 1 = n$ is odd. Then the transform $f \longrightarrow \widehat{f}$ is inverted by the formula*

$$c_{p,q} f = \big(\square_q^{(n-1)/2} \widehat{f}\big)^\vee, \qquad f \in \mathcal{D}(\boldsymbol{G}(p,n)),$$

where

$$c_{p,q} = (-4\pi)^{\frac{1}{2}(n-1)} \frac{\Gamma(\frac{1}{2}(p+1))\Gamma(\frac{1}{2}(q+1))}{(\Gamma(\frac{1}{2}))^2}.$$

Remark. For $p = 0$ and for $q = 0$ this reduces to the inversion formulas in Theorem 2.1 and 2.3. The idea of the proof is to express $f \longrightarrow \widehat{f}$ as a composition of the usual Radon transform on $\boldsymbol{R}^{p+1}$ and the dual transform on $\boldsymbol{R}^{q+1}$. For this reason the case when p and q are both even is technically a little simpler and in the proof below we shall restrict ourselves to this case.

Recall the orthogonal decomposition $\boldsymbol{R}^n = x_o + \boldsymbol{R}e_{p+1} + \xi_o$. Let $\boldsymbol{G}(p, \xi_o^{\perp})$ denote the space of p-planes in $\xi_o^{\perp} = \boldsymbol{R}e_{p+1} + x_o$. With $f \in \mathcal{D}(X)$ consider the transform

$$(41) \qquad (R_p f)(\ell) = \int_{\xi_o} f(w + \ell)dw, \qquad \ell \in \boldsymbol{G}(p, \xi_o^{\perp}).$$

Then, if $y \in \xi_o^{\perp}$, we have by (35),

$$\widehat{f}(y + \xi_o) = \int_{K_q} \int_{\xi_o} f(y + w + k_q \cdot x_o)dw\, dk_q = \int_{K_q} (R_p f)(y + k_q \cdot x_o)dk_q$$

so

$$(42) \qquad \widehat{f}(y + \xi_o) = (R_p f)^{\text{dual}}(y), \qquad y \in \xi_o^{\perp},$$

where the superscript 'dual' refers to the dual Radon transform on the $(p+1)$-dimensional space $\xi_o^{\perp}$. Since $R_p f \in \mathcal{D}(\boldsymbol{G}(p, \xi_o^{\perp}))$ we can use Theorem 2.3 to invert (42). This introduces the Laplacian $L_{\xi_o^{\perp}}$. However, by (37), we have (with $\xi_o^{\perp}$ identified with the fiber $\pi_q^{-1}(\xi_o)$),

$$(43) \qquad L_{\xi_o^{\perp}}(\widehat{f}|\xi_o^{\perp}) = \Box_q(\widehat{f})|\xi_o^{\perp}$$

so Theorem 2.3 implies

$$(44) \qquad \int_{x_o} (\Box_q^{p/2}\widehat{f})(y + z + \xi_o)dz = c'(R_p f)(y + x_o), \qquad y \in \xi_o^{\perp},$$

where $c' = (-4\pi)^{p/2}\Gamma(\frac{1}{2}(p+1))/\Gamma(\frac{1}{2})$. Now we apply $(L_{\xi_o^{\perp}})^{\frac{q}{2}}$ to (44), and after the differentiation put $y = re_{p+1}$. On the left hand side we use (43) and on the right hand side we note that

$$(R_p f)(y_1 e_1 \cdots + y_{p+1}e_{p+1} + x_o) = (R_p f)(y_{p+1}e_{p+1} + x_o).$$

This gives the formula

$$(45)\qquad \int_{x_o} (\square_q^{(n-1)/2}\widehat{f})(re_{p+1}+z+\xi_o)dz = c'\frac{d^q}{dr^q}\big(R_pf(re_{p+1}+x_o)\big).$$

For $k_p \in K_p$ consider the twisted transform $R_p^{k_p} : f \longrightarrow (R_pf^{k_p^{-1}})^{k_p}$. By (41) we have for $\ell' \in \boldsymbol{G}(p, k_p \cdot \xi_o^{\perp})$,

$$(R_p^{k_p}f)(\ell') = (R_pf^{k_p^{-1}})(k_p^{-1}\cdot \ell') = \int_{\xi_o} f(k_p(w+k_p^{-1}\cdot \ell'))dw$$

so

$$(46)\qquad (R_p^{k_p}f)(\ell') = \int_{k_p\cdot\xi_o} f(z+\ell')dz, \qquad \ell' \in \boldsymbol{G}(p, k_p\cdot \xi_o^{\perp}).$$

Thus we get from (45), replacing ξ_o by $k_p \cdot \xi_o$,

$$(47)\quad \int_{x_o} (\square_q^{(n-1)/2}\widehat{f})(k_p\cdot re_{p+1}+z+k_p\cdot\xi_o)dz = c'\frac{d^q}{dr^q}(R_p^{k_p}f(k_p\cdot re_{p+1}+x_o)).$$

Finally, we define the function $Sf \in \mathcal{D}(\boldsymbol{G}(q, x_o^{\perp}))$ by

$$(48)\qquad (Sf)(\ell'') = \int_{\ell''} f(w''+x_o)dw'', \qquad \ell'' \in \boldsymbol{G}(q, x_o^{\perp}).$$

This is just the ordinary Radon transform of the restriction $f|\pi_p^{-1}(x_o)$ of f to the fiber $\pi_p^{-1}(x_o) = x_o^{\perp}$. By the inversion formula in Theorem 2.1 we have

$$(49)\qquad c''f(x_o) = \int_{K_p} (\square^{q/2}Sf)(k_p\cdot\xi_o)dk_p, \qquad \square = \frac{d^2}{dr^2},$$

with $c'' = (-4\pi)^{q/2}\Gamma(\frac{1}{2}(q+1))/\Gamma(\frac{1}{2})$.

On the other hand, if $y \in \xi_o^{\perp}$

$$(50)\ (R_p^{k_p}f)(k_p\cdot(y+x_o)) = \int_{\xi_o} f(k_p\{z+(y+x_o)\})dz = (Sf)(k_p\cdot(y+\xi_o)).$$

We put here $y = re_{p+1}$ and apply $\square^{q/2} = \frac{d^q}{dr^q}$ to both sides. Then (49) and (47) give the desired result.

3. The Totally Geodesic Radon Transform on Constant Curvature Spaces.

A. The Generalized Transforms.

Let X be a complete n-dimensional simply connected Riemannian manifold of constant curvature $\kappa \neq 0$. We normalize the metric such that $\kappa = \pm 1$ so X is either the sphere $\boldsymbol{S}^n$ with the metric induced from $\boldsymbol{R}^n$ or the hyperbolic space $\boldsymbol{H}^n$, that is the unit ball $|x| < 1$ in $\boldsymbol{R}^n$ with the Riemannian structure

$$ds^2 = 4\frac{dx_1^2 + \cdots + dx_n^2}{(1 - x_1^2 - \cdots - x_n^2)^2},$$

of constant curvature -1. For a fixed integer k $(1 \leq k \leq n-1)$ let Ξ be the set of totally geodesic submanifolds of X of dimension k. The *totally geodesic Radon transform* $f \longrightarrow \widehat{f}$ is defined by

$$\widehat{f}(\xi) = \int_{\xi} f(x)dm(x), \qquad \xi \in \Xi, \tag{51}$$

where dm is the Riemannian measure on ξ, and the dual transform $\varphi \longrightarrow \overset{\vee}{\varphi}$ is defined by

$$\overset{\vee}{\varphi}(x) = \int_{x \in \xi} \varphi(\xi)d\mu(\xi), \tag{52}$$

where μ is the unique measure on the compact space $\{\xi \in \Xi : x \in \xi\}$, invariant under all rotations around x, and having total measure 1.

In [GGA], Ch. I, §4 we have inverted the transform (51) for k even and proved a support theorem for it in the case $X = \boldsymbol{H}^n$ (loc. cit. Theorem 4.2). We shall now complete these results and here the abstract Radon transform from No. 1 (with the more general incidence definition) will play a useful role.

Let $I(X)$ denote the group of isometries of X. Then G, the identity component of $I(X)$, acts transitively on X and on Ξ. We fix an origin $o \in X$ (say $o = 0$ for $\boldsymbol{H}^n$, the north pole for $\boldsymbol{S}^n$) and let K denote the isotropy subgroup of G at o. Fix $p \in \boldsymbol{R}^+$, fix $\xi_p \in \Xi$ at distance $d(o, \xi_p) = p$ from o and let H_p be the subgroup of G mapping ξ_p into itself. Then we have

$$X = G/K, \qquad \Xi = G/H_p. \tag{53}$$

Proposition 3.11. *The point $x \in X$ and the manifold $\xi \in \Xi$ are incident if and only if $d(x, \xi) = p$, d denoting distance.*

Proof. Let $x = gK$, $\xi = \gamma H_p$. Then $gK \cap \gamma H_p \neq \emptyset \Longrightarrow \gamma^{-1}gK \cap H_p \neq \emptyset \Longrightarrow \gamma^{-1}g \cdot o = h_p \cdot o$ for some $h_p \Longrightarrow d(\gamma^{-1}g \cdot o, \xi_p) = p$ so $d(x, \xi) = p$. On the other hand, if $d(x, \xi) = p$ so $d(o, g^{-1}\gamma \cdot \xi_p) = p$ then by the transitivity, $g^{-1}\gamma \cdot \xi_p = k \cdot \xi_p$ for some $k \in K$. Hence $g^{-1}\gamma h = k$ for some $h \in H_p$ so x and ξ are incident.

The abstract Radon transform for the double fibration

$$\begin{array}{ccc} & G/(K \cap H_p) & \\ \swarrow & & \searrow \\ G/K & & G/H_p \end{array}$$

now becomes (because of Proposition 3.11)

$$(54) \qquad \widehat{f}_p(\xi) = \int\limits_{d(x,\xi)=p} f(x)dm(x), \qquad \check{\varphi}_p(x) = \int\limits_{d(x,\xi)=p} \varphi(\xi)d\mu(\xi).$$

Here dm is the Riemannian measure and $d\mu$ the invariant average over the set $\check{x}$. Note that this normalization of the measures may not agree with that of (5) (6) since we have not yet normalized the Haar measures on G, K, H_p and $K \cap H_p$. The subscript p in (54) is added for emphasis; note that $\widehat{f}_o = \widehat{f}$, $\check{\varphi}_o = \check{\varphi}$ in (51) and (52).

For $X = \boldsymbol{S}^2$, $\widehat{f}_p$ is the integral of f over a circle on $\boldsymbol{S}^2$ of length $2\pi \cos p$, $\check{\varphi}_p(s)$ is the average of φ over the set of great circles at distance p from s. For $X = \boldsymbol{H}^2$, $\widehat{f}_p(\xi)$ is the integral of f over the "equidistant curve" from ξ in the sense of non-Euclidean geometry (cf. [GGA], Introduction, Exercise B. 8).

While we are here just interested in inverting the transform $f \longrightarrow \widehat{f}$ we shall see that the more general transform $\varphi \longrightarrow \check{\varphi}_p$ will enter in the inversion formula.

B. The Inversion Formulas.

We shall now prove an inversion formula for the transform (51). Consider a pair $x \in X, \xi \in \Xi$ at distance p, i.e. $d(x, \xi) = p$. Choose $g \in G$ such that $g \cdot o = x$ and let dk denote the normalized Haar measure on K. The family $kg^{-1} \cdot \xi$ $(k \in K)$ constitutes the set of elements in Ξ incident to o.

Thus

$$(55)\qquad \begin{aligned}(\widehat{f})^{\vee}_p(g\cdot o) &= \int_K \widehat{f}(gkg^{-1}\cdot\xi)dk = \int_K dk \int_\xi f(gkg^{-1}\cdot y)dm(y)\\ &= \int_\xi \Big(\int_K f(gkg^{-1}\cdot y)dk\Big)dm(y) = \int_\xi (M^{d(x,y)}f)(g\cdot o)dm(y),\end{aligned}$$

where, as in Ch. II, §2, No. 4, $(M^r f)(z)$ is the average of f on the sphere $S_r(z)$. Let x_o be a point on ξ at the minimum distance p from x and let

$$(56)\qquad r = d(x_o, y) \qquad q = d(x,y).$$

Since ξ is totally geodesic, $d(x_o, y)$ equals the distance between x_o and y in ξ. The last integrand $(M^{d(x,y)}f)(x)$ above is constant in y on each sphere in ξ with center x_o.

Consider first the case $X = \boldsymbol{H}^n$. We use geodesic polar coordinates in ξ with center x_o and note that ξ is itself a hyperbolic space of curvature -1. Then (55) implies

$$(57)\qquad (\widehat{f})^{\vee}_p(x) = \Omega_k \int_0^\infty (M^q f)(x)\sinh^{k-1} r\,dr, \qquad \text{for } f(x) = 0(e^{-kd(o,x)}),$$

where Ω_k is the area of the unit sphere in $\boldsymbol{R}^k$ and q is given by (56).

We use now the cosine formula on the non-Euclidean right-angled triangle (xx_oy). By (56) and $p = d(x_o, x)$ we then have

$$(58)\qquad \cosh q = \cosh p\ \cosh r.$$

Now fix x and put

$$(59)\qquad F(\cosh q) = (M^q f)(x), \qquad \widehat{F}(\cosh p) = (\widehat{f})^{\vee}_p(x).$$

Then by (57)–(58),

$$\widehat{F}(\cosh p) = \Omega_k \int_0^\infty F(\cosh p\ \cosh r)(\sinh r)^{k-1}dr.$$

Writing here $t = \cosh p$, $s = \cosh r$ this becomes

$$(60)\qquad \widehat{F}(t) = \Omega_k \int_1^\infty F(ts)(s^2-1)^{\frac{k}{2}-1}ds.$$

Here we substitute $u = (ts)^{-1}$ and then put $v = t^{-1}$. Then (60) becomes

$$v^{-1}\widehat{F}(v^{-1}) = \Omega_k \int_0^v \{F(u^{-1})u^{-k}\}(v^2 - u^2)^{\frac{k}{2}-1} du. \tag{61}$$

We write this in the form

$$H(v) = \int_0^v G(u)(v^2 - u^2)^{\frac{1}{2}(k-2)} du$$

which is inverted by (see e.g. [GGA], Ch. I, §2)

$$G(u) = \frac{2^{k+1}}{(k-1)!}\frac{\Omega_k}{\Omega_{k+1}} u \left(\frac{d}{d(u^2)}\right)^k \int_0^u (u^2 - t^2)^{\frac{k}{2}-1} tH(t)dt. \tag{62}$$

Hence (61) implies

$$F(u^{-1})u^{-k} = c\, u \left(\frac{d}{d(u^2)}\right)^k \int_o^u (u^2 - v^2)^{\frac{k}{2}-1} \widehat{F}(v^{-1})dv, \tag{63}$$

where $c^{-1} = (k-1)!\Omega_{k+1}/2^{k+1}$. Let $\ell m\ v$ denote the positive solution q to the equation $\cosh q = v^{-1}$ $(v \leq 1)$. Putting $q = 0, u = 1$ we deduce the following result from (59) and (63).

Theorem 3.12. *The k-dimensional totally geodesic Radon transform $f \longrightarrow \widehat{f}$ on the hyperbolic space $X = \boldsymbol{H}^n$ is inverted by*

$$f(x) = c\left[\left(\frac{d}{d(u^2)}\right)^k \int_o^u \left((\widehat{f})^\vee_{\ell m\ v}\right)(x)(u^2 - v^2)^{\frac{k}{2}-1} dv\right]_{u=1}, f(x) = 0(e^{-kd(o,x)}),$$

where

$$c = \frac{2^{k+1}}{(k-1)!\Omega_{k+1}}, \quad \ell m\ v = \cosh^{-1}(v^{-1}). \tag{64}$$

We consider now the remaining case $X = \boldsymbol{S}^n$. If f is odd, i.e. $f(-x) = -f(x)$, then $\widehat{f} \equiv 0$, so we assume f even, i.e. $f(-x) = f(x)$. Again, we start with the formula

$$(\widehat{f})^\vee_p(x) = \int_\xi (M^{d(x,y)} f)(x) dm(y)$$

and use geodesic polar coordiantes in ξ with center x_o. Using the evenness of f we get in place of (57),

$$(\widehat{f})_p^\vee(x) = 2\Omega_k \int_o^{\frac{\pi}{2}} (M^q f)(x) \sin^{k-1} r \, dr. \tag{65}$$

By spherical trigonometry used on the right-angled triangle (xx_oy),

$$\cos q = \cos p \ \cos r. \tag{66}$$

Again we fix x and put

$$F(\cos q) = (M^q f)(x), \quad \widehat{F}(\cos p) = (\widehat{f})_p^\vee(x).$$

Then, using (66),

$$\widehat{F}(\cos p) = 2\Omega_k \int_o^{\frac{\pi}{2}} F(\cos p \ \cos r)(\sin r)^{k-1} dr, \tag{67}$$

which on putting $v = \cos p, \ u = v \cos r$ becomes

$$v^{k-1}\widehat{F}(v) = 2\Omega_k \int_o^v F(u)(v^2 - u^2)^{\frac{k}{2}-1} du. \tag{68}$$

Using (62) again this inverts to

$$F(u) = \tfrac{c}{2} u \left(\frac{d}{d(u^2)}\right)^k \int_o^u (u^2 - v^2)^{\frac{k}{2}-1} v^k \widehat{F}(v) dv,$$

where c is the same as in (64). This proves the following counterpart to Theorem 3.12.

Theorem 3.13. *The k-dimensional totally geodesic Radon transform $f \longrightarrow \widehat{f}$ on S^n is for even f inverted by*

$$f(x) = \tfrac{c}{2} \left[\left(\frac{d}{d(u^2)}\right)^k \int_o^u \left((\widehat{f})^\vee_{\cos^{-1} v}\right)(x) v^k (u^2 - v^2)^{\frac{k}{2}-1} dv \right]_{u=1},$$

with c as in (64).

The inversion formulas in these two theorems have a simple geometric interpretation: $(\widehat{f})^\vee_p(x)$ is the average of the integrals of f over the k-dimensional totally geodesic submanifolds of X tangent to $S_p(x)$.

The formulas show clearly the significance of the parity of k. If k is even the differentiations in the inversion formulas can be carried out and the integral sign disappears. This is consistent with the inversion formulas for k even proved in [GGA], Ch. I, §4, for $X = \boldsymbol{H}^n$ and $X = \boldsymbol{S}^n$, respectively,

$$Cf = [L + (k-1)(n-k)][L + (k-3)(n-k+2)] \dots [L + 1(n-2)]((\widehat{f})^\vee),$$
$$2Cf = [L - (k-1)(n-k)][L - (k-3)(n-k+2)] \dots [L - 1(n-2)]((\widehat{f})^\vee),$$

where f is even in the second case and

$$C = \frac{\Gamma(\frac{n}{2})}{\Gamma(\frac{n-k}{2})}(-4\pi)^{\frac{k}{2}}.$$

Example. The simplest case is $X = \boldsymbol{H}^3, k = 2$. The formula above is

$$f = -\frac{1}{2\pi}(L+1)((\widehat{f})^\vee), \tag{69}$$

whereas Theorem 3.12 gives

$$f(x) = \tfrac{2}{\pi}\left[\left(\frac{d}{d(u^2)}\right)^2 \int_0^u (\widehat{f})^\vee_{\ell m\ v}(x)\, dv\right]_{u=1}.$$

Now $d/d(u^2) = (2u)^{-1}\, d/du$ so this reduces to

$$f(x) = -\tfrac{1}{2\pi}(\widehat{f})^\vee(x) + \tfrac{1}{2\pi}\left[\tfrac{d}{du}(\widehat{f})^\vee_{\ell m\ u}\right]_{u=1}. \tag{70}$$

Comparing (69) and (70) we get

$$\left[\tfrac{d}{du}(\widehat{f})^\vee_{\ell m\ u}\right]_{u=1} = -L((\widehat{f})^\vee).$$

Remark. In [GGA], Theorem 4.11, Ch. I it is stated that the antipodal transform $f \longrightarrow \widehat{f}$ for a compact two-point homogeneous space X maps $\mathcal{E}(X)$ onto $\mathcal{E}(\Xi)$. This is proved *loc. cit.* page 175 except for the case $X = \boldsymbol{P}^n(\boldsymbol{R})$ (n even) where the proof only shows that the range of the map $f \longrightarrow \widehat{f}$

is dense. To verify the surjectivity statement in full we observe by *loc. cit.* page 163 that in terms of the spherical harmonics expansion(*loc. cit.* p. 21)

$$f = \sum_{k,m} a_{km} S_{km} \qquad (k \text{ even})$$

we have

$$\widehat{f} = \sum_{k,m} a_{km} \varphi_k(\tfrac{\pi}{2}) S_{km} \qquad (k \text{ even}).$$

By *loc. cit.* page 23 we have

$$\varphi_{2k}(\tfrac{\pi}{2}) = \tfrac{\Omega_{n-1}}{\Omega_n}(-1)^k \int_0^\pi (\cos\varphi)^{2k}(\sin\varphi)^{n-2} d\varphi \sim k^{-\frac{(n-1)}{2}}.$$

The characterization of the sequences $\{a_{km}\}$ as those which are rapidly decreasing in k ([GGA], Introduction, Theorem 3.4) thus shows that the mapping $f \longrightarrow \widehat{f}$ is surjective.

C. Support Theorems.

While the inversion formulas above determine f in terms of $\widehat{f}$ these formulas do not directly describe the support of f in terms of the support of $\widehat{f}$. The reason is that these formulas involve the operation $\varphi \longrightarrow \overset{\vee}{\varphi}$. Nevertheless, the conclusion

$$(71) \qquad \widehat{f}(\xi) = 0 \quad \text{for } d(0,\xi) > A \Longrightarrow f(x) = 0 \quad \text{for } d(0,x) > A,$$

which is one of the principal problems in §1, No. 2, is desirable, for example for applications. See [GGA], p. 131 for applications to tomography, and Ch. V, Lemma 1.5, of this book for applications to differential equations.

In the case $k < n-1$ we shall now use the injectivity of the k-dimensional Radon transform on $X_0 (X_0 = \boldsymbol{R}^{n-1},\ \boldsymbol{H}^{n-1},\ \boldsymbol{S}^{n-1})$ to conclude (71) for the Radon transform on $X (X = \boldsymbol{R}^n,\ \boldsymbol{H}^n,\ \boldsymbol{S}^n)$.

For $X = \boldsymbol{R}^n$ and $X = \boldsymbol{H}^n$ the support theorems read as follows ([GGA], p. 105, 153), taking into account (6)§1 used on a $(k+1)$-plane in $\boldsymbol{R}^n$ disjoint from $B_A(0)$ and Theorem 3.12 used on a $(k+1)$-dimensional totally geodesic submanifold of $\boldsymbol{H}^n$, disjoint from $B_A(0)$.

Theorem 3.14. *Let $f \in C(\boldsymbol{R}^n)$ satisfy the condition:*

For each $m \in \boldsymbol{Z}^+$ the function

$$(72) \qquad x \longrightarrow f(x)|x|^m \quad \textit{is bounded}.$$

Then (71) holds.

In the case $k < n-1$ assumption (72) can be replaced by the weaker assumption

(72′) $\qquad f$ *is integrable over each* $(k+1)$*-plane in* $\boldsymbol{R}^n$.

Theorem 3.15. *Let $f \in C(\boldsymbol{H}_n)$ satisfy the decay condition:*

For each $m \in \boldsymbol{Z}^+$ the function

$$x \longrightarrow f(x)e^{m\, d(0,x)} \qquad \textit{is bounded.}$$

Then (71) holds.

In the case $k < n-1$ the dccay condition is only required for $m = k+1$.

For the sphere we have a similar result. Let N denote the north pole on $\boldsymbol{S}^n$, B a closed spherical cap with center N and radius A, and B' the symmetric one with center at the south pole S.

Theorem 3.16. *Let f be an even function on $\boldsymbol{S}^n$ satisfying the condition:*

For each $m \subset \boldsymbol{Z}^+$, the function

$$s \longrightarrow f(s)(\cos d(N,s))^{-m} \qquad \textit{is bounded.} \tag{73}$$

Then

$$\widehat{f}(\xi) = 0 \quad \textit{for} \quad \xi \cap B = \emptyset \tag{74}$$

implies

$$f(s) = 0 \quad \textit{for} \quad s \notin B \cup B'.$$

In the case $k < n-1$ condition (73) can be dropped.

Proof. In the case $k < n-1$ this follows from the injectivity of the Radon transform used on a totally geodesic $\boldsymbol{S}^{k+1} \subset \boldsymbol{S}^n$ disjoint from B. Thus we assume $k = n-1$ and for simplicity we take $n = 2$. We derive the result from Theorem 3.14 by the central projection μ of the lower hemisphere onto $\boldsymbol{R}^2$ (See figure); this map is given by $\mu(s) = x$, s being the intersection of the line $o'x$ with the lower hemisphere. Then $\mu(\xi) = \ell$ where ξ is the semicircle with diameter AB passing through s and ℓ is the line through x

parallel to AB. Let φ and θ denote the lengths of the arcs SM and Ms, respectively. The plane $o'So''$ is perpendicular to ℓ and intersects ξ in M. If p denotes the distance from o'' to x (with sign) and $q = |So''|$ we have for $f \in C(\boldsymbol{S}^2)$

$$\widehat{f}(\xi) = \int_{\xi} f(s)d\theta = \int_{\ell} (f \circ \mu^{-1})(x)\frac{d\theta}{dp}dp.$$

Also

$$\tan\varphi = q, \quad \tan\theta = p/(1+q^2)^{\frac{1}{2}} \tag{75}$$

so

$$\frac{dp}{d\theta} = (1+q^2)^{\frac{1}{2}}(1+\tan^2\theta) = (1+p^2+q^2)(1+q^2)^{-\frac{1}{2}},$$

whence

$$\frac{dp}{d\theta} = (1+|x|^2)\cos\varphi.$$

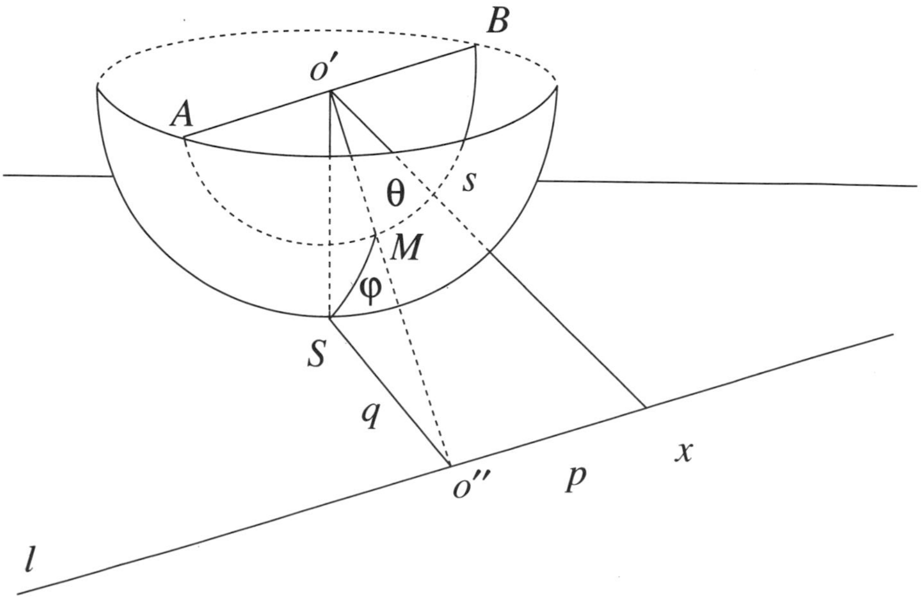

Since $\varphi = d(S, \xi)$ this proves

$$\cos d(S,\xi)\widehat{f}(\xi) = \int_{\ell} (f \circ \mu^{-1})(x)(1+|x|^2)^{-1} dm(x). \tag{76}$$

The triangle $o'Sx$ shows that

$$\tan d(S,s) = |x| \tag{77}$$

so condition (73) for f implies condition (72) for $f \circ \mu^{-1}$. Now assuming (73) and (74) we have by Theorem 3.14 and (76) that $f \circ \mu^{-1}(x) = 0$ for $|x| > \tan A$ so by (77)

$$f(s) = 0 \quad \text{for} \quad d(S,s) > A \tag{78}$$

as desired.

4. The d-plane Transform.

Let d be an integer, $0 < d < n$, and consider the d-plane transform defined by

$$\widehat{f}(\xi) = \int_{\xi} f(x)dm(x), \qquad f \in \mathcal{D}(\boldsymbol{R}^n), \tag{79}$$

ξ being a d-dimensional plane with the Euclidean measure dm. The dual transform $\varphi \longrightarrow \overset{\vee}{\varphi}$ is defined by

$$\overset{\vee}{\varphi}(x) = \int_{x \in \xi} \varphi(\xi)d\mu(\xi), \qquad \varphi \in \mathcal{E}(\boldsymbol{G}(d,n)), \tag{80}$$

where $d\mu$ is the measure on the set of d-planes through x which is invariant under the rotations around x and $\mu\{\xi : x \in \xi\} = 1$. Without proofs we shall report on the range question (Problem B in §1) for these transforms. The most natural range questions are: How are the ranges

$$\mathcal{D}(\boldsymbol{R}^n)^\wedge \qquad \text{and} \qquad \mathcal{E}(\boldsymbol{G}(d,n))^\vee \tag{81}$$

characterized? This is answered in Corollary 3.17 and Theorems 3.20, 3.22 below. In the process we determine also the range $\mathcal{E}'(\boldsymbol{R}^n)^\wedge$ (Theorem 3.21). In addition we consider the space $\mathcal{S}(\boldsymbol{R}^n)$ of rapidly decreasing functions on $\boldsymbol{R}^n$. As in [GGA], Ch. I, §2, we view $\boldsymbol{G}(d,n)$ as a fiber bundle over

$\boldsymbol{G}_{d,n}$, the manifold of d-dimensional subspaces of $\boldsymbol{R}^n$, the projection π : $\boldsymbol{G}(d,n) \longrightarrow \boldsymbol{G}_{d,n}$ being the map which to $\xi \in \boldsymbol{G}(d,n)$ assigns the parallel plane $\sigma = \pi(\xi)$ through the origin. The fiber $\pi^{-1}(\sigma)$ is naturally identified with the orthogonal complement $\sigma^{\perp}$. One can then in a natural way define the space $\mathcal{S}(\boldsymbol{G}(d,n))$ of rapidly decreasing functions on $\boldsymbol{G}(d,n)$ (Gonzalez [1984] [1987]).

For the case $d = n-1$ the ranges (81) are described in Theorems 2.4 and 2.5. The first theorem includes the statement

$$f \in \mathcal{S}(\boldsymbol{R}^n), \qquad \widehat{f} \in \mathcal{D}(\boldsymbol{G}(n-1,n)) \Longrightarrow f \in \mathcal{D}(\boldsymbol{R}^n) \tag{82}$$

or more succinctly,

$$\widehat{\mathcal{S}} \cap \mathcal{D} = \widehat{\mathcal{D}}. \tag{83}$$

Theorem 2.4 (ii) extends to the case $d \leq n-1$ as follows. With $\xi \in \boldsymbol{G}(d,n)$ put $\sigma = \pi(\xi)$ and identify $\pi^{-1}(\sigma)$ with $\sigma^{\perp}$ as above. Putting $x'' = \sigma^{\perp} \cap \xi$ we write

$$\xi = (\sigma, x'').$$

Then (79) takes the form

$$\widehat{f}(\sigma, x'') = \int_{\sigma} f(x' + x'')\,dx'. \tag{84}$$

Now we define the space $\mathcal{D}_H(\boldsymbol{G}(d,n))$ as the set of smooth functions $\varphi(\xi) = \varphi_{\sigma}(x'')$ on $\boldsymbol{G}(d,n)$ of compact support satisfying the following moment condition:

(H) *For each $k \in \boldsymbol{Z}^+$ there exists a homogeneous k^{th} degree polynomial P_k on $\boldsymbol{R}^n$ such that for each $\sigma \in \boldsymbol{G}_{d,n}$ the polynomial*

$$P_{\sigma,k}(u'') = \int_{\sigma^{\perp}} \varphi_{\sigma}(x'')(x'', u'')^k\,dx'', \qquad u'' \in \sigma^{\perp},$$

coincides with the restriction $P_k|\sigma^{\perp}$.

Theorem 2.4 (ii) now has the following consequence ([GGA], Ch. I, §2).

Corollary 3.17. *The d-dimensional Radon transform is a bijection of $\mathcal{D}(\boldsymbol{R}^n)$ onto $\mathcal{D}_H(\boldsymbol{G}(d,n))$.*

A somewhat more complicated L^2-version of this result was given by Solmon [1976]. Theorem 2.4 (i) shows that this corollary extends from $\mathcal{D}(\boldsymbol{R}^n)$

to $\mathcal{S}(\boldsymbol{R}^n)$ provided $d = n - 1$; however if $d < n - 1$ it was shown by an example in Gonzalez [1984] that the moment condition (H) is not sufficient to characterize the image $\mathcal{S}(\boldsymbol{R}^n)^\wedge$.

To explain the description of the range $\mathcal{S}(\boldsymbol{R}^n)^\wedge$ for $d < n - 1$ consider first the case $d = 1$, $n = 3$ where (79) is the X-ray transform in $\boldsymbol{R}^3$. Writing a line ξ in the form:

$$\xi : \qquad x_1 = \alpha_1 t + \beta_1,\ x_2 = \alpha_2 t + \beta_2,\ x_3 = t$$

so that

$$\widehat{f}(\xi) = (\alpha_1^2 + \alpha_2^2 + 1)^{\frac{1}{2}} \int_{\boldsymbol{R}} f(\alpha_1 t + \beta_1, \alpha_2 t + \beta_2, t)\,dt.$$

Then a simple differentiation shows that

$$\left(\frac{\partial^2}{\partial\alpha_1\partial\beta_2} - \frac{\partial^2}{\partial\alpha_2\partial\beta_1}\right)\left((\alpha_1^2 + \alpha_2^2 + 1)^{-\frac{1}{2}}\widehat{f}\right) = 0, \tag{85}$$

and John showed in [1938] that, roughly speaking, the range is given by the null-space of the "ultrahyperbolic operator" in (85). An extension of John's result to the d-plane transform in $\boldsymbol{R}^n$ appears in Gelfand, Gindikin and Graev [1982]. Richter in [1986] (and to some extent Grinberg in [1987]) gave a complete proof and used this to give a more conceptual characterization of the range $\mathcal{S}(\boldsymbol{R}^n)^\wedge$. To describe Richter's result consider the natural action of the Lie group $\boldsymbol{M}(n)$ or $\boldsymbol{R}^n$ and on $\boldsymbol{G}(d,n)$ which induces natural homomorphisms

$$\lambda :\ \boldsymbol{D}(\boldsymbol{M}(n)) \longrightarrow \boldsymbol{E}(\boldsymbol{R}^n),$$
$$\Lambda :\ \boldsymbol{D}(\boldsymbol{M}(n)) \longrightarrow \boldsymbol{E}(\boldsymbol{G}(d,n)),$$

where for a manifold M, $\boldsymbol{E}(M)$ denotes the algebra of all differential operators on M and if M is a Lie group, $\boldsymbol{D}(M) \subset \boldsymbol{E}(M)$ is the subalgebra of left-invariant differential operators on M. (In [GGA], Ch. II, §5, No. 7 these homomorphisms are denoted ν.) Since $f \longrightarrow \widehat{f}$ and $\varphi \longrightarrow \check{\varphi}$ commute with the action of $\boldsymbol{M}(n)$ we have for $P \in \boldsymbol{D}(\boldsymbol{M}(n))$,

$$(\lambda(P)f)^\wedge = \Lambda(P)\widehat{f}, \qquad (\Lambda(P)\varphi)^\vee = \lambda(P)\check{\varphi} \tag{86}$$

With the usual imbedding of $\boldsymbol{M}(n)$ into $\boldsymbol{GL}(n+1,\boldsymbol{R})$ (see (21), this section) the Lie algebra of $\boldsymbol{M}(n)$ has a basis given by the matrices $Z_i = E_{i,n+1}$ ($1 \leq i \leq n$) and $X_{ij} = E_{ij} - E_{ji}$($1 \leq i < j \leq n$). Viewing them as left-invariant vector fields on $\boldsymbol{M}(n)$ we define $V_{ij\ell} \in \boldsymbol{D}(\boldsymbol{M}(n))$ by

$$V_{ij\ell} = Z_i X_{j\ell} + Z_j X_{\ell i} + Z_\ell X_{ij}, \quad 1 \leq i < j < \ell \leq n. \tag{87}$$

Then it is not hard to see (e.g. [GGA], Ch. II, §5, (76)) that

$$\lambda(X_{ij}) = x_i \frac{\partial}{\partial x_j} - x_j \frac{\partial}{\partial x_i}, \qquad \lambda(Z_i) = -\frac{\partial}{\partial x_i}.$$

Consequently, $\lambda(V_{ij\ell}) = 0$ so by (86), $\Lambda(V_{ij\ell})\widehat{f} = 0$. Richter's theorem is the converse.

Theorem 3.18. *Let $1 \le d \le n-2$. Then the image of $\mathcal{S}(\boldsymbol{R}^n)$ under $f \longrightarrow \widehat{f}$ is given by*

$$\mathcal{S}(\boldsymbol{R}^n)^\wedge = \{\varphi \in \mathcal{S}(\boldsymbol{G}(d,n)) : \qquad \Lambda(V_{ij\ell})\varphi = 0 \qquad 1 \le i < j < \ell \le n\}$$

In the case $d = n-1$ the map $\varphi \longrightarrow \overset{\vee}{\varphi}$ is one-to-one on $\mathcal{S}(\boldsymbol{G}(n-1,n))$ (cf. Theorem 2.5) so by (86), $\Lambda(V_{ij\ell}) = 0$. Thus Theorem 3.18 fails for $d = n-1$ and instead one has Theorem 2.4 (i).

While Richter's proof of Theorem 3.18 proceeded via the coordinate-generalization of (85), Gonzalez in [1991] found a more direct conceptual proof of it. He proved also in [1990b] the following interesting refinement. Let

$$U = \sum_{i<j<\ell} V_{ij\ell}^2.$$

Then $U \in \boldsymbol{D}(\boldsymbol{M}(n))$ is a bi-invariant operator on $\boldsymbol{M}(n)$ so, in contrast to $\Lambda(V_{ij\ell})$, the differential operator $\Lambda(U)$ on $\boldsymbol{G}(d,n)$ is invariant under the action of $\boldsymbol{M}(n)$. In terms of the notation of Exercises **B2** and **B5**, the operator U is given by

$$U = D_{P_2} - \tfrac{1}{6}\binom{n-1}{2} D_{P_1}.$$

Theorem 3.19. *Let $1 \le d \le n-2$. Then the image of $\mathcal{S}(\boldsymbol{R}^n)$ under $f \longrightarrow \widehat{f}$ is given by*

$$\mathcal{S}(\boldsymbol{R}^n)^\wedge = \{\varphi \in \mathcal{S}(\boldsymbol{G}(d,n)) : \qquad \Lambda(U)\varphi = 0\}.$$

On the other hand, from the support theorem for the d-plane transform ([GGA], Ch. I, §2, No. 6) we know that

$$f \in \mathcal{S}(\boldsymbol{R}^n), \qquad \widehat{f}(\xi) = 0 \qquad \text{for } d(o,\xi) > A$$

implies

$$f(x) = 0 \qquad \text{for} \qquad |x| > A. \tag{88}$$

Taking Theorem 3.19 into account, (88) implies the following result.

Theorem 3.20. *Let $1 \leq d \leq n-2$. Then the image $\mathcal{D}(\boldsymbol{R}^n)^\wedge$ under the d-plane transform is characterized by*

$$\mathcal{D}(\boldsymbol{R}^n)^\wedge = \{\varphi \in \mathcal{D}(\boldsymbol{G}(d,n)) : \qquad \Lambda(U)\varphi = 0\}.$$

In particular, for smooth functions φ on $\boldsymbol{G}(d,n)$ of compact support $(d \leq n-2)$ the moment condition (H) is equivalent to the differential equation $\Lambda(U)\varphi = 0$. For $\varphi \in \mathcal{S}(\boldsymbol{G}(d,n))$ this is no longer true as shown by Gonzalez' example quoted above. We recall also that Kurusa has in [1991a] proved Theorem 3.20 (with $\Lambda(U)$ replaced by the analog of (85)) directly from Corollary 3.17.

For constant curvature analogs of Theorems 3.19–3.20 see Berenstein and Tarabusi [1992], Gonzalez [1993], Grinberg [1985] and Kakehi [1992], [1993].

We consider next the question of describing the range $\mathcal{E}'(\boldsymbol{R}^n)^\wedge$ under the d-plane transform as defined by (12). Again, the answer is quite different in the two cases $d = n-1$ and $d \leq n-2$.

Theorem 3.21.

(i) *In the case $d = n-1$,*

$$\mathcal{E}'(\boldsymbol{R}^n)^\wedge = \{\sigma \in \mathcal{E}'(\boldsymbol{G}(n-1,n)) : \qquad \sigma(\mathcal{N}) = 0\},$$

where $\mathcal{N}$ is the kernel of the dual transform (described in Theorem 2.5).

(ii) *In the case $d < n-1$,*

$$\mathcal{E}'(\boldsymbol{R}^n)^\wedge = \{\sigma \in \mathcal{E}'(\boldsymbol{G}(d,n)) : \qquad \Lambda(V_{ij\ell})\sigma = 0 \text{ for all } i,j,k\}.$$

Part (i) is Theorem 2.6(i). Part (ii) is proved in Gonzalez [1991]. In both cases we conclude that

$$\mathcal{E}'(\boldsymbol{R}^n)^\wedge \text{ is closed in } \mathcal{E}'(\boldsymbol{G}(d,n)) \qquad (1 \leq d \leq n-1). \tag{89}$$

We now use Theorem 2.15 on the map $\alpha : \varphi \longrightarrow \check{\varphi}$. Its adjoint ${}^t\alpha$ is $\sigma \longrightarrow \widehat{\sigma}$ and it is one-to-one (cf. the proof of Theorem 2.21 in [GGA], Ch. I) so by (89) we conclude the following result (cf. Theorem 2.5(iii)).

Theorem 3.22. *For each d $(1 \leq d \leq n-1)$.*

$$\mathcal{E}(\boldsymbol{G}(d,n))^{\vee} = \mathcal{E}(\boldsymbol{R}^n).$$

Since $\Lambda(D)\widehat{f} = 0$ for $\lambda(D) = 0$ it is of interest to state another result by Richter [1989].

Theorem 3.23. *The kernel of λ coincides with the two-sided ideal in $\boldsymbol{D}(\boldsymbol{M}(n))$ generated by the operators V_{ijk} in (87) and the operators*

$$X_{ij}X_{\ell m} - X_{i\ell}X_{jm} + X_{im}X_{j\ell} \qquad 1 \leq i < j < \ell < m \leq n.$$

Because of the second formula in (86) if P is in the kernel of λ then the space $\Lambda(P)\mathcal{E}(\Xi)$ belongs to the kernel of the dual transform $\varphi \longrightarrow \overset{\vee}{\varphi}$. According to Gonzalez [1991] the following stronger result holds.

Theorem 3.24. *The kernel of $\varphi \longrightarrow \overset{\vee}{\varphi}$ is the closed subspace generated by the spaces*

$$\Lambda(V_{ij\ell})\mathcal{E}(\Xi) \qquad (1 \leq i < j < \ell \leq n).$$

5. Analogy with the Poisson Integral.

It is of some interest to compare Theorem 3.19 to the classical Poisson integral. The theorem describes the range of the Radon transform

$$\widehat{f}(gH) = \int_{H/L} f(ghK)\,dh_L \tag{90}$$

for the following two homogeneous spaces

$$\boldsymbol{R}^n = G/K = \boldsymbol{M}(n)/\boldsymbol{O}(n), \qquad \boldsymbol{G}(d,n) = G/H = \boldsymbol{M}(n)/\boldsymbol{M}(d) \times \boldsymbol{O}(n-d)$$

for which $L = K \cap H = \boldsymbol{O}(d) \times \boldsymbol{O}(n-d)$.

Since $d \leq n-2$, the dimension of the "target space" $\boldsymbol{G}(d,n)$ is larger than $\dim \boldsymbol{R}^n$ so, although the transform (90) is one-to-one, it is natural to expect the range $\mathcal{D}(\boldsymbol{R}^n)^\wedge$ to be "small" compared to $\mathcal{D}(\boldsymbol{G}(d,n))$; in this case it is in fact cut down to the null-space of a certain differential operator.

Now consider the Poisson integral

$$(PF)(z) = \int_B \frac{1-|z|^2}{|z-b|^2} F(b)\,db, \qquad |z| < 1, \tag{91}$$

where B is the unit circle with normalized measure db. The unit disk D as well as the boundary B are both homogeneous spaces of the group $G = \boldsymbol{SU}(1,1)$, i.e.,

$$B = G/H, \qquad D = G/K, \tag{92}$$

where H and K are the isotropy groups of the points $(1,0)$ and $(0,0)$, respectively. We put $L = H \cap K$. Furthermore, the Poisson kernel $(1-|z|^2)/|z-b|^2$ is the Jacobian of the action of G on B :

$$\frac{d(g^{-1}\cdot b)}{db} = P(g\cdot o, b)$$

([GGA], p. 45). Thus

$$(PF)(g\cdot o) = \int_B P(g\cdot o, b)F(b)db = \int_B F(b)\frac{d(g^{-1}b)}{db}db,$$

which equals

$$\int_B F(g\cdot b)db = \int_{K/L} F(gkH)dk_L.$$

This shows that the Poisson integral (91) is just the Radon transform $F \longrightarrow \widehat{F}$ for the homogeneous spaces (92). Recall here the remark following Proposition 3.1.

The characterization of the functions PF as solutions of the Laplace equation, say

$$P\big(L^\infty(B)\big) = \{\psi \in L^\infty(D) : L\psi = 0\}, \tag{93}$$

is therefore a very close analog to Theorem 3.19.

The analogy goes further as we now explain in the case $d=1$, $n=3$ so $\boldsymbol{G}(d,n)$ is the space of lines in $\boldsymbol{R}^3$. Gauss' mean value theorem for harmonic functions asserts the equivalence

$$L\psi = 0 \Longleftrightarrow \int_S \psi(s)dw(s) = \psi(s_o),$$

if S is a circle with center s_o and dw the normalized invariant measure. Taking Theorem 3.19 into account, the analog for $\boldsymbol{G}(1,3)$ is another result from John [1938].

Theorem 3.25. *A function φ in $\mathcal{S}(\boldsymbol{G}(1,3))$ satisfies $\Lambda(U)\varphi = 0$ if and only if it has the following mean value property. For each hyperboloid of revolution H of one sheet the mean values of φ over the two families of generating lines of H are equal. (The variable of integration is the polar angle in the equatorial plane of H).*

John proves this in [1938] by transforming the differential equation (85) to the ultrahyperbolic equation

$$\frac{\partial^2}{\partial x_1^2} + \frac{\partial^2 u}{\partial x_2^2} - \frac{\partial^2 u}{\partial x_3^2} - \frac{\partial^2 u}{\partial x_4^2} = 0$$

on which he then uses Ásgeirsson's mean value theorem ([GGA], Ch. II, §5, No. 6).

EXERCISES AND FURTHER RESULTS

A. Radon Transform on $\boldsymbol{R}^n$.

1. Let $\widetilde{\mathcal{N}}$ denote the kernel of the dual transform $\varphi \longrightarrow \overset{\vee}{\varphi}$ mapping $C(\boldsymbol{P}^n)$ into $C(\boldsymbol{R}^n)$. In the topology of uniform convergence on compact sets, the space $\widetilde{\mathcal{N}}$ is the closed subspace of $C(\boldsymbol{P}^n)$ generated by

$$E_k \otimes p^\ell, \qquad (k - \ell > 0 \quad \text{and even }).$$

Here E_k is the eigenspace of the Laplacian $L_{\boldsymbol{S}^{n-1}}$ for eigenvalue $-k(k+n-2)$.

2. Let $Y_k \in E_k$ and let $\varphi \in \mathcal{S}(\boldsymbol{R})$ satisfy $\varphi(-p) = (-1)^k\varphi(p)$. Let $Y_k \otimes \varphi$ denote the function on $\boldsymbol{P}^n$ defined by $(Y_k \otimes \varphi)(w,p) = \varphi(p)Y_k(w)$. Prove that

$$(Y_k \otimes \varphi)^\vee(rw) = c_o r^k Y_k(w) \int\limits_{\boldsymbol{S}^{n+2k-1}} \varphi^{(k)}(r\eta_1)d\eta$$

where $\varphi^{(k)} = d^k\varphi/dp^k$ and $c_o^{-1} = \Omega_n(2\pi)^k$.

3. *The Legendre polynomial $P_k(n, \cos\theta)$ for $n = 2$* (cf. §2, No. 3). Prove that

$$P_k(2, \cos\theta) = \cos\ k\theta$$

(The polynomial $s \longrightarrow P_k(2,s)$ is the so called Tschebyscheff polynomial).

4. Let $K \subset \boldsymbol{O}(n)$ be a closed subgroup which is transitive on the unit sphere $\boldsymbol{S}^{n-1}$ and put $G = \boldsymbol{R}^n \cdot K$. Let $0 < d < n$, let ξ_o be a fixed d-plane in $\boldsymbol{R}^n$ and put $\Xi = G \cdot \xi_o$. Put

$$(1) \qquad \widehat{f}(\xi) = \int_{\xi} f(x)\, dm(x), \qquad f \in \mathcal{S}(\boldsymbol{R}^n)$$

where dm is the Euclidean measure on ξ and

$$\overset{\vee}{\varphi}(x) = \int_{\xi \ni x} \varphi(\xi) d\mu(\xi), \qquad \varphi \in C(\Xi),$$

where $d\mu$ is the normalized invariant measure on the compact subset of $\xi \in \Xi$ passing through x. Show that the inversion formula (Fuglede [1958], Helgason [1959, 1984a]) for the $d-plane$ transform

$$(2) \qquad cf = (-L)^{\frac{d}{2}}((\widehat{f})^{\vee}), \qquad f \in \mathcal{S}(\boldsymbol{R}^n),$$

$\left(c = \frac{\Gamma(n/2)}{\Gamma((n-d)/2)}(4\pi)^{\frac{d}{2}}\right)$ remains valid for the above restricted Radon transform with identical proof (remark by Grinberg [1985]). Examples:

(i) $K = \boldsymbol{U}(n)$ acting on $\boldsymbol{C}^n = \boldsymbol{R}^{2n}$. Taking ξ_o as the complex k-plane spanned by the first k coordiante vectors, $\Xi = G \cdot \xi_o$ becomes the set of all complex k-planes and (2) takes the form

$$(2') \qquad c_o f = (-L)^k((\widehat{f})^{\vee}) \qquad \left(c_o = \frac{\Gamma(n)}{\Gamma(n-k)}(4\pi)^k\right).$$

(ii) $K = \boldsymbol{U}(n) \subset \boldsymbol{O}(2n)$ acting on $\boldsymbol{R}^{2n}$ and $\xi_o = \boldsymbol{R}^n \times 0$. Show that $K \cdot \xi_o$ consists of all the subspaces of $\boldsymbol{R}^{2n}$ which are Lagrangian, that is are maximal isotropic under the skew-symmetric form

$$\{(x, y), (u, v)\} = x \cdot v - y \cdot u \qquad x, y \in \boldsymbol{R}^n$$

where $\cdot$ is the inner product on $\boldsymbol{R}^n$.

5. Let $X = \boldsymbol{R}^{2n+1}$ and put σ_A for the set of hyperplanes ξ in X for which $d(0, \xi) = A$. Then show that

$$T \in \mathcal{E}'(X), \quad \operatorname{supp}(\widehat{T}) \subset \sigma_A \Longrightarrow \operatorname{supp}(T) \subset S_A(0).$$

6. With convolution $*$ on $\boldsymbol{P}^n$ defined as in (7) §2 show that if $f \in \mathcal{D}(\boldsymbol{R}^n)$, $\varphi \in \mathcal{D}(\boldsymbol{P}^n)$ then $\overset{\vee}{\varphi} * f = (\varphi * \widehat{f})^{\vee}$.

B. Homogeneous Spaces; Grassmann Manifolds.

1. Let G be a group, K and H two subgroups and $L = K \cap H$. The points $x = gK$ in $X = G/K$ and $\xi = \gamma H$ in $\Xi = G/H$ are said to be *incident* if as cosets of G they are not disjoint. Prove the following statements:

(i) $G/L = \{(x, \xi) \in X \times \Xi : x, \xi \text{ incident}\}$. This is Proposition 3.1.

(ii) The property $KHK = G$ is equivalent to the property: Any two $x_1, x_2 \in X$ are incident to some $\xi \in \Xi$. Similar statement for $HKH = G$.

(iii) The condition

$$HK \cap KH = K \cup H$$

is equivalent to the property:

For any two $x_1 \neq x_2$ in X there is at most one $\xi \in \Xi$ incident to both. In particular, this property is equivalent to:

For any two $\xi_1 \neq \xi_2$ in Ξ there is at most one $x \in X$ incident to both.

2*. Let G/H be a reductive homogeneous space ([GGA], Ch. II). If $\mathfrak{g}$ and $\mathfrak{h}$ denote the Lie algebras of G and H, respectively, let $\mathfrak{m} \subset \mathfrak{g}$ be a subspace satisfying

$$\mathfrak{g} = \mathfrak{h} \oplus \mathfrak{m}, \qquad Ad_G(h)\mathfrak{m} \subset \mathfrak{m} \quad (h \in H).$$

Let $I(\mathfrak{m})$ denote the algebra of $Ad_G(H)$-invariant elements in the symmetric algebra $S(\mathfrak{m})$. If $Z_1, \ldots, Z_s$ is any basis of $\mathfrak{m}$ and for $P \in I(\mathfrak{m})$ we define the differential operator D_P on G/H by

$$(D_P f)(gH) = \left\{ P\left(\frac{\partial}{\partial t_1}, \ldots, \frac{\partial}{\partial t_s}\right) f(g \exp(\Sigma t_i Z_i)H) \right\}_{(t)=0}$$

the mapping $P \longrightarrow D_P$ is a basis-independent linear bijection of $I(\mathfrak{m})$ onto $\boldsymbol{D}(G/H)$, the algebra of G-invariant differential operators on G/H ([GGA], Theorem 4.9, Ch. II).

3. In the notation of Exercise **B2** suppose $P_1, \ldots P_\ell$ is a system of generators of $I(\mathfrak{m})$. Then the $D_{P_i} (1 \leq i \leq \ell)$ generate $\boldsymbol{D}(G/H)$. If the P_i are algebraically independent then so are the D_{P_i}.

4*. Consider the Grassmann manifold

$$X = \boldsymbol{G}(p, n) = G/H_p = \boldsymbol{M}(n)/\boldsymbol{O}(n-p) \times \boldsymbol{M}(p)$$

from §3, No. 2, the group H_p being given by (24) §3. Show that G/H_p is reductive (even symmetric) with a complementary subspace $\mathfrak{m} \subset \mathfrak{g}$ given by the matrices

$$S = \begin{pmatrix} 0 & Y & 0 \\ -{}^tY & 0 & W \\ 0 & 0 & 0 \end{pmatrix} \qquad \begin{array}{l} Y \text{ any } p\times(n-p) \text{ matrix} \\ {}^tY \text{ its transpose} \\ W \subset \boldsymbol{R}^{n-p} \end{array}$$

and the action of $Ad(h_p)$ in (23) given by

$$Ad(h_p)S = \begin{pmatrix} 0 & aYa^{-1} & 0 \\ -b^tYa^{-1} & 0 & b\cdot W + b^tYa^{-1}\cdot V \\ 0 & 0 & 0 \end{pmatrix}.$$

The matrices

$$\begin{aligned} X_{ij} &= E_{ip+j} - E_{p+ji} \quad (1 \le i \le p,\ 1 \le j \le n-p) \\ U_k &= E_{p+k,n+1} \quad (1 \le k \le n-p) \end{aligned}$$

form a basis of $\mathfrak{m}$. The algebra $I(\mathfrak{m})$ and thus the algebra $\boldsymbol{D}(G/H_p)$ of G-invariant differential operators on $\boldsymbol{G}(p,n)$ is given by the following result:

Consider the $(p+1) \times (n-p)$-matrix with vector entries

$$A = \begin{pmatrix} U_1, \dots, U_{n-p} \\ X_{11}, \dots, X_{1n-p} \\ X_{p1}, \dots, X_{pn-p} \end{pmatrix}.$$

For $1 \le k \le \min(p+1, n-p)$ let $P_k \in S(\mathfrak{m})$ be the sum of the squares of the $k \times k$-minors of A having vectors U_j in the first row:

$$P_k = \sum_{\substack{1\le i_1<\cdots<i_k\le n-p \\ 1\le j_1<\cdots<j_{k-1}\le p}} det^2 \begin{pmatrix} U_{i_1} \dots U_{i_k} \\ X_{j_1 i_1} \dots X_{j_1 i_k} \\ \vdots \\ X_{j_{k-1} i_1} \dots X_{j_{k-1} i_k} \end{pmatrix}.$$

Then the polynomials P_k are algebraically independent generators of $I(\mathfrak{m})$ (Gonzalez-Helgason [1986]).

5*. Let G be a connected Lie group with Lie algebra $\mathfrak{g}$, $I(\mathfrak{g} \subset S(\mathfrak{g})$ the subalgebra of $Ad(G)$-invariants in the symmetric algebra $S(\mathfrak{g})$. Let $\boldsymbol{D}(G)$ denote the algebra of left-invariant differential operators on G, $\boldsymbol{Z}(G)$ the subalgebra of bi-invariant differential operators. If $Z_1, \ldots, Z_s$ is a basis of $\mathfrak{g}$ and we to $P \in S(\mathfrak{g})$ associate the differential operator D_P on G by

$$(D_P f)(g) = \left\{ P\left(\frac{\partial}{\partial t_1}, \ldots, \frac{\partial}{\partial t_s}\right) f(g \exp(\Sigma t_i Z_i)) \right\}_{(t)=0}$$

then the mapping $P \longrightarrow D_P$ is a bijection of $S(\mathfrak{g})$ onto $\boldsymbol{D}(G)$ (independent of the basis) mapping $I(\mathfrak{g})$ onto $\boldsymbol{Z}(G)$. (Gelfand [1950], Harish-Chandra [1953], [GGA], Ch. II.)

6*. In this exercise we describe the algebra $\boldsymbol{Z}(G)$ of bi-invariant differential operators on the group $G = \boldsymbol{M}(n)$. By (21) a basis of $\mathfrak{g}$ is given by

$$\begin{aligned} X_{ij} &= E_{ij} - E_{ji} \quad (1 \leq i < j \leq n), \\ U_k &= E_{k,n+1} \quad (1 \leq k \leq n). \end{aligned}$$

Consider the $(n+1) \times (n+1)$ skew-symmetric matrix with vector entries

$$B = \begin{pmatrix} 0, & X_{12}, \ldots, X_{1n}, U_1 \\ -X_{12}, & 0, \ldots, X_{2n}, U_2 \\ & \\ & \\ -U_1, -U_2, & \ldots, -U_n, \ 0 \end{pmatrix}.$$

For $1 \leq j \leq [\frac{1}{2}(n+1)]$ let $Q_j \in S(\mathfrak{g})$ be the sum of the $2j \times 2j$ skew-symmetric minors of B having vectors U_k in the last row and column.

Then the Q_j are algebraically independent generators of $I(\mathfrak{g})$ and by Exercise **B3** the D_{Q_j} are algebraically independent generators of $\boldsymbol{Z}(G)$ (Gonzalez [1988]).

7.* With the notation of Exercises **B2** and **B5** consider the homomorphism $\mu : \boldsymbol{Z}(G) \longrightarrow \boldsymbol{D}(G/H)$ defined by restricting a $D \in \boldsymbol{Z}(G)$ to functions which are right-invariant under H. The map μ is surjective if G/H is a classical symmetric space but may fail to be surjective if G is an exceptional group (Helgason, [1992a]).

Using Exercises **B4** and **B6** show that $\mu : \boldsymbol{Z}(G) \longrightarrow \boldsymbol{D}(G/H_p)$ is surjective in the case

$$G/H_p = \boldsymbol{M}(n)/\boldsymbol{O}(n-p) \times \boldsymbol{M}(p)$$

8. For the dual Grassmann manifolds G/H_p and G/H_q $(p + q = n - 1)$ let $D \in \boldsymbol{Z}(G)$ and let $D_p \in \boldsymbol{D}(G/H_p)$, $D_q \in \boldsymbol{D}(G/H_q)$ correspond to D under μ in **B7**. Then

$$(D_p f)^\wedge = D_q \widehat{f}$$

generalizes Prop. 3.9 to all invariant differential operators on G/H_p and G/H_q.

9*. (Cormack and Quinto [1980]). Given $f \in C(\boldsymbol{R}^n)$ let

$$(Sf)(y) = (M^{|y|/2} f)(y/2),$$

the mean value of f over a sphere with center $y/2$ passing through the origin.

(i) For $w \in \boldsymbol{S}^{n-1}, p \in \boldsymbol{R}$ put

$$\varphi(w,p) = f(pw)|p|^{n-2}.$$

Then

$$(Sf)(y) = \left(\frac{2}{|y|}\right)^{n-2} \overset{\vee}{\varphi}(y),$$

where $\overset{\vee}{\varphi}$ is the dual transform of φ (§2, (5)).

(ii) Let $f \in C^\infty(\boldsymbol{R}^n)$ and $A \geq 0$. If $(Sf)(y) = 0$ for $|y| \leq A$ then $f(y) = 0$ for $|y| \leq A$.

(iii) Let E denote the set of restrictions of the entire functions on $\boldsymbol{C}^n$ to $\boldsymbol{R}^n$. Then $f \longrightarrow Sf$ is a bijection of E onto itself.

10. Show that if A and B are arbitrary $n \times n$ matrices then

$$(n^2-1) \int_{U(n)} |\operatorname{Tr}(AV) Tr(BV)|^2 dV =$$

$$\operatorname{Tr}(AA^*)\operatorname{Tr}(BB^*) + \operatorname{Tr}(AB)\operatorname{Tr}(A^*B^*) - \frac{1}{n}\{\operatorname{Tr}(ABB^*A^*) + \operatorname{Tr}(BAA^*B^*)\}$$

11. Let X be the flat torus T^2 and Ξ the set of closed geodesics. Prove that the corresponding Radon transform is injective on $\mathcal{E}(X)$ (Strichartz [1982]).

12. Let U/K be a compact symmetric space which is not a sphere. Let $f \in \mathcal{E}(U/K)$ be such that $\int_\gamma f = 0$ for all closed geodesics γ in U/K. Then $f \equiv 0$.

13. As in (92) §3 let $D = G/K$ be the open unit disk, $G = \boldsymbol{SU}(1,1)$ and $B = G/H$ the boundary. Show for the corresponding incidence that

$$\widehat{b} = D \qquad (b \in B), \qquad \overset{\vee}{d} = B \qquad (d \in D).$$

NOTES

This chapter is to some extent a continuation of Chapter I in [GGA] whose historical notes should therefore supplement those below, particularly as regards earlier work on the Radon transform.

The inversion of the Radon transform $f \longrightarrow \widehat{f}$ on $\boldsymbol{R}^n$ goes back to Radon [1917] and John [1955]. Their inversion formulas are different in the cases n odd and n even. The unified version in Theorem 2.1 which emphasizes the dual transform $\varphi \longrightarrow \overset{\vee}{\varphi}$ is from Helgason [1965a]. The inversion of $\varphi \longrightarrow \overset{\vee}{\varphi}$ on $\mathcal{S}^*(P^n)$ in Theorem 2.2 is also from there but the improved version in Theorem 2.3 (Cor. 2.14) was proved by Gonzalez [1984] for n odd and by Solmon [1987] in general. The proof in the text, via Proposition 2.12, follows a substantial simplification due to Madych and Solmon [1988] where several refinements are also given. The range theorems for $\mathcal{D}(\boldsymbol{R}^n)$ and $\mathcal{E}'(\boldsymbol{R}^n)$ (Theorem 2.4 and 2.6) are from Helgason [1964b], [1965a] and [1983a], respectively. The latter implies that the range $\mathcal{E}'(\boldsymbol{R}^n)^\wedge$ is closed which, as Hertle [1984] observed, implies immediately that $\mathcal{E}(\boldsymbol{P}^n)^\vee = \mathcal{E}(\boldsymbol{R}^n)$ (Theorem 2.5(iii)). Hertle also proved that the Radon transform $f \longrightarrow \widehat{f}$ is not a homeomorphism of $\mathcal{D}(\boldsymbol{R}^n)$ onto the subspace $\mathcal{D}(\boldsymbol{R}^n)^\wedge$ of $\mathcal{D}(\boldsymbol{P}^n)$. This subspace is thus an example "occurring in nature" of a closed subspace of an LF-space which is not an LF-space.

The determination of the kernel $\mathcal{N}$ of the dual transform (Theorem 2.5) by means of a K-finite decomposition is sketched in Helgason [1984a] (p. 226, 562) and here carried out in detail in §2. A result of this type as well as Proposition 2.11 is given by Ludwig [1966].

The Radon transform and its dual for the double fibration (§3, No.1)

$$\begin{array}{ccc} & Z = G/K \cap H & \\ \swarrow & & \searrow \\ X = G/K & & \Xi = G/H \end{array}$$

was initiated in Helgason [1966a] where the adjointness relation (8)§3 is proved. Once Haar measures on G and on the closed subgroups K and H have been fixed the definitions of $f \longrightarrow \widehat{f}$ and $\varphi \longrightarrow \overset{\vee}{\varphi}$ are completely canonical. A further generalization, de-emphasizing G but postulating properties of the entering measures, was given in Gelfand, Graev and Shapiro [1969] and further developed for example in Guillemin-Sternberg [1977]. Our definition uses the concept of *incidence* for $X = G/K$ and $\Xi = G/H$ which goes back to Chern [1942]. Even when the elements of X can be viewed as subsets of Ξ and vice versa it turns out to be advantageous (cf. Theorems 3.12 and Theorem 3.13) not to restrict incidence to the naive one $x \in \xi$. Proposition 3.1 was remarked to me by Gaillard. This group-theoretic view of incidence is also important in the inversion theorem on Grassmann manifolds (Theorem 3.10), due to Gonzalez [1984, 1987]. The earlier version of the inversion formula (Helgason [1965a]), while it looks the same, was less satisfactory, because of convergence difficulties coming from the use of the incidence $x \cap \xi \neq \emptyset$ (rather than $x \cap \xi \neq \emptyset$, $x \perp \xi$). The inversion formula for the Radon transform on $\boldsymbol{H}^n$ and $\boldsymbol{S}^n$ was given by the author in [1959] (the less elegant case of k odd was also proved then (and the method used in [1964b] for the support theorem for $\boldsymbol{H}^n$) but not published until [1990] as Theorems 3.12 and 3.13 when it was realized that these formulas fit into the general incidence framework). See also Semyanistyi [1961] for another method for $k = n - 1$ and Berenstein-Tarabusi [1991] and Kurusa [1991b] for other methods for general k.

Support theorems for the Radon transform on $\boldsymbol{R}^n$ and $\boldsymbol{H}^n$ are given in Helgason [1964b], [1965a] and [1980b]. See Boman [1991] and Boman and Quinto [1987] for interesting generalizations. For $\boldsymbol{S}^n$ such support theorems were proved by Quinto [1983] and in the stronger form of Theorem 3.16, by Kurusa [1992]. A related argument is indicated by Quinto [1987], §3.

CHAPTER II

A DUALITY FOR SYMMETRIC SPACES

This chapter deals with an important example of homogeneous spaces in duality, namely a symmetric space $X = G/K$ of the noncompact type and the space $\Xi = G/MN$ of horocycles in X. (See Theorem 1.1 below and the notation accompanying it). These spaces

$$X = G/K, \qquad \Xi = G/MN,$$

show many analogies; some are reminiscent of analogies between $\boldsymbol{R}^n$ and the space $\boldsymbol{P}^n$ of hyperplanes but others are more subtle because of the richer structure of X (see Prop. 1.7 for an example). When viewed in terms of Ξ, the Bruhat decomposition of G becomes an analog of the Cartan decomposition of G (and X), see §1.

In §2 we compare the algebras $\boldsymbol{D}(G/K)$ and $\boldsymbol{D}(G/MN)$ of invariant differential operators on X and Ξ respectively, discuss their joint eigenfunctions and corresponding eigenspace representations. A more detailed study comes later (Ch. III, §6 and Ch. VI, §3).

The Radon transform on X assigns to a function f on X a function $\widehat{f}$ on Ξ by integrating f over each horocycle ξ in X. In §3 we prove the inversion formula and the Plancherel formula for this Radon transform. The inversion formula is local (that is given by a differential operator) exactly when G has all its Cartan subgroups conjugate. (For the Radon transform on $\boldsymbol{R}^n$ relative to hyperplanes the inversion formula is local exactly if n is odd.)

The range problems for the Radon transform and its dual are postponed until Chapter IV after the necessary tools from Fourier transform theory on X have been developed.

The analogy between X and Ξ leads to the consideration of the analogs on Ξ to the spherical functions on X. These are the conical distributions (and conical functions) on Ξ. These are constructed by analytic continuation. They are studied in §4–§6 and again in Chapter VI §3. They play various roles in representation theory, as extreme weight vectors, as intertwining operators and as special representation coefficients. While the conical distributions in general relate to infinite-dimensional representations the conical functions correspond to finite-dimensional representations.

§1. THE SPACE OF HOROCYCLES.

1. Definition and Coset Representation.

As often before, let D denote the non-Euclidean disk $|z| < 1$ with the Riemannian metric $ds^2 = (1-x^2-y^2)^{-2}(dx^2+dy^2)$. As explained in [GGA], Introduction, §4, the geometric analog for D of hyperplanes in $\mathbf{R}^n$ are the *horocycles*, that is the circles in D tangential to the boundary $B: \ |z| = 1$. Writing D in the form

$$D = \mathbf{SU}(1,1)/\mathbf{SO}(2)$$

(cf. [GGA], Introduction, §4, (5)) we shall now describe the horocycles group-theoretically; this will suggest the generalization to a symmetric space of the noncompact type.

We recall that the Lie algebra $\mathfrak{su}(1,1)$ of $\mathbf{SU}(1,1)$ is given by

$$\mathfrak{su}(1,1) = \left\{ \begin{pmatrix} i\alpha & \beta \\ \overline{\beta} & -i\alpha \end{pmatrix} : \alpha \in \mathbf{R}, \beta \in \mathbf{C} \right\}$$

and an Iwasawa decomposition by

$$\mathfrak{su}(1,1) = \mathfrak{g}_o = \mathfrak{k}_o + \mathfrak{a}_o + \mathfrak{n}_o,$$

where

$$\mathfrak{k}_o = \mathbf{R}\begin{pmatrix} i & 0 \\ 0 & -i \end{pmatrix}, \qquad \mathfrak{a}_o = \mathbf{R}\begin{pmatrix} 0 & 1 \\ 1 & 0 \end{pmatrix}, \qquad \mathfrak{n}_o = \mathbf{R}\begin{pmatrix} i & -i \\ i & -i \end{pmatrix}.$$

The subgroup $N_o = \exp \mathfrak{n}_o$ of $G_o = \mathbf{SU}(1,1)$ equals

$$N_o = \left\{ \begin{pmatrix} 1+in & -in \\ in & 1-in \end{pmatrix} : \ n \in \mathbf{R} \right\}$$

and the horocycle with $0 \leq x < 1$ as diameter equals the orbit $\xi_o = N_o \cdot 0$. Any other horocycle ξ can be written in the form $ka \cdot \xi_o$ ($k \in K_o = \exp \mathfrak{k}_o, a \in A_o = \exp \mathfrak{a}_o$). But then $\xi = kaN_o(ka)^{-1} \cdot (ka \cdot 0)$ so ξ is an orbit of a group conjugate to N_o. Conversely, if $g \in G_o, z \in D$, and we write $g^{-1} \cdot z = na \cdot 0$ then the orbit $gN_og^{-1} \cdot z$ is the horocycle $ga \cdot \xi_o$.

Generalizing this situation let G be a connected semisimple Lie group with finite center, $X = G/K$ the associated symmetric space where K is a maximal compact subgroup of G. Let $G = KAN$ be an Iwasawa decomposition as defined for example in [DS], Ch. IX, §1. As usual, let M denote the centralizer of A in K, let o be the origin $\{K\}$ in X, and ξ_o the orbit $N \cdot o$.

Definition. A *horocycle* in X is any orbit $N' \cdot x$ where $x \in X$ and N' is a subgroup of G conjugate to N.

The choice of Iwasawa decomposition is immaterial here since all such decompositions are conjugate (cf.[DS], Ch. IX, Exercise A5).

Theorem 1.1.

(i) *Each horocycle is a closed submanifold of* X.

(ii) *The group* G *acts transitively on the set of horocycles in* X. *The subgroup of* G *which maps the horocycle* ξ_o *into itself equals* MN.

For the proof and subsequent developments we recall some standard notation. Let $\mathfrak{g}, \mathfrak{k}, \mathfrak{a}, \mathfrak{n}$ and $\mathfrak{m}$ denote the Lie algebras of the groups G, K, A, N, and M above, and let Σ denote the set of roots of $\mathfrak{g}$ with respect to $\mathfrak{a}$ (also called the restricted roots). For $\alpha \in \Sigma$ let $\mathfrak{g}_\alpha$ denote the corresponding root space. Let $m_\alpha = \dim(\mathfrak{g}_\alpha)$, the *multiplicity* of α. Let $\mathfrak{a}^+$ denote the Weyl chamber corresponding to $\mathfrak{n}$, i.e. $\mathfrak{n} = \Sigma_{\alpha \in \Sigma^+} \mathfrak{g}_\alpha$, where Σ^+ is the set of $\alpha \in \Sigma$ which are positive on $\mathfrak{a}^+$. As usual, we put $2\rho = \sum_{\alpha > 0} m_\alpha \alpha$, the sum of the positive roots with multiplicity. Let $\mathfrak{p}$ be the orthogonal complement of $\mathfrak{k}$ in $\mathfrak{g}$ with respect to the Killing form B of $\mathfrak{g}$ and let θ be the Cartan involution corresponding to the decomposition $\mathfrak{g} = \mathfrak{k} + \mathfrak{p}$. We usually give X the Riemannian structure induced by the Killing form B restricted to $\mathfrak{p}$. Given λ in the complex dual $\mathfrak{a}_C^*$ of $\mathfrak{a}$ we define A_λ in the complexification $\mathfrak{a}^C$ of $\mathfrak{a}$ by $B(A_\lambda, H) = \lambda(H), (H \in \mathfrak{a})$. We also put $\mathfrak{a}_+^* = \{\lambda \in \mathfrak{a}_C^* : A_\lambda \in \mathfrak{a}^+\}$. We often write $\langle , \rangle$ for B as well as the bilinear form which B induces on $\mathfrak{a}^c$ and on $\mathfrak{a}_c^*$.

By [DS], Ch. II, Prop. 4.4 and Ch. VI, Exercise B.2 we have the following results.

Lemma 1.2. *The horocycle* $N \cdot o$ *is a closed submanifold of* X *and is orthogonal to the manifold* $A \cdot o$ *at* o.

Lemma 1.3. $(N\theta N) \cap K = \{e\}$.

We can now prove Theorem 1.1. Let $g, h \in G$ and consider the orbits $gNg^{-1}h \cdot o$ and $N \cdot o$. Writing $h^{-1}g = kan$ and using $aNa^{-1} = N$ we find that $hkN \cdot o = gNg^{-1}h \cdot o$ so G acts transitively. Now Part (i) is clear.

Now $mNm^{-1} \subset N$ for each $m \in M$ so MN is a group which maps $N \cdot o$ into itself. On the other hand, let $g \in G$ be such that $gN \cdot o = N \cdot o$. Writing $g = kan$ we conclude that $kann_1 = k_1$ for some $n_1 \in N, k_1 \in K$. Hence $a = e$ so $kN \cdot o = N \cdot o$. Since the action of k on the tangent space X_o preserves orthogonality, Lemma 1.2 implies $Ad_G(k)\mathfrak{a} = \mathfrak{a}$, that is k lies in M', the normalizer of A in K. As usual, let W denote the Weyl group M'/M and w or $|W|$ its order. Suppose $k \notin M$. Then the Weyl group element $s = kM$ does not leave $\mathfrak{a}^+$ invariant so given $H_s \in s^{-1}\mathfrak{a}^+$ there exists an $\alpha \in \Sigma^+$ such that $\alpha(H_s) < 0$. Select $Y \neq 0$ in $\mathfrak{g}_\alpha$. Since $\alpha(s^{-1}H) < 0$ for $H \in \mathfrak{a}^+$ we have $Ad(k)Y \in \theta\mathfrak{n}$ so $kNk^{-1} \cap \theta(N) \neq \{e\}$. But $kN \cdot o = N \cdot o$ so $kNk^{-1} \subset NK$ and now Lemma 1.3 gives a contradiction. Hence $k \in M$ and the theorem is proved.

Remark. The group $Ad_G(M)$ is the centralizer of $Ad_G(A)$ in $Ad_G(K)$ and is therefore independent of the choice of G (with Lie algebra $\mathfrak{g}$).

In fact, if $Ad(k)Ad(a) = Ad(a)Ad(k)$ then $ka = akz$ with z in center of G which lies in K. Applying θ and eliminating z we get $ka^{-2} = a^{-2}k$, verifying the remark.

Definition. The set of horocycles in X with the differentiable structure of G/MN will be called the *dual space* of X and denoted by Ξ.

Remark. The manifolds X and Ξ have the same dimension and show many analogies reminiscent of the duality between points and hyperplanes in $\boldsymbol{R}^n$.

Note however that in contrast to the space $\boldsymbol{P}^n$ of hyperplanes π, the space Ξ of horocycles $\xi \subset D$ is not symmetric (Exercise **A1**). If $p \in \pi$, $q \in \xi$ the geodesic symmetries s_p of $\boldsymbol{R}^n$ and s_q of X satisfy

$$s_p \cdot \pi = \pi \qquad \text{(with opposite orientation);}$$
$$s_q \cdot \xi = \text{ horocycle tangent to } \xi \text{ at } q.$$

(See figure relating to Proposition 1.12.) Thus the analogy between $\boldsymbol{P}^n$ and Ξ would be closer if the hyperplanes were taken with orientation.

2. The Isotropy Actions for X and for Ξ.

We shall now develop some group-theoretic analogies for X and Ξ. Part (i) in the next result is just the classical "polar coordinate" representation for a symmetric space ([DS], Ch. IX, §1).

Proposition 1.4.

(i) *The mapping* $\varphi : (kM, a) \longrightarrow kaK$ *is a differentiable mapping of* $(K/M) \times A$ *onto* X. *If* $A^+ = \exp \mathfrak{a}^+$ *the restriction of* φ *to* $K/M \times A^+$ *is a diffeomorphism onto the regular set* $X' \subset X$; *also* $X = K\overline{A^+} \cdot o$.

(ii) *The mapping* $\psi : (kM, a) \longrightarrow kaMN$ *is a diffeomorphism of* $(K/M) \times A$ *onto* Ξ.

It is clear from Theorem 1.1 that ψ is surjective. If $k_1, k_2 \in K$, $a_1, a_2 \in A$ such that $k_1a_1N \cdot o = k_2a_2N \cdot o$ then $(k_2a_2)^{-1}k_1a_1 \in MN$ so $k_1M = k_2M$ and $a_1 = a_2$. Thus ψ is injective.

Next let $\mathfrak{l}$ denote the orthogonal complement of $\mathfrak{m}$ in $\mathfrak{k}$ and consider the natural maps

$$(1) \qquad p_o : K \longrightarrow K/M, \qquad p : G \longrightarrow G/K, \qquad \pi : G \longrightarrow G/MN.$$

The differentials dp_o and $d\pi$ map $\mathfrak{l}$ and $\mathfrak{l}+\mathfrak{a}$, respectively, isomorphically onto the tangent spaces $(K/M)_{p_o(e)}$ and $(G/MN)_{\pi(e)}$. Let $a \in A, k \in K, H \in \mathfrak{a}, L \in \mathfrak{l}$. Then for $t \in \boldsymbol{R}$,

$$\psi(k(\exp tL)M, a) = ka\exp(Ad(a^{-1})tL)MN$$
$$\psi(kM, a\exp tH) = ka(\exp tH)MN$$

so, $\tau(x)$ denoting translation by x,

$$d\psi(d\tau(k)dp_o(L), d\tau(a)H) = d\tau(ka)(d\pi(Ad(a^{-1})L + H).$$

The right hand side is $\neq 0$ unless $Ad(a^{-1})L + H \in \mathfrak{m} + \mathfrak{n}$. But this happens only if $L = H = 0$ so ψ is regular. This proves the proposition.

Continuing the analogy between X and Ξ we shall describe the double coset spaces $K\backslash G/K$ and $MN\backslash G/MN$. The first is the orbit space of K acting on G/K and the second is the orbit space of MN acting on G/MN.

Proposition 1.5. *The following relations are natural identifications:*

(i) $K\backslash G/K = A/W$ *(orbit space of* W *on* A*).*

(ii) $MN\backslash G/MN = A \times W$.

Part (i) is a restatement of [DS], Ch. IX, Theorem 1.1, taking Ch. VII, Theorem 2.22 into account. Part (ii) amounts to the fact that each $g \in G$ can be written (with obvious notations) $g = m_1n_1m'am_2n_2$ where $a \in A$ is unique and $m' \in M'$ is unique mod M. The existence and the uniqueness of m' is clear from the Bruhat decomposition ([DS], Ch. IX, Theorem 1.4) but it remains to prove the uniqueness of $a \in A$. This is contained in the following lemma.

Lemma 1.6. *Let* $\mathfrak{a}^+$ *and* $(\mathfrak{a}^+)'$ *be two Weyl chambers in* $\mathfrak{a}$. *Let* $G = KAN$ *and* $G = KAN'$ *be the corresponding Iwasawa decompositions. Then*

$$(2) \qquad (NN') \cap MA = \{e\}.$$

Proof. In the case $(\mathfrak{a}^+)' = -\mathfrak{a}^+$ we have $N' = \theta N$ and (2) is contained in [DS], Ch. VI, Exercise B2. The general case will follow if we can prove

$$N' \subset N\theta N. \tag{3}$$

This however is contained in [GGA], Ch. IV, §6 (28).

We restate the results above in terms of the isotropy action. Since $A/W = \overline{A^+}$, $A'/W = A^+$ we can legitimately write

$$X = K \cdot (A/W) \cdot o, \tag{4}$$

$$X' = K \cdot (A'/W) \cdot o. \tag{5}$$

For $s \in W$ let m_s be a representative of s in M' and ξ_s the horocycle $m_s MN$. Let Ξ_s denote the orbit $MNA \cdot \xi_s$ so by Prop. 1.5,

$$\Xi = \bigcup_{s \in W} \Xi_s \qquad \text{(disjoint union)}\ . \tag{6}$$

The proof of Prop. 1.5 showed that given $\xi \in \Xi_s$ there exists a unique $a(\xi) \in A$ such that

$$\xi = mna(\xi) \cdot \xi_s$$

with $m \in M, n \in N$.

The unique summand in (6) of maximum dimension is Ξ_{s^*} where s^* is the Weyl group element which interchanges $\mathfrak{a}^+$ and $-\mathfrak{a}^+$. We put for simplicity

$$m^* = m_{s^*},\ \xi^* = \xi_{s^*},\ \Xi^* = \Xi_{s^*}$$

and observe that $\xi_e = \xi_o$. Writing $(A \times W) \cdot \xi_o$ for the union of the submanifolds $A \cdot \xi_s$ $(s \in W)$ we have in analogy with (4) and (5)

$$\Xi = MN \cdot (A \times W) \cdot \xi_o \tag{7}$$

$$\Xi^* = MN \cdot (A \cdot \xi^*). \tag{8}$$

By a Weyl chamber in $\mathfrak{p}$ we understand a Weyl chamber in some maximal abelian subspace of $\mathfrak{p}$. Then K acts transitively on the set of all Weyl chambers in $\mathfrak{p}$ and the subgroup which leaves $\mathfrak{a}^+$ invariant equals M. Now by Prop. 1.4 each $\xi \in \Xi$ can be written $\xi = kaMN$ where $kM \in K/M$ and $a \in A$ are unique. We say the Weyl chamber kM is *normal* to ξ and call

$\log a$ the *composite distance* from o to ξ. More generally, if $x = g_1 K \in X, \xi = g_2 MN \in \Xi$ we write $\langle x, \xi \rangle = H(g_1^{-1} g_2)$ and call it the *composite distance from x to ξ*. Here $H(g) \in \mathfrak{a}$ is given by $g \in K \exp H(g) N$. It is easy to see that $\langle x, \xi \rangle$ is well-defined and invariant under the action of G.

Remark. $B = K/M$ is identified with the set of all Weyl chambers in $\mathfrak{p}$.

Proposition 1.7.

(i) *Given $x \in X, b \in B$ there exists a unique horocycle passing through x with normal b.*

(ii) *Given a horocycle ξ and a point $x \in \xi$ there exist exactly $|W|$ distinct horocycles passing through x with tangent space at x equal to ξ_x.*

Proof. (i) If $x = gK, b = kM$ the horocycle $\xi(x, b) = k \exp(-H(g^{-1}k)) \cdot \xi_o$ is the unique horocycle in question. For (ii) we may assume $x = o$, $\xi = N \cdot o$. Each Weyl chamber $\mathfrak{a}_i^+ \subset \mathfrak{a}$ gives rise to a group N_i for which the Iwasawa decomposition $G = KAN_i$ holds. If ξ' is any horocycle through o then $\xi' = kN \cdot o$ $(k \in K)$. If in addition ξ and ξ' have the same tangent space at o then k maps this tangent space, and hence its orthogonal complement, into itself. By Lemma 1.2, $Ad(k)\mathfrak{a} \subset \mathfrak{a}$ so $k \in M'$ and $\xi' = N_i \cdot o$ for some N_i above. To see that two different N_i give different orbits of o suppose $N_i \cdot o = N_j \cdot o$. Then $N_i K = N_j K$ which by Lemma 1.3 and (3) implies $N_i = N_j$.

Remark. For $x = o, \xi = \xi_o$ the $|W|$ horocycles indicated in (ii) are the horocycles $\xi_s = m_s \cdot \xi_o$.

Proposition 1.8. *Let $\xi_1, \xi_2 \in \Xi$. The following conditions are equivalent:*

(i) *ξ_1 and ξ_2 have the same normal.*

(ii) *There exists a geodesic γ perpendicular to both ξ_1 and ξ_2 such that $\tau\xi_1 = \xi_2$ for a transvection τ* ([DS], Ch. IV §3) *along γ.*

(iii) *ξ_1 and ξ_2 are orbits of the same group $g^{-1}Ng$.*

Proof. (i) $\Longrightarrow$ (ii). Let ξ_o denote the origin $N \cdot o$ in Ξ. We have $\xi_1 = k_1 a_1 \cdot \xi_o$, $\xi_2 = k_2 a_2 \cdot \xi_o$ and with assumption (i) we can take $k_1 = k_2$. Since $a_1 \cdot \xi_o$ and $a_2 \cdot \xi_o$ have the property in (ii) so do ξ_1 and ξ_2. For (ii) $\Longrightarrow$ (iii) select $g \in G$ such that $g \cdot \xi_2 = \xi_o$. Then $g \cdot \xi_1 = \tau' \cdot \xi_o$ for a transvection τ' along a geodesic γ' perpendicular to $g \cdot \xi_1$ and ξ_o. If γ' intersects ξ_o at $n \cdot o$ then $n^{-1} \cdot \gamma'$ is a geodesic in $A \cdot o$ so for some $a \in A, n^{-1} g \cdot \xi_1 = a \cdot \xi_o$.

Thus $\xi_1 = g^{-1}Ng \cdot (g^{-1}a \cdot o)$ and $\xi_2 = g^{-1}Ng \cdot (g^{-1} \cdot o)$ so (iii) is proved.

Finally, for (iii) $\Longrightarrow$ (i) we may assume that ξ_1 and ξ_2 are orbits of the group $N^k = kNk^{-1}$ for some fixed $k \in K$. Since $G = N^k A^k K$ each orbit of the group N^k intersects $A^k \cdot o$. Thus for suitable $a_1, a_2 \in A$ we have $\xi_i = N^k k a_i \cdot o$ $(i = 1, 2)$ so ξ_1 and ξ_2 have the same normal kM.

Definition. Two horocycles ξ_1 and ξ_2 with the properties of Proposition 1.8 will be called *parallel horocycles.*

We shall generally use the notation $\overline{N}$ for the group θN, whose Lie algebra is $\overline{\mathfrak{n}} = \theta \mathfrak{n}$.

3. Geodesics in the Horocycle Space.

Although the coset space G/MN is not symmetric in general, not even reductive (see Exercise **A1**) it does possess curves resembling geodesics. We shall now define these.

If $H \in \mathfrak{a}$ the geodesic $t \longrightarrow \exp tH \cdot o$ in X is said to be regular if the vector H is regular. More generally, a geodesic γ in X is said to be *regular* if its stabilizer $\{g \in G : \ g \cdot \gamma = \gamma\}$ has minimum dimension. By a *flat* in X we shall mean a totally geodesic submanifold of X whose curvature tensor vanishes identically. As shown by Cartan, the *maximal flats* in X are of the form $gA \cdot o$ $(g \in G)$ (cf. [DS], Ch. V, §6).

Let γ be a regular geodesic in X. Pick any point x on γ, and consider the isotropy subgroup K_x of G at x, its Lie algebra $\mathfrak{k}_x \subset \mathfrak{g}$, the orthogonal complement $\mathfrak{p}_x$ of $\mathfrak{k}_x$ in $\mathfrak{g}$ and the isomorphism $d\pi_x : \mathfrak{p}_x \longrightarrow X_x$ induced by the natural mapping $\pi_x : g \longrightarrow g \cdot x$ of G onto X. Let $d_x \in \mathfrak{p}_x$ be the vector such that $(d\pi_x)_e(d_x)$ is the tangent vector to the geodesic at x. The Iwasawa group N_x corresponding to the unique Weyl chamber (in the centralizer of d_x in $\mathfrak{p}_x$) which contains d_x is given by

$$N_x = \left\{ z \in G : \lim_{t \to +\infty} \exp(-td_x) z \exp(td_x) = e \right\}$$

(cf. [DS], Ch. VI, Exercise B2). The group N_x is independent of the choice of x on γ. To see this let $y \in \gamma$ be arbitrary. In order to prove $N_x = N_y$ we may assume $x = o$, $d_x = d_o \in \mathfrak{a}^+$ so $N_x = N$. Then $y = \exp(sd_o) \cdot o \quad (s \in \boldsymbol{R})$, $K_y = \exp(sd_o) K \exp(-sd_o)$ and $\mathfrak{p}_y = Ad(\exp(sd_o))\mathfrak{p}$. In particular $d_o \in \mathfrak{p}_y$. Also $\pi_y(\exp(td_o)) = \exp(t+s)d_o \cdot o$ so $(d\pi_y)_e(d_o)$ equals the tangent vector to γ at y. Since $d_o \in \mathfrak{p}_y$ the definition of d_y implies $d_o = d_y$, whence $N = N_y$. Thus we write N_γ instead of N_x. Putting

$$\Gamma_\gamma(t) = N_\gamma \cdot \gamma(t) \qquad (t \in \boldsymbol{R})$$

we call the curve Γ_γ a *geodesic in the horocycle space* $\Xi = G/MN$.

Lemma 1.9. *For* $g \in G, t \in \boldsymbol{R}$ *we have*

$$g \cdot \Gamma_\gamma(t) = \Gamma_{g\cdot\gamma(t)}$$

so the action of G *on* Ξ *permutes the geodesics.*

Proof. We have

$$g \cdot \Gamma_\gamma(t) = gN_\gamma \cdot \gamma(t) = gN_\gamma g^{-1} \cdot ((g \cdot \gamma)(t)). \tag{9}$$

If $x_1 = g \cdot x$ then $K_{x_1} = gK_x g^{-1}$, $\mathfrak{k}_{x_1} = Ad(g)\mathfrak{k}$, $\mathfrak{p}_{x_1} = Ad(g)\mathfrak{p}_x$ and

$$\pi_{x_1}(\exp\ tAd(g)d_x) = g \exp\ td_x \cdot x$$

so $(d\pi_{x_1})_e(Ad(g)d_x)$ equals the tangent vector to $g \cdot \gamma$ at x_1. Thus

$$N_{g\cdot\gamma} = gN_\gamma g^{-1} \tag{10}$$

which by (9) proves the lemma.

Corollary 1.10. *Any two horocycles* $\Gamma_\gamma(t_1)$ *and* $\Gamma_\gamma(t_2)$ *on the curve* Γ_γ *correspond under a transvection of* X *along* γ.

Because of Lemma 1.9 we may assume $\gamma(t) = \exp tH \cdot o$ $(H \in \mathfrak{a}^+)$ in which case the statement is obvious.

We shall now consider the problem of connecting two horocycles by means of a geodesic in Ξ.

Proposition 1.11. *Let* $\xi_1, \xi_2 \in \Xi$. *Then there exists a maximal flat* F *perpendicular to both* ξ_1 *and* ξ_2.

Proof. We may take $\xi_2 = \xi_o$. Writing $\xi_1 = k_1 a_1 \cdot \xi_o$ $(k_1 \in K, a_1 \in A)$ and writing by the Bruhat decomposition $k_1 = nam'm_1n_1$ we have $\xi_1 = nam' \cdot \xi_o$. The flat $F = nA \cdot o$ then has the stated property.

Remark. If ξ_1, ξ_2 in Prop. 1.11 are parallel there are infinitely many F with the stated property. In fact, if $\xi_2 = \xi_o$, then all the maximal flats $F_n = nA \cdot o \quad (n \in N)$ will qualify.

If m' is an arbitrary fixed element in M' it is convenient to refer to the isometry $x \longrightarrow m' \cdot x$ of X as a *reflection* in o. More generally, the isometry $x \longrightarrow gm'g^{-1} \cdot x$ will be called a reflection in $g \cdot o$ for $g \in NA$.

Proposition 1.12. *Let* $\xi_1, \xi_2 \in \Xi$. *Then*

$$\xi_1 = s\tau\xi_2,$$

where τ *is a transvection along a geodesic* γ *perpendicular to both* ξ_1 *and* ξ_2, *and* s *is a reflection in the point* $\xi_1 \cap \gamma$. *Finally* ξ_1 *and* $\tau\xi_2$ *have the same tangent space at this point.*

Proof. As before we can take $\xi_2 = \xi_o$ and $\xi_1 = nam' \cdot \xi_o \quad (n \in N, a \in A, m' \in M')$. Put $a_t = \exp(t \log a)$ and let γ be the geodesic $na_t \cdot o \quad (t \in \boldsymbol{R})$. The transvection $\tau = \tau(na)$ along γ maps ξ_o into $na \cdot \xi_o$ which under the reflection $s : \; x \longrightarrow nam'(na)^{-1} \cdot x$ gets mapped onto ξ_1. Finally we note that ξ_1 and $\tau\xi_2$ have the same tangent space at the point $na \cdot o$; in fact the flat $nA \cdot o$ is perpendicular to both. (See figure.)

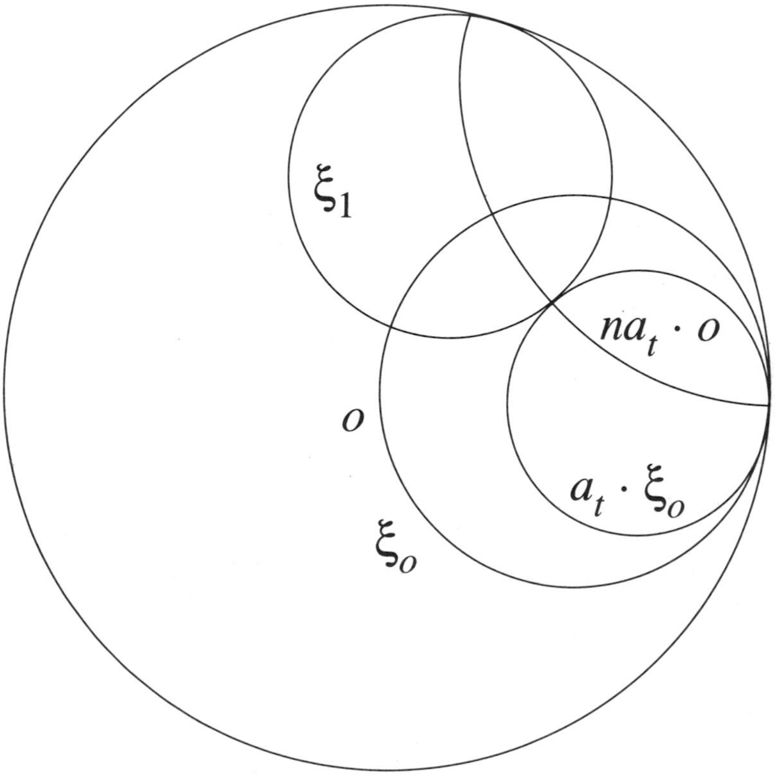

A submanifold Ξ_1 of Ξ will be called *totally geodesic* if each geodesic in Ξ which is tangential to Ξ_1 at some point lies completely in Ξ_1.

Now fix an element $H^o \in \overline{\mathfrak{a}^+}$ and let $X' = G'/K'$ be the corresponding *boundary component* as constructed in [GGA], Ch. IV, §8, No. 3. (Note that in [GGA], p. 461, line 4 from the bottom, $\mathfrak{a}^+$ should be $\overline{\mathfrak{a}^+}$ and in (9), p. 462 the symbol $\mathfrak{a}' \cap \mathfrak{a}^+$ should be omitted.) We recall the construction of X'. Let

$$\Sigma^o = \{\alpha \in \Sigma : \ \alpha(H^o) = 0\}, \tag{11}$$

let $\mathfrak{g}'$ denote the (semisimple) subalgebra of $\mathfrak{g}$ generated by all $\mathfrak{g}_\alpha (\alpha \in \Sigma^o)$ and put

$$\mathfrak{k}' = \mathfrak{g}' \cap \mathfrak{k}, \ \mathfrak{a}' = \mathfrak{g}' \cap \mathfrak{a}, \ \mathfrak{n}' = \mathfrak{g}' \cap \mathfrak{n}, \ \mathfrak{m}' = \mathfrak{g}' \cap \mathfrak{m}. \tag{12}$$

Putting

$$\mathfrak{a}^o = \{H \in \mathfrak{a} : \ \alpha(H) = 0 \text{ for } \alpha \in \Sigma^o\} \tag{13}$$

we see that

$$\mathfrak{a} = \mathfrak{a}' \oplus \mathfrak{a}^o \tag{14}$$

is an orthogonal decomposition with respect to the Killing form of $\mathfrak{g}$. The restrictions $\alpha' = \alpha|\mathfrak{a}'$ $(\alpha \in \Sigma^o)$ form the set Σ' of restricted roots of $\mathfrak{g}'$ with respect to $\mathfrak{a}'$. The subset

$$(\mathfrak{a}')^+ = \{H \in \mathfrak{a}' : \ \alpha(H) > 0 \qquad \text{for} \qquad \alpha \in \Sigma^o \cap \Sigma^+\} \tag{15}$$

is nonempty: if $H \in \mathfrak{a}^+$, its orthogonal projection $H' \in \mathfrak{a}'$ (relative to (14)) is contained in $(\mathfrak{a}')^+$. Also $(\mathfrak{a}')^+$ is a maximal connected subset of the subset $(\mathfrak{a}')'$ of $\mathfrak{a}'$ where all $\alpha \in \Sigma'$ are $\neq 0$. Thus $(\mathfrak{a}')^+$ is a Weyl chamber for Σ'. Let $(\Sigma')^+$ denote the corresponding set of positive roots. We have the compatible Iwasawa decompositions,

$$\mathfrak{g} = \mathfrak{k} + \mathfrak{a} + \mathfrak{n}, \qquad \mathfrak{g}' = \mathfrak{k}' + \mathfrak{a}' + \mathfrak{n}', \tag{16}$$

and

$$\mathfrak{a}' = \sum_{\alpha \in \Sigma^o} \boldsymbol{R} A_\alpha, \qquad \mathfrak{n}' = \sum_{\alpha \in \Sigma^o \cap \Sigma^+} \mathfrak{g}_\alpha. \tag{17}$$

Let G', K', A', N' denote the analytic subgroups of G corresponding to the algebras $\mathfrak{g}', \mathfrak{k}', \mathfrak{a}', \mathfrak{n}'$ above. Then (*loc. cit.* p. 463)

$$G' = K'A'N' \tag{18}$$

and

$$K' = G' \cap K, \qquad A' = G' \cap A, \qquad N' = G' \cap N. \tag{19}$$

If $g \in G'$ then $Ad_G(g)$ fixes $\mathfrak{a}^o$ pointwise; hence the centralizer M' of $\mathfrak{a}'$ in K' is contained in M, i.e. $M' \subset M$. On the other hand, $M \cap K' \subset M'$ so

$$M' = G' \cap M. \tag{20}$$

We split the set Σ^+ of positive roots by means of Σ^o and put

$$\Delta' = \Sigma^o \cap \Sigma^+, \qquad \Delta'' = \Sigma^+ - \Sigma^o \cap \Sigma^+, \tag{21}$$

$$\bar{\mathfrak{n}}' = \sum_{\alpha \in \Delta'} \mathfrak{g}_{-\alpha}, \qquad \bar{\mathfrak{n}}'' = \sum_{\alpha \in \Delta''} \mathfrak{g}_{-\alpha}. \tag{22}$$

Then $\bar{\mathfrak{n}}'$ and $\bar{\mathfrak{n}}''$ are Lie algebras whose direct sum is $\bar{\mathfrak{n}}$. Let $\overline{N}'$ and $\overline{N}''$ denote the corresponding analytic subgroups of G. Then $\mathfrak{g}'$ normalizes $\bar{\mathfrak{n}}''$ and G' normalizes $\overline{N}''$.

Proposition 1.13. *Let X' be a boundary component of X corresponding to a vector $H^o \in \overline{\mathfrak{a}^+}$. Let Ξ and Ξ' be the dual spaces of X and X', respectively, and let Ξ^o denote the orbit $G' \cdot \xi_o$ in Ξ. Then*

(i) *Ξ^o is totally geodesic in Ξ.*

(ii) *The mapping $\xi \longrightarrow \xi \cap X'$ is a bijection of Ξ^o onto Ξ'.*

Proof. Since $G' \cap MN = M'N'$ the mapping

$$g' \cdot \xi_o \longrightarrow g'M'N' \tag{23}$$

is a well-defined bijection of Ξ^o onto Ξ'. Since the Iwasawa decompositions $G = NAK$, $G' = N'A'K'$ are compatible we find that $(N \cdot o) \cap X' = N' \cdot o$ so $(g' \cdot \xi_o) \cap X' = g'N' \cdot o$ so the map (23) is just the map $\xi \longrightarrow \xi \cap X'$. This proves (ii).

For (i) we first consider a geodesic Γ_γ in Ξ tangential to Ξ^o at ξ_o. By Cor. 1.10 there exists a $Z \in \mathfrak{p}$ such that $\Gamma_\gamma(t) = \exp(tZ) \cdot \xi_o$, $(t \in \boldsymbol{R})$. The tangent space to $\Xi = G/MN$ at ξ_o is canonically identified with the factor space $\mathfrak{g}/(\mathfrak{m}+\mathfrak{n})$; the assumption about Γ_γ therefore means that there exists a vector $Y \in \mathfrak{g}'$ such that $Z - Y \in \mathfrak{m} + \mathfrak{n}$. Since $\mathfrak{g}' = \mathfrak{n}' + \mathfrak{m}' + \mathfrak{a}' + \theta\mathfrak{n}'$, $\mathfrak{m}' = \mathfrak{m} \cap \mathfrak{k}'$ we may assume that $Y \in \mathfrak{a}' + \theta\mathfrak{n}'$. Let $\mathfrak{q}$ (resp. $\mathfrak{q}'$) denote the orthogonal complement of $\mathfrak{a}$ in $\mathfrak{p}$ (resp. of $\mathfrak{a}'$ in $\mathfrak{p}'$). Then the mapping $L \longrightarrow L - \theta L$

is a bijection of $\mathfrak{n}$ onto $\mathfrak{q}$ mapping $\mathfrak{n}'$ onto $\mathfrak{q}'$. Let $X_1, \ldots, X_d$ be a basis of $\mathfrak{n}$ such that $X_1, \ldots, X_c$ span $\mathfrak{n}'$. Then for certain $H \in \mathfrak{a}$, $H' \in \mathfrak{a}'$ we have

$$Z = H + \sum_1^d c_i(X_i - \theta X_i), \qquad Y = H' + \sum_1^c b_j \theta X_j$$

for $c_i, b_j \in \boldsymbol{R}$. The relation $Z - Y \in \mathfrak{m} + \mathfrak{n}$ then implies $H = H'$, $c_i = 0$ for $c < i \leq d$ so $Z \in \mathfrak{p}'$. Thus Γ_γ lies in Ξ^o.

Now Lemma 1.9 reduces the general case to the special one considered so the proposition is proved.

§2. INVARIANT DIFFERENTIAL OPERATORS.

1. The Isomorphisms.

As in §1 let G be a connected semisimple Lie group with finite center, $X = G/K$ the associated symmetric space and $\Xi = G/MN$ the horocycle space for X. Let $\boldsymbol{D}(X)$ denote the algebra of differential operators D on X which are invariant under the action of G (cf. Ch. I, §3, No. 1). Similarly we consider the algebra $\boldsymbol{D}(\Xi)$ of G-invariant differential operators on Ξ. We also have to consider the algebra $\boldsymbol{D}(A)$ of differential operators on A which are invariant under all translations (A is abelian) as well as the subalgebra $\boldsymbol{D}_W(A) \subset \boldsymbol{D}(A)$ of invariants under the action of W on A.

When N acts on X the orbits are transversal to the submanifold $A \cdot o$ in the sense of [GGA], Ch. II §3 (29). Thus if D is a differential operator on X we can define the radial part $\Delta_N(D)$ which is a differential operator on A such that $(Df)(a \cdot o) = (\Delta_N(D)\overline{f})(a)$ for each N-invariant function f on $X, \overline{f}$ denoting $f|A \cdot o$ ([GGA], Ch. II, §3). If μ is a linear function on $\mathfrak{a}$ the function e^μ on A is defined by $e^\mu(a) = e^{\mu(\log a)}$.

Let $\boldsymbol{D}(G)$ denote the algebra of all differential operators on G invariant under all left translations and $\boldsymbol{D}_K(G) \subset \boldsymbol{D}(G)$ the subalgebra of operators invariant under all right translations by K. Since G/K is a reductive homogeneous space, $\boldsymbol{D}(X)$ is isomorphic to the algebra $\boldsymbol{D}_K(G)/\boldsymbol{D}_K(G) \cap \boldsymbol{D}(G)\mathfrak{k}$ (Helgason [1959]) which on the other hand is isomorphic to $\boldsymbol{D}_W(A)$ (Harish-Chandra [1958a]). This leads to the following result.

Theorem 2.1. *The mapping*

$$\Gamma: \; D \longrightarrow e^{-\rho}\Delta_N(D) \circ e^\rho \tag{1}$$

is an isomorphism of $\boldsymbol{D}(X)$ *onto* $\boldsymbol{D}_W(A)$.

For a proof of the above results see [GGA], Ch. II, §§4–5. We shall now obtain an analogous description of the algebra $\boldsymbol{D}(\Xi)$.

Because of Prop. 1.4, Ξ can be viewed as a trivial fibre bundle with base K/M and the fibre over kM given by $F_k = \{kaMN : a \in A\}$. If $g \in G$ let $k(g)$, $a(g)$, $n(g)$ denote the components in the Iwasawa decomposition $G = KAN$. Since

$$gkaMN = k(gk)a(gk)aMN \tag{2}$$

we see that the action of g on Ξ maps the fibre F_k onto the fibre $F_{k(gk)}$ inducing the translation

$$kaMN \longrightarrow k(gk)a(gk)aMN \tag{3}$$

on the fibres. Thus if U belongs to $\boldsymbol{D}(A)$ the operator D_U on $\mathcal{E}(\Xi)$ given by

$$(D_U\varphi)(kaMN) = U_a(\varphi(kaMN)) \tag{4}$$

belongs to $\boldsymbol{D}(\Xi)$. It turns out that the entire algebra $\boldsymbol{D}(\Xi)$ arises in this fashion.

Theorem 2.2. *The mapping $U \longrightarrow D_U$ is an isomorphism of $\boldsymbol{D}(A)$ onto $\boldsymbol{D}(\Xi)$. In particular, $\boldsymbol{D}(\Xi)$ is commutative.*

In preparation for the proof, put $H = MN$. If G/H were a reductive coset space we know from [GGA], Ch. II, Theorem 4.6 that the algebra $\boldsymbol{D}(\Xi) = \boldsymbol{D}(G/H)$ is determined by the symmetric algebra invariants for H acting on the tangent space $(G/H)_{\xi_o}$. Although G/H is not reductive in general we shall nevertheless determine $\boldsymbol{D}(G/H)$ by means of the mentioned invariants.

Consider the direct decomposition

$$\mathfrak{g} = (\mathfrak{m} + \mathfrak{n}) + (\mathfrak{l} + \mathfrak{a}), \tag{5}$$

where, as before, $\mathfrak{l}$ is the orthogonal complement of $\mathfrak{m}$ in $\mathfrak{k}$. Let σ be the projection of $\mathfrak{g}$ on $\mathfrak{l} + \mathfrak{a}$ given by (5). We know ([GGA], Ch. II, §4 (15)) that the natural mapping

$$\pi : G \longrightarrow G/H, \qquad d\pi_e : \mathfrak{g} \longrightarrow (G/H)_{\xi_o}$$

satisfies

$$\big(d\pi \circ Ad_G(h)\big)(Y) = \big(d\tau(h)_{\xi_o} \circ d\pi\big)(Y), \qquad h \in H,\ Y \in \mathfrak{g}.$$

Since $Ad_G(h)(Y)-(\sigma\circ Ad_G(h))(Y)$ is annihilated by $d\pi$ it follows that under the isomorphism

$$d\pi : \mathfrak{l}+\mathfrak{a} \longrightarrow (G/H)_{\xi_o}$$

the endomorphism $\sigma \circ Ad_G(h)$ of $\mathfrak{l}+\mathfrak{a}$ corresponds to $d\tau(h)_{\xi_o}$.

Lemma 2.3. *Let $I(\mathfrak{l}+\mathfrak{a})$ denote the set of polynomials in the symmetric algebra $S(\mathfrak{l}+\mathfrak{a})$ which are invariant under the action of the group $\sigma\circ Ad_G(MN)$. Then*

$$I(\mathfrak{l}+\mathfrak{a}) = S(\mathfrak{a}).$$

Proof. It will be convenient to complexify the situation. Thus let $\mathfrak{g}^C$ be the complexification of $\mathfrak{g}$ and if $\mathfrak{b}\subset\mathfrak{g}$ is any subspace let $\mathfrak{b}^C$ be the complex subspace of $\mathfrak{g}^C$ generated by $\mathfrak{b}$. We shall prove that $S(\mathfrak{a}^C)$ is the set of $\sigma\circ Ad_G(MN)$-invariants in $S(\mathfrak{l}^C+\mathfrak{a}^C)$. For $X\in\mathfrak{m}^C+\mathfrak{n}^C$ let $d(X)$ denote the derivation of $S(\mathfrak{l}^C+\mathfrak{a}^{C})$ extending the endomorphism $\sigma\circ ad_{\mathfrak{g}}(X)$ of $\mathfrak{l}^C+\mathfrak{a}^C$. We shall then prove the stronger result that if $q\in S(\mathfrak{l}^C+\mathfrak{a}^C)$ such that

$$d(X)q=0 \qquad \text{for } X\in\mathfrak{n}^C \tag{6}$$

then $q\in S(\mathfrak{a}^C)$. We extend $\mathfrak{a}$ to a maximal abelian subalgebra $\mathfrak{h}$ of $\mathfrak{g}$. Then $\mathfrak{h}$ is invariant under θ and $\mathfrak{h}^C$ is a Cartan subalgebra of $\mathfrak{g}^C$ ([DS], Ch. VI, §3). We order the set Δ of roots of $\mathfrak{g}^C$ with respect to $\mathfrak{h}^C$ in such a way that $\alpha\in\Delta$ is >0 whenever its restriction $\overline{\alpha}=\alpha|\mathfrak{a}$ is >0. Let $P_+=\{\alpha\in\Delta:\ \overline{\alpha}>0\}$ and let $\alpha_1>\alpha_2>\cdots>\alpha_p$ be the members of P_+ in decreasing order. In terms of the root spaces $(\mathfrak{g}^C)^\alpha$ we have (loc. cit)

$$\mathfrak{n}^C=\sum_{\alpha\in P_+}(\mathfrak{g}^C)^\alpha. \tag{7}$$

For each α select a nonzero vector $X_\alpha\in(\mathfrak{g}^C)^\alpha$ and put $E_\alpha=X_\alpha+\theta X_\alpha$. Then

$$\mathfrak{l}^C=\sum_{\alpha\in P_+}CE_\alpha. \tag{8}$$

If $\alpha\in\Delta$ let $H_\alpha\in\mathfrak{h}^C$ be determined by $B(H,H_\alpha)=\alpha(H)$ and put $H'_\alpha=\sigma(H_\alpha)$. If $\alpha\in P_+$ then $H'_\alpha\neq 0$; otherwise $H_\alpha\in\mathfrak{h}^C\cap\mathfrak{k}^C$ and $\overline{\alpha}=0$. If α^θ denotes the root $H\longrightarrow\alpha(\theta H)$ then $\alpha\longrightarrow-\alpha^\theta$ is a permutation of P_+. Let $\alpha,\beta\in P_+$. Then

$$d(X_\beta)(E_\alpha)=\sigma\big([X_\beta,X_\alpha+\theta X_\alpha]\big)=\sigma\big([X_\beta,\theta X_\alpha]\big)=c\,\sigma\big([X_\beta,X_{\alpha^\theta}]\big),$$

where $c \neq 0$. Since $\left[(\mathfrak{g}^{\boldsymbol{C}})^{\alpha}, (\mathfrak{g}^{\boldsymbol{C}})^{-\alpha}\right] = \boldsymbol{C}H_\alpha$ we have

$$d(X_\beta)E_\alpha = \begin{cases} 0 & \text{if } \beta > -\alpha^\theta \\ C_\beta H'_\beta & \text{if } \beta = -\alpha^\theta. \end{cases}$$

Here $C_\beta \neq 0$. Now suppose $q \in S(\mathfrak{l}^{\boldsymbol{C}} + \mathfrak{a}^{\boldsymbol{C}})$ satisfies (6). We write q as a polynomial in $E_{\alpha_1}, \ldots E_{\alpha_p}$ whose coefficients are polynomials in a fixed basis $H_1, \ldots, H_\ell$ of $\mathfrak{a}$:

$$q = \sum_{(n)} A_{(n)}(H_1, \ldots, H_\ell) E_{\alpha_1}^{n_1} \ldots E_{\alpha_p}^{n_p} \tag{9}$$

The derivation $d(X_{\alpha_1})$ annihilates each H_i and all $E_{\alpha_1}, \ldots, E_{\alpha_p}$ except one, say E_{α_j}, for which $-\alpha_j^\theta = \alpha_1$. Consequently,

$$0 = d(X_{\alpha_1})q = C_{\alpha_1} H'_{\alpha_1} \sum_{(n)}{}' A_{(n)}(H_1, \ldots, H_\ell) n_j E_{\alpha_1}^{n_1} \ldots E_{\alpha_j}^{n_j - 1} \ldots E_{\alpha_p}^{n_p}$$

where $\sum'$ ranges over the terms in (9) for which $n_j > 0$. Since $C_{\alpha_1} H'_{\alpha_1} \neq 0$ this last equation shows that $\sum'$ is an empty sum, in other words, E_{α_j} does not occur in (9). Applying $d(X_{\alpha_2})$ to (9) we see similarly that $E_{\alpha_k}\left(-\alpha_k^\theta = \alpha_2\right)$ does not occur in (9) etc. Since $\alpha \longrightarrow -\alpha^\theta$ is a permutation of P_+ this proves that q in (6) belongs to $S(\mathfrak{a})$; this proves the lemma because the converse inclusion is obvious.

For the proof of Theorem 2.2 let $H_{\ell+1}, \ldots, H_{\ell+p}$ be any basis of $\mathfrak{l}$. For $\epsilon > 0$ sufficiently small the mapping

$$\left(t_1, \ldots, t_{p+\ell}\right) \longrightarrow \exp\left(\sum_1^{p+\ell} t_i H_i\right) H$$

is a diffeomorphism of the cube $|t_i| < \epsilon$ onto an open neighborhood V of the origin ξ_o in G/H. The inverse is a local coordinate system on V and the coordinate vector fields $\partial_i = \frac{\partial}{\partial t_i}$ satisfy

$$\left(\frac{\partial}{\partial t_i}\right)_o = d\pi(H_i) \qquad (1 \leq i \leq \ell + p) \tag{10}$$

with π as in (4) §1. If $D \in \boldsymbol{D}(\Xi)$ there exists a unique polynomial P in $p+\ell$ variables such that for all $\varphi \in \mathcal{E}(\Xi)$,

$$(D\varphi)(\xi_o) = \left\{P(\frac{\partial}{\partial t_1}, \ldots, \frac{\partial}{\partial t_{p+\ell}}) \varphi(\exp(\sum_i t_i H_i) H)\right\}_{(t)=0}. \tag{11}$$

In the decomposition of P into monomials let P^* be the sum of the terms of highest degree. For $h \in H$ we write the mapping $\tau(h) : gH \longrightarrow hgH$ in coordinates

$$\tau(h)\exp\big(\Sigma t_i H_i\big)H = \exp\big(\Sigma s_i H_i\big)H.$$

Writing $F\big(t_1, \ldots, t_{p+\ell}\big) = \varphi\big(\exp\big(\Sigma t_i H_i\big)H\big)$ the invariance $D\big(\varphi \circ \tau(h)\big)(\xi_o) = (D\varphi)(\xi_o)$ implies

$$\left\{P\left(\frac{\partial}{\partial t_1}, \ldots, \frac{\partial}{\partial t_{p+\ell}}\right)\big(F(t_1, \ldots, t_{p+\ell}) - F\big(s_1, \ldots, s_{p+\ell}\big)\big)\right\}_{(t)=0} = 0 \tag{12}$$

Although the coordinate change $\big(t_1, \ldots, t_{p+\ell}\big) \longrightarrow \big(s_1, \ldots, s_{p+\ell}\big)$ is not linear the highest degree term P^* transforms linearly. In other words,

$$\big\{P^*\big(\partial_1, \ldots, \partial_{p+\ell}\big)F(s_1, \ldots, s_{p+\ell})\big\}_{(t)=0}$$

is a linear combination

$$\sum_\alpha a_\alpha \big(D^\alpha F\big)(0) \tag{13}$$

of derivatives of F at 0 and the sum of the highest degree monomial in (13) is obtained from P^* by the linear transformation given by the Jacobian matrix

$$\left(\frac{\partial s_i}{\partial t_j}\right)_{(t)=0.} \tag{14}$$

The function F in (12) being arbitrary we can isolate the highest degree derivatives in this equation and deduce that P^* is invariant under (14), in other words, P^* is invariant under the linear isotropy group $\big\{d\tau(h)_{\xi_o} : h \in H\big\}$. Because of Lemma 2.3 this implies that P^* is a polynomial in the derivatives $\partial_1, \ldots, \partial_\ell$ alone. The operator D_{P^*} defined by (4) then satisfies

$$\big(D_{P^*}\varphi\big)(\xi_o) = \big\{P^*\big(\partial_1, \ldots, \partial_\ell\big)\varphi\big(\exp\big(t_1 H_1 + \cdots + t_\ell H_\ell\big)H\big)\big\}_{(t)=0}.$$

Then $D - D_{P^*}$ has lower order than D at ξ_o, hence everywhere by the invariance. By induction we get $D = D_U$ for some $U \in \boldsymbol{D}(A)$. This concludes the proof of Theorem 2.2, the injectivity of the map $U \longrightarrow D_U$ being obvious.

2. Radial Part Interpretation.

Let us now consider the decompositions

$$X' = K \cdot \big(A^+ \cdot o\big) \tag{15}$$

(16) $$\Xi^* = MN \cdot \big(A \cdot \xi^*\big)$$

from (5), (8) §1. These are decompositions of open dense subsets of X and Ξ, respectively, into orbits of the isotropy groups with explicit cross sections $A^+ \cdot o$ and $A \cdot \xi^*$. The isotropy group orbits intersect these cross sections transversally so radial parts ([GGA], Ch. II, §3) of differential operators are well defined. Because of Theorems 2.1 and 2.2 we have two surjective isomorphisms

(17) $$\Gamma : \boldsymbol{D}(X) \longrightarrow \boldsymbol{D}_W(A),$$

(18) $$\widehat{\Gamma} : \boldsymbol{D}(\Xi) \longrightarrow \boldsymbol{D}(A),$$

where $\widehat{\Gamma}(D_U) = U$. These have interpretations in terms of the decompositions (15) and (16).

Proposition 2.4. *Let $D \in \boldsymbol{D}(X)$ and let $\Delta_K(D)$ be the radial part of D for the action of K on X' with transversal manifold $A^+ \cdot o$. Then*

(19) $$\Delta_K(D) = e^{-\rho}\Gamma(D) \circ e^{\rho} + E',$$

where E' has degree less than that of D and where

$$\lim_{t \to +\infty} E'_{tH} = 0 \textit{ for any } H \in \mathfrak{a}^+,$$

E'_{tH} being the operator E' with its coefficients evaluated at tH.

This is a consequence of a more explicit result in Harish-Chandra [1958a], Lemma 26, or [GGA], Ch. II, Prop. 5.23.

Proposition 2.5. *Let $D \in \boldsymbol{D}(\Xi)$ and let $\Delta_{MN}(D)$ be the radial part of D for the action of MN on Ξ^* with transversal manifold $A \cdot \xi^*$ (which we identify with A via the map $a \longrightarrow a \cdot \xi^*$). Then*

(20) $$\Delta_{MN}(D) = \widehat{\Gamma}(D)^{s^*},$$

the image of $\widehat{\Gamma}(D)$ under the involutive diffeomorphism $s^* : a \longrightarrow m^* a (m^*)^{-1}$ of A.

Proof. If φ is MN-invariant and we put $a^* = s^*(a)$ we have by (4)

$$\begin{aligned}\big(\Delta_{MN}(D)\varphi\big)\big(a^* \cdot \xi^*\big) &= \big(D\varphi\big)\big(m^* a \cdot \xi_o\big)\\ &= \widehat{\Gamma}(D)_a\big(\varphi(m^* a \cdot \xi_o)\big) = \widehat{\Gamma}(D)_a\big(\varphi(a^* \cdot \xi^*)\big)\end{aligned}$$

and this proves (20).

3. Joint Eigenspaces and Eigenspace Representations.

Let L be a Lie group and H a closed subgroup. Let $\boldsymbol{D}(L/H)$ denote the algebra of differential operators D on L/H which are invariant under all translations $\tau(l) : xH \longrightarrow lxH$ of L/H. Let $\chi : \boldsymbol{D}(L/H) \longrightarrow \boldsymbol{C}$ be a homomorphism and denote by E_χ and $\mathcal{D}'_\chi$ the *joint eigenspaces*

$$E_\chi = \{f \in \mathcal{E}(L/H) : \; Df = \chi(D)f \quad \text{for} \quad D \in \boldsymbol{D}(L/H)\}$$

$$\mathcal{D}'_\chi = \{T \in \mathcal{D}'(L/H) : \; DT = \chi(D)T \quad \text{for} \quad D \in \boldsymbol{D}(L/H)\}.$$

For the latter definition it is assumed that L/H has an L-invariant measure dl_H and that DT is defined by $(DT)(f) = T(D^*f)$ where the adjoint is taken relative to dl_H ([GGA], Ch. II, §2, No.3). The spaces E_χ and $\mathcal{D}'_\chi$ are given the topologies induced by those of $\mathcal{E}(L/H)$ and $\mathcal{D}'(L/H)$. The natural representation T_χ of L on E_χ, i.e. $\big(T_\chi(l)f\big)(xH) = f\big(l^{-1}xH\big)$ is called an *eigenspace representation*. Similarly we have an eigenspace representation of L on each $\mathcal{D}'_\chi$.

In order to have a rich supply of spaces $E_\chi, \mathcal{D}'_\chi$ the commutativity of $\boldsymbol{D}(L/H)$ is a desirable property. It is therefore useful to remark that L/H may be representable as another coset space L'/H' such that $\boldsymbol{D}(L'/H')$ is commutative although $\boldsymbol{D}(L/H)$ is not. Consider for example the nilpotent group N in the Iwasawa decomposition $G = KAN$. It can be viewed in various ways:

$$N = N \times N/\Delta N = MN/M$$

where ΔN is the diagonal in $N \times N$ and M the centralizer of A in K. While $\boldsymbol{D}(N)$ is abelian only if G/K is a hyperbolic space, $\boldsymbol{D}(N \times N/\Delta N)$ is always abelian ([GGA], Ch. II, Theorem 5.7) and $\boldsymbol{D}(MN/M)$ is abelian provided the symmetric space G/K has rank one (cf. Exercise **A7**).

Another method for dealing with the noncommutativity of $\boldsymbol{D}(L/H)$ is to replace it with a commutative subalgebra (see Ch. VI, §1, No. 2).

Now we apply these considerations to the symmetric space $X = G/K$ and the horocycle space $\Xi = G/MN$. Here $\boldsymbol{D}(G/K)$ and $\boldsymbol{D}(G/MN)$ are both commutative. Since the algebra $\boldsymbol{D}(G/K)$ contains elliptic operators, for example the Laplace-Beltrami operator, a joint eigendistribution would by classical regularity theorems be a C^∞ function (even analytic function). Thus $E_\chi = \mathcal{D}'_\chi$ for $\boldsymbol{D}(G/K)$.

Using the isomorphism (17) and (18) we can parameterize the set of these joint eigenspaces in a convenient fashion. For $\lambda \in \mathfrak{a}_c^*$ consider the following joint eigenspaces

$$\mathcal{E}_\lambda(X) = \{f \in \mathcal{E}(X) : \ Df = \Gamma(D)(i\lambda)f \text{ for } D \in \boldsymbol{D}(X)\} \tag{21}$$

$$\mathcal{D}'_\lambda(\Xi) = \{\Psi \in \mathcal{D}'(\Xi) : \ D\Psi = \widehat{\Gamma}(D)(i\lambda - \rho)\Psi \ \text{ for } \ D \in \boldsymbol{D}(\Xi), \tag{22}$$

and put

$$\mathcal{E}_\lambda(\Xi) = \mathcal{E}(\Xi) \cap \mathcal{D}'_\lambda(\Xi).$$

Proposition 2.6. (i) *The spaces $\mathcal{E}_\lambda(X)$ and $\mathcal{D}'_\lambda(\Xi)$ for $\lambda \in \mathfrak{a}_c^*$ constitute all the joint eigenspaces of the algebras $\boldsymbol{D}(X)$ and $\boldsymbol{D}(\Xi)$, respectively.*

(ii) *The functions f in $\mathcal{E}_\lambda(X)$ are characterized by the functional equation*

$$\int_K f(gk \cdot x)dk = \varphi_\lambda(x) f(g \cdot o), \qquad x \in X, g \in G,$$

where φ_λ is the unique spherical function in $\mathcal{E}_\lambda(X)$.

(iii) *The functions ψ in $\mathcal{E}_\lambda(\Xi)$ are characterized by the functional equation*

$$\psi(ga \cdot \xi_o) = \psi(g \cdot \xi_o) e^{(i\lambda - \rho)(\log a)}, \qquad g \in G, \ a \in A.$$

Proof. Because of (17) and [GGA], Lemma 3.11, Ch. III, each homomorphism of $\boldsymbol{D}(X)$ into $\boldsymbol{C}$ has the form $D \longrightarrow \Gamma(D)(i\lambda)$ for some $\lambda \in \mathfrak{a}_c^*$. For $\boldsymbol{D}(\Xi)$ the similar statement follows from (18) which also implies the functional equation for ψ. The functional equation for f is a special case of Prop. 2.4, Ch. IV in [GGA]. Note that $\mathcal{E}_\lambda(X) = \mathcal{E}_{s\lambda}(X)$ for each $s \in W$.

These joint eigenspaces $\mathcal{E}_\lambda(X)$ and $\mathcal{D}'_\lambda(\Xi)$ will be of major interest later on, particularly as natural representation spaces for the group G. Let T_λ and τ_λ denote the natural representations of G on $\mathcal{E}_\lambda(X)$ and $\mathcal{E}_\lambda(\Xi)$, respectively, i.e.

$$\big(T_\lambda(g)f\big)(x) = f(g^{-1} \cdot x), \qquad \big(\tau_\lambda(g)\psi\big)(\xi) = \psi(g^{-1} \cdot \xi)$$

and let $\widetilde{\tau}_\lambda$ denote the corresponding representation of G on $\mathcal{D}'_\lambda(\Xi)$, i.e.

$$\widetilde{\tau}_\lambda(g)\Psi = \Psi^{\tau(g)} \qquad \Psi \in \mathcal{D}'_\lambda(\Xi).$$

Here $\mathcal{D}'(\Xi)$ is, as usual, given the strong topology (cf. Ch. I, §2, No. 6).

4. The Mean Value Operators.

Let G be a connected Lie group and K a compact subgroup. *We do not assume in this subsection that G/K is symmetric.* Let $\pi : \ G \longrightarrow G/K$ be the natural mapping and put $o = \pi(e)$. For each $x \in G$ we associate a linear operator $M^x : C(G/K) \longrightarrow C(G/K)$ defined by

$$(M^x f)(gK) = \int_K f(gkx \cdot o)dk, \qquad g \in G, \tag{23}$$

where dk is the normalized Haar measure on K. If G/K is two-point homogeneous, $(M^x f)(p)$ is the average of f over the sphere $S_{d(o,x\cdot o)}(p)$.

The isotropy subgroup of G at gK is gKg^{-1} and the set $\{gkx\cdot o : k \in K\}$ in (23) is just the orbit $gKg^{-1} \cdot (gx \cdot o)$ of the point $gx \cdot o$ under this isotropy subgroup. Thus (23) is well defined even if f is not defined on all of G/K. Also

$$(M^x f)(gK) = \int_K \widetilde{f}(gkx)dk \tag{24}$$

if $\widetilde{f} = f \circ \pi$ and $M^{kxk'} = M^x$ for $k, k' \in K$. From [GGA], II, Cor. 4.10 and III, Theorem 1.10 we know that $\boldsymbol{D}(G/K)$ is generated by finitely many operators, say $\Delta_1, \ldots \Delta_\ell$ (and the identity). We can assume that the Δ_i annihilate the constants ("have no constant term").

Let $\mathfrak{g}$ and $\mathfrak{k}$ denote the Lie algebras of G and K, respectively, and let $\mathfrak{p}$ denote a fixed subspace of $\mathfrak{g}$ such that $\mathfrak{g} = \mathfrak{k} + \mathfrak{p}$ (direct sum) and $Ad_G(K)\mathfrak{p} \subset \mathfrak{p}$. Select a positive definite quadratic form on $\mathfrak{p}$, invariant under $Ad_G(K)$. This gives rise to a G-invariant Riemannian structure on G/K.

Theorem 2.7. *Let $p \in G/K$ and f an analytic function on a neighborhood U of p. Then there exists a ball $B_\delta(0) \subset \mathfrak{p}$ and a neighborhood $V \subset U$ of p with the property: for each $X \in B_\delta(0)$ there exists a sequence of polynomials $P_1, P_2, \ldots$ without constant term such that for $q \in V$,*

$$(M^{\exp X} f)(q) = f(q) + \sum_{m=1}^{\infty} (P_m(\Delta_1, \ldots, \Delta_\ell)f)(q). \tag{25}$$

Here P_m is given by

$$\left(\int_K (Ad(k)\widetilde{X})^m dk\right)\widetilde{f} = m!(P_m(\Delta_1, \ldots, \Delta_\ell)f)^{\sim},$$

$\widetilde{X}$ being the left invariant vector field such that $\widetilde{X}_e = X$ and, by definition, $Ad(k)\widetilde{X} = (Ad(k)X)^{\sim}$.

Proof. Let B be a submanifold of G which π maps diffeomorphically onto a neighborhood $W \subset U$ of p. For each $q \in W$ let g_q be the element in B for which $\pi(g_q) = q$. The mapping

$$(t_1, \ldots, t_r) \longrightarrow \pi(g_p \exp(t_1 X_1 + \cdots + t_r X_r))$$

is a diffeomorphism of an open ball in $\boldsymbol{R}^r$ onto an open neighborhood W_p of p. The inverse mapping, say φ_p, is a coordinate system on W_p. For each $q \in W$ put

$$\varphi_q = \varphi_p \circ \tau\big(g_p g_q^{-1}\big), \qquad W_q = \tau\big(g_q g_p^{-1}\big) W_p.$$

Then the pair $\big(\varphi_q, W_q\big)$ is a local chart on G/K containing q. It is clear that W_q is "close to" W_p if q is close to p. We can therefore select an open neighborhood V' of p so small that for each $q \in V'$, the neighborhood W_q contains V' and such that the function $f \circ \varphi_p^{-1}$ has a Taylor series expansion

$$\big(f \circ \varphi_p^{-1}\big)\big(t_1, \ldots, t_r\big) = \sum a_{n_1 \ldots n_r}(p) t_1^{n_1} \ldots t_r^{n_r}$$

absolutely convergent on $\varphi_p(V')$. The coordinates $t_i(\varphi_p(s))$ and $t_j(\varphi_q(s))$ are connected by

$$\begin{aligned} &\pi\big(g_p \exp\big(t_1(\varphi_p(s))X_1 + \cdots + t_r(\varphi_p(s))X_r\big)\big) \\ &= \pi\big(g_q \exp\big(t_1(\varphi_q(s))X_1 + \cdots + t_r(\varphi_q(s))X_r\big)\big) \end{aligned}$$

for $s \in W_q \cap W_p$. Shrinking V' further we can assume that for each $q \in V'$

$$t_i(\varphi_p(s)) = P_q^i(t_1(\varphi_q(s)), \ldots, t_r(\varphi_q(s))) \qquad (1 \leq i \leq r)$$

where $P_q^1, \ldots, P_q^r$ are power series absolutely convergent for $s \in V'$. The Taylor series for $f \circ \varphi_q^{-1}$ around the origin is obtained by substituting the expressions for $t_i(\varphi_p(s))$ into the Taylor series for $f \circ \varphi_p^{-1}$. We can therefore shrink V' once more so that for each $q \in V'$ the Taylor expansion for $f \circ \varphi_q^{-1}$,

$$f \circ \varphi_q^{-1} = \sum a_{n_1 \ldots n_r}(q) t_1^{n_1} \ldots t_r^{n_r},$$

is absolutely convergent on $\varphi_q(V')$. Since $\varphi_q(V') = \varphi_p(\tau(g_p g_q^{-1})V')$ we can select an open neighborhood V of p, $p \in V \subset V'$, such that the set $\bigcap_{q \in V} \varphi_q(V')$ contains an open ball $\Sigma_i t_i^2 < \delta^2$ in $\boldsymbol{R}^r$. We may assume that $g_p K g_p^{-1} \cdot V \subset V$ and that V has compact closure contained in V'. Now let $X \in B_\delta(0) \subset \mathfrak{p}$ and $q \in V$. Then

$$\big[M^{\exp X} f\big](q) = \int_K \widetilde{f}\big(g_q k \exp X k^{-1}\big) dk = \int_K \widetilde{f}\big(g_q \exp Ad(k)X\big) dk.$$

Putting $Ad(k)X = t_1X_1 + \cdots + t_rX_r$ we have by [GGA], II, Theorem 4.3

$$\begin{aligned}&\left[\left(Ad(k)\widetilde{X}\right)^m \widetilde{f}\right](g_q)\\ &= \left[\left(t_1\frac{\partial}{\partial x_1} + \cdots + t_r\frac{\partial}{\partial x_r}\right)^m \widetilde{f}(g_q\exp(x_1X_1 + \cdots + x_rX_r))\right](0)\\ &= \left[\left(t_1\frac{\partial}{\partial x_1} + \cdots + t_r\frac{\partial}{\partial x_r}\right)^m (f\circ\varphi_q^{-1})(x_1,\ldots,x_r)\right](0).\end{aligned}$$

Since $t_1^2 + \cdots + t_r^2 < \delta^2$ and since the Taylor series for $f \circ \varphi_q^{-1}$ converges in the ball $t_1^2 + \cdots + t_r^2 < \delta^2$, we have

$$\int_K \widetilde{f}(g_q \exp Ad(k)X)dk = \int_K \sum_0^\infty \frac{1}{m!}\left[\left(Ad(k)\widetilde{X}\right)^m \widetilde{f}\right](g_q)dk.$$

Due to the uniform convergence of the Taylor series, the summation and integration can be interchanged. For $m > 0$, the differential operator

$$D_X = \int_K (Ad(k)\widetilde{X})^m dk = \int_K Ad(k)\cdot \widetilde{X}^m dk$$

belongs to $\boldsymbol{D}_K(G)$. This operator gives rise to an operator in $\boldsymbol{D}(G/K)$ which can be written $m!P_m(\Delta_1,\ldots,\Delta_\ell)$ where P_m is a polynomial without constant term. Since

$$D_X\widetilde{F} = (m!P_m(\Delta_1,\ldots,\Delta_\ell)F)^\sim$$

for $F \in C^\infty(G/K)$, we obtain

$$[M^{\exp X}f](q) = f(q) + \sum_1^\infty [P_m(\Delta_1\ldots,\Delta_\ell)f](q),$$

and this proves the theorem.

Corollary 2.8. *Let φ be a spherical function on G/K and $\lambda_1,\ldots\lambda_\ell$ given by $\Delta_i\varphi = \lambda_i\varphi$ $(1 \le i \le \ell)$. Then for x sufficiently close to e in G*

$$\widetilde{\varphi}(x) = 1 + \sum_{m=1}^\infty P_m(\lambda_1,\ldots,\lambda_\ell) \tag{26}$$

and on analytic functions

$$M^x = I + \sum_{m=1}^\infty P_m(\Delta_1,\ldots\Delta_\ell). \tag{27}$$

In fact, the spherical function φ is analytic and by (24), $M^x\varphi = \widetilde{\varphi}(x)\varphi$.

Example. Consider the plane $\boldsymbol{R}^2$ as a homogeneous space $\boldsymbol{R}^2 = \boldsymbol{M}_o(\boldsymbol{R}^2)/\boldsymbol{SO}(2)$ ([GGA], IV, §2, No. 3). Then the function

$$\varphi(x,y) = J_o\big((-\lambda)^{\frac{1}{2}}(x^2+y^2)^{\frac{1}{2}}\big) = \sum_{k=0}^{\infty} \frac{1}{2^{2k}(k!)^2}\lambda^k(x^2+y^2)^k$$

is a spherical function on $\boldsymbol{R}^2$ and $L_{\boldsymbol{R}^2}\varphi = \lambda\varphi$. Then (27) becomes the well-known expansion

$$M^r = J_o\big(\sqrt{-L}\,r\big) = \sum_{k=0}^{\infty} \frac{1}{2^{2k}(k!)^2}L^k r^{2k} \tag{28}$$

for the operator of spherical average ([GGA], I, §2).

Now suppose G is unimodular with Haar measure dg. The convolution

$$(f*g)(x) = \int_G f(xy^{-1})g(y)dy = \int_G f(y)g(y^{-1}x)dy \tag{29}$$

is well-defined for $f, g \in C(G)$ if at least one of them has compact support. The subspace $C_c(G)$ is closed under convolution and the subspace $C_c^\natural(G)$ of the functions $f \in C_c(G)$ which are bi-invariant under K is a subalgebra.

Theorem 2.9. *The following properties of G/K are equivalent:*

(i) *The algebra $\boldsymbol{D}(G/K)$ of invariant differential operators on G/K is commutative.*

(ii) *The convolution algebra $C_c^\natural(G)$ is commutative.*

Proof. First we prove (i) $\Longrightarrow$ (ii). Let $f \in C_c(G/K)$ be analytic in a relatively compact neighborhood C_o of o. Then by Theorem 2.7,

$$\big(M^xM^yf\big)(o) = \big(M^yM^xf\big)(o) \tag{30}$$

for all $x, y \in G$ near e. Taking f K-invariant (30) implies

$$\int_K F(xky)dk = \int_K F(ykx)dk \tag{31}$$

for $F \in C_c^\natural(G)$ analytic on a connected relatively compact neighborhood C of e. By analyticity (31) holds for all $x, y \in G$ for which $xKy \cup yKx \subset C$.

Now given a compact subset V of G the functions $H \in C_c(G)$ analytic on a neighborhood of V form a dense subset of $C(V)$. (Compare Ch. V, Lemma 2.9.) Thus (31) holds for all $x, y \in G$ and all $F \in C_c^\natural(G)$. Replacing in (31) y by y^{-1} and integrating against $\varphi(y)$, $\varphi \in C_c^\natural(G)$, we obtain the desired relation

$$F * \varphi = \varphi * F. \tag{32}$$

Conversely, suppose (ii) holds and let $\boldsymbol{D}_K(G)$ be as in subsection No. 1 (without G assumed semisimple). Let $f, g \in \mathcal{D}^\natural(G)$, the convolution algebra of K-bi-invariant functions in $\mathcal{D}(G)$ and let $D, E \subset \boldsymbol{D}_K(G)$. Since D is left invariant, (ii) implies

$$D(f * g) = f * Dg = Dg * f,$$

so

$$DE(f * g) = D(Eg * f) = Eg * Df = Df * Eg,$$
$$ED(f * g) = ED(g * f) = Eg * Df.$$

Thus E and D commute on the subset $\mathcal{D}^\natural \times \mathcal{D}^\natural$ of $\mathcal{D}^\natural$, hence by density, on all of $\mathcal{D}^\natural$.

Now let $F \in \mathcal{D}(G), F(gk) \equiv F(g)$ and consider

$$F^*(x, y) = (M^x F)(y).$$

Let the subscript 1 denote differentiation with respect to the first variable. Since F^* is bi-invariant under K in the first variable we have

$$(D_1 F^*)(x, y) = \int_K (DF)(ykx)dk = (DF)^*(x, y)$$

so

$$(DF)(y) = (D_1 F^*)(e, y).$$

Hence

$$\begin{aligned}((DE)F)(y) &= (D_1(EF)^*)(e, y) = (D_1 E_1 F^*)(e, y)\\ &= (E_1 D_1 F^*)(e, y) = (EDF)(y)\end{aligned}$$

so (i) holds.

§3. THE RADON TRANSFORM AND ITS DUAL.

1. Measure-theoretic Preliminaries.

Before discussing the main subject of this section in detail it will be convenient to establish some conventions about the normalization of invariant measures on the groups and homogeneous spaces considered in §§1–2.

The Killing form B induces Euclidean measures on A, its Lie algebra $\mathfrak{a}$ and the dual space $\mathfrak{a}^*$. If $\ell = \dim(A)$ we multiply these measures by the factor $(2\pi)^{-\ell/2}$ and thereby obtain invariant measures da, dH and $d\lambda$ on $A, \mathfrak{a}$ and $\mathfrak{a}^*$ respectively. This normalization has the property that the Fourier transform

(1) $$f^*(\lambda) = \int_A f(a)e^{-i\lambda(\log a)}da, \qquad \lambda \in \mathfrak{a}^*$$

is inverted without a multiplicative constant, i.e.

(2) $$f(a) = \int_{\mathfrak{a}^*} f^*(\lambda)e^{i\lambda(\log a)}d\lambda, \quad f \in \mathcal{D}(A).$$

We normalize the Haar measures dk and dm on the compact groups K and M, respectively, such that the total measure is 1. The Haar measures dn and $d\overline{n}$ on the nilpotent groups N and $\overline{N} = \theta N$ are normalized such that

(3) $$\theta(dn) = d\overline{n}, \quad \int_{\overline{N}} e^{-2\rho(H(\overline{n}))}d\overline{n} = 1,$$

(cf. [GGA], Ch. IV, §6). Then we normalize the Haar measure dg on G such that

(4) $$\begin{aligned}\int_G f(g)dg &= \int_{KAN} f(kan)e^{2\rho(\log a)}dk\ da\ dn \\ &= \int_{NAK} f(nak)e^{-2\rho(\log a)}dn\ da\ dk\end{aligned}$$

(loc. cit., Ch. I, §5). If $m \in M$ the mapping $n \longrightarrow mnm^{-1}$ is an automorphism of N mapping dn into a multiple of dn. Since M is compact, dn is actually preserved and this in turn implies that the product measure $dm\ dn$ is a bi-invariant measure on NM.

If U is a Lie group and P a closed subgroup, with left invariant measures du and dp, respectively, the U-invariant measure du_P on U/P (when it exists) will be normalized by

$$\int_U f(u)du = \int_{U/P}\left(\int_P f(up)dp\right)du_P. \tag{5}$$

In particular, we have a K-invariant measure dk_M on K/M of total measure 1. We shall also use the notation

$$dx = dg_K, \quad d\xi = dg_{MN}$$

for the invariant measures on $X = G/K$ and $\Xi = G/MN$ respectively. By uniqueness, dx is a constant multiple of the measure on X induced by the Riemannian structure on X given by the Killing form B.

Lemma 3.1. *For each $\varphi \in \mathcal{D}(\Xi)$*

$$\int_\Xi \varphi(\xi)d\xi = \int_{K/M\times A} \varphi(ka\cdot\xi_o)e^{2\rho(\log a)}da\,dk_M$$

$$\int_\Xi \varphi(\xi)d\xi = \int_{AN} \varphi(an\cdot\xi^*)da\,dn$$

$$\int_\Xi \varphi(\xi)d\xi = \int_{\overline{N}A} \varphi(\overline{n}a\cdot\xi_o)e^{2\rho(\log a)}d\overline{n}\,da.$$

Proof. Because of the diffeomorphism in Prop. 1.4 (ii) there exists an analytic function D on $K/M \times A$ such that

$$\int_\Xi \varphi(\xi)d\xi = \int_{K/M\times A} \varphi(ka\cdot\xi_o)D(kM,a)da\,dk_M. \tag{6}$$

Since $d\xi$ is K-invariant we see by replacing φ by $\varphi^{\tau(k)}$ that D is constant in the first variable so we write $D(a)$ instead of $D(kM,a)$. Let $\alpha \in \mathcal{D}(A)$, $\beta \in \mathcal{D}(N)$ be arbitrary and define $F \in \mathcal{D}(G)$, $\varphi \in \mathcal{D}(\Xi)$ by

$$F(kan) = \alpha(a)\beta(n), \qquad \varphi(gMN) = \int_{MN} F(gmn)dm\,dn.$$

Then

$$\int_{\Xi} \varphi(\xi)d\xi = \int_{G} F(g)dg = \int_{KAN} F(kan)e^{2\rho(\log a)}dkdadn$$
$$= \int_{A} \alpha(a)e^{2\rho(\log a)}da \int_{N} \beta(n)dn.$$

On the other hand, by (6),

$$\int_{\Xi} \varphi(\xi)d\xi = \int_{A} \alpha(a)D(a)da \int_{N} \beta(n)dn$$

so the first relation of the lemma follows. For the second let S denote the group NA. Then $s \longrightarrow s \cdot \xi^*$ is a bijection of S onto Ξ^*. The measure on S which hereby corresponds to $d\xi$ is a left invariant Haar measure on S. But the measure

$$F \longrightarrow \int_{AN} F(an)da\, dn$$

is a left invariant Haar measure on S so the desired formula holds with a constant factor, say c, on the right. To evaluate c we note that

$$\int_{\Xi} \varphi(\xi)d\xi = \int_{\Xi} \varphi(m^* \cdot \xi)d\xi = c \int_{NA} \varphi(m^* anm^* \cdot \xi_o)da\, dn.$$

Writing $\overline{n} = (m^*)^{-1}nm^*$ and recalling that $\theta(dn) = d\overline{n}$, $s^*(da) = da$ this integral becomes

$$c \int_{\overline{N}A} \varphi(a\overline{n}MN)da\, d\overline{n} = c \int_{\overline{N}A} \varphi(\overline{n}aMN)e^{2\rho(\log a)}dad\overline{n}$$

the last relation coming from [GGA], Ch. I, Cor. 5.2. Here we put $\overline{n} = k(\overline{n}) \exp H(\overline{n})n_1$ so the last integral is

$$c \int_{\overline{N}A} \varphi(k(\overline{n})aMN)e^{2\rho(\log a)}e^{-2\rho(H(\overline{n}))}d\overline{n}da$$

which by the formula

$$\int_{K/M} F(kM)dk_M = \int_{\overline{N}} F(k(\overline{n})M)e^{-2\rho(H(\overline{n}))}d\overline{n}$$

(e.g. [GGA], Ch. I) reduces to

$$c \int_{(K/M)\times A} \varphi(kaMN)e^{2\rho(\log a)}dk_M da.$$

Comparing with the first formula of the lemma we get $c = 1$ as desired.

2. Integral Transforms and Differential Operators.

Let $X = G/K$, $\Xi = G/MN$ be as above and note that each $\xi \in \Xi$, as a submanifold of X, inherits a Riemannian structure (and hence a measure) from that of X. The mapping $n \longrightarrow n \cdot o$ is a diffeomorphism of N onto the horocycle ξ_o. We carry the measure dn over on ξ_o. This gives us a measure ds on ξ_o, which is invariant under the action of MN. Thus we get by translation by elements of G a well defined measure ds on each horocycle ξ; by uniqueness this is a multiple of the Riemannian measure on ξ by a constant factor, independent of ξ.

The *Radon transform* (or the *horocycle transform*) of a function f on X is defined by

$$\widehat{f}(\xi) = \int_{\xi} f(x)\, ds(x) \tag{7}$$

for all $\xi \in \Xi$ for which this integral exists. Since ξ is a closed submanifold of X the integral (7) exists for all ξ if f has compact support.

Along with (7) the *dual transform* $\varphi \longrightarrow \check{\varphi}$ is defined for each continuous φ on Ξ by

$$\check{\varphi}(x) = \int_{\check{x}} \varphi(\xi)d\mu(\xi), \tag{8}$$

where $\check{x}$ is the set of horocycles passing through x. The isotropy subgroup of G at x acts transitively on $\check{x}$ and $d\mu$ is the corresponding invariant measure normalized such that $\mu(\check{x}) = 1$. The spaces $X = G/K$ and $\Xi = G/MN$ are homogeneous spaces in duality in the sense of Ch. I, §3 and satisfy the transversality condition of Lemma 3.2, Ch. I (cf. Exercise **A1**). Furthermore,

$$\int_{\Xi} \widehat{f}(\xi)\varphi(\xi)d\xi = \int_{X} f(x)\check{\varphi}(x)dx. \tag{9}$$

For convenience we write the integral transforms (7) and (8) in group-theoretic terms:

$$\widehat{f}(g \cdot \xi_o) = \int_N f(gn \cdot o)dn, \quad \overset{\vee}{\varphi}(g \cdot o) = \int_K \varphi(gk \cdot \xi_o)dk. \tag{10}$$

The following injectivity result provides a good motivation for studying these transforms.

Theorem 3.2. *Let $f \in L^1(X)$. Then $\widehat{f}(\xi)$ exists for almost all $\xi \in \Xi$. If $\widehat{f}(\xi) = 0$ for almost all ξ then $f(x) = 0$ for almost all x.*

Proof. Define $F \in L^1(G)$ by $F(g) = f(g \cdot o)$. By the Fubini theorem,

$$\begin{aligned} \infty > \int_G |F(g)|dg &= \int_{KAN} |F(kan)|e^{2\rho(\log a)}dk\, da\, dn \\ &= \int_{KA} e^{2\rho(\log a)} \int_N |f(kan \cdot o)|dn\, dk\, da \\ &= \int_{(K/M)\times A} e^{2\rho(\log a)}\left(\int_N |f(kan \cdot o)|dn\right)dk_M da \end{aligned}$$

so for almost all $(kM, a) \in (K/M) \times A$,

$$\int_N |f(kan \cdot o)|dn < \infty.$$

Thus by Prop. 1.4, $\widehat{f}(\xi)$ exists for almost all $\xi \in \Xi$.

Next let $L^\natural(G)$ denote the space of functions in $L^1(G)$ which are bi-invariant under K. For $\lambda \in \mathfrak{a}_C^*$ we consider the spherical function

$$\varphi_\lambda(g) = \int_K e^{(i\lambda+\rho)(A(kg))}dk, \tag{11}$$

where $\exp A(g)$ ($A(g) \in \mathfrak{a}$) is the A-component of g in the Iwasawa decomposition $G = NAK$. Let $\lambda = \xi + i\eta$ $(\xi, \eta \in \mathfrak{a}^*)$ be such that φ_λ is bounded. The set of $\mu \in \mathfrak{a}_C^*$ for which φ_μ is bounded is the tube over the convex hull of the points $s\rho$ ($s \in W$) ([GGA], Ch. IV, §8). In particular, $\varphi_{i\eta}$ is bounded, so if $F \in L^\natural(G)$

$$\int_K \int_G |F(g)e^{(i\lambda+\rho)(A(kg))}|dgdk = \int_G |F(g)|\varphi_{i\eta}(g)dg < \infty.$$

Thus the inner integral is finite for almost all k, hence for all $k \in K$ because of the bi-invariance of F under K. Thus using (4) and the Fubini theorem on $G = NAK$ we get for φ_λ bounded,

$$\int_G F(g)\varphi_\lambda(g)dg = \int_G F(g)e^{(i\lambda+\rho)(A(g))}dg = \int_A \varphi_F(a)e^{i\lambda(\log a)}da$$

where

$$\varphi_F(a) = e^{-\rho(\log a)} \int_N F(na)dn.$$

Now suppose $f \in L^1(X)$ satisfies $\widehat{f}(\xi) = 0$ for almost all $\xi \in \Xi$. Let $\varphi \in L^\natural(G) \cap \mathcal{D}(G)$ and consider the function

$$F_o(g) = \int_G \varphi(h)f(h^{-1}g \cdot o)dh,$$

which clearly belongs to $L^\natural(G)$ and satisfies

$$\int_N F_o(an)dn = \int_G \varphi(h)\widehat{f}(h^{-1}a \cdot \xi_o)dh = 0$$

for all a. Thus $\varphi_{F_o} \equiv 0$ so

$$\int_G F_o(g)\varphi_\lambda(g)dg = 0 \qquad \text{for } \varphi_\lambda \text{ bounded.}$$

By the semisimplicity of the Banach algebra $L^\natural(G)$ ([GGA], Ch. IV, Prop. 3.8) this implies $F_o \equiv 0$. Letting $\varphi \longrightarrow \delta$, the delta distribution at e we get $f = 0$ in $L^1(X)$.

The injectivity of $f \longrightarrow \widehat{f}$ being established we shall now investigate how the transforms (10) intertwine with the invariant differential operators on X and Ξ. In Chapter IV when more tools have been developed we shall determine the ranges $\mathcal{D}(X)^\wedge$ and $\mathcal{E}(\Xi)^\vee$ as well as the kernel of $\varphi \longrightarrow \check{\varphi}$.

Note that the mapping $D \longrightarrow e^\rho D \circ e^{-\rho}$ is an automorphism of $\boldsymbol{D}(A)$ (extension of the mapping $H \longrightarrow H - \rho(H)$ $(H \in \mathfrak{a})$ to the symmetric algebra $S(\mathfrak{a})$) so we can define a mapping

$$D \longrightarrow \widehat{D} \qquad \text{of } \boldsymbol{D}(X) \text{ into } \boldsymbol{D}(\Xi)$$

by the condition

$$\widehat{\Gamma}(\widehat{D}) = e^{-\rho}\Gamma(D) \circ e^{\rho}. \tag{12}$$

Equivalently we can write (by (18) §2)

$$\widehat{D} = D_{e^{-\rho}\Gamma(D)\circ e^{\rho}}. \tag{13}$$

Theorem 3.3. *For the isomorphism*

$$D \longrightarrow \widehat{D} = D_{e^{-\rho}\Gamma(D)\circ e^{\rho}}$$

of $\boldsymbol{D}(X)$ *into* $\boldsymbol{D}(\Xi)$ *we have*

$$(Df)^{\wedge} = \widehat{D}\widehat{f} \qquad\qquad f \in \mathcal{D}(X)$$
$$(\widehat{D}\varphi)^{\vee} = D\check{\varphi} \qquad\qquad \varphi \in \mathcal{E}(\Xi).$$

This result is of course closely related to Prop. 3.4 in Ch. I. However it is not contained in it since the operators in $\boldsymbol{D}(X)$ are not always induced by the bi-invariant operators on G (cf. Ch. III, Theorem 10.3).

For the proof we recall that

$$\int_N f(an \cdot o)dn = e^{-2\rho(\log a)} \int_N f(na \cdot o)dn \tag{14}$$

([GGA], Ch. I, Cor. 5.2). With $A(g) \in \mathfrak{a}$ as before we put

$$r(g \cdot o) = e^{2\rho(A(g))},$$
$$f^*(g \cdot o) = e^{-2\rho(A(g))} \int_N f(ng \cdot o)dn.$$

Both of these are N-invariant functions on X and by (14),

$$\int_N (Df)(na \cdot o)dn = (D(rf^*))(a \cdot o)$$
$$= \Delta_N(D)_a\Big(e^{2\rho(\log a)}\widehat{f}(a \cdot \xi_o)\Big).$$

Replacing f by Df in (14) this gives

$$(Df)^{\wedge}(a \cdot \xi_o) = e^{-2\rho(\log a)}\Delta_N(D)_a\Big(e^{2\rho(\log a)}\widehat{f}(a \cdot \xi_o)\Big),$$

and by (1) §2 and (12), (13) this expression reduces to

$$\widehat{\Gamma}(\widehat{D})_a\big(\widehat{f}(a\cdot\xi_o)\big) = (\widehat{D}\widehat{f})(a\cdot\xi_o).$$

Since $D, \widehat{D}$ and $f \longrightarrow \widehat{f}$ are G-invariant this proves the first formula of the theorem.

For the second formula we need the following lemma. Here the adjoint of a differential operator is indicated by the superscript $*$; it will be clear from the context which volume element is used in the definition of the adjoint.

Lemma 3.4. *Let $U \longrightarrow D_U$ be the isomorphism of $\mathbf{D}(A)$ onto $\mathbf{D}(\Xi)$ defined by* (4) *§2. Then*

$$(D_U)^* = D_{e^{-2\rho}U^* \circ e^{2\rho}}.$$

For the proof define U' by $(D_U)^* = D_{U'}$. Then by (4) §2 and Lemma 3.1,

$$\int\limits_{K/M\times A} U_a\big(\varphi(ka\cdot\xi_o)\big)\psi(ka\cdot\xi_o)e^{2\rho(\log a)}\,dk_M\,da$$

$$= \int\limits_{K/M\times A} \varphi(ka\cdot\xi_o)e^{2\rho(\log a)}U'_a(\psi(ka\cdot\xi_o))\,dk_M\,da.$$

Let $\alpha, \beta \in \mathcal{D}(A)$ and put $\varphi(ka\cdot\xi_o) = \alpha(a)$, $\psi(kM, a) = \beta(a)$. Then the formula above shows $U^*(e^{2\rho}\beta) = e^{2\rho}U'(\beta)$ so $U' = e^{-2\rho}U^* \circ e^{2\rho}$ as claimed.

To conclude the proof of Theorem 3.3 we invoke the formula

$$\Gamma(D^*) = \Gamma(D)^*, \qquad D \in \mathbf{D}(X), \tag{15}$$

from [GGA], Ch. II, Lemma 5.21. We have, using (9),

$$\int\limits_X (\widehat{D}\varphi)^\vee(x)f(x)\,dx = \int\limits_\Xi (\widehat{D}\varphi)(\xi)\widehat{f}(\xi)\,d\xi = \int\limits_\Xi \varphi(\xi)(\widehat{D})^*\widehat{f}(\xi)\,d\xi.$$

Since the taking of adjoints changes the order of factors we have, using (13),

$$(\widehat{D})^* = D_V,$$

where

$$V = e^{-2\rho}\big(e^{-\rho}\Gamma(D)\circ e^{\rho}\big)^* e^{2\rho} = e^{-2\rho}e^{\rho}\Gamma(D)^* \circ e^{-\rho}\circ e^{2\rho}$$

so by (15),

$$(\widehat{D})^* = (D^*)^\wedge. \tag{16}$$

Introducing this in the integral above it becomes

$$\int_\Xi \varphi(\xi)(D^*)^\wedge \widehat{f}(\xi)d\xi = \int_\Xi \varphi(\xi)(D^*f)^\wedge(\xi)d\xi$$

$$= \int_X \overset{\vee}{\varphi}(x)(D^*f)(x)dx = \int_X (D\overset{\vee}{\varphi})(x)f(x)dx.$$

Since $f \in \mathcal{D}(X)$ is arbitrary this concludes the proof of the theorem.

The first formula of Theorem 3.3 will be quite valuable on a later occasion because it shows how the Radon transform converts the unwieldy differential operator D into the operator $\widehat{D}$ which is in fact a differential operator of constant coefficients.

3. The Inversion Formula and the Plancherel Formula for the Radon Transform.

We shall now show how a function f on X can be recovered explicitly from its Radon (horocycle) transform. There is a certain similarity with the Radon (hyperplane) transform in $\boldsymbol{R}^n$. This was inverted by means of a differential operator when n is odd, otherwise by means of a pseudodifferential operator. For X a similar dichotomy is present. However, it does not depend on $\dim X$ but rather on the number of conjugacy classes of Cartan subgroups of G : the horocycle transform on X is inverted by means of a differential operator exactly when all the Cartan subgroups of G are conjugate. Otherwise it is inverted by means of a pseudo-differential operator.

To begin with we recall Harish-Chandra's inversion formula for the spherical transform on G [1958b]; a simpler proof is given in [GGA], Ch. IV, §7 and Exercise C4. Let $\mathcal{D}^\natural(G)$ and $\mathcal{E}^\natural(G)$ be the space of functions $F \in \mathcal{D}(G)$ (respectively $F \in \mathcal{E}(G)$) which are bi-invariant under K. The *spherical transform* on G is defined by

$$\widetilde{F}(\lambda) = \int_G F(g)\varphi_{-\lambda}(g)dg, \qquad F \in \mathcal{D}^\natural(G) \tag{17}$$

where φ_λ is the spherical function defined above ($\lambda \in \mathfrak{a}^*$). This is inverted by Harish-Chandra's $\boldsymbol{c}$-function. This function is given by

$$\lim_{t\to+\infty} e^{(-i\lambda+\rho)(tH)}\varphi_\lambda(\exp tH) = \int_{\overline{N}} e^{-(i\lambda+\rho)(H(\overline{n}))}d\overline{n} = \boldsymbol{c}(\lambda) \tag{18}$$

for $H \in \mathfrak{a}^+$ arbitrary, and $\mathrm{Re}(i\lambda) \in \mathfrak{a}^*_+$, this being also the range of λ for which the integral converges absolutely. The function $\mathbf{c}(\lambda)$ extends to the meromorphic function

(18′)

$$\mathbf{c}(\lambda) = c_0 \prod_{\alpha \in \Sigma_o^+} \frac{2^{-\langle i\lambda, \alpha_o \rangle}\Gamma(\langle i\lambda, \alpha_o\rangle)}{\Gamma(\frac{1}{2}(\frac{1}{2}m_\alpha + 1 + \langle i\lambda, \alpha_o\rangle))\Gamma(\frac{1}{2}(\frac{1}{2}m_\alpha + m_{2\alpha} + \langle i\lambda, \alpha_o\rangle))}$$

on $\mathfrak{a}^*_{\boldsymbol{C}}$. Here Σ_o^+ is the set of positive indivisible roots, α_o is the normalized root $\alpha/\langle \alpha, \alpha\rangle$ and the constant c_o is given by the condition $\mathbf{c}(-i\rho) = 1$. Recall that w denotes the order of W.

The inversion formula indicated is

$$F(g) = w^{-1} \int_{\mathfrak{a}^*} \widetilde{F}(\lambda)\varphi_\lambda(g)|\mathbf{c}(\lambda)|^{-2} d\lambda, \quad F \in \mathcal{D}^\natural(G). \tag{19}$$

Remark. The limit formula (18) compares the spherical function φ_λ on A^+ with the plane wave function $e^{i\lambda-\rho}$ on A^+ (somewhat in the spirit of *scattering theory*). Here it is illuminating to note that by Prop. 2.4 the constant-coefficient operator $e^{-\rho}\Gamma(D) \circ e^\rho$ is a kind of a limit of the operator $\Delta_K(D)$ as we go to ∞ in the Weyl chamber A^+; furthermore, φ_λ satisfies the differential equations

$$\Delta_K(D)\varphi_\lambda = \Gamma(D)(i\lambda)\varphi_\lambda, \quad D \in \boldsymbol{D}(G/K)$$

while the plane wave $e^{i\lambda-\rho}$ satisfies the "limit equations"

$$(e^{-\rho}\Gamma(D) \circ e^\rho)(e^{i\lambda-\rho}) = \Gamma(D)(i\lambda)e^{i\lambda-\rho}.$$

We shall now prove an inversion formula for the Radon transform $f \longrightarrow \widehat{f}$ on $\mathcal{D}(X)$. The spaces $A, \mathfrak{a}$ and $\mathfrak{a}^*$ being metric vector spaces, the spaces $\mathcal{S}(A)$ $\mathcal{S}(\mathfrak{a})$ and $\mathcal{S}(\mathfrak{a}^*)$ of rapidly decreasing functions are well defined. We define the space $\mathcal{S}(\Xi)$ of rapidly decreasing functions on $\Xi = K/M \times A$ as follows. The negative of the Killing form B restricted to $\mathfrak{l}$ (orthogonal complement of $\mathfrak{m}$ in $\mathfrak{k}$) induces a K-invariant Riemannian structure on K/M. Let $L_{K/M}$ and L_A denote the Laplace-Beltrami operators on K/M and A, respectively, and let $\mathcal{S}(\Xi)$ denote the set of $\varphi \in \mathcal{E}(\Xi)$ (with $\varphi(kM, a) = \varphi(ka \cdot \xi_o)$) such that for each polynomial $P(x, y)$ and each $\ell \in \boldsymbol{Z}^+$,

$$\sup_{kM,a} \mid (P(L_{K/M}, L_A)\varphi)(kM, a)(1 + |a|^\ell) \mid < \infty. \tag{20}$$

Here $|a|$ stands for $B(\log a, \log a)^{\frac{1}{2}}$.

We now need some simple facts about the $\boldsymbol{c}$-function. Let π be the product of the positive indivisible roots considered as a function on $\mathfrak{a}_C^*$, i.e.

$$\pi(\lambda) = \prod_{\alpha \in \Sigma_o^+} \langle \lambda, \alpha \rangle.$$

Definition. A smooth function on a Euclidean space is said to have *slow growth* if each of its derivatives is bounded by a polynomial.

Lemma 3.5. *The functions*

$$\lambda \longrightarrow \boldsymbol{c}(\lambda)\pi(\lambda), \qquad \lambda \longrightarrow \big(\boldsymbol{c}(\lambda)\pi(\lambda)\big)^{-1} \qquad (\lambda \in \mathfrak{a}^*)$$

have slow growth. Hence the function

$$\lambda \longrightarrow \boldsymbol{c}(\lambda)^{-1} \qquad\qquad (\lambda \in \mathfrak{a}^*)$$

has slow growth and is a tempered distribution on $\mathfrak{a}^*$.

Proof. As observed in the proof of Prop. 7.2, Ch. IV in [GGA] we have

$$\boldsymbol{c}(\lambda)^{-1} = c_o^{-1} \prod_{\alpha \in \Sigma_o^+} C_\alpha(z_\alpha),$$

where $z_\alpha = \langle i\lambda, \alpha_o \rangle$ and

$$C_\alpha(z) = 2\pi^{\frac{1}{2}} \frac{\Gamma(\frac{1}{4}m_{2\alpha} + \frac{1}{2} + \frac{1}{2}z)\Gamma(\frac{1}{4}m_\alpha + \frac{1}{2}m_{2\alpha} + \frac{1}{2}z)}{\Gamma(\frac{1}{2}z + \frac{1}{2})\Gamma(\frac{1}{2}z)}.$$

Using $\Gamma(z)z = \Gamma(z+1)$ we obtain both $\boldsymbol{c}(\lambda)\pi(\lambda)$ and $\big(\boldsymbol{c}(\lambda)\pi(\lambda)\big)^{-1}$ as a product of one-variable functions of the type

$$s(x) = \frac{\Gamma(a + ix)}{\Gamma(b + ix)}, \qquad x \in \boldsymbol{R}. \tag{21}$$

where $a, b > 0$ are constants. It suffices therefore to prove the following elementary result.

Lemma 3.6. *The function* $s(x)$ *given by* (21) *has each of its derivatives bounded by a polynomial.*

Proof. Using the formula

$$\frac{\Gamma'(z)}{\Gamma(z)} = \sum_{m=1}^{\infty} \Big(\frac{1}{m} - \frac{1}{m+z}\Big) - \gamma - \frac{1}{z},$$

where γ is Euler's constant, we find that

$$-i\frac{s'(x)}{s(x)} = \frac{1}{b+ix} - \frac{1}{a+ix} + \sum_1^\infty\left(\frac{1}{n+b+ix} - \frac{1}{n+a+ix}\right)$$

so

$$s'(x) = s(x)t(x), \tag{22}$$

where $t(x)$ is a function each of whose derivatives is bounded. On the other hand,

$$\lim_{|x|\to\infty} \mid \Gamma(c+ix) \mid \exp(\tfrac{1}{2}\pi|x|)|x|^{\frac{1}{2}-c} = (2\pi)^{\frac{1}{2}}$$

so $s(x)$ is bounded by a polynomial. Using (22) the lemma follows by induction.

Because of Lemma 3.5 the multiplication by $\mathbf{c}(\lambda)^{-1}$ maps $\mathcal{S}(\mathfrak{a}^*)$ into itself. Using the Euclidean Fourier transform

$$F^*(\lambda) = \int_A F(a)e^{-i\lambda(\log a)}da, \qquad F \in \mathcal{S}(A),$$

we define the linear operators j and $\overline{j}$ of $\mathcal{S}(A)$ into itself by the requirements

$$(jF)^*(\lambda) = \mathbf{c}(\lambda)^{-1}F^*(\lambda), \tag{23}$$

$$(\overline{j}F)^*(\lambda) = \mathbf{c}(-\lambda)^{-1}F^*(\lambda). \tag{24}$$

The operator j commutes with translation on A; actually, it is a convolution with a tempered distribution. We can then define the endomorphisms J and $\overline{J}$ of $\mathcal{S}(\Xi)$ by

$$(J\varphi)(kaMN) = j_a(\varphi(kaMN)),$$

$$(\overline{J}\varphi)(kaMN) = \overline{j}_a(\varphi(kaMN)).$$

Let e^ρ denote the function on $\Xi = (K/M) \times A$ given by $e^\rho(kM, a) = e^{\rho(\log a)}$ and define the space $\mathcal{S}_\rho(\Xi)$ by

$$\mathcal{S}_\rho(\Xi) = \{\varphi \in \mathcal{E}(\Xi) : e^\rho\varphi \in \mathcal{S}(\Xi)\}.$$

Then the operators $\Lambda = e^{-\rho}J \circ e^\rho$, $\overline{\Lambda} = e^{-\rho}\overline{J} \circ e^\rho$ map the space $\mathcal{S}_\rho(\Xi)$ into itself. It is now convenient to observe the following fact.

Lemma 3.7. *The spaces $\mathcal{S}(\Xi), \mathcal{S}_\rho(\Xi)$ as well as the operators $J, \overline{J}, \Lambda$ and $\overline{\Lambda}$ are invariant under the action of G on Ξ.*

This is clear from the fact which is evident from (2) §2 that when Ξ is viewed as a fibre bundle over K/M the action of G on Ξ is fibre-preserving, inducing a translation on each fibre, combined with the fact that j is invariant under translations on A.

For $F \in \mathcal{E}(A)$ let F' denote the function $a \longrightarrow F(a^{-1})$ and for an operator D define D' by $D'F = (DF')'$. Also denote complex conjugate by "conj".

Lemma 3.8. *We have*

(i) $\overline{j} = j'$ *on* A;

(ii) $\Lambda(\operatorname{conj}(\varphi)) = \operatorname{conj}(\Lambda\varphi)$;

(iii) $\overline{\Lambda}$ *equals the adjoint of* Λ *in the sense that for* $\psi_1, \psi_2 \in \mathcal{D}(\Xi)$,

$$\int_\Xi \psi_1(\xi)\overline{\Lambda}\psi_2(\xi)d\xi = \int_\Xi \Lambda\psi_1(\xi)\psi_2(\xi)d\xi.$$

Proof. Part (i) is immediate from the definition taking the formula $\operatorname{conj} \mathbf{c}(\lambda) = \mathbf{c}(-\lambda)$ $(\lambda \in \mathfrak{a}^*)$ into account. Next note that $F \in \mathcal{S}(A)$ is real-valued if and only if $F^*(-\lambda) = \operatorname{conj}(F^*(\lambda))$ for $\lambda \in \mathfrak{a}^*$. Thus j maps real-valued functions into real-valued functions so (ii) follows. For (iii) put $\varphi_i(kM, a) = e^{\rho(\log a)}\psi_i(ka \cdot \xi_o)$ and

$$h(a) = \int_K \int_A \varphi_1(kM, b)\varphi_2(kM, ab)dkdb.$$

Then by the translation-invariance of j

$$(jh)(a) = \int_K \int_A \varphi_1(kM, b)(J\varphi_2)(kM, ab)dkdb. \tag{25}$$

Also

$$h'(a) = h(a^{-1}) = \int_K \int_A \varphi_1(kM, ab)\varphi_2(kM, b)dkdb$$

so

$$(\overline{j}h')(a) = \int_K \int_A (\overline{J}\varphi_1)(kM, ab)\varphi_2(kM, b)dkdb. \tag{26}$$

But by (i), $(jh)(e) = (\bar{j}h')(e)$ so we can equate the right hand sides of (25) and (26) for $a = e$. Substituting the definitions of φ_i we get (iii).

Theorem 3.9. *Let $f_1, f_2 \in \mathcal{D}(X)$. Then*

$$w \int_X f_1(x) f_2(x) dx = \int_\Xi (\Lambda \widehat{f_1})(\xi)(\Lambda \widehat{f_2})(\xi) d\xi.$$

For the proof we consider a modified convolution product $F_1 \# F_2$ (non-associative) of two functions $F_1, F_2 \in C_c(G)$. This is defined by

$$(F_1 \# F_2)(g) = \int_G F_1(h) F_2(hg) dh$$

and because of the right invariance of dh this is just $(F_1' * F_2)(g)$. Also if $f_1, f_2 \in C_c(X)$ we define the function $f_1 \# f_2$ on X by

$$(f_1 \# f_2)(g \cdot o) = (F_1 \# F_2)(g)$$

if $F_i(g) = f_i(g \cdot o)$ $(i = 1, 2; g \in G)$.

Lemma 3.10. *Let $f_1, f_2 \in C_c(X)$. Then $f_1 \# f_2$ is K-invariant and its Radon transform satisfies*

$$(f_1 \# f_2)^\wedge(a \cdot \xi_o) = \int_K \int_A e^{2\rho(\log b)} \widehat{f_1}(kb \cdot \xi_o) \widehat{f_2}(kba \cdot \xi_o) dk\, db.$$

Proof. The K-invariance is obvious and

$$(f_1 \# f_2)^\wedge(a \cdot \xi_o) = \int_N (F_1 \# F_2)(an) dn = \int_G F_1(g) \widehat{f_2}(ga \cdot \xi_o) dg$$

$$= \int_{KAN} F_1(kbn) \widehat{f_2}(kbna \cdot \xi_o) e^{2\rho(\log b)} dk\, db\, dn.$$

Since $\widehat{f_2}(kbna \cdot \xi_o) = \widehat{f_2}(kba \cdot \xi_o)$ the lemma follows.

Now if $F \in \mathcal{D}^\natural(G)$ we know from the spherical transform theory that

$$\widetilde{F}(\lambda) = \int_A \varphi_F(a) e^{-i\lambda(\log a)} da, \tag{27}$$

where

$$\varphi_F(a) = e^{\rho(\log a)} \int_N F(an)dn \tag{28}$$

(see e.g. [GGA], Ch. IV, §7). Since $\mathbf{c}(-\lambda) = \text{conj}(\mathbf{c}(\lambda))$ on $\mathfrak{a}^*$ we have by (23) and (24)

$$(j\bar{j}\varphi_F)^*(\lambda) = |\mathbf{c}(\lambda)|^{-2}\widetilde{F}(\lambda).$$

Thus by the Euclidean Fourier inversion formula together with (19) we deduce

$$F(e) = w^{-1}(j\bar{j}\varphi_F)(e). \tag{29}$$

Let $f_1, f_2 \in \mathcal{D}(X)$, $F_i(g) = f_i(g \cdot o)$ $(i = 1, 2)$ so by Lemma 3.10,

$$\varphi_{F_1 \# F_2}(a) = e^{\rho(\log a)}(f_1 \# f_2)^\wedge(a \cdot \xi_o)$$

$$= \int_K \int_A e^{\rho(\log b)} \widehat{f}_1(kb \cdot \xi_o) e^{\rho(\log(ab))} \widehat{f}_2(kba \cdot \xi_o) dk db.$$

Viewing $\varphi_i = e^\rho \widehat{f}_i$ as a function on $K/M \times A$ we obtain

$$\varphi_{F_1 \# F_2}(a) = \int_K \int_A \varphi_1(kM, b)\varphi_2(kM, ab) db\, dk, \tag{30}$$

which is just the expression $h(a)$ in (25). We have

$$(\bar{j}h')(a) = (jh)(a^{-1}) = \int_K \int_A \varphi_1(kM, ab)(J\varphi_2)(kM, b) dk\, db$$

and then

$$(j\bar{j}h')(a) = \int_{K \times A} (J\varphi_1)(kM, ab)(J\varphi_2)(kM, b) dk\, db. \tag{31}$$

Note that $(j\bar{j}h')' = \bar{j}jh$ and put $a = e$. Then

$$(\bar{j}j\varphi_{F_1 \# F_2})(e) = \int_{K \times A} (J\varphi_1)(kM, b)(J\varphi_2)(kM, b) dk\, db.$$

Here we introduce the operator Λ and combine the result with Lemma 3.1 and (29). Then Theorem 3.9 follows.

Corollary 3.11. *For each* $f \in \mathcal{D}(X)$,

$$w \int_X |f(x)|^2 dx = \int_\Xi |(\Lambda \widehat{f})(\xi)|^2 d\xi.$$

This is immediate from Lemma 3.8 and Theorem 3.9.

For the inversion formula in its best form we require an analog for X of the Darboux equation in $\boldsymbol{R}^n$ ([GGA], Ch. I, Lemma 2.14).

Theorem 3.12. *Let* $f \in \mathcal{E}(X)$ *and define* F *on* $X \times X$ *by*

$$F(gK, hK) = \int_K f(gkh \cdot o)dk.$$

Then for each $D \in \boldsymbol{D}(X)$,

$$D_1 F = D_2 F$$

where the subscripts indicate action on the first and second variable, respectively.

Remark. In the notation of §2, No. 4 we have

$$F(gK, hK) = (M^h f)(gK).$$

Proof. For $f \in \mathcal{E}(X)$ let $\widetilde{f} \in \mathcal{E}(G)$ denote the lift given by $\widetilde{f}(g) = f(g \cdot o)$. Let $\varphi \in \mathcal{D}(X)$. Then the convolution $f \times \varphi$ is the function on X given by

$$(f \times \varphi)(g \cdot o) = (\widetilde{f} * \widetilde{\varphi})(g), \tag{32}$$

and we have ([GGA], Ch. II §5),

$$D(f \times \varphi) = Df \times \varphi = f \times D\varphi, \qquad D \in \boldsymbol{D}(X), \tag{33}$$

We also note that

$$(D_2 F)(gK, hK) = \int_K (Df)(gkh \cdot o)dk. \tag{34}$$

Let $\varphi \in \mathcal{D}(X)$ be K-invariant. Then by a simple computation

$$\int_G F(gK, hK)\widetilde{\varphi}(h^{-1})dh = (f \times \varphi)(g \cdot o)$$

so by (33) and (34),

$$\int_G (D_1 F)(gK, hK)\widetilde{\varphi}(h^{-1})dh = D(f \times \varphi)(g \cdot o)$$

$$= \big((Df) \times \varphi\big)(g \cdot o) = \int_G (D_2 F)(gK, hK)\widetilde{\varphi}(h^{-1})dh.$$

Since the functions $h \longrightarrow (D_i F)(gK, hK)$ are bi-invariant under K and $\widetilde{\varphi}$ arbitrary in $\mathcal{D}^\natural(G)$ this relation between $D_1 F$ and $D_2 F$ proves the theorem.

Theorem 3.13.

(i) *The Radon transform has the following inversion formula*

$$f = w^{-1}\big(\Lambda\overline{\Lambda}\widehat{f}\big)^\vee, \qquad f \in \mathcal{D}(X).$$

(ii) Suppose $\mathfrak{g}$ has all its Cartan subalgebras conjugate. Then $j\overline{j}$ is a differential operator in $\boldsymbol{D}_W(A)$. Let $\square$ be the operator in $\boldsymbol{D}(X)$ such that $\Gamma(\square) = j\overline{j}$. Then the inversion is given by

$$f = w^{-1}\,\square((\widehat{f})^\vee), \qquad f \in \mathcal{D}(X).$$

Part (i) follows from Theorem 3.9 if we take Lemma 3.8 and the adjoint relation (9) into account.

For Part (ii) we recall that $\boldsymbol{c}(\lambda)^{-1}$ is now a polynomial and $\boldsymbol{c}(\lambda)\boldsymbol{c}(-\lambda)$ Weyl group invariant so $j\overline{j} \in \boldsymbol{D}_W(A)$. Consider the K-invariant function

$$f_o(x) = \int_K f(k \cdot x)dk$$

and the integral (cf. (28))

$$\varphi_{f_o}(a) = e^{\rho(\log a)} \int_N f_o(an \cdot o)dn = e^{\rho(\log a)}\widehat{f_o}(a \cdot \xi_o).$$

Using Theorem 3.3, (13), and (4) §2 we deduce

$$\varphi_{Df_o} = \Gamma(D)\varphi_{f_o}. \tag{35}$$

Let $\square$ be the operator in $\boldsymbol{D}(X)$ for which $\Gamma(\square) = j\bar{j}$. From (29) we obtain

$$f(o) = w^{-1}\varphi_{\square f_o}(e) = w^{-1}(\square f_o)^{\wedge}(\xi_o)$$

so

$$f(o) = w^{-1} \int\limits_{K\times N} (\square f)(kn \cdot o)dk\, dn.$$

Let $g \in G$ and replace here f by $f^{\tau(g)^{-1}}$. Since $\square$ is G-invariant we get

$$f(g \cdot o) = w^{-1} \int\limits_{K\times N} (\square f)(gkn \cdot o)dkdn. \tag{36}$$

But by the Darboux equation (Theorem 3.12) we have with some abuse of notation,

$$\int\limits_K (\square f)(gk \cdot x)dk = \square_{gK}\Big(\int\limits_K f(gk \cdot x)dk\Big)$$

so the integral in (36) equals

$$\int\limits_N dn \int\limits_K (\square f)(gkn \cdot o)dk = \int\limits_N \Big(\square_{gK}\Big(\int\limits_K f(gkn \cdot o)dk\Big)\Big)dn$$

$$= \square_{gK}\Big(\int\limits_{K\times N} f(gkn \cdot o)dk\, dn\Big) = \big(\square(\widehat{f})^{\vee}\big)(g \cdot o).$$

This concludes the proof.

The next natural step would be to characterize the range $\mathcal{D}(X)^{\wedge}$ intrinsically and to prove a support theorem to the effect that if a function $f \in \mathcal{E}(X)$ is "rapidly decreasing" and $\widehat{f}(\xi) = 0$ for all ξ disjoint from a ball $B \subset X$ then $f \equiv 0$ on $X - B$. The Euclidean analogs are proved in [GGA], Ch. I, §2. These theorems will only be proved much later because they require more refined tools. We shall then see that the support theorem combined with the intertwining property in Theorem 3.3 has important applications to differential equations.

4. The Poisson Transform.

As in §1, No. 1 let $\xi(x,b)$ denote the horocycle passing through the point $x \in X$ with *normal* $b \in B = K/M$. We denote by $A(x,b) \in \mathfrak{a}$ the *composite distance* from the origin o to $\xi(x,b)$. This notion is the symmetric space analog of the scalar product (y,ω) where $y,\omega \in \boldsymbol{R}^n$ and $|\omega| = 1$. In fact, (y,ω) represents the distance (with sign) of the origin in $\boldsymbol{R}^n$ to the hyperplane through y with normal vector ω.

The vector-valued inner product $A(x,b)$ generalizes the non-Euclidean inner product $\langle\ ,\ \rangle$ considered in [GGA], Introd. §4, No. 4. It has a simple expression in terms of the Iwasawa decompositions $G = KAN$, $G = NAK$. If $g \in G$, let $H(g) \in \mathfrak{a}, A(g) \in \mathfrak{a}$ be determined by $g = k_1 \exp H(g) n_1 = n_2 \exp A(g) k_2$ with $k_1, k_2 \in K$, $n_1, n_2 \in N$. Then

$$A(g) = -H(g^{-1}). \tag{37}$$

If $x = gK$, $b = kM$ we see that since $g^{-1}k \in K \exp H(g^{-1}k)N$ the point gK lies on the horocycle $k\exp(-H(g^{-1}k)) \cdot \xi_o$. Thus we have by (37)

$$A(gK, kM) = A(k^{-1}g) = -H(g^{-1}k). \tag{38}$$

The following result is the analog of the obvious Euclidean result that for each $\mu \in \boldsymbol{C}$, the function $y \longrightarrow e^{i\mu(y,\omega)}$ is an eigenfunction of the Laplacian on $\boldsymbol{R}^n$.

Proposition 3.14. *Let $\lambda \in \mathfrak{a}^*_{\boldsymbol{C}}, b \in B$. Then the function*

$$e_{\lambda,b} : x \longrightarrow e^{(i\lambda+\rho)(A(x,b))}$$

is a joint eigenfunction of $\boldsymbol{D}(X)$. In fact, $e_{\lambda,b}$ belongs to the joint eigenspace $\mathcal{E}_\lambda(X)$ (from (21) *§2).*

Proof. Let first $b = b_o = eM$. Then e_{λ,b_o} is N-invariant and its restriction, say $\bar{e}$, to $A \cdot o$ is $e_{\lambda,b_o}(a \cdot o) = e^{(i\lambda+\rho)(\log a)}$. By Theorem 2.1, if $D \in \boldsymbol{D}(X)$,

$$(De_{\lambda,b_o})\,|_{A\cdot o} = \Delta_N(D)\bar{e} = \left(e^{\rho}\Gamma(D) \circ e^{-\rho}\right)\bar{e} = \Gamma(D)(i\lambda)\bar{e}$$

so

$$De_{\lambda,b_o} = \Gamma(D)(i\lambda)e_{\lambda,b_o} \tag{39}$$

both sides of the equation being N-invariant.

More generally, if $b = kM$ then $e_{\lambda,b}(x) = e_{\lambda,b_o}(k^{-1} \cdot x)$ so by the K-invariance of D and (39),

$$De_{\lambda,b} = \Gamma(D)(i\lambda)e_{\lambda,b}. \tag{40}$$

Corollary 3.15. *We have*

$$\Gamma(D)(\rho) = 0$$

for every $D \in \boldsymbol{D}(G/K)$ without constant term (annihilating the constants).

In fact, the spherical function φ_λ equals

$$\varphi_\lambda(g) = \int_B e^{(i\lambda+\rho)(A(gK,b))}db$$

and satisfies $\varphi_{s\lambda} = \varphi_\lambda$ for $s \in W$. By (40) we have $D\varphi_\lambda = \Gamma(D)(i\lambda)\varphi_\lambda$. Since $\varphi_\lambda \equiv 1$ for $i\lambda = \pm\rho$ the corollary follows.

Because of this corollary, for each $b \in B$ and each $s \in W$ the function

$$u(x) = e^{(s\rho+\rho)(A(x,b))}$$

is *harmonic,* that is $Du = 0$ for each $D \in \boldsymbol{D}(G/K)$ annihilating the constants. The function

$$P(x,b) = e^{2\rho(A(x,b))} \tag{41}$$

is called the *Poisson kernel.*

If $F \in C(B)$ its *Poisson transform* $\mathcal{P}_\lambda F$ is defined by

$$(\mathcal{P}_\lambda F)(x) = \int_B e^{(i\lambda+\rho)(A(x,b))}F(b)db. \tag{42}$$

It is obvious from (40) that $\mathcal{P}_\lambda$ maps $C(B)$ into the joint eigenspace $\mathcal{E}_\lambda(X)$. We shall see later that the *injectivity question* for

$$\mathcal{P}_\lambda : \ C(B) \longrightarrow \mathcal{E}_\lambda(X)$$

is of considerable importance. Here we shall prove a preliminary result in this direction.

Theorem 3.16. *Suppose* $\lambda \in \mathfrak{a}_c^*$ *satisfies*

$$\operatorname{Re}(\langle i\lambda, \alpha\rangle) > 0 \quad \textit{for } \alpha \in \Sigma^+.$$

Then if $H \in \mathfrak{a}^+$, $a_t = \exp tH$,

$$\lim_{t\to+\infty} e^{(-i\lambda+\rho)(tH)} \int_B e^{(i\lambda+\rho)(A(a_t\cdot o, b))} F(b)db = \mathbf{c}(\lambda)F(b_o)$$

where b_o *is the origin* eM *in* $B = K/M$ *and* $F \in C(B)$.

Remark. It is of interest to observe that the limit above is independent of the choice of H in Weyl chamber $\mathfrak{a}^+$.

Proof. As in No. 1 we transfer the integration from K/M to $\overline{N}$ by which $dk_M = e^{-2\rho(H(\overline{n}))}d\overline{n}$. Since

$$A(a\cdot o, k(\overline{n})M) = -H(a^{-1}k(\overline{n})) = H(\overline{n}) - H(a^{-1}\overline{n}a) + \log a$$

the left hand side above becomes

$$e^{2\rho(tH)} \int_{\overline{N}} e^{-(i\lambda+\rho)(H(a_t^{-1}\overline{n}a_t))} e^{(i\lambda-\rho)(H(\overline{n}))} F(k(\overline{n})M)d\overline{n},$$

which by the substitution $\overline{n} \longrightarrow a_t\overline{n}a_t^{-1}$ becomes

$$\int_{\overline{N}} e^{(i\lambda-\rho)(H(a_t\overline{n}a_t^{-1}))} e^{-(i\lambda+\rho)(H(\overline{n}))} F(k(a_t\overline{n}a_t^{-1})M)d\overline{n}. \tag{43}$$

Since $\overline{n} = \exp X$, $X \in \Sigma_{\alpha>o}\mathfrak{g}_{-\alpha}$ it is easily seen that $a_t\overline{n}a_t^{-1} \longrightarrow e$ as $t \longrightarrow +\infty$ so formally our integral (43) converges to

$$\int_{\overline{N}} e^{-(i\lambda+\rho)(H(\overline{n}))} F(eM)d\overline{n} = \mathbf{c}(\lambda)F(b_o).$$

However, we must here justify going to the limit under the integral sign and here this justification is somewhat delicate. For this let $\lambda = \xi + i\eta$ so our assumption on λ amounts to $-A_\eta \in \mathfrak{a}^+$. Now it was proved by Harish-Chandra [1958a] (see also [GGA], Ch. IV, §6) that

$$B(H, H(\overline{n})) \geq 0, \ B(H, H(\overline{n}) - H(a_t\overline{n}a_t^{-1})) \geq 0 \qquad \text{for } H \in \mathfrak{a}^+.$$

Thus if we choose $\epsilon > 0$ such that

$$0 < \epsilon < 1, \qquad A_\rho + \epsilon A_\eta \in \mathfrak{a}^+$$

and put $\kappa = \sup |F|$ the integrand in (43) is majorized by

$$\begin{aligned}
&e^{-(\eta+\rho)(H(a_t\overline{n}a_t^{-1}))}e^{(\eta-\rho)(H(\overline{n}))}\kappa \\
&\leq e^{(-\eta+\epsilon\eta)(H(a_t\overline{n}a_t^{-1}))}e^{(\eta-\rho)(H(\overline{n})}\kappa \\
&\leq e^{(-\eta+\epsilon\eta)(H(\overline{n}))}e^{(\eta-\rho)(H(\overline{n}))}\kappa = e^{(\epsilon\eta-\rho)(H(\overline{n}))}\kappa
\end{aligned}$$

which is integrable over $\overline{N}$. This justifies letting $t \longrightarrow +\infty$ under the integral sign in (43) and proves the theorem.

The following result relates the Poisson transform $\mathcal{P}_\lambda$ to the dual transform $\varphi \longrightarrow \overset{\vee}{\varphi}$.

Proposition 3.17. *Let $F \in \mathcal{E}(B)$ and define φ on Ξ by*

$$\varphi(ka \cdot \xi_o) = e^{(i\lambda-\rho)(\log a)}F(kM).$$

Then $\varphi \in \mathcal{E}_\lambda(\Xi)$ and

$$\mathcal{P}_\lambda F = \overset{\vee}{\varphi}.$$

Proof. That $\varphi \in \mathcal{E}_\lambda(\Xi)$ is obvious from Proposition 2.6 and the formula $\mathcal{P}_\lambda F = \overset{\vee}{\varphi}$ is a special case of Theorem 5.9 later in this chapter.

5. The Dual Transform and the Poisson Kernel.

We shall now express the transforms $f \longrightarrow \widehat{f}$, $\varphi \longrightarrow \overset{\vee}{\varphi}$ more explicitly in terms of the Iwasawa decompositions

(44) $$g = k(g)\exp H(g)n_1 = n_2 \exp A(g)u(g), \qquad k(g),\, u(g) \in K.$$

We put

(45) $$k_g = k(gk), \qquad g(kM) = k_gM,$$

expressing the action of g on $K/M = G/MAN$. Since A normalizes N we have

(46) $$H(ghk) = H(gk_h) + H(hk) \qquad g, h \in G$$

so

$$A(g\cdot x, g(b)) = A(x,b) + A(g\cdot o, g(b)). \tag{47}$$

Harish-Chandra's formula

$$\int_K F(k_{g^{-1}})dk = \int_K F(h)e^{-2\rho(H(gk))}dk \tag{48}$$

(cf. [GGA], Ch. I, Lemma 5.19) amounts to

$$T_g^*(dk) = e^{-2\rho(H(gk))}dk, \tag{49}$$

for the pull-back of the map $T_g : k \longleftrightarrow k_g$ of K onto K. This can be restated for the Jacobian of the map $kM \longrightarrow g(kM)$

$$\frac{d(g(b))}{db} = e^{-2\rho(H(gk))} \tag{50}$$

so by (38)

$$\frac{d(g^{-1}(b))}{db} = e^{2\rho(A(g\cdot o,b))}, \tag{51}$$

the Poisson kernel.

The Radon transform $f \longrightarrow \widehat{f}$ is given by

$$\widehat{f}(g\cdot\xi_o) = \int_N f(gn\cdot o)dn, \tag{52}$$

so by (4) and (14) we obtain

$$\int_{G/K} f(g\cdot o)dg_K = \int_A\int_N f(an\cdot o)da\,dn = \int_A \widehat{f}(a\cdot\xi_o)da. \tag{53}$$

For the dual transform we have, by definition,

$$\overset{\vee}{\varphi}(g\cdot o) = \int_K \varphi(gk\cdot\xi_o)dk. \tag{54}$$

Here we change the variable from k to $k_{g^{-1}}$ and use the relations

$$k(gk_{g^{-1}}) = k, \quad H(gk_{g^{-1}}) = -H(g^{-1}k) = A(g\cdot o, kM)$$

and

(55) $$dk_{g^{-1}} = e^{-2\rho(H(g^{-1}k))}dk,$$

the last being a restatement of (49). Then the right hand side of (54) equals

$$\int_K \varphi(gk_{g^{-1}} \cdot \xi_o)dk_{g^{-1}} = \int_K \varphi(k\exp(-H(g^{-1}k)) \cdot \xi_o)e^{-2\rho(H(g^{-1}k))}dk$$

so we have proved the following result.

Theorem 3.18. *The dual transform* $\varphi \longrightarrow \overset{\vee}{\varphi}$ *is given by*

(56) $$\overset{\vee}{\varphi}(x) = \int_{K/M} \varphi(k\exp A(x,kM) \cdot \xi_o)e^{2\rho(A(x,kM))}dk_M.$$

Thus the natural measure $\mu = \mu_x$ on $\overset{\vee}{x}$ in (8) is given in terms of the Poisson kernel:

(57) $$d\mu_x = P(x,b)db.$$

For the non-Euclidean disk

$$D = \boldsymbol{SU}(1,1)/\boldsymbol{SO}(2)$$

the metric is conformal to the Euclidean metric. Thus the invariant measure μ_x in (57) is just the usual angle measure. Equation (57) therefore amounts to the following fact (see figure).

The origin $\xi_o = N\cdot 0$ in the horocycle space is the circle with the segment $(0,1)$ as diameter. For $x \in D$ let $\xi(x, e^{i\varphi})$ be the horocycle through x and $e^{i\varphi}$. Then by (57) the angle θ between ξ_o and $\xi(x, e^{i\varphi})$ satisfies

(58) $$\frac{d\theta}{d\varphi} = P(x, e^{i\varphi}) = \frac{1-|x|^2}{|x-e^{i\varphi}|^2} = e^{2\langle x, e^{i\varphi}\rangle},$$

where $\langle x, e^{i\varphi}\rangle$ is the non-Euclidean distance from 0 to $\xi(x, e^{i\varphi})$ (see [GGA], Introduction, §4).

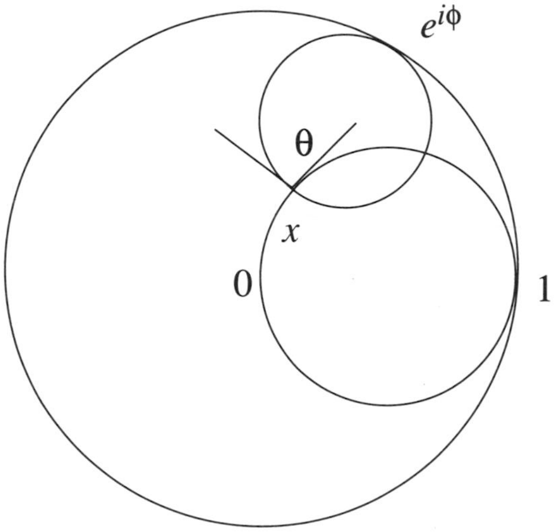

§4. FINITE-DIMENSIONAL SPHERICAL AND CONICAL REPRESENTATIONS.

1. Conical Distributions. Elementary Properties.

Our emphasis on the analogies between the symmetric space $X = G/K$ and the horocycle space $\Xi = G/MN$ suggests considering the analogs for Ξ of the spherical functions on X.

We recall that a *spherical function* on X is a complex-valued function $\varphi \not\equiv 0$ on X which is 1) K-invariant; 2) eigenfunction of each $D \in \boldsymbol{D}(X)$. For such a function we have $\varphi(o) \neq 0$ so it is customary to normalize φ by $\varphi(o) = 1$.

Definition. A distribution Ψ on Ξ is said to be *conical* if

(i) $\Psi^{\tau(mn)} = \Psi \qquad m \in M, n \in N$

(ii) For each $D \in \boldsymbol{D}(\Xi)$, ψ is an eigendistribution of D.

In (i), $\tau(mn)$ denotes the translation $gMN \longrightarrow mngMN$ and the action of it on distributions is that explained in Ch. I, §1, No. 1. A *conical function* is defined in the same way. The passage from functions to distributions preserves the analogy with spherical functions. In fact, since the Laplace-Beltrami operator L_X belongs to $\boldsymbol{D}(X)$ a "spherical distribution" would automatically be a spherical function, being an eigenfunction of an elliptic operator with analytic coefficients. The spherical functions on X are given by Harish-Chandra's formula

$$\varphi_\lambda(g) = \int_K e^{(i\lambda+\rho)(A(kg))}dk, \qquad g \in G,$$

λ running through $\mathfrak{a}_c^*$. Also $\varphi_\mu = \varphi_\lambda$ if and only if $\lambda \in W \cdot \mu$ so the set of spherical functions is parametrized by $\mathfrak{a}_c^*/W$. Remembering the identifications

$$K\backslash G/K \approx A/W, \tag{1}$$

$$MN\backslash G/MN \approx A \times W, \tag{2}$$

from §1, No. 2 we are led to the following guess.

Conjecture. The set of conical distributions is "parametrized" by $\mathfrak{a}_c^* \times W$.

We shall see later (cf. Theorem 5.16) that this conjecture is true at least for generic eigenvalues; it is also true without exception for X of rank one (cf. Theorems 6.18 and 6.21).

A representation T of G on a locally convex vector space V is said to be *spherical* if V contains a vector $v \neq 0$ (called a *spherical vector*) fixed under $T(K)$. Each spherical function φ can be written

$$\varphi(g \cdot o) = \langle T(g^{-1})v, v' \rangle, \tag{3}$$

where T is an irreducible spherical representation with contragredient T' and v and v' are $T(K)$ and $T'(K)$ fixed vectors, respectively ([GGA], Ch. IV, Theorem 4.5). A representation π of G on a locally convex vector space V is said to be *conical* if V contains a vector $v \neq 0$ (called a *conical vector*) fixed under all of $\pi(MN)$. We shall establish several analogs of (3) for conical distributions and conical representations (cf. Theorem 3.10 in Ch. VI). It will then turn out that *conical functions* will correspond to the *finite-dimensional* conical representations.

For the study of the conical distributions we use the decomposition

$$\Xi = \bigcup_{s \in W} \Xi_s, \qquad \Xi_s = MNA \cdot \xi_s \tag{4}$$

from (6) §1. The open orbit is $\Xi^* = MNA \cdot \xi^*$ where $\xi^* = \xi_{s^*}$, s^* being the Weyl group element exchanging $\mathfrak{a}^+$ and $-\mathfrak{a}^+$; the other orbits have lower dimension. Let $\xi \in \Xi^*$. Then we know from Prop. 1.5 and its proof that there exists a unique element $a(\xi) \in A$ such that

$$\xi = mna(\xi) \cdot \xi^* \qquad m \in M, n \in N. \tag{5}$$

Consider now the isomorphism $U \longrightarrow D_U$ of $\boldsymbol{D}(A)$ onto $\boldsymbol{D}(\Xi)$ as given by (4) §2,

$$\big(D_U\varphi\big)(ka \cdot \xi_o) = U_a\big(\varphi(ka \cdot \xi_o)\big) \qquad \varphi \in \mathcal{D}(\Xi). \tag{6}$$

If $\mu \in \mathfrak{a}_c^*$ we extend μ to a homomorphism of $\boldsymbol{D}(A)$ into $\boldsymbol{C}$; by the above we get a homomorphism $\chi_\mu : \boldsymbol{D}(\Xi) \longrightarrow \boldsymbol{C}$ given by

$$\chi_\mu(D_U) = \mu(U).$$

Every homomorphism of $\boldsymbol{D}(\Xi)$ into $\boldsymbol{C}$ is obtained in this way.

Theorem 4.1. *Let Ψ be a conical distribution on Ξ and let $\lambda \in \mathfrak{a}_C^*$ be determined such that*

$$D\Psi = \chi_{i\lambda-\rho}(D)\Psi \qquad D \in \boldsymbol{D}(\Xi). \tag{7}$$

Then there exists a function $\psi_\lambda \in \mathcal{E}(\Xi^*)$ such that $\Psi = \psi_\lambda$ on Ξ^*, that is

$$\Psi(\varphi) = \int_{\Xi^*} \psi_\lambda(\xi)\varphi(\xi)d\xi \quad \text{for} \quad \varphi \in \mathcal{D}(\Xi^*). \tag{8}$$

This function ψ_λ is given by

$$\psi_\lambda(\xi) = \psi_\lambda(mna(\xi) \cdot \xi^*) = e^{(is^*\lambda+\rho)(\log a(\xi))}\psi_\lambda(\xi^*), \qquad \xi \in \Xi^*. \tag{9}$$

On the other hand, let $\lambda \in \mathfrak{a}_C^*$ and let $\psi_\lambda \in \mathcal{E}(\Xi^*)$ be defined by (9) *and $\psi_\lambda(\xi^*) \neq 0$. Then ψ_λ is locally integrable on Ξ if and only if*

$$Re(\langle i\lambda, \alpha\rangle) > 0 \qquad \alpha \in \Sigma^+, \tag{10}$$

(Re= real part). If (10) *is satisfied then ψ_λ as a distribution on Ξ is a conical distribution.*

For the proof we consider for each $a \in A$ the diffeomorphism $\sigma(a)$: $gMN \longrightarrow gaMN$ of Ξ onto itself. It is well defined since a normalizes MN. Using Lemma 3.1 we see quickly

$$\int_\Xi \varphi(\xi)d\xi = e^{2\rho(\log a)} \int_\Xi \varphi(\sigma(a)\xi)d\xi. \tag{11}$$

Let $U \in \boldsymbol{D}(A)$ and put $D = D_U$. Then if $\varphi \in \mathcal{D}(\Xi)$, $g \in G$, $a \in A$,

$$(D\varphi)(ga \cdot \xi_o) = D\varphi^{\tau(g^{-1})}(a \cdot \xi_o) = U_a\big(\varphi^{\tau(g^{-1})}(a \cdot \xi_o)\big)$$

so

$$(D\varphi)(\xi) = \Big\{U_a\big(\varphi^{\sigma(a^{-1})}(\xi)\big)\Big\}_{a=e}. \tag{12}$$

We now generalize (12) to distributions.

Lemma 4.2. *Let $\Psi \in \mathcal{D}'(\Xi), U \in \boldsymbol{D}(A)$ and for each $a \in A$ define the functional $U_a\big(\Psi^{\sigma(a)}\big)$ by*

$$U_a\big(\Psi^{\sigma(a)}\big)(\varphi) = U_a\big(\Psi^{\sigma(a)}(\varphi)\big), \qquad \varphi \in \mathcal{D}(\Xi).$$

Then $U_a\big(\Psi^{\sigma(a)}\big)$ is a distribution on Ξ and

$$D^*\Psi = \Big\{U_a\big(\Psi^{\sigma(a)}\big)\Big\}_{a=e} \tag{13}$$

if $D = D_U$ and D^* its adjoint.

Proof. Define $\Phi \in \mathcal{E}(A \times \Xi)$ by

$$\Phi(a, \xi) = \varphi(\sigma(a)\xi)$$

and if $a_o \in A$ let u denote the distribution

$$u(F) = \big(UF\big)(a_o), \qquad F \in \mathcal{D}(A).$$

If $\Psi \in \mathcal{D}'(\Xi)$ the tensor product $u \otimes \Psi$ is a distribution on $A \times \Xi$ satisfying

$$(u \otimes \Psi)_{(a,\xi)}\big(f(a, \xi)\big) = u_a\big(\Psi_\xi(f(a, \xi))\big) = \Psi_\xi\big(u_a(f(a, \xi))\big) \tag{14}$$

for $f \in \mathcal{D}(A \times \Xi)$ (Schwartz [1966], Ch. IV). The support of $u \otimes \Psi$ is contained in $\{a_o\} \times \Xi$ which intersects the support of Φ in a compact set. Hence $(u \otimes \Psi)(\Phi)$ can be defined as $(u \otimes \Psi)(\alpha\Phi)$ where α is an arbitrary function in $\mathcal{D}(A \times \Xi)$ which is identically 1 on a neighborhood of the compact set indicated. Two such α give the same result. Then by (14),

$$u_a\big(\Psi_\xi\big(\alpha(a, \xi)\varphi(\sigma(a)\xi)\big)\big) = \Psi_\xi\big(u_a(\alpha(a, \xi)\varphi(\sigma(a)\xi))\big). \tag{15}$$

If a is sufficiently close to a_o then $\alpha(a,\xi) = 1$ on a neighborhood of the support of the function $\xi \longrightarrow \varphi(\sigma(a)\xi))$. For such a we have

$$\Psi_\xi(\alpha(a,\xi)\varphi(\sigma(a)\xi)) = \Psi_\xi(\varphi(\sigma(a)\xi))$$

so by (15), since $\operatorname{supp}(u) \subset \{a_o\}$,

$$u_a\left(\Psi_\xi(\varphi^{\sigma(a^{-1})}(\xi)\right) = \Psi_\xi\left(u_a(\varphi(\sigma(a)\xi)\right),$$

which means

$$\left\{U_a\left(\Psi(\varphi^{\sigma(a^{-1})})\right)\right\}_{a=a_o} = \Psi_\xi\left(\left\{U_a(\varphi^{\sigma(a^{-1})}(\xi))\right\}_{a=a_o}\right).$$

Combining this with (12) we obtain

$$\Psi(D\varphi) = \left\{U_a\left(\Psi(\varphi^{\sigma(a^{-1})})\right)\right\}_{a=e}. \tag{16}$$

It follows that $\{U_a\left(\Psi^{\sigma(a)}\right)\}_{a=e}$ is a distribution on Ξ and by (16)

$$D^*\Psi = \left\{U_a\left(\Psi^{\sigma(a)}\right)\right\}_{a=e}$$

as claimed. That $U_b(\Psi^{\sigma(b)})$ is a distribution for each b follows by using (16) with Ψ replaced by $\Psi^{\sigma(b)}$. This proves the lemma.

We can now rephrase condition (ii) in the definition of a conical distribution.

Proposition 4.3. *A distribution Ψ on Ξ is an eigendistribution of each $D \in \boldsymbol{D}(\Xi)$ if and only if for each $a \in A$, $\Psi^{\sigma(a)}$ is a constant multiple of Ψ. Moreover, if $\lambda \in \mathfrak{a}_c^*$ then for each $\Psi \in \mathcal{D}'(\Xi)$ the condition*

$$D\Psi = \chi_{i\lambda-\rho}(D)\Psi \qquad D \in \boldsymbol{D}(\Xi) \tag{17}$$

is equivalent to

$$\Psi^{\sigma(a)} = e^{-(i\lambda+\rho)(\log a)}\Psi, \qquad a \in A. \tag{18}$$

Proof. Suppose first (17) holds. Then by (13)

$$\{U_a\left(\Psi^{\sigma(a)}\right)\}_{a=e} = \chi_{i\lambda-\rho}(D^*)\Psi.$$

Fix $b \in A$, $\varphi \in \mathcal{D}(\Xi)$ and apply both sides to $\varphi^{\sigma(b^{-1})}$. Since by definition

$$U_a\big(\Psi^{\sigma(a)}\big)\big(\varphi^{\sigma(b^{-1})}\big) = U_a\big(\Psi^{ab}(\varphi)\big)$$

we obtain

$$\big\{U_a\big(\Psi^{\sigma(ab)}(\varphi)\big)\big\}_{a=e} = \chi_{i\lambda-\rho}(D^*)\Psi^{\sigma(b)}(\varphi).$$

Thus the function $t_\varphi : b \longrightarrow \Psi^{\sigma(b)}(\varphi)$ on A satisfies

$$Ut_\varphi = \chi_{i\lambda-\rho}(D^*)t_\varphi, \qquad U \in \boldsymbol{D}(A).$$

It follows that

$$\Psi^{\sigma(b)}(\varphi) = \Psi(\varphi)e^{\mu(\log b)}$$

where $\mu \in \mathfrak{a}_C^*$ is given by

$$\chi_{i\lambda-\rho}(D^*) = \chi_\mu(D) \qquad D \in \boldsymbol{D}(\Xi).$$

By Lemma 3.4

$$\chi_{i\lambda-\rho}(D^*) = \chi_{i\lambda-\rho}\big(D_{e^{-2\rho}U^*\circ e^{2\rho}}\big).$$

The automorphism $U \longrightarrow e^{-2\rho}U \circ e^{2\rho}$ of $\boldsymbol{D}(A)$ is the same as the automorphism induced by $H \longrightarrow H + 2\rho(H)$ $(H \in \mathfrak{a})$. Also if $\nu \in \mathfrak{a}_C^*$,

$$\chi_\nu\big(H + 2\rho(H)\big) = \chi_{\nu+2\rho}(H), \quad H^* = -H$$

so $\chi_{i\lambda-\rho}(D^*)$ equals

$$\chi_{i\lambda+\rho}\big(D_{U^*}\big) = \chi_{-i\lambda-\rho}(D_U).$$

Thus $\mu = -i\lambda - \rho$, proving (18). The converse follows by reversing the above steps.

Passing now to the proof of Theorem 4.1 suppose $\Psi \in \mathcal{D}'(\Xi)$ is conical. Let $\lambda \in \mathfrak{a}_C^*$ be such that (7) holds. We define a smooth function α on Ξ^* by the formula

$$\alpha(\xi) = e^{(-is^*\lambda+\rho)(\log a(\xi))}, \qquad \xi \in \Xi^*.$$

Then

$$\alpha^{\sigma(a)} = e^{(i\lambda+\rho)(\log a)}\alpha.$$

Combining this with (18) and the invariance of Ψ under N the restriction $\Psi^* = \Psi \mid \mathcal{D}(\Xi^*)$ satisfies

$$(19) \qquad \big(\alpha\Psi^*\big)^{\tau(n)\sigma(a)} = \alpha\Psi^*, \qquad n \in N, a \in A.$$

Consider the *direct* product $N \times A$ whose Haar measure is $dn\, da$. The mapping $(n, a) \longrightarrow nm^*a \cdot \xi_o$ is a diffeomorphism of $N \times A$ onto Ξ^* (Lemma 1.3). Under this diffeomorphism $\alpha\Psi^*$ corresponds to a distribution on $N \times A$ which by (19) is invariant under all left translations (since A is abelian) and therefore is a constant multiple of $dn\, da$. (Convolving with a function it is easy to see that a left-invariant distribution on a Lie group is Haar measure.) It follows that

$$\big(\alpha\Psi^*\big)(\varphi) = c \int\limits_{N \times A} \varphi(nm^*a \cdot \xi_o)dn\, da, \quad \varphi \in \mathcal{D}(\Xi^*),$$

where c is a constant. Since s^* preserves da we deduce from Lemma 3.1 and (14) §3 that

$$(\alpha\Psi^*)(\varphi) = c \int\limits_{\Xi} \varphi(\xi)\beta(\xi)d\xi,$$

where $\beta(\xi) = e^{2\rho(\log a(\xi))}$. Since $\frac{1}{\alpha} \in \mathcal{E}(\Xi^*)$ this proves that Ψ^* is a C^∞ function and given by (9).

On the other hand, let ψ_λ be defined by (9) with $\psi_\lambda(\xi^*) = 1$. The local integrability of ψ_λ is equivalent to

$$(20) \qquad \int\limits_{\Xi} |\psi_\lambda(\xi)\varphi(\xi)|d\xi < \infty, \qquad \varphi \in \mathcal{D}(\Xi).$$

Under the diffeomorphism $kM \longrightarrow kMAN$ of K/M onto G/MAN the set $\{kM : k \cdot \xi_o \in \Xi^*\}$ corresponds to *the* open orbit $MANm^*MAN$ of MAN on G/MAN. Thus the complement

$$\big\{kM : \; k \cdot \xi_o \in \Xi - \Xi^*\big\}$$

is a union of submanifolds of K/M of lower dimensions. Thus the integral

$$(21) \qquad \int\limits_{K/M} |\psi_\lambda(k \cdot \xi_o)|dk_M$$

makes sense and because of Lemma 3.1 it is finite if and only if (20) holds. Transferring the integral to $\overline{N}$ (via the map $\overline{n} \longrightarrow k(\overline{n})M$, see §3, No.1) we see that ψ_λ is locally integrable on Ξ if and only if

$$\int\limits_{\overline{N}} \big|\; \psi_\lambda\big(m^*k(\overline{n}) \cdot \xi_o\big) \;\big|\; e^{-2\rho(H(\overline{n}))}d\overline{n} < \infty.$$

On the other hand,

$$\psi_\lambda(m^*k(\overline{n})\cdot\xi_o) = e^{(-i\lambda+\rho)(H(\overline{n}))}$$

so the local integrability of ψ_λ is equivalent to the integral

$$\int_{\overline{N}} e^{-(i\lambda+\rho)(H(\overline{n}))}d\overline{n}$$

being absolutely convergent. The integral represents Harish-Chandra's **c**-function and as shown by Gindikin-Karpelevič its absolute convergence is equivalent to (10) (cf. [GGA], Ch. IV, §6).

Finally, if ψ_λ as defined by (9) is locally integrable on Ξ then the distribution $\varphi \longrightarrow \int_{\Xi}\varphi(\xi)\psi_\lambda(\xi)d\xi$ is MN-invariant. Using Proposition 4.3 we see that it is a conical distribution. This concludes the proof of Theorem 4.1.

Considering the definition of χ_μ we see from §2, (22) that the space $\mathcal{D}'_\lambda = \mathcal{D}'_\lambda(\Xi)$ is given by

$$\mathcal{D}'_\lambda(\Xi) = \{\Psi \in \mathcal{D}'(\Xi) : D\Psi = \chi_{i\lambda-\rho}(D)\Psi \text{ for } D \in \mathbf{D}(\Xi)\}.$$

Proposition 4.4. *The space $\mathcal{D}'_\lambda(\Xi)$ consists precisely of the distributions Ψ on Ξ given by*

$$\Psi(\psi) = \int_{K/M}\Big(\int_A \psi(ka\cdot\xi_o)e^{(i\lambda+\rho)(\log a)}da\Big)dS(kM), \tag{22}$$

where S is an arbitrary distribution on K/M.

Proof. Because of (18) it is obvious that Ψ as defined by (22) is a member of $\mathcal{D}'_\lambda(\Xi)$. On the other hand, suppose $\Phi \in \mathcal{D}'_\lambda$. Fix $\varphi \in \mathcal{E}(K/M)$. If $\alpha \in \mathcal{D}(A)$ we denote the function $(kM, a) \longrightarrow \varphi(kM)\alpha(a)$ by $\varphi\otimes\alpha$. Viewing it as a function on Ξ (Proposition 1.4) we consider the distribution

$$\tau_\varphi :\ \alpha \longrightarrow \Phi(\varphi\otimes\alpha), \qquad \alpha \in \mathcal{D}(A).$$

If $R(b)$ denotes the right translation $a \longrightarrow ab$ on A we find

$$(\tau_\varphi)^{R(b)}(\alpha) = \Phi^{\sigma(b)}(\varphi\otimes\alpha)$$

so by (18)

$$(\tau_\varphi)^{R(b)} = e^{-(i\lambda+\rho)(\log b)}\tau_\varphi.$$

This implies that τ_φ is a constant multiple of a character of A, more precisely,

$$\Phi(\varphi \otimes \alpha) = \tau_\varphi(\alpha) = c_\varphi \int_A e^{(i\lambda+\rho)(\log a)}\alpha(a)da \tag{23}$$

for all $\alpha \in \mathcal{D}(A)$, c_φ being a constant. Fix $\alpha_o \in \mathcal{D}(A)$ such that the function $e^{(i\lambda+\rho)(\log a)}\alpha_o(a)$ has integral 1 over A. Then we see that the map $\varphi \longrightarrow c_\varphi = \Phi(\varphi \otimes \alpha_o)$ is a distribution S on K/M and by (23)

$$\Phi(\varphi \otimes \alpha) = \int_{K/M} \Big(\int_A (\varphi \otimes \alpha)(ka \cdot \xi_o) e^{(i\lambda+\rho)(\log a)} da\Big) dS(kM)$$

for $\varphi \in \mathcal{E}(K/M)$, $\alpha \in \mathcal{D}(A)$. The corresponding functions $\varphi \otimes \alpha$ span a dense subspace of $\mathcal{D}((K/M) \times A)$ (Schwartz [1966], Ch. IV) so the proposition follows.

Proposition 4.5. *Let $\Psi \in \mathcal{D}'_\lambda$ be a conical distribution and $\Phi \in \mathcal{D}'_\mu$ a conical function. Then $\Phi\Psi$ is a conical distribution in $\mathcal{D}'_{\mu+\lambda+i\rho}$.*

Proof. Because of Proposition 4.3 we just have to verify the appropriate homogeneity under $\sigma(a)$, remembering that $\sigma(a)$ does not preserve the measure $d\xi$ but $(d\xi)^{\sigma(a)} = e^{-2\rho(\log a)}d\xi$. Since

$$\Psi^{\sigma(a)} = e^{-(i\lambda+\rho)(\log a)}\Psi, \qquad (\Phi d\xi)^{\sigma(a)} = e^{-(i\mu+\rho)(\log a)}\Phi d\xi$$

we have for Φ as a function

$$\Phi^{\sigma(a)} = e^{-(i\mu-\rho)(\log a)}\Phi$$

and

$$(\Phi\Psi)^{\sigma(a)} = e^{(-i\mu+\rho-i\lambda-\rho)(\log a)}\Phi\Psi$$

proving the result.

2. Conical Functions and Finite-dimensional Representations.

In this subsection we determine the conical functions and establish their intimate connection with an important class of representations. We continue use of the representation of $\xi \in \Xi^*$,

$$\xi = mna(\xi) \cdot \xi^*, \qquad \xi \in \Xi^*.$$

Theorem 4.6. *Let ψ be a conical function on Ξ. Then*

$$\psi(\xi) = e^{-\mu(\log a(\xi))}\psi(\xi^*), \qquad \xi \in \Xi^*, \tag{24}$$

where $\mu \in \mathfrak{a}_c^*$ satisfies

$$\frac{\langle \mu, \alpha \rangle}{\langle \alpha, \alpha \rangle} \in \boldsymbol{Z}^+ \qquad \text{for } \alpha \in \Sigma^+. \tag{25}$$

Conversely, if $\mu \in \mathfrak{a}_C^*$ satisfies (25) *then the function ψ defined by* (24) *(with $\psi(\xi^*)$ arbitrary) extends to a conical function on Ξ.*

Before the proof we state additional results. Let ζ be any function on Ξ, E_ζ the smallest vector space over $\boldsymbol{C}$ containing all the G-translates $\zeta^{\tau(g)}$ of ζ and π_ζ the natural representation of G on E_ζ, i.e. $(\pi_\zeta(g)\psi)(\xi) = \psi(g^{-1} \cdot \xi)$ for $g \in G, \xi \in \Xi, \psi \in E_\zeta$. For *weights* and related terms see No. 6 in this section.

Theorem 4.7.

(i) *The mapping $\psi \longrightarrow \pi_\psi$ is a bijection of the set of conical functions $\not\equiv 0$ on Ξ (proportional ψ identified) onto the set of finite-dimensional irreducible conical representations of G (equivalent representations identified). Also*

$$\psi(g \cdot \xi_o) = \langle \pi_\psi(g^{-1})\mathbf{e}, \mathbf{e}' \rangle \qquad (g \in G) \tag{26}$$

where $\mathbf{e}$ and $\mathbf{e}'$, respectively, are contained in the (one-dimensional) highest weight spaces of π_ψ and of its contragredient representation.

(ii) *An irreducible representation π is conical if and only if its highest weight vector is $\pi(M)$-fixed.*

(iii) *Let π be an irreducible representation with conical vector $\mathbf{e}$. Then $\mathbf{e}$ is a highest weight vector.*

Formula (26) is of course a close analog to formula (3) for the spherical function so far as finite-dimensional representations are concerned. Later we shall extend this correspondence in a modified fashion to infinite-dimensional representations as well.

Theorem 4.8.

(i) *Let ψ be a conical function and μ given by* (24). *Extend μ to a linear form $\widetilde{\mu}$ on a Cartan subalgebra $\mathfrak{h}^C$ of $\mathfrak{g}^C$ taking $\widetilde{\mu} = 0$ on a Cartan subalgebra $\mathfrak{h}_\mathfrak{m}$ of $\mathfrak{m}$. Then $\widetilde{\mu}$ is the highest weight of the (irreducible) representation π_ψ.*

(ii) *The highest weights of the finite-dimensional irreducible conical representations are the linear forms* $\widetilde{\mu}$ *on* $\mathfrak{h}^{\boldsymbol{C}}$ *characterized by*

$$\frac{\langle\mu,\alpha\rangle}{\langle\alpha,\alpha\rangle} \in \boldsymbol{Z}^+ \textit{ for } \alpha \in \Sigma^+; \qquad \widetilde{\mu}(\mathfrak{h}_{\mathfrak{m}}) = 0 \tag{27}$$

μ *denoting the restriction* $\widetilde{\mu}|\mathfrak{a}$.

Definition. μ is called the *highest restricted weight* of π_ψ. (See No. 6 in this §.)

It is convenient to prove these three theorems together. The idea is as follows. First we determine the conical functions in the case $G = \boldsymbol{SL}(2, \boldsymbol{R})$. For a general G and Ξ we construct for each $\alpha \in \Sigma^+$ a subgroup $G_\alpha \subset G$ locally isomorphic to $\boldsymbol{SL}(2, \boldsymbol{R})$ and a corresponding horocycle space $\Xi^\alpha \subset \Xi$. If ψ is a conical function on Ξ a suitable modification of it has a restriction to Ξ^α which is a nonzero conical function and one arrives at the integrality condition (25). On the other hand, if (25) is satisfied, we can construct a certain finite-dimensional representation π_μ of G whose highest weight $\widetilde{\mu}$ satisfies (27). Using the structure of the group of components of M we prove that the function $g \longrightarrow \langle\pi_\mu(g^{-1})e, e^*\rangle$ defines a conical function on Ξ.

Now let $G = \boldsymbol{SL}(2, \boldsymbol{R})$ with the standard Iwasawa decomposition $G = KAN$ where $K = \boldsymbol{SO}(2)$ and

$$A = \left\{a = \begin{pmatrix} d & 0 \\ 0 & d^{-1} \end{pmatrix} : d > 0 \text{ in } \boldsymbol{R}\right\}, \qquad N = \left\{n = \begin{pmatrix} 1 & x \\ 0 & 1 \end{pmatrix} : x \in \boldsymbol{R}\right\}.$$

Lemma 4.9.

(i) *The conical functions on* G/MN $(G = \boldsymbol{SL}(2, \boldsymbol{R}))$ *are the constant multiples of the functions*

$$\psi\left(\begin{pmatrix} x_{11} & x_{12} \\ x_{21} & x_{22} \end{pmatrix} \cdot \xi_o\right) = (x_{21})^{2n} \qquad n \in \boldsymbol{Z}^+. \tag{28}$$

(ii) *Let* G *be a connected Lie group with finite center, locally isomorphic to* $\boldsymbol{SL}(2, \boldsymbol{R})$. *Then Theorem* 4.6 *holds for* G/MN.

Proof. (i) Here $M = \{\epsilon I : \epsilon = \pm 1\}$ and for m^* we can use $\begin{pmatrix} 0 & 1 \\ -1 & 0 \end{pmatrix}$. Suppose $\psi \not\equiv 0$ is a conical function. Then by Theorem 4.1 there exists a $\mu \in \mathfrak{a}^*_{\boldsymbol{C}}$ such that

$$\psi(\xi) = e^{-\mu(\log a(\xi))}\psi(\xi^*), \quad \xi \in \Xi^*.$$

Here Σ^+ consists of just one element α, and $\rho = \frac{1}{2}\alpha$ satisfies $\rho\begin{pmatrix} 1 & 0 \\ 0 & -1 \end{pmatrix} = 1$. Let $z \in \boldsymbol{C}$ be such that $\mu = z\rho$. Then

$$\psi(\begin{pmatrix} \epsilon & 0 \\ 0 & \epsilon \end{pmatrix}\begin{pmatrix} 1 & x \\ 0 & 1 \end{pmatrix}\begin{pmatrix} d & 0 \\ 0 & d^{-1} \end{pmatrix}\begin{pmatrix} 0 & 1 \\ -1 & 0 \end{pmatrix} \cdot \xi_o) = \psi(\xi^*)e^{-z\log d}.$$

Thus on the set Ξ^* the function ψ is given by

$$\psi(\begin{pmatrix} x_{11} & x_{12} \\ x_{21} & x_{22} \end{pmatrix} \cdot \xi_o) = \psi(\xi^*)|x_{21}|^z.$$

Since $\psi \in \mathcal{E}(\Xi)$ we conclude $z \in 2\boldsymbol{Z}^+$ so (28) holds. Conversely, if $n \in \boldsymbol{Z}^+$ formula (28) gives a well defined C^∞ function on G/MN. Also

$$\psi(mn_1ga \cdot \xi_o) = \psi(g \cdot \xi_o)e^{2n\rho(\log a)} \tag{29}$$

so by Proposition 4.3 it is a conical function. Note that the linear forms $\mu = z\rho$ above are characterized by

$$\frac{\langle \mu, \alpha\rangle}{\langle \alpha, \alpha\rangle} \in \boldsymbol{Z}^+, \qquad \alpha \in \Sigma^+,$$

so Theorem 4.6 is proved for $G = \boldsymbol{SL}(2, \boldsymbol{R})$.

For the second part of Lemma 4.9 it is convenient to prove first the following general fact.

Lemma 4.10. *Let G be a connected semisimple Lie group with finite center and K, A, M determined as before. Let G_o be another group covered by G and K_o, A_o, M_o determined similarly. If $p : G \longrightarrow G_o$ is the covering map, then*

$$p(K) = K_o, \qquad p(A) = A_o, \qquad p(M) = M_o.$$

Proof. The first relations are obvious and so is the inclusion $p(M) \subset M_o$. Suppose $m_o \in M_o$. Then $m_oa_o = a_om_o$, $a_o \in A_o$. Select $k \in K$ such that $p(k) = m_o$. Then for each $a \in A$, $ka = akz$ where z belongs to the kernel of p. Then z belongs to the center of G so $z = k_o \in K$. Applying the Cartan involution θ to the relation $ka = akk_o$ and eliminating k_o we see that $ka^{-2} = a^{-2}k$ so since $a \in A$ was arbitrary, $k \in M$ as desired.

For Lemma 4.9 (ii) we first apply Lemma 4.10 to $G = \boldsymbol{SL}(2, \boldsymbol{R})$ and $G_o = Ad(G)$. We conclude that $M_o = \{e\}$. Then we deduce from Lemma 4.10 that if G is locally isomorphic to $\boldsymbol{SL}(2, \boldsymbol{R})$ then M equals the center of G (so its action on G/MN is trivial). We conclude that under the natural bijection $G/MN \longrightarrow G_o/M_oN_o$ the isotropy actions correspond so the conical functions are the same. Now Part (ii) of Lemma 4.9 is obvious.

We can now prove the first half of Theorem 4.6. Let $\alpha \in \Sigma^+$ and select $X_\alpha \neq 0$ in the root space $\mathfrak{g}_\alpha$ (§1, No. 1). Then the space

$$\mathfrak{g}^\alpha = \boldsymbol{R}A_\alpha + \boldsymbol{R}X_\alpha + \boldsymbol{R}\theta X_\alpha$$

is a subalgebra of $\mathfrak{g}$ isomorphic to $\mathfrak{sl}(2, \boldsymbol{R})$. Let $G^\alpha \subset G$ denote the corresponding analytic subgroup. If $\mathfrak{k}^\alpha = \mathfrak{k} \cap \mathfrak{g}^\alpha, \mathfrak{p}^\alpha = \mathfrak{p} \cap \mathfrak{g}^\alpha$ the decomposition $\mathfrak{g}^\alpha = \mathfrak{k}^\alpha + \mathfrak{p}^\alpha$ is a Cartan decomposition of $\mathfrak{g}^\alpha$. The space $\mathfrak{a}^\alpha = \boldsymbol{R}A_\alpha$ is maximal abelian in $\mathfrak{p}^\alpha$ and $(\boldsymbol{R}^+ - (0))A_\alpha$ is a Weyl chamber $(\mathfrak{a}^\alpha)^+$. In the usual manner (§1, No. 1) we form the subgroups $K^\alpha, A^\alpha, M^\alpha, N^\alpha, \overline{N}^\alpha$ of G^α and the subalgebras $\mathfrak{m}^\alpha$, $\mathfrak{n}^\alpha$ and $\overline{\mathfrak{n}}^\alpha$. We have then $K^\alpha = G^\alpha \cap K, N^\alpha = G^\alpha \cap N, A^\alpha = G^\alpha \cap A$. Since M^α and M may not be connected the relation $M^\alpha = G^\alpha \cap M$ is less obvious. However, if $g \in M^\alpha$ then $Ad(g)$ fixes each element of the hyperplane $\alpha(H) = 0$ in $\mathfrak{a}$ (since $[H, \mathfrak{g}^\alpha] = 0$) as well as the normal line $\mathfrak{a}^\alpha$. Thus $g \in M$. This proves the following result.

Lemma 4.11. *The mapping $gM^\alpha N^\alpha \longrightarrow gMN$ is an injection of $\Xi^\alpha (= G^\alpha/M^\alpha N^\alpha)$ into Ξ.*

Let $\langle\ ,\ \rangle$ and $\langle\ ,\ \rangle_\alpha$ denote the bilinear forms on $\mathfrak{a}_C^*$ and $(\mathfrak{a}^\alpha)_C^*$, respectively, induced by the Killing forms of $\mathfrak{g}$ and $\mathfrak{g}^\alpha$. If $\mu \in \mathfrak{a}_c^*$ let $\overline{\mu}$ denote the restriction to $\mathfrak{a}^\alpha$. Since $\overline{\mu}$ is proportional to $\overline{\alpha}$ and $\langle \mu, \alpha \rangle = \overline{\mu}(A_\alpha)$ we have

$$\frac{\langle \mu, \alpha \rangle}{\langle \alpha, \alpha \rangle} = \frac{\langle \overline{\mu}, \overline{\alpha} \rangle_\alpha}{\langle \overline{\alpha}, \overline{\alpha} \rangle_\alpha}. \tag{30}$$

Suppose now $\psi \not\equiv 0$ is a conical function on Ξ. Then by Theorem 4.1 there exists a $\mu \in \mathfrak{a}_c^*$ such that (24) holds say with $\psi(\xi^*) = 1$. Let $m_\alpha^* \in K^\alpha$ be such that $Ad_{G^\alpha}(m_\alpha^*)$ induces the nontrivial Weyl group element on $\mathfrak{a}^\alpha$. Viewing Ξ^α as a subset of Ξ (Lemma 4.11) we consider the function ψ_α on Ξ^α given by

$$\psi_\alpha(g^\alpha M^\alpha N^\alpha) = \psi\big((m^*)^{-1} m_\alpha^* g^\alpha MN\big), \quad g^\alpha \in G^\alpha.$$

Then

$$\psi_\alpha\big(m_\alpha n_\alpha a_\alpha m_\alpha^* M^\alpha N^\alpha\big) = \psi\big((m^*)^{-1} a_\alpha^{-1} MN\big) = e^{(s^*\mu)(\log a_\alpha)},$$

where $(s\mu)(H) = \mu(s^{-1}H)$. Since we know that ψ_α is smooth on Ξ^α this formula shows that ψ_α is a conical function. Then Lemma 4.9(ii) and (30) imply that

$$\frac{\langle -s^*\mu, \alpha \rangle}{\langle \alpha, \alpha \rangle} \in \boldsymbol{Z}^+.$$

Since $-s^*$ permutes the set Σ^+, we have proved that any conical function satisfies (24) and (25).

For the converse of Theorem 4.6 suppose μ satisfies (25). We extend it to the linear form $\widetilde{\mu}$ on the Cartan subalgebra $\mathfrak{h}^C = (\mathfrak{h}_{\mathrm{m}})^C \oplus \mathfrak{a}^C$ of $\mathfrak{g}^C$ by condition (27). As in §2 we order the set Δ of roots of $\mathfrak{g}^C$ with respect to $\mathfrak{h}^C$ in such a way that $\beta \in \Delta$ is > 0 whenever its restriction $\overline{\beta} = \beta|\mathfrak{a}$ is > 0. We know from [DS], Ch. VII, Lemma 8.4 that

$$\langle \beta, \beta \rangle = m\langle \overline{\beta}, \overline{\beta} \rangle \qquad \text{where } m = 1, 2, \text{ or } 4.$$

Moreover, if $m = 4$ then $2\overline{\beta} \in \Sigma^+$. Writing

$$2\frac{\langle \widetilde{\mu}, \beta \rangle}{\langle \beta, \beta \rangle} = \begin{cases} 0 & \overline{\beta} = 0 \\ 2m^{-1}\langle \mu, \overline{\beta} \rangle / \langle \overline{\beta}, \overline{\beta} \rangle & 2\overline{\beta} \notin \Sigma^+ \\ 4m^{-1}\langle \mu, 2\overline{\beta} \rangle / \langle 2\overline{\beta}, 2\overline{\beta} \rangle, & 2\overline{\beta} \in \Sigma^+ \end{cases}$$

we see that the right hand side belongs to $\boldsymbol{Z}^+$. Thus if $\widetilde{G}^C$ is the simply connected Lie group with Lie algebra $\mathfrak{g}^C$ there exists a finite-dimensional irreducible representation π of $\widetilde{G}^C$ (and $\mathfrak{g}^C$) on a complex vector space E such that π has highest weight $\widetilde{\mu}$. Let $\boldsymbol{e} \in E$ be a highest weight vector $\neq 0$. Let $\widetilde{G}, \widetilde{K}, \widetilde{A}, \widetilde{N}$ denote the analytic subgroups of $\widetilde{G}^C$ corresponding to $\mathfrak{g}, \mathfrak{k}, \mathfrak{a}, \mathfrak{n}$, respectively, let $\widetilde{M}$ denote the centralizer of $\widetilde{A}$ in $\widetilde{K}$, and $\widetilde{M_o}$ its identity component.

We shall now prove that π restricted to $\widetilde{G}$ is a conical representation. The algebra $\mathfrak{m}^C$ is spanned by $\mathfrak{h}^C \cap \mathfrak{k}^C$ and the root vectors $X_\beta, X_{-\beta}$ where $\beta \in \Delta$, $\beta > 0$ and $\overline{\beta} = 0$. If $H \in \mathfrak{h}^C \cap \mathfrak{k}^C$ then $\pi(\exp H)\boldsymbol{e} = e^{\widetilde{\mu}(H)}\boldsymbol{e} = \boldsymbol{e}$. Since $\widetilde{\mu} + \beta$ is not a weight of $\pi, \pi(X_\beta)\boldsymbol{e} = 0$. Secondly, $\pi(X_{-\beta})\boldsymbol{e} = 0$; in fact, since $\langle \widetilde{\mu}, \beta \rangle = 0$, the linear form $\widetilde{\mu} - \beta$ is the image of $\widetilde{\mu} + \beta$ under the Weyl reflection s_β and thus is not a weight. This proves $\pi(m)\boldsymbol{e} = \boldsymbol{e}$ for all $m \in \widetilde{M_0}$. For $\widetilde{M}$ itself we use the fact that $\widetilde{M} = \widetilde{M_0}\exp(i\mathfrak{a}) \cap \widetilde{K}$ ([DS], p. 435) and the fact that

$$\left\{ H \in i\mathfrak{a} : \quad \exp H \in \widetilde{K} \right\} \tag{31}$$

consists of the integral linear combinations of the vectors $2\pi i \; A_\alpha / \langle \alpha, \alpha \rangle$ $\alpha \in \Sigma^+$. The condition (25) then implies that $\pi(\exp H)\boldsymbol{e} = e^{\mu(H)}\boldsymbol{e} = \boldsymbol{e}$ for each H in the set (31). Thus $\pi(m)\boldsymbol{e} = \boldsymbol{e}$ for all $m \in \widetilde{M}$ so, since obviously $\pi(n)\boldsymbol{e} = \boldsymbol{e}$, $\pi|\widetilde{G}$ is an irreducible conical representation of $\widetilde{G}$. By Schur's lemma, the center of $\widetilde{G}$ is mapped into scalars by $\widetilde{\pi}$. This center is contained in $\widetilde{M}$ and $\pi(m)\boldsymbol{e} = \boldsymbol{e}$ for $m \in \widetilde{M}$; hence π is the identity on the center and we get an irreducible representation π_μ of G on E by putting

$$\pi_\mu(g) = \pi(\widetilde{g}) \tag{32}$$

if $g \in G$ and $\widetilde{g} \in \widetilde{G}$ are related by $Ad_G(g) = Ad_{\widetilde{G}}(\widetilde{g})$. Since M and $\widetilde{M}$ have the same images in the adjoint group (Lemma 4.10) (32) implies that

$$\pi_\mu(mn)\boldsymbol{e} = \boldsymbol{e} \qquad (m \in M, n \in N)$$

so π_μ is conical.

Let $(\pi_\mu)'$ denote the contragredient representation on the dual space E' of E and let $\boldsymbol{e}'$ be a nonzero highest weight vector. Consider the function

$$F(g) = \langle \pi_\mu(g^{-1})\boldsymbol{e}, \boldsymbol{e}' \rangle \qquad g \in G.$$

It is clear that $F(mngn') = F(g)$ $(n' \in N)$. Hence we have for all $m \in M, n \in N$,

$$F(gmn) = F(g), \qquad g \in MNAm^*N, \tag{33}$$

but since this set of g is dense in G, (33) holds for all $g \in G$. We can then define $\psi \in \mathcal{E}(\Xi)$ such that

$$F(g) = \psi(g \cdot \xi_0)$$

and then

$$\psi(mna \cdot \xi^*) = F(am^*) = e^{-\mu(\log a)}\psi(\xi^*)$$

so ψ is conical. Also $\psi \not\equiv 0$ since the elements $\pi_\mu(g^{-1})\boldsymbol{e}$ span E. This completes the proof of Theorem 4.6.

As for Theorem 4.8(i) we have, starting with μ satisfying (25), constructed the irreducible conical representation π_μ and shown that the function ψ defined on Ξ^* by (24) extends smoothly to Ξ and

$$\psi(g \cdot \xi_0) = \langle \pi_\mu(g^{-1})\boldsymbol{e}, \boldsymbol{e}' \rangle, \qquad g \in G.$$

This relation shows that E_ψ is finite-dimensional. Consider the mapping

$$v \longrightarrow \Psi_v, \qquad \Psi_v(g \cdot \xi_0) = \langle \pi_\mu(g^{-1})v, \boldsymbol{e}' \rangle \tag{34}$$

of E into E_ψ. It is injective by the irreducibility of π_μ. The surjectivity follows from the relation

$$\Psi_{\pi_\mu(h)\boldsymbol{e}} = \psi^{\tau(h)},$$

which also shows that (34) sets up an equivalence between π_μ and π_ψ.

For surjectivity in Theorem 4.7(i) let π_o be any finite-dimensional irreducible conical representation of G on a complex vector space E_o. Let $\boldsymbol{e}_o \neq 0$

be a vector fixed under $\pi_o(MN)$ and $e'_o \neq 0$ a vector in the highest weight space for π'_o. Again we consider

$$F^*(g) = \langle \pi_o(g^{-1})e_o, e'_o \rangle, \qquad g \in G.$$

By $\pi_o(MN)e_o = e_o$, $\pi'_o(A)e'_o \subset \boldsymbol{R}e'_o$, $F^*(g) = \psi^*(g \cdot \xi_o)$ where ψ^* is a conical function on Ξ; just as for (34) we see that the representation π_{ψ^*} is equivalent to π_o. For the injectivity in (i) we remark that if ψ_1 and ψ_2 are conical functions such that π_{ψ_1} and π_{ψ_2} are equivalent then ψ_1 and ψ_2 are proportional. In fact this is clear from (26), the highest weight spaces being one-dimensional. If μ is the highest restricted weight of π_{ψ^*} then by Theorem 4.8(i), ψ^* satisfies (24) so

$$\psi^*(mnam^* \cdot \xi_o) = e^{-\mu(\log a)}\psi^*(\xi^*).$$

This implies $\psi^*(bg \cdot \xi_o) = e^{-\mu(\log b)}\psi^*(g \cdot \xi_o)$ for all $b \in A$, $g \in G$. Hence

$$\langle \pi_o(b^{-1})e_o, \pi'_o(g)e'_o \rangle = e^{-\mu(\log b)}\langle e_o, \pi'_o(g)e'_o \rangle$$

which by the irreducibility of π'_o implies that e_o is a highest weight vector. This proves Theorem 4.7(ii) and (iii). Finally Theorem 4.8(ii) results from Theorem 4.7(i) and Theorem 4.8(i).

3. The Finite-dimensional Spherical Representations.

We shall now show that the representations we have been discussing have another characterization. We retain our usual assumption that G is any connected semisimple Lie group with finite center.

Theorem 4.12. *Let π be a complex irreducible finite-dimensional representation of G. Then π is spherical if and only if it is conical. The highest weights of these representations are thus characterized by* (27).

Proof. Suppose first π is spherical. Let $v \neq 0$ be a spherical vector in the corresponding representation space E and let π' be the contravariant representation on the dual space E'. Then π' is irreducible. Select $v'_o \in E'$ such that $\langle v, v'_o \rangle = 1$. Then the vector $v' = \int_K \pi'(k)v'_o dk$, is a spherical vector $\neq 0$ for π'. Consider the function on G/K given by

$$\varphi(gK) = \langle \pi(g^{-1})v, v' \rangle, \qquad g \in G. \tag{35}$$

The G-translates of φ span a finite-dimensional function space E_φ on G/K. The mapping

$$u \longrightarrow F_u \qquad \text{where} \qquad F_u(gK) = \langle \pi(g^{-1})u, v' \rangle \tag{36}$$

is a bijection of E onto E_φ setting up the equivalence of π with the natural representation π_φ of G on E_φ. Let $\psi \in E_\varphi$ be in the highest weight space, $\psi \neq 0$. Then $\pi_\varphi(n)\psi = \psi$ and by $G = NAK$

$$\psi(nmg \cdot o) = \psi(nmn_1a_1k_1 \cdot o) = \psi(g \cdot o)$$

so π_φ, and hence π, is conical.

On the other hand suppose π is conical. By Theorem 4.7, π is equivalent to π_ψ where ψ is a conical function. It then suffices to prove that the vector

$$\psi_o = \int_{K/M} \pi_\psi(k)\psi \, dk_M$$

is $\neq 0$. However,

$$\begin{aligned}\psi_o(\xi_o) &= \int_{K/M} \psi(m^*k \cdot \xi_o)dk_M \\ &= \int_{\overline{N}} \psi(m^*k(\overline{n}) \cdot \xi_o)e^{-2\rho(H(\overline{n}))}d\overline{n} = \psi(\xi^*) \int_{\overline{N}} e^{(s^*\mu-2\rho)(H(\overline{n}))}d\overline{n} \neq 0\end{aligned}$$

since

$$m^*k(\overline{n}) \cdot \xi_o = m^*\overline{n}\exp(-H(\overline{n})) \cdot \xi_o = m^*\overline{n}m^{*-1}\exp(-s^*(H(\overline{n}))) \cdot \xi^*.$$

Corollary 4.13. *Let π_μ be an irreducible finite-dimensional conical representation with highest restricted weight μ. Then $(\pi_\mu)'$ is also conical and equivalent to $\pi_{-s^*\mu}$.*

In fact, $(\pi_\mu)'$ is conical by Theorem 4.12. Let ψ' denote the corresponding conical function and μ' the corresponding highest restricted weight. Then

$$\psi'(g \cdot \xi_o) = \langle \pi'_\mu(g^{-1})e', e\rangle = \langle \pi_\mu(g)e, e'\rangle = \psi(g^{-1} \cdot \xi_o)$$

so

$$\begin{aligned}\psi(\exp H\ m^* \cdot \xi_o) &= \psi'((m^*)^{-1}\exp(-H) \cdot \xi_o) \\ &= \psi'(\exp(-s^*H)m^* \cdot \xi_o).\end{aligned}$$

Thus

$$e^{-\mu(H)}\psi(\xi^*) = e^{\mu'(s^*H)}\psi'(\xi^*),$$

proving the corollary.

4. Conical Models and Spherical Models.

We shall now describe somewhat different models of the irreducible finite-dimensional conical (spherical) representations.

As before, $\mathcal{D}'_\lambda(\Xi)$ consists of the distributions on Ξ satisfying (18). From the description in Prop. 4.4 it is clear that if $\Psi \in \mathcal{D}'_\lambda(\Xi)$ is G-finite (or even just K-finite) it is given by a smooth function.

Definition. For $\lambda \in \mathfrak{a}_C^*$ let $\mathcal{D}'_{\lambda,G} = \mathcal{D}'_{\lambda,G}(\Xi)$ denote the space of G-finite elements in $\mathcal{D}'_\lambda(\Xi)$.

Since $(d\xi)^{\sigma(a)} = e^{-2\rho(\log a)}d\xi$ we note the following result.

Lemma 4.14. *$\mathcal{D}'_{\lambda,G}$ consists of the functions $\varphi \in \mathcal{E}(\Xi)$ such that*

(i) *φ is G-finite;*

(ii) $\varphi(ga \cdot \xi_o) = e^{(i\lambda-\rho)(\log a)}\varphi(g \cdot \xi_o)$.

These spaces now form the models indicated.

Proposition 4.15. *With $\mathcal{D}'_{\lambda,G}$ as above we have*

$$\mathcal{D}'_{\lambda,G} \neq \{0\} \quad \textit{if and only if} \quad \frac{\langle i\lambda - \rho, \alpha\rangle}{\langle \alpha, \alpha\rangle} \in \boldsymbol{Z}^+, \alpha \in \Sigma^+. \tag{37}$$

In this case, $\mathcal{D}'_{\lambda,G}$ is finite-dimensional and irreducible under G. As λ runs through the elements of $\mathfrak{a}_C^*$ satisfying (37) the spaces $\mathcal{D}'_{\lambda,G}$ form the models of all the inequivalent finite-dimensional conical representations of G. The highest restricted weight on $\mathcal{D}'_{\lambda,G}$ is $-s^*(i\lambda - \rho)$.

Proof. Let ψ be a nonzero element in $\mathcal{D}'_{\lambda,G}$ and E_ψ the (finite-dimensional) span of the G-translates of ψ. By weight theory there exists a nonzero vector $\psi_o \in E_\psi$ which is N-invariant. The decomposition (6) §1 then shows that ψ_o is also M-invariant so ψ_o is conical. Using Lemma 4.14 (ii) for $g = m^*$ and the fact that $-s^*$ permutes Σ^+ condition (37) follows (in both directions). We have

$$\mathcal{D}'_{\lambda,G} \supset E_\psi \supset E_{\psi_o},$$

where E_{ψ_o} is spanned by the G-translates of ψ_o. Since ψ_o is (up to a constant factor) the only conical function in $\mathcal{D}'_{\lambda,G}$ and since the representation of G on E_ψ is completely reducible we deduce $E_\psi = E_{\psi_o}$. Thus each $\psi \in \mathcal{D}'_{\lambda.G}$ belongs to E_{ψ_o} and the proposition follows.

We now state and prove the counterpart for X of this last result. It is even easier to prove.

Recall that the isomorphism $\Gamma: \boldsymbol{D}(X) \longrightarrow \boldsymbol{D}_W(A)$ (Theorem 2.1) and the spherical function φ_λ ((11) §3) are related by

$$D\varphi_\lambda = \Gamma(D)(i\lambda)\varphi_\lambda, \qquad \lambda \in \mathfrak{a}^*_{\boldsymbol{C}}. \tag{38}$$

(cf. [GGA], Ch.II, §5). We introduce the *joint eigenspace*

$$\mathcal{E}_\lambda(X) = \{f \in \mathcal{E}(X) : \ Df = \Gamma(D)(i\lambda)f \quad \text{for } D \in \boldsymbol{D}(X)\} \tag{39}$$

which by (17) is the analog of $\mathcal{D}'_\lambda(\Xi)$. Let $\mathcal{E}_{\lambda,G} = \mathcal{E}_{\lambda,G}(X)$ denote the space of G-finite elements in $\mathcal{E}_\lambda(X)$. Note that $\mathcal{E}_{s\lambda} = \mathcal{E}_\lambda$ and $\mathcal{E}_{s\lambda,G} = \mathcal{E}_{\lambda,G}$ for $s \in W$.

Proposition 4.16. *With $\mathcal{E}_{\lambda,G}$ as above we have $\mathcal{E}_{\lambda,G} \neq \{0\}$ if and only if*

$$\text{For some } s \in W, \qquad \frac{\langle -is\lambda - \rho, \alpha\rangle}{\langle \alpha, \alpha\rangle} \in \boldsymbol{Z}^+, \ \alpha \in \Sigma^+. \tag{40}$$

In this case, $\mathcal{E}_{\lambda,G}$ is finite-dimensional and irreducible under G. All finite-dimensional spherical representation of G arise in this fashion. Also $-is\lambda-\rho$ is the highest restricted weight.

Proof. Suppose $\mathcal{E}_{\lambda,G} \neq \{0\}$, let φ be a nonzero element, and E_φ the space spanned by the G-translates of φ. Since each irreducible subspace of E_φ contains a K-invariant $\neq 0$, and since $\boldsymbol{C}\varphi_\lambda$ is the subspace of K-invariants in $\mathcal{E}_\lambda(X)$ we see that E_φ is irreducible. Since φ was arbitrary, $\mathcal{E}_{\lambda,G}$ equals E_φ and is thus finite-dimensional and irreducible. Let $\psi \neq 0$ in $\mathcal{E}_{\lambda,G}$ be a highest weight vector and μ the highest restricted weight. Then

$$\psi(ngK) = \psi(gK), \ \psi(a^{-1}gK) = e^{\mu(\log a)}\psi(gK).$$

This implies $\psi(K) \neq 0$ and $\psi(aK) = e^{-\mu(\log a)}\psi(K)$. Let $D \in \boldsymbol{D}(X)$ and let $\Delta_N(D)$ denote the radial part of D for N acting on X. Then

$$(D\psi)(a\cdot o) = (\Delta_N(D)\overline{\psi})(a\cdot o), \qquad \text{where } \overline{\psi} = \psi | A\cdot o. \tag{41}$$

By Theorem 2.1 we have $\Delta_N(D) = e^\rho\Gamma(D)\circ e^{-\rho}$ so (41) implies

$$\Gamma(D)(i\lambda)e^{-\mu(\log a)} = \Gamma(D)(-\rho-\mu)e^{-\mu(\log a)}.$$

It follows ([GGA], Ch. III, §3) that $i\lambda$ and $-\rho-\mu$ are conjugate under W so $\mu = -is\lambda - \rho$ for some $s \in W$. Thus (40) follows from criterion (25) for conical representations.

On the other hand suppose (40) holds. Then putting $\mu = -is\lambda - \rho$, μ is the highest restricted weight for a finite-dimensional irreducible conical (spherical) representation π. This is equivalent to the representation π_φ on E_φ from the proof of Theorem 4.12 where φ is the function (35). The function

$$\psi(gK) = e^{-\mu(A(g))} = e^{(is\lambda+\rho)(A(g))} \tag{42}$$

is a highest weight vector and by the uniqueness of K-fixed vectors in E_φ,

$$\varphi(gK) = \int_K e^{(is\lambda+\rho)(A(kg))}dk = \varphi_\lambda(gK) \tag{43}$$

so $\mathcal{E}_{\lambda,G} \neq 0$.

5. Simultaneous Euclidean Imbeddings of X and of Ξ Horocycles as Plane Sections.

Let $G = \boldsymbol{SL}(2, \boldsymbol{R})$ and Ad its adjoint representation on $\mathfrak{g} = \mathfrak{sl}(2, \boldsymbol{R})$. The vector $X_1 = \left(\begin{smallmatrix} 0 & 1 \\ -1 & 0 \end{smallmatrix}\right)$ in $\mathfrak{g}$ is fixed under $Ad(K)$ (where $K = \boldsymbol{SO}(2)$) so the representation Ad is spherical. Also the vector $E_1 = \left(\begin{smallmatrix} 0 & 1 \\ 0 & 0 \end{smallmatrix}\right)$ is fixed under $Ad(MN)$ where

$$MN = \left\{ \begin{pmatrix} \epsilon & x \\ 0 & \epsilon \end{pmatrix} : \epsilon = \pm 1, x \in \boldsymbol{R} \right\}.$$

Let B denote the Killing form of $\mathfrak{g}$. As shown in [GGA], Ch. V, Ex. A11,

(i) *The map $gK \longrightarrow Ad(g)X_1$ is a bijection of G/K onto the upper half of the quadric*

$$B(X, X) = B(X_1, X_1) \qquad \textit{in } \mathfrak{g}.$$

(ii) *The map $gMN \longrightarrow Ad(g)E_1$ is a bijection of the horocycle space G/MN onto the upper half of the null cone*

$$B(Z, Z) = 0 \qquad \textit{in } \mathfrak{g}.$$

(iii) *The horocycles in G/K are given by the plane sections $B(X, Z) = -1$, Z being a fixed vector in the null cone above.*

We note also that the geodesics in the quadric above are the nonempty sections with 2-planes through the origin (compare[GGA], Ch. I, §4, No. 1). Thus a function on this quadric can be determined by its integrals over plane sections through the origin([GGA] Ch. I, Cor. 4.3) and also by integrals over plane sections parallel to some vector in the null cone(Theorem 3.13, this chapter).

We shall now generalize properties (i), (ii), and (iii) to

$$X = G/K, \qquad \Xi = G/MN$$

where, as before, G is an arbitrary connected semisimple Lie group with finite center. If $\ell = \operatorname{rank} G/K$ we consider linear forms $\mu_1, \ldots, \mu_\ell$ on $\mathfrak{a}$ each satisfying the integrality condition (25) and let π_j $(1 \leq j \leq \ell)$ denote the conical irreducible representation of G with highest restricted weight μ_j on the complex finite-dimensional vector space V_j. Let $\boldsymbol{e}_j \in V_j$ be a highest weight vector and let $v_j \in V_j$ be a K-fixed vector. Let π be the direct sum representation $\pi = \pi_1 \dot{+} \ldots \dot{+} \pi_\ell$ of G on the direct sum $V = \oplus_j V_j$ and put

$$\boldsymbol{e} = \oplus_j \boldsymbol{e}_j, \qquad \boldsymbol{v} = \oplus_j v_j. \tag{44}$$

Theorem 4.17. *Assume the linear form $\mu_1, \ldots, \mu_\ell \in \mathfrak{a}^*$ satisfy the integrality condition* (25) *and are linearly independent. Let π be the representation of G on V defined above and $\boldsymbol{e}$ and $\boldsymbol{v}$ the vectors defined by* (44). *Then the maps*

$$gK \longrightarrow \pi(g)\boldsymbol{v}, \qquad gMN \longrightarrow \pi(g)\boldsymbol{e}$$
$$X \longrightarrow V, \qquad \Xi \longrightarrow V$$

are injective and imbed X and Ξ as G-orbits in V.

Proof. Fix π_j and as in the proof of Theorem 4.12 select v_j' in the dual of V_j fixed under K in the contragredient representation π_j' satisfying $\langle v_j, v_j' \rangle = 1$. Put

$$\varphi_j(gK) = \langle \pi_j(g^{-1})v_j, v_j' \rangle \tag{45}$$

and let E_j denote the finite-dimensional subspace of $\mathcal{E}(X)$ spanned by the G-translates of φ_j. Let

$$F_v(gK) = \langle \pi_j(g^{-1})v, v_j' \rangle \qquad v \in V_j$$

and consider again the bijection

$$v \longrightarrow F_v \tag{46}$$

setting up the equivalence between π_j and the natural representation π_{φ_j} of G on E_j. Since the spaces E_j $(1 \leq j \leq \ell)$ consist of representation coefficients of inequivalent representations the sum $E = \Sigma_j E_j$ is direct.

The representation π is realized on it with spherical and conical vectors respectively given by

$$\varphi(gK) = \sum_j \varphi_j(gK) = \sum_j \langle \pi_j(g^{-1})v_j, v_j' \rangle \tag{47}$$

$$\psi(gK) = \sum_j e^{-\mu_j(A(g))} \tag{48}$$

(cf. (42)). To establish Theorem 4.17 it remains to prove that under the action of G on E the isotropy group of φ is K and the isotropy group of ψ is MN.

Dealing first with φ let L be the isotropy subgroup

$$L = \{\ell \in G : \ \pi(\ell)\varphi = \varphi\}.$$

Because of the directness of the sum $E = \Sigma_j E_j$ we have $\varphi_j^{\tau(\ell)} = \varphi_j$ for each $\ell \in L$ and each j. Let j be fixed and in accordance with (42) and (43) select $\lambda \in \mathfrak{a}_c^*$ such that

$$\varphi_j(gK) = \varphi_\lambda(g) = \int_K e^{(i\lambda+\rho)(A(kg))} dk \tag{49}$$

and $\mu_j = -(i\lambda+\rho)$. We shall now use our symmetry identity for the spherical function φ_λ ([GGA], Ch. IV, Lemma 4.4),

$$\varphi_\lambda(h^{-1}g) = \int_K e^{(-i\lambda+\rho)(A(kh))} e^{(i\lambda+\rho)(A(kg))} dk. \tag{50}$$

Put $\nu = -s^*\mu_j = is^*\lambda - \rho$ and note that ν satisfies the integrality condition (25) and that $\varphi_\lambda = \varphi_{s^*\lambda}$. Using this in (50) the identity $\varphi_j^{\tau(\ell)} = \varphi_j$ takes the form

$$\int_K e^{\nu+2\rho(A(kg))} e^{-\nu(A(k\ell))} dk = \int_K e^{(\nu+2\rho)(A(kg))} dk.$$

Since $\langle \nu + \rho, \alpha \rangle > 0$ we can take the limit in Theorem 3.16 and deduce that

$$\mathbf{c}(-i(\nu+\rho)) e^{-\nu(A(\ell))} = \mathbf{c}(-i(\nu+\rho)),$$

whence $\nu(A(\ell)) = 0$. Since this conclusion holds for each j $(1 \le j \le \ell)$ and since the μ_j are linearly independent, we conclude

$$A(\ell) = 0 \qquad \ell \in L. \tag{51}$$

Of course $K \subset L$, but suppose now ℓ_o is an element of L not in K. We decompose according to the Cartan decomposition $G = KAK$, $\ell_o = k_1 a_o k_2$ with $a_o \neq e$ in A. The identity $\varphi(\ell_o gK) \equiv \varphi(gK)$ together with the K-invariance of φ implies $a_o \in L$ contradicting (51). This proves the first half of Theorem 4.17.

It remains to prove that if the function ψ in (48) satisfies $\psi^{\tau(h)} = \psi$ then $h \in MN$. Since the sum $E = \Sigma_j E_j$ is direct we can deduce from (48) that for each j

$$e^{\mu_j(H(gh))} = e^{\mu_j(H(g))} \qquad g \in G. \tag{52}$$

Thus it remains to prove that $H(gh) = H(g)$ for all $g \in G$ implies $h \in MN$. We prove a stronger result.

Lemma 4.18. *Let* $h \in G$.

(i) *If* $H(nh) = 0$ *for all* $n \in N$ *then* $h \in MN$.

(ii) *If* $H(ah) = H(a)$ *for all* $a \in A$ *then* $h \in MN$.

Proof. (i) $H(nh) = 0$ implies $nh = k_1 n_1$ so $h^{-1}n^{-1} = n_1^{-1}k_1^{-1}$. Thus h^{-1} maps the horocycle ξ_o into itself so, by Theorem 1.1, $h \in MN$.

(ii) Writing $h = k_o a_o n_o$ we first note that $a_o = e$ so $H(ak_o) = H(a)$. Consider the Bruhat decomposition,

$$G = \bigcup_{s \in W} B m_s B, \qquad B = MAN, \tag{53}$$

([DS], IX, §1) where m_s is a representative of the Weyl group element s. Put $N^s = m_s N m_s^{-1}$. Then we have the diffeomorphic decomposition

$$N^{s^{-1}} = (\overline{N} \cap N^{s^{-1}}) \cdot (N \cap N^{s^{-1}}) \tag{54}$$

([GGA], IV, §6 (28)). Now (53) implies

$$G/B = \bigcup_{s \in W} N m_s \cdot \omega, \tag{55}$$

where ω is the origin in G/B. Putting $\overline{N}_s = \overline{N} \cap N^{s^{-1}}$ we can write $Nm_s \cdot \omega = m_s \overline{N}_s \cdot \omega$. Decomposing $\overline{n}_s \in \overline{N}_s$ as $\overline{n}_s = k(\overline{n}_s)a(\overline{n}_s)n(\overline{n}_s)$ we deduce

$$Nm_s \cdot \omega = \{m_s k(\overline{n}_s) \cdot \omega : \ \overline{n}_s \in \overline{N}_s\}.$$

Applying the diffeomorphism $kM \longrightarrow kMAN$ of K/M onto G/B to (55) we thus obtain

$$K/M = \bigcup_{s\in W} m_s k(\overline{N}_s)M. \tag{56}$$

Hence $k_o = m_s k(\overline{n}_s)m$ for some $s \in W$ so putting $b = m_s^{-1} a m_s$ we have $H(bk(\overline{n}_s)) = H(m_s b k(\overline{n}_s)m) = H(ak_o) = H(a) = s \log b$. However

$$H(bk(\overline{n}_s)) = H(b\overline{n}_s b^{-1}) - H(\overline{n}_s) + \log b$$

so

$$H(b\overline{n}_s b^{-1}) - H(\overline{n}_s) = s\log b - \log b. \tag{57}$$

Let $b = \exp tH$ $(H \in \mathfrak{a}^+)$ and let $t \longrightarrow +\infty$. The left hand side converges to $-H(\overline{n}_s)$ whereas the right hand side has no limit unless $s = e$. Thus $k_o \in M$ and the lemma and Theorem 4.17 are proved.

Corollary 4.19. *In the embedding of Theorem 4.17, the horocycle space* Ξ *is the cone*

$$\{\pi(k)(s_1 e_1 + \cdots + s_\ell e_\ell) : \quad \textit{all } s_i > 0,\ k \in K\}. \tag{58}$$

The geodesics in the horocycle space are given by the curves

$$t \longrightarrow \pi(k)\big(s_1^t e_1 + \cdots + s_\ell^t e_\ell\big).$$

In particular, if $\operatorname{rank}(G/K) = 1$ *and* V *is given a* $\pi(K)$*-invariant inner product,* Ξ *is a circular cone.*

In fact, writing $g = kan$ we have

$$\pi(g)e = \pi(ka)e = \pi(k)\big(e^{\mu_1(\log a)}e_1 + \cdots + e^{\mu_\ell(\log a)}e_\ell\big)$$

and since the μ_i form a basis of $\mathfrak{a}^*$ we can for given $s_1, \ldots, s_\ell > 0$ find a unique $a \in A$ such that $s_i = e^{\mu_i(\log a)} (1 \leq i \leq \ell)$. Moreover, under our imbedding the geodesic $k \exp tHN \cdot o$ in Ξ is mapped into the curve

$$t \longrightarrow \pi(k)\big(e^{t\mu_1(H)}e_1 + \cdots + e^{t\mu_\ell(H)}e_\ell\big).$$

Consider now the contragredient representations $\pi_1', \ldots \pi_\ell'$ and corresponding highest weight vectors $e_1', \ldots e_\ell'$. Then the π_i' are also conical (Cor. 4.13) so the vectors e_i' are fixed under M. Let W_ℓ denote the annihilator

$$\{w \in V : \langle w, e_i'\rangle = 0 \quad \text{for } 1 \leq i \leq \ell\}.$$

We can now generalize the description of the horocycles in the hyperbolic plane given at the beginning of this subsection.

Theorem 4.20. *With the embedding $X = \pi(G)\boldsymbol{v} \subset V$ from Theorem 4.17 the horocycle $\pi(N)\boldsymbol{v}$ is the intersection of X with the plane $\boldsymbol{v} + W_\ell$.*

Proof. First we remark that $\langle \boldsymbol{v}, e_i' \rangle \neq 0$ for each i. In fact by Cor. 4.13,

$$\langle \pi(nak)\boldsymbol{v}, e_i' \rangle = e^{\nu(\log a)} \langle \boldsymbol{v}, e_i' \rangle, \qquad \nu = -s^* \mu_i, \tag{59}$$

and now the statement follows from the irreducibility of π_i. Next we note that for each i, $\langle \pi(N)\boldsymbol{v}, e_i' \rangle = \langle \boldsymbol{v}, e_i' \rangle$ so the horocycle $\pi(N)\boldsymbol{v}$ is contained in the intersection $X \cap (\boldsymbol{v} + W_\ell)$. On the other hand suppose $x \in X \cap (\boldsymbol{v} + W_\ell)$. Then for some $n \in N, a \in A, x = \pi(na)\boldsymbol{v}$ so

$$\langle \pi(na)\boldsymbol{v}, e_i' \rangle = \langle \boldsymbol{v}, e_i' \rangle \qquad \text{for } 1 \leq i \leq \ell. \tag{60}$$

Thus by (59), $(s^*\mu_i)(\log a) = 0$ so $a = e$ whence $x \in \pi(N)\boldsymbol{v}$ as claimed.

6. Restricted Weights.

Let π be a representation of $\mathfrak{g}$ on a finite-dimensional vector space V over $\boldsymbol{C}$. This induces a representation of $\mathfrak{g}^{\boldsymbol{C}}$ and of the simply connected complexification $G^{\boldsymbol{C}}$ on V. Let G and U be the analytic subgroups of $G^{\boldsymbol{C}}$ corresponding to $\mathfrak{g}$ and $\mathfrak{u} = \mathfrak{k} + i\mathfrak{p}$ and consider again the subgroups $K, A, N, \overline{N}$ and M with Lie algebras $\mathfrak{k}, \mathfrak{a}, \mathfrak{n}, \overline{\mathfrak{n}}$ and $\mathfrak{m}$. The choice of $\mathfrak{n}$ corresponds to an ordering of the dual space $\mathfrak{a}^*$. We choose a compatible ordering of the dual of the space $\mathfrak{h}_{\boldsymbol{R}} = i(\mathfrak{h} \cap \mathfrak{k}) + \mathfrak{a}$ on which all roots of $\mathfrak{g}^{\boldsymbol{C}}$ with respect to $\mathfrak{h}^{\boldsymbol{C}} = (\mathfrak{h}_{\boldsymbol{R}})^{\boldsymbol{C}}$ are real-valued.

We choose an inner product $\langle\ ,\ \rangle$ on V with respect to which $\pi(U)$ consists of unitary operators, and put $\|v\|^2 = \langle v, v \rangle$. If $*$ denotes the adjoint operation,

$$\pi(T)^* = -\pi(T) \qquad (T \in \mathfrak{k}), \qquad \pi(X)^* = \pi(X) \qquad (X \in \mathfrak{p}) \tag{61}$$

so

$$\pi(Z)^* = -\pi(\theta Z) \qquad Z \in \mathfrak{g} \tag{62}$$

and

$$\pi(g)^* = \pi(\theta g^{-1}) \qquad g \in G. \tag{63}$$

Definition. (i) A $\lambda \in (\mathfrak{h}^{\boldsymbol{C}})^*$ is called a *weight* of π if there exists a $v \neq 0$ in V such that

$$\pi(H)v = \lambda(H)v, \qquad (H \in \mathfrak{h}^{\boldsymbol{C}}).$$

(ii) A $\mu \in (\mathfrak{a}^{\boldsymbol{C}})^*$ is called a *restricted weight* if there exists a $w \neq 0$ in V such that

$$\pi(H)w = \mu(H)w, \qquad (H \in \mathfrak{a}).$$

The vector v (resp. w) is called a *weight vector* (resp. *restricted weight vector*).

It is clear from (61) that weights and restricted weights are real-valued on $\mathfrak{h}_{\boldsymbol{R}}$ and $\mathfrak{a}$, respectively.

Proposition 4.21. *The restricted weights coincide with the restrictions of the weights to* $\mathfrak{a}$.

Proof. It suffices to prove that each restricted weight μ_o is a restriction $\lambda_o|\mathfrak{a}$ where λ_o is a weight. But if $v \neq 0$ such that $\pi(H)v = \mu_o(H)v$ $(H \in \mathfrak{a})$ and we diagonalize the commutative family $\pi(\mathfrak{h}_{\boldsymbol{R}})$ of self-adjoint operators we see that $v = \Sigma v_\lambda$ where v_λ satisfies $\pi(H)v_\lambda = \lambda(H)v_\lambda$ $(H \in \mathfrak{h}_{\boldsymbol{R}})$. This proves the result.

Let $\Delta = \Delta(\mathfrak{g}^{\boldsymbol{C}}, \mathfrak{h}^{\boldsymbol{C}})$ the set of roots, let $\Sigma = \Sigma(\mathfrak{g}, \mathfrak{a})$, the set of restricted roots and let $\Delta_o = \{\alpha \in \Delta : \ \overline{\alpha} = 0\}$. Let $B = B(\Delta)$ be a basis of Δ, put $B_o = B \cap \Delta_o$ and let $\overline{B} \subset \Sigma$ be the set of restrictions of $B - B_o$ to $\mathfrak{a}$. Then Δ_o is a root system with basis B_o and $\overline{B}$ is a basis of the root system Σ (cf. e.g. [DS], Ch. X, Exercise F7). While the cardinality of $B - B_o$ is in general larger than $\ell = \dim \mathfrak{a}$ we can select $\alpha_1, \ldots, \alpha_\ell \in B - B_o$ such that $\overline{B} = \{\overline{\alpha}_1, \ldots, \overline{\alpha}_\ell\}$.

Let $\Lambda(\pi)$ (resp. $\overline{\Lambda}(\pi)$) denote the set of weights (resp. restricted weights) of π.

Proposition 4.22. *Suppose π is a finite-dimensional irreducible representation of $\mathfrak{g}$ on V. Let λ denote its highest weight and put $\mu = \lambda|\mathfrak{a} = \overline{\lambda}$. Then*

(i) *Each restricted weight $\nu \in \overline{\Lambda}(\pi)$ has the form*

$$\nu = \mu - \sum_1^\ell m_i \overline{\alpha}_i, \qquad m_i \in \boldsymbol{Z}^+.$$

(ii) *Each $\nu \in \overline{\Lambda}(\pi)$ belongs to the convex hull $C(\mu)$ of the orbit $W \cdot \mu$.*

Proof. Part (i) is an immediate consequence of the corresponding statement for $\Lambda(\pi)$ if we take the facts above about $\overline{B}, B$ and B_o into account. For Part (ii) note that by ([DS], VII, Prop. 8.10) each $s \in W$ permutes $\overline{\Lambda}(\pi)$. Select $s \in W$ such that $s\nu \in C\ell(\mathfrak{a}_+^*)$, $C\ell$ denoting closure. Then by (i)

$$s\nu = \mu - \sum_1^\ell m_i\overline{\alpha}_i. \tag{64}$$

On the other hand, the closure C^+ of $\mathfrak{a}^+$, the dual cone

$${}^+C = \{H \in \mathfrak{a} : \langle H, H'\rangle \geq 0 \qquad \text{for } H' \in C^+\}$$

and the negative ${}^-C = -{}^+C$ satisfy for each $H \in C^+$

$$C(H) = \bigcup_{\sigma \in W} \sigma\{C^+ \cup (H + {}^-C)\} = \bigcap_{\sigma \in W} \sigma(H + {}^-C), \tag{65}$$

(cf. [GGA], IV, Lemma 8.3). Using the first identity on the vector $H = A_\lambda$ we see since $A_{s\nu} \in C^+$ that $s\nu \in C(\mu)$ as claimed.

We now restate Theorem 4.8 and 4.12 in terms of restricted weights. We recall that since the center of G lies in M our assumption that G be contained in the simply connected complexification $G^{\boldsymbol{C}}$ is immaterial for the spherical representations.

As usual, let Σ_o be the set of indivisible roots in Σ, and $\Sigma_0^+ = \Sigma_o \cap \Sigma^+$. Let Σ_* be the set of unmultipliable roots in Σ and $\Sigma_*^+ = \Sigma_* \cap \Sigma^+$. Then Σ_* is a root system and if $\gamma_1, \ldots, \gamma_\ell$ is any basis of Σ a basis $\beta_1, \ldots, \beta_\ell$ of Σ_* can be obtained by putting $\beta_i = \gamma_i$ if $2\gamma_i \notin \Sigma$, $\beta_i = 2\gamma_i$ if $2\gamma_i \in \Sigma$ ([DS], pp. 456 and 475).

Proposition 4.23. *With an arbitrary basis $\beta_1, \ldots, \beta_\ell$ of the system Σ_* of unmultipliable roots in Σ let $\mu_1, \ldots, \mu_\ell \in \mathfrak{a}^*$ be determined by*

$$\frac{\langle \mu_i, \beta_j\rangle}{\langle \beta_j, \beta_j\rangle} = \delta_{ij}, \qquad 1 \leq i, j \leq \ell.$$

Let $\lambda \in \mathfrak{a}^*$. Then λ is the highest restricted weight for an irreducible spherical representation of G if and only if

$$\lambda = \sum_{i=1}^\ell m_i\mu_i, \qquad m_i \in \boldsymbol{Z}^+.$$

Proof. For $\beta \in \Sigma_*$ put $\beta^* = \beta/\langle \beta, \beta\rangle$. Then the lattice $\sum_1^\ell \boldsymbol{Z}\beta_j^*$ is invariant under each simple reflection s_{β_k} (since Σ_* is a root system so

$2\langle\beta_k,\beta_j^*\rangle \in \boldsymbol{Z}$) hence invariant under W. Since each $\beta \in \Sigma_*$ equals $s\beta_k$ for some $s \in W$, $1 \leq k \leq \ell$, we see that $\sum_1^\ell \boldsymbol{Z}\beta_j^* = \Sigma_\beta \boldsymbol{Z}\beta^*$. Thus

$$\langle\lambda,\beta_j^*\rangle \in \boldsymbol{Z}^+ \qquad (1 \leq j \leq \ell) \Longleftrightarrow \langle\lambda,\beta^*\rangle \in \boldsymbol{Z}^+ \qquad (\beta \in \Sigma_*^+)$$

and this last condition is clearly equivalent to

$$\frac{\langle\lambda,\alpha\rangle}{\langle\alpha,\alpha\rangle} \in \boldsymbol{Z}^+ \qquad \text{for all } \alpha \in \Sigma^+.$$

Thus the proposition is a restatement of (27).

7. The Component $H(\overline{n})$.

In [GGA], IV, §6, No. 3 we proved Harish-Chandra's result that $H(\overline{n}) - H(a\overline{n}a^{-1}) \in {}^+C$ for $a \in \exp \mathfrak{a}^+$ and consequently $H(\overline{n}) \in {}^+C$ for each $\overline{n} \in \overline{N}$. For G/K of rank one we have in Ch. II, Theorem 6.1 a simple explicit formula for $H(\overline{n})$ from which its behavior for large $\overline{n}$ are readily read.

In this subsection we deal with the case of arbitrary rank and shall prove some results which relate $H(\overline{n})$ to the Cartan- and Bruhat decomposition. It will be convenient to use the notation $a \longrightarrow +\infty$ to mean $\lim \alpha(\log a) = \infty$ for each $\alpha \in \Sigma^+$. Also $|Z|^2 = -B(Z,\theta Z)$ for $Z \in \mathfrak{g}$.

Proposition 4.24. *Fix $\overline{n}$ and for $a \in A$ let $a' \in C\ell(A^+)$ be the unique element for which $\overline{n}a \in Ka'K$. Then*

$$\lim_{a\longrightarrow\infty} |\log a' - \log a - H(\overline{n})| = 0. \tag{66}$$

Remark. For G/K of rank 1 (66) shows that the distance between the sphere through $\overline{n}a\cdot o$ with center o and the horocycle $Na\cdot o$ is approximately $|H(\overline{n})|$ for large a.

Proof. Let π be an irreducible representation of G on V of highest weight Λ and put $\nu = \Lambda|\mathfrak{a}$. Let

$$V = V_\nu + \sum_{\mu<\nu} V_\mu = V_\nu + W$$

be the decomposition into weight subspaces. Let $v_1, \ldots v_r$ be an orthonormal basis of V_ν and $w_1, \ldots, w_p$ an orthonormal basis of W consisting of restricted weight vectors, of restricted weights $\mu_1, \ldots, \mu_p$, respectively. If L is a linear transformation of V with expression ℓ_{ij} in the basis $(v_1, \ldots, w_p)$ we put $\|L\|^2 = \Sigma_{ij} |\ell_{ij}|^2$. Then, $\| \; \|$ also denoting the norm on V,

$$\|\pi(a')\|^2 = \|\pi(\overline{n}a)\|^2 = \sum_1^r \|\pi(\overline{n}a)v_i\|^2 + \sum_1^p \|\pi(\overline{n}a)w_j\|^2$$

so

$$re^{2\nu(\log a')} + \sum_1^p e^{2\mu_i(\log a')} =$$

$$e^{2\nu(\log a)} \sum_1^r \|\pi(\overline{n})v_i\|^2 + \sum_1^p e^{2\mu_i(\log a)} \|\pi(\overline{n})w_i\|^2.$$

Decompose $\overline{n}$ according to $G = KAN$ and note that $\pi(n)v_i = v_i$ for all i. Hence $\|\pi(\overline{n})v_i\| = e^{\nu(H(\overline{n}))}$. Putting $H'(a) = \log a' - \log a - H(\overline{n})$ the equation above can be written

$$r + \sum_1^p \|\pi(\overline{n})w_j\|^2 e^{(2\mu_i - 2\nu)(\log a) - 2\nu(H(\overline{n}))} =$$

$$e^{2\nu(H'(a))} \left\{ r + \sum_1^p e^{(2\mu_i - 2\nu)(\log a')} \right\}.$$

Let $a \longrightarrow +\infty$. Then we get

$$r = \lim_{a \longrightarrow +\infty} e^{2\nu(H'(a))} \{r + \sum_1^p e^{(2\mu_i - 2\nu)(\log a')}\} \geq \limsup_{a \to \infty} \left(re^{2\nu(H'(a))}\right)$$

so $\limsup \nu(H'(a)) \leq 0$. On the other hand, $\overline{n}a = k'a'k$ so $H(\overline{n}a) = H(\overline{n}) + \log a = H(a'k)$. Combining this with $\nu(\log a') \geq \nu(H(a'k))$ ([GGA], IV, Lemma 6.5) we have $\nu(H'(a)) \geq 0$. Thus $\lim_{a \to +\infty} \nu(H'(a)) = 0$ for each ν so by Prop. 4.23, $H'(a) \longrightarrow 0$ as claimed.

For further reference we note here an additional result. Let $v \in V$ be a unit vector of weight Λ. Then if $X \in \mathfrak{g}_{-\alpha} (\alpha \in \Sigma^+)$ we have $\pi(X)v \in V_{\nu - \alpha}$. Hence $\pi(\overline{n})v - v \in W$ for each $\overline{n} \in \overline{N}$ so

$$\pi(\overline{n}^a)v = v + \sum_1^p a_i w_i$$

where, because of (63),

$$a_i = \langle \pi(\overline{n}^a)v, w_i \rangle = e^{(\mu_i - \nu)(\log a)} \langle \pi(\overline{n})v, w_i \rangle.$$

Decomposing $\overline{n}^a$ according to $G = KAN$ we deduce (taking norms)

$$e^{2\nu(H(\overline{n}^a))} = 1 + \sum_1^p e^{(2\mu_i - 2\nu)(\log a)} |\langle \pi(\overline{n})v, w_i \rangle|^2. \tag{67}$$

Consider now the decomposition (53). As usual let s^* denote the Weyl group element which interchanges $\mathfrak{a}^+$ and $-\mathfrak{a}^+$. We put $m^* = m_{s^*}$. Multiplying (53) on the left by m^* we can write

$$G = \overline{N}MAN \cup \overline{N}m_{s_1}MAN \cup \ldots. \tag{68}$$

Here just the first summand is an open submanifold of G and the components of an element $g \in \overline{N}MAN$ are uniquely determined by g. We express this by writing

$$g = \overline{n}(g)m(g) \exp B(g) n_B(g), \qquad B(g) \in \mathfrak{a}.$$

Proposition 4.25. *Fix $\overline{n} \in \overline{N}$. Then*

$$\lim_{a \to +\infty} \big(H(a^{-1}\overline{n}a) - \log a + s^* \log a\big) = \begin{cases} B(m^*\overline{n}) & \text{if } m^*\overline{n} \in \overline{N}MAN \\ -\infty & \text{else} \end{cases}.$$

Remark. $\lim_{a \to +\infty} f(a) = -\infty$ means that for each $\alpha \in \Sigma^+$,

$$\alpha(f(a)) \longrightarrow -\infty.$$

Proof. If $g \in \overline{N}MAN$ then

$$B(g) = \lim_{b \to +\infty} H(bgb^{-1}).$$

We put here $g = m^*\overline{n}$ and note that $a \longrightarrow +\infty \Longleftrightarrow m^*a^{-1}(m^*)^{-1} \longrightarrow +\infty$. Then

$$\begin{aligned} B(m^*\overline{n}) &= \lim_{a \to +\infty} H(m^*a^{-1}(m^*)^{-1}m^*\overline{n}m^*a(m^*)^{-1}) \\ &= \lim_{a \to +\infty} (H(a^{-1}\overline{n}) + s^* \log a) \end{aligned}$$

as claimed. If $m^*\overline{n} \notin \overline{N}MAN$ then by (68) $m^*\overline{n} = \overline{n}_1 m_s m a_1 n_1, (s \neq e)$. The expression above is

$$H(a^{-1}\overline{n}) + s^* \log a = H(m^* a^{-1}(m^*)^{-1}\overline{n}_1 m_s) + \log a_1 + s^* \log a$$

which, putting $b = m^* a^{-1}(m^*)^{-1}$, becomes

$$\begin{aligned} &H(m_s^{-1}(b\overline{n}_1 b^{-1})m_s m_s^{-1} b m_s) + \log a_1 - \log b \\ &= H(m_s^{-1}(b\overline{n}_1 b^{-1})m_s) + s^{-1} \log b - \log b + \log a_1. \end{aligned}$$

Since $a \longrightarrow +\infty$ is equivalent to $b \longrightarrow +\infty$ this has limit $0 + (-\infty) + \log a_1$, whence the second relation above.

Remarks. 1. For the case $G = \boldsymbol{SL}(n, \boldsymbol{R})$ where $\overline{N}$ is the group of unipotent lower-triangular matrices and $m^* = E_{n1} + \cdots + E_{1n}$, $m^*\overline{N}$ consists of the matrices in $\overline{N}$ turned 90° clockwise. As shown in [DS], IX, Exercise A2, $m^*\overline{n}$ lies in $\overline{N}MAN$ if all the minors

$$\det\Big(((m^*\overline{n})_{\ell m})_{1 \leq \ell, m \leq i}\Big) \quad \text{for } 1 \leq i \leq n \text{ are } \neq 0.$$

2. In general the set $S = \{\overline{n} \in \overline{N} : m^*\overline{n} \in \overline{N}MAN\}$ is open and dense in $\overline{N}$. In fact, the map $\varphi : (\overline{n}, m, a, n) \longrightarrow m^*\overline{n}man$ is a diffeomorphism of $\overline{N} \times M \times A \times N$ onto an open dense subset of G. The pre-image $\varphi^{-1}(\overline{N}MAN)$ is open and dense in $\overline{N} \times M \times A \times N$ but equals $S \times M \times A \times N$.

3. Looking at Theorem 6.1 later in this chapter we see how the limit operation in Prop. 4.25 picks out the terms in the polynomial

$$e^{2\alpha(H(\overline{n}))} = (1 + c|X|^2)^2 + 4c|Y|^2$$

which have the highest degree of homogeneity relative to the automorphism $\overline{n} \longrightarrow \overline{n}^a$ of $\overline{N}$.

§5. CONICAL DISTRIBUTIONS.

1. The Construction of $\Psi'_{\lambda,s}$.

Again let G be a noncompact connected semisimple Lie group with finite center, $X = G/K$ the associated symmetric space and $\Xi = G/MN$ the space of horocycles in X. We shall now take up the problem of determining all the conical distributions.

As a consequence of Prop. 4.3 the problem of determining the conical distributions on Ξ amounts to finding the MN-invariants in $\mathcal{D}'(\Xi)$ which are "homogeneous" under the translations $\sigma(a) : \; gMN \longrightarrow gaMN$ of Ξ. For this it will be convenient to use the decomposition

$$\Xi = \bigcup_{s\in W} \Xi_s \tag{1}$$

from §1, where

$$\Xi_s = MNA \cdot \xi_s, \qquad \xi_s = m_s MN.$$

In fact, writing $a^s = m_s a m_s^{-1}$ so

$$mnbm_s aMN = mna^s bm_s MN$$

we see that the action of $\tau(mn)\sigma(a)$ on Ξ_s coincides with the action of $\tau(mna^s)$. After defining a certain natural measure ν on Ξ_s wc are going to look for functions ψ on Ξ_s with the property that the functional

$$\Psi : \; \varphi \longrightarrow \int_{\Xi_s} \varphi(\xi)\psi(\xi)d\nu(\xi) \tag{2}$$

should be MN-invariant and satisfy

$$\Psi^{\sigma(a)} = e^{-(i\lambda+\rho)(\log a)}\Psi \tag{3}$$

in accordance with (9) §4.

To define ν consider the groups

$$\begin{aligned} N^s &= m_s N m_s^{-1} & \overline{N}^s &= m_s \overline{N} m_s^{-1} \\ N'_s &= N \cap \overline{N}^s & \overline{N}_s &= m_s^{-1} N'_s m_s = \overline{N} \cap N^{s^{-1}}, \end{aligned}$$

whose Lie algebras, respectively, are given by

$$\begin{aligned} \mathfrak{n}^s &= \sum_{s^{-1}\beta>0} \mathfrak{g}_\beta, & \overline{\mathfrak{n}}^s &= \sum_{s^{-1}\beta<0} \mathfrak{g}_\beta, \\ \mathfrak{n}'_s &= \sum_{\beta>0, s^{-1}\beta<0} \mathfrak{g}_\beta, & \overline{\mathfrak{n}}_s &= \sum_{\beta<0, s\beta>0} \mathfrak{g}_\beta. \end{aligned}$$

Then $\mathfrak{n} = (\mathfrak{n} \cap \mathfrak{n}^s) + \mathfrak{n}'_s$ and in analogy with §4 (54),

$$N = (N \cap N^s)N'_s = N'_s(N \cap N^s)$$

in the sense that the product maps are diffeomorphisms under which the Haar measures correspond (cf. [GGA], Ch. IV §6, No. 6 for the simple justification). Thus if we choose a Haar measure dn'_s on N'_s it corresponds via the identification $N/N \cap N^s = N'_s$ to an N-invariant measure on $N/N \cap N^s$. From Prop. 1.5 it is clear that when MNA acts on Ξ_s the isotropy subgroup of ξ_s is $M(N \cap N^s)$. Thus the map $(n'_s, a) \longrightarrow n'_s a \cdot \xi_s$ is a diffeomorphism of $N'_s \times A$ onto Ξ_s and the remarks above show that the measure $dn'_s da$ hereby corresponds to an MN-invariant measure $d\nu$ on Ξ_s. The requirements (2) and (3) now lead us to define a function ψ on Ξ_s by

$$\psi(n'_s a \cdot \xi_s) = \psi(a \cdot \xi_s) = e^{(is\lambda + s\rho)(\log a)}\psi(\xi_s) \tag{4}$$

and a functional $\Psi'_{\lambda,s}$ by

$$\Psi'_{\lambda,s}(\varphi) = \int_{\Xi_s} \varphi(\xi) e^{(is\lambda+s\rho)(\log a(\xi))} d\nu(\xi) \tag{5}$$

on the set of $\varphi \in \mathcal{D}(\Xi)$ for which the integral converges absolutely. Here $a(\xi) \in A$ is defined by $\xi = mna(\xi) \cdot \xi_s$ $(m \in M, n \in N)$ as in §1. Since Ξ_s is not in general closed in Ξ (in fact only for $s = e$) the absolute convergence of (5) is not automatic.

Theorem 5.1. *The integral defining $\Psi'_{\lambda,s}$ is absolutely convergent for all $\varphi \in \mathcal{D}(\Xi)$ if and only if*

$$Re(\langle \alpha, i\lambda \rangle) > 0 \qquad \text{for } \alpha \in \Sigma^+ \cap s^{-1}\Sigma^-, \tag{6}$$

Re denoting real part. If (6) holds, then $\Psi'_{\lambda,s}$, as defined by (5), is a conical distribution in $\mathcal{D}'_\lambda(\Xi)$.

Proof. Let $\mu = is\lambda + s\rho$ so by definition,

$$\Psi'_{\lambda,s}(\varphi) = \int_{N'_s \times A} \varphi(n'_s a m_s \cdot \xi_o) e^{\mu(\log a)} dn'_s da.$$

The mapping $n'_s \longrightarrow m_s^{-1} n'_s m_s$ is an isomorphism of N'_s onto $\overline{N}_s$. Let $d\overline{n}_s$ denote the Haar measure on $\overline{N}_s$ which hereby corresponds to dn'_s. Then putting $a^{s^{-1}} = m_s^{-1} a m_s$ our integral becomes

$$\int_{\overline{N}_s \times A} \varphi(m_s \overline{n}_s a^{s^{-1}} \cdot \xi_o) e^{\mu(\log a)} d\overline{n}_s da$$

so with the Iwasawa decomposition $g = k(g) \exp H(g) n(g)$,

$$\Psi'_{\lambda,s}(\varphi) = \int\limits_{\overline{N}_s \times A} \varphi(m_s k(\overline{n}_s) \exp H(\overline{n}_s) a \cdot \xi_o) e^{(i\lambda+\rho)(\log a)} d\overline{n}_s da, \tag{7}$$

which we write as

$$\int\limits_{\overline{N}_s} e^{-(i\lambda+\rho)(H(\overline{n}_s))} d\overline{n}_s \int\limits_{A} \varphi(m_s k(\overline{n}_s) a \cdot \xi_o) e^{(i\lambda+\rho)(\log a)} da.$$

Here the integral over A is absolutely convergent for each $\varphi \in \mathcal{D}(\Xi)$. Taking φ K-invariant we deduce that (5) is absolutely convergent for each $\varphi \in \mathcal{D}(\Xi)$ if and only if the integral

$$\int\limits_{\overline{N}_s} e^{-(i\lambda+\rho)(H(\overline{n}_s))} d\overline{n}_s \tag{8}$$

is absolutely convergent. By the criterion of Gindikin-Karpelevič ([GGA], IV, Theorem 6.13) this happens exactly when (6) holds.

Assuming (6) holds we can define the functional $S'_{\lambda,s}$ on $\mathcal{E}(K/M)$ by

$$S'_{\lambda,s}(F) = \int\limits_{\overline{N}_s} F(m_s k(\overline{n}_s) M) e^{-(i\lambda+\rho)(H(\overline{n}_s))} d\overline{n}_s. \tag{9}$$

Then $S'_{\lambda,s}$ is a distribution and by the computation above

$$\Psi'_{\lambda,s}(\varphi) = \int\limits_{K/M} \Big(\int\limits_{A} \varphi(ka \cdot \xi_o) e^{(i\lambda+\rho)(\log a)} da \Big) dS'_{\lambda,s}(kM). \tag{10}$$

According to Prop. 4.4 we then have $\Psi'_{\lambda,s} \in \mathcal{D}'_\lambda(\Xi)$ as claimed.

The mapping $\lambda \longrightarrow \Psi'_{\lambda,s}$ is a holomorphic function on the tube (6) with values in $\mathcal{D}'(\Xi)$. We shall now extend it to a meromorphic distribution-valued function on the complex space $\mathfrak{a}^*_C$. For this we carry out the analogous continuation of the function $\lambda \longrightarrow S'_{\lambda,s}$.

We observe that m_s normalizes M and that $k(m\overline{n}_s m^{-1}) = mk(\overline{n}_s)m^{-1}$. This shows that

(i) $S'_{\lambda,s}$ is M-invariant

(ii) $S'_{\lambda,s}$ does not depend on the choice of the representative m_s of s.

2. The Reduction to Rank One.

The convolution of distributions on K induces a convolution of distributions on K/M as follows. If $t_1, t_2 \in \mathcal{D}'(K)$ the convolution $t_1 * t_2$ is defined by

$$(t_1 * t_2)(f) = \int\limits_{K \times K} f(k_1 k_2) dt_1(k_1) dt_2(k_2),$$

for $f \in \mathcal{D}(K)$. Let $\dot{f} \in \mathcal{D}(K/M)$ be defined by

$$\dot{f}(kM) = \int\limits_M f(km) dm$$

and if $S \in \mathcal{D}'(K/M)$ define $\widetilde{S} \in \mathcal{D}'(K)$ by

$$\widetilde{S}(f) = S(\dot{f}).$$

Then $\widetilde{F} = F \circ \pi$ if $F \in \mathcal{D}(K/M)$ and π is the natural mapping of K onto K/M. If $S_1, S_2 \in \mathcal{D}'(K/M)$ we define the *convolution* $S_1 \times S_2 \in \mathcal{D}'(K/M)$ by

$$(S_1 \times S_2)(F) = (\widetilde{S}_1 * \widetilde{S}_2)(\widetilde{F}).$$

Lemma 5.2. *Let $S, T \in \mathcal{D}'(K/M)$ and suppose T is invariant under the action of M. Then*

$$(S \times T)(F) = \int\limits_{K/M} \Big(\int\limits_{K/M} F^{\tau(k^{-1})}(\ell M) dT(\ell M) \Big) dS(kM).$$

Proof. The inner integral, say $g(k)$, satisfies $g(km) \equiv g(k)$ because of the M-invariance of T. The right hand side therefore equals

$$\int\limits_K \Big(\int\limits_K \widetilde{F}(k\ell) d\widetilde{T}(\ell) \Big) d\widetilde{S}(k)$$

which is $(\widetilde{S} * \widetilde{T})(\widetilde{F}) = (S \times T)(F)$.

Each $s \in W$ can be written $s = s_{\alpha_1} \ldots s_{\alpha_p}$ where each α_i is a simple restricted root and s_{α_i} the corresponding reflection. This expression for s is said to be *reduced* if p is as small as possible. As usual, Σ_o^+ denotes the set of positive indivisible roots, $\Sigma_o^- = -\Sigma_o^+$, and we put

$$\Sigma_s^+ = \Sigma_o^+ \cap s^{-1} \Sigma_o^-.$$

Theorem 5.3. *Let $s \in W$ and $s = s_{\alpha_1} \dots s_{\alpha_p}$ any reduced expression of s. Put*

$$s^{(q)} = s_{\alpha_{q+1}} \dots s_{\alpha_p}, \qquad s^{(p)} = e.$$

Then for a suitable normalization of the Haar measures on $\overline{N}_{s_{\alpha_i}}$ (independent of λ)

$$S'_{\lambda,s} = S'_{s^{(1)}\lambda, s_{\alpha_1}} \times \dots \times S'_{s^{(p)}\lambda, s_{\alpha_p}}$$

for all λ in the tube

$$Re(\langle \alpha, i\lambda \rangle) > 0 \qquad \text{for } \alpha \in \Sigma_s^+.$$

Proof. For the proof we need some facts about the action of W on the root system (see e.g. [GGA], IV§6, No. 5). In particular, we have by *loc. cit.* (25)

$$\Sigma_s^+ = \Sigma_{s^{(1)}}^+ \cup \{\beta\}, \tag{11}$$

where $\beta = -s^{-1}\alpha_1$.

Now we use the "rank-one reduction" method on the root β (*loc. cit.* No. 6). Let $\mathfrak{g}_{(\beta)}$ denote the subalgebra of $\mathfrak{g}$ generated by the root spaces $\mathfrak{g}_{-\beta}$ and $\mathfrak{g}_\beta$ and let G_β denote the analytic subgroup of G with Lie algebra $\mathfrak{g}_{(\beta)}$. Then G_β is semisimple of real rank one and we have the Cartan and Iwasawa decompositions

$$\begin{gathered} \mathfrak{g}_{(\beta)} = \mathfrak{k}_\beta + \mathfrak{p}_\beta, \qquad \mathfrak{g}_{(\beta)} = \mathfrak{k}_\beta + \mathfrak{a}_\beta + \mathfrak{n}_\beta \\ G_\beta = K_\beta A_{(\beta)} N_\beta, \end{gathered} \tag{12}$$

where $\mathfrak{k}_\beta, \mathfrak{p}_\beta, \mathfrak{a}_\beta = \boldsymbol{R}A_\beta, \mathfrak{n}_\beta$ are obtained from $\mathfrak{g}_{(\beta)}$ by intersection with $\mathfrak{k}, \mathfrak{p}, \mathfrak{a}, \mathfrak{n}$, respectively and the groups $K_\beta = K \cap G_\beta$, $A_{(\beta)} = A \cap G_\beta$, $N_\beta = N \cap G_\beta$ are connected. Also $Ad_G(G_\beta)$ is closed in $Ad_G(G)$ ([DS], IX, proof of Lemma 2.2) and since G_β is the identity component of $Ad_G^{-1}(Ad_G(G_\beta))$ it is a closed subgroup of G. Thus K_β is compact and G_β has finite center. Note also that M_β, the centralizer of A_β in K_β, satisfies $M_\beta = M \cap G_\beta$ (loc. cit. Lemma 2.3). Since $[\mathfrak{g}_\beta, \mathfrak{g}_\beta] = \mathfrak{g}_{2\beta}$ the group N_β has Lie algebra $\mathfrak{g}_\beta + \mathfrak{g}_{2\beta}$. We write $\overline{\mathfrak{n}}_\beta$ for the Lie algebra of $\overline{N}_\beta = \theta N_\beta$ and put $P = MAN$, $P_\beta = M_\beta A_\beta N_\beta$. Taking the negatives and doubles in (11) we have

$$\overline{\mathfrak{n}}_s = \overline{\mathfrak{n}}_{s'} + \overline{\mathfrak{n}}_\beta, \tag{13}$$

where for simplicity we write s' for $s^{(1)}$. This implies (cf. §4, (54))

$$\overline{N}_s = \overline{N}_{s'}\overline{N}_\beta = \overline{N}_\beta \overline{N}_{s'} \tag{14}$$

in the sense that the product maps are diffeomorphisms under which $d\overline{n}_s$ corresponds to $d\overline{n}_{s'}d\overline{n}_\beta$. Identifying K/M and G/P we have by (9) and (14),

$$S'_{\lambda,s}(F) = \int_{\overline{N}_s} F(m_s\overline{n}_sP)e^{-(i\lambda+\rho)(H(\overline{n}_s))}d\overline{n}_s$$

$$= \int_{\overline{N}_\beta\overline{N}_{s'}} F(m_s\overline{n}_\beta\overline{n}_{s'}P)e^{-(i\lambda+\rho)(H(\overline{n}_\beta\overline{n}_{s'}))}d\overline{n}_\beta d\overline{n}_{s'},$$

which by writing $\overline{n}_\beta = k_\beta a_\beta n_\beta$, by (12), becomes

$$\int_{\overline{N}_\beta\overline{N}_{s'}} F(m_sk_\beta a_\beta n_\beta\overline{n}_{s'}P)e^{-(i\lambda+\rho)(H(a_\beta n_\beta\overline{n}_{s'}))}d\overline{n}_\beta d\overline{n}_{s'}.$$

Now by definition, $\overline{N}_{s'} = N^{(s')^{-1}} \cap \overline{N}$ so by §4, (54),

$$N^{(s')^{-1}} = \overline{N}_{s'} \cdot (N^{(s')^{-1}} \cap N) \tag{15}$$

and the Haar measures correspond under the product map. Consequently,

$$\overline{N}_{s'} = N^{(s')^{-1}}/(N^{(s')^{-1}} \cap N) \tag{16}$$

and the Haar measure $d\overline{n}_{s'}$ corresponds to the invariant measure on the right hand side. Now N_β normalizes $N^{(s')^{-1}}$ (since $s'\beta = \alpha_1 > 0$) so by (15) we have a decomposition

$$n_\beta\overline{n}_{s'}n_\beta^{-1} = (\overline{n}_{s'})^*n. \tag{17}$$

The mapping $\overline{n}_{s'} \longrightarrow (\overline{n}_{s'})^*$ of $\overline{N}_{s'}$ onto itself corresponds via (16) to the diffeomorphism of the quotient space (16) induced by the conjugation by n_β on $N^{(s')^{-1}}$. Since $Ad_L(n_\beta)$ is unipotent (both for $L = N^{(s')^{-1}}$ and $L = N^{(s')^{-1}} \cap N$) the conjugation by n_β preserves the Haar measures so the mapping $\overline{n}_{s'} \longrightarrow (\overline{n}_{s'})^*$ is measure preserving. Consequently, n_β can be omitted in the last integral. Now we use the substitution $\overline{n}_{s'} \longrightarrow a_\beta\overline{n}_{s'}a_\beta^{-1}$. Clearly

$$\frac{d(a_\beta\overline{n}_{s'}a_\beta^{-1})}{d\overline{n}_{s'}} = e^{2\vartheta(\log a_\beta)},$$

where $2\vartheta = \sum_{\alpha<o,s'\alpha>o} \alpha$. We can write

$$-2\rho = \sum_{\alpha<o,s'\alpha>o} \alpha + \sum_{\alpha<o,s'\alpha<o} \alpha$$

$$2(s')^{-1}\rho = \sum_{\alpha<o,s'\alpha>o} \alpha + \sum_{\alpha>o,s'\alpha>o} \alpha$$

so by adding we find $2\vartheta = (s')^{-1}\rho - \rho$. Hence our integral above becomes

$$\int\limits_{\overline{N}_\beta \overline{N}_{s'}} F(m_s k_\beta \overline{n}_{s'} P) e^{-(i\lambda+\rho)(H(\overline{n}_{s'}))} e^{-(i\lambda+(s')^{-1}\rho)(\log a_\beta)} d\overline{n}_\beta d\overline{n}_{s'}.$$

Now we wish to express this in terms of the root α_1. We construct the group G_{α_1} and the appropriate subgroups $K_{\alpha_1}, A_{\alpha_1}, N_{\alpha_1}\overline{N}_{\alpha_1}$, and M_{α_1}. Since $s'\beta = \alpha_1$ these groups are obtained from the previous ones by conjugating by $m_{s'}$. Thus if $\overline{n}_{\alpha_1} = m_{s'}\overline{n}_\beta m_{s'}^{-1}$ the decomposition

$$\overline{n}_{\alpha_1} = k_{\alpha_1} a_{\alpha_1} n_{\alpha_1}$$

according to $G_{\alpha_1} = K_{\alpha_1} A_{\alpha_1} N_{\alpha_1}$ is obtained from $\overline{n}_\beta = k_\beta a_\beta n_\beta$ by conjugating the factors by $m_{s'}$. Since $m_s m_{s'}^{-1} = m m_{s_{\alpha_1}}$ (with $m \in M$) our integral becomes (with $d\overline{n}_{\alpha_1}$ corresponding to $d\overline{n}_\beta$)

$$\int\limits_{\overline{N}_{\alpha_1} \overline{N}_{s'}} F(m m_{s_{\alpha_1}} k_{\alpha_1}(\overline{n}_{\alpha_1}) m_{s'} \overline{n}_{s'} P) e^{-(i\lambda+\rho)(H(\overline{n}_{s'}))} e^{-(is'\lambda+\rho)(\log a_{\alpha_1})} d\overline{n}_{\alpha_1} d\overline{n}_{s'}.$$

Here we can delete the factor m since $S'_{\lambda,s}$ is M-invariant. Putting $k = m_{s_{\alpha_1}} k(\overline{n}_{\alpha_1})$ this integral is

$$\int\limits_{\overline{N}_{\alpha_1}} S'_{\lambda,s'}(F^{\tau(k^{-1})}) e^{-(is'\lambda+\rho)(H(\overline{n}_{\alpha_1}))} d\overline{n}_{\alpha_1}.$$

Since α_1 is simple the symmetry s_{α_1} permutes $\Sigma^+ - \{\alpha_1, 2\alpha_1\}$. It follows that the group $\overline{N}_{\alpha_1}$ coincides with $\overline{N}_{s_{\alpha_1}}$. But then it follows from (9) and Lemma 5.2 that the last integral is just

$$\left(S'_{s'\lambda,s_{\alpha_1}} \times S'_{\lambda,s'}\right)(F).$$

Now Theorem 5.3 follows by iteration because the expression for $s^{(q)}$ in the theorem is necessarily reduced.

We shall now relate the convolution factors in Theorem 5.3 to conical distributions for rank one spaces.

Suppose now $\alpha \in \Sigma^+$ simple and $\mu \in \mathfrak{a}^*_C$. As remarked above $\overline{N}_\alpha = \overline{N}_{s_\alpha}$ so

$$S'_{\mu,s_\alpha}(F) = \int\limits_{\overline{N}_\alpha} F(m_{s_\alpha}\overline{n}_\alpha P) e^{-(i\mu+\rho)(H(\overline{n}_\alpha))} d\overline{n}_\alpha \tag{18}$$

and by Theorem 5.1 and (10) this converges absolutely for all $F \in \mathcal{E}(G/P)$ if and only if $\mathrm{Re}(\langle i\mu, \alpha\rangle) > 0$. We construct the rank-one group G_α in the same manner as G_β above. Select $m_\alpha^* \in K_\alpha$ such that $Ad_{G_\alpha}(m_\alpha^*) = -1$ on $\mathfrak{a}_\alpha$. Since $Ad_G(m_{s_\alpha})$ and each element in the group $Ad_G(G_\alpha)$ fix the hyperplane $\alpha = 0$ in $\mathfrak{a}$ elementwise we have $m_{s_\alpha}^{-1} m_\alpha^* \in M$. Since $P_\alpha = M_\alpha A_\alpha N_\alpha = G_\alpha \cap P$ the manifold G_α/P_α is imbedded as a submanifold of G/P. Let $\overline{\mu}$ denote the restriction $\mu|\mathfrak{a}_\alpha$ and consider the distribution $T_{\overline{\mu},s_\alpha}$ on G_α/P_α defined by

$$(19) \qquad T_{\overline{\mu},s_\alpha}(f) = \int\limits_{\overline{N}_\alpha} f(m_\alpha^* \overline{n}_\alpha P_\alpha) e^{-(i\mu+\rho)(H(\overline{n}_\alpha))} d\overline{n}_\alpha$$

for $f \in \mathcal{E}(G_\alpha/P_\alpha)$. According to Theorem 5.1, the absolute convergence of (19) for all f is equivalent to $\mathrm{Re}(\langle i\overline{\mu} + \overline{\rho} - \rho_\alpha, \overline{\alpha}\rangle) > 0$ where ρ_α is the ρ-function for G_α/K_α. However, since α is simple, $\langle \overline{\rho}, \overline{\alpha}\rangle = \langle \rho_\alpha, \overline{\alpha}\rangle$ ([GGA], IV, §6, No. 6) so (19) is absolutely convergent if and only if $\mathrm{Re}(\langle i\overline{\mu}, \overline{\alpha}\rangle) > 0$. Moreover, since $m_{s_\alpha}^{-1} m_\alpha^* \in M$ and S'_{μ,s_α} M-invariant,

$$(20) \qquad S'_{\mu,s_\alpha}(F) = T_{\overline{\mu},s_\alpha}(\overline{F}), \qquad F \in \mathcal{E}(G/P),$$

where $\overline{F}$ is the restriction of F to G_α/P_α. Theorem 5.3 and (20) reduce the problem of meromorphic continuation of the function $\lambda \longrightarrow \Psi'_{\lambda,s}$ to a rank-one problem.

3. The analytic continuation of $\Psi_{\lambda,s}$.

It turns out that the function $\lambda \longrightarrow S'_{\lambda,s}$ has the same singularities as the partial $\boldsymbol{c}$-function

$$(21) \qquad \boldsymbol{c}_s(\lambda) = \int\limits_{\overline{N}_s} e^{-(i\lambda+\rho)(H(\overline{n}_s))} d\overline{n}_s.$$

This is not altogether unexpected since by (9), $\boldsymbol{c}_s(\lambda) = S'_{\lambda,s}(1)$. According to the Gindikin-Karpelevič product formula

$$(21') \qquad \boldsymbol{c}_s(\lambda) = \prod_{\alpha \in \Sigma_s^+} \boldsymbol{c}_\alpha(\lambda_\alpha),$$

([GGA], IV, §6, No. 6) where $\boldsymbol{c}_\alpha$ is the $\boldsymbol{c}$-function for the rank-one space G_α/K_α and λ_α is the restriction of λ to $\mathfrak{a}_\alpha$. The formula holds for a specific normalization of $d\overline{n}_s$.

If we now express c_α by means of (18′) in §3 we obtain a formula for c_s as a fraction where both sides are products of Γ-factors. It turns out that for us both the numerator and denominator have their individual significance. If $\langle\ \rangle_\alpha$ is the Killing form for G_α then since $\mathfrak{a}_\alpha$ is one-dimensional

$$\left\langle i\lambda_\alpha, \frac{\alpha_\alpha}{\langle\alpha_\alpha, \alpha_\alpha\rangle_\alpha}\right\rangle_\alpha = \frac{\langle i\lambda, \alpha\rangle}{\langle\alpha, \alpha\rangle}.$$

Thus we obtain from (18′) in §3 and (21′)

$$c_s(\lambda) = a_s d_s(\lambda) e_s(\lambda), \tag{22}$$

where

$$d_s(\lambda) = \prod_{\alpha\in\Sigma_s^+} \Gamma(\langle i\lambda, \alpha_o\rangle) 2^{-\langle i\lambda, \alpha_o\rangle}, \tag{23}$$

$$e_s^{-1}(\lambda) = \prod_{\alpha\in\Sigma_s^+} \Gamma\left(\tfrac{1}{2}\left(\tfrac{m_\alpha}{2} + 1 + \langle i\lambda, \alpha_o\rangle\right)\right) \Gamma\left(\tfrac{1}{2}\left(\tfrac{m_\alpha}{2} + m_{2\alpha} + \langle i\lambda, \alpha_o\rangle\right)\right). \tag{24}$$

Here $\Sigma_s^+ = \Sigma_o^+ \cap s^{-1}\Sigma_o^-$, $\quad \alpha_o = \alpha/\langle\alpha, \alpha\rangle$ and the constant a_s is given by

$$a_s = \prod_{\alpha\in\Sigma_s^+} 2^{(m_\alpha/2)+m_{2\alpha}} \Gamma\left(\tfrac{1}{2}(m_\alpha + m_{2\alpha} + 1)\right). \tag{25}$$

In other words, d_s and e_s^{-1} are the numerator and denominator in the formula (22) for c_s. Note that while both involve the restricted root structure for G, d_s does not involve the multiplicities. Also $s\Sigma_s^+ = -\Sigma_{s^{-1}}^+$ so $a_s = a_{s^{-1}}$.

Theorem 5.4. *Let $s \in W$. The mapping*

$$\lambda \longrightarrow d_s(\lambda)^{-1}\Psi'_{\lambda,s}$$

extends from the tube (6) to a holomorphic function, denoted $\Psi_{\lambda,s}$, on $\mathfrak{a}_C^*$ with values in $\mathcal{D}'(\Xi)$. For each $\lambda \in \mathfrak{a}_C^*$, $\Psi_{\lambda,s}$ is a conical distribution in $\mathcal{D}'_\lambda(\Xi)$.

Let $S_{\lambda,s}$ correspond to $\Psi_{\lambda,s}$ in the sense of Prop. 4.4.

Proof. First we assume rank $G/K = 1$ so W has order 2. If $s = e$, then $\overline{N}_s = \{e\}$ and

$$\Psi'_{\lambda,e}(\varphi) = c\int_A \varphi(a\cdot\xi_o) e^{(i\lambda+\rho)(\log a)}\, da,$$

where c is a constant so our statement is obvious. Consider therefore $s \neq e$ in W. With α the unique element in Σ_o^+ we have

$$\boldsymbol{d}_s(\lambda) = \Gamma(\langle i\lambda, \alpha_o\rangle)2^{-\langle i\lambda, \alpha_o\rangle}.$$

This is meromorphic on $\mathfrak{a}_C^*$ with simple poles given by

$$(26) \qquad -\langle i\lambda, \alpha_o\rangle \in \boldsymbol{Z}^+.$$

For $\mathrm{Re}(\langle i\lambda, \alpha\rangle) > 0$ we have

$$S'_{\lambda,s}(F) = \int_{\overline{N}} F(m^*k(\overline{n})M)e^{-(i\lambda+\rho)(H(\overline{n}))}d\overline{n},$$

where $m^* \in K$ is a representative of s. Let $z \in \boldsymbol{C}$ be determined by $\lambda = z\rho$ and let ψ be the function on $K/M - \{m^*M\}$ given by

$$\psi(k(\overline{n})M) = e^{\rho(H(\overline{n}))}.$$

Transferring the integral to K/M (§3, No. 1) it becomes (since $(m^*)^2 \in M$ and ψ M-invariant),

$$\int_{K/M} F(m^*kM)\psi(kM)^{-iz+1}dk_M = \int_{K/M} F(kM)\psi(m^*kM)^{-iz+1}dk_M.$$

Now as proved in [DS], IX, §2 the components of a $g \in \overline{N}MAN$ are uniquely determined by g so we write

$$(27) \qquad g = \overline{n}(g)m(g)\exp B(g)n_B(g).$$

Transferring the last integral to $\overline{N}$ we then obtain

$$(28) \qquad S'_{\lambda,s}(F) = \int_{\overline{N}} F(k(\overline{n})M)e^{-(i\lambda-\rho)(H(\overline{n}(m^*\overline{n})))}e^{-2\rho(H(\overline{n}))}d\overline{n}.$$

The kernel in this integral can be determined explicitly by means of the following formulas from [DS], IX, Theorem 3.8. Let $|Z|^2 = -B(Z, \theta Z)$ and $\overline{n} = \exp(X+Y)$ where $X \in \mathfrak{g}_{-\alpha}$ $Y \in \mathfrak{g}_{-2\alpha}$. Then

$$(29) \qquad e^{\rho(H(\overline{n}))} = \{(1+c|X|^2)^2 + 4c|Y|^2\}^p,$$

$$(30) \qquad e^{\rho(B(m^*\overline{n}))} = \{c^2|X|^4 + 4c|Y|^2\}^p,$$

where $4p = m_\alpha + 2m_{2\alpha}$, $c^{-1} = 4(m_\alpha + 4m_{2\alpha})$, m_α and $m_{2\alpha}$ being the appropriate multplicities. Writing down formula (27) for $g = m^*\overline{n}$ ($\overline{n} \neq e$) we see that

$$H(\overline{n}) = H(\overline{n}(m^*\overline{n})) + B(m^*\overline{n}). \tag{31}$$

The integral (28) therefore becomes

$$\int_{\mathfrak{g}_{-\alpha}} \int_{\mathfrak{g}_{-2\alpha}} F^*(X,Y) \left[1 + \frac{1 + 2c|X|^2}{c^2|X|^4 + 4c|Y|^2}\right]^{p(1-iz)} dX dY, \tag{32}$$

where F^* is an integrable function on $\mathfrak{g}_{-\alpha} \times \mathfrak{g}_{-2\alpha}$ relative to the Euclidean measure $dXdY$. In order to extend this to a meromorphic function of z we may assume F^* of compact support; in fact, if we write $F^* = F' + F''$ where $F' \equiv 0$ near 0, F'' smooth of compact support, the contribution from F' will be holomorphic for all $z \in \boldsymbol{C}$.

Assuming first $\mathfrak{g}_{-2\alpha} \neq \{0\}$ we use polar coordinates in each of the spaces $\mathfrak{g}_{-\alpha}$ and $\mathfrak{g}_{-2\alpha}$ and write $r = c^{\frac{1}{2}}|X|$, $s = 2c^{\frac{1}{2}}|Y|$. Then the integral becomes

$$\int_0^\infty \int_0^\infty F_1(r,s) \left(1 + \frac{1+2r^2}{r^4+s^2}\right)^{p(1-iz)} r^{m_\alpha - 1} s^{m_{2\alpha}-1} dr ds,$$

where $F_1(r,s)$ is essentially the average of F^* over the product $S^r \times S^s$ of spheres in $\mathfrak{g}_{-\alpha}$ and $\mathfrak{g}_{-2\alpha}$, respectively. Next we put $x = r^2, y = s$ and then use polar coordinates (ρ, θ) in the quadrant $x \geq 0$, $y \geq 0$. Then our expression becomes

$$\int_0^\infty \varphi_z(\rho)\rho^{2piz-1} d\rho, \qquad Re(iz) > 0, \tag{33}$$

where

$$\varphi_z(\rho) = \tfrac{1}{2} \int_0^{\frac{\pi}{2}} F_1\left((\rho\cos\theta)^{\frac{1}{2}}, \rho\sin\theta\right) \left(1 + 2\rho\cos\theta + \rho^2\right)^{p(1-iz)} g(\theta) d\theta, \tag{34}$$

with $g(\theta) = (\cos\theta)^{(m_\alpha/2)-1}(\sin\theta)^{m_{2\alpha}-1}$. The function $F_1(r,s)$ extends to an even C^∞ function of r and s; thus we can allow ρ to take negative values and see from (34) that φ_z is smooth for $-1 < \rho < \infty$. Moreover, φ_z and its derivatives in this interval depend holomorphically on z. It follows (see e.g. [GGA], I, §2, No. 8) that the integral (33) extends to a meromorphic function of z in the plane $\boldsymbol{C}$ with at most simple poles when

$$-2piz \in \boldsymbol{Z}^{+}. \tag{35}$$

In the case when $\mathfrak{g}_{-2\alpha} = \{0\}$ we still get expression (33) but this time F_1 is an even function of a single variable and $\varphi_z(\rho)$ is the function

$$\tfrac{1}{2}F_1(\rho^{\frac{1}{2}})(1+\rho)^{2p(1-iz)}.$$

Thus the possible poles of (33) are still given by (35). Since this condition coincides with (26) we have obtained the desired extension of the mapping $\lambda \longrightarrow \boldsymbol{d}_s(\lambda)^{-1}\Psi'_{\lambda,s}$ to a holomorphic mapping from $\mathfrak{a}^*_{\boldsymbol{C}}$ to $\mathcal{D}'(\Xi)$ in the rank-one case.

For G/K of arbitrary rank we use the product formula in Theorem 5.3. Consider the j th factor $S'_{s^{(j)}\lambda, s_{\alpha_j}}$ in the decomposition of $S'_{\lambda,s}$. We construct the symmetric space $G_{\alpha_j}/K_{\alpha_j}$ corresponding to the simple root α_j and let $\langle\ ,\ \rangle_j$ denote the associated Killing form. According to what we just proved, for each $f \in \mathcal{E}(G_{\alpha_j}/P_{\alpha_j})$, the function

$$\nu \longrightarrow \Gamma(\langle i\nu, (\overline{\alpha}_j)_o\rangle_j)^{-1}T_{\nu,s_{\alpha_j}}(f)$$

is holomorphic on $(\mathfrak{a}_{\alpha_j})^*_{\boldsymbol{C}}$. (Here and below, bar denotes restriction to $\mathfrak{a}_{\alpha_j}$). Thus by (20) and Theorem 5.3 the mapping

$$\lambda \longrightarrow \prod_{1\leq j\leq p} \Gamma\left(\langle i(s^{(j)}\lambda)^{-}, (\overline{\alpha}_j)_o\rangle_j\right)^{-1} S'_{\lambda,s} \tag{36}$$

is holomorphic on the entire space $\mathfrak{a}^*_{\boldsymbol{C}}$ with values in $\mathcal{D}'(K/M)$. Since $\mathfrak{a}_{\alpha_j}$ is one-dimensional the argument for Γ in (36) can be simplified to

$$\langle is^{(j)}\lambda, (\alpha_j)_o\rangle. \tag{37}$$

Now by iteration of (11) we have

$$\Sigma_s^+ = \{\beta_1, \ldots, \beta_p\}, \tag{38}$$

where $\beta_{q+1} = -{s^{(q)}}^{-1}\alpha_{q+1}$ (cf. [GGA], IV, Corollary 6.11). Since

$${s^{(j)}}^{-1}\alpha_j = -s_{\alpha_p}\ldots s_{\alpha_{j+1}}s_{\alpha_j}(\alpha_j) = -{s^{(j-1)}}^{-1}\alpha_j = \beta_j \tag{39}$$

the expression in (37) equals $\langle i\lambda, (\beta_j)_o\rangle$ so by (38) the product of the factors in (36) equals the Γ-factor in $\boldsymbol{d}_s(\lambda)^{-1}$. This concludes the proof of Theorem 5.4.

Corollary 5.5. *With the notation of Theorem 5.3 we have*

$$d_s(\lambda) = d_{s_{\alpha_1}}(s^{(1)}\lambda)\dots d_{s_{\alpha_p}}(s^{(p)}\lambda) \tag{40}$$

for all λ where $d_s(\lambda)$ is holomorphic and

$$S_{\lambda,s} = S_{s^{(1)}\lambda,s_{\alpha_1}} \times \cdots \times S_{s^{(p)}\lambda,s_{\alpha_p}} \tag{41}$$

for all $\lambda \in \mathfrak{a}_C^*$.

In fact (40) follows directly from (39) and (38) and now (41) follows from Theorem 5.3.

For the problem of determining all the conical distributions we are now led to the question whether the constructed $\Psi_{\lambda,s}$ are actually $\neq 0$ or equivalently, $S_{\lambda,s} \neq 0$. Prop. 5.7 below shows that the factors in (41) are all $\neq 0$. However, this does not guarantee that the convolution product $S_{\lambda,s}$ is $\neq 0$ (in contrast to the well-known fact that the convolution algebra $\mathcal{E}'(\boldsymbol{R}^n)$ has no zero divisors.)

Proposition 5.6. *Let $s \in W$ and $\lambda \in \mathfrak{a}_C^*$ such that $d_s(\lambda)^{-1} \neq 0$. Then $\Psi_{\lambda,s} \neq 0$.*

Proof. We write (9) in the form

$$S'_{\lambda,s}(F) = \int_{\overline{N}_s} F(m_s\overline{n}_sP)e^{-(i\lambda+\rho)(H(\overline{n}_s))}d\overline{n}_s. \tag{42}$$

Let $g \in \mathcal{D}(\overline{N}_s)$ and extend g to a $\widetilde{g} \in \mathcal{D}(\overline{N})$. Since the map $\overline{n} \longrightarrow \overline{n}P$ is a diffeomorphism of $\overline{N}$ onto an open set in G/P there exists a function $F \in \mathcal{D}(G/P)$ such that $F(m_s\overline{n}) = \widetilde{g}(\overline{n})$ for $\overline{n} \in \overline{N}$. Then

$$S'_{\lambda,s}(F) = \int_{\overline{N}_s} g(\overline{n}_s)e^{-(i\lambda+\rho)(H(\overline{n}_s))}d\overline{n}_s \tag{43}$$

and this function of λ extends from the tube (6) to a holomorphic function on $\mathfrak{a}_C^*$. By uniqueness of holomorphic continuation it coincides with $d_s(\lambda)S_{\lambda,s}(F)$. Now if $\Psi_{\lambda,s} = 0$ at some λ for which $d_s(\lambda)^{-1} \neq 0$ we deduce that the expression (43) vanishes for all $g \in \mathcal{D}(\overline{N}_s)$ which is impossible.

Proposition 5.7. *If α is a simple root then $\Psi_{\lambda,s_\alpha} \neq 0$ for all $\lambda \in \mathfrak{a}_C^*$.*

Proof. As stated before, $\Sigma_{s_\alpha}^+$ consists of the element α alone and as remarked in connection with (37),

$$\Gamma(\langle is_\alpha\lambda, \alpha_o\rangle) = \Gamma(\langle i(s_\alpha\lambda)^-, (\overline{\alpha})_o\rangle_\alpha)$$

so by (20)

$$S_{\lambda,s_\alpha}(F) = \boldsymbol{d}_{s_\alpha}(\overline{\lambda})^{-1}T_{\overline{\lambda},s_\alpha}(\overline{F}). \tag{44}$$

Since G_α/P_α is a closed submanifold of G/P carrying the induced topology, each $\varphi \in \mathcal{E}(G_\alpha/P_\alpha)$ is the restriction of some $F \in \mathcal{E}(G/P)$ (see e.g. [DS], I, Exercise D5). Hence if $\Psi_{\lambda,s_\alpha} = 0$, equation (44) implies

$$\boldsymbol{d}_{s_\alpha}(\overline{\lambda})^{-1}T_{\overline{\lambda},s_\alpha} = 0.$$

This means that for the proof we may take G/K of rank one. Because of Prop. 5.6 it suffices to consider the λ which are poles of $\boldsymbol{d}_{s_\alpha}$ and to prove that if $F \in \mathcal{E}(K/M)$ is suitably chosen the integral (33) as a function of z has a nonvanishing residue at the point $2piz = -m$ ($m \in \boldsymbol{Z}^+$). By the well-known formula for this residue (cf. [GGA], I, §2, (71) this amounts to

$$\left(\frac{d^m\varphi_z}{d\rho^m}\right)_{\rho=0} \neq 0.$$

However, this is immediate if we choose F such that for $X \in \mathfrak{g}_{-\alpha}, Y \in \mathfrak{g}_{-2\alpha}$ sufficiently small, F^* in (32) has the form $F^*(X,Y) = |X|^{2m}$.

Corollary 5.8. *For the value $\lambda = 0$ we have*

$$S_{o,s} = c\ \delta_{eM}, \qquad s \in W,$$

where $c \neq 0$ is a constant and δ_{eM} is the delta distribution at the origin in K/M.

In fact, the residue of (33) at $z = 0$ is $\varphi_o(0)$ so each factor in (44) is the delta distribution up to a nonzero constant factor.

Heuristic Remarks. The orbit space descriptions

$$K\backslash G/K \approx A/W, \qquad MN\backslash G/MN \approx A \times W$$

in §4, No. 1 suggested the conjecture that the set of conical distributions should be naturally parametrized by $\mathfrak{a}_C^* \times W$. Let us now re-examine this conjecture in the light of Theorem 5.3 and Corollary 5.5.

The residues of the function $\lambda \longrightarrow \Psi'_{\lambda,s}$, that is the values of $\Psi_{\lambda,s}$ at the singular points $\lambda = \lambda_o$ (where $\boldsymbol{d}_s(\lambda)^{-1} = 0$) have an interesting geometric interpretation. The closure of the orbit Ξ_s is known to consist of Ξ_s together with the orbits Ξ_t where the $t \in W$ are obtained by deleting arbitrarily factors from the reduced expression $s = s_{\alpha_1} \ldots s_{\alpha_p}$. The value $\Psi_{\lambda_o,s}$ then consists of a linear combination of certain transversal derivatives of the various $\Psi_{\lambda,t}$ constructed for these other orbits. This is in fact suggested by the convolution formula in Theorem 5.3 where at a singularity, some of the factors become certain derivatives of the delta function at the origin of K/M. During the proof of Theorem 5.4 we saw in fact that in the rank-one case the values of $\Psi_{\lambda,s}$ at the poles of $\boldsymbol{d}_s(\lambda)$ are certain transversal derivatives of $\Psi_{\lambda,e}$. We shall see in the next section that for G/K of rank one this holomorphic continuation procedure gives us all the conical distributions except for the case $\lambda = 0$. For this particular value of λ, $\mathcal{D}'_\lambda(\Xi)$ contains in addition to $\Psi_{o,e}$ a conical distribution Ψ_o which is constructed more in the spirit of a Cauchy principal value rather than by holomorphic continuation.

For the purpose of determining all the conical distributions we now invoke the dual transform $\Psi \longrightarrow \check{\Psi}$ of $\mathcal{D}'(\Xi)$ into $\mathcal{D}'(X)$ (Ch. I, Cor. 3.4) as well as the joint eigenspaces $\mathcal{E}_\lambda(X), \mathcal{D}'_\lambda(\Xi)$ of the algebras $\boldsymbol{D}(X)$, $\boldsymbol{D}(\Xi)$, respectively, described in this chapter, §2, No. 3.

Theorem 5.9. *Let $\lambda \in \mathfrak{a}^*_C$. Then the dual transform $\Psi \in \mathcal{D}'(\Xi) \longrightarrow \check{\Psi} \in \mathcal{D}'(X)$ maps $\mathcal{D}'_\lambda(\Xi)$ into $\mathcal{E}_\lambda(X)$.*

More precisely, if $\Psi \in \mathcal{D}'_\lambda(\Xi)$ is written

$$\Psi(\varphi) = \int_{K/M} \left(\int_A \varphi(ka \cdot \xi_o) e^{(i\lambda+\rho)(\log a)} da \right) dS(kM),$$

(Prop. 4.4) then

$$\check{\Psi}(x) = \int_{K/M} e^{(i\lambda+\rho)(A(x,kM))} dS(kM),$$

with $A(x, kM)$ as in §3, No. 4.

Proof. Let $f \in \mathcal{D}(X)$. Then the Radon transform $\widehat{f}$ satisfies

$$\widehat{f}(ka \cdot \xi_o) = \int_N f(kan \cdot o) dn$$

and

$$\int_X f(x)dx = \int_{N\times A} f(an\cdot o)dn\,da \tag{45}$$

(cf. §3, (4) and (14)). Hence

$$\begin{aligned}\int_A \widehat{f}(a\cdot\xi_o)e^{(i\lambda+\rho)(\log a)}da &= \int_A e^{(i\lambda+\rho)(\log a)}\int_N f(an\cdot o)dn\,da\\ &= \int_X e^{(i\lambda+\rho)(A(x,eM))}f(x)dx.\end{aligned}$$

Replacing here f by $f^{\tau(k^{-1})}$ we obtain

$$\begin{aligned}\overset{\vee}{\Psi}(f) = \Psi(\widehat{f}) &= \int_{K/M}\Big(\int_A \widehat{f}(ka\cdot\xi_o)e^{(i\lambda+\rho)(\log a)}da\Big)dS(kM)\\ &= \int_{K/M}\Big(\int_X e^{(i\lambda+\rho)(A(x,kM))}f(x)dx\Big)dS(kM).\end{aligned}$$

In the inner integration X can be replaced by a relatively compact open set (independent of $k\in K$). Then by the "Fubini theorem" for distributions (Schwartz [1966], Ch. VI, §3) the two integration signs above can be exchanged so we get

$$\overset{\vee}{\Psi}(f) = \int_X f(x)\Big(\int_{K/M} e^{(i\lambda+\rho)(A(x,kM))}dS(kM)\Big)dx$$

proving the stated formula for $\overset{\vee}{\Psi}(x)$. Finally $\overset{\vee}{\Psi}\in\mathcal{E}_\lambda(X)$ by Proposition 3.14.

Theorem 5.10. *For each $\lambda\in\mathfrak{a}^*_C$ and each $s\in W$ the function $(\Psi_{\lambda,s})^\vee$ is given by*

$$(\Psi_{\lambda,s})^\vee(na\cdot o) = a_s\boldsymbol{e}_s(\lambda)e^{(is\lambda+\rho)(\log a)}, \qquad n\in N, a\in A,$$

where the constant $a_s\neq 0$ is given by (25). Equivalently,

$$\int_B e^{(i\lambda+\rho)(A(x,b))}dS_{\lambda,s}(b) \equiv a_s\boldsymbol{e}_s(\lambda)\int_B e^{(is\lambda+\rho)(A(x,b))}d\delta_o(b)$$

where δ_o is the delta-distribution on $B = K/M$, at the origin eM.

Proof. Since the map $\Psi \longrightarrow \check{\Psi}$ commutes with the G-action and since $\Psi'_{\lambda,s}$ is N-invariant we have

$$(\Psi'_{\lambda,s})^{\vee}(na \cdot o) = (\Psi'_{\lambda,s})^{\vee}(a \cdot o).$$

Now we use (10), (9) and Theorem 5.9. Since $A(a \cdot o, kM) = -H(a^{-1}k)$ we obtain

$$(\Psi'_{\lambda,s})^{\vee}(a \cdot o) = \int_{\overline{N}_s} e^{-(i\lambda+\rho)(H(a^{-1}m_s k(\overline{n}_s))} e^{-(i\lambda+\rho)(H(\overline{n}_s))} d\overline{n}_s.$$

We put $b = m_s^{-1}a^{-1}m_s$ and then use the obvious formula

$$H(bk(\overline{n})) = \log b + H(b\overline{n}b^{-1}) - H(\overline{n}).$$

Then

$$\left(\Psi'_{\lambda,s}\right)^{\vee}(a \cdot o) = e^{-(i\lambda+\rho)(\log b)} \int_{\overline{N}_s} e^{-(i\lambda+\rho)(H(b\overline{n}_s b^{-1}))} d\overline{n}_s$$

and as we saw during the proof of Theorem 5.3

$$\frac{d(b\overline{n}_s b^{-1})}{d\overline{n}_s} = e^{(s^{-1}\rho-\rho)(\log b)} = e^{(-\rho+s\rho)(\log a)}.$$

Hence

$$\left(\Psi'_{\lambda,s}\right)^{\vee}(a \cdot o) = e^{(is\lambda+s\rho)(\log a)} e^{(\rho-s\rho)(\log a)} \int_{\overline{N}_s} e^{-(i\lambda+\rho)(H(\overline{n}_s))} d\overline{n}_s,$$

which by (22) equals

$$a_s \boldsymbol{d}_s(\lambda) \boldsymbol{e}_s(\lambda) e^{(is\lambda+\rho)(\log a)}.$$

Now we divide by $\boldsymbol{d}_s(\lambda)$ and use analytic continuation for all $\lambda \in \mathfrak{a}_C^*$.

Corollary 5.11. Let $s \in W$ and $\lambda \in \mathfrak{a}_C^*$ a point where $\boldsymbol{e}_s(\lambda) \neq 0$. Then $\Psi_{\lambda,s} \neq 0$.

This result and Prop. 5.6 show that the only possible $\lambda \in \mathfrak{a}_C^*$ for which $\Psi_{\lambda,s}$ might be 0 are the common singularities of the numerator and denumerator of $\boldsymbol{c}_s(\lambda)$. We shall now see by an example that $\Psi_{\lambda,s}$ can indeed vanish for some λ.

Example. $G = \boldsymbol{SL}(3, \boldsymbol{R})$. Here are two simple restricted roots α_1, α_2 and the corresponding positive Weyl chamber $\mathfrak{a}^+$ is a $60°$ sector of the plane. The Weyl group element $s = s_{\alpha_1} s_{\alpha_2} s_{\alpha_1}$ maps $\mathfrak{a}^+$ to its negative and this expression for s is reduced. Also $\Sigma_s^+ = \Sigma_o^+$ consists of α_1, α_2 and $\alpha_1 + \alpha_2$. Now we select $\lambda \in \mathfrak{a}_C^*$ such that both $\boldsymbol{d}_s(\lambda)^{-1}$ and $\boldsymbol{e}_s(\lambda)$ vanish. For this we determine λ by

$$\langle i\lambda, (\alpha_1)_o \rangle = -\langle i\lambda, (\alpha_2)_o \rangle = \tfrac{1}{2}. \tag{46}$$

Then $\langle i\lambda, \alpha_1 + \alpha_2 \rangle = 0$ so $\boldsymbol{d}_s(\lambda)^{-1} = 0$ and $\boldsymbol{e}_s(\lambda) = 0$ since $m_{\alpha_i} = 1$, $m_{2\alpha_i} = 0$. Actually, $3i\lambda = \alpha_1 - \alpha_2$ and in the notation of Theorem 5.3,

$$s^{(1)} = s_{\alpha_2} s_{\alpha_1}, \qquad s^{(2)} = s_{\alpha_1}, \qquad s_{\alpha_1}(\alpha_2) = s_{\alpha_2}(\alpha_1) = \alpha_1 + \alpha_2. \tag{47}$$

By (41) we also have

$$S_{\lambda,s} = S_{s^{(1)}\lambda, s_{\alpha_1}} \times S_{s^{(2)}\lambda, s_{\alpha_2}} \times S_{\lambda, s_{\alpha_1}}. \tag{48}$$

Here we claim that up to a constant nonzero factor

(i) $S_{s^{(2)}\lambda, s_{\alpha_2}} = \delta_{eM}$

(ii) $S_{\lambda, s_{\alpha_1}}$ is the characteristic function of the orbit $K_{\alpha_1} \cdot eM$ on K/M.

For (i) we just observe that $s^{(2)}\lambda = s_{\alpha_1}\lambda$ vanishes identically on $\mathfrak{a}_{\alpha_2}$ (since $s_{\alpha_1}(\mathfrak{a}_{\alpha_2}) = \mathfrak{a}_{\alpha_1+\alpha_2}$ and $\lambda(\mathfrak{a}_{\alpha_1+\alpha_2}) = 0$).

For (ii) we note that the restrictions of ρ and $i\lambda$ are given by

$$\rho_1 = \rho|\mathfrak{a}_{\alpha_1} = \frac{\langle \rho, \alpha_1 \rangle}{\langle \alpha_1, \alpha_1 \rangle}\alpha_1 = \tfrac{1}{2}\alpha_1 \qquad i\lambda|\mathfrak{a}_{\alpha_1} = \frac{\langle i\lambda, \alpha_1 \rangle}{\langle \alpha_1, \alpha_1 \rangle}\alpha_1 = \tfrac{1}{2}\alpha_1$$

so, up to a constant (since $m_{s_{\alpha_1}}$ can be chosen in K_{α_1})

$$S_{\lambda, s_{\alpha_1}}(F) = \int_{\overline{N}_{\alpha_1}} F(m_{s_{\alpha_1}} k(\overline{n}_{\alpha_1}) M_{\alpha_1}) e^{-2\rho_1(H(\overline{n}_{\alpha_1}))} d\overline{n}_{\alpha_1}$$

$$= \int_{K_{\alpha_1}/M_{\alpha_1}} F(kM_{\alpha_1}) dk_{M_{\alpha_1}}.$$

Using (i) and (ii) we can compute $S_{\lambda,s}(F)$. By (i) the middle factor in (48) can be omitted; also, combining (ii) and Lemma 5.2 we see that

$$S_{\lambda,s}(F) = \int_{K/M} \left(\int_{K_{\alpha_1}/M_{\alpha_1}} F^{\tau(k^{-1})}(\ell M_{\alpha_1}) d\ell_{M_{\alpha_1}} \right) dS_{s^{(1)}\lambda, s_{\alpha_1}}(kM).$$

Here $S_{s^{(1)}\lambda,s_{\alpha_1}}$ has support in the orbit $K_{\alpha_1} \cdot eM \subset K/M$ so in the integral above k is restricted to K_{α_1}. But then by the invariance of $dk_{M_{\alpha_1}}$, k can be removed altogether and we obtain

$$S_{\lambda,s}(F) = S_{s^{(1)}\lambda,s_{\alpha_1}}(1) \cdot \int_{K_{\alpha_1}/M_{\alpha_1}} F(\ell M_{\alpha_1}) d\ell_{M_{\alpha_1}}) d\ell_{M_{\alpha_1}}$$

Here the first factor equals $\boldsymbol{e}_{s_{\alpha_1}}(s^{(1)}\lambda)$ which is given by (24). Using (46) and (47) we find that this factor vanishes so indeed $S_{\lambda,s}(F) = 0$. Summarizing, we have obtained the following result.

Proposition 5.12. *Let $G = \boldsymbol{SL}(3, \boldsymbol{R})$, let α_1, α_2 be the simple restricted roots and s^* the Weyl group element which maps the positive Weyl chamber into its negative. Let $\lambda \in \mathfrak{a}_C^*$ be determined by*

$$\frac{\langle i\lambda, \alpha_1\rangle}{\langle \alpha_1, \alpha_1\rangle} = \frac{1}{2}, \qquad \frac{\langle i\lambda, \alpha_2\rangle}{\langle \alpha_2, \alpha_2\rangle} = -\frac{1}{2}.$$

Then

$$\Psi_{\lambda,s^*} = 0.$$

4. The Determination of the Conical Distributions.

Because of Theorem 5.9 it is natural to introduce the following terminology.

Definition. The element $\lambda \in \mathfrak{a}_C^*$ is said to be *simple* if the dual transform $\Psi \longrightarrow \check{\Psi}$ is injective on $\mathcal{D}'_\lambda(\Xi)$.

By Theorem 5.9, λ is simple if and only if the Poisson transform

(49) $$\mathcal{P}_\lambda : \; S \in \mathcal{D}'(B) \longrightarrow f \in \mathcal{E}_\lambda(X)$$

given by

$$f(x) = \int_B e^{(i\lambda+\rho)(A(x,b))} dS(b)$$

is injective. Here $B = K/M$ as usual.

Lemma 5.13. *Let $\lambda \in \mathfrak{a}_C^*$. Then λ is simple if and only if the mapping*

(50) $$F \in \mathcal{E}(B) \longrightarrow f \in \mathcal{E}_\lambda(X)$$

given by

$$f(x) = \int_B e^{(i\lambda+\rho)(A(x,b))} F(b)db$$

is injective.

Proof. It suffices to prove the "if" part. Assume that $S \in \mathcal{D}'(B)$ satisfies

$$\int_B e^{(i\lambda+\rho)(A(x,b))} dS(b) = 0 \qquad \text{for all } x.$$

If $F \in \mathcal{E}(K)$ it follows that

$$\int_K \int_B e^{(i\lambda+\rho)(A(k^{-1}\cdot x,b))} dS(b)F(k)dk \equiv 0$$

so if $\widetilde{S}$ denotes the lift of S to K (cf. Ch. I, §2, No. 5)

$$\int_K \left(\int_K e^{(i\lambda+\rho)(A(x,kk_o M))} d\widetilde{S}(k_o) \right) F(k)dk \equiv 0.$$

Hence the function $F_1 \in \mathcal{E}(B)$ defined by

$$F_1(kM) = \int_K F(kk_o^{-1}) d\widetilde{S}(k_o)$$

satisfies

$$\int_B e^{(i\lambda+\rho)(A(x,b))} F_1(b)db \equiv 0.$$

Since the map (50) is assumed injective, $F_1 \equiv 0$ so since F is arbitrary, $S = 0$.

Corollary 5.14. *If* $Re(\langle i\lambda, \alpha\rangle) > 0$ *for each* $\alpha \in \Sigma^+$ *then* λ *is simple.*

This is immediate from Lemma 5.13 and Theorem 3.16 since $\boldsymbol{c}(\lambda) \neq 0$ under the assumption on λ.

For the determination of the conical distributions we first recall that each conical distribution lies in the space $\mathcal{D}'_\lambda(\Xi)$ for some $\lambda \in \mathfrak{a}^*_C$. We also recall that $\lambda \in \mathfrak{a}^*_C$ is *regular* if $\langle \lambda, \alpha\rangle \neq 0$ for all $\alpha \in \Sigma$ or equivalently $s\lambda \neq \lambda$ for all $s \neq e$ in W (see for example [GGA], III, Exercise 2).

Theorem 5.15. *Suppose λ satisfies the conditions*

(i) λ is regular;

(ii) λ is simple;

(iii) $\boldsymbol{e}_{s*}(\lambda) \neq 0$.

Then the linear combinations

$$\sum_{s \in W} c_s \Psi_{\lambda,s} \qquad c_s \in \boldsymbol{C}$$

constitute all the conical distributions in $\mathcal{D}'_\lambda(\Xi)$.

Proof. Suppose $\Psi \in \mathcal{D}'_\lambda(\Xi)$ is a conical distribution. Since the dual transform $\Phi \longrightarrow \check{\Phi}$ commutes with the G-action, $\check{\Psi}$ is by Theorem 5.9 an N-invariant function f in the joint eigenspace $\mathcal{E}_\lambda(X)$. Because of Theorem 2.1 and the definition of $\mathcal{E}_\lambda(X)$ the restriction $\overline{f} = f|A \cdot o$ satisfies

$$(e^\rho \Gamma(D) \circ e^{-\rho})(\overline{f}) = \Gamma(D)(i\lambda) f, \qquad D \in \boldsymbol{D}(X).$$

Since the homomorphism Γ in Theorem 2.1 is surjective we have

$$L(e^{-\rho}\overline{f}) = L(i\lambda)(e^{-\rho}\overline{f}) \tag{51}$$

for each $L \in \boldsymbol{D}_W(A)$, where on the right L is viewed as a polynomial function on $\mathfrak{a}^*_{\boldsymbol{C}}$. The functions $a \longrightarrow e^{is\lambda(\log a)}$ on A constitute $|W|$ linearly independent solutions to (51), $|W|$ denoting the order of the Weyl group. On the other hand, the dimension of the solution space to (51) equals $|W|$ ([GGA], Ch. III, Theorem 3.13 (of Harish-Chandra and Steinberg)). This proves that $\check{\Psi}$ is a linear combination

$$\check{\Psi}(na \cdot o) = \sum_{s \in W} c_s e^{(is\lambda + \rho)(\log a)}$$

and now by Theorem 5.10, the simplicity of λ, and the condition $\boldsymbol{e}_{s^*}(\lambda) \neq 0$ imply that

$$\Psi = \sum_{s \in W} c_s \Psi_{\lambda,s}$$

as claimed.

We shall prove later(Ch. III,§4) that conditions (ii) and (iii) are equivalent. Anticipating this result and writing $\pi(\lambda) = \Pi_\alpha \langle \lambda, \alpha \rangle$ ($\alpha \in \Sigma_o^+$) as in §3 we can restate Theorem 5.15 as follows.

Theorem 5.16. *Suppose $\lambda \in \mathfrak{a}_C^*$ such that $\pi(\lambda)\boldsymbol{e}_{s^*}(\lambda) \neq 0$. Then the linear combinations*

$$\sum_{s \in W} c_s \Psi_{\lambda,s}, \qquad c_s \in \boldsymbol{C},$$

constitute all the conical distributions in $\mathcal{D}'_\lambda(\Xi)$.

To a large extent this establishes the conjecture formulated in §4, No. 1 that $\mathfrak{a}_C^* \times W$ should parametrize the set of conical distributions.

Theorem 5.16 is stated here for orientation; however, since it uses the criterion $\boldsymbol{e}_{s^*}(\lambda) \neq 0 \iff \lambda$ simple it will not be used until this criterion has been proved.

We shall now derive an important consequence of Theorem 5.10.

Theorem 5.17. *For all $\lambda \in \mathfrak{a}_C^*$ and all $s \in W$,*

$$S_{\lambda,s} \times S_{s\lambda,s^{-1}} = a_s^2 \boldsymbol{e}_s(\lambda)\boldsymbol{e}_s(-\lambda)\delta_o, \tag{52}$$

where δ_o is the delta-distribution on $B = K/M$ and a_s is given by (25).

Proof. Let us apply the left hand side of (52) to $e^{(is\lambda+\rho)(A(x,b))}$. Then by Lemma 5.2

$$\int_B e^{(is\lambda+\rho)(A(x,b))} d\big(S_{\lambda,s} \times S_{s\lambda,s^{-1}}\big)(b) =$$

$$\int_{K/M} \Bigg(\int_{K/M} e^{(is\lambda+\rho)(A(x,k\ell M))} dS_{s\lambda,s^{-1}}(\ell M)\Bigg)\, dS_{\lambda,s}(kM).$$

In the inner integral we write $A(x, k\ell M) = A(k^{-1} \cdot x, \ell M)$ and use Theorem 5.10. Then our expression becomes (since $s^{-1}(s\lambda) = \lambda$ and $a_s = a_{s^{-1}}$),

$$a_s \boldsymbol{e}_{s^{-1}}(s\lambda) \int_{K/M} e^{(i\lambda+\rho)(A(x,kM))} dS_{\lambda,s}(kM)$$

which again by Theorem 5.10 is

$$a_s^2 \boldsymbol{e}_{s^{-1}}(s\lambda)\boldsymbol{e}_s(\lambda) \int_B e^{(is\lambda+\rho)(A(x,b))} d\delta_o(b)$$

Here we can use the easy formula $\boldsymbol{e}_{s^{-1}}(s\lambda) = \boldsymbol{e}_s(-\lambda)$ and if $s\lambda$ is simple we can conclude from the formula proved that (52) holds whenever $s\lambda$ is simple.

Thus by Corollary 5.14, (52) holds on an open subset of $\mathfrak{a}_C^*$, hence on all of $\mathfrak{a}_C^*$ by analytic continuation.

We conclude this section with another identity for the distributions $S_{\lambda,s}$ which will also be useful later.

Let S be an M-invariant distribution on K/M. Then its lift $\widetilde{S}$ to K given by

$$\widetilde{S}(f) = S(\dot{f}), \qquad \dot{f}(kM) = \int_M f(km)dm$$

is bi-invariant under M and so is the image $\widetilde{S}'$ of $\widetilde{S}$ under the diffeomorphism $k \longrightarrow k^{-1}$. Thus we can define $S^* \in \mathcal{D}'(K/M)$ by

$$\widetilde{S^*} = \widetilde{S}'. \tag{53}$$

Thus

$$S^*(F) = \int_K \widetilde{F}(k^{-1})d\widetilde{S}(k) = \widetilde{S}(\widetilde{F}').$$

Theorem 5.18. *For all $\lambda \in \mathfrak{a}_c^*$ and all $s \in W$*

$$(S_{-\lambda,s})^* = S_{s\lambda,s^{-1}}. \tag{54}$$

Proof. The method is to verify the identity for the rank-one case and then use the product formula and (20) to reduce the general case to the rank-one case.

For the rank-one case we anticipate Proposition 6.4 in the next section (for which an independent proof is given). Writing S for $S'_{\lambda,s}$ in (9) we then have by Prop. 6.4,

$$\begin{aligned}\widetilde{S}(\widetilde{F}) &= \int_{\overline{N}} \widetilde{F}(m^*k(\overline{n}_o))e^{-(i\lambda+\rho)(H(\overline{n}_o))}d\overline{n}_o \\ &= \int_{\overline{N}} \widetilde{F}(k(\overline{n})^{-1}m^*)e^{-(i\lambda+\rho)(H(n))}d\overline{n}\end{aligned}$$

so

$$S(F) = \widetilde{S}(\widetilde{F}) = \widetilde{S}'(\widetilde{F}) = S^*(F).$$

Since $\boldsymbol{d}_s(-\lambda) = \boldsymbol{d}_s(s\lambda)$ this proves $(S_{-\lambda,s})^* = S_{-\lambda,s}$ verifying (54) in this case.

Next we prove (54) in the case when s is the reflection in a simple root α. In this case $s = s^{-1}$ and (20) implies

$$S_{s\lambda,s} = S_{-\lambda,s} \tag{55}$$

so equation (54) amounts to $(S_{-\lambda,s})^* = S_{-\lambda,s}$. To verify this we write (20) in the form

$$S(F) = T(\overline{F}) \qquad \text{for} \qquad F \in \mathcal{E}(K/M), \qquad \overline{F} = F|(K_\alpha/M_\alpha). \tag{56}$$

We shall prove

$$S^*(F) = T^*(\overline{F}), \tag{57}$$

which, since we just proved $T^* = T$, would imply $S^* = S$. Since S^* is M-invariant and T^* M_α-invariant it suffices to verify (57) for M-invariant F. Then $\widetilde{F'} = \widetilde{F^*}$ where F^* is also an M-invariant function on K/M. We have of course

$$\widetilde{S}(\widetilde{F}) = S(F) = T(\overline{F}) = \widetilde{T}(\widetilde{\overline{F}}), \tag{58}$$

where the tildes denote both lifts to K and to K_α. Thus

$$S^*(F) = \widetilde{S}(\widetilde{F}') = \widetilde{S}(\widetilde{F^*}) = T(\overline{F^*}) = \widetilde{T}((\overline{F^*})^\sim). \tag{59}$$

However,

$$(\overline{F^*})^\sim(k_\alpha) = \overline{F^*}(k_\alpha M_\alpha) = F^*(k_\alpha M) = \\ \widetilde{F}(k_\alpha^{-1}) = (\overline{F})^\sim(k_\alpha^{-1}) = ((\overline{F})^\sim)'(k_\alpha)$$

so the last term in (59) equals

$$\widetilde{T}'((\overline{F})^\sim) = T^*(\overline{F}),$$

proving (57). This proves (54) if s is a reflection in a simple root.

Now we need a simple lemma.

Lemma 5.19. *Let $S, T \in \mathcal{D}'(K/M)$ both be M-invariant. Then*

$$(S \times T)^* = T^* \times S^*.$$

Proof. We have

$$((S \times T)^*)^\sim = ((S \times T)^\sim)' = (\widetilde{S} * \widetilde{T})' = \\ \widetilde{T}' * \widetilde{S}' = (\widetilde{T^*}) * (\widetilde{S^*}) = (T^* \times S^*)^\sim.$$

For the proof of Theorem 5.18 we use the product decomposition (41); for simplicity we write s_i for s_{α_i}. Then by the lemma and the first part of the proof

$$(S_{-\lambda,s})^* = S_{-s^{(p)}\lambda,s_p} \times \cdots \times S_{-s^{(1)}\lambda,s_1}$$

and writing $t = s^{-1} = s_p \cdots s_1$

$$S_{s\lambda,s^{-1}} = S_{t^{(1)}s\lambda,s_p} \times \cdots \times S_{t^{(p)}s\lambda,s_1}.$$

Taking (55) into account, the qth factor in the first expression is

$$S_{s_{p-q+1}s^{(p-q+1)}\lambda,s_{p-q+1}}.$$

The qth factor in the second expression is

$$S_{t^{(q)}s\lambda,s_{p-q+1}}.$$

On the other hand,

$$s_{p-q+1}s^{(p-q+1)} = s_{p-q+1}\cdots s_p,$$
$$t^{(q)}s = (s_{p-q}\cdots s_1)(s_1\cdots s_p) = s_{p-q+1}\cdots s_p.$$

Thus the corresponding factors coincide so (54) is proved.

§6. SOME RANK-ONE RESULTS.

1. Component Computations.

We still consider a connected noncompact semisimple Lie group G with finite center and preserve the notation G, K, A, N, $\overline{N}$, M, M', θ and $\mathfrak{g}, \mathfrak{k}, \mathfrak{a}, \mathfrak{n}, \overline{\mathfrak{n}}\ \mathfrak{m}, \theta$, and $\mathfrak{p}$ from the last section. In the present section we assume that G/K has rank one. Here a useful tool is the $\boldsymbol{SU}(2,1)$-reduction ([DS], Chapter IX, §3). Consider the three decompositions

$$G = K\overline{A^+}K$$
$$G = KAN$$
$$G = \overline{N}MAN \cup m^*MAN.$$

Let $g \in G$ and suppose we want to compute the components of g according to these decompositions. Sometimes g can be imbedded in a subgroup $G^* \subset G$ locally isomorphic to $\boldsymbol{SU}(2,1)$ such that the Cartan, Iwasawa and Bruhat decompositions of G^* are compatible with the ones above. Then the components of g can be found by computing within the group $\boldsymbol{SU}(2,1)$. The following result ([DS], IX, §3 and [GGA] IV, Ex. C3) is proved in this way and in the present section we encounter further applications of the method. According to the decompositions above we write

$$g = k_1 \exp A^+(g) k_2, \qquad g = k(g) \exp H(g) n(g),$$
$$g = \overline{n}(g) m(g) \exp B(g) n_B(g), \qquad g \in \overline{N}MAN,$$

where $A^+(g) \in \overline{\mathfrak{a}^+}$ and $H(g), B(g) \in \mathfrak{a}$. Recall that Σ^+ consist of α (and possibly 2α).

Writing

$$m^*\overline{n} = \overline{n}(m^*\overline{n}) m \exp B(m^*\overline{n}) n_1$$
$$= k \exp H(\overline{n}(m^*\overline{n})) \exp B(m^*\overline{n}) n_2$$

we see that

$$H(\overline{n}) = H(\overline{n}(m^*\overline{n})) + B(m^*\overline{n})$$

Theorem 6.1. *We put* $|Z|^2 = -B(Z, \theta Z)$ *for* $Z \in \mathfrak{g}$. *If* $\overline{n} \in \overline{N}$ *we write* $\overline{n} = \exp(X+Y)$ $(X \in \mathfrak{g}_{-\alpha}, Y \in \mathfrak{g}_{-2\alpha})$ *then for* $a \in A$

(i)
$$\cosh^2(\alpha(A^+(\overline{n}a))) = \left[\cosh(\alpha(\log a)) + \tfrac{1}{2} c\, e^{\alpha(\log a)} |X|^2\right]^2 + c\, e^{2\alpha(\log a)} |Y|^2$$

(ii)
$$e^{\rho(H(\overline{n}))} = \left[(1 + c|X|^2)^2 + 4c|Y|^2\right]^{\frac{1}{4}(m_\alpha + 2m_{2\alpha})}$$

(iii)
$$e^{\rho(B(m^*\overline{n}))} = \left[c^2|X|^4 + 4c|Y|^2\right]^{\frac{1}{4}(m_\alpha + 2m_{2\alpha})}.$$

Here m_α *and* $m_{2\alpha}$ *are the multiplicities of the roots* α *and* 2α *and* $c^{-1} = 4(m_\alpha + 4m_{2\alpha})$.

As a consequence we can write down the Poisson kernel (§3, No. 4) and the conical functions explicitly in the rank one case.

Corollary 6.2. *Let* $\overline{n} = \exp(X+Y)$ $(X \in \mathfrak{g}_{-\alpha}$, $Y \in \mathfrak{g}_{-2\alpha})$ *and* $a_t = \exp tH$ *where* $\alpha(H) = 1$.

(i) *The Poisson kernel*

$$P(x, b) = e^{2\rho(A(x,b))}$$

is given by

$$P(a_tK, k(\overline{n})M) = \left[\frac{e^{2t}((1+c|X|^2)^2+4c|Y|^2)}{(1+c|e^tX|^2)^2+4c|e^{2t}Y|^2}\right]^{\langle\rho,\alpha_o\rangle}$$

where $\alpha_o = \alpha/\langle\alpha,\alpha\rangle$.

(ii) *If* $\mu \in \mathfrak{a}_c^*$ *the function*

$$\psi(\xi) = e^{-\mu(\log a(\xi))} \qquad \xi \in \Xi^*$$

from Theorem 4.1–4.6 is given by

$$\psi(\overline{n}a_t \cdot \xi_o) = \left[e^{2t}(c^2|X|^4 + 4c|Y|^2)\right]^{\frac{1}{2}\langle\mu,\alpha_o\rangle}$$

and

$$\log a(\overline{n}a \cdot \xi_o) = -\log a - B(m^*\overline{n}), \qquad a \in A, \overline{n} \in \overline{N}.$$

For (i) we just observe that if $a \in A, \overline{n} \in \overline{N}$, then

$$P(aK, k(\overline{n})M) = e^{2\rho(\log a)}e^{2\rho(H(\overline{n}))}e^{-2\rho(H(a^{-1}\overline{n}a))}.$$

For (ii) we have $\psi(mnam^* \cdot \xi_o) = e^{-\mu(\log a)}$ so moving m^* to the left,

$$\psi(m^*\overline{n}a \cdot \xi_o) = e^{\mu(\log a)}.$$

Now we write

$$m^*\overline{n} = \overline{n}(m^*\overline{n})m_1 \exp B(m^*\overline{n})n_1$$

so

$$\psi(\overline{n}(m^*\overline{n})a \cdot \xi_o) = e^{\mu(\log a - B(m^*\overline{n}))}.$$

Here we put $\overline{n} = \overline{n}(m^*\overline{n}_1)$ and note that since m^* can be chosen such that $(m^*)^2 \in Z$ (Prop. 6.3 below) we have

$$\overline{n}(m^*\overline{n}(m^*\overline{n}_1)) = \overline{n}_1,$$
$$B(m^*\overline{n}(m^*\overline{n}_1)) = -B(m^*\overline{n}_1).$$

Consequently,

$$\psi(\overline{n}a_t \cdot \xi_o) = e^{\mu(tH + B(m^*\overline{n}))}$$

and (ii) follows.

2. The Inversion of $\overline{N}$.

Since the group $G_o = \boldsymbol{SU}(2,1)$ will remain a useful tool we start with some notational conventions for this case. As in [DS], IX, §3 all general concepts connected with this group will be given the subscript o. The automorphism $\theta_o : \ g \longrightarrow I_{2,1} g I_{2,1}$ where

(1) $$I_{2,1} = \begin{pmatrix} -1 & 0 & 0 \\ 0 & -1 & 0 \\ 0 & 0 & 1 \end{pmatrix}$$

is a Cartan involution and $\mathfrak{k}_o$, the fixed point set, consists of the matrices

(2) $$\begin{pmatrix} A & & 0 \\ & & 0 \\ 0 & 0 & -Tr(A) \end{pmatrix} \qquad \text{A skew-hermitian}$$

and $\mathfrak{p}_o$ of the matrices

(3) $$\begin{pmatrix} 0 & 0 & z_1 \\ 0 & 0 & z_2 \\ \overline{z}_1 & \overline{z}_2 & 0 \end{pmatrix} \qquad (z_1, z_2 \in \boldsymbol{C}).$$

For $\mathfrak{a}_o$ and $\mathfrak{a}_o^+$ we take

(4) $$\mathfrak{a}_o = \boldsymbol{R}H_o, \qquad \mathfrak{a}_o^+ = \{tH_o : \ t > 0\},$$

where

(5) $$H_o = \begin{pmatrix} 0 & 0 & 1 \\ 0 & 0 & 0 \\ 1 & 0 & 0 \end{pmatrix}.$$

The restricted roots are $\pm\alpha_o, \pm 2\alpha_o$ where

$$\alpha_o(H_o) = 1.$$

The centralizer $\mathfrak{m}_o$ of $\mathfrak{a}_o$ in $\mathfrak{k}_o$ is $\mathfrak{m}_o = \boldsymbol{R}T_o$ where

(6) $$T_o = \begin{pmatrix} i & 0 & 0 \\ 0 & -2i & 0 \\ 0 & 0 & i \end{pmatrix}.$$

The root spaces $\mathfrak{g}_{\alpha_o}$ and $\mathfrak{g}_{-\alpha_o}$ are the sets of matrices

(7) $$\begin{pmatrix} 0 & z & 0 \\ -\overline{z} & 0 & \overline{z} \\ 0 & z & 0 \end{pmatrix} \text{ for } \mathfrak{g}_{\alpha_o}, \qquad \begin{pmatrix} 0 & z & 0 \\ -\overline{z} & 0 & -\overline{z} \\ 0 & -z & 0 \end{pmatrix} \text{ for } \mathfrak{g}_{-\alpha_o}$$

with $z \in \boldsymbol{C}$. The root spaces $\mathfrak{g}_{2\alpha_o}, \mathfrak{g}_{-2\alpha_o}$ are the sets of matrices

$$(8) \qquad \begin{pmatrix} it & 0 & -it \\ 0 & 0 & 0 \\ it & 0 & -it \end{pmatrix} \text{ for } \mathfrak{g}_{2\alpha_o}, \qquad \begin{pmatrix} it & 0 & it \\ 0 & 0 & 0 \\ -it & 0 & -it \end{pmatrix} \text{ for } \mathfrak{g}_{-2\alpha_o}$$

where $t \in \boldsymbol{R}$. Thus $m_{\alpha_o} = 2$, $m_{2\alpha_o} = 1$ and $\rho = 2\alpha_o$.

Proposition 6.3. *Suppose the symmetric space G/K has rank one. Then an element $m^* \in K$ can be chosen such that $Ad_G(m^*) = -I$ on $\mathfrak{a}$ and $(m^*)^2 \in Z$, the center of G.*

Proof. For $G_o = \boldsymbol{SU}(2,1)$ this is clear since the element (1) above can be chosen as m^*. Next consider the universal covering group $\widetilde{G}_o = \boldsymbol{SU}(2,1)^\sim$, $\widetilde{K}_o$ the analytic subgroup with Lie algebra $\mathfrak{k}_o$, and $\widetilde{M}_o$ the centralizer of $\mathfrak{a}_o$ in $\widetilde{K}_o$. Since $\widetilde{K}_o$ covers K_o we can select $\widetilde{m}^* \in \widetilde{K}_o$ over m^*. Then $Ad_{\widetilde{G}_o}(\widetilde{m}^*)$ and $Ad_{G_o}(m^*)$ agree on $\mathfrak{g}_o$. Thus

$$(9) \qquad Ad_{\widetilde{G}_o}(\widetilde{m}^*) = -I \text{ on } \mathfrak{a}_o, \qquad Ad_{\widetilde{G}_o}(\widetilde{m}^*) = I \text{ on } \mathfrak{m}_o,$$

and $(\widetilde{m}^*)^2$ is an element m in the one-dimensional group $\widetilde{M}_o$ which is connected since $\widetilde{K}_o/\widetilde{M}_o$ is a sphere ([DS], IX, Exercise B1). We can thus write $m = t^2$ $(t \in \widetilde{M}_o)$ and note by (9) that $\widetilde{m}^*$ commutes elementwise with $\widetilde{M}_o$. The element $m_o^* = t^{-1}\widetilde{m}^*$ then has the desired properties

$$(10) \qquad Ad_{\widetilde{G}_o}(m_o^*) = -I \text{ on } \mathfrak{a}_o, \qquad (m_0^*)^2 = e.$$

a) If G is locally isomorphic to $\boldsymbol{SU}(2,1)$ it is covered by $\widetilde{G}_o$ and we can use for m^* the image of m_o^* under the covering map. This will satisfy

$$(11) \qquad Ad_G(m^*) = -I \text{ on } \mathfrak{a}_o, \qquad (m^*)^2 = e.$$

Next consider the case $G = \boldsymbol{SO}_o(1,n)$ the identity component of $\boldsymbol{SO}(1,n)$. Here K is the set of matrices

$$\begin{vmatrix} 1 & 0 \\ 0 & u \end{vmatrix}, \qquad u \in \boldsymbol{SO}(n).$$

For $\mathfrak{a}_o$ we can take $\boldsymbol{R}(E_{12} + E_{21})$ and then the diagonal matrix

$$m^* = E_{11} - E_{22} - E_{33} + E_{44} + \cdots + E_{n+1n+1}$$

satisfies (11). The image $m_o^* = Ad_G(m^*)$ in $G_o = Ad_G(G)$ will also satisfy (11).

b) If G is locally isomorphic to $\boldsymbol{SO}_o(1,n)$ then it is a covering group of G_o. If $m^* \in G$ is mapped onto m_o^* by the covering map then

$$Ad_G(m^*) = -I \text{ on } \mathfrak{a}_o, \quad (m^*)^2 \in Z.$$

Consider now the case of general G. If Σ^+ consists of two elements then the $\boldsymbol{SU}(2,1)$ reduction ([DS], Chapter IX, §3) yields a subgroup $G^* \subset G$ locally isomorphic to $\boldsymbol{SU}(2,1)$ and we are through by case a) above. If Σ^+ consists of just one element, G is locally isomorphic to $\boldsymbol{SO}_o(1,n)$ so we are through by case b) above.

Remark. For $G = \boldsymbol{SO}(1,n)$ and for the cases when Σ^+ has two elements, m^* could be chosen such that $(m^*)^2 = e$. For $G = \boldsymbol{SL}(2,\boldsymbol{R})$ this is not possible because $(m^*)^2 = -I$ for each choice of m^*.

We recall that the set $\overline{N}MAN$ is a dense subset of G and if $g \in \overline{N}MAN$ there exists a unique element $\overline{n}(g) \in \overline{N}$ such that $g \in \overline{n}(g)MAN$.

Definition. Let m^* be as in Prop. 6.3. Then the mapping

$$J: \quad \overline{n} \longrightarrow \overline{n}(m^*\overline{n}) \tag{12}$$

of $\overline{N} - \{e\}$ onto itself is called the *inversion* of $\overline{N}$.

Remarks. Consider the diffeomorphism $\overline{n} \longrightarrow k(\overline{n})M$ of $\overline{N}$ onto $K/M - \{m^*M\}$. Writing

$$\begin{aligned} m^*\overline{n} &= \overline{n}(m^*\overline{n})man \\ &= k(\overline{n}(m^*\overline{n}))a_1n_1man = k(\overline{n}(m^*\overline{n}))ma_2n_2 \end{aligned}$$

we see that

$$k(\overline{n}(m^*\overline{n}))M = k(m^*\overline{n})M = m^*k(\overline{n})M.$$

Thus, under the map $\overline{n} \longrightarrow k(\overline{n})M$, J corresponds to the involution

$$kM \longrightarrow m^*kM \tag{13}$$

of K/M. Thus J is indeed an involution. Note that it is not completely canonical since m^* is not quite uniquely determined by the properties of Proposition 6.3.

Example. Let $G_o = \boldsymbol{SU}(2,1)$ and let m^* be the matrix (1). Then the inversion

$$J : \overline{n} \longrightarrow \overline{n}(m^*\overline{n})$$

is given as follows: If (by (7) and (8))

$$(14)\quad \overline{n} = \exp\begin{pmatrix} it & z & it \\ -\overline{z} & 0 & -\overline{z} \\ -it & -z & -it \end{pmatrix}, \qquad \overline{n}(m^*\overline{n}) = \exp\begin{pmatrix} iu & w & iu \\ -\overline{w} & 0 & -\overline{w} \\ -iu & -w & -iu \end{pmatrix}$$

then

$$(15)\qquad w = \frac{-z}{|z|^2 + 2it}, \qquad u = \frac{-t}{|z|^4 + 4t^2}$$

(cf. [DS], Chapter IX, §3). The mapping $kM_o \longrightarrow Ad(k)H_o$ identifies K_o/M_o with a sphere $S_{r_o}(0)$ in $\mathfrak{p}$ and then the involution (13) is just the *antipodal mapping.* We shall also encounter another interesting involution on $\overline{N}$.

Proposition 6.4. *Let $m^* \in M'$ be any element inducing the nontrivial Weyl group element. Given $\overline{n} \in \overline{N}$ there exists a unique $\overline{n}_o \in \overline{N}$ such that*

$$(16)\qquad m^*k(\overline{n}_o)M = k(\overline{n})^{-1}m^*M.$$

The map $j : \overline{n} \longrightarrow \overline{n}_o$ is an involutive diffeomorphism of $\overline{N}$ onto itself, preserving $d\overline{n}$, and

$$(17)\qquad H(\overline{n}_o) = H(\overline{n}).$$

In the case $G = \boldsymbol{SU}(2,1)$ the involution $\overline{n} \longrightarrow \overline{n}_o$ is (in terms of (14)) given by

$$(18)\quad t_o = -t, \qquad z_o = -z\left(\frac{1 + |z|^2 + 2it}{1 + |z|^2 - 2it}\right)^{\frac{1}{2}} \qquad \left(\textit{Here } \left(\frac{a}{\overline{a}}\right)^{\frac{1}{2}} = \frac{a}{|a|}\right).$$

Proof. We have seen that the map $\overline{n}_o \longrightarrow m^*k(\overline{n}_o)M$ is a diffeomorphism of $\overline{N}$ onto $K/M - \{M\}$. We now claim that the map $\overline{n} \longrightarrow k(\overline{n})^{-1}m^*M$ is also a diffeomorphism of $\overline{N}$ onto $K/M - \{M\}$. For this, note first that for $m \in M$, $k(\overline{n}^m) = (k(\overline{n}))^m$, superscript denoting conjugation. For surjectivity let $k \notin M$ and select $\overline{n} \in \overline{N}$ such that $m^*k(\overline{n})M = k^{-1}M$. Then for some $m \in M$,

$$km^* = mk(\overline{n})^{-1} = (k(\overline{n}^m))^{-1}m$$

so $kM = k(\overline{n}^m)^{-1}m^*M$ as desired. The injectivity follows in the same way, the first mapping being injective. The first mapping being a diffeomorphism the same is now clear for the second. Next we prove formula (17). For this we first observe that the validity of (17) is independent of how m^* is chosen mod M.

We now use the $\boldsymbol{SU}(2,1)$ reduction. Let α (and possibly 2α) be the positive restricted roots so we have

$$\overline{\mathfrak{n}} = \mathfrak{g}_{-\alpha} + \mathfrak{g}_{-2\alpha}.$$

Then $\overline{n} = \exp X \exp Y$ $(X \in \mathfrak{g}_{-\alpha}, Y \in \mathfrak{g}_{-2\alpha})$. Suppose $Y \neq 0$. Then the analytic subgroup $G^* \subset G$ whose Lie algebra $\mathfrak{g}^*$ is generated by $X, Y, \theta X, \theta Y$ is locally isomorphic to $\boldsymbol{SU}(2,1)$. By the remark above we may take m^* inside G^* and by (11) we can take it such that $(m^*)^2 = e$. Since the groups G^* and G have compatible Iwasawa and Bruhat decompositions ([DS], Chapter IX, Lemma 3.7) it suffices to prove (17) for the case $G_o = \boldsymbol{SU}(2,1)$. We have

$$\overline{n} = k(\overline{n}) \exp H(\overline{n}) u \tag{19}$$

and if according to (7) and (8)

$$\overline{n} = \exp \begin{pmatrix} it & z & it \\ -\overline{z} & 0 & -\overline{z} \\ -it & -z & -it \end{pmatrix}, \qquad u = \exp \begin{pmatrix} i\tau & \zeta & -i\tau \\ -\overline{\zeta} & 0 & \overline{\zeta} \\ i\tau & \zeta & -i\tau \end{pmatrix},$$

then ([DS], Chapter IX, Lemma 3.9),

$$\zeta = \frac{-z}{1 + |z|^2 - 2it}, \qquad \tau = \frac{-t}{(1 + |z|^2)^2 + 4t^2}, \tag{20}$$

and

$$e^{\rho_o(H(\overline{n}))} = (1 + |z|^2)^2 + 4t^2. \tag{21}$$

Choosing m^* as the matrix (1) we have, using (19),

$$\begin{aligned} k(\overline{n})^{-1}m^* &= m^*m^*k(\overline{n})^{-1}m^* \\ &= m^* \exp(-H(\overline{n}))\theta_o(u)\theta_o(\overline{n})^{-1} \end{aligned}$$

and consequently

$$k(\overline{n})^{-1}m^* = m^*k[\exp(-H(\overline{n}))\theta_o(u)\exp(H(\overline{n}))],$$

whence

$$\overline{n}_o = \theta_o(u)^{\exp(-H(\overline{n}))}. \tag{22}$$

Now

$$\theta_o(u) = \exp\begin{pmatrix} i\tau & \zeta & i\tau \\ -\overline{\zeta} & 0 & -\overline{\zeta} \\ -i\tau & -\zeta & -i\tau \end{pmatrix} \tag{23}$$

and by *loc. cit.*

$$H(\overline{n}) = sH_o, \qquad s = \tfrac{1}{2}\rho_o(H(\overline{n})) \tag{24}$$

$$\exp(-H(\overline{n})) = \begin{pmatrix} \cosh s & 0 & -\sinh s \\ 0 & 1 & 0 \\ -\sinh s & 0 & \cosh s \end{pmatrix}.$$

Since $(\exp X)^g = \exp(X^g)$ we find

$$\log \overline{n}_o = (\log \theta_o(u))^{\exp(-H(\overline{n}))}$$

so writing

$$\overline{n}_o = \exp\begin{pmatrix} it_o & z_o & it_o \\ -\overline{z}_o & 0 & -\overline{z}_o \\ -it_o & -z_o & -it_o \end{pmatrix},$$

t_o and z_o are readily computed in terms of τ and ζ. Combining with (20) we get the formulas for the coordinate expression of the map $\overline{n} \longrightarrow \overline{n}_o$,

$$t_o = -t, \quad z_o = -z\frac{((1+|z|^2)^2 + 4t^2)^{\frac{1}{2}}}{1+|z|^2 - 2it}.$$

This proves (18). We also conclude that $|t_o| = |t|$, $|z_o| = |z|$ so (17) follows from (21) because $\overline{n} \in \overline{N}$ was arbitrary. Similarly, it suffices to prove the involutiveness $j(j(\overline{n})) = \overline{n}$ for $\overline{n} \in \boldsymbol{SU}(2,1)$. However (18) implies (with obvious notation)

$$(t_o)_o = -t_o = t,$$

$$(z_o)_o = -z_o\left(\frac{1+|z_o|^2 + 2it_o}{1+|z_o|^2 - 2it_o}\right)^{\frac{1}{2}} = z$$

as desired. Finally we must prove that j preserves $d\overline{n}$, that is if $j^*(d\overline{n}_o) = \varphi(\overline{n})d\overline{n}$ then $\varphi \equiv 1$. Again it suffices to show $\varphi(\overline{n}) = 1$ for $\overline{n} \in \overline{N} \cap G^*$, in other words it suffices to prove the statement $\varphi \equiv 1$ in case our group is $\boldsymbol{SU}(2,1)$. This amounts to proving that the map $(t,z) \longrightarrow (t_o, z_o)$ has Jacobian 1. Writing $z = re^{i\theta}$, $z_o = re^{i\psi}$ the map has the coordinate expression

$$(t, r, \theta) \longrightarrow (-t, r, \psi),$$

where $\psi = \theta +$ Arctan $(2t/1+r^2)$. This map preserves the measure $dt\, r\, dr\, d\theta$ which is proportional to $d\bar{n}$. This concludes the proof, the case $G = \boldsymbol{SO}_o(1,n)$ being trivial.

3. The Simplicity Criterion.

As mentioned in §5, No.4 it will turn out that an element $\lambda \in \mathfrak{a}_C^*$ is simple if and only if $\boldsymbol{e}_{s^*}(\lambda) \neq 0$. In this subsection we shall prove this for the case when G/K has rank one; later, with more tools at our disposal, we shall deal with the general case.

Theorem 6.5. *(G/K of rank one.) Let $\lambda \in \mathfrak{a}_c^*$ and let s denote the non-trivial Weyl group element. Then*

λ is simple if and only if $\boldsymbol{e}_s(\lambda) \neq 0$.

Since $\Psi_{\lambda,s} \neq 0$ for all $\lambda \in \mathfrak{a}_c^*$ (Prop. 5.7) it is clear from Theorem 5.10 that λ is not simple if $\boldsymbol{e}_s(\lambda) = 0$. For the converse we first prove the following result.

Lemma 6.6. *Let $\lambda \in \mathfrak{a}_c^*$ be real-valued, that is,*

$$Re(i\lambda) = 0. \tag{25}$$

Then λ is simple.

Proof. Let $s \in \boldsymbol{R}$ be determined by $\lambda = s\rho$. If λ is not simple there would exist an $F \in \mathcal{E}(B)$ such that

$$\int_B e^{(i\lambda+\rho)(A(x,b))} F(b)db = 0, \quad x \in X. \tag{26}$$

We may assume $F(eM) \neq 0$. As we have seen (§3, No. 4) this can be written

(27) $$\int_{\overline{N}} e^{-(i\lambda+\rho)(H(v^a)+H(v))}F(k(v)M)e^{-2\rho(H(v))}dv = 0.$$

Here we use the formula for $e^{\rho(H(\overline{n}))}$ in Theorem 6.1 taking $\overline{n} = \exp(X_1+Y_1)$. To simplify put $X = c^{\frac{1}{2}}X_1$, $Y = 2c^{\frac{1}{2}}Y_1$, $F(X,Y) = F(k(\exp(X_1+Y_1))M)$. Take $H \in \mathfrak{a}^+$ such that $\alpha(H) = 1$ and put $P = \frac{1}{4}(m_\alpha + 2m_{2\alpha})$. Then if $a = \exp tH$,

$$\begin{aligned} e^{\rho(H(\overline{n}^a))} &= \left[\left(1+e^{-2t}|X|^2\right)^2 + e^{-4t}|Y|^2\right]^P \\ &= e^{-4Pt}\left[\left(e^{2t}+|X|^2\right)^2 + |Y|^2\right]^P. \end{aligned}$$

Hence (27) takes the form

$$\int \left\{\frac{(1+|X|^2)^2+|Y|^2}{(e^{2t}+|X|^2)^2+|Y|^2}\right\}^{P(1+is)} F(X,Y)\frac{dX dY}{[(1+|X|^2)^2+|Y|^2]^{2P}} = 0.$$

Now we take d/dt under the integral sign and obtain

$$\int \frac{[(1+|X|^2)^2+|Y|^2]^{P(is-1)}}{[(e^{2t}+|X|^2)^2+|Y|^2]^{P(is+1)+1}}(e^{2t}+|X|^2)F(X,Y)dX\, dY = 0.$$

The differentiated integral converges uniformly for t in any compact interval so the differentiation is indeed allowed. Here we make the substitution $X = e^tS$, $Y = e^{2t}T$. This gives

(28) $$\int \frac{[(1+e^{2t}|S|^2)^2+e^{4t}|T|^2]^{P(is-1)}}{[(1+|S|^2)^2+|T|^2]^{P(is+1)+1}}(1+|S|^2)F(e^tS, e^{2t}T)\, dS\, dT = 0.$$

The integrand is bounded by a constant times

$$\left[\left(1+|S|^2\right)^2+|T|^2\right]^{-P-\frac{1}{2}},$$

which equals $e^{-(\rho+\epsilon\rho)(H(v))}$(where $\epsilon = 1/2P$) and is thus integrable ([GGA], Chapter IV, §6, No. 2). Thus we can let $t \longrightarrow -\infty$ under the integral sign in (28) and deduce

(29) $$F(eM)\cdot \int_{\mathfrak{g}_{-\alpha}\times\mathfrak{g}_{-2\alpha}} (1+|S|^2)[(1+|S|^2)^2+|T|^2]^{-Q}dS\, dT = 0,$$

where $Q = P(is+1)+1$. It just remains to verify that the integral in (29) is $\neq 0$. Taking polar coordinates r_α and $r_{2\alpha}$ in $\mathfrak{g}_{-\alpha}$ and $\mathfrak{g}_{-2\alpha}$ and then taking $r = r_\alpha^2, \rho = r_{2\alpha}^2$ our integral is a constant multiple of

$$\int_o^\infty \int_o^\infty (1+r)[(1+r)^2+\rho]^{-Q} r^{\frac{1}{2}m_\alpha - 1} \rho^{\frac{1}{2}m_{2\alpha}-1} dr\, d\rho.$$

Here we put $\rho = (1+r)^2\sigma$ where by the integral reduces to

$$\int_o^\infty r^{\frac{1}{2}m_\alpha - 1}(1+r)^{1-2Q+m_{2\alpha}} dr \int_o^\infty (1+\sigma)^{-Q}\sigma^{\frac{1}{2}m_{2\alpha}-1} d\sigma$$

$$= \frac{\Gamma(\frac{1}{2}m_{2\alpha})\Gamma(2Q - m_{2\alpha} - 1 - \frac{1}{2}m_{2\alpha})}{\Gamma(2Q - m_{2\alpha} - 1)} \frac{\Gamma(\frac{1}{2}m_{2\alpha})\Gamma(Q - \frac{1}{2}m_{2\alpha})}{\Gamma(Q)},$$

which is $\neq 0$ for $s \in \boldsymbol{R}$. Hence we get the contradiction $F(eM) = 0$, which proves the lemma.

For Theorem 6.5 it now remains to prove that if $\boldsymbol{e}_s(\lambda) \neq 0$ then λ is simple. Suppose λ not simple. Then, by Lemma 6.6 and Corollary 5.14, $-\lambda$ is simple so by the first part of the proof $\boldsymbol{e}_s(-\lambda) \neq 0$. Now we use the formula in Theorem 5.17 which here implies

$$S_{-\lambda,s} \times S_{\lambda,s} = a_s^2 \boldsymbol{e}_s(\lambda)\boldsymbol{e}_s(-\lambda)\delta_o.$$

Taking convolution with any $F \in \mathcal{E}(B)$ on the left we obtain

$$F_1 \times S_{\lambda,s} = a_s^2 \boldsymbol{e}_s(\lambda)\boldsymbol{e}_s(-\lambda)F, \tag{30}$$

where $F_1 = F \times S_{-\lambda,s}$. On the other hand, by Lemma 5.2 and Theorem 5.10

$$\int_B e^{(i\lambda+\rho)(A(x,b))}(F_1 \times S_{\lambda,s})(b)db = a_s \boldsymbol{e}_s(\lambda) \int_B e^{(-i\lambda+\rho)(A(x,b))} F_1(b)db. \tag{31}$$

Since λ is not simple there exists an $F \not\equiv 0$ in $\mathcal{E}(B)$ such that

$$\int_B e^{(i\lambda+\rho)(A(x,b))}F(b)db = 0.$$

Then (30)–(31) imply $F_1 \not\equiv 0$ and

$$\int_B e^{(-i\lambda+\rho)(A(x,b))}F_1(b)db = 0,$$

contradicting the simplicity of $-\lambda$. This concludes the proof of Theorem 6.5.

4. THE ALGEBRA $\boldsymbol{D}(K/M)$.

We keep the rank-one assumption on G/K and shall now describe the algebra $\boldsymbol{D}(K/M)$ of K-invariant differential operators on K/M. We consider again the direct decomposition

$$\mathfrak{k} = \mathfrak{m} + \mathfrak{l}, \tag{32}$$

where $\mathfrak{l}$ and $\mathfrak{m}$ are orthogonal with respect to the Killing form B of $\mathfrak{g}$, which we shall also denote $\langle , \rangle$. For $\beta \in \mathfrak{a}^*$ we put

$$\mathfrak{l}_\beta = \{T \in \mathfrak{l} : \ (\mathrm{ad}H)^2 T = \beta(H)^2 T \text{ for } H \in \mathfrak{a}\}$$

and have then the further decomposition

$$\mathfrak{l} = \mathfrak{l}_\alpha + \mathfrak{l}_{2\alpha}. \tag{33}$$

Let $T_1, \dots, T_{m_\alpha}$ and $S_1, \dots, S_{m_{2\alpha}}$ be bases of $\mathfrak{l}_\alpha$ and $\mathfrak{l}_{2\alpha}$, respectively, orthonormal with respect to $-B$. Viewing these as left invariant vector fields on K we put

$$\omega_\alpha = \sum_{i=1}^{m_\alpha} T_i^2, \qquad \omega_{2\alpha} = \sum_{j=1}^{m_{2\alpha}} S_j^2. \tag{34}$$

These elements of the algebra $\boldsymbol{D}(K)$ of left invariant differential operators on K are right invariant under M. They induce elements $\Delta_\alpha, \Delta_{2\alpha} \in \boldsymbol{D}(K/M)$ by

$$\omega_\alpha(f \circ \pi) = (\Delta_\alpha f) \circ \pi, \quad \omega_{2\alpha}(f \circ \pi) = (\Delta_{2\alpha} f) \circ \pi,$$

$\pi : \ K \longrightarrow K/M$ being the natural mapping. According to [GGA], II, Theorem 4.9, we have with $\partial_i = \partial/\partial t_i$,

$$(\Delta_\alpha f)(\pi(g)) = \left[\sum_i \partial_i^2 \left((f \circ \pi)(g \exp \Sigma t_i T_i) \right) \right]_{t_i = 0} \tag{35}$$

and similarly for $\Delta_{2\alpha}$ (with S_j). The commutation relations

$$[\mathfrak{m} + \mathfrak{l}_{2\alpha}, \mathfrak{l}_\alpha] \subset \mathfrak{l}_\alpha, \quad [\mathfrak{l}_\alpha, \mathfrak{l}_\alpha] \subset \mathfrak{m} + \mathfrak{l}_{2\alpha}$$
$$[\mathfrak{m} + \mathfrak{l}_{2\alpha}, \mathfrak{m} + \mathfrak{l}_{2\alpha}] \subset \mathfrak{m} + \mathfrak{l}_{2\alpha}$$

([DS], VII, §11) show that if S is the analytic subgroup of K with Lie algebra $\mathfrak{m} + \mathfrak{l}_{2\alpha}$ then K/S is symmetric. The element ω_α is invariant under $Ad_K(S)$ so each S_j commutes with ω_α (cf. [GGA], II, Lemma 4.4). In particular, $\omega_{2\alpha}$ and ω_α, and therefore also $\Delta_{2\alpha}$ and Δ_α, commute.

Theorem 6.7. *(G/K of rank one). The algebra $\mathbf{D}(K/M)$ is commutative with ≤ 2 generators. In addition:*

(i) *If $\dim \mathfrak{l}_{2\alpha} > 1$ then $\mathbf{D}(K/M)$ is generated by Δ_α and $\Delta_{2\alpha}$.*

(ii) *If $\dim \mathfrak{l}_{2\alpha} = 1$ then $\mathbf{D}(K/M)$ is generated by Δ_α and S_1.*

(iii) *If $\dim \mathfrak{l}_{2\alpha} = 0$, $\dim \mathfrak{l}_\alpha > 1$ then $\mathbf{D}(K/M)$ is generated by Δ_α.*

(iv) *If $\dim \mathfrak{l}_\alpha = 1$ (so $\mathfrak{l}_{2\alpha} = 0$) then $\mathbf{D}(K/M)$ is generated by T_1.*

Proof. (i) Let $I((\mathfrak{l}_\alpha + \mathfrak{l}_{2\alpha})^*)$ denote the algebra of $Ad_K(M)$-invariant polynomial functions on $\mathfrak{l}_\alpha + \mathfrak{l}_{2\alpha}$. Because of Exercise **D3**, each $p \in I((\mathfrak{l}_\alpha + \mathfrak{l}_{2\alpha})^*)$ is constant on each product $S_{r_1}(0) \times S_{r_2}(0)$ where $S_{r_i}(0)$ is a sphere in $\mathfrak{l}_{i\alpha}$. Hence if $Z = \Sigma_i t_i T_i + \Sigma_j s_j S_j$ then

$$p(Z) = f((\Sigma_i t_i^2)^{\frac{1}{2}}, (\Sigma_j s_j^2)^{\frac{1}{2}}),$$

where $f(x, y)$ is defined for $x \geq 0, y \geq 0$. By the invariance of p

$$f(|t|, |s|) = p(\epsilon t T_1, \delta s S_1)$$

if $\epsilon^2 = \delta^2 = 1$ so $f(x, y)$ is a polynomial in x^2 and y^2. Hence each $Ad_K(M)$-invariant in the symmetric algebra $S(\mathfrak{l}_\alpha + \mathfrak{l}_{2\alpha})$ is a polynomial in $\Sigma_i T_i^2$ and $\Sigma_j S_j^2$. Thus, by [GGA], II, Cor. 4.10 and (35) above, each $D \in \mathbf{D}(K/M)$ is a polynomial in Δ_α and $\Delta_{2\alpha}$, proving (i). For case (ii) we use Exercise **D2** and the remark that here $Ad_K(M)$ acts trivially on $\mathfrak{l}_{2\alpha}$ ([DS], IX, Lemma 3.5).

From Theorem 2.9 we can now deduce the following result.

Corollary 6.8. *The convolution algebra $C^\natural(K)$ of M-bi-invariant continuous functions on K is commutative.*

5. An Additional Conical Distribution for $\boldsymbol{\lambda} = 0$.

In view of Theorems 5.15 and 6.5 we have determined all the conical distributions in $\mathcal{D}'_\lambda(\Xi)$ provided $\pi(\lambda)\boldsymbol{e}_s(\lambda) \neq 0$. We shall now sketch the method of Hu[1973], [1975] for dealing with this discrete set of λ for which $\pi(\lambda)\boldsymbol{e}_s(\lambda) = 0$.

Proposition 6.9. *A conical distribution Ψ is completely determined by its restriction to the open set $\overline{N}A \cdot \xi_o$.*

Proof. Because of the homogeneity condition

$$\Psi^{\tau(mn)\sigma(a)} = e^{-(i\lambda+\rho)(\log a)}\Psi$$

(cf. Prop. 4.3) the support of Ψ is invariant under the action $g \cdot \xi_o \longrightarrow mnga \cdot \xi_o$ of MNA on Ξ. If $\Psi | \overline{N}A \cdot \xi_o = 0$ then $\mathrm{supp}(\Psi) \subset m^* A \cdot \xi_o$ which however is impossible because of the invariance mentioned. This proves the proposition.

As Cor. 5.8 shows the analytic continuations $\Psi_{\lambda,s}$ and $\Psi_{\lambda,e}$ produce proportional conical distributions for $\lambda = 0$. We shall now construct an additional conical distribution in $\mathcal{D}'_o$. Let $Y_1, \dots, Y_p$ and $Z_1, \dots, Z_q$ be bases of $\mathfrak{g}_{-\alpha}$ and $\mathfrak{g}_{-2\alpha}$, respectively, orthonormal with respect to $-B(X, \theta Y)$. Fix $H \in \mathfrak{a}$ such that $\alpha(H) = 1$. We know from Theorem 4.1 that each conical distribution in $\mathcal{D}'_\lambda$ coincides on the open set $\Xi^* = MNA \cdot \xi^*$ with a constant multiple of the function ψ_λ given by

$$\psi_\lambda(mna \cdot \xi^*) = e^{(-i\lambda+\rho)(\log a)}. \tag{36}$$

Proposition 6.10. *For $\varphi \in \mathcal{D}(\Xi)$ let $\varphi_o \in \mathcal{D}(\Xi)$ be determined by*

$$\varphi_o(ka \cdot \xi_o) = \varphi(a \cdot \xi_o) \qquad k \in K, a \in A,$$

and ψ_o defined by (36). Define the functional Ψ_o by

$$\Psi_o(\varphi) = \int_\Xi (\varphi(\xi) - \varphi_o(\xi))\psi_o(\xi) d\xi, \quad \varphi \in \mathcal{D}(\Xi).$$

Then Ψ_o is a conical distribution in $\mathcal{D}'_o(\Xi)$ and not proportional to $\Psi_{o,e}$.

Proof. We first show that Ψ_o is well defined. Consider the coordinate system

$$\exp(\Sigma y_i Y_i) \exp(\Sigma z_j Z_j)(\exp tH) a \cdot \xi_o \longrightarrow (y_1, \dots, y_p, z_1, \dots z_q, t)$$

around $a \cdot \xi_o$. Since $\varphi - \varphi_o = 0$ on Ξ_e we have for small y_i, z_j, t

$$(\varphi - \varphi_o)(y, z, t) = \sum_i y_i f_i(y, z, t) + \sum_j z_j h_j(y, z, t),$$

where the f_i, h_j are smooth in y, z, t. Consider now the conical function

$$F(\xi) = e^{-2\alpha(\log a(\xi))}, \qquad \xi \in \Xi^*,$$

which by Corollary 6.2 is given by

$$F(y,z,t) = e^{2\alpha(\log a)} e^{2t} \left[c^2(\Sigma y_i^2)^2 + 4c(\Sigma z_j^2)\right].$$

Thus if $r > 0$ is sufficiently small $(\varphi - \varphi_o)(\xi)/F(\xi)^r$ is a continuous function on Ξ of compact support. By Theorem 4.1, $e^{(-2r\alpha+\rho)(\log a(\xi))}$ is locally integrable so the integral

$$\int_\Xi (\varphi(\xi) - \varphi_o(\xi))\psi_o(\xi)d\xi$$

$$\int_\Xi (\varphi - \varphi_o)(\xi)F(\xi)^{-r}e^{(-2r\alpha+\rho)(\log a(\xi))}d\xi$$

exists. Thus Ψ_o is a well defined functional on $\mathcal{D}(\Xi)$.

Next we define $\Psi_j \in \mathcal{D}'(\Xi)$ for $j = 1, 2, \dots$ by

$$\Psi_j(\varphi) = \int_\Xi (\varphi - \varphi_o)(\xi)e^{(\rho-\alpha/j)(\log a(\xi))}d\xi. \tag{37}$$

Since $\psi_o(\xi) = e^{\rho(\log a(\xi))}$ and the function

$$F(\xi)^{\frac{1}{2j}} = e^{-\frac{\alpha}{j}(\log a(\xi))}$$

is bounded on the support of $\varphi - \varphi_o$ uniformly in j we have by the dominated convergence theorem,

$$\Psi_o(\varphi) = \lim_{j\to\infty} \Psi_j(\varphi), \qquad \varphi \in \mathcal{D}(\Xi). \tag{38}$$

Now by (18) in Proposition 4.3 it is obvious that $\Psi_j \in \mathcal{D}'_{-\alpha/j}(\Xi)$ so by (38) and Schwartz [1966] Ch. III, Th. XIII, $\Psi_o \in \mathcal{D}'_o(\Xi)$. On the other hand, the function $\eta(\xi) = e^{(\rho-\alpha/j)(\log a(\xi))}$ is locally integrable (Theorem 4.1) so

$$\Psi_j(\varphi) = \int_\Xi \varphi(\xi)\eta(\xi)d\xi - \int_A \varphi(a\cdot\xi_o) \int_{K/M} \eta(ka\cdot\xi_o)e^{2\rho(\log a)}dk_M da$$

and now the MN-invariance of Ψ_j is obvious. Now (38) shows that Ψ_o is a conical distribution in $\mathcal{D}'_o(\Xi)$. Note that Ψ_j is a linear combination of $\Psi_{-\alpha/j,s}$ and $\Psi_{-\alpha/j,e}$.

6. Conical Distributions for the Exceptional λ.

Because of Prop. 6.9 we identify a conical distribution Ψ with its restriction to the open set $\overline{N}A \cdot \xi_o$ which we now identify with $\overline{N} \times A$. Let $\Psi \in \mathcal{D}'_\lambda$ be conical with support in $A \cdot \xi_o$. By Exercise **D1**, $\Psi = D\Psi_{\lambda,o}$ where D is a differential operator on Ξ transversal to $A \cdot \xi_o$ and in fact a polynomial in the vector fields on Ξ generated by $\overline{N}$. For which $\lambda \in \mathfrak{a}_c^*$ do such D (with order $D > 0$) occur? We shall now address this question.

Lemma 6.11. *There exist bases $Y_o, Y_1, \ldots, Y_{p-1}$ and $Z_1, \ldots, Z_q$ of $\mathfrak{g}_{-\alpha}$ and $\mathfrak{g}_{-2\alpha}$, respectively such that*

(i) $-\langle Y_i, \theta Y_j \rangle = 2\delta_{ij}\langle \alpha, \alpha \rangle$,

(ii) $-\langle Z_k, \theta Z_\ell \rangle = 2\delta_{k\ell}\langle \alpha, \alpha \rangle$,

(iii) $[Y_o, Y_j] = 0 \qquad q+1 \leq j \leq p-1$

(iv) $[Z_i, \theta Y_o] = 2Y_i \qquad 1 \leq i \leq q$.

Proof. We verify this by means of the classification and must therefore consider the following cases (notation as in [DS], X, §6, No. 2)

a) $(\mathfrak{g}, \mathfrak{k}) = (\mathfrak{f}_{4(20)}, \mathfrak{so}(9))$; b) $\mathfrak{g} = \mathfrak{so}(n, 1)$;

c) $\mathfrak{g} = \mathfrak{su}(n, 1)$; d) $\mathfrak{g} = \mathfrak{sp}(n, 1)$.

In case a) we have $p = m_\alpha = 8, q = m_{2\alpha} = 7$ so condition (iii) is vacuous. Now select

$$Y_o \in \mathfrak{g}_{-\alpha}, \qquad Z_1, \ldots, Z_7 \in \mathfrak{g}_{-2\alpha}, \quad H \in \mathfrak{a}$$

such that (ii) holds and

$$-\langle Y_o, \theta Y_o \rangle = 2/\langle \alpha, \alpha \rangle, \quad \alpha(H) = 1.$$

Then $\langle H, H \rangle = 1/\langle \alpha, \alpha \rangle$, $[Y_o, \theta Y_o] = 2H$, $[Z_i, \theta Z_i] = 4H$, $[Z_i, \theta Z_j] \in \mathfrak{m}$, $i \neq j$. Let

$$Y_i = \tfrac{1}{2}[Z_i, \theta Y_o] \quad i = 1, \ldots, 7.$$

Then if $1 \leq i \leq 7$,

$$-\langle Y_i, \theta Y_j \rangle = -\tfrac{1}{4}\langle [Z_i, \theta Y_o], [\theta Z_j, Y_o] \rangle,$$

which by the Killing form invariance becomes

$$-\tfrac{1}{4}\langle [Z_i, \theta Z_j], [\theta Y_o, Y_o] \rangle = 2\delta_{ij}/\langle \alpha, \alpha \rangle.$$

Similarly $\langle Y_o, \theta Y_i \rangle = 0$ so the lemma is verified for this case.

b) $\mathfrak{g} = \mathfrak{so}(n,1)$. Here $\mathfrak{g}_{-2\alpha} = 0, \mathfrak{g}_{-\alpha}$ abelian, so the lemma is trivial.

c) Here

$$\mathfrak{g} = \mathfrak{su}(n,1) = \left\{ \begin{pmatrix} A & {}^t\overline{V} \\ V & -Tr(A) \end{pmatrix} \middle| \begin{array}{c} A \text{ skew-Hermitian} \\ V \in \boldsymbol{C}^n \end{array} \right\}$$

and $\theta X = -{}^t\overline{X}$, $\langle X, Y\rangle = 2(n+1)\mathrm{Tr}(XY)$. The vector $H = E_{1\,n+1} + E_{n+1\,1}$ satisfies $\alpha(H) = 1$ for the indivisible restricted root $\alpha > 0$ and then

$$\mathfrak{g}_{-\alpha} = \{V_\alpha : \ V \in \boldsymbol{C}^{n-1}\},$$

where

$$V_\alpha = \begin{pmatrix} 0 & -\overline{V} & 0 \\ {}^tV & 0 & {}^tV \\ 0 & \overline{V} & 0 \end{pmatrix}. \tag{39}$$

$$\mathfrak{g}_{-2\alpha} = \{Z_t = it(-E_{11} - E_{1n+1} + E_{n+1\ 1} + E_{n+1\ n+1}) : \ t \in \boldsymbol{R}\}$$

(see [DS], IX, §3). Let E_i be the ith coordinate vector in $\boldsymbol{R}^{n-1}$. Then we take Z_1 as Z_t for $t = 1$ and put

$$Y_0, Y_1, \ldots, Y_{2n-3} = (E_1)_\alpha, (iE_1)_\alpha, \ldots, (iE_{n-1})_\alpha.$$

Then since $\langle \alpha, \alpha\rangle = 1/(4(n+1))$ the relations of the lemma are readily verified.

d) Here we use the set $\boldsymbol{H}$ of quaternions

$$\mathfrak{g} = \mathfrak{sp}(n,1) = \left\{ \begin{pmatrix} A & {}^t\overline{V} \\ V & -TrA \end{pmatrix} \middle| \begin{array}{c} A \text{ skew Hermitian, entries} \\ \text{in } \boldsymbol{H}, V \subset \boldsymbol{H}^n \end{array} \right\}.$$

Here $\overline{q} = a - b\boldsymbol{i} - c\boldsymbol{j} - d\boldsymbol{k}$ for a quaternion $q = a + b\boldsymbol{i} + c\boldsymbol{j} + d\boldsymbol{k}$. Again the Cartan involution θ is given by $\theta X = -{}^t\overline{X}$ and $\langle X, Y\rangle = 2(n+1)\mathrm{Tr}(XY)$ ([DS], III, §8). Let E_{ij}, H, α, E_i be defined as above. Then

$$\mathfrak{g}_{-\alpha} = \{V_\alpha : \ V \in \boldsymbol{H}^{n-1}\}$$

with V_α as in (39) and

$$\mathfrak{g}_{-2\alpha} = \boldsymbol{R}Z(\boldsymbol{i}) + \boldsymbol{R}Z(\boldsymbol{j}) + \boldsymbol{R}Z(\boldsymbol{k}),$$

where

$$Z(\epsilon) = \epsilon(-E_{11} - E_{1n+1} + E_{n+1\ 1} + E_{n+1\ n+1}), \qquad \epsilon = \boldsymbol{i}, \boldsymbol{j}, \boldsymbol{k}.$$

Put

$$Z_1 = Z(\boldsymbol{i}), Z_2 = Z(\boldsymbol{j}), Z_3 = Z(\boldsymbol{k})$$

$$Y_0 = (E_1)_\alpha, \qquad Y_1 = (\boldsymbol{i}E_1)_\alpha,\ Y_2 = (\boldsymbol{j}E_1)_\alpha, \qquad Y_3 = (\boldsymbol{k}E_1)_\alpha$$

and choose the other Y_m to be

$$(\epsilon E_\ell)_\alpha, \epsilon = 1, \boldsymbol{i}, \boldsymbol{j}, \boldsymbol{k}, \quad 2 \leq \ell \leq n-1.$$

Then the conditions of the lemma are satisfied.

Corollary 6.12. *Let* $T_i = [\theta Y_o, Y_i]$, $\quad 1 \leq i \leq p-1$. *Then*

(i) $T_i \in \mathfrak{m}, \quad 1 \leq i \leq p-1.$

(ii) $[T_i, Y_i] = -2Y_o, \quad q+1 \leq i \leq p-1.$

(iii) $[T_i, Y_i] = -6Y_o, \quad 1 \leq i \leq q.$

In fact $T_i \in \mathfrak{m}$ follows from $[\theta Y_o, Y_i] \subset \mathfrak{a} + \mathfrak{m}$ and part (i) of the lemma together with [DS], Ch. IX, §1, (8). The remaining statements follow from parts (iii), (iv) of the lemma, combined with the Jacobi identity.

Consider now the natural representation ν of G on $\mathcal{E}(\Xi)$ given by

$$\nu(g)f = f^{\tau(g)},$$

where $\tau(g)(xMN) = gxMN$. This induces a homomorphism ν of $\mathfrak{g}$ into the Lie algebra of vector fields on Ξ :

$$(\nu(X)f)(\xi) = \left\{\frac{d}{dt} f(\exp(-tX)\cdot\xi)\right\}_{t=0}. \tag{40}$$

As usual, ν extends to a homomorphism of the algebra $\boldsymbol{D}(G)$ into the algebra $\boldsymbol{E}(\Xi)$ of all differential operators on Ξ. For simplicity we shall write X instead of $\nu(X)$.

Lemma 6.13. *Let* $\Delta = \sum_{i=0}^{n-1} Y_i^2$, $\quad X_o = \theta Y_o$. *Then we have (with* $[A, B] = AB - BA$ *in* $\boldsymbol{D}(G)$,

(i) $[X_o, Y_o] = -2H, [X_o, Y_i] = T_i, 1 \leq i \leq p-1;$

$[X_o, Z_j] = -2Y_j, \quad 1 \leq j \leq q.$

(ii) $[T_i, \Delta] = 0, \quad 1 \leq i \leq p.$

(iii) $[H, \Delta^k] = -2k\Delta^k.$

(iv) $[X_o, \Delta] = 2Y_o(-2H - p - 2q + 2) + 2\sum_{i=1}^{p-1} Y_i T_i$

Proof. Part (i) was already proved; (ii) is clear since $Ad_G(M)$ preserves $\langle\ ,\ \rangle$; (iii) is straightforward since $Y_i \in \mathfrak{g}_{-\alpha}$ and (iv) comes from (i) and Corollary 6.12.

Proposition 6.14. *Assume the multiplicity $p = m_\alpha$ is > 1. Suppose a polynomial Q in Y_i $(0 < i \leq p-1)$ and Z_j $(1 \leq j \leq q)$ is an M-invariant differential operator. Then Q is a (commutative) polynomial in Δ and the Z_j.*

The proof is the same as that of Theorem 6.7.

Lemma 6.15. *Consider the differential operator*

$$D = \sum_{k, 2i_o + i_1 + \cdots + i_q = \ell} a_{k, i_o, i_1, \ldots, i_q} \Delta^{i_o} Y_k Z_1^{i_1} \ldots Z_q^{i_q} + P$$

where P is a polynomial (in the Y_i and Z_j) of degree $\leq \ell$ and the coefficients $a_{k,(i)}$ are scalars. If $(D\varphi)(e) = 0$ for all $\varphi \in \mathcal{D}(\overline{N})$ then each coefficient $a_{k,(i)} = 0$.

Proof. Consider the coordinate system

$$\exp\Big(\sum_i y_i Y_i + \Sigma_j z_j Z_j\Big) \longrightarrow (y_o, \ldots, z_q)$$

and let $X, Z \in \overline{\mathfrak{n}}$. Since $[\overline{\mathfrak{n}}, [\overline{\mathfrak{n}}, \overline{\mathfrak{n}}]] = 0$ we have

$$\begin{aligned} (X\varphi)(\exp Z) &= \left\{\frac{d}{dt}\varphi(\exp(-tX)\exp Z)\right\}_{t=0} \\ &= \left\{\frac{d}{dt}\varphi\big(\exp(Z + t(-X - \tfrac{1}{2}[X, Z]))\big)\right\}_{t=0}. \end{aligned} \tag{41}$$

Taking $X = Z_j$ this gives since $[Z_j, Z] = 0$,

$$(Z_j\varphi)(y_o, \ldots, z_q) = -\frac{\partial}{\partial z_j}\varphi(y_o, \ldots, z_q).$$

Next we take $X = Y_i$ and obtain

$$(Y_i\varphi)(y_o, \ldots, z_q) = -\Big(\frac{\partial}{\partial y_i} + \tfrac{1}{2}\sum_k y_k[Y_i, Y_k]\Big)\varphi(y_o, \ldots, z_q).$$

Hence

$$(\Delta^i Y_k Z_1^{i_1} \dots Z_q^{i_q} \varphi)(e) =$$
$$\left\{\left(\sum_r \frac{\partial^2}{\partial y_r^2}\right)^i \left(-\frac{\partial}{\partial y_k}\right)\left(-\frac{\partial}{\partial z_1}\right)^{i_1} \cdots \left(\frac{\partial}{\partial z_q}\right)^{i_q} \varphi\right\}(0) + (D_1\varphi)(0)$$

where D_1 is a polynomial in $\frac{\partial}{\partial y_i}$, $\frac{\partial}{\partial z_j}$ of degree less than that of the first term. Now the statement of the lemma follows by choosing φ suitably.

Definition. For $\lambda \in \mathfrak{a}_C^*$ let $\mathrm{Con}(\mathcal{D}'_\lambda)$ (respectively $\mathrm{Con}(\mathcal{D}'_{\lambda,e})$) denote the set of conical distributions in $\mathcal{D}'_\lambda$ (respectively the set of conical distributions in $\mathcal{D}'_\lambda$ with support in $\Xi_e = A \cdot \xi_o$).

Proposition 6.16. *Assume again* $p > 1$. *Suppose* $\mathrm{Con}(\mathcal{D}'_{\lambda,e}) \neq \boldsymbol{C}\Psi_{\lambda,e}$. *Then*

(i) $\lambda = i\ell\alpha$, $\ell \in \boldsymbol{Z}^+ - (0)$.

(ii) *Each* $\Psi \in \mathrm{Con}(\mathcal{D}'_{\lambda,e})$ *has the form*

$$\Psi = (c_1 D + c_2)\Psi_{\lambda,e},$$

where $D = \Delta^\ell + D_1, D_1$ *a polynomial in the* Y_i *and* Z_j *of degree* $< 2\ell$ *and* $c_1, c_2 \in \boldsymbol{C}$.

(iii) $\mathrm{Con}(\mathcal{D}'_{\lambda,e}) = \boldsymbol{C}\Psi_{\lambda,s} + \boldsymbol{C}\Psi_{\lambda,e}$.

Remark. We saw in §5, No. 3, particularly (28)–(33), how in the rank-one case the residues of the function $\lambda \longrightarrow \Psi'_{\lambda,s}$ are related to the residues of the distribution-valued function $\alpha \longrightarrow x_+^\alpha$ ([GGA], I, §2). Since these residues are given by derivatives of the delta function it is a priori clear that the values of $\Psi_{\lambda,s}$ at the poles of $\boldsymbol{d}_s(\lambda)$, i.e. $\lambda \in i\boldsymbol{Z}^+\alpha$, are certain transversal derivatives of $\Psi_{\lambda,e}$. The main point of Prop. 6.16 is that in this manner we obtain all conical distributions with support in Ξ_e.

Proof. Suppose $\Psi \in \mathrm{Con}(\mathcal{D}'_{\lambda,e})$ is not a multiple of $\Psi_{\lambda,e}$. Then by Exercise **D1** we have for a suitable polynomial Q',

$$\Psi = Q'(Y_o, \dots, Y_{p-1}, Z_1, \dots, Z_q)\Psi_{\lambda,e}. \tag{42}$$

Both Ψ and $\Psi_{\lambda,e}$ are MN-invariant so by the compactness of M we can take $Q'(Y_o, \dots, Z_q)$ in (42) invariant under $Ad_G(M)$, hence by Prop. 6.14 of the form $Q(\Delta, Z_1, \dots, Z_q)$. Recall now that if $Z \in \mathfrak{g}$, $(Z\Psi)(\varphi) = \Psi(Z^*\varphi)$ and this adjoint Z^* equals $-Z$ (since $d\xi$ is G-invariant). Now if $X \in \mathfrak{m} + \mathfrak{n}$ we have

$$X\Psi = 0, \qquad X\Psi_{\lambda,e} = 0. \tag{43}$$

Hence the element $X_o = \theta Y_o \in \mathfrak{g}_\alpha$ satisfies

$$[X_o, Q]\Psi_{\lambda,e} = X_o(Q\Psi_{\lambda,e}) - Q(X_o\Psi_{\lambda,o}) = 0 \tag{44}$$

and of course

$$H\Psi_{\lambda,e} = -(i\lambda + \rho)(H)\Psi_{\lambda,e}. \tag{45}$$

If we now write (44) out explicitly using the preceding bracket relations we will be led to a condition on λ. We obtain in this way

$$\begin{aligned}
[X_o, \Delta^k]\Psi_{\lambda,e} &= \sum_{j=1}^{k} \Delta^{j-1}[X_o, \Delta]\Delta^{k-j}\Psi_{\lambda,e} \\
&= 2\sum_{j=1}^{k} \Delta^{j-1}Y_o(-2H - p - 2q + 2)\Delta^{k-j}\Psi_{\lambda,e} \\
&= 2\sum_{j=1}^{k} \Delta^{j-1}Y_o\Delta^{k-j}(4(k-j) + 2 + 2i\lambda(H))\Psi_{\lambda,e} \\
&= 4\sum_{j=1}^{k} \Delta^{k-1}Y_o\,(2(k-j) + 1 + i\lambda(H))\,\Psi_{\lambda,e} + P\Psi_{\lambda,e}
\end{aligned}$$

where P is a polynomial in the Y_i and Z_j of degree $< 2k-1$. Suppose now the top degree terms of Q are

$$a_{i_oi_1\cdots_q}\Delta^{i_o}Z_1^{i_1}\dots Z_q^{i_q}, \qquad 2i_o + i_1 + \cdots + i_q = N.$$

We use the formula just proved and Lemma 6.13 to calculate the leading terms in $[X_o, Q]\Psi_{\lambda,e}$. Note in particular that $[X_o, \Delta^k]\Psi_{\lambda,e}$ involves derivatives of order $\leq 2k-1$. Therefore

$$\begin{aligned}
&[X_o, Q]\Psi_{\lambda,e} = \\
&\Big(\sum_{(i)} a_{i_oi_1\dots i_q}\sum_{k\geq 1}\Delta^{i_o}Z_1^{i_1}\dots[X_o, Z_k^{i_k}]\dots Z_q^{i_q} + P_1\Big)\Psi_{\lambda,e} \\
&= -2\sum_{(i)} a_{i_oi_1\dots i_q}\Big(\sum_{k\geq 1} i_k\Delta^{i_o}Y_kZ_1^{i_1}\dots Z_k^{i_k-1}\dots Z_q^{i_q}\Big)\Psi_{\lambda,e} + P_1\Psi_{\lambda,e}
\end{aligned}$$

where P_1 is a polynomial in the Y_i and Z_j of degree $< N$. Choose $g \in \mathcal{D}(A)$ with $\Psi_{\lambda,e}(g) = 1$ and for any $f \in \mathcal{D}(\overline{N})$ denote the function $(\overline{n}, a) \longrightarrow f(\overline{n})g(a)$ by $f \otimes g$. We have

$$\begin{aligned} 0 &= [X_o, Q](\Psi_{\lambda,e})(f \otimes g) \\ &= \Big[\Big(-2 \sum_{(i)} a_{i_o i_1 \dots i_q} \sum_{k \geq 1} i_k \Delta^{i_o} Y_k Z_1^{i_1} \dots Z_k^{i_k - 1} \dots Z_q^{i_q}\Big) f\Big](e) + (P_1 f)(e). \end{aligned}$$

Because of Lemma 6.15 (and the appearance of i_k as coefficients) we deduce that

$$a_{i_o i_1 \dots i_q} = 0 \qquad \text{if} \qquad 2i_o + i_1 + \dots + i_q = N, \quad i_1 + \dots + i_q > 0.$$

This means that the highest degree term in Q has the form $c\Delta^\ell$ $(2\ell = N, c \neq 0)$. Consider now the terms in Q of degree $2\ell - 1$:

$$b_{i_o i_1 \dots i_q} \Delta^{i_o} Z_1^{i_1} \dots Z_q^{i_q}, \quad 2i_o + i_1 + \dots + i_q = 2\ell - 1.$$

Then

$$\begin{aligned} [X_o, Q]\Psi_{\lambda,e} &= c[X_o, \Delta^\ell]\Psi_{\lambda,e} \\ &+ \Big(\sum_{(i)} b_{i_o i_1 \dots i_q} \sum_{k} \Delta^{i_o} Z_1^{i_1} \dots [X_o, Z_k^{i_k}] \dots Z_q^{i_q} + P_2\Big)\Psi_{\lambda,e} \\ &= 4c \sum_{j=1}^{\ell} \Delta^{\ell-1} Y_o (2(\ell - j) + 1 + i\lambda(H))\Psi_{\lambda,e} \\ &- 2\Big(\sum_{(i)} b_{i_o i_1 \dots i_q} \sum_{k \geq 1} i_k \Delta^{i_o} Y_k Z_1^{i_1} \dots Z_k^{i_k - 1} \dots Z_q^{i_q}\Big)\Psi_{\lambda,e} + P_3 \Psi_{\lambda,e}, \end{aligned}$$

where P_2 and P_3 are polynomials of degree $< 2\ell - 1$ in Y_i, Z_j. The left hand side being 0 we deduce again from Lemma 6.15 that

$$\sum_{j=1}^{\ell} \{2(\ell - j) + 1 + i\lambda(H)\} = 0,$$

which gives $i\lambda(H) = -\ell$ so $\lambda = i\ell\alpha$, proving (i).

In the case $\lambda = i\ell\alpha$, ℓ an integer > 0, $\Psi_{\lambda,s}$ is an element in $\mathrm{con}(\mathcal{D}'_{\lambda,e})$ not proportional to $\Psi_{\lambda,e}$. Hence by the above, $\Psi_{\lambda,s} = c\, D\Psi_{\lambda,e}$, where $c \neq 0, D = \Delta^\ell + D_1$ where D_1 is a polynomial in Y_i, Z_j of degree $< 2\ell$. For any $\Psi \in \mathrm{Con}(\mathcal{D}'_{\lambda,e})$, $\lambda = i\ell\alpha$, we have $\Psi = (c_1 \Delta^\ell + D_2)\Psi_{\lambda,e}$ so

$$\Psi - \frac{c_1}{c}\Psi_{\lambda,s} = (D_2 - c_1 D_1)\Psi_{\lambda,e}.$$

However, the degree of $D_2 - c_1 D_1$ is $< 2\ell$ so $\Psi - \frac{c_1}{c}\Psi_{\lambda,s}$ must be proportional to $\Psi_{\lambda,e}$ so (ii) and (iii) are proved.

For the case $p = 1$ we have $\mathfrak{g}_{-2\alpha} = 0$, $\mathfrak{g}_{-\alpha} = \boldsymbol{R}Y_0$ and the action of M on $\mathfrak{g}_{-\alpha}$ is trivial. Now if $\Psi \in$ Con $(\mathcal{D}'_{\lambda,e})$, (42) takes the form

$$\Psi = Q(Y_o)\Psi_{\lambda,e}. \tag{46}$$

Let ℓ denote the degree of Q such that $Q(Y_o) = \sum_{k=0}^{\ell} a_k Y_o^k$ with $a_\ell \neq 0$. Then

$$\begin{aligned}[][X_o, Y_o^k] &= \sum_{j=1}^{k} Y_o^{k-j}[X_o, Y_o]Y_o^{j-1} \\ &= \sum_{j=1}^{k} Y_o^{k-j}(-2H)Y_o^{j-1} = 2\sum_{j=1}^{k} Y_o^{k-j}Y_o^{j-1}(-H + j - 1)\end{aligned}$$

since $HY_o^j = Y_o^j H - jY_o^j$. Hence by (46)

$$\begin{aligned}[][X_o, Y_o^k]\Psi_{\lambda,e} &= 2\sum_{j=1}^{k} Y_o^{k-1}((i\lambda + \rho)(H) + k - 1)\Psi_{\lambda,e} \\ &= 2\left\{k(i\lambda(H) + \tfrac{1}{2}) + \tfrac{k}{2}(k-1)\right\} Y_o^{k-1}\Psi_{\lambda,e}.\end{aligned} \tag{47}$$

Now

$$[X_o, Q(Y_o)]\Psi_{\lambda,e} = X_o\Psi - Q(Y_o)X_o\Psi_{\lambda,e} = 0$$

so by (47) we have for $0 \leq k \leq \ell$.

$$a_k\left\{k(i\lambda(H) + \tfrac{1}{2}) + \tfrac{k}{2}(k-1)\right\} = 0. \tag{48}$$

Since $a_\ell \neq 0$ we deduce from (48) that

$$\lambda = \frac{i}{2}\ell\alpha \tag{49}$$

and then $a_k = 0$ for $0 < k < \ell$. On the other hand, if (49) holds then (47) shows that $Y_o^\ell\Psi_{\lambda,e}$ is a conical distribution. This proves the following result.

Proposition 6.17. *Assume* $p = \dim \mathfrak{g}_\alpha = 1$. *If* $\mathrm{Con}(\mathcal{D}'_{\lambda,e}) \neq \boldsymbol{C}\Psi_{\lambda,e}$ *then*

$$\lambda = \frac{i}{2}\ell\alpha, \quad \ell \in \boldsymbol{Z}^+ - (0).$$

If $\lambda = \frac{i}{2}\ell\alpha$ $(\ell \in \boldsymbol{Z}^+ - (0))$ *then* $Y_o^\ell \Psi_{\lambda,e}$ *and* $\Psi_{\lambda,e}$ *span* $\operatorname{Con}(\mathcal{D}'_{\lambda,e})$.

Remark. In the case $p = 1, X$ is the hyperbolic plane and the full isometry group $I(X)$ has two components. If we replace G by $I(X)$ then the new M contains an element m_o such that $Ad(m_o)Y_o = -Y_o$. Then the Q in (46), being M-invariant, will only appear for ℓ even and only those appear above. With this definition of G Proposition 6.16 would remain valid for $p = 1$.

Using Theorem 6.5 and Proposition 6.16 we can now extend the determination of $\operatorname{Con}(\mathcal{D}'_\lambda)$ in Theorem 5.15 to all $\lambda \in \mathfrak{a}^*_{\boldsymbol{C}}$.

Theorem 6.18. (Hu) *Suppose G/K has rank one and assume $p > 1$. Then*

$$\operatorname{Con}(\mathcal{D}'_\lambda) = \boldsymbol{C}\Psi_{\lambda,s} + \boldsymbol{C}\Psi_{\lambda,e} \qquad \text{if } \lambda \neq 0;$$
$$\operatorname{Con}(\mathcal{D}'_o) = \boldsymbol{C}\Psi_o + \boldsymbol{C}\Psi_{o,e},$$

so $\dim \operatorname{Con}(\mathcal{D}'_\lambda) = 2$ *for each* $\lambda \in \mathfrak{a}^*_{\boldsymbol{C}}$.

Proof. Assume first that $\lambda \notin i\boldsymbol{Z}^+\alpha$, in particular $\lambda \neq 0$. Then $\boldsymbol{d}_s(\lambda)^{-1} \neq 0$ and by Proposition 5.7, $\Psi_{\lambda,s} \neq 0$. Since

$$(\Psi_{\lambda,s})^{\tau(a)} = e^{(i\lambda-\rho)(\log a)}\Psi_{\lambda,s}$$
$$(\Psi_{\lambda,e})^{\tau(a)} = e^{(-i\lambda-\rho)(\log a)}\Psi_{\lambda,e}$$

it is clear that $\Psi_{\lambda,s}$ is not proportional to $\Psi_{\lambda,e}$. Now Prop. 6.16 shows that $\Psi_{\lambda,s} \notin \mathcal{D}'_{\lambda,e}$. Hence by Theorem 4.1, for any $\Psi \in \operatorname{Con}(\mathcal{D}'_\lambda)$ there exists a $c \in \boldsymbol{C}$ such that $\Psi - c\Psi_{\lambda,s}$ has support in Ξ_e, i.e. $\Psi - c\Psi_{\lambda,s} \in \operatorname{Con}(\mathcal{D}'_{\lambda,e})$, which by our assumption on λ equals $\boldsymbol{C}\Psi_{\lambda,e}$ (Prop. 6.16). This proves that

$$\operatorname{Con}(\mathcal{D}'_\lambda) = \boldsymbol{C}\Psi_{\lambda,s} + \boldsymbol{C}\Psi_{\lambda,e}, \qquad \lambda \notin i(\boldsymbol{Z}^+ - 0)\alpha. \tag{50}$$

Next consider the case $\lambda = 0$. Here $\Psi_o \notin \operatorname{Con}(\mathcal{D}'_{o,e})$ so again if $\Psi \in \operatorname{Con}(\mathcal{D}'_o)$ there exists a constant $c \in \boldsymbol{C}$ such that $\Psi - c\Psi_o \in \operatorname{Con}(\mathcal{D}'_{o,e})$ which by Prop. 6.16 equals $\boldsymbol{C}\Psi_{o,e}$. This settles the case $\lambda = 0$. It remains to prove the following result.

Lemma 6.19. *If* $\lambda \in i(\boldsymbol{Z}^+ - 0)\alpha$ *then*

$$\operatorname{Con}(\mathcal{D}'_\lambda) = \operatorname{Con}(\mathcal{D}'_{\lambda,e}).$$

Proof. Here it is convenient to exploit the simple fact that the product of a conical function and a conical distribution is another conical distribution (Proposition 4.5). Consider the conical function

$$\psi(\xi) = e^{-\epsilon\alpha(\log a(\xi))}, \qquad \xi \in \Xi^*,$$

where $\epsilon = 1$ if $q = 0$, $\epsilon = 2$ if $q > 0$. If $\Psi \in \mathrm{Con}(\mathcal{D}'_{i\ell\alpha}) - \mathrm{Con}(\mathcal{D}'_{i\ell\alpha,e})$ then by Theorem 4.1,

$$(\Psi|\Xi^*)(\xi) = c\, e^{(\ell\alpha+\rho)(\log a(\xi))}, \qquad c \neq 0$$

so

$$\psi\Psi \in \mathrm{Con}(\mathcal{D}'_{i(\ell-\epsilon)\alpha}) - \mathrm{Con}(\mathcal{D}'_{i(\ell-\epsilon)\alpha,e}).$$

Hence

$$\text{(51)} \qquad \mathrm{Con}(\mathcal{D}'_{i(\ell-\epsilon)\alpha}) = \mathrm{Con}(\mathcal{D}'_{i(\ell-\epsilon)\alpha,e}) \Longrightarrow \mathrm{Con}(\mathcal{D}'_{i\ell\alpha}) = \mathrm{Con}(\mathcal{D}'_{i\ell\alpha,e}).$$

We now invoke Theorem 5.15 and the simplicity criterion in Theorem 6.5. We have the following:

a) Assume $q = 0$. If $p \neq 2$ it is easy to see that $\boldsymbol{e}_s(i\alpha) \neq 0$. Consequently,

$$\mathrm{Con}(\mathcal{D}'_{i\alpha}) = \boldsymbol{C}\Psi_{i\alpha,s} + \boldsymbol{C}\Psi_{i\alpha,e} = \mathrm{Con}(\mathcal{D}'_{i\alpha,e})$$

so by (51)

$$\mathrm{Con}(\mathcal{D}'_{i\ell\alpha}) = \mathrm{Con}(\mathcal{D}'_{i\ell\alpha,e}),\ \ell \in \boldsymbol{Z}^+ - (0).$$

b) Assume $q \geq 1$. Again $\boldsymbol{e}_s(i\alpha) \neq 0$ and

$$\text{(52)} \qquad \mathrm{Con}(\mathcal{D}'_{i\alpha}) = \mathrm{Con}(\mathcal{D}'_{i\alpha,e}).$$

Furthermore if $p \neq 2$ then $\boldsymbol{e}_s(2i\alpha) \neq 0$ and

$$\text{(53)} \qquad \mathrm{Con}(\mathcal{D}'_{2i\alpha}) = \mathrm{Con}(\mathcal{D}'_{2i\alpha,e}).$$

From (51), (52) and (53) we now deduce

$$\mathrm{Con}(\mathcal{D}'_{i\ell\alpha}) = \mathrm{Con}(\mathcal{D}'_{i\ell\alpha,e}), \quad \ell \in \boldsymbol{Z}^+ - 0$$

for this case too. This proves Lemma 6.19 (and therefore Theorem 6.18) except for the cases when $p = 2$. Here $q = 0$ or $q = 1$. For Lemma 6.19 and Theorem 6.18 it just remains to prove the following special result.

Proposition 6.20. *Let α be the positive indivisible root.*

(i) *If* $G = \boldsymbol{SO}_o(1,3)$ *then* $\mathrm{Con}(\mathcal{D}'_{i\alpha}) = \mathrm{Con}(\mathcal{D}'_{i\alpha,e})$

(ii) *If* $G = \boldsymbol{SU}(1,2)$ *then* $\mathrm{Con}(\mathcal{D}'_{2i\alpha}) = \mathrm{Con}(\mathcal{D}'_{2i\alpha,e})$

This result being rather special we relegate its proof to a series of exercises at the end of this chapter following Hu[1973].

For the case left over from Theorem 6.18 we have the following result.

Theorem 6.21. *Suppose* $p = 1$ *so* G/K *is the hyperbolic plane. Then*

(i) *If* $\lambda \neq i(\ell - \frac{1}{2})\alpha, \quad \ell \in \boldsymbol{Z}^+$, *we have*

$$\begin{aligned} \mathrm{Con}(\mathcal{D}'_\lambda) &= \boldsymbol{C}\Psi_{\lambda,s} + \boldsymbol{C}\Psi_{\lambda,e}, \qquad \lambda \neq 0; \\ \mathrm{Con}(\mathcal{D}'_o) &= \boldsymbol{C}\Psi_o + \boldsymbol{C}\Psi_{o,e}. \end{aligned}$$

(ii) *If* $\lambda = i(\ell - \frac{1}{2})\alpha, \ell \in \boldsymbol{Z}^+$ *then*

$$\mathrm{Con}(\mathcal{D}'_\lambda) = \boldsymbol{C}\Psi_{\lambda,s} + \boldsymbol{C}Y_o^{2\ell-1}\Psi_{\lambda,e} + \boldsymbol{C}\Psi_{\lambda,e}.$$

Proof. Part (i) is clear from Theorem 6.5 and Proposition 6.16. For (ii) we have by the proof of Theorem 6.18 that if $\Psi \in \mathrm{Con}(\mathcal{D}'_\lambda)$ then for some $c \in \boldsymbol{C}$ $\Psi - c\Psi_{\lambda,s}$ has support in Ξ_e. Then (ii) follows from the second part of Proposition 6.17.

Remark. Note that in case (ii), $\mathrm{Con}(\mathcal{D}'_\lambda)$ is 3-dimensional, otherwise it has dimension 2. As remarked after Proposition 6.17, if we replace G by $I(X)$ then the distributions $Y_o^{2\ell-1}\Psi_{\lambda,e}$ are no longer conical so $\dim \mathrm{Con}(\mathcal{D}'_\lambda) = 2$. On the other hand this enlargement of G has no effect in the cases $p > 1$. Thus one can say in view of Theorems 5.16, 6.18 and 6.21 that the conjecture from §4, No. 1 that the set of conical distributions is parametrized by $\mathfrak{a}_c^* \times W$ is for the most part verified for general G/K and completely verified for G/K of rank one.

It is interesting that this complication with the case $X = \boldsymbol{H}^2$ comes up again in another context (see remarks before Lemma 3.9 in Chapter III).

EXERCISES AND FURTHER RESULTS

A. The Spaces $X = G/K$ and $\Xi = G/MN$.

1. Show that axioms (i) (ii) for homogeneous spaces in duality (Ch. I, §3) and the properties of Ch. I, Lemma 3.2 are satisfied for

$$X = G/K, \qquad \Xi = G/MN.$$

Show that G/MN is not in general reductive.

2. If $f \in \mathcal{D}^{\natural}(G)$ then

$$\int_{NA} f(an)\,dn\,da = \int_{NA} f(an)e^{2\rho(\log a)}\,da\,dn.$$

3. If G is connected and locally isomorphic to $\boldsymbol{SL}(2, \boldsymbol{R})$ then M equals the center of G.

4*. (Mostow) In the case of the complex ball

$$B^{+} = \left\{z \in \boldsymbol{C}^{n} : \ |z_1|^2 + \cdots + |z_n|^2 < 1\right\}$$

viewed as a symmetric space of $\boldsymbol{SU}(n,1)$ ([DS], X, Exercise D1) the horocycles (which are certain Euclidean ellipsoids) are limits of spheres (which are also certain Euclidean ellipsoids).

5*. Let $X = G/K$ be as usual and let $\mathcal{F}(X)$ denote the space of continuous functions on X decreasing faster than any function $x \longrightarrow e^{-kd(o,x)}$ $(k \geq 0)$, d denoting the distance. Let $\nu : \boldsymbol{D}(G) \longrightarrow \boldsymbol{E}(X)$ denote the homomorphism induced by the action of G on X ([GGA], II, §5, No. 7) and let

$$\mathcal{F}^{\infty}(X) = \{f \in \mathcal{E}(X) : \nu(D)f \in \mathcal{F}(X) \quad \text{for all } D \in \boldsymbol{D}(G)\}.$$

Suppose X has rank one, $G = KAN$ an Iwasawa decomposition, $a_t = \exp tH$ where H belongs to the positive Weyl chamber corresponding to N. The *interior* of the horocycle $N{\cdot}o$ is by definition the set $\bigcup_{t>0} Na_t{\cdot}o$. By transitivity of G on the set of geodesics γ this defines the interior of any horocycle. Let $\widehat{f}$ denote the X-ray transform of a function f on X.

Let ξ be a horocycle in X and suppose $f \in \mathcal{F}^{\infty}(X)$ satisfies

$$\widehat{f}(\gamma) = 0,$$

whenever the geodesic γ is disjoint from ξ. Then

$$f(x) = 0 \quad \text{for} \quad x \notin \text{interior of } \xi.$$

(cf. Helgason [1987]).

6*. For the usual positive definite form $(X,Y) \longrightarrow -B(X,\theta Y)$ on $\mathfrak{g}$ let $X_1,\ldots,X_{m_\alpha}$ be an orthonormal basis of the restricted root space $\mathfrak{g}_\alpha$ ($\alpha \in \Sigma^+$), and let $L_\alpha = \widetilde{X}_1^2 + \cdots + \widetilde{X}_{m_\alpha}^2$ where $\widetilde{X}_i$ is the left invariant vector field on N such that $(\widetilde{X}_i)_e = X_i$. We view L_α as a differential operator on X by $(L_\alpha f)(na\cdot o) = (L_\alpha)_n(f(na\cdot o))$. Prove that the Laplacian L_X on X is given by

$$(L_X f)(na\cdot o) = \left[\left(e^\rho L_{A^o} e^{-\rho} - \langle\rho,\rho\rangle + 2\sum_{\alpha\in\Sigma^+} e^{2\alpha} L_\alpha\right) f\right](na\cdot o)$$

(cf. Anderson [1979], Williams [1985], Faraut [1982b] and [GGA], p. 267).

7. If the symmetric space G/K has rank one then $\boldsymbol{D}(MN/M)$ is commutative.

8. As in §6 (where G/K has rank one) let $J : \overline{n} \longrightarrow \overline{n}(m^*\overline{n})$ be the inversion on $\overline{N} - \{e\}$. Show that

$$\frac{d(J\overline{n})}{d\overline{n}} = e^{-2\rho(B(m^*\overline{n}))},$$

which is explicitly given by Theorem 6.1.

9. Let $G_o = \boldsymbol{SU}(2,1)$ and let the notation be as in §6, No. 2.

(i) Show that Ad_{G_o} is a spherical representation and that

$$\boldsymbol{v} = \begin{pmatrix} i & 0 & 0 \\ 0 & i & 0 \\ 0 & 0 & -2i \end{pmatrix} \quad \text{and } \boldsymbol{e} = \begin{pmatrix} i & 0 & -i \\ 0 & 0 & 0 \\ i & 0 & -i \end{pmatrix}$$

are a spherical and conical vector, respectively.

(ii) Consider the symmetric space $X_o = Ad(G_o)\boldsymbol{v}$, and the horocycle space $\Xi_o = Ad(G_o)\boldsymbol{e}$ both imbedded in the Lie algebra $\mathfrak{g}_o = \mathfrak{su}(2,1)$. Consider the vector

$$H_o = \begin{pmatrix} 0 & 0 & 1 \\ 0 & 0 & 0 \\ 1 & 0 & 0 \end{pmatrix},$$

which spans $\mathfrak{a}_o$ and put $a_t = \exp tH_o$. The curves $Ad(a_t)\boldsymbol{v}$ and $Ad(a_t)\boldsymbol{e}$ are geodesics in X_o and Ξ_o, respectively. Show that $t \longrightarrow Ad(a_t)\boldsymbol{v}$ is a part of the intersection of X_o with a certain 2-plane through $\boldsymbol{v}$. It is a hyperbola and the curve $t \longrightarrow Ad(a_t)\boldsymbol{e}$ is a straight line parallel to one of the asymptotes.

10*. (Generalization) Let G/K be an irreducible Hermitian symmetric space. Let $v \neq 0$ be in the (one-dimensional) center of the Lie algebra of K and consider the imbedding

$$gK \longrightarrow Ad(g)v$$

of G/K into $\mathfrak{g}$. Show that each geodesic in G/K is a part of the intersection of G/K with a certain 2-plane in $\mathfrak{g}$.

B. The Conical Functions for $G = \boldsymbol{SL}(n+1, \boldsymbol{R})$.

With the notation of [GGA], V, Exercise A9 let $\alpha_i = e_i - e_{i+1}$ be the simple roots, whereby $\omega_r = e_1 + \cdots + e_r$ $(1 \leq r \leq n)$ are the fundamental weights (for the representations of $\boldsymbol{SL}(n+1, \boldsymbol{C})$.

(i) Show that the highest weights of the irreducible conical representations of $\boldsymbol{SL}(n+1, \boldsymbol{R})$ are the elements in the set

$$\sum_{i=1}^{n} \boldsymbol{Z}^+(2\omega_i)$$

(ii) Let ψ be a conical function and μ the highest weight of the corresponding representation π_ψ. Let

$$-s^*\mu = \sum_1^n k_i(2\omega_i), \qquad k_i \in \boldsymbol{Z}^+,$$

where s^* is as in §1.

Then

$$\psi(g \cdot \xi_o) = \psi(\xi^*) D_1(g)^{2k_1} D_2(g)^{2k_2} \ldots D_n(g)^{2k_n},$$

where $D_i(g)$ is the $i \times i$ minor in $g = (g_{k\ell})$ with $g_{n+1,1}$ in the lower left hand corner.

C. Hyperbolic Space; Inversion and Support Theorem.

1. The Inversion Formula for the Radon Transform on the Hyperbolic Space.

Let $\boldsymbol{H}^n$ be the hyperbolic space

$$\{x \in \boldsymbol{R}^n : \qquad |x| < 1\}$$

with the Riemannian structure

$$ds^2 = \frac{dx_1^2 + \cdots + dx_n^2}{(1 - x_1^2 - \cdots - x_n^2)^2}$$

whose sectional curvature is -4. Show that the horocycles are the $(n-1)$-spheres tangential to the boundary $|x| = 1$.

(i) Show that the induced metric on each horocycle is flat.

(ii) Let $\xi_o = N \cdot o$, $q \in \xi_o$ and $r = d(o, q)$ the distance between o and q in $\boldsymbol{H}^n$. Let $\rho = d'(o, q)$ be the distance between o and q in ξ_o. Show that

$$\rho = \sinh r$$

(cf. [GGA], p. 36) and that in the notation of (10) §3,

$$(\widehat{f})^\vee(p) = \Omega_{n-1} \int_o^\infty (M^r f)(p) \rho^{n-2} d\rho$$

where Ω_{n-1} is the area of the unit sphere in $\boldsymbol{R}^{n-1}$ and $(M^r f)(p)$ is the average of f on $S_r(p)$.

(iii) Let R and R^* denote the Radon transform and its dual for $\boldsymbol{H}^n$, where horocycles are replaced by totally geodesic hypersurfaces. Then (cf. Helgason [1959] or [GGA], Ch. I, with $\boldsymbol{H}^n$ of curvature -4),

$$R^*(Rf)(p) = \Omega_{n-1} 2^{2-n} \int_o^\infty (M^r f)(p) \mathrm{sh}^{n-2}(2r) dr.$$

Deduce that since $S_r(p)$ has area $2^{1-n}\mathrm{sh}^{n-1}(2r)\Omega_n$ we have

$$R^*(Rf)(p) = 2\frac{\Omega_{n-1}}{\Omega_n} \int_{\boldsymbol{H}^n} f(q) \mathrm{sh}^{-1}(2(d(p, q)) dq$$

(compare [GGA], Ch. I §2 (17)).

(iv) Deduce from (ii) and (iii) that if $n = 3$ then

$$(\widehat{f})^\vee = R^*(Rf)$$

and that if $n = 2$ then

$$(\widehat{f})^\vee(p) = \frac{1}{\pi}\int_{\boldsymbol{H}^2} f(q)\mathrm{sh}^{-1}(d(p,q))dq.$$

(v) Show that the radial part of the Laplacian L is

$$\Delta(L) = \frac{d^2}{dr^2} + 2(n-1)\coth(2r)\frac{d}{dr}.$$

(vi) Using the Darboux equation show that if $k \in \boldsymbol{Z}^+$ is even

$$(L + (k+2)(2n-k-4))\int_0^\infty \mathrm{sh}^k r\ \ \mathrm{sh}(2r)(M^r f)(p)dr$$

$$= -k(n-k-2)\int_0^\infty \mathrm{sh}^{k-2} r\,\mathrm{sh}(2r)(M^r f)(p)dr.$$

If $k = 0$ the right hand side should be replaced by $-2(n-2)f(p)$.

(vii) Iterating (vi) show that for n odd,

$$\big(L + (n-1)(n-1)\big)\big(L + (n-3)(n+1)\big)\dots\big(L + 2(2n-4)\big)\big(\widehat{f}\big)^\vee = cf$$

where c is a (computable) constant. As clear *a priori* from (iv) if $n = 3$ the formula coincides with that of [GGA] Ch. I, Theorem 4.5 for totally geodesic hypersurfaces.

2*. Support theorem for the horocycle transform on $\boldsymbol{H}^n$. Let d denote distance function on $\boldsymbol{H}^n$ and B a fixed ball in $\boldsymbol{H}^n$. Let f be a continuous function on $\boldsymbol{H}^n$ satisfying the conditions

(i) For each $m \in \boldsymbol{Z}^+$, $f(x)e^{md(o,x)}$ is bounded.

(ii) $\widehat{f}(\xi) = 0$ whenever the horocycle ξ lies outside B.

Then

$$f(x) = 0 \qquad \text{for} \qquad x \notin B.$$

D. Conical Distributions.

1. Let V be a manifold and W a closed submanifold carrying the induced topology, $X_1, \dots, X_p, Y_1, \dots, Y_q$ vector fields on V such that for each $v \in V$, $(X_i)_v, (Y_j)_v$ form a basis of V_v and for each $w \in W$ the $(Y_j)_w$ belong to W_w. Let T be a distribution on V having support in W. Then there exists unique distributions $T_{n_1 \dots n_p}$ on W such that T is a locally finite sum

$$T = \sum_n X_1^{n_1} \dots X_p^{n_p} (\widetilde{T}_{n_1} \dots n_p)$$

where the tilde denotes the extension of a distribution on W to V (Schwartz [1966], Bruhat [1956], p. 107).

2*. Let $\alpha \in \Sigma_0^+$. If $\dim \mathfrak{g}_{j\alpha} > 1$ then $\operatorname{Ad} M$ act transitively on the unit sphere in $\mathfrak{g}_{j\alpha}$ (relative to the form $B_\theta(X, Y) = -B(X, \theta X)$.)

3*. Assume $\Sigma^+ = \{\alpha, 2\alpha\}$ and that $\dim \mathfrak{g}_{2\alpha} > 1$. Let $X \in \mathfrak{g}_\alpha$ and M_o the centralizer of X in M. Then $\operatorname{Ad}(M_o)$ acts transitively on the unit sphere in $\mathfrak{g}_{2\alpha}$ (for B_θ) (cf. Kostant [1975a], Ch. II).

4. Let G/K have rank one. By Proposition 6.9 any $\Psi \in \operatorname{Con}(\mathcal{D}'_\lambda)$ is determined by its restriction to the open set $\overline{N}A \cdot \xi_o$. Prove that given any $\Psi \in \operatorname{Con}(\mathcal{D}'_\lambda)$ there exists a unique $S_\Psi \in \mathcal{D}'(\overline{N})$ such that for all $\varphi \in \mathcal{D}(\overline{N}A \cdot \xi_o)$,

$$\Psi(\varphi) = \int_{\overline{N}} \left(\int_A \varphi(\overline{n}a \cdot \xi_o) e^{(i\lambda+\rho)(\log a)} da \right) dS_\Psi(\overline{n}).$$

We write this in the form

$$\Psi = S_\Psi \otimes T_\lambda,$$

$T_\lambda \in \mathcal{D}'(A)$ being defined by

$$T_\lambda(F) = \int_A F(a) e^{(i\lambda+\rho)(\log a)} da.$$

Show that with Ψ_o as in Proposition 6.10

(a) $S_{\Psi_{\lambda,e}} = \delta$ (delta function at $e \in \overline{N}$).

(b) $S_{\Psi_o}(f) = \int_{\overline{N}} \left[f(\overline{n}) - f(e) e^{-\rho(H(\overline{n}))} \right] e^{-\rho(B(m^* \overline{n}))} d\overline{n}$.

5. (Proposition 6.20 for $G = \boldsymbol{SO}_o(1,3)$.) Here

$$\mathfrak{g} = \left\{ \begin{pmatrix} 0 & V \\ {}^tV & A \end{pmatrix} : \; V \in \boldsymbol{R}^3, \qquad A \quad 3 \times 3 \text{ skew symmetric} \right\}$$

The Cartan involution is $X \longrightarrow -{}^tX$ and we put

$$H = E_{12} + E_{21}, \qquad X = E_{13} + E_{31} + E_{23} - E_{32}, \qquad T = E_{34} - E_{43}$$
$$Y = -E_{13} - E_{31} + E_{23} - E_{32}, \qquad Z = -E_{14} - E_{41} + E_{24} - E_{42}.$$

Then $\alpha(H) = 1$, $\rho = \alpha$, $\mathfrak{g}_{-\alpha} = \boldsymbol{R}Y + \boldsymbol{R}Z$, $X \in \mathfrak{g}_\alpha$ and $\mathfrak{m} = \boldsymbol{R}T$. Suppose $\operatorname{Con}(\mathcal{D}'_{i\alpha}) \neq \operatorname{Con}(\mathcal{D}'_{i\alpha,e})$.

(i) Show that there is a $\Psi \in \operatorname{Con}(\mathcal{D}'_{i\alpha})$ such that if $\varphi \in \mathcal{D}(\Xi^*)$

$$\Psi(\varphi) = \int\limits_{\overline{N} \times A \cdot \xi_o} \varphi(\overline{n}a \cdot \xi_o) e^{-2\alpha(B(m^*\overline{n}))} d\overline{n} da.$$

(ii) Prove that with S_Ψ as in Problem **D4**,

$$S_\Psi(f) = \int\limits_{\boldsymbol{R}^2} f(\exp(uY + vZ))(u^2 + v^2)^{-2} du\, dv$$

if $f \in \mathcal{D}(\overline{N} - \{e\})$.

(iii) Consider for $f \in \mathcal{D}(\overline{N})$ the Taylor polynomial (with $f(u,v) = f(\exp(uY + vZ))$)

$$(Pf)(u,v) = uf_u(0,0) + vf_v(0,0) + \tfrac{1}{2}u^2 f_{uu}(0,0) + uvf_{uv}(0,0) + \tfrac{1}{2}v^2 f_{vv}(0,0)$$

and define $S \in \mathcal{D}'(\overline{N})$ by

$$S(f) = \int \left[f(u,v) - f(0,0) - \{(Pf)(u,v)/(1 + u^2 + v^2)\} \right] (u^2 + v^2)^{-2} dudv.$$

Show that if $\Delta = \frac{\partial^2}{\partial u^2} + \frac{\partial^2}{\partial v^2}$ then

$$S_\Psi = S + D\delta$$

where δ is the delta function at e and D is a polynomial in Δ.

(iv) Using

$$F\Psi \in \operatorname{Con}(\mathcal{D}'_0) = \boldsymbol{C}\Psi_0 + \boldsymbol{C}\Psi_{0,e}$$

where F is the conical function

$$F(\xi) = e^{-\alpha(\log a(\xi))} \qquad \xi \in \Xi^*,$$

show that D in (iii) equals $c\Delta + c'$ $(c, c' \in \boldsymbol{C})$ so

$$\Psi = (S + (c\Delta + c')\delta) \otimes T_{i\alpha}.$$

Since $\delta \otimes T_{i\alpha} = \Psi_{i\alpha,e}$ is conical we may assume Ψ in (i) of the form

$$\Psi = (S + c\,\Delta\delta) \otimes T_{i\alpha}.$$

(v) Show that in the decomposition $\overline{N}MAN$

$$\overline{n}(\exp sX \exp(uY + vZ)) = \exp(xY + yZ)$$

where

$$x = (u - s(u^2+v^2))/(1 + s^2(u^2+v^2) - 2su)$$
$$y = v/(1 + s^2(u^2+v^2) - 2su).$$

(vi) Let $n = \exp sX \in N$. By the invariance of Ψ,

$$h(s) = \Psi(\varphi^{\tau(n^{-1})}) - \Psi(\varphi) = 0, \quad s \in \boldsymbol{R}.$$

Choose $g \in \mathcal{D}(A)$ such that $T_{i\alpha}(g) = 1$ and $f \in \mathcal{D}(\overline{N})$ such that $f(u,v) = u$ near the origin. Show that for $\varphi(\overline{n}a \cdot \xi_o) = f(\overline{n})g(a)$,

$$h'(0) \neq 0,$$

contradicting the existence of Ψ.

6. (Proposition 6.20 for $G = \boldsymbol{SU}(2,1)$.) Here

$$\mathfrak{g} = \left\{ \begin{pmatrix} Z & {}^tV \\ \overline{V} & -Tr(Z) \end{pmatrix} : \ V \in \boldsymbol{C}^2, \ Z \quad 2 \times 2 \text{ skew Hermitian.} \right\}$$

The Cartan involution is $X \longrightarrow -{}^t\overline{X}$.

As in §6 No. 2, (5), consider the subspace

$$\mathfrak{a} = \boldsymbol{R}H, \qquad H = E_{13} + E_{31}$$

whereby the positive restricted indivisible root α is given by $\alpha(H) = 1$. Also $\rho = 2\alpha$. Then

$$\mathfrak{g}_\alpha = \boldsymbol{R}X_1 + \boldsymbol{R}X_2, \qquad \mathfrak{g}_{2\alpha} = \boldsymbol{R}X_3$$
$$\mathfrak{g}_{-\alpha} = \boldsymbol{R}Y_1 + \boldsymbol{R}Y_2, \qquad \mathfrak{g}_{-2\alpha} = \boldsymbol{R}Y_3$$

where

$$X_1 = \begin{pmatrix} 0 & 1 & 0 \\ -1 & 0 & 1 \\ 0 & 1 & 0 \end{pmatrix}, \qquad X_2 = \begin{pmatrix} 0 & i & 0 \\ i & 0 & -i \\ 0 & i & 0 \end{pmatrix}, \qquad X_3 = \begin{pmatrix} i & 0 & -i \\ 0 & 0 & 0 \\ i & 0 & -i \end{pmatrix}$$

and $Y_j = \theta X_j$ $(j = 1, 2, 3)$. Suppose now

$$\mathrm{Con}(\mathcal{D}'_{2i\alpha}) \neq \mathrm{Con}(\mathcal{D}'_{2i\alpha,e}).$$

(i) Show that there is a $\Psi \in \mathrm{Con}(\mathcal{D}'_{2i\alpha})$ such that if $\varphi \in \mathcal{D}(\Xi^*)$,

$$\Psi(\varphi) = \int\limits_{\overline{N}A} \varphi(\overline{n}a \cdot \xi_o) e^{-4\alpha(B(m^*\overline{n}))} d\overline{n}\, da.$$

Prove that if $\overline{n} = \exp(uY_1 + vY_2 + \frac{1}{2}wY_3)$ then

$$\Psi(\varphi) = \int \varphi(\exp(uY_1 + vY_2 + \tfrac{1}{2}wY_3)a \cdot \xi_o)((u^2 + v^2)^2 + w^2)^{-2} du\, dv\, dw$$

(ii) Consider for $f \in \mathcal{D}(\overline{N})$, $f(u, v, w) = f(\exp(uY_1 + vY_2 + \frac{1}{2}wY_3)$ the Taylor polynomial $P(f)$ at $(u, v, w) = 0$ (with $f(0, 0, 0)$ removed) up to fourth order (giving u and v degree 1, w degree 2). Define $S \in \mathcal{D}'(\overline{N})$ by

$$S(f) = \int \left[f(u, v, w) - f(0, 0, 0) - \{P(f)/(1 + (u^2 + v^2)^2 + w^2)^3\}\right] d\mu$$

where $d\mu = \left[(u^2 + v^2)^2 + w^2\right]^2 du\, dv\, dw$. Show that, writing $\Psi = S_\Psi \otimes T_{2i\alpha}$ according to Exercise **D4**, we have

$$S_\Psi = S + D\delta$$

where D is a polynomial in

$$\Delta = \frac{\partial^2}{\partial u^2} + \frac{\partial^2}{\partial v^2} \quad \text{and} \quad \frac{\partial}{\partial w}$$

(iii) Consider the conical function on Ξ given by

$$F(\xi) = e^{-2\alpha(\log a(\xi))} \qquad \xi \in \Xi^*.$$

Show that

$$F(D\delta \otimes T_{2i\alpha}) = c\Psi_{0,e} \qquad c \in \boldsymbol{C}$$

and deduce that

$$D = a_1\Delta^2 + a_2\Delta + a_3\Delta\frac{\partial}{\partial w} + a_4\frac{\partial^2}{\partial w^2} + a_5.$$

(iv) With $n = \exp sX_1$, $\overline{n} = \exp(uY_1 + vY_2 + \frac{1}{2}wY_3)$ let $\overline{n}(n\overline{n}) = \exp(xY_1 + yY_2 + \frac{1}{2}zY_3)$. Show that

$$x = \left[(1 - 2su + s^2(u^2+v^2))(u - s(u^2+v^2)) - (v - sw)(2sv - s^2w)\right] J$$
$$y = \left[(1 - 2su + s^2(u^2+v^2))(v - sw) + (u - s(u^2+v^2))(2sv - s^2w)\right] J$$
$$z = \left[(1 - 2su + s^2(u^2+v^2))w + (u^2+v^2)(2sv - s^2w)\right] J$$

and

$$(x^2+y^2)^2 + z^2 = ((u^2+v^2)^2 + w^2)J$$

where

$$J^{-1} = (1 - 2su + s^2)(u^2+v^2)^2 + (2sv - s^2w)^2$$

(v) Choose $f \in \mathcal{D}(\overline{N})$ such that $f(u,v,w) = u^3$ near the origin. Let

$$f^s(\overline{n}) = f(\overline{n}(\exp sX_1 \cdot \overline{n})).$$

By the N-invarinace of Ψ,

$$h(s) = (S + D\delta)(f^s) - (S + D\delta)(f^s) \equiv 0.$$

Show however, by direct computation, that

$$h'(0) \neq 0$$

contradicting the existence of Ψ.

E*. The Heisenberg Group $\boldsymbol{H}_n$.

This group is the set $\boldsymbol{C}^n \times \boldsymbol{R}$ with the multiplication

$$(z,t)(z',t') = (z + z', t + t' + Im(z^*z')),$$

where $z = \sum_{i=1}^{n} z_i E_{i1}$ (column vector), $z^* = \sum_{i=1}^{n} \overline{z}_i E_{1i}$ (row vector). We put $|z| = (|z_1|^2 + \cdots + |z_n|^2)^{\frac{1}{2}} = (z^*z)^{\frac{1}{2}}$.

1. The group $\boldsymbol{H}_n$ can be realized as the group of matrices

$$\begin{pmatrix} 1 + \frac{1}{2}|z|^2 + it & z^* & -\frac{1}{2}|z|^2 - it \\ z & I_n & z \\ \frac{1}{2}|z|^2 + it & z^* & 1 - \frac{1}{2}|z|^2 - it \end{pmatrix},$$

with Lie algebra $\mathfrak{h}_n \subset \mathfrak{sl}(n+2, \boldsymbol{C})$ given by

$$\begin{pmatrix} it & z^* & -it \\ z & 0 & -z \\ it & z^* & -it \end{pmatrix}, z \in \boldsymbol{C}^n, \quad t \in \boldsymbol{R}.$$

Show that $\boldsymbol{H}_n$ is isomorphic to the group $\overline{N}$ in the Iwasawa decomposition of $\boldsymbol{SU}(1, n+1)$ (cf. [DS], Ch. IX, §3). Show also that if $U \in \boldsymbol{U}(n)$ the mapping $(z, t) \longrightarrow (Uz, t)$ is an automorphism of $\boldsymbol{H}_n$.

2. For $X \in \mathfrak{h}_n$ let $\widetilde{X}$ denote the corresponding left invariant vector field on $\boldsymbol{H}_n$. In the basis X_j, $Y_j (1 \le i \le n), T$ of $\mathfrak{h}_n$ given by

$$X_j = E_{1j+1} + E_{j+11} + E_{n+2j+1} - E_{j+1n+2}$$
$$Y_j = -iE_{1j+1} + iE_{j+11} - iE_{n+2j+1} - iE_{j+1n+2}$$
$$T = E_{11} - E_{1n+2} + E_{n+21} - E_{n+2n+2},$$

we have

$$\widetilde{X}_j = \frac{\partial}{\partial x_j} - y_j \frac{\partial}{\partial t}, \qquad \widetilde{Y}_j = \frac{\partial}{\partial y_j} + x_j \frac{\partial}{\partial t}, \qquad \widetilde{T} = \frac{\partial}{\partial t}.$$

Then the *sub-Laplacian*

$$L_o = \sum_{j=1}^{n} (\widetilde{X}_j^2 + \widetilde{Y}_j^2)$$

equals

$$L_o = \sum_{j=1}^{n} \left[\frac{\partial^2}{\partial x_j^2} + \frac{\partial^2}{\partial y_j^2} + 2x_j \frac{\partial^2}{\partial y_j \partial t} - 2y_j \frac{\partial^2}{\partial x_j \partial t} + (x_j^2 + y_j^2) \frac{\partial^2}{\partial t^2} \right].$$

The operators L_o and $\widetilde{T}$ generate the subalgebra of $\boldsymbol{D}(H_n)$ of operators invariant under the action of $\boldsymbol{U}(n)$ in **1**. Compare the algebra $\boldsymbol{D}(MN/M)$ in [GAA], Exercise B10 in Ch. IV.

3. Let

$$Q(z, t) = |z|^4 + 4t^2, \qquad (z, t) \in \boldsymbol{H}_n,$$

which is essentially a power of $e^{\rho(B(m^* \overline{n}))}$ from Theorem 6.1. (The mapping J in **A8** is a kind of an inversion in the unit sphere in this norm.) Show that

$$\|(z, t)\| = Q(z, t)^{\frac{1}{4}}$$

has the homogeneity

$$\|s(z,t)\| = |s|\|(z,t)\| \qquad\qquad s \in \boldsymbol{R}$$

and that, defining for $g_1, g_2 \in \boldsymbol{H}_n$

$$d(g_1, g_2) = \|g_2^{-1} g_1\|,$$

d is a left-invariant metric on $\boldsymbol{H}_n$.

4. In $\boldsymbol{R}^n$ the function $Q_o(X) = |X|^2$ satisfies

$$L_{\boldsymbol{R}^n} Q_o^\lambda = 2\lambda(2\lambda + n - 2) Q_o^{\lambda-1}.$$

Prove that the function Q in **E3** satisfies

$$L_o Q(z,t)^\lambda = 8\lambda(2\lambda + n)|z|^2 Q(z,w)^{\lambda-1}$$

and

$$(L_o^2 + 4(2\lambda+n)^2 \frac{\partial^2}{\partial t^2}) Q(z,t)^\lambda = 64\lambda^2(2\lambda+n)(2\lambda+n-1) Q(z,t)^{\lambda-1}.$$

5. In $\boldsymbol{R}^n (n > 2)$ the function

$$G_o(X) = \frac{1}{(2-n)\Omega_n} |X|^{2-n}$$

is a fundamental solution of $L_{\boldsymbol{R}^n}$,

$$L_{\boldsymbol{R}^n} G_o = \delta_o.$$

For $\boldsymbol{H}_n$ the function

$$G(z,t) = -c_n(|z|^4 + 4t^2)^{-\frac{1}{2}n}, \qquad c_n = \frac{1}{\Omega_n^2} \frac{2^{n-1}}{\pi},$$

satisfies

$$L_o G = \delta.$$

6. The Kelvin transformation in $\boldsymbol{R}^n$ is defined by

$$(Kf)(X) = |X|^{2-n} f(\frac{X}{|X|^2}).$$

Then, if $L = L_{\boldsymbol{R}^n}$,

$$(LKf)(X) = |X|^{-4}(KLf)(X), \qquad f \in \mathcal{D}(\boldsymbol{R}^n - (0)).$$

For $\boldsymbol{H}_n$ one defines similarly, using the inversion J in §6, (12) (15),

$$(Kf)(z,t) = (|z|^4 + 4t^2)^{-\frac{n}{2}} f\left(\frac{-z}{|z|^2 + 2it}, \frac{-t}{|z|^4 + 4t^2}\right).$$

Then

$$(L_o Kf)(z,t) = (|z|^4 + 4t^2)^{-1}(KL_o f)(z,t).$$

NOTES

This chapter is for the most part based on the author's papers [1963], [1964b], [1965b], [1966a] and [1970a].

§§1–2. Horocycles are classical objects for hyperbolic spaces $\boldsymbol{H}^n$ and are considered for any homogeneous space G/C (G complex semisimple, C any closed subgroup) in Gelfand and Graev [1959], [1964] as orbits of the conjugates of the nilpotent group N. The study of the horocycle space Ξ for a general symmetric space $X = G/K$ of the noncompact type, the identifications $\Xi = G/MN$, $MN\backslash\Xi = A \times W$ and $\boldsymbol{D}(G/MN) = \boldsymbol{D}(A)$ (Theorems 1.1 and 2.2 and Prop. 1.5) are from the author's paper [1963] mentioned above, whereas the discussion of the geodesics in Ξ and the identification of Ξ' for a boundary component $X' \subset X$ (Prop. 1.13) are from [1970a]. The method will also be useful in Ch. III, §4.

As a generalization of the space of spherical harmonics of a given degree, the construction of the general eigenspace representations for any homogeneous space L/H was proposed by the author in [1970b]. Some cases will be studied in detail in Ch. III, §6, §8, and in Ch. VI where the role of the commutativity of $\boldsymbol{D}(L/H)$ will be made clear. The characterization of the joint eigenfunctions on X and on Ξ in Proposition 2.6 is from Helgason [1962a] and [1970a]. The expansion of the mean-value operator (Theorem 2.7) generalizing the so-called Pizetti's formula in $\boldsymbol{R}^n$ (cf. (28)§2 and John [1955], p. 88) and the commutativity equivalence (Theorem 2.9) are from the author's paper [1959] and book [1984a] p. 485, respectively. A more precise form of Cor. 2.8 is given in Rouvière [1991a], Theorem 4.2.

§3. The injectivity of the Radon transform on $L^1(X)$ (Theorem 3.2) is proved in Helgason [1970a]. It rests on the boundedness criterion for the spherical function φ_λ proved by Helgason and Johnson [1969]. It was proved by Gelfand and Naimark [1948] for $\boldsymbol{SL}(n, \boldsymbol{C})$ and by Harish-Chandra [1954] for complex semisimple Lie groups G that a function on G can be determined in terms of its integrals over the conjugacy classes and their translates. It is pointed out in Gelfand and Graev [1959] how this implies that a function of compact support on the symmetric spaces G/K (for G complex) can be determined in terms of its integrals over the horocycles in G/K. This proof does not extend to real groups G. The inversion formula for the general symmetric space G/K as well as the Plancherel formula for the Radon transform is from the author's papers [1963], [1964b], [1965b] and [1970a]. Several examples were worked out in Gindikin and Karpelevič [1964] and [1981]. For further work on the explicit inversion of the Radon transform of K-invariant functions (the "Abel Transform") see Meeney [1986] and Beerends [1987], [1988]. A natural extension of the Radon transform to sections of vector bundles over X is carried out in Branson, Ólafsson and Schlichtkrull [1993b]. The analog of the Darboux equation (Theorem 3.12) was proved by the author in [1959]; the simplified proof in the text (using (33) §3) was kindly communicated to me by Koornwinder. Prop. 3.14 on the eigenfunctions $e_{\lambda,b}$ was observed in Karpelevič [1965] §17.10 and in Helgason [1965b]. Theorem 3.16 is from Harish-Chandra [1958a] in the case when $F \equiv 1$ and for general F in Helgason [1970a], p. 129. More general versions are given by Vogan [1981] (Lemma 4.2.15) and Wallach [1988] (Theorem 5.3.4) in connection with Langlands' classification.

The connection between $\varphi \longrightarrow \overset{\vee}{\varphi}$ and the Poisson kernel is discussed in the author's paper [1966a]; relation (50) is from Furstenberg [1963].

§4. The theory of conical functions, conical distributions and their connection with representation theory is developed in Helgason [1966a] and [1970a]. In particular, the identity of the finite-dimensional conical representations and spherical representations is first stated and proved in these papers. Generalizations were given by Schlichtkrull [1984a] and Johnson [1987a]. The motivation of conical distribution was the analogy with spherical functions based on the duality between X and Ξ, which for example suggests the conjecture in §4, No. 1. While the analogy is very close so far as conical functions are concerned (cf. Theorems 4.7, 4.12, and 4.17 and Propositions 4.15–4.16) the conical distributions offer more far-fetched analogies like Theorem 3.10 in Ch. VI. The imbedding results in No. 5 are from Helgason [1991]. In the proof of Lemma 4.18 (i) and Proposition 4.25 I have used some simplifying remarks by Schlichtkrull who also observed the fact in Remark 2 at the end of No. 5. Proposition 4.24 is due to Harish-Chandra [1958b].

§5. The construction of the conical distributions (Theorems 5.3, 5.4 and 5.16) is from Helgason [1970a]; the proof of Theorem 5.3 utilizes, among other things, a method of Schiffmann [1971], which in turn is an extension of the reduction technique of Gindikin and Karpelevič in [1962]. An interesting relationship between conical vectors and Whittaker vectors (representation space vectors transforming under N according to a character) is established in Goodman-Wallach [1980]. Thereby our determination of the conical distributions (Theorem 5.16) has applications to Whittaker vector theory. An extension of the conical distribution theory to p-adic groups has been carried out by Williams [1978].

§**6.** The $\boldsymbol{SU}(2,1)$-reduction and the ensuing component computation in No. 1–2 is from the author's paper [1970a]. Parts (ii) and (iii) in Theorem 6.1 were independently obtained by Schiffmann [1971], using different methods not involving $\boldsymbol{SU}(2,1)$. The inversion J from No. 2 was introduced independently in these two papers. It has later been used in analysis on the Heisenberg group see e.g. Korányi [1982b]. The simplicity criterion is from Helgason [1970a], Ch. IV using Theorem 5.17; the case $\lambda \neq 0, \lambda \in \mathfrak{a}^*$, in Lemma 6.6 was originally done using methods from Bruhat [1956], combined with Theorem 5.17, and the case $\lambda = 0$ by a method proposed by Coifman. The proof of Lemma 6.6 in the text (which does both cases simultaneously) is due to Dadok [1976]. Proposition 6.4 is from the author's paper [1973a]. As observed by Lepowsky [1975], and Flensted-Jensen and Koornwinder [1979a], Theorem 6.7 is a simple consequence of Kostant's theorem in Exercise **D3**. The results from No. 5–6 determining the conical distributions in the rank-one case not covered by Theorem 5.16 are due to Hu ([1973] and [1975]). Related algebraic results were obtained by Lepowsky [1975]. For the case $G = \boldsymbol{SO}_0(n,1)$ see also Harzallah [1975] and Strasburger [1984] and for a certain extension to $\boldsymbol{O}(p,q)$, Faraut and Harzallah [1984]. See also Farah and Kamoun [1990] for further work in this area.

CHAPTER III

THE FOURIER TRANSFORM ON A SYMMETRIC SPACE.

In this chapter we define a Fourier transform $f \longrightarrow \widetilde{f}$ on a symmetric space of the noncompact type and investigate its operational properties.

There are two natural viewpoints towards the definition of a Fourier transform on $X = G/K$:

(i) Since the Bessel transform

$$F(r) \longrightarrow \widetilde{F}(\lambda) = \int_0^\infty F(r) J_0(r\lambda) r\,dr$$

is the Fourier transform on $\boldsymbol{R}^2$ specialized to radial functions and since the symmetric space analog of the Bessel transform is the spherical transform

$$F(a) \longrightarrow \widetilde{F}(\lambda) = \int_A F(a) \varphi_{-\lambda}(a) \delta(a)\,da$$

the Fourier transform $f \longrightarrow \widetilde{f}$ on X should reduce to the spherical transform when specialized to K-invariant functions.

(ii) In $\boldsymbol{R}^n$ the Fourier transform is the one-dimensional Fourier transform of the Radon transform (Ch. I, §2, (1)). By analogy, the Fourier transform on $X = G/K$ should be the flat Fourier transform (along A) of the Radon transform.

Fortunately, both of these requirements (i) and (ii) are satisfied by the same definition (4) §1 of $f \longrightarrow \widetilde{f}$ (see §5, No. 3). This Fourier transform converts functions f on X to functions $\widetilde{f}$ on $\mathfrak{a}^* \times K/M$. The inversion formula for this transform comes from combining the inversion of the spherical transform on G with a certain surprising symmetry property of the spherical function (Theorem 1.1). The Plancherel formula equates the L^2 norm of f on X with the L^2 norm of $\widetilde{f}$on $\mathfrak{a}_+^* \times K/M$ relative to a certain explicit measure. Since $A^+ \times K/M \sim X$ via the "polar coordinate" decomposition of X the symmetric space can be considered as self-dual under the Fourier transform just as is the case for Euclidean space $\boldsymbol{R}^n$.

Using series expansions for the Eisenstein integrals we obtain a Paley-Wiener type theorem for this Fourier transform, that is an intrinsic characterization of the range $\mathcal{D}(X)^\sim$. In order to get a characterization with intrinsic topologies on both sides it is more effective to specialize to functions on X of a given K-type. This in turn gives an explicit representation of all K-finite joint eigenfunctions of the invariant differential operators on X as Poisson transforms of K-finite functions on

K/M. Coupled with the explicit simplicity criterion in §4 (the injectivity criterion for the Poisson transform) the topological Paley-Wiener type theorem also gives an explicit irreducibility criterion for the eigenspace representations on $X = G/K$.

The analog of this theory for the tangent space X_0 is developed in §§7–8. Because this case is in some sense degenerate, the Eisenstein integral theory is no longer as useful so the methods employed for X do not have workable analogs for X_0. Instead we work out the theory for the flat space by considering it as a limiting case of the curved space X. While the resulting Paley-Wiener theorems (Theorems 7.7 and 5.11) show certain analogies there is an important difference which emerges in the definition of $\mathcal{H}^\delta$ in (50) §5 and the definition of $\mathcal{K}^\delta$ in (29) §7. In the first case the divisibility is a consequence of the Weyl group invariance, in the second case it is not. The simplicity criteria (Theorem 4.4 and 8.2) and the description of the K-finite joint eigenfunctions (Theorems 6.1 and 8.5) also show certain differences.

In §9 we show how the compact case and the noncompact case can be viewed as analytic continuations of each other and in §10 we use the Fourier analysis on X to relate the G-invariant differential operators on X to the bi-invariant differential operators on G.

In §11 we specialize to the case of G/K of rank one and work out an explicit formula for the Eisenstein integral and put the Paley Wiener type theorem in a more concrete form.

In §12 we develop the spherical transform theory further than in [GGA], Ch. IV, using functional analytic methods and positive definite functions.

§1. THE INVERSION AND THE PLANCHEREL FORMULA.

1. The Symmetry of the Spherical Function.

Again let G be a connected semisimple Lie group with finite center, $X = G/K$ the associated symmetric space. We shall preserve the notation from Chapter II, in particular the measure normalizations from §3, No. 1.

Consider first the Fourier transform on $\boldsymbol{R}^n$ in polar coordinate form

$$\widetilde{F}(\lambda\omega) = \int_{\boldsymbol{R}^n} F(x) e^{-i\lambda(x,\omega)}\,dx, \qquad |\omega| = 1, \lambda \in \boldsymbol{R}. \tag{1}$$

Geometrically, the scalar product (x, ω) represents the (signed) distance from the origin to the hyperplane passing through x, having unit normal ω. As pointed out in Chapter II, §3, No. 4, the symmetric space analog of this scalar product is the vector $A(x, b) \in \mathfrak{a}$ which is the composite distance from the origin $o = \{K\}$ in X to the horocycle $\xi(x,b)$ through $x \in X$ with normal $b \in B = K/M$. In terms of the Iwasawa decompositions $G = NAK = KAN$ we have written

$$g \in N \exp A(g)K, \quad g \in K \exp H(g)N, \tag{2}$$

with $A(g), H(g) \in \mathfrak{a}$ uniquely determined and then

$$A(g) = -H(g^{-1}), \quad A(gK, kM) = A(k^{-1}g). \tag{3}$$

By the formal geometric analogy between (x,ω) and $A(x,b)$ we are led to the following definition.

Definition. If f is a function on X the *Fourier transform* $\widetilde{f}$ is defined by

$$\widetilde{f}(\lambda, b) = \int_X f(x)e^{(-i\lambda+\rho)(A(x,b))}\,dx \tag{4}$$

for all $\lambda \in \mathfrak{a}^*_{\boldsymbol{C}}$, $b \in B$ for which this integral exists.

Analogies. The definition (4) comes from the formal analogy between (x,ω) and $A(x,b)$; the ρ in (4) is put there for formal technical reasons. However there is also a more conceptual analogy between (1) and (4). For $\lambda \in \boldsymbol{C} - \{0\}$ the functions

$$e_{\pm\lambda,\omega} : x \longrightarrow e^{\pm i\lambda(x,\omega)} \qquad x \in \boldsymbol{R}^n$$

span precisely the space of eigenfunctions of the Laplacian L which are constant on hyperplanes with normal ω with eigenvalue $-\lambda^2$:

$$Le_{\pm\lambda,\omega} = -\lambda^2 e_{\pm\lambda,\omega}. \tag{5}$$

On the other hand the functions

$$e_{s\lambda,b} : x \longrightarrow e^{(is\lambda+\rho)(A(x,b))} \qquad x \in X$$

for $s \in W$, are constant on each horocycle with normal b, and satisfy the system

$$De_{s\lambda,b} = \Gamma(D)(i\lambda)e_{s\lambda,b}, \qquad D \in \boldsymbol{D}(G/K). \tag{6}$$

Also because of [GGA], Cor. 5.19, Ch. II and Theorem 3.13, Ch. III, we know that if λ is regular, the functions $e_{s\lambda,b}$ form a basis of the space of functions f which satisfy the system (6) and are constant on each horocycle with normal b.

While the relationship between $\widetilde{F}(-\lambda\omega)$ and $\widetilde{F}(\lambda\omega)$ is rather trivial, the relationship between the various $\widetilde{f}(s\lambda, b)$ $(s \in W)$ is of considerable interest as will be seen later (see Ch. VI, Corollary 3.9).

If f is K-invariant we can in (2) replace f by $f^{\tau(k)}$. Using $A(k \cdot x, b) = A(x, k^{-1}b)$ we see that the right hand side of (4) is independent of b. Integrating it over b we see from the spherical function formula $\varphi_\lambda(gK) = \int e^{(i\lambda+\rho)(A(kg))}dk$ that if f is K-invariant, $\widetilde{f}(\lambda, b) = \int f(x)\varphi_{-\lambda}(x)dx$, the spherical transform of f. Thus the Fourier transform (4) is an extension of the spherical transform from K-invariant functions to "arbitrary" functions on X.

Some of the simpler properties of this transform rely on the following formula which we refer to as "*the symmetry of the spherical function* φ_λ."

Theorem 1.1. *For each* $\lambda \in \mathfrak{a}_C^*, g, h \in G$,

$$\varphi_\lambda(g^{-1}h) = \int_K e^{(i\lambda+\rho)(A(kh))}e^{(-i\lambda+\rho)(A(kg))}dk. \tag{7}$$

Proof. Writing $g = n_1 \exp A(g)u(g)$ $(u(g) \in K)$ we have $kgh = n \exp A(kg)u(kg)h$ so, since A normalizes N

$$A(kgh) = A(kg) + A(u(kg)h). \tag{8}$$

Hence

$$\varphi_\lambda(g^{-1}h) = \int_K e^{(i\lambda+\rho)(A(kg^{-1}))}e^{(i\lambda+\rho)(A(u(kg^{-1})h))}dk.$$

Here we substitute the relation

$$A(kg^{-1}) = -A(u(kg^{-1})g),$$

which follows from (8). The relations

$$g = n_1 \exp A(g)u(g) = k(g)\exp H(g)n_2 \qquad k(g) \in K$$

imply $k(g) = u(g^{-1})^{-1}$, $A(g) = -H(g^{-1})$. Harish-Chandra's formula

$$\int_K F(k(g^{-1}k))dk = \int_K F(k)e^{-2\rho(H(gk))}dk$$

([GGA], Ch. I, Lemma 5.19) can thus be stated (replacing $F(k)$ by $F(k^{-1})$),

$$\int_K F(u(kg^{-1}))dk = \int_K F(k)e^{2\rho(A(kg))}dk.$$

Now the formula above for $\varphi_\lambda(g^{-1}h)$ gives the desired formula (7).

Consider now the convolution $\times$ on G/K given by

$$(f_1 \times f_2) \circ \pi = (f_1 \circ \pi) * (f_2 \circ \pi) \tag{9}$$

$\pi : G \longrightarrow G/K$ being the natural mapping.

Lemma 1.2. *If $f \in \mathcal{D}(X)$ and $\lambda \in \mathfrak{a}_c^*$ then*

$$(f \times \varphi_\lambda)(x) = \int_B e^{(i\lambda+\rho)(A(x,b))}\widetilde{f}(\lambda, b)db.$$

Proof. We have by the definition

$$(f \times \varphi_\lambda)(x) = \int_G f(g \cdot o)\varphi_\lambda(g^{-1} \cdot x)dg \tag{10}$$

so the lemma follows from (7).

Theorem 1.3. *For each $f \in \mathcal{D}(X)$,*

$$f(x) = w^{-1} \int_{\mathfrak{a}^*} \int_B e^{(i\lambda+\rho)(A(x,b))}\widetilde{f}(\lambda, b)|\mathbf{c}(\lambda)|^{-2}d\lambda\, db, \qquad x \in X,$$

where w is the order of the Weyl group.

Proof. Fix $g \in G$ and define f_1 on G by

$$f_1(g_1) = \int_K f(gkg_1 \cdot o)dk.$$

Then f_1 is a K-bi-invariant function in $\mathcal{D}(G)$ so Harish-Chandra's inversion formula for the spherical transform applies. This gives

$$f_1(e) = w^{-1} \int_{\mathfrak{a}^*} \widetilde{f_1}(\lambda)|\mathbf{c}(\lambda)|^{-2}d\lambda$$

([GGA], IV, Theorem 7.5 and Exercise C4). But

$$\widetilde{f_1}(\lambda) = \int_G \Big(\int_K f(gkg_1 \cdot o)dk\, \varphi_{-\lambda}(g_1)\Big)dg_1$$

$$= \int_G f(gg_1 \cdot o)\varphi_{-\lambda}(g_1)dg_1 = \int_G f(gg_1 \cdot o)\varphi_\lambda(g_1^{-1})dg_1$$

by (7) for $h = e$; hence by Lemma 1.2,

$$\widetilde{f_1}(\lambda) = \int_B e^{(i\lambda+\rho)(A(g\cdot o,b))}\widetilde{f}(\lambda, b)db.$$

Since $f_1(e) = f(g \cdot o)$ the theorem is proved.

Since the convolution $\times$ on X is not commutative the Fourier transform does not convert it into multiplication. However, this does hold if the second factor is K-invariant.

Lemma 1.4. *Let $f, \varphi \in C_c(X)$ and suppose φ is K-invariant. Then*

$$(f \times \varphi)^\sim(\lambda, b) = \widetilde{f}(\lambda, b)\widetilde{\varphi}(\lambda), \tag{11}$$

where $\widetilde{\varphi}$ is the spherical transform of φ.

Proof. Using (8) we have

$$(f \times \varphi)^\sim(\lambda, kM) = \int_G \Big(\int_G f(g \cdot o)\varphi(g^{-1}h \cdot o)dg\Big) \quad e^{(-i\lambda+\rho)(A(hK,kM))}dh$$

$$= \int_G \int_G f(g \cdot o)\varphi(h \cdot o)e^{(-i\lambda+\rho)(A(u(k^{-1}g)h))}e^{(-i\lambda+\rho)(A(gK,kM)}dg\, dh$$

$$= \widetilde{f}(\lambda, kM)\widetilde{\varphi}(\lambda),$$

the K-invariance of φ justifying the last identity.

Remark. Since $\varphi \times f$ is K-invariant the analog of (11) can not hold for it.

While the inversion formula for $f \longrightarrow \widetilde{f}$ depends only on the symmetry property (7) the full Plancherel theorem requires a criterion for simplicity. Proposition 1.7, although rather crude, is sufficient. However, the analog of the Paley-Wiener theorem (a characterization of $\mathcal{D}(X)^{\sim}$) will require more preparation and more refined tools.

2. The Plancherel Formula

In order to state the Plancherel formula we recall the notation $\mathfrak{a}_+^*$ for the preimage

$$\mathfrak{a}_+^* = \{\lambda \in \mathfrak{a}^* : A_\lambda \in \mathfrak{a}^+\}$$

of the positive Weyl chamber $\mathfrak{a}^+$. Here $B(A_\lambda, H) = \lambda(H)$ for $H \in \mathfrak{a}$.

Theorem 1.5. *The Fourier transform $f(x) \longrightarrow \widetilde{f}(\lambda, b)$ defined by (4) extends to an isometry of $L^2(X)$ onto $L^2(\mathfrak{a}_+^* \times B)$ (with the measure $|\mathbf{c}(\lambda)|^{-2} d\lambda\, db$ on $\mathfrak{a}_+^* \times B$). Moreover,*

$$\int_X f_1(x)\overline{f_2(x)}dx = \frac{1}{w}\int_{\mathfrak{a}^* \times B} \widetilde{f}_1(\lambda, b)\overline{\widetilde{f}_2(\lambda, b)}|\mathbf{c}(\lambda)|^{-2} d\lambda\, db.$$

Proof. If $\lambda \in \mathfrak{a}^*$ we deduce from Lemma 1.2 that

$$\int_X (f_1 \times \varphi_\lambda)(x)\overline{f_2(x)}dx = \int_B \widetilde{f}_1(\lambda, b)\overline{\widetilde{f}_2(\lambda, b)}db. \tag{12}$$

We multiply both sides of (12) by $|\mathbf{c}(\lambda)|^{-2}$ and integrate over $\mathfrak{a}^*$. Using Lemma 1.2 and the inversion formula we obtain the stated formula. Since $\varphi_{s\lambda} = \varphi_\lambda (s \in W)$ the right hand side of (12) is W-invariant; hence

$$\int_X |f(x)|^2 dx = \int_{\mathfrak{a}_+^* \times B} |\widetilde{f}(\lambda, b)|^2 |\mathbf{c}(\lambda)|^{-2} d\lambda\, db. \tag{13}$$

For the surjectivity statement we need the following results about simplicity.

Proposition 1.6. *Suppose $-\lambda \in \mathfrak{a}_c^*$ is simple. Then the space of functions*

$$b \longrightarrow \widetilde{f}(\lambda, b) \tag{14}$$

as f runs through $\mathcal{D}(X)$ is dense in $L^2(B)$.

Proof. Suppose $F \in L^2(B)$ is orthogonal to all the functions (14), that is,

$$\int_B \widetilde{f}(\lambda, b)\overline{F(b)}db = 0, \qquad f \in \mathcal{D}(X).$$

Then

$$\int_X f(x)\left(\int_B e^{(-i\lambda+\rho)(A(x,b))}\overline{F(b)}db\right)dx = 0$$

for all $f \in \mathcal{D}(X)$. Thus the inner integral vanishes identically in x so the simplicity of $-\lambda$ implies $F \equiv 0$.

Proposition 1.7. *There exists an analytic function Ψ on the real space $\mathfrak{a}^*$ such that $\Psi(0) \neq 0$ and such that*

(15) $$\Psi(\lambda) \neq 0 \qquad \textit{implies} \qquad \lambda \textit{ simple.}$$

Proof. If λ is not simple there exists an $F \not\equiv 0$ in $\mathcal{E}(B)$ such that

(16) $$\int_B e^{(i\lambda+\rho)(A(x,b))}F(b)db = 0 \qquad \text{for } x \in X.$$

Here we may assume $F(eM) \neq 0$. As in Theorem 3.16, Ch. II we transfer the integral from B to $\overline{N}$ by the map $\overline{n} \longrightarrow k(\overline{n})M$. Then we get for each $a \in A$,

(17) $$\int_{\overline{N}} e^{-(i\lambda+\rho)(H(a\overline{n}a^{-1}))}e^{(i\lambda-\rho)(H(\overline{n}))}F(k(\overline{n})M)d\overline{n} = 0.$$

In the proof of the theorem quoted we made the substitution $\overline{n} \longrightarrow \overline{n}^a$ and then let $a \longrightarrow +\infty$ (in A^+) under the integral sign. For $Re(A_{i\lambda}) \in \mathfrak{a}^+$ this could be justified but for $\lambda \in \mathfrak{a}^*$ the process is more complicated since $\int_{\overline{N}} e^{-\rho(H(\overline{n}))}d\overline{n} = \infty$.

Our new approach comes from looking at the behavior of $H(\overline{n}^a)$ for $a \longrightarrow -\infty$ established in Proposition 4.25, Ch. II. Since $A_\rho \in \mathfrak{a}^+$ and since by their definition in Proposition 4.23, Ch. II, the A_{μ_i} lie on the edges of the Weyl chamber we have $\rho = \sum_1^\ell a_i\mu_i$ with $a_i > 0$. We would like to have such a relation with the μ_i replaced by linearly independent highest restricted weights ν_i of spherical irreducible representations π_i of G, such that in addition, $A_{\nu_i} \in \mathfrak{a}^+$. Let I be the $\ell \times \ell$ identity matrix and E the $\ell \times \ell$ matrix with 1 in each entry. Fix $n \in \boldsymbol{Z}^+$. Viewing μ as a column vector with coordinates $\mu_1, \ldots, \mu_\ell$ we define the column vector $\nu = (\nu_1, \ldots, \nu_\ell)$ by

(18) $$\nu = (nI + E)\mu.$$

Then each ν_i satisfies the integrality condition of Prop. 4.23, Ch. II and in addition, $A_{\nu_i} \in \mathfrak{a}^+$. The matrix equation (18) is inverted by

$$\mu = \left(\frac{1}{n}I - \frac{1}{n(n+\ell)}E\right)\nu \tag{19}$$

and the equation $\rho = \sum_1^\ell a_i\mu_i$ becomes

$$\rho = \sum_1^\ell b_i\nu_i, \qquad nb_i = a_i - \frac{1}{(n+\ell)}\sum_j a_j. \tag{20}$$

Thus, taking n large enough we can ensure that each $b_i > 0$.

Going back to (17) we write

$$\rho + i\lambda = \sum_{i=1}^\ell \lambda_i(2\nu_i) \qquad \lambda_i \in \boldsymbol{C} \tag{21}$$

so

$$e^{(\rho+i\lambda)(H(\overline{n}))} = \prod_{i=1}^\ell \left(e^{2\nu_i(H(\overline{n}))}\right)^{\lambda_i}. \tag{22}$$

According to (67) in §4, Ch. II, we can write

$$e^{2\nu_i(H(\overline{n}))} = 1 + \sum_\mu h^i_\mu(\overline{n}), \qquad h^i_\mu(\overline{n}^a) = e^{(2\mu-2\nu_i)(\log a)}h^i_\mu(\overline{n}), \tag{23}$$

where μ runs over the restricted weights $\neq \nu_i$ of the representation π_i. Motivated by Proposition 4.25, Ch. II, we multiply (17) by $e^{-(i\lambda+\rho)(\log a - s^*\log a)}$. For simplicity we write u for $\overline{n}$; we fix $H \in \mathfrak{a}^+$ and put $a = a_t = \exp tH$. Then by (17) we have

$$\int_{\overline{N}} e^{-(i\lambda+\rho)(H(u^{a_t})+tH-ts^*H)}e^{(i\lambda-\rho)(H(u))}F(k(u)M)du = 0. \tag{24}$$

Now using (21)–(23) we find that

$$e^{-(i\lambda+\rho)(H(u^{a_t})+tH-ts^*H)} =$$

$$\prod_{i=1}^\ell \left\{e^{(2\nu_i-2s^*\nu_i)(tH)} + \sum_{\mu\neq\nu_i,s^*\nu_i} e^{(2\mu-2s^*\nu_i)(tH)}h^i_\mu(u) + h^i_{s^*\nu_i}(u)\right\}^{-\lambda_i}. \tag{25}$$

Here we have separated out the term $h^i_{s^*\nu_i}$, of weight $\mu = s^*\nu_i$. The important feature is that this term has a coefficient independent of t and can thus be eliminated by differentiation with respect to t. Thus we apply d/dt to (24).

Let $H^i_t(u)$ denote the expression inside $\{\ \}$ in (25) and put $K^i_t(u) = (d/dt)H^i_t(u)$. Differentiating (24) we get

(26)

$$\int_{\overline{N}} e^{-(i\lambda+\rho)(H(u^{a_t})+tH-ts^*H)} \sum_{i=1}^{\ell} \lambda_i \frac{K^i_t(u)}{H^i_t(u)} e^{(i\lambda-\rho)(H(u))} F(k(u)M)du = 0.$$

This integral converges uniformly for t in any compact interval so the differentiation under the integral sign is permissible. In (26) we can cancel the constant factor $e^{-(i\lambda+\rho)(tH-ts^*H)}$ and then we make the substitution $v = u^{a_t}$, $dv = e^{-2\rho(tH)}du$. We have

$$K^i_t(u) = e^{(2\nu_i-2s^*\nu_i)(tH)}(2\nu_i - 2s^*\nu_i)(H)+ \sum_{\mu\neq\nu_i,s^*\nu_i} e^{(2\mu-2s^*\nu_i)(tH)}(2\mu - 2s^*\nu_i)(H)h^i_\mu(u)$$

so by (23)

$$K^i_t(v^{a-t}) = e^{(2\nu_i-2s^*\nu_i)(tH)}\big[(2\nu_i-2s^*\nu_i)(H)+ \sum_{\mu\neq\nu_i,s^*\nu_i} (2\mu-2s^*\nu_i)(H)h^i_\mu(v)\big],$$

which we write as

$$K^i_t(v^{a-t}) = e^{(2\nu_i-2s^*\nu_i)(tH)}k^i_H(v).$$

Also, by (23)

$$\begin{aligned} H^i_t(v^{a-t}) &= e^{(2\nu_i-2s^*\nu_i)(tH)}\big[1 + \sum_{\mu\neq\nu_i,s^*\nu_i} h^i_\mu(v) + h^i_{s^*\nu_i}(v)\big] \\ &= e^{(2\nu_i-2s^*\nu_i)(tH)}e^{2\nu_i(H(v))} \end{aligned}$$

so (26) implies

(27)

$$\int_{\overline{N}} e^{-(i\lambda+\rho)(H(v))}e^{(i\lambda-\rho)(H(v^{a-t}))} \sum_{i=1}^{\ell} \lambda_i e^{-2\nu_i(H(v))}k^i_H(v)F(k(v^{a-t})M)dv = 0.$$

Here we would like to let $t \longrightarrow -\infty$ under the integral sign because then $F(k(v^{a-t})M) \longrightarrow F(eM)$. This limit operation will now be justified by the dominated convergence theorem. We have $\rho(H(\overline{n})) \geq 0$ for $\overline{n} \in \overline{N}$ so in (27),

$$\left| e^{-(i\lambda+\rho)(H(v))}e^{(i\lambda-\rho)(H(v^{a-t}))} \right| \leq e^{-\rho(H(v))}.$$

We also recall ([GGA], IV, Theorem 6.13) that for $\nu \in \mathfrak{a}^*$,

$$\int_{\overline{N}} e^{-(\rho+\nu)(H(\overline{n}))} d\overline{n} < \infty$$

if and only if $A_\nu \in \mathfrak{a}^+$. In particular, for $0 < \epsilon < 1$, $1 \leq i \leq \ell$,

$$\int_{\overline{N}} e^{-(\rho+2\epsilon\nu_i)(H(\overline{n}))} d\overline{n} < \infty.$$

Because of (23) it will therefore suffice to show that for $\mu \neq \nu_i, s^*\nu_i$ and ϵ sufficiently small, the ratio

$$\frac{h^i_\mu(v)}{(1 + h^i_{s^*\nu_i}(v))^{1-\epsilon}} \tag{28}$$

is bounded for $v \in \overline{N}$. To see this we note that by Proposition 4.22, Ch. II,

$$s^*\mu = \nu_i - \sum_1^\ell m_i \overline{\alpha}_i$$

so

$$\mu = s^*\nu_i + \sum_1^\ell m_i\beta_i, \qquad \beta_i = -s^*\overline{\alpha}_i. \tag{29}$$

Of course (28) is bounded in a neighborhood C of e in $\overline{N}$. But by (23) and (29) we have

$$\frac{h^i_\mu(v^a)}{(h^i_{s^*\nu_i}(v^a))^{1-\epsilon}} = \frac{h^i_\mu(v)}{(h^i_{s^*\nu_i}(v))^{1-\epsilon}} e^{2(\Sigma m_i\beta_i - \epsilon\nu_i')(\log a)},$$

where $\nu_i' = \nu_i - s^*\nu_i$. With $a^{-1} \in A^+$ we see that the bound on C extends to C^a, hence to $\overline{N}$.

This justifies taking $t \longrightarrow -\infty$ under the integral sign in (27). Since $F(eM) \neq 0$ we conclude that for each $H \in \mathfrak{a}^+$

$$\sum_{j=1}^{\ell} \lambda_j \int_{\overline{N}} e^{-(i\lambda+\rho+2\nu_j)(H(v))} \Big(\nu_j'(H) + \sum_{\mu\neq\nu_j, s^*\nu_j} (\mu - s^*\nu_j)(H) h^j_\mu(v)\Big) dv = 0. \tag{30}$$

Here $\nu_j'(H) > 0$ and $(\mu - s^*\nu_j)(H) > 0$ by (29).

First note that if $\lambda = 0$ then by (20) and (21) each $\lambda_j > 0$ so (30) is impossible. This shows that the point $\lambda = 0$ is simple. More generally, we express each ν_j' and $\mu - s^*\nu_j$ in (30) in terms of the canonical basis $\mu_1, \dots \mu_\ell$ in Prop. 4.23, Ch. II. Then (30) becomes a relation of the form $\sum_{j=1}^{\ell} c_j(\lambda)\mu_j(H) = 0 \quad (H \in \mathfrak{a}^+)$ and Prop. 1.7 is proved with the function

$$\Psi(\lambda) = \sum_{j=1}^{\ell} c_j(\lambda)\overline{c_j(\lambda)}.$$

Coming back to the surjectivity question in Theorem 1.5 suppose $F \in L^2(\mathfrak{a}_+^* \times B)$ is orthogonal to the range $\mathcal{D}(X)^\sim$, i.e.

$$\int_{\mathfrak{a}_+^* \times B} \widetilde{f}(\lambda, b)F(\lambda, b)|\boldsymbol{c}(\lambda)|^{-2} d\lambda db = 0, \quad f \in \mathcal{D}(X). \tag{31}$$

Let $\mathcal{D}^\natural(X)$ and $\mathcal{E}^\natural(X)$ denote the space of K-invariant functions in $\mathcal{D}(X)$ and in $\mathcal{E}(X)$, respectively. Then the space of spherical transforms $\mathcal{D}^\natural(X)^\sim$ is a uniformly dense subspace of the W-invariant continuous functions on $\mathfrak{a}^*$ vanishing at ∞ ([GGA], Ch. IV, §7, No. 2). Because of Lemma 1.4, (31) implies

$$\int_{\mathfrak{a}_+^* \times B} \widetilde{f}(\lambda, b)\widetilde{\varphi}(\lambda)F(\lambda, b)|\boldsymbol{c}(\lambda)|^{-2} d\lambda \, db = 0, \quad \varphi \in \mathcal{D}^\natural(X).$$

By the density mentioned this holds with $\widetilde{\varphi}$ replaced by an arbitrary function $\psi \in C_c(\mathfrak{a}_+^*)$. Hence there exists a null set $N_f \subset \mathfrak{a}_+^*$ (depending on f) such that if $\lambda \notin N_f$ then

$$\int_B \widetilde{f}(\lambda, b)F(\lambda, b)db = 0. \tag{32}$$

Since by assumption $b \longrightarrow F(\lambda, b)$ is in $L^2(B)$ for almost all $\lambda \in \mathfrak{a}_+^*$ we may assume this for $\lambda \notin N_f$. For each $n \in \boldsymbol{Z}^+$ let $\varphi_n \in \mathcal{D}(X)$ be the smoothed-out characteristic function of the ball $B_n(o) \subset X$ of radius n, i.e. $\varphi_n(x) = 1$ for $d(o, x) < n - \frac{1}{n}$, $\varphi_n(x) = 0$ for $d(o, x) > n + \frac{1}{n}$ and $0 \leq \varphi_n(x) \leq 1$ for all x. Fix a global coordinate system on X and let $\mathfrak{M}$ denote the (countable) set of all functions on X of the form $\varphi_n(x)p(x)$ where p is a polynomial in the coordinates with rational coefficients. Let $N = \bigcup_{f \in \mathfrak{M}} N_f$ so (32) holds for all

$f \in \mathfrak{M}$ and all $\lambda \notin N$. If $f \in \mathcal{D}(X)$ we can choose $n \in \boldsymbol{Z}^+$ such that $f = \varphi_n f$ and then a suitable subsequence $f_k = \varphi_n f_k$ $(k = 1, 2, \dots)$ from $\mathfrak{M}$ converges uniformly to f. Then for each $\lambda \in \mathfrak{a}^*$, $\widetilde{f_k}(\lambda, b)$ converges to $\widetilde{f}(\lambda, b)$ uniformly on B. Thus (32) holds for all $f \in \mathcal{D}(X)$ and $\lambda \notin N$. Now by Proposition 1.7 there exists a null set $N_o \subset \mathfrak{a}^*_+$ such that each $\lambda \in \mathfrak{a}^*_+ - N_o$ is simple. For $\lambda \in \mathfrak{a}^*$ the simplicity of λ is equivalent to that of $-\lambda$. Thus by (32) and Proposition 1.6 we deduce

$$\int_B F(\lambda, b) P(b) db = 0 \quad \text{for all} \quad P \in L^2(B)$$

provided $\lambda \in \mathfrak{a}^*_+ - N_o - N$. Hence $\int_E F(\lambda, b) d\lambda\, db = 0$ for each rectangle $E \subset (\mathfrak{a}^*_+ - N_o - N) \times B$ whence $F \equiv 0$ almost everywhere on $\mathfrak{a}^*_+ \times B$.

Remark. The "polar coordinate" mapping $(kM, a) \longrightarrow kaK$ is a bijection of $(K/M) \times A^+$ onto the regular set $X' \subset X$ whose complement has measure 0 ([DS], Ch. IX, Cor. 1.2). Thus Theorem 1.5 shows that just as in the Euclidean space $\boldsymbol{R}^n$:

The symmetric space X is self-dual under the Fourier transform.

§2. GENERALIZED SPHERICAL FUNCTIONS (EISENSTEIN INTEGRALS).

1. Reduction to Zonal Spherical Functions.

As usual, let $\widehat{K}$ denote the set of equivalence classes of unitary irreducible representations of K; for each $\delta \in \widehat{K}$ let V_δ be a vector space (with inner product $\langle\ ,\ \rangle$) on which a representation of class δ is realized; let such a representation also be denoted by δ. As in [GGA], Ch. V, §3 let $\widehat{K}_M$ denote the set of elements $\delta \in \widehat{K}$ for which the subspace

$$V_\delta^M = \{v \in V_\delta : \ \delta(m)v = v \ \text{ for } \ m \in M\}$$

is $\neq 0$. If $\delta \in \widehat{K}_M$ the contragredient representation $\check{\delta}$ also belongs to $\widehat{K}_M$. We put

$$d(\delta) = \dim V_\delta, \ \ell(\delta) = \dim V_\delta^M.$$

Definition. For $\delta \in \widehat{K}_M, \lambda \in \mathfrak{a}^*_{\mathbf{c}}$ the function

$$\Phi_{\lambda,\delta}(x) = \int_K e^{(i\lambda+\rho)(A(x,kM))}\delta(k)dk, \quad x \in X,$$

is called the *generalized spherical function* of class δ. The term *Eisenstein integral* is also used.

The function $\Phi_{\lambda,\delta}$ maps X into $\mathrm{Hom}(V_\delta, V_\delta)$ the space of linear maps of V_δ into itself. It satisfies the conditions

$$\Phi_{\lambda,\delta}(k \cdot x) = \delta(k)\Phi_{\lambda,\delta}(x), \quad \Phi_{\lambda,\delta}(x)\delta(m) = \Phi_{\lambda,\delta}(x) \tag{1}$$

$$D\Phi_{\lambda,\delta} = \Gamma(D)(i\lambda)\Phi_{\lambda,\delta}, \quad D \in \boldsymbol{D}(G/K) \tag{2}$$

by Ch. II, Prop. 3.14. If $\delta(k) = 1$ $(k \in K)$ then $\Phi_{\lambda,\delta}$ reduces to the (zonal) spherical function φ_λ. We shall now prove a formula relating the general $\Phi_{\lambda,\delta}$ to the special φ_λ. As before we use φ_λ both for the zonal spherical function on X and for the corresponding K-bi-invariant function on G.

Theorem 2.1. *Fix $v, w \in V_\delta$ and assume $-\lambda$ is simple. Then there exists a right-invariant differential operator D on G such that*

$$\langle v, \Phi_{-\bar{\lambda},\delta}(gK)w\rangle = (D\varphi_\lambda)(g), \qquad g \in G.$$

Proof. Consider the Poisson transform $F \longrightarrow f$ given by

$$f(x) = \int_B e^{(i\lambda+\rho)(A(x,b))} F(b)\,db. \tag{3}$$

It is clear from Schwarz's inequality that the mapping $F \longrightarrow f$ is continuous from $L^2(B)$ to $\mathcal{E}(X)$ with the customary topology (cf. Ch. I, §1, No. 1).

Lemma 2.2. *If $-\lambda$ is simple the functions*

$$b \longrightarrow \sum_j a_j e^{(-i\lambda+\rho)(A(g_j\cdot o,b))}, \qquad a_j \in \boldsymbol{C}, \qquad g_j \in G, \tag{4}$$

form a dense subspace of $L^2(B)$.

This is immediate from the definition because no $F \neq 0$ in $L^2(B)$ can be orthogonal to all the functions (4).

Lemma 2.3. *Suppose $-\lambda$ is simple. Then the closed subspace $\mathcal{E}_{(\lambda)}(X) \subset \mathcal{E}(X)$ generated by the G-translates of φ_λ contains all f in (3) with $F \in L^2(B)$.*

For this we use the symmetry of φ_λ as expressed in Theorem 1.1. Given $F \in L^2(B)$ we choose a sequence of functions (4) converging to F. By the continuity mentioned above the corresponding Poisson transforms will converge to f, proving the lemma.

For the proof of Theorem 2.1 consider the natural representation T of G on $\mathcal{E}_{(\lambda)}(X)$. This T is irreducible, quasisimple and K-finite ([GGA], Ch. IV, Theorem 4.5). Let dT denote the differential of T representing $\mathfrak{g}$ on the space $\mathcal{E}^{\infty}_{(\lambda)}$ of K-finite vectors in $\mathcal{E}_{(\lambda)}(X)$. Let $\boldsymbol{D}(G)$ denote the algebra of left-invariant differential operators on G. Then we have

$$\boldsymbol{D}(G) = \oplus_{\delta \in \widehat{K}} \boldsymbol{D}_\delta(G), \tag{5}$$

where $\boldsymbol{D}_\delta(G)$ is the set of elements in $\boldsymbol{D}(G)$ which under $Ad_G(K)$ transform according to δ. Then as shown in the proof of the cited theorem, (with no assumption on λ)

$$dT(\boldsymbol{D}_\delta(G))\varphi_\lambda = \mathcal{E}_{(\lambda),\delta}, \tag{6}$$

the space of K-finite vectors in $\mathcal{E}_{(\lambda)}(X)$ of type δ. If $v_1, \ldots, v_{d(\delta)}$ is a basis of V_δ and $w \in V_\delta$ then the function

$$f_i(x) = \langle v_i, \Phi_{-\bar{\lambda},\delta}(x) w \rangle$$

satisfies

$$f_i^{\tau(k)} = \sum_{p=1}^{d(\delta)} \delta(k)_{pi} f_p$$

so by Lemma 2.3 the function $x \longrightarrow \langle v, \Phi_{-\bar{\lambda},\delta}(x) w\rangle$ does belong to $\mathcal{E}_{(\lambda),\delta}$.

Recall now that if $X_1, \ldots, X_r \in \mathfrak{g}$, $f \in \mathcal{E}_{(\lambda)}(X)$, then

$$(dT(X_1 \cdots X_r)f)(x) = \left\{ \frac{\partial^r}{\partial t_1 \cdots \partial t_r} f(\exp(-t_r X_r) \cdots \exp(-t_1 X_1) \cdot x) \right\}_{t=0}. \tag{7}$$

Viewing the members of $\mathcal{E}_{(\lambda)}(X)$ as functions on G, (7) shows that the members of $dT(\boldsymbol{D}(G))$ are right invariant differential operators on G. Now (6) shows the existence of the desired operator D.

We shall now put (6) in a more explicit form by invoking some results about the structure of $\boldsymbol{D}(G)$. Let $I(\mathfrak{p}^*)$ denote the algebra of K-invariant polynomial functions on $\mathfrak{p}$ and $H(\mathfrak{p}^*)$ the space of corresponding harmonic polynomials ([GGA], Ch. III, §1). Then (*loc. cit.*)

$$S(\mathfrak{p}^*) = I(\mathfrak{p}^*)H(\mathfrak{p}^*), \tag{8}$$

$S(\mathfrak{p}^*)$ denoting the algebra of polynomial functions on $\mathfrak{p}$. We shall now need the following result of Kostant and Rallis [1971] (cf. [GGA], III, Theorem 5.4).

Theorem 2.4. *If $H_o \in \mathfrak{a}$ is regular then each $h \in H(\mathfrak{p}^*)$ is determined by its values on the orbit $Ad_G(K) \cdot H_o$.*

The result quoted actually states this with K replaced by K^θ, the group of fixed points of θ extended to the (complex) adjoint group of $\mathfrak{g}^C$. However, K^θ and its identity component $(K^\theta)_o$ have the same orbits of points in $\mathfrak{a}^c$ ([GGA], Ch. III, Lemma 4.2). Thus, if $h(k \cdot H_o) = 0$ for all $k \in K$ this holds for all $k \in K^\theta$ as desired.

Proposition 2.5. *Suppose $H_o \in \mathfrak{a}$ is regular. Then the restriction mapping*

$$h \longrightarrow h\big|_{Ad_G(K)H_o} \tag{9}$$

is a bijection of $H(\mathfrak{p}^*)$ onto the space of K-finite functions in $\mathcal{E}(K/M)$.

Proof. The mapping

$$kM \longrightarrow Ad(k)H_o \tag{10}$$

is a bijection of K/M onto the orbit $Ad_G(K)H_o$ ([GGA], III, Exercise 2). On the other hand (*loc. cit.* Introduction, Exercise A1) each K-finite function on $K/M = Ad_G(K)H_o$ is the restriction of a polynomial $p \in S(\mathfrak{p}^*)$ which by (8) can be taken harmonic. The injectivity of (9) comes from Theorem 2.4.

If $\delta \in \widehat{K}_M$ let H_δ (resp. $\mathcal{E}_\delta(K/M)$) denote the space of $h \in H(\mathfrak{p}^*)$ (resp. $f \in \mathcal{E}(K/M)$) of type δ. Then (9) maps H_δ onto $\mathcal{E}_\delta(K/M)$. Fix an orthonormal basis

$$v_1, \ldots, v_{d(\delta)} \qquad \text{of} \qquad V_\delta \tag{11}$$

such that

$$v_1, \ldots, v_{\ell(\delta)} \quad \text{span} \quad V_\delta^M. \tag{12}$$

Then the functions

$$kM \longrightarrow \langle v_j, \delta(k)v_i\rangle \quad 1 \leq j \leq d(\delta), 1 \leq i \leq \ell(\delta)$$

form a basis of $\mathcal{E}_\delta(K/M)$ and

$$\mathcal{E}_\delta(K/M) = \oplus_{i=1}^{\ell(\delta)} \mathcal{E}_{\delta,i}(K/M), \tag{13}$$

where $\mathcal{E}_{\delta,i}(K/M)$ is the space of functions

$$F_{v,i}(kM) = \langle v, \delta(k)v_i \rangle, \qquad v \in V_\delta. \tag{14}$$

The map $v \longrightarrow F_{v,i}$ is an isomorphism of V_δ onto $\mathcal{E}_{\delta,i}(K/M)$ commuting with the action of K.

Let $S(\mathfrak{g})$ and $S(\mathfrak{p})$ denote the (complex) symmetric algebras over $\mathfrak{g}$ and $\mathfrak{p}$, respectively. Let $P \longrightarrow P^*$ denote the symmetrization mapping of $S(\mathfrak{g})$ onto $\boldsymbol{D}(G)$. Identifying $S(\mathfrak{p}^*)$ and $S(\mathfrak{p})$ via the Killing form of $\mathfrak{g}$, $I(\mathfrak{p}^*)$, $H(\mathfrak{p}^*)$ and H_δ become subspaces of $S(\mathfrak{g})$. Let I^*, H^* and H_δ^* denote their respective images under the symmetrization. By a simple induction ([GGA], Ch. II, Lemma 4.7) we have

$$\boldsymbol{D}(G) = \boldsymbol{D}(G)\mathfrak{k} \oplus S(\mathfrak{p})^*. \tag{15}$$

A similar proof, using (8) gives

$$\boldsymbol{D}(G) = \boldsymbol{D}(G)\mathfrak{k} + H^*I^* = \boldsymbol{D}(G)\mathfrak{k} + I^*H^* \tag{16}$$

as follows. Let $p \in S(\mathfrak{p})$ be homogeneous and decompose it according to (8), $p = \Sigma_k h_k i_k$, where all h_k, i_k are homogeneous. Since the monomials $(X_1^*)^{k_1} \cdots (X_n^*)^{k_n}(X_1, \ldots, X_n$ being a basis of $\mathfrak{g}$, compatible with $\mathfrak{g} = \mathfrak{k} + \mathfrak{p})$ form a basis of $\boldsymbol{D}(G)$ we see that if $q = p^* - \Sigma_k h_k^* i_k^*$, then $\deg q < \deg p$. We decompose q according to (15) and repeat the argument for the component in $S(\mathfrak{p})^*$ etc.

Now we can state (6) in a stronger form, namely

$$dT(H_\delta^*)\varphi_\lambda = \mathcal{E}_{(\lambda),\delta}. \tag{17}$$

This is clear from (16) since $dT(\boldsymbol{D}(G)\mathfrak{k})$ annihilates φ_λ and φ_λ is an eigenfunction of each element in $dT(I^*)$.

Because of the Iwasawa decomposition $\mathfrak{g} = \mathfrak{a}+\mathfrak{n}+\mathfrak{k}$ we have the following direct decomposition of the universal enveloping algebra $\boldsymbol{D}(G)$,

$$\boldsymbol{D}(G) = \boldsymbol{D}(A) \oplus (\mathfrak{n}\boldsymbol{D}(G) + \boldsymbol{D}(G)\mathfrak{k}) \tag{18}$$

(cf. [GGA], II, Lemma 5.14). Let $D \longrightarrow q^D$ denote the corresponding projection of $\boldsymbol{D}(G)$ onto $\boldsymbol{D}(A)$. For $\lambda \in \mathfrak{a}_c^*$ consider the function

$$\zeta_\lambda(g) = e^{(-i\lambda+\rho)(A(g))}, \qquad g \in G \tag{19}$$

with $A(g)$ as in (2) §1. This function satisfies

(20) $$\zeta_\lambda(ngk) \equiv \zeta_\lambda(g), \qquad \zeta_\lambda(mg) \equiv \zeta_\lambda(g), \qquad m \in M$$

and as a result of the first relation,

(21) $$(D\zeta_\lambda)(e) = (q^D\zeta_\lambda)(e) = q^D(\rho - i\lambda).$$

Because of (13) and the K-isomorphism of H_δ onto $\mathcal{E}_\delta(K/M)$ the vector space $E_\delta = \mathrm{Hom}_K(V_\delta, H_\delta^*)$ of linear maps $V_\delta \longrightarrow H_\delta^*$ commuting with the actions of K (i.e. δ and Ad) has dimension $\ell(\delta)$.

Let $\check{\delta}$ denote the contragredient representation of K on the dual space $V_\delta' = V_{\check{\delta}}$. For $\lambda \in \mathfrak{a}_c^*$ consider the linear map

$$Q^\delta(\lambda) : \; E_\delta \longrightarrow V_{\check{\delta}}^M$$

given by

(22) $$(Q^\delta(\lambda)(\epsilon))(v) = q^{\epsilon(v)}(\rho - i\lambda), \qquad \epsilon \in E_\delta, v \in V_\delta.$$

The right hand side is indeed invariant under $v \longrightarrow \delta(m)v$ by virtue of (20)–(21).

Let $\epsilon_1, \ldots, \epsilon_{\ell(\delta)}$ be any basis of E_δ and as before, $v_1, \ldots, v_{\ell(\delta)}$ an orthonormal basis of V_δ^M. We shall often find it convenient to represent $Q^\delta(\lambda)$ by the matrix

(22′) $$Q^\delta(\lambda)_{ij} = q^{\epsilon_j(v_i)}(\rho - i\lambda), \qquad 1 \leq i, j \leq \ell(\delta).$$

Remark. If we change the basis (ϵ_j) to another one (η_k) where $\eta_k = \Sigma a_{jk}\epsilon_j$ then the matrix $Q^\delta(\lambda)$ changes to $Q^\delta(\lambda)A$ where $A = (a_{jk})$.

Using once more the symmetry of φ_λ (Theorem 1.1) we can put (17) into more explicit form.

Under the left regular representation of G on $\mathcal{E}(G)$ ([GGA], IV, §1) each $D \in \boldsymbol{D}(G)$ is represented by a right invariant differential operator $\tau(D)$ such that $D \longrightarrow \tau(D)$ is a homomorphism and if $X_1, \ldots, X_r \in \mathfrak{g}$ then

(23)
$$(\tau(X_1 \cdots X_r)f)(g) = \left\{\frac{\partial^r}{\partial t_1 \cdots \partial t_r} f(\exp(-t_rX_r) \cdots \exp(-t_1X_1)g)\right\}_{t=0}.$$

We note that $\tau(X)^{L_g} = \tau(\mathrm{Ad}(g)X)$ and more generally,

$$\tau(D)^{L_g} = \tau(Ad(g)D), \quad D^{R_{g^{-1}}} = \mathrm{Ad}(g)D, \quad D \in \boldsymbol{D}(G), g \in G. \tag{24}$$

Theorem 2.6. *For each* $\lambda \in \mathfrak{a}_c^*$, $v \in V_\delta$, $1 \leq j \leq \ell(\delta)$,

$$(\tau(\epsilon_j(v))\varphi_\lambda)(g) = \sum_{i=1}^{\ell(\delta)} Q^\delta(\lambda)_{ij} \int_K e^{(i\lambda+\rho)(A(gK,kM))} \langle v, \delta(k)v_i \rangle dk.$$

Proof. By Theorem 1.1,

$$\varphi_\lambda(h^{-1}g) = \int_K \zeta_{-\lambda}(k^{-1}g)\zeta_\lambda(k^{-1}h)dk$$

so, putting $h = \exp t_1 X_1 \ldots \exp t_r X_r$ in (23),

$$(\tau(X_1 \cdots X_r)\varphi_\lambda)(g) = \int_K \zeta_{-\lambda}(k^{-1}g)\left((X_1 \cdots X_r)\zeta_\lambda\right)(k^{-1})dk, \tag{25}$$

where on the right, $X_1 \cdots X_r$ is a member of $\boldsymbol{D}(G)$. Using (20) we get

$$(\tau(\epsilon_j(v))\varphi_\lambda)(g) = \int_K \zeta_{-\lambda}(k^{-1}g)\left(\epsilon_j(v)^{R_k}\zeta_\lambda\right)(e)dk. \tag{26}$$

But

$$\epsilon_j(v)^{R_k} = \mathrm{Ad}(k^{-1})(\epsilon_j(v)) = \epsilon_j(\delta(k^{-1})v) = \epsilon_j\left(\sum_1^{d(\delta)} a_p v_p\right)$$

and if $m \in M$ then by (20),

$$\begin{aligned}(\epsilon_j(v_p)\zeta_\lambda)(e) &= (\epsilon_j(v_p)\zeta_\lambda^{L_m})(e) = (\epsilon_j(v_p)\zeta_\lambda)(m^{-1}) \\ &= (\epsilon_j(v_p)^{R_m}\zeta_\lambda)(e) = (\epsilon_j(\delta(m^{-1})v_p)\zeta_\lambda)(e).\end{aligned}$$

Integrating over M we see that this expression vanishes for $p > \ell(\delta)$. Hence

$$(\epsilon_j(v)^{R_k}\zeta_\lambda)(e) = \sum_{i=1}^{\ell(\delta)} a_i(\epsilon_j(v_i)\zeta_\lambda)(e) = \sum_{i=1}^{\ell(\delta)} a_i Q^\delta(\lambda)_{ij}.$$

Since $a_i = \langle v, \delta(k)v_i \rangle$ this proves the theorem.

Remark. For each $\delta \in \widehat{K}_M$, $\det(Q^\delta(\lambda)) \not\equiv 0$. This is clear from Cor. 5.14 in Ch. II together with Lemma 2.11 below whose proof fits in at this stage.

2. The Expansion of $\Phi_{\lambda,\delta}$.

We shall now use Theorem 2.1 to obtain an expansion of $\Phi_{\lambda,\delta}$ on $A^+ \cdot o$ generalizing Harish-Chandra's expansion of φ_λ in terms of exponentials ([GGA], Ch. IV, §5). For this we recall some essential notation.

For $\alpha \in \Sigma^+$ let as in [DS], Ch. VII, §11,

$$\mathfrak{k}_\alpha = \{T \in \mathfrak{k} : \ (\operatorname{ad} H)^2 T = \alpha(H)^2 T \quad \text{for } H \in \mathfrak{a}\} \tag{27}$$

and $T_1^\alpha, \ldots, T_{m_\alpha}^\alpha$ a basis of $\mathfrak{k}_\alpha$, orthonormal with respect to the negative of the Killing form $\langle , \rangle$ of $\mathfrak{g}$. Let

$$\omega_\alpha = \sum_{i=1}^{m_\alpha} T_i^\alpha \cdot T_i^\alpha, \tag{28}$$

where the T_i^α are viewed as left-invariant differential operators on G. Then ω_α is bi-invariant under M. Let as usual $\alpha_1, \ldots, \alpha_\ell$ denote the simple restricted roots and Λ the set of integral linear combinations

$$\Lambda = \{n_1\alpha_1 + \cdots + n_\ell \alpha_\ell : \ n_j \in \boldsymbol{Z}^+\}. \tag{29}$$

Theorem 2.7. *There exist certain meromorphic functions $\boldsymbol{C}_s (s \in W)$ on $\mathfrak{a}_c^*$ and rational functions $\boldsymbol{\Gamma}_\mu (\mu \in \Lambda)$ on $\mathfrak{a}_c^*$ all with values in* $\operatorname{Hom}(V_\delta^M, V_\delta^M)$ *such that for $H \in \mathfrak{a}^+$, $v \in V_\delta^M$,*

$$\Phi_{\lambda,\delta}(\exp H \cdot o)v = \sum_{s \in W, \mu \in \Lambda} e^{(is\lambda - \rho - \mu)(H)} \boldsymbol{\Gamma}_\mu(s\lambda) \boldsymbol{C}_s(\lambda) v. \tag{30}$$

Here the functions $\boldsymbol{\Gamma}_\mu$ are given by the recursion formulas

$$\begin{aligned} &(\langle \mu, \mu \rangle - 2i\langle \mu, \lambda \rangle)\boldsymbol{\Gamma}_\mu(\lambda) \\ &= 2 \sum_{\alpha \in \Sigma^+} m_\alpha \sum_{k \geq 1} \{\langle \alpha, \mu - 2k\alpha + \rho \rangle - i\langle \alpha, \lambda \rangle\} \boldsymbol{\Gamma}_{\mu - 2k\alpha}(\lambda) \\ &- 4 \sum_{\alpha \in \Sigma^+} \sum_{k \geq 1} k\, \delta(\omega_\alpha) \boldsymbol{\Gamma}_{\mu - 2k\alpha}(\lambda), \qquad \boldsymbol{\Gamma}_0 \equiv I. \end{aligned} \tag{31}$$

Here $\delta(\omega_\alpha)$ is the differential of δ applied to ω_α. In these sums k runs over the set of integers ≥ 1 for which $\mu - 2k\alpha \in \Lambda$.

Remark. The expansion (30) holds for λ in the connected open dense subset of $\mathfrak{a}_c^*$ where all the $\boldsymbol{\Gamma}_\mu(s\lambda)$ and $\boldsymbol{C}_s(\lambda)$ are holomorphic. This set can be read off from (31) and formulas for $\boldsymbol{C}_s(\lambda)$ proved later (see Proposition 2.9 and Ch. II, Theorem 5.4).

We now embark on the proof of Theorem 2.7. As in [GGA], II, §5, let $\mathcal{R}$ denote the ring of functions on $A^+ \cdot o$ which can be expanded in an absolutely convergent series on $A^+ \cdot o$ with zero constant term,

$$f(\exp H \cdot o) = \sum_{\mu \in \Lambda} a_\mu e^{-\mu(H)}, \quad a_\mu \in \boldsymbol{C}, a_o = 0. \tag{32}$$

Lemma 2.8. *Let D be a right-invariant differential operator on G. Then there exist operators E_A, $E_i \in \boldsymbol{D}(A)$ and functions $g_i \in \mathcal{R}$ such that on A^+,*

$$\overline{Df} = (E_A + \sum_i g_i E_i)\overline{f}, \qquad f \in \mathcal{E}^\natural(G), \tag{33}$$

the bar denoting restriction to A.

Remark. In the language of radial parts of differential operators ([GGA], II, §3, No.3) Lemma 2.8 states that $E_A + \Sigma g_i E_i$ is the radial part of D for the left and right action of $K \times K$ on G:

$$(k_1, k_2) : \ g \longrightarrow k_1 g k_2^{-1}$$

with A^+ as transversal submanifold.

Proof of Lemma 2.8. We consider again the decomposition of $\boldsymbol{D}(G)$ given by (18). Let $\boldsymbol{D}^q(G)$ denote the set of $E \in \boldsymbol{D}(G)$ of degree $\leq q$; define $\boldsymbol{D}^q(A)$ similarly. If $E \in \boldsymbol{D}^q(G)$ and if (X_i) is a basis of $\mathfrak{n}$ compatible with the decomposition $\mathfrak{n} = \sum_{\alpha \in \Sigma^+} \mathfrak{g}_\alpha$ of $\mathfrak{n}$ into root subspaces $\mathfrak{g}_\alpha$ we can select $D_i \in \boldsymbol{D}^{q-1}(G)$ such that

$$E - E'_A - \sum_i X_i D_i \in \boldsymbol{D}(G)\mathfrak{k}, \tag{34}$$

where E'_A is the $\boldsymbol{D}(A)$-component of E according to (18). Let $X_i \in \mathfrak{g}_\alpha$, $X_i = Y_i + Z_i$ ($Y_i \in \mathfrak{p}, Z_i \in \mathfrak{k}$) and fix $h = \exp H$ in A' (the set of regular elements in A). Then

$$\begin{aligned} 2\,\mathrm{Ad}(h^{-1})Z_i &= \mathrm{Ad}(h^{-1})X_i + \theta\,\mathrm{Ad}(h)X_i \\ &= e^{-\alpha(H)}(Y_i + Z_i) + e^{\alpha(H)}(-Y_i + Z_i) \end{aligned}$$

so

$$Y_i = (\coth \alpha(H))Z_i - (\sinh \alpha(H))^{-1}\,\mathrm{Ad}(h^{-1})Z_i.$$

Consequently,

$$X_i \equiv g_\alpha(h^{-1})Z_i \qquad \mathrm{mod}\ (\mathrm{Ad}(h^{-1})\mathfrak{k}), \tag{35}$$

where $g_\alpha(h) = 1 - \coth \alpha(H)$. We substitute this into (34) and replace Z_iD_i by $D_iZ_i + [Z_i, D_i]$. Here $D_iZ_i \in \boldsymbol{D}(G)\mathfrak{k}$ and $[Z_i, D_i] \in \boldsymbol{D}^{q-1}(G)$. Repeating this process with E replaced by $[Z_i, D_i]$ etc. we obtain by induction elements g_i in the ring generated by the g_α and operators $E_i' \in \boldsymbol{D}^{q-1}(A)$ such that

$$E - E_A' - \sum_i g_i(h^{-1})E_i' \in (\mathrm{Ad}(h^{-1})\mathfrak{k})\boldsymbol{D}(G) + \boldsymbol{D}(G)\mathfrak{k}. \tag{36}$$

Now each element

$$C \in (\mathrm{Ad}(h^{-1})\mathfrak{k})\boldsymbol{D}(G) + \boldsymbol{D}(G)\mathfrak{k}$$

satisfies

$$(Cf)(h) = 0 \qquad \text{if} \quad f \in \mathcal{E}^\natural(G) \tag{37}$$

(see [GGA], II, §5, No.2). Let us now use (36) on the operator $E = D^\vee$, the image of the right-invariant operator D under the diffeomorphism $\vee$: $x \longrightarrow x^{-1}$ of G. Then by (36)–(37) we have

$$(D^\vee f)(h) = ((E_A' + \sum_i g_i(h^{-1})E_i')f)(h), \qquad f \in \mathcal{E}^\natural(G). \tag{38}$$

Now $(D^\vee f)(h) = (D\check{f})(h^{-1})$ and $(E_i'f)(h) = (E_i'f)^\vee(h^{-1}) = (E_i\check{f})(h^{-1})$ where $E_i = (E_i')^\vee$. Putting similarly $E_A = (E_A')^\vee$ we deduce from (38)

$$(D\check{f})(h^{-1}) = ((E_A + \sum_i g_i(h^{-1})E_i)\check{f})(h^{-1}). \tag{39}$$

Here we take $h^{-1} \in A^+$ and note that the restriction of g_i to A^+ belongs to $\mathcal{R}$. Then (39) proves the lemma.

Now Harish-Chandra's expansion for φ_λ has the form ([GGA], IV, Theorem 5.5)

$$\varphi_\lambda(\exp H) = \sum_{s\in W,\mu\in\Lambda} e^{(is\lambda-\rho-\mu)(H)}\Gamma_\mu(s\lambda)\mathbf{c}(s\lambda), \qquad H \in \mathfrak{a}^+, \tag{40}$$

where $\mathbf{c}$ is the $\mathbf{c}$-function and the Γ_μ are certain rational functions on $\mathfrak{a}^*_{\mathbf{c}}$. Also each g_i is given by an absolutely convergent series

$$g_i(\exp H) = \sum_{\mu\in\Lambda,\mu\neq 0} \gamma_{i\mu}e^{-\mu(H)}, \quad H \in \mathfrak{a}^+, \tag{41}$$

and now we use Lemma 2.8 with D taken as the operator in Theorem 2.1; its dependence on λ is clear from Theorem 2.6, in fact

$$\int_K e^{(i\lambda+\rho)(A(gK,kM))}\langle v, \delta(k)v_i\rangle dk = \sum_{j=1}^{\ell(\delta)} P^\delta(\lambda)_{ji}(\tau(\epsilon_j(v))\varphi_\lambda)(g), \tag{42}$$

where the matrix $P^\delta(\lambda)$ is the inverse of $Q^\delta(\lambda)$. We now make temporarily the following assumptions on λ :

(i) $\det Q^\delta(\lambda) \neq 0,\ \det \overset{\vee}{Q^\delta}(\lambda) \neq 0$,

(ii) $\mathbf{c}$ holomorphic at each $s\lambda$, $\qquad s \in W$.

(iii) If $s, \sigma \in W,\ \mu, \nu \in \Lambda$ then

$$is\lambda - \mu \neq i\sigma\lambda - \nu \text{ unless } s = \sigma, \mu = \nu.$$

(iv) $\langle \mu, \mu\rangle - 2i\langle \mu, s\lambda\rangle \neq 0$ for $\mu \in \Lambda - \{0\}, s \in W$.

The last assumption guarantees that each Γ_μ is holomorphic at each $s\lambda$ (*loc. cit*). Thus λ is restricted to a certain open, dense connected set $\mathfrak{a}^*_C(\delta)$ in the space $\mathfrak{a}^*_C$. Now we take $h = \exp H$ ($H \in \mathfrak{a}^+$) and observe that $\Phi_{\lambda,\delta}(h\cdot o)$ maps V_δ^M into itself. Using (40)–(42) and Lemma 2.8 we get an expansion valid for $v, w \in V_\delta^M$

$$\langle v, \Phi_{-\overline{\lambda},\delta}(h\cdot o)w\rangle = \sum_{s\in W,\mu\in\Lambda} B_{s,\mu}(\lambda; v, w)e^{(is\lambda-\mu-\rho)(H)} \tag{43}$$

where by (iii) the coefficients $B_{s,\mu}(\lambda; v, w)$ are uniquely determined. Since the left hand side of (43) is bilinear in v, w the uniqueness implies the same for each coefficient so we have

$$B_{s,\mu}(\lambda; v, w) = \langle B_{s,\mu}(\lambda)v, w\rangle, \tag{44}$$

where $B_{s,\mu}(\lambda) \in \operatorname{Hom}(V_\delta^M, V_\delta^M)$. Also (40) and (42) imply that $B_{s,\mu}$ is holomorphic in $\mathfrak{a}_c^*(\delta)$. In the basis (12) we put $v = v_i$, $w = v_j$. Then by (43),

$$(45) \qquad \int_K e^{(i\lambda+\rho)(A(h\cdot o,kM))}\overline{\delta(k)_{ij}}dk = \sum_{s\in W,\mu\in\Lambda} \langle B_{s,\mu}(\lambda)v_i, v_j\rangle e^{(is\lambda-\mu-\rho)(H)}.$$

Since the conjugates of $\delta(k)_{ij}$ are the matrix elements of the contragredient $\check{\delta}(k)$ ([GGA], Ch. IV, §1) we get by replacing in (45) δ by $\check{\delta}$, the expansion

$$(46) \qquad \Phi_{\lambda,\delta}(\exp H\cdot o)v = \sum_{s\in W,\mu\in\Lambda} e^{(is\lambda-\mu-\rho)(H)}A_{s,\mu}(\lambda)v,$$

where $v \in V_\delta^M$ and $A_{s,\mu}(\lambda) \in \operatorname{Hom}(V_\delta^M, V_\delta^M)$ (the transpose of $B_{s,\mu}(\lambda)$). Now that we know that an expansion of the form (30) exists, its coefficients are computed by just using the fact that $\Phi_{\lambda,\delta}$ is an eigenfunction of the Laplace-Beltrami operator $L = L_X$ of X, namely

$$L\Phi_{\lambda,\delta} = -(\langle\lambda,\lambda\rangle + \langle\rho,\rho\rangle)\Phi_{\lambda,\delta}$$

(§1, (6) and [GGA], Ch. II, Cor. 5.20). We now use the polar coordinate form of the Laplacian

$$(Lf)(h\cdot o) = \big[(L_A + \sum_{\alpha\in\Sigma^+} m_\alpha(\coth\alpha)A_\alpha))\overline{f}\big](h\cdot o)$$
$$+ \sum_{\alpha\in\Sigma^+}(\sinh\alpha(\log h))^{-2}((\operatorname{Ad}(h^{-1})\omega_\alpha)\widetilde{f})(h),$$

where L_A is the Laplacian on $A\cdot o$, $\langle A_\alpha, H\rangle = \alpha(H)$ for $H \in \mathfrak{a}$, $\overline{f}$ denotes the restriction of f to $A\cdot o$ and $\widetilde{f}$ is the lift of f to G, i.e. $\widetilde{f}(g) = f(g\cdot o)$ (cf. [GGA], II, Theorem 5.24). Next we note that

$$((\operatorname{Ad}(h^{-1})\omega_\alpha)\widetilde{f})(h) = (\omega_\alpha^{R(h)}\widetilde{f})(h) = [\omega_\alpha(\widetilde{f}^{R(h^{-1})})](e)$$

and the function $F = (\Phi_{\lambda,\delta})^{R(h^{-1})}$ satisfies for $T \in \mathfrak{k}$,

$$(T^2F)(e) = \left\{\frac{d^2}{dt^2}F(\exp tT)\right\}_{t=0} = \left\{\frac{d^2}{dt^2}\delta(\exp tT)\right\}_{t=0} F(e) = \delta(T)^2F(e).$$

Hence the restriction $\overline{\Phi_{\lambda,\delta}}$ of $\Phi_{\lambda,\delta}$ to $A^+\cdot o$ satisfies

$$\left\{L_A + \sum_{\alpha\in\Sigma^+} m_\alpha(\coth\alpha)A_\alpha + \sum_{\alpha\in\Sigma^+}(\sinh\alpha)^{-2}\delta(\omega_\alpha) + \langle\lambda,\lambda\rangle + \langle\rho,\rho\rangle\right\}\overline{\Phi}_{\lambda,\delta} = 0.$$

Here we plug in expansion (46), the formula

$$\coth \alpha = 1 + 2\sum_{k\geq 1} e^{-2k\alpha}$$

and the differentiated formula

$$(\sinh \alpha)^{-2} = 4\sum_{k\geq 1} ke^{-2k\alpha}.$$

Identifying the coefficients we find that for each $s \in W$, $A_{s,\mu}(\lambda)$ satisfies the recursion formula for $\mathbf{\Gamma}_\mu(s\lambda)$ given by (31). Multiplying (31) by $A_{s,o}(\lambda)$ on the *right* we thus deduce,

$$A_{s,\mu}(\lambda) = \mathbf{\Gamma}_\mu(s\lambda)A_{s,o}(\lambda) \tag{47}$$

under the additional assumption on λ that the coefficient to $\mathbf{\Gamma}_\mu(s\lambda)$ in (31) should always be $\neq 0$; this however is guaranteed by condition (iv) above.

We now put

$$\boldsymbol{C}_s(\lambda) = A_{s,o}(\lambda), \qquad s \in W. \tag{48}$$

The rationality of the function $\mathbf{\Gamma}_\mu$ is obvious from (31). For the meromorphy of $\boldsymbol{C}_s(\lambda)$ we write (46) in the form

$$e^{\rho(\log h)}\Phi_{\lambda,\delta}(h\cdot o)v = \sum_{s\in W} \Phi(s\lambda : h)\boldsymbol{C}_s(\lambda)v \tag{49}$$

for $v \in V_\delta^M$, $h \in A^+$, $\lambda \in \mathfrak{a}_{\boldsymbol{c}}^*(\delta)$, where

$$\Phi(\lambda : h) = e^{i\lambda(\log h)}\sum_{\mu\in\Lambda} \mathbf{\Gamma}_\mu(\lambda)e^{-\mu(\log h)}.$$

Assuming in addition

$$Re(\langle i\lambda, \alpha\rangle) > 0 \qquad \text{for} \quad \alpha \in \Sigma^+ \tag{50}$$

we know that for $s \neq e$ in W, $e^{(is\lambda - i\lambda)(t\log h)} \longrightarrow 0$ for $t \longrightarrow +\infty$. Thus we get by applying the limit theorem (Ch. II, Theorem 3.16) to (49)

$$\boldsymbol{c}(\lambda)v = \boldsymbol{C}_e(\lambda)v, \qquad v \in V_\delta^M. \tag{51}$$

For any $s \in W$ we fix a representative $m_s \in M'$ and as in Chapter II, §1, put

$$\overline{N}_s = \overline{N} \cap m_s^{-1}Nm_s.$$

The partial $\boldsymbol{c}$-function is given by

$$\boldsymbol{c}_s(\lambda) = \int_{\overline{N}_s} e^{-(i\lambda+\rho)(H(\overline{n}_s))} d\overline{n}_s \tag{52}$$

and is explicitly evaluated by the formulas (22)–(25) in Ch. II, §5. Now we shall prove a formula for $\boldsymbol{C}_s(\lambda)$.

Proposition 2.9. *Let $s \in W$. Then $\boldsymbol{C}_{s^{-1}}$ extends to a meromorphic function on $\mathfrak{a}_{\mathbf{c}}^*$ with values in $\operatorname{Hom}(V_\delta^M, V_\delta^M)$ by the formula*

$$\boldsymbol{C}_{s^{-1}}(s\lambda)v = \frac{\boldsymbol{c}(\lambda)}{\boldsymbol{c}_s(\lambda)} \int_{\overline{N}_s} e^{-(i\lambda+\rho)(H(\overline{n}_s))} \delta(k(\overline{n}_s)^{-1} m_s^{-1})\, d\overline{n}_s\, v,$$

where $v \in V_\delta^M$ is arbitrary. Both sides are holomorphic in a region containing the set where $\lambda \longrightarrow \boldsymbol{c}(\lambda)/\boldsymbol{e}_s(\lambda)$ is holomorphic.

Proof. We recall from Chapter II, Theorem 5.10 that

$$e^{(is\lambda+\rho)(A(a\cdot o, kM))} = a_s^{-1} \boldsymbol{e}_s(\lambda)^{-1} \int_B e^{(i\lambda+\rho)(A(a\cdot o, k\cdot b))} dS_{\lambda,s}(b). \tag{53}$$

We lift $S_{\lambda,s}$ to $\widetilde{S}_{\lambda,s} \in \mathcal{D}'(K)$ as in Ch. II, §5, No. 2, multiply (53) by $\delta(k)$ and integrate over K. We then find for $a \in A^+$

$$\begin{aligned} &a_s^{-1}\boldsymbol{e}_s(\lambda)^{-1} \int_K \int_K e^{(i\lambda+\rho)(A(a\cdot o, k\ell))} \delta(k) v\, d\widetilde{S}_{\lambda,s}(\ell)\, dk \\ &= \int_K e^{(is\lambda+\rho)(A(a\cdot o, kM))} \delta(k) v\, dk = \sum_{\sigma\in W, \mu\in\Lambda} e^{(i\sigma\lambda-\mu-\rho)(\log a)} \boldsymbol{\Gamma}_\mu(\sigma\lambda) \boldsymbol{C}_{\sigma s^{-1}}(s\lambda) v. \end{aligned}$$

The left hand side here is, if $\chi(k) = \delta(k)v$,

$$a_s^{-1}\boldsymbol{e}_s(\lambda)^{-1} \int_K e^{(i\lambda+\rho)(A(a\cdot o, kM))} \chi * \widetilde{S}_{\lambda,s}(k) dk.$$

Multiplying by $e^{(-i\lambda+\rho)(\log a)}$ and using Ch. II, Theorem 3.16, we get for $\operatorname{Re}(\langle i\lambda, \alpha\rangle) > 0$ ($\alpha \in \Sigma^+$) the formula

$$a_s^{-1}\boldsymbol{e}_s(\lambda)^{-1} (\chi * \widetilde{S}_{\lambda,s})(e) \boldsymbol{c}(\lambda) = \boldsymbol{C}_{s^{-1}}(s\lambda) v,$$

which gives the stated formula for $C_{s^{-1}}(s\lambda)$. According to Chapter II, §5, formula (9) and Theorem 5.4, this formula gives a meromorphic continuation of $C_{s^{-1}}$ to $\mathfrak{a}_c^*$. Since we already knew (30) for an open set of λ it holds by analytic continuation on the set of λ for which both sides are holomorphic. This set can be read off from (31) above and Chapter II, Theorem 5.4 and (9) §5. This proves Proposition 2.9 and thus concludes the proof of Theorem 2.7.

While the zonal spherical function φ_λ is W-invariant in λ the Eisenstein integral transforms under W in a nontrivial but computable fashion.

Notation. Given $\delta \in \widehat{K}_M$ and $\lambda \in \mathfrak{a}_c^*$ we put

$$(54)\qquad \begin{aligned} A(\lambda, m_s) &= \int_{\overline{N}_s} \delta(m_s k(\overline{n}_s)) e^{-(i\lambda+\rho)(H(\overline{n}_s))} d\overline{n}_s, \\ B(\lambda, m_s) &= c_s(\lambda)^{-1} A(\lambda, m_s) \end{aligned}$$

$$(55)\qquad \Gamma_{m_s,\lambda} = c_s(\lambda)^{-1} \int_{\overline{N}_s} \delta(k(\overline{n}_s)^{-1} m_s^{-1}) e^{-(i\lambda+\rho)(H(\overline{n}_s))} d\overline{n}_s.$$

All three are meromorphic functions on $\mathfrak{a}_c^*$ with values in $\operatorname{Hom}(V_\delta, V_\delta)$. Since $k(\overline{n}_s)^m = k(\overline{n}_s^m)$ it is easy to see that $A(\lambda, m_s)$ and its adjoint $A(\lambda, m_s)^*$ leave V_δ^M invariant; consequently,

$$(A(\lambda, m_s)|V_\delta^M)^* = A(\lambda, m_s)^*|V_\delta^M.$$

On V_δ^M we have $A(\lambda, m_s m) = A(\lambda, m_s)$ etc. so we often write $A(\lambda, s)$, $B(\lambda, s)$ and $\Gamma_{s,\lambda}$ for the restrictions of these operators to V_δ^M.

Theorem 2.10. *The generalized spherical function $\Phi_{\lambda,\delta}$ satisfies the functional equations*

$$\Phi_{s\lambda,\delta}(x) = \Phi_{\lambda,\delta}(x)\Gamma_{m_s,\lambda}.$$

Moreover,

$$\Gamma_{s,\lambda} v = \frac{C_{s^{-1}}(s\lambda)}{c(\lambda)} v \qquad v \in V_\delta^M.$$

Proof. Replacing in (53) k by $k_o^{-1}k$ we see that $a \cdot o$ can be replaced by any $x \in X$. Hence

$$\begin{aligned}
&\int_K e^{(is\lambda+\rho)(A(x,kM))}\delta(k)dk \\
&= a_s^{-1}e_s(\lambda)^{-1}\int_K\int_K e^{(i\lambda+\rho)(A(x,k\ell M))}\delta(k)dk\, d\widetilde{S}_{\lambda,s}(\ell) \\
&= a_s^{-1}e_s(\lambda)^{-1}\int_K\int_K e^{(i\lambda+\rho)(A(x,kM))}\delta(k)dk\, \delta(\ell^{-1})d\widetilde{S}_{\lambda,s}(\ell) \\
&= \left(\int_K e^{(i\lambda+\rho)(A(x,kM))}\delta(k)dk\right)\Gamma_{m_s,\lambda}
\end{aligned}$$

since $\boldsymbol{c}_s(\lambda) = a_s\boldsymbol{d}_s(\lambda)\boldsymbol{e}_s(\lambda)$.

3. Simplicity (Preliminary Results).

We recall that $\lambda \in \mathfrak{a}_{\mathbf{c}}^*$ is *simple* if the Poisson transform $F \longrightarrow f$ given by

$$f(x) = \int_B e^{(i\lambda+\rho)(A(x,b))}F(b)db, \qquad F \in \mathcal{E}(B), \tag{56}$$

is one-to-one. In Chapter II, Cor. 5.14 we have a convenient sufficient condition for simplicity. This was proved by studying the behavior of f at infinity. We shall now prove a preliminary necessary and sufficient condition for simplicity; in contrast to the previous method this utilizes the fact that by the analyticity of f, its identical vanishing is equivalent to the vanishing of all its derivatives at the origin.

The non-simplicity of $-\lambda$ is equivalent to the existence of a $F \not\equiv 0$ in $\mathcal{E}(B)$ such that

$$\int_B e^{(-i\lambda+\rho)(A(x,b))}F(b)db = 0, \qquad x \in X. \tag{57}$$

By (3) §1 and (19) this is equivalent to

$$\int_K \zeta_\lambda(k^{-1}g)F(kM)dk \equiv 0. \tag{58}$$

Because of the analyticity of ζ_λ this in turn is equivalent to

$$\left\{ D_g \left(\int_K \zeta_\lambda(k^{-1}g)F(kM)dk \right) \right\}_{g=e} = 0, \qquad D \in \mathbf{D}(G). \tag{59}$$

Consider now the decomposition (16). If $D \in \mathbf{D}(G)\mathfrak{k}$ then $D\zeta_\lambda \equiv 0$. Also if $D \in I^*$ then by (24) D is bi-invariant under K so

$$(D\zeta_\lambda)(k^{-1}) = (D\zeta_\lambda)(e) = \text{ constant in } k$$

so (59) gives no additional information for $D \in I^*$. Thus it suffices to look at (59) for $D \in H^*$.

Lemma 2.11. *Let $\lambda \in \mathfrak{a}_{\mathbf{c}}^*$. Then $-\lambda$ is simple if and only if*

$$\det(Q^\delta(\lambda)) \neq 0 \qquad \textit{for all } \delta \in \widehat{K}_M.$$

Proof. The nonsimplicity of $-\lambda$ is equivalent to (57) above. Thus we may assume $F(\{M\}) \neq 0$. Replacing x in (57) by $m \cdot x$ amounts to replacing F by $F^{\tau(m^{-1})}$. Integrating this over M we see that we may assume F to be M-invariant. We now assume this in the equivalent relations (59) where as we have seen it suffices to consider $D \in H^*$. Since $H^* = \oplus_{\delta \in \widehat{K}_M} H_\delta^*$ and, since H_δ^* is spanned by the $\epsilon_j(V_\delta)$, equation (59) is equivalent to

$$\int_K (\epsilon_j(v_i)\zeta_\lambda)(k^{-1})F(kM)dk = 0 \tag{60}$$

for all $\delta \in \widehat{K}_M$, $1 \leq i \leq d(\delta)$, $1 \leq j \leq \ell(\delta)$. In connection with (26) we saw that

$$(\epsilon_j(v_i)\zeta_\lambda)(k^{-1}) = (\epsilon_j(v_i)^{R(k)}\zeta_\lambda)(e) = \begin{cases} 0 & \text{if } i > \ell(\delta) \\ \sum_{p=1}^{\ell(\delta)} \langle v_i, \delta(k)v_p\rangle Q^\delta(\lambda)_{pj} & \text{if } i \leq \ell(\delta) \end{cases}$$

so (60) is equivalent to

$$\sum_{p=1}^{\ell(\delta)} Q^\delta(\lambda)_{pj} \int_K \langle v_i, \delta(k)v_p\rangle F(kM)dk = 0 \tag{61}$$

for $1 \le i, j \le \ell(\delta)$, $\delta \in \widehat{K}_M$. The M-invariance of F implies that (61) holds also for $\ell(\delta) < i \le d(\delta)$. However, the functions $kM \longrightarrow \langle v_i, \delta(k)v_p \rangle$ span $\mathcal{E}(B)$ so (61) amounts to $\det(Q^\delta(\lambda)) = 0$ for some δ. This completes the proof of the lemma.

We can now restate Theorem 2.1, taking (42) and Lemma 2.11 into account.

Theorem 2.12. *Let $\lambda \in \mathfrak{a}_c^*$ and assume $-\lambda$ is simple. Let the matrix $P^\delta(\lambda)$ be the inverse of $Q^\delta(\lambda)$. Then for $v \in V_\delta$ and $1 \le i \le \ell(\delta)$,*

$$\int_{K/M} e^{(i\lambda+\rho)(A(gK,kM))} \langle v, \delta(k)v_i \rangle dk = \sum_{j=1}^{\ell(\delta)} P^\delta(\lambda)_{ji} (\tau(\epsilon_j(v))\varphi_\lambda)(g).$$

§3. THE Q^δ-MATRICES.

1. The K-finite Functions in $\mathcal{E}_\lambda(\Xi)$.

For $\lambda \in \mathfrak{a}_c^*$ let $\mathcal{E}_\lambda(\Xi)_K$ denote the space of K-finite functions in $\mathcal{E}_\lambda(\Xi)$ and consider the K-fixed vector $\eta_\lambda \in \mathcal{E}_\lambda(\Xi)$ given by

$$\eta_\lambda(g \cdot \xi_o) = e^{(i\lambda-\rho)(H(g))}, \qquad g \in G. \tag{1}$$

We also think of η_λ as a function on G itself. We begin with a result which shows how η_λ generates $\mathcal{E}_\lambda(\Xi)_K$. For $D \in \boldsymbol{D}(G)$ let $\tau(D)$ denote the right-invariant differential operator on G defined by (23) §2. Because of (24) §2 and the fact that each $D \in \boldsymbol{D}(G)$ is K-finite relative to the action of $\mathrm{Ad}_G(K)$ on $\boldsymbol{D}(G)$ it is obvious that

$$\tau(\boldsymbol{D}(G))\eta_\lambda \subset \mathcal{E}_\lambda(\Xi)_K. \tag{2}$$

With $H^* \subset \boldsymbol{D}(G)$ as in §2 consider the map

$$T_\lambda : \ H^* \longrightarrow \mathcal{E}_\lambda(\Xi)_K \tag{3}$$

given by

$$T_\lambda(u) = \tau(u)\eta_\lambda. \tag{4}$$

Then

$$T_\lambda(\mathrm{Ad}(k)u)(\xi) = T_\lambda(u)(k^{-1} \cdot \xi). \tag{5}$$

Theorem 3.1. *Assume $-\lambda$ simple. Then*

$$\tau(\boldsymbol{D}(G))\eta_\lambda = \mathcal{E}_\lambda(\Xi)_K, \tag{6}$$

and more precisely, the map T_λ is a bijection of H^* onto $\mathcal{E}_\lambda(\Xi)_K$.

Remark. Formula (6) is an analog of Theorem 2.1. Note however the difference that Theorem 2.1 does not state that $\tau(\boldsymbol{D}(G))\varphi_\lambda$ constitutes the space $\mathcal{E}_\lambda(X)_K$ of K-finite elements in $\mathcal{E}_\lambda(X)$. While this is true (for $-\lambda$ simple) this will require more refined tools (see Theorem 6.1).

Proof of Theorem 3.1. First we observe that $\mathcal{E}_\lambda(\Xi)$ consists of the functions

$$F \otimes e^{i\lambda-\rho} : \ (kM, a) \longrightarrow F(kM)e^{(i\lambda-\rho)(\log a)}, \tag{7}$$

where $F \in \mathcal{E}(B)$; here $F \otimes e^{i\lambda-\rho}$ is K-finite if and only if F is K-finite. Let $\mathcal{E}_\lambda(\Xi)(\delta)$ denote the space of K-finite functions in $\mathcal{E}_\lambda(\Xi)$ of type $\delta \in \widehat{K}_M$. By the above remark,

$$\dim \mathcal{E}_\lambda(\Xi)(\delta) < \infty. \tag{8}$$

With ζ_λ as in (19) §2 we have $\eta_\lambda(g) = \zeta_\lambda(g^{-1})$ which we write $\eta_\lambda = \zeta_\lambda^\vee$ so

$$\tau(D)\eta_\lambda = (D\zeta_\lambda)^\vee, \qquad D \in \boldsymbol{D}(G). \tag{9}$$

We have seen ((60) §2) that if $-\lambda$ is simple no $F \neq 0$ in $\mathcal{E}(B)$ is orthogonal to all the functions $k \longrightarrow (D\zeta_\lambda)^\vee(k)$. Thus by the finite-dimensionality in (8) we have

$$\tau(\boldsymbol{D}_\delta(G))\eta_\lambda = \mathcal{E}_\lambda(\Xi)(\delta), \tag{10}$$

which proves (6). Also by (10), T_λ maps H_δ^* into $\mathcal{E}_\lambda(\Xi)(\delta)$. Now if $j \in I^*$ ((16) §2) then $\tau(j)\eta_\lambda$ is K-invariant and hence a multiple of η_λ. Since $\tau(\boldsymbol{D}(G)\mathfrak{k})$ annihilates η_λ we deduce from (16) §2 and (6) that $T_\lambda(H^*) = \mathcal{E}_\lambda(\Xi)_K$. But by Prop. 2.5, H_δ^* and $\mathcal{E}_\lambda(\Xi)(\delta)$ have the same dimension so now the bijectivity of T_λ follows.

2. Connections with Harmonic Polynomials.

We now go back to the notation of Ch. II, §5, No. 1. In particular, we consider for $s \in W$ the subgroup $\overline{N}_s = \overline{N} \cap m_s^{-1} N m_s$ and put

$$(I'_{\lambda,s}\psi)(g \cdot \xi_o) = \int_{\overline{N}_s} \psi(g m_s \overline{n}_s \cdot \xi_o) d\overline{n}_s, \quad \psi \in \mathcal{E}_\lambda(\Xi). \tag{11}$$

This is well defined because we saw (*loc. cit.*) that the measure $d\overline{n}_s$ on $\overline{N}_s = N^{s^{-1}}/(N^{s^{-1}} \cap N)$ is $N^{s^{-1}}$-invariant. For convergence we observe the following result.

Lemma 3.2. *With* $*$ *defined as in Ch. II, §5, (53) we have*

$$I'_{\lambda,s}(F \otimes e^{i\lambda-\rho}) = (F \times (S'_{-\lambda,s})^*) \otimes e^{is\lambda-\rho} \tag{12}$$

and (11) converges absolutely if

$$\operatorname{Re}(\langle -i\lambda, \alpha \rangle) > 0 \qquad \text{for} \quad \alpha \in \Sigma^+ \cap s^{-1}\Sigma^-. \tag{13}$$

Proof. Putting $b = m_s^{-1} a m_s$ we have for $\psi = F \otimes e^{i\lambda-\rho}$

$$\int_{\overline{N}_s} \psi(k a m_s \overline{n}_s \cdot \xi_o) d\overline{n}_s = \int_{\overline{N}_s} \psi(k m_s b \overline{n}_s b^{-1} b \cdot \xi_o) d\overline{n}_s.$$

Now

$$\frac{d(b\overline{n}_s b^{-1})}{d\overline{n}_s} = e^{(s^{-1}\rho-\rho)(\log b)} \tag{14}$$

(Ch. II, proof of Theorem 5.10) so since ψ is in $\mathcal{E}_\lambda(\Xi)$ the integral becomes

$$\int_{\overline{N}_s} \psi(k m_s \overline{n}_s \cdot \xi_o) d\overline{n}_s \quad e^{(\rho-s^{-1}\rho)(\log b)} e^{(i\lambda-\rho)(\log b)}$$

$$= \int_{\overline{N}_s} \psi(k m_s k(\overline{n}_s) \cdot \xi_o) e^{(i\lambda-\rho)(H(\overline{n}_s))} d\overline{n}_s \, e^{(is\lambda-\rho)(\log a)}$$

and now the lemma follows from Ch. II, §5 (9) and Theorem 5.1.

Corollary 3.3. *$I'_{\lambda,s}$ maps $\mathcal{E}_\lambda(\Xi)$ into $\mathcal{E}_{s\lambda}(\Xi)$, commutes with the action of G and*

$$I'_{\lambda,s}\eta_\lambda = \boldsymbol{c}_s(-\lambda)\eta_{s\lambda}.$$

More generally, if $w \in V_\delta, v \in V_\delta^M$, then

$$I'_{\lambda,s}\left(e^{(i\lambda-\rho)(\log a)}\langle w, \delta(k)v\rangle\right) = e^{(is\lambda-\rho)(\log a)}\boldsymbol{c}_s(-\lambda)\langle w, \delta(k)(\Gamma_{s,-\lambda})^* v\rangle.$$

In fact, if $F(kM) = \langle w, \delta(k)v\rangle$ then

$$I'_{\lambda,s}(F \otimes e^{i\lambda-\rho})(k \cdot \xi_o) = \int_{\overline{N}_s} e^{(i\lambda-\rho)(H(\overline{n}_s))}\langle w, \delta(k)\delta(m_s k(\overline{n}_s))v\rangle d\overline{n}_s$$

and now the result follows from (55) in §2.

The result above suggests considering the normalized operator

$$I_{\lambda,s} = \boldsymbol{c}_s(-\lambda)^{-1} I'_{\lambda,s}, \tag{15}$$

the factor being well-defined under condition (13).

Corollary 3.4. *Assume condition (13). Then the diagram*

$$\begin{array}{ccc} \mathcal{E}_\lambda(\Xi)_K & \xrightarrow{I_{\lambda,s}} & \mathcal{E}_{s\lambda}(\Xi)_K \\ T_\lambda \nwarrow & & \nearrow T_{s\lambda} \\ & H^* & \end{array}$$

is commutative, i.e., $I_{\lambda,s}T_\lambda(u) = T_{s\lambda}(u)$.

In fact, if $u \in H^*$ then, since $I'_{\lambda,s}$ commutes with the action of G on Ξ,

$$I_{\lambda,s}T_\lambda(u) = I_{\lambda,s}\tau(u)\eta_\lambda = \tau(u)I_{\lambda,s}\eta_\lambda = \tau(u)\eta_{s\lambda}$$

proving $I_{\lambda,s}T_\lambda(u) = T_{s\lambda}(u)$.

Since by (21) §2 and (9) above,

$$(T_\lambda(u))(\xi_o) = q^u(\rho - i\lambda)$$

the corollaries above show that there is a connection between the operator $\Gamma_{s,\lambda}$ and the Q^δ-polynomials. To get this in explicit form we consider again the vector space

$$E_\delta = \mathrm{Hom}_K(V_\delta, H_\delta^*)$$

which, as remarked in §2, has dimension $\ell(\delta)$.

Let $v \longrightarrow v'$ be the conjugate-linear map of V_δ onto $V_{\check{\delta}}$ such that $v'(w) = \langle w, v\rangle$. Since $(A^*v)' = {}^tA(v')$, Corollary 3.3 can be written

$$I_{\lambda,s}(e^{i\lambda-\rho} \otimes F) = e^{is\lambda-\rho} \otimes F_\lambda, \tag{16}$$

where

$$F(kM) = v'(\delta(k^{-1})w), \qquad F_\lambda(kM) = ({}^t\Gamma_{s,-\lambda}(v'))(\delta(k^{-1})w).$$

Let $\epsilon \in E_\delta$ and $w \in V_\delta$. Since by (22) §2 and (19)

$$\begin{aligned} T_\lambda(\epsilon(w))(ka \cdot \xi_o) &= T_\lambda(\epsilon(\delta(k^{-1})w))(a \cdot \xi_o) \\ &= e^{(i\lambda-\rho)(\log a)}(Q^\delta(\lambda)\epsilon)(\delta(k^{-1})w), \end{aligned}$$

Corollary 3.4 and (16) implies (taking $v' = Q^\delta(\lambda)(\epsilon)$),

$$\left({}^t\Gamma_{s,-\lambda}\right) Q^\delta(\lambda)(\epsilon) = Q^\delta(s\lambda)(\epsilon). \tag{17}$$

Here we replace δ by $\check{\delta}$ and recall that $(\check{\delta}(k^{-1})(v'))(w) = v'(\delta(k)w)$. In the notation of (54)–(55) of §2 we have

$$(\Gamma_{s,\lambda})^{\check{\delta}} = {}^t\left((B(\lambda,s)^\delta\right), \tag{18}$$

the superscripts δ and $\check{\delta}$ having their obvious meaning. Taking transpose again in (18), we obtain the following result from (17).

Theorem 3.5. *Let $\delta \in \widehat{K}_M$. Then the restriction of $B(\lambda, m_s)$ to V_δ^M is given by*

$$B(\lambda, s) = Q^{\check{\delta}}(-s\lambda)Q^{\check{\delta}}(-\lambda)^{-1}, \qquad s \in W.$$

We shall need a simple relationship between $Q^{\check{\delta}}(\lambda)$ and $Q^\delta(\lambda)$ which is best expressed in the matrix form ((22)′ §2)

$$Q^\delta(\lambda)_{ij} = q^{\epsilon_j(v_i)}(\rho - i\lambda), \tag{19}$$

where $v_1, \ldots, v_{\ell(\delta)}$ is an orthonormal basis of V_δ^M and $\epsilon_1, \ldots, \epsilon_{\ell(\delta)}$ any basis of E_δ. Corresponding to the direct decompositions

$$\begin{aligned} \mathfrak{a}_{\mathbf{c}}^* &= \mathfrak{a}^* + i\mathfrak{a}^* \\ \boldsymbol{D}(G) &= \boldsymbol{D}(G)_{\boldsymbol{R}} + i\boldsymbol{D}(G)_{\boldsymbol{R}}, \end{aligned}$$

where $\boldsymbol{D}(G)_{\boldsymbol{R}}$ is the set of operators in $\boldsymbol{D}(G)$ with real coefficients, consider the conjugations

$$\lambda \longrightarrow \overline{\lambda} \qquad (\lambda \in \mathfrak{a}_{\boldsymbol{c}}^*), \qquad D \longrightarrow \overline{D} \qquad (D \in \boldsymbol{D}(G)).$$

It is clear that $(H^*)^- = H^*$.

Lemma 3.6. *For* $\epsilon \in E_\delta = \mathrm{Hom}_K(V_\delta, H_\delta^*)$ *define the map* $\epsilon' : V_{\check{\delta}} \longrightarrow H^*$ *by*

$$\epsilon'(v') = \epsilon(v)^- \qquad v \in V_\delta.$$

Then $\epsilon' \in E_{\check{\delta}}$, *and relative to the bases* $(v_i), (v_i'), (\epsilon_j), (\epsilon_j')$ *we have*

$$Q^{\check{\delta}}(\lambda)_{ij} = (Q^\delta(-\overline{\lambda})_{ij})^-. \tag{20}$$

Proof. First we prove that

$$(H_\delta^*)^- = H_{\check{\delta}}^*. \tag{21}$$

For this let $h_1, \dots, h_{\ell(\delta)}$ be a basis of a K-invariant irreducible subspace of H_δ^*. Then, dot denoting the adjoint action,

$$k \cdot h_j = \sum_j \delta(k)_{ij} h_i. \tag{22}$$

Since the adjoint action on $\boldsymbol{D}(G)$ commutes with the conjugation $D \longrightarrow \overline{D}$ we have

$$k \cdot \overline{h}_j = (k \cdot h_j)^- = \sum_i (\delta(k)_{ij})^- \overline{h}_i = \sum_i (\check{\delta}(k))_{ij} \overline{h}_i \tag{23}$$

so $\overline{h}_j \in H_{\check{\delta}}^*$, proving (21). Next we have (e.g. [GGA], p. 393)

$$\begin{aligned} \epsilon'(\check{\delta}(k)v') &= \epsilon'((\delta(k)v)') = (\epsilon(\delta(k)v))^- \\ &= (k \cdot \epsilon(v))^- = k \cdot (\epsilon(v))^- \end{aligned}$$

which by (21) belongs to $H_{\check{\delta}}^*$. Thus $\epsilon \longrightarrow \epsilon'$ is a complex-linear bijection of E_δ onto $E_{\check{\delta}}$. In the bases $(v_i'), (\epsilon_j')$

$$Q^{\check{\delta}}(\lambda)_{ij} = q^{\epsilon_j'(v_i')}(\rho - i\lambda).$$

Since $q^{\overline{u}} = (q^u)^-$ we thus have

$$Q^{\check{\delta}}(\lambda)_{ij} = \left(q^{\epsilon_j(v_i)}\right)^- (\rho - i\lambda) = \left(q^{\epsilon_j(v_i)}(\rho + i\overline{\lambda})\right)^-,$$

proving (20).

3. A Product Formula for $\det Q^\delta(\lambda)$ (Preliminary Version).

With $\delta \in \widehat{K}_M$ and $s \in W$ we consider again the endomorphism $A(\lambda, s)$ of V_δ^M given by

$$A(\lambda, s)v = \int_{\overline{N}_s} e^{-(i\lambda+\rho)(H(\overline{n}_s))}\delta(m_s k(\overline{n}_s))v\, d\overline{n}_s, \tag{24}$$

absolute convergence being guaranteed for

$$\mathrm{Re}(\langle \alpha, i\lambda\rangle) > 0 \qquad \text{for} \quad \alpha \in \Sigma^+ \cap s^{-1}\Sigma^-. \tag{25}$$

Let $w \in V_\delta$ and put $F(kM) = \langle \delta(k)v, w\rangle$. Then

$$\langle A(\lambda, s)v, w\rangle = S'_{\lambda,s}(F), \tag{26}$$

so by Ch. II, Theorem 5.3, and Lemma 5.2,

$$A(\lambda, s) = A(s^{(1)}\lambda, s_{\alpha_1})A(s^{(2)}\lambda, s_{\alpha_2})\dots A(s^{(p)}\lambda, s_{\alpha_p}), \tag{27}$$

where $s = s_{\alpha_1}\dots s_{\alpha_p}$ is a reduced expression of s. For each $\beta \in \Sigma^+$ such that $\frac{1}{2}\beta \notin \Sigma^+$ we consider the rank-one symmetric space G_β/K_β as in Ch. II §5, No. 2 together with the subgroups $A_{(\beta)} = A \cap G_\beta$, $N_\beta = N \cap G_\beta$, and $M_\beta = M \cap G_\beta$. For δ the trivial one-dimensional representation, (27) becomes the product formula

$$\boldsymbol{c}_s(\lambda) = \prod_{j=1}^{p} \boldsymbol{c}_{\alpha_j}(\lambda_j), \tag{28}$$

where $\lambda_j = (s^{(j)}\lambda)|\mathfrak{a}_{\alpha_j}$, and $\boldsymbol{c}_{\alpha_j}$ is the $\boldsymbol{c}$-function for $G_{\alpha_j}/K_{\alpha_j}$. (Here we have normalized the Haar measure on each $\overline{N}_{s_{\alpha_j}}$ such that $\exp(-2\rho(H(\overline{n}_{s_{\alpha_j}}))$ has integral 1).

We shall now combine (27) with Theorem 3.5 to calculate the determinant of $B(\lambda, m_s)|V_\delta^M$ in two ways. Note first that since $\mathrm{Ad}_G(M)$ leaves each root space $\mathfrak{g}_\beta$ invariant, the group M normalizes G_β and K_β. Hence $K_\beta M$ is a group and since $M \cap K_\beta = M_\beta$ we have

$$K_\beta M/M = K_\beta/M_\beta. \tag{29}$$

Lemma 3.7. *Let*

$$C^\natural(K_\beta M) = \{f \in C(K_\beta M) : \ f \ \ \textit{bi-invariant under } M\}$$
$$C^\natural(K_\beta) = \{F \in C(K_\beta) : \ F \ \ \textit{bi-invariant under } M_\beta\}$$

and let

$$f^\natural = f|K_\beta \qquad (= \textit{ restriction to } K_\beta).$$

Then the convolution algebras $C^\natural(K_\beta M)$ *and* $C^\natural(K_\beta)$ *are commutative and the map*

$$f \longrightarrow f^\natural \tag{30}$$

is an isomorphism of $C^\natural(K_\beta M)$ *into* $C^\natural(K_\beta)$.

Proof. We know from Cor. 6.8 in Ch. II that $C^\natural(K_\beta)$ is commutative. Under the diffeomorphism $k_\beta M \longrightarrow k_\beta M_\beta$ of $K_\beta M/M$ onto K_β/M_β the actions of K_β correspond. Hence the algebra $\boldsymbol{D}(K_\beta M/M)$ is a subalgebra of $\boldsymbol{D}(K_\beta/M_\beta)$ which by Theorem 6.6, Ch. II is commutative. Hence $\boldsymbol{D}(K_\beta M/M)$ is commutative so, by Theorem 2.9, Ch. II, the convolution algebra $C^\natural(K_\beta M)$ is commutative.

If $h \in C(K_\beta M)$ the function $(k, m) \longrightarrow h(km)$ on $K_\beta \times M$ is continuous and the functional

$$h \longrightarrow \int_{K_\beta} dk \left(\int_M h(km)dm \right) \tag{31}$$

is a Haar measure on $K_\beta M$. Hence if $f, g \in C^\natural(K_\beta M), \ell \in K_\beta$

$$(f * g)^\natural(\ell) = \int_{K_\beta}\int_M f(\ell(km)^{-1})g(km)dkdm = \int_{K_\beta} g(k)dk \int_M f(\ell mk^{-1})dm.$$

If f is a spherical function for the pair $(K_\beta M, M)$ the integral over M equals $f(\ell)f(k^{-1})$ so

$$(f * g)^\natural(\ell) = f(\ell) \int_{K_\beta} g(k)f(k^{-1})dk = f(\ell) \int_{K_\beta} g^\natural(k)f^\natural(k^{-1})dk,$$

which by Exercise **D2** in this chapter equals $(f^\natural * g^\natural)(\ell)$. Thus we obtain

$$(f * g)^\natural = f^\natural * g^\natural \qquad f, g \in C^\natural(K_\beta M) \tag{32}$$

by decomposing $f \in C^\natural(K_\beta M)$ into spherical functions. The injectivity of $f \longrightarrow f^\natural$ being obvious the lemma is proved.

Lemma 3.8. *Suppose* $\dim K_\beta > 1$. *Then the map* $\tau : M_\beta k M_\beta \longrightarrow MkM$ *is a bijection between the double coset spaces*

$$M_\beta \backslash K_\beta / M_\beta \text{ and } M\backslash(K_\beta M)/M,$$

and the mapping (30) is surjective. Moreover, an irreducible representation π *of* $K_\beta M$ *with an* M*-fixed vector remains irreducible when restricted to* K_β *and all* M_β*-spherical irreducible representation of* K_β *are obtained in this way.*

Proof. The surjectivity of τ is obvious; to prove injectivity we must prove that if $mkm' = k'$ for some $m, m' \in M$, $k, k' \in K_\beta$ then $m_\beta k m'_\beta = k'$ for some $m_\beta, m'_\beta \in M_\beta$. Now $K_\beta M/M$ and K_β/M_β have tangent space $\mathfrak{l}_\beta + \mathfrak{l}_{2\beta}$ at the origin (Ch. II, §6, No. 4) so

$$kM = \exp(T_1 + T_2)M \qquad T_1 \in \mathfrak{l}_\beta,\ T_2 \in \mathfrak{l}_{2\beta}. \tag{33}$$

Since $mkM = mkm^{-1}M$ we see that

$$mkM = \exp(S_1 + S_2)M, \quad S_1 \in \mathfrak{l}_\beta,\ S_2 \in \mathfrak{l}_{2\beta}. \tag{34}$$

If $\dim \mathfrak{l}_{2\beta} > 1$ or if $\dim \mathfrak{l}_{2\beta} = 0$ and $\dim \mathfrak{l}_\beta > 1$ we know from Ch. II, Exercises **D2, 3** that for some $m_\beta \in M_\beta$, $S_i = \mathrm{Ad}(m_\beta)T_i$ $(i = 1, 2)$. Thus $mkM = m_\beta kM$ so $mk = m_\beta k m''$ $(m'' \in M)$. Multiplying by m' we get $k' = m_\beta k m'' m'$ so by $K_\beta \cap M = M_\beta$ we get $k' = m_\beta k m'_\beta$ as desired.

In the case $\dim \mathfrak{l}_{2\beta} = 1$ we claim that $\mathrm{Ad}_G(M)$ restricted to $\mathfrak{l}_{2\beta}$ is the identity. By Remark following Lemma 1.3 in Ch. II, $\mathrm{Ad}_G(M)$ is independent of the choice of G so we take G inside the simply connected complexification $G^{\mathbf{c}}$. Then $M = M_0(\exp(i\mathfrak{a}) \cap K)$ ([DS], Ch. IX, Exercise A3), M_0 denoting the identity component of M. By $\dim \mathfrak{l}_{2\beta} = 1, \mathrm{Ad}_G(M_o)|\mathfrak{l}_{2\beta} = $ identity. On the other hand, if $H \in i\mathfrak{a}$ is such that $\exp H \in K$ then $\mathrm{Ad}(\exp H)\mathfrak{l}_{2\beta} \subset \mathfrak{k}$, $(\mathrm{ad}\, H)^{2n-1}\mathfrak{l}_{2\beta} \subset i\mathfrak{p}$ so $e^{\mathrm{ad}\, H}|\mathfrak{l}_{2\beta} = \cosh(\mathrm{ad}\, H)|\mathfrak{l}_{2\beta}$ and $\mathrm{Ad}(\exp H)|\mathfrak{l}_{2\beta}$ is scalar multiplication by $\cosh 2\beta(H)$. But by [DS], Ch. VII, Theorem 8.5, H is an integral linear combination of the vectors $2\pi i A_\alpha/\langle\alpha, \alpha\rangle$ $(\alpha \in \Sigma)$ so since $2\langle\beta, \alpha\rangle/\langle\alpha, \alpha\rangle \in \boldsymbol{Z}$, we have $e^{2\beta(H)} = 1$. Thus in (34) we have $S_2 = T_2$ and by the Exercise **D2** quoted, $S_1 = \mathrm{Ad}(m_\beta)T_1$ and now $m_\beta k m'_\beta = k'$ follows as before.

For the surjectivity of the map (30) let $F \in C^\natural(K_\beta)$. Define f on $K_\beta M$ by

$$f(k_\beta m) = F(k_\beta).$$

If $k_\beta m = k'_\beta m'$ the injectivity of τ implies that $f(k_\beta m) = f(k'_\beta m')$ so f is well-defined. Also, if $m' \in M$

$$f(m' k_\beta m) = f(k'_\beta m' m) = F(k'_\beta)$$

where $k'_\beta = m'k_\beta(m')^{-1}$. By the injectivity of τ, $F(k'_\beta) = F(k_\beta)$. Thus $f \in C^\natural(K_\beta M)$ and $f|K_\beta = F$.

Finally, we consider the representation π (without invoking the assumption $\dim K_\beta > 1$ for the time being). Since $C^\natural(K_\beta M)$ and $C^\natural(K_\beta)$ are both commutative convolution algebras, the irreducible spherical representations for the pair $(K_\beta M, M)$ and the pair (K_β, M_β) are by [GGA], Theorem 3.4, Ch. IV, given by the $K_\beta M$-translates (respectively, K_β-translates) of the functions on $K_\beta M/M = K_\beta/M_\beta$ which are spherical for the pair $(K_\beta M, M)$ (respectively the pair (K_β, M_β)).

For $\dim K_\beta > 1$ the convolution algebras $C^\natural(K_\beta M)$ and $C^\natural(K_\beta)$ coincide, hence by [GGA], Theorem 3.3 in Ch. IV, the spherical functions φ on $K_\beta M/M = K_\beta/M_\beta$ are the same for both pairs $(K_\beta M, M)$ and (K_β, M_β). Since φ is M-invariant the span of its $K_\beta M$-translates is the same as the span of its K_β-translates. Thus the restriction $\pi|K_\beta$ is just the natural representation of K_β on the span of translates of a spherical function on K_β/M_β, hence is irreducible. Since φ can be an arbitrary spherical function on K_β/M_β the last statement of the lemma is obvious.

We must now consider the case $\dim K_\beta = 1$. This case causes a certain complication because the circle is a homogeneous space both of the circle group $\boldsymbol{SO}(2)$ and of the noncommutative group $\boldsymbol{O}(2)$ and in these two cases the spherical functions are $\theta \longrightarrow e^{i\ell\theta}$ and $\theta \longrightarrow \cos \ell\theta$, respectively.

Lemma 3.9. *Suppose* $\dim K_\beta = 1$. *Then we have two cases*

(a) $mkm^{-1} = k$ *for all* $k \in K_\beta$, $m \in M$.

(b) $mkm^{-1} = k^{-1}$ *for all* $k \in K_\beta$, *some* $m \in M$.

In Case (a) an M-spherical irreducible representation π of $K_\beta M$ remains irreducible when restricted to K_β.

In Case (b) an M-spherical irreducible representation $\pi \not\equiv I$ of $K_\beta M$ decomposes when restricted to K_β into two irreducible one-dimensional components π_o and its contragredient $\check{\pi}_o$.

In both cases all M_β-spherical irreducible representations of K_β are so obtained (by $\pi|K_\beta$ in Case (a) and by π_o in Case (b).)

Proof. In Case (a) the map τ in Lemma 3.8 is still bijective and as a result the mapping $f \longrightarrow f^\natural$ is still surjective. Thus the irreducibility of $\pi|K_\beta$ follows as before and all M_β-spherical irreducible representations of K_β are so obtained.

In Case (b) we must compare M-spherical and M_β-spherical functions on the circle $K_\beta M/M = K_\beta/M_\beta$. The action of M_β on K_β/M_β being trivial the M_β-spherical functions are just the characters χ of K_β/M_β. On the other hand, $mkM = mk^{-1}M$ for a suitable m for which $\chi(mkm^{-1}) = \chi(k^{-1}) = \overline{\chi}(k)$ so the M-spherical functions are $\varphi = \frac{1}{2}(\chi+\overline{\chi})$. Hence $\varphi^{\tau(k)} = \frac{1}{2}(\chi(k^{-1})\chi + \chi(k)\overline{\chi})$, so if $\chi \not\equiv 1$ the M-spherical irreducible representation π given by φ has the properties stated.

Lemma 3.10. *With $\delta \in \widehat{K}_M$ arbitrary, let V denote the $K_\beta M$-invariant subspace of V_δ generated by V_δ^M and $V = \bigoplus_{i=1}^{\ell} V_i$ a decomposition into $K_\beta M$-irreducible subspaces. Then $\ell = \ell(\delta)$, the dimension of V_δ^M, and* $\dim(V_\delta^M \cap V_i) = 1$ *for each i.*

Proof. Let $V' = \bigoplus_{i\in I} V_i$ be the sum of the spaces V_i having a nonzero $\delta(M)$-fixed vector. If $v \in V_\delta^M$ and $v = \sum_1^{\ell} v_i \quad (v_i \in V_i)$ then by the M-invariance of each V_i, each v_i is $\delta(M)$-fixed so $v_i \in V_\delta^M$. Thus $V_\delta^M \subset V'$ whence $V \subset V'$. Conversely if $i \in I$ then $V_\delta^M \cap V_i \neq 0$ and V_i is spanned by $\delta(K_\beta M)(V_\delta^M \cap V_i)$ so $V' \subset V$. Thus $V' = V$ and $\dim(V_\delta^M \cap V_i) > 0$ for $1 \leq i \leq \ell$. Since the convolution algebra $C^\natural(K_\beta M)$ is commutative we have by [GGA], Ch. V, Theorem 3.5, that

$$\dim(V_\delta^M \cap V_i) = 1 \qquad \text{for} \quad 1 \leq i \leq \ell. \tag{35}$$

This proves the lemma.

Fixing a unit vector v_i in each $V_\delta \cap V_i$ we have

$$V_\delta^M = \sum_{i=1}^{\ell(\delta)} \boldsymbol{C} v_i. \tag{36}$$

Let $\delta_i(1 \leq i \leq \ell(\delta))$ denote the representation of $K_\beta M$ on V_i defined by the lemma.

Coming back to the decomposition (27) fix j $(1 \leq j \leq p)$, take $\beta = \alpha_j$ and let δ_i denote the (irreducible) representation of $K_{\alpha_j}M$ on V_i given by δ. The operator $A(s^{(j)}\lambda, m_{s_{\alpha_j}})$ maps $V_i \cap V_\delta^M$ into itself by multiplication with the scalar

$$\int_{\overline{N}_{\alpha_j}} \langle \delta_i(m_{s_{\alpha_j}} k(\overline{n}_{\alpha_j}))v_i, v_i\rangle e^{-(i\lambda_j+\rho_{\alpha_j})(H(\overline{n}_{\alpha_j}))} d\overline{n}_{\alpha_j}, \tag{37}$$

where we have used the fact that the restriction $\rho|\mathfrak{a}_{\alpha_j}$ equals the ρ-function ρ_{α_j} for the space $G_{\alpha_j}/K_{\alpha_j}$ (cf. [GGA], Ch. IV, §6 (34)). Let $m_{\alpha_j} \in M'_{\alpha_j}$ (the normalizer of $\mathfrak{a}_{\alpha_j}$ in K_{α_j}) be such that $\mathrm{Ad}(m_{\alpha_j}) = -1$ on $\mathfrak{a}_{\alpha_j}$. Then (Ch. II, §5, No. 2) $m_{\alpha_j} = mm_{s_{\alpha_j}}$ with $m \in M$ so in (37) we can replace $m_{s_{\alpha_j}}$ by m_{α_j}. In order to compute (37) using Theorem 3.5 on G_{α_j} we must first inquire about the irreducibility of δ_i restricted to K_{α_j}. This is answered by Lemmas 3.8–3.9. The irreducibility can only fail in Case (b) in Lemma 3.9 and for this case Theorem 3.5 does not apply directly.

Lemma 3.11. *Suppose* $\dim K_{\alpha_j} = 1$ *and that*

$$\delta_i|K_{\alpha_j} = \delta_{i,o} \oplus \check{\delta}_{i,o}$$

according to Lemma 3.9. Let $\mathbf{e}$ *and* $\check{\mathbf{e}}$ *be unit vectors in the representation spaces of* $\delta_{i,o}$ *and* $\check{\delta}_{i,o}$, *respectively. Then the integral (37) equals*

$$(38) \qquad \int_{\overline{N}_{\alpha_j}} \langle \delta_{i,o}(m_{\alpha_j}k(\overline{n}))\mathbf{e}, \mathbf{e}\rangle e^{-(i\lambda_j+\rho_{\alpha_j})(H(\overline{n}))}\,d\overline{n},$$

which also equals

$$(39) \qquad \int_{\overline{N}_{\alpha_j}} \langle \check{\delta}_{i,o}(m_{\alpha_j}k(\overline{n}))\check{\mathbf{e}}, \check{\mathbf{e}}\rangle e^{-(i\lambda_j+\rho_{\alpha_j})(H(\overline{n}))}\,d\overline{n}.$$

Proof. Since $2\varphi = \chi + \overline{\chi}$ in Lemma 3.9 we can normalize $\mathbf{e}$ and $\check{\mathbf{e}}$ such that $\sqrt{2}v_i = \mathbf{e} + \check{\mathbf{e}}$. Thus it suffices to prove that expressions (38) and (39) are equal. By Theorem 3.5 these expressions are, up to the factor $\mathbf{c}_{\alpha_j}(\lambda_j)$, given respectively by,

$$(40) \qquad \frac{Q^{\check{\delta}_{i,o}}(\lambda_j)}{Q^{\check{\delta}_{i,o}}(-\lambda_j)} \quad \text{and} \quad \frac{Q^{\delta_{i,o}}(\lambda_j)}{Q^{\delta_{i,o}}(-\lambda_j)}.$$

Using now Theorem 2.10, equation (17) and [GGA], Prop. 4.12 p. 50, we see that the first expression (40) has the form $r(i\lambda_j)$ where r is the ratio of two polynomials with real coefficients. In particular, $\overline{r(i\lambda_j)} = r(-i\overline{\lambda}_j)$. Combining this with (20) we see that the two expressions in (40) actually coincide.

Let $\delta(i,j)$ $(1 \leq i \leq \ell(\delta), 1 \leq j \leq p)$ denote the representation δ_i of K_{α_j} defined above with the proviso that in the case when $\delta_i|K_{\alpha_j}$ is not irreducible, $\delta(i,j)$ stands for the component $\delta_{i,o}$. Then (37) represents a diagonalization of the linear transformation

$$\mathbf{c}_{\alpha_j}(\lambda_j)^{-1}A(s^{(j)}\lambda, s_{\alpha_j}) \tag{41}$$

with eigenvalues

$$Q_{\alpha_j}^{\delta(i,j)^\vee}(\lambda_j)Q_{\alpha_j}^{\delta(i,j)^\vee}(-\lambda_j)^{-1}, \tag{42}$$

where the subscript α_j indicates the Q-polynomial for the space $G_{\alpha_j}/K_{\alpha_j}$. Then Theorem 3.5, (27), and (28) provide two ways of calculating $\det(B(\lambda,s)|V_\delta^M)$, yielding the following formula

$$\det \overset{\vee}{Q}{}^{\delta}(s\lambda)\det \overset{\vee}{Q}{}^{\delta}(\lambda)^{-1} = \prod_{\substack{1\leq i\leq \ell(\delta)\\ 1\leq j\leq p}} Q_{\alpha_j}^{\delta(i,j)^\vee}(-\lambda_j)Q_{\alpha_j}^{\delta(i,j)^\vee}(\lambda_j)^{-1}. \tag{43}$$

Now we recall from Ch. II, §5, No. 3 that

$$\Sigma_s^+ = \Sigma_o^+ \cap s^{-1}\Sigma_o^- = \{\beta_1, \dots, \beta_p\} \text{ where } \alpha_j = s^{(j)}\beta_j. \tag{44}$$

Then

$$\lambda_j = s^{(j)}\lambda|\mathfrak{a}_{\alpha_j} = (\lambda|\mathfrak{a}_{\beta_j}) \circ (s^{(j)})^{-1} = \lambda^j \circ (s^{(j)})^{-1} \tag{45}$$

where λ^j is defined by this last relation. If we now conjugate the group G_{α_j} by $m_{s(j)}^{-1}$ we get G_{β_j} and the corresponding subgroups $K_{\beta_j}, A_{\beta_j}, N_{\beta_j}, \overline{N}_{\beta_j}$ and M_{β_j}. We also get the representation

$$k_{\beta_j} \longrightarrow \delta(i,j)(m_{s(j)}k_{\beta_j}m_{s(j)}^{-1}) \qquad \text{of } K_{\beta_j} \tag{46}$$

which we also denote by $\delta(i,j)$. The multiplicities of $\beta_j, 2\beta_j$ being the same as those of $\alpha_j, 2\alpha_j$ we have

$$\rho_{\alpha_j}(H(\overline{n}_{\alpha_j})) = (s^{(j)^{-1}}\rho_{\alpha_j})(H(\overline{n}_{\beta_j})) = \rho_{\beta_j}(H(\overline{n}_{\beta_j})). \tag{47}$$

Combining this with (45) we obtain

$$\mathbf{c}_{\alpha_j}(\lambda_j) = \mathbf{c}_{\beta_j}(\lambda^j). \tag{48}$$

Putting these remarks together and writing $\lambda_\beta = \lambda|\mathfrak{a}_\beta$, $\delta(i,\beta_j) = \delta(i,j)$ we obtain

$$(49)\quad \det \overset{\vee}{Q}{}^{\delta}(s\lambda) \det \overset{\vee}{Q}{}^{\delta}(\lambda)^{-1} \prod_{\beta\in\Sigma_s^+, 1\le i\le \ell(\delta)} Q_\beta^{\delta(i,\beta)^\vee}(-\lambda_\beta) Q_\beta^{\delta(i,\beta)^\vee}(\lambda_\beta)^{-1}.$$

Here we take for s the "maximal" Weyl group element s^*. Then $\Sigma_s^+ = \Sigma_o^+$ so clearing denominators we obtain from (49)

$$\det \overset{\vee}{Q}{}^{\delta}(s^*\lambda) \prod_{\beta\in\Sigma_o^+, 1\le i\le \ell(\delta)} Q_\beta^{\delta(i,\beta)^\vee}(\lambda_\beta)$$
$$= \det \overset{\vee}{Q}{}^{\delta}(\lambda) \prod_{\beta\in\Sigma_o^+, 1\le i\le \ell(\delta)} Q_\beta^{\delta(i,\beta)^\vee}(-\lambda_\beta).$$

By the simplicity criterion (Ch. II, Theorem 6.5), we know that either λ_β or $-\lambda_\beta$ is simple. Thus by Lemma 2.11, the two products above (over $\beta \in \Sigma_o^+, 1 \le i \le \ell(\delta)$) are relatively prime so we deduce that

$$(50)\qquad \det \overset{\vee}{Q}{}^{\delta}(\lambda) = P_\delta(\lambda) \prod_{\beta\in\Sigma_o^+, 1\le i\le \ell(\delta)} Q_\beta^{\delta(i,\beta)^\vee}(\lambda_\beta),$$

where $P_\delta(\lambda)$ is a polynomial. The following result is now crucial.

Lemma 3.12. *The polynomial P_δ in (50) is Weyl group invariant.*

Proof. Fix a simple root γ and let s be the reflection $s = s_\gamma$. Then s permutes $\Sigma_o^+ - \{\gamma\}$ so

$$\det \overset{\vee}{Q}{}^{\delta}(s\lambda) = P_\delta(s\lambda) \prod_{1\le i\le \ell(\delta)} Q_\gamma^{\delta(i,\gamma)^\vee}(-\lambda_\gamma) \prod_{\beta\in\Sigma_o^+-(\gamma), 1\le i\le \ell(\delta)} Q_\beta^{\delta(i,\beta)^\vee}(\lambda_\beta).$$

Here we divide by the expression in (50) and compare the result with formula (49) for $s = s_\lambda$. In this case $\Sigma_s^+ = (\gamma)$ and as a result we get $P_\delta(s\lambda) = P_\delta(\lambda)$. Since the simple reflections generate W the lemma is proved.

§4. THE SIMPLICITY CRITERION.

In this section we shall establish the necessary and sufficient condition $e_{s^*}(\lambda) \neq 0$ for simplicity of a $\lambda \in \mathfrak{a}_c^*$. We begin by proving a sharpened form of Corollary 5.14 in Ch. II.

Proposition 4.1. *Let $\lambda \in \mathfrak{a}_c^*$ be such that*

$$\text{Re}(\langle i\lambda, \alpha \rangle) \geq 0 \text{ for } \alpha \in \Sigma^+. \tag{1}$$

Then λ is simple.

Proof. We proceed by induction on the rank ℓ, the case $\ell = 1$ being proved by Corollary 5.14 and Lemma 6.6 in Ch. II. The proof uses ideas similar to those used in proving the criterion of the boundedness of the spherical function in [GGA], IV, §8, No. 3 and combines these methods with the product formula (50) in §3.

We recall first (Ch. II, §5, No. 4) that λ is simple if and only if the map $\psi \longrightarrow \overset{\vee}{\psi}$ from $\mathcal{E}_\lambda(\Xi)$ to $\mathcal{E}_\lambda(X)$ is injective. Recall that $\mathcal{E}_\lambda(\Xi)$ denotes $\mathcal{E}(\Xi) \cap \mathcal{D}'_\lambda(\Xi)$ and its elements are thus characterized by

$$\psi(ga \cdot \xi_o) = \psi(g \cdot \xi_o) e^{(i\lambda - \rho)(\log a)}. \tag{2}$$

Explicitly, λ is simple if and only if

$$\psi \in \mathcal{E}_\lambda(\Xi), \qquad \int_K \psi(gk \cdot \xi_o) dk \equiv 0 \Longrightarrow \psi \equiv 0. \tag{3}$$

By translation invariance we can reformulate this as follows:
(4)

$$\lambda \in \mathfrak{a}_c^* \text{ is simple if and only if } \psi \in \mathcal{E}_\lambda(\Xi), \int_K \psi(gk \cdot \xi_o) dk \equiv 0 \Longrightarrow \psi(\xi_o) = 0.$$

Let $\lambda \in \mathfrak{a}_c^*$ satisfy (1). We write $\lambda = \xi + i\eta$ so $i\lambda = i\xi - \eta$. Then the element $H^o = A_{-\eta}$ belongs to $\overline{\mathfrak{a}^+}$.

If $H^o \neq 0$ we associate with it the boundary component $X' = G'/K'$ defined in Ch. II, §1, No. 3 and use the notation there associated with X'. On the groups $\overline{N}'$ and $\overline{N}''$ we select Haar measures dv and dw, respectively, such that under the diffeomorphism $(v, w) \longrightarrow vw$ of $\overline{N}' \times \overline{N}''$ onto $\overline{N}$ the product measure $dv\, dw$ corresponds to $d\overline{n}$ (cf. [GGA], p. 465). Let $2\rho'$ and $2\rho''$ denote the sum of the roots (with multiplicity) in Δ' and Δ'', respectively. As in *loc. cit.* p. 466, fix $H \in \mathfrak{a}$ such that

$$\alpha(H) = 0 \text{ for } \alpha \in \Delta'; \qquad \alpha(H) > 0 \text{ for } \alpha \in \Delta''$$

and put $a_t = \exp tH \quad (t \in \boldsymbol{R})$.

We assume now

$$(5) \qquad \psi \in \mathcal{E}_\lambda(\Xi), \qquad \int_K \psi(gk \cdot \xi_o)dk = 0, \qquad g \in G.$$

Here we put $g = a_t$ and transfer the integration from K/M to $\overline{N}$ via the map $\overline{n} \longrightarrow k(\overline{n})M$ for which $dk_M = e^{-2\rho(H(\overline{n}))}d\overline{n}$. Then

$$(6) \qquad \int_{\overline{N}} \psi(a_t k(\overline{n}) \cdot \xi_o)e^{-2\rho(H(\overline{n}))}d\overline{n} = 0$$

and since $a_t k(\overline{n}) \cdot \xi_o = a_t \overline{n} \exp(-H(\overline{n})) \cdot \xi_o$, we find

$$(7) \qquad \int_{\overline{N}} \psi(a_t \overline{n} a_t^{-1} \cdot \xi_o)e^{-(i\lambda+\rho)(H(\overline{n}))}d\overline{n} = 0.$$

Here we plug in $\overline{n} = vw$, $d\overline{n} = dv\ dw$ from above and get

$$(8) \qquad \int_{\overline{N}' \times \overline{N}''} \psi(a_t vw a_t^{-1} \cdot \xi_o)e^{-(i\lambda+\rho)(H(vw))}dv\ dw = 0.$$

Now we use the decomposition $v = k'a'n'$ from (18), Ch. II, §1, and invoke the fact that G' commutes elementwise with a_t. Thus our integral is

$$(9) \qquad \int \psi(k' a_t a' n' w a_t^{-1} \cdot \xi_o)e^{-(i\lambda+\rho)(H(a'n'w))}dv\ dw = 0.$$

Now G' normalizes $\overline{N}''$ and by the unipotency of $\mathrm{Ad}(n')$, the automorphism $w \longrightarrow n'w(n')^{-1}$ of $\overline{N}''$ preserves the measure dw. Thus n' can be removed in (9). Next note that

$$\frac{d(a'w(a')^{-1})}{dw} = e^{-2\rho''(\log a')}$$

so (9) becomes

$$(10) \qquad \int_{\overline{N}' \times \overline{N}''} \psi(k' a_t w a_t^{-1} \cdot \xi_o)K_\lambda(a', w)dvdw = 0$$

where, because of (2),

$$\begin{aligned} K_\lambda(a', w) &= e^{(i\lambda-\rho)(\log a')}e^{-(i\lambda+\rho)(H(w))}e^{-(i\lambda+\rho)(\log a')}e^{2\rho''(\log a')} \\ &= e^{-2\rho'(H(v))}e^{-(i\lambda+\rho)(H(w))}. \end{aligned}$$

Using (2) again we thus get

$$\int_{\overline{N}'} e^{-2\rho'(H(v))}dv \int_{\overline{N}''} \psi(k(v)k(w^{a_t})\cdot\xi_o)e^{(i\lambda-\rho)(H(a_t w a_t^{-1}))}e^{-(i\lambda+\rho)(H(w))}dw = 0.$$

As shown in [GGA], p. 466 the integrand is bounded by a constant multiple of

$$e^{-2\rho'(H(v))}e^{(\epsilon\eta-\rho)(H(w))},$$

where ϵ is such that $0 < \epsilon < 1$, $\rho + \epsilon\eta \in \mathfrak{a}^*_+$. This being integrable over $\overline{N}' \times \overline{N}''$ we can let $t \longrightarrow +\infty$; then $w^{a_t} \longrightarrow e$ and as shown *loc. cit.* the inner integral has a nonzero limit so we deduce

$$\int_{\overline{N}'} \psi(k(v)\cdot\xi_o)e^{-2\rho'(H(v))}dv = 0. \tag{11}$$

Now by [GGA], p. 462 the positive roots of $\mathfrak{g}'$ with respect to $\mathfrak{a}'$ are the restrictions $\alpha' = \alpha|\mathfrak{a}'$ $(\alpha \in \Sigma^o \cap \Sigma^+)$. Thus $2\rho'|\mathfrak{a}'$ is the sum of these restrictions with multiplicity so (11) amounts to

$$\int_{K'} \psi(k\cdot\xi_o)dk = 0. \tag{12}$$

Applying this argument (starting with (5) to a translated function $\psi\circ\tau(g')$ $(g' \in G')$ we have proved

$$\int_{K'} \psi(g'k\cdot\xi_o)dk = 0. \tag{13}$$

By Ch. II, Proposition 1.13, the horocycle space Ξ' is identified with the set of intersections of X' with the members of the orbit $G'\cdot\xi_o$. Also if $\lambda' = \lambda|\mathfrak{a}'$, $\psi' = \psi|\Xi'$, we have by (2) and (13)

$$\psi' \in \mathcal{E}_{\lambda'}(\Xi'), \quad \int_{K'} \psi'(g'k\cdot\xi_o)dk = 0, \quad g' \in G'. \tag{14}$$

Thus we have proved the implication

$$(5) \Longrightarrow (14) \text{ if } (1) \text{ holds and } \mathrm{Re}(i\lambda) \neq 0.$$

The stage is now set for proving Proposition 4.1 by induction. Since this will involve the product formula (50) §3 the proof will be clearest if we proceed through the steps $\ell = 1, 2, 3\ldots$. The case $\ell = 1$ being settled let $\ell = 2$ and assume (1). We treat the cases $\mathrm{Re}(i\lambda) \neq 0$ and $\mathrm{Re}(i\lambda) = 0$ separately.

a) If $\mathrm{Re}(i\lambda) \neq 0$ we use criterion (4). Since (5) implies that (14) holds for the rank-one space X' and since $\mathrm{Re}(\langle i\lambda', \alpha'\rangle) \geq 0$ for $\alpha' \in (\Sigma')^+$ (by Ch. II, §1 (14)) we have λ' simple so by (4), $\psi'(\xi_o \cap X') = 0$ whence $\psi(\xi_o) = 0$ and λ is simple.

This shows by Lemma 2.11 and (50) §3, that if $\mathrm{Re}(i\lambda) \neq 0$ and if λ satisfies (1) then $P^\delta(-\lambda) \neq 0$ for all δ. Combining this with the Weyl group invariance of P_δ we deduce that the zero set of P_δ is contained in the real space $\mathfrak{a}^*$. This implies that P_δ is constant (otherwise $P_\delta(\lambda_1, \lambda_2) = 0$ has a λ_2 solution for each λ_1). This constant is $\neq 0$ for each δ.

b) If $\mathrm{Re}(i\lambda) = 0$ we use the product formula (50) in §3. Because of the simplicity of each real λ in the rank-one case the right hand side of (50) is $\neq 0$ for $\mathrm{Re}(i\lambda) = 0$ (now that we know that $P_\delta \neq 0$ in case $\ell = 2$) so again by Lemma 2.11, λ is simple.

For $\ell = 3$ we proceed in the same way. We assume (1) and wish to prove (4). If $\mathrm{Re}(i\lambda) \neq 0$ then X' has rank 1 or 2. If $\mathrm{Re}(i\lambda') \neq 0$ then since (5) $\Longrightarrow$(14) the previous case gives $\psi'(\xi_o') = 0$ so (4) follows. If $\mathrm{Re}(i\lambda') = 0$ the previous case shows λ' simple so again since (5) $\Longrightarrow$ (14) we deduce (4). Thus if λ satisfies (1) and if $\mathrm{Re}(i\lambda) \neq 0$ then λ is simple.

This conclusion, combined with Lemma 2.11, and the Weyl group invariance of P_δ implies that its zero set lies in $\mathfrak{a}^*$ so as before, P_δ is a constant. Thus by Lemma 6.6 (Ch. II) the right hand side of (50) §3 is $\neq 0$ for λ real so again by Lemma 2.11, if $\mathrm{Re}(i\lambda) = 0$ then λ is simple. This concludes the case $\ell = 3$.

This induction process proves Proposition 4.1.

As a corollary of the proof we get the full product formula.

Theorem 4.2. *The product formula (50) in §3 holds with P_δ a nonzero constant.*

In order to use the product formula (50) effectively we need the following result.

Proposition 4.3. *Fix $\beta \in \Sigma_o^+$; as δ runs through $\widehat{K}_M$, the representations $\delta(i, \beta)$ $(1 \leq i \leq \ell(\delta))$ run through all of $(\widehat{K}_\beta)_{M_\beta}$.*

Proof. Let $\pi \in (\widehat{K}_\beta)_{M_\beta}$. Because of Proposition 2.5 in the easy rank-one case we can realize π on the space $H_\pi(\mathfrak{p}_\beta^*)$ of harmonic polynomial functions on $\mathfrak{p}_\beta$ of type π, via the adjoint action of K_β on $\mathfrak{p}_\beta$. Let $v \neq 0$ in $H_\pi(\mathfrak{p}_\beta^*)$ be a $\pi(M_\beta)$-fixed vector. Now $\mathrm{Ad}_G(M)$ leaves $\mathfrak{p}_\beta$ invariant and induces an action on $S(\mathfrak{p}_\beta^*)$. Except for Case (b) in Lemma 3.9 v is fixed under this action of $\mathrm{Ad}_G(M)$ and $H_\pi(\mathfrak{p}_\beta^*)$ thus M-invariant and we have an irreducible M-spherical representation $\tilde{\pi}$ of $K_\beta M$ on $H_\pi(\mathfrak{p}_\beta^*)$. In Case (b) in Lemma 3.9 the harmonic polynomials are linear combinations of $(x+iy)^\ell$ and $(x-iy)^\ell$ ($\ell \in \boldsymbol{Z}$) so we have either

$$H_\pi(\mathfrak{p}_\beta^*) = \boldsymbol{C}(x+iy)^\ell, \quad H_{\check{\pi}}(\mathfrak{p}_\beta^*) = \boldsymbol{C}(x-iy)^\ell$$

or

$$H_\pi(\mathfrak{p}_\beta^*) = \boldsymbol{C}(x-iy)^\ell, \quad H_{\check{\pi}}(\mathfrak{p}_\beta^*) = \boldsymbol{C}(x+iy)^\ell,$$

where $\ell \in \boldsymbol{Z}$ is fixed. In either case $\pi \oplus \check{\pi}$ extends to an irreducible representation $\tilde{\pi}$ of $K_\beta M$ with the $\tilde{\pi}(M)$-fixed vector $v = (x+iy)^\ell + (x-iy)^\ell$.

In any case $H(\mathfrak{p}_\beta^*)$ is spanned by the powers of the nilpotent elements in $\mathfrak{p}_\beta^{\boldsymbol{C}}$ ([GGA], Ch. III, Cor. 1.9 and p. 18.) Now a nilpotent element X in $\mathfrak{p}_\beta^{\boldsymbol{C}}$ is a nilpotent element in $\mathfrak{g}^{\boldsymbol{C}}$. In fact, as shown in [GGA], p. 370 there exists an element $T \in \mathfrak{k}^{\boldsymbol{C}}$ such that the orbit $\exp(\boldsymbol{C}T)\cdot X$ has 0 in its closure so the K-invariant polynomials on $\mathfrak{p}^{\boldsymbol{C}}$ without constant term vanish on X so by *loc. cit.* Theorem 4.3, X is nilpotent in $\mathfrak{g}^{\boldsymbol{C}}$. Thus $H(\mathfrak{p}_\beta^*) \subset H(\mathfrak{p}^*)$. Let W be the span of the K-orbit of v (above) in $H(\mathfrak{p}^*)$. Let $W = \bigoplus_\delta V_\delta$ be its decomposition into K-irreducibles and put $w = \Sigma_\delta w_\delta$ for $w \in W$. Fix δ such that the component v_δ of v in V_δ is $\neq 0$. Then $v_\delta \in V_\delta^M$. Construct $V = \bigoplus_{i=1}^{\ell(\delta)} V_i$ as in Lemma 3.10 and fix i such that the V_i-component $v_{\delta,i}$ of v_δ is $\neq 0$. Then the correspondence $J:\ \tilde{\pi}(k_\beta m)v \longrightarrow \delta_i(k_\beta m)v_{\delta,i}$ gives a well-defined equivalence between $\tilde{\pi}$ and δ_i. In fact, suppose $\sum_j a_j\tilde{\pi}(k_j)v = 0$ for some $a_j \in \boldsymbol{C}$, $k_j \in K_\beta M$. Then $\Sigma_j a_j\delta(k_j)v_\delta = 0$ whence $\sum_j a_j\delta_i(k_j)v_{\delta,i} = 0$ so the map is well-defined. Restricting to K_β we get the desired equivalence between π and $\delta(i,\beta)$.

Theorem 4.4. *Let $\lambda \in \mathfrak{a}_c^*$. Then*

$$\lambda \text{ is simple if and only if } \boldsymbol{e}_{s^*}(\lambda) \neq 0.$$

Proof. Because of Proposition 4.3 and Lemma 2.11, λ is simple if and only if for each $\beta \in \Sigma_o^+$, the restriction λ_β is simple. But by Ch. II, Theorem

6.5 λ_β is simple if and only if $\boldsymbol{e}^\beta(\lambda_\beta) \neq 0$, where $\boldsymbol{e}^\beta$ is the $\boldsymbol{e}_{s^*}$-function for G_β/K_β and since

$$\boldsymbol{e}_{s^*}(\lambda) = \text{const.} \prod_{\beta \in \Sigma_o^+} \boldsymbol{e}^\beta(\lambda_\beta),$$

the theorem follows.

Let us restate this in more direct terms.

Theorem 4.5. *Let $\lambda \in \mathfrak{a}_c^*$ and define the Poisson transform $\mathcal{P}_\lambda$ by*

$$(\mathcal{P}_\lambda F)(x) = \int_B e^{(i\lambda+\rho)(A(x,b))} F(b)\,db, \qquad F \in \mathcal{E}(B).$$

Then $\mathcal{P}_\lambda$ is injective if and only if $\boldsymbol{e}_{s^*}(\lambda) \neq 0$.

§5. THE PALEY-WIENER THEOREM FOR THE FOURIER TRANSFORM ON $X = G/K$.

In this section we shall give an explicit description of the range $\mathcal{D}(X)^\sim$ of $\mathcal{D}(X)$ under the Fourier transform. Then we specialize the result to functions which are K-finite of a given type and obtain thereby a stronger topological version of the result. We recall the definition (§1, No. 1)

$$\widetilde{f}(\lambda, b) = \int_X f(x) e^{(-i\lambda+\rho)(A(x,b))}\,dx \tag{1}$$

of the Fourier transform $f \longrightarrow \widetilde{f}$ on X. In order to describe the range $\mathcal{D}(X)^\sim$ we introduce the following definition.

Definition. A C^∞ function $\psi(\lambda, b)$ on $\mathfrak{a}_c^* \times B$, holomorphic in λ is called a *holomorphic function of uniform exponential type* if there exists a constant $R \geq 0$ such that for each $N \in \boldsymbol{Z}^+$,

$$\sup_{\lambda \in \mathfrak{a}_c^*, b \in B} e^{-R|\operatorname{Im}\lambda|}(1 + |\lambda|)^N |\psi(\lambda, b)| < \infty. \tag{2}$$

Here $\operatorname{Im}\lambda = \eta$ if $\lambda = \xi + i\eta\,(\xi, \eta \in \mathfrak{a}^*)$ and $|\lambda| = (|\xi|^2 + |\eta|^2)^{\frac{1}{2}}$.

The space of functions ψ satisfying (2) will be denoted $\mathcal{H}^R(\mathfrak{a}^* \times B)$ and we denote by $\mathcal{H}(\mathfrak{a}^* \times B)$ their union over all $R > 0$. Also $\mathcal{H}(\mathfrak{a}^* \times B)_W$ denotes the space of functions $\psi \in \mathcal{H}(\mathfrak{a}^* \times B)$ satisfying

$$\int_B e^{(is\lambda+\rho)(A(x,b))} \psi(s\lambda, b)\,db = \int_B e^{(i\lambda+\rho)(A(x,b))} \psi(\lambda, b)\,db \tag{3}$$

for $s \in W$, $\lambda \in \mathfrak{a}_c^*$, $x \in X$. The space of W-invariant holomorphic functions $\lambda \longrightarrow \psi(\lambda)$ (independent of b) satisfying (2) is denoted $\mathcal{H}_W(\mathfrak{a}_c^*)$.

Theorem 5.1. *The Fourier transform* $f(x) \longrightarrow \widetilde{f}(\lambda, b)$ *is a bijection of* $\mathcal{D}(X)$ *onto* $\mathcal{H}(\mathfrak{a}^* \times B)_W$. *Moreover,* $\psi = \widetilde{f}$ *satisfies (2) if and only if* $\operatorname{supp}(f) \subset C\ell(B_R(o))$.

Writing $\psi_\lambda(b) = \psi(\lambda, b)$ condition (3) can be written

$$\mathcal{P}_{s\lambda}(\psi_{s\lambda}) = \mathcal{P}_\lambda(\psi_\lambda), \qquad s \in W, \tag{4}$$

in terms of the Poisson transform (Ch. II, §3, No. 4).

1. Estimates of the Γ-coefficients.

For $\delta \in \widehat{K}_M$ we consider now the Eisenstein integral

$$\Phi_{\lambda,\delta}(x) = \int_K e^{(i\lambda+\rho)(A(x,kM))}\delta(k)\,dk, \tag{5}$$

which for $x = \exp H \cdot o$ $(H \in \mathfrak{a}^+)$ and $v \in V_\delta^M$ has the expansion

$$e^{\rho(\log h)}\Phi_{\lambda,\delta}(h \cdot o)v = \sum_{s \in W} \Phi(s\lambda : h)\boldsymbol{C}_s(\lambda)v \quad (h \in A^+) \tag{6}$$

where

$$\Phi(\lambda : h) = e^{i\lambda(\log h)} \sum_{\mu \in \Lambda} \boldsymbol{\Gamma}_\mu(\lambda) e^{-\mu(\log h)}. \tag{7}$$

This was established in §2, No. 2. We saw also that the recursion formula for $\boldsymbol{\Gamma}_\mu$ (§2, (31), is equivalent to the differential equation

$$\Big\{L_A + \sum_{\alpha \in \Sigma^+} m_\alpha(\coth\alpha)A_\alpha + \sum_{\alpha\in\Sigma^+}(\sinh\alpha)^{-2}\delta(\omega_\alpha) + \langle\lambda,\lambda\rangle + \langle\rho,\rho\rangle\Big\}\Phi_\lambda = 0, \tag{8}$$

where $\Phi_\lambda(h) = e^{-\rho(\log h)}\Phi(\lambda : h)$.

Let Δ be defined on A by

$$\Delta(\exp H) = c \prod_{\alpha \in \Sigma^+} (\sinh\alpha(H))^{m_\alpha} \qquad (H \in \mathfrak{a}) \tag{9}$$

the constant c being determined such that

$$\int_G f(g)dg = \int_K \int_K \int_{A^+} f(k_1ak_2)\Delta(a)dk_1dadk_2. \tag{10}$$

(Usually we denote the right hand side of (9) by δ but this would conflict with our present use of δ). As pointed out in [GGA], Ch. II, Prop. 3.9 and Theorem 3.7,

$$L_A + \sum_{\alpha\in\Sigma^+} m_\alpha(\coth\alpha)A_\alpha = \Delta^{-\frac{1}{2}}L_A \circ \Delta^{\frac{1}{2}} - d, \tag{11}$$

where $d = \Delta^{-\frac{1}{2}}L_A(\Delta^{\frac{1}{2}})$. Combining this with (8) we see that the function $\Psi_\lambda = \Delta^{\frac{1}{2}}\Phi_\lambda$ on A^+ with values in $\mathrm{Hom}(V_\delta^M, V_\delta^M)$ satisfies the differential equation

$$L_A\Psi_\lambda - d\,\Psi_\lambda + \sum_{\alpha\in\Sigma^+} (\sinh\alpha)^{-2}\delta(\omega_\alpha)\Psi_\lambda = (-\langle\lambda,\lambda\rangle - \langle\rho,\rho\rangle)\Psi_\lambda. \tag{12}$$

We now expand the coefficients in terms of the descending exponentials $e^{-\mu}(\mu \in \Lambda)$. We have on $\mathfrak{a}^+$,

$$\Delta(\exp H) = \text{const. } e^{2\rho(H)}\prod(1-e^{-2\alpha(H)})^{m_\alpha}$$

so

$$\Delta^{-\frac{1}{2}}(\exp H) = e^{-\rho(H)}\sum_{\mu\in\Lambda} b_\mu e^{-\mu(H)},$$

$$\Delta^{\frac{1}{2}}(\exp H) = e^{\rho(H)}\sum_{\mu\in\Lambda} c_\mu e^{-\mu(H)},$$

$$d(\exp H) = \langle\rho,\rho\rangle + \sum_{\mu\in\Lambda'} d_\mu e^{-\mu(H)}, \qquad \Lambda' = \Lambda - \{0\}$$

(since $\lim_{H\to+\infty} d(\exp H) = \langle\rho,\rho\rangle$). Also

$$(\sinh\alpha(H))^{-2} = \sum_{\mu\in\Lambda'} e_\mu^\alpha e^{-\mu(H)}$$

so

$$\sum_{\alpha\in\Sigma^+}(\sinh\alpha)^{-2}(H)\delta(\omega_\alpha) = \sum_{\mu\in\Lambda'} D_\mu e^{-\mu(H)},$$

where $D_\mu = \sum_{\alpha\in\Sigma^+} e_\mu^\alpha\delta(\omega_\alpha)$. All these coefficients b_μ, c_μ, d_μ and e_μ^α have at most polynomial growth in μ. Combining with (7) we get

$$\Psi_\lambda(\exp H) = \sum_{\mu\in\Lambda} A_\mu(\lambda)e^{(i\lambda-\mu)(H)}, \quad H \in \mathfrak{a}^+, \tag{13}$$

where $A_\mu(\lambda) \in \mathrm{Hom}(V_\delta^M, V_\delta^M)$ is given by

$$A_\mu(\lambda) = \sum_{\nu, \mu-\nu \in \Lambda} \mathbf{\Gamma}_{\mu-\nu}(\lambda) c_\nu. \tag{14}$$

Now we substitute formula (13) for Ψ_λ into (12) and use the relation

$$(L_A \Psi_\lambda)(\exp H \cdot o) = \sum_{\mu \in \Lambda} \langle i\lambda - \mu, i\lambda - \mu \rangle A_\mu(\lambda) e^{(i\lambda-\mu)(H)}.$$

Equating the coefficients to $e^{-\mu}$ we obtain the recursion formula

$$\{\langle \mu, \mu \rangle - 2i\langle \lambda, \mu \rangle\} A_\mu(\lambda) = \sum_{0 \neq \nu, \mu-\nu \in \Lambda} (d_\nu - D_\nu) A_{\mu-\nu}(\lambda). \tag{15}$$

For a linear transformation $A : V_\delta^M \longrightarrow V_\delta^M$ we denote by $\|A\|$ its Hilbert-Schmidt norm $(Tr(AA^*))^{\frac{1}{2}}$. Then $\|AB\| \leq \|A\| \, \|B\|$.

Lemma 5.2. *Let $H \in \mathfrak{a}^+$. Then for a suitable constant K_H we have*

$$\|\mathbf{\Gamma}_\mu(\lambda)\| \leq K_H e^{\mu(H)} \qquad \mu \in \Lambda, \qquad \lambda \in \mathfrak{a}^* + i\mathfrak{a}^*_+.$$

Proof. If $\mu \in \Lambda, \mu = m_1\alpha_1 + \cdots + m_\ell \alpha_\ell$, put $m(\mu) = m_1 + \cdots + m_\ell$. Then by (15) and assumption on λ,

$$\|A_\mu(\lambda)\| \leq c\, m(\mu)^{-1} \sum_{0 \neq \nu, \mu-\nu \in \Lambda} (|d_\nu| + \|D_\nu\|) \|A_{\mu-\nu}(\lambda)\|, \tag{16}$$

where c is a constant. Since $|d_\nu|$ and $\|D_\nu\|$ have at most polynomial growth in ν there exists an integer N_o such that

$$c \sum_{\nu \in \Lambda} (|d_\nu| + \|D_\nu\|) e^{-\nu(H)} \leq N_o.$$

Then we select a constant K'_H such that

$$\|A_\nu(\lambda)\| \leq K'_H e^{\nu(H)} \tag{17}$$

for $\nu \in \Lambda, m(\nu) \leq N_o$, $\lambda \in \mathfrak{a}^* + i\mathfrak{a}^*_+$. This can be done since $A_o(\lambda) =$ const. and (16) shows inductively that each $\|A_\nu(\lambda)\|$ is bounded in the indicated range of λ. We shall now prove by induction that (17) holds for all $\nu \in \Lambda$. Let

$N > N_o$ be an integer and suppose (17) holds for all $\nu \in \Lambda$ with $m(\nu) < N$. Let $\mu \in \Lambda$ with $m(\mu) = N$. Then by the above inequalities

$$\|A_\mu(\lambda)\| \leq c\,N^{-1} \sum_{\nu\neq 0, \mu-\nu\in\Lambda} (|d_\nu| + \|D_\nu\|) K'_H e^{(\mu-\nu)(H)} \leq K'_H e^{\mu(H)}$$

proving (17) for all $\nu \in \Lambda$. Finally $\Phi_\lambda = \Delta^{-\frac{1}{2}}\Psi_\lambda$ implies

$$\mathbf{\Gamma}_\mu(\lambda) = \sum_{\nu,\mu-\nu\in\Lambda} A_{\mu-\nu}(\lambda) b_\nu$$

so, since b_ν has at most polynomial growth in ν, the estimate (17) implies a similar one for $\mathbf{\Gamma}_\mu(\lambda)$.

2. Some Identities for $\boldsymbol{C}_s$.

Another tool needed for the Paley-Wiener theorem is the following identity for the linear transformations $\boldsymbol{C}_s$ from Proposition 2.9.

Theorem 5.3. *Let $\delta \in \widehat{K}_M$ and let $s \in W$ be arbitrary. Then the linear transformation $\boldsymbol{C}_{s^{-1}} : V_\delta^M \longrightarrow V_\delta^M$ satisfies the identity*

$$\frac{\boldsymbol{C}_{s^{-1}}(-s\lambda)}{\mathbf{c}(-\lambda)} \frac{\boldsymbol{C}_{s^{-1}}(-s\overline{\lambda})^*}{\mathbf{c}(\lambda)} = I, \tag{18}$$

where $*$ denotes adjoint and I the identity operator.

Consider again the endomorphism $A(\lambda, s)$ of V_δ^M,

$$A(\lambda, s)v = \int_{\overline{N}_s} e^{-(i\lambda+\rho)(H(\overline{n}_s))}\delta(m_s k(\overline{n}_s))v\, d\overline{n}_s, \quad v \in V_\delta^M \tag{19}$$

whose adjoint satisfies

$$A(\overline{\lambda}, s)^* v = \int_{\overline{N}_s} e^{(i\lambda-\rho)(H(\overline{n}_s))}\delta(k(\overline{n}_s)^{-1} m_s^{-1})v\, d\overline{n}_s, \quad v \in V_\delta^M.$$

By Proposition 2.9, we have on V_δ^M,

$$\frac{A(\overline{\lambda}, s)^*}{\mathbf{c}_s(-\lambda)} = \frac{\boldsymbol{C}_{s^{-1}}(-s\lambda)}{\mathbf{c}(-\lambda)}, \qquad \frac{A(\lambda, s)}{\mathbf{c}_s(\lambda)} = \frac{\boldsymbol{C}_{s^{-1}}(-s\overline{\lambda})^*}{\mathbf{c}(\lambda)},$$

where we have also used $(\boldsymbol{c}_s(\overline{\lambda}))^- = \boldsymbol{c}_s(-\lambda)$. Thus (18) amounts to

$$\frac{A(\overline{\lambda}, s)^*}{\boldsymbol{c}_s(-\lambda)} \frac{A(\lambda, s)}{\boldsymbol{c}_s(\lambda)} = I \qquad \text{on } V_\delta^M. \tag{20}$$

We consider first the rank one case. Here we have Kostant's result, $\dim V_\delta^M = 1$ (Cor. 6.8, Ch. II and [GGA], Theorem 3.5, Ch. V). Again let m^* denote a representative in K of the nontrivial Weyl group element. We start with a simple lemma.

Lemma 5.4. *Suppose $X = G/K$ has rank one and let $v \in V_\delta^M$. Then*

$$\int_{\overline{N}} \langle \delta(k(\overline{n})^{-1}m^*)v, v\rangle e^{-(i\lambda+\rho)(H(\overline{n}))} d\overline{n}$$
$$= \int_{\overline{N}} \langle \delta(m^* k(\overline{n}))v, v\rangle e^{-(i\lambda+\rho)(H(\overline{n}))} d\overline{n}.$$

Proof. In the notation of Proposition 6.4, Ch. II, the first integral equals

$$\int_{\overline{N}} \langle \delta(m^* k(\overline{n}_o))v, v\rangle e^{-(i\lambda+\rho)(H(\overline{n}))} d\overline{n},$$

which since $H(\overline{n}_o) = H(\overline{n})$, $d\overline{n} = d\overline{n}_o$, equals the second integral.

Remark. Another proof comes from remarking that both integrals are meromorphic in λ and (taking complex conjugations) coincide for $\lambda \in i\mathfrak{a}^*$.

Still assuming rank $X = 1$ we have $s\lambda = -\lambda$ so Theorem 2.10 gives $\Gamma_{s,\lambda}\Gamma_{s,-\lambda} = 1$ so

$$\boldsymbol{c}(\lambda)^{-1}\boldsymbol{C}_s(-\lambda)\boldsymbol{c}(-\lambda)^{-1}\boldsymbol{C}_s(\lambda) = 1,$$

while Lemma 5.4 amounts to

$$\boldsymbol{C}_s(-\lambda) = \boldsymbol{C}_s(\overline{\lambda})^- = \boldsymbol{C}_s(\overline{\lambda})^*.$$

This proves Theorem 5.3 in the rank one case.

For the general case we consider again the decomposition

$$A(\lambda, s) = A(s^{(1)}\lambda, s_{\alpha_1}) \cdots A(s^{(p)}\lambda, s_{\alpha_p}) \qquad \text{on } V_\delta^M \tag{21}$$

from (27) §3. Then

$$A(\overline{\lambda}, s)^* = A(s^{(p)}\overline{\lambda}, s_{\alpha_p})^* \cdots A(s^{(1)}\overline{\lambda}, s_{\alpha_1})^* \quad \text{on} \quad V_\delta^M.$$

Thus Theorem 5.3 will follow if we prove for each j

$$\frac{A(s^{(j)}\overline{\lambda}, s_{\alpha_j})^*}{c_{\alpha_j}(-\lambda_j)} \frac{A(s^{(j)}\lambda, s_{\alpha_j})}{c_{\alpha_j}(\lambda_j)} = I \qquad \text{on } V_\delta^M. \tag{22}$$

Here $\lambda_j = s^{(j)}\lambda|\mathfrak{a}_{\alpha_j}$ as before. Now the decomposition (36) §3 gives a diagonalization of $A(s^{(j)}\lambda, s_{\alpha_j})$ on V_δ^M with diagonal elements given by (37) §3. Thus (22) follows from the rank one case, already settled. This proves Theorem 5.3.

Remark. The identity (20) can also be written

$$B(\overline{\lambda}, s)^* B(\lambda, s) = I \qquad \text{on } V_\delta^M. \tag{23}$$

In particular, if $\lambda \in \mathfrak{a}^*$, $B(\lambda, s)$ is unitary on V_δ^M.

3. The Fourier Transform and the Radon Transform. K-types.

If f is a function on X its Radon transform $\widehat{f}$ and its Fourier transform $\widetilde{f}$ are respectively defined by

$$\widehat{f}(\xi) = \int_\xi f(x)ds(x), \qquad \widetilde{f}(\lambda, b) = \int_X f(x)e^{(-i\lambda+\rho)(A(x,b))}dx. \tag{24}$$

Here ξ is any horocycle in X, and the measures ds and dx are normalized as in Ch. II, §3, No. 1. These transforms are defined for those ξ, λ, and b for which the integrals (24) converge absolutely.

Now if $b = kM$ then $A(x, kM) = A(k^{-1}x, eM)$ so

$$\widetilde{f}(\lambda, kM) = \int_X f(k \cdot x)e^{(-i\lambda+\rho)(A(x,eM))}dx$$

If $x = an \cdot o$ then $A(x, e) = -H(n^{-1}a^{-1}) = H(a)$ so by

$$\int_X f(x)dx = \int_{AN} f(an \cdot o)da\, dn$$

we have, if ξ_o is the horocycle $N \cdot o$,

$$\widetilde{f}(\lambda, kM) = \int_A \widehat{f}(ka \cdot \xi_o) e^{(-i\lambda+\rho)(\log a)} da. \tag{25}$$

Thus in analogy with the Euclidean case (Ch. I, §2, (6)) **the Fourier transform on X is the A-Euclidean Fourier transform of the Radon transform.**

Let δ be an irreducible unitary representation of K on a vector space V_δ of dimension $d(\delta)$. For $f \in \mathcal{E}(X)$ we put

$$f^\delta(x) = d(\delta) \int_K f(k \cdot x) \delta(k^{-1}) dk. \tag{26}$$

Then f^δ is a C^∞ map from X to $\mathrm{Hom}(V_\delta, V_\delta)$ satisfying

$$f^\delta(k \cdot x) = \delta(k) f^\delta(x). \tag{27}$$

If $f^\delta \not\equiv 0$ then (27) shows that $f^\delta(a \cdot o) \neq 0$ for some $a \in A$ so again by (27) the space V_δ^M of $\delta(M)$-fixed vectors is $\neq 0$.

By analogy with (24) we now define

$$\widehat{f^\delta}(\xi) = \int_\xi f^\delta(x) ds(x), \quad \widetilde{f^\delta}(\lambda, b) = \int_X f^\delta(x) e^{(-i\lambda+\rho)(A(x,b))} dx.$$

Putting $\widehat{f^\delta}(a) = \widehat{f^\delta}(a \cdot \xi_o),)$ $\widetilde{f^\delta}(\lambda) = \widetilde{f^\delta}(\lambda, eM)$ we have

$$\widetilde{f^\delta}(\lambda, kM) = \delta(k)\widetilde{f^\delta}(\lambda), \qquad \widehat{f^\delta}(ka \cdot \xi_o) = \delta(k)\widehat{f^\delta}(a \cdot \xi_o). \tag{28}$$

Then we conclude the following result from Theorem 1.3.

Corollary 5.5. *Let $f \in \mathcal{D}(X)$ and $\delta \in \widehat{K}_M$. Then, with Δ as in (10),*

$$\widetilde{f^\delta}(\lambda) = \int_{A^+} \Phi_{\overline{\lambda},\delta}(a \cdot o)^* f^\delta(a \cdot o) \Delta(a) da$$

and

$$f^\delta(a \cdot o) = w^{-1} \int_{\mathfrak{a}^*} \Phi_{\lambda,\delta}(a \cdot o) \widetilde{f^\delta}(\lambda) |c(\lambda)|^{-2} d\lambda.$$

For this we just have to remark that by (5),

$$\Phi_{\overline{\lambda},\delta}(x)^* = \int_K e^{(-i\lambda+\rho)(A(x,kM))}\delta(k^{-1})dk. \tag{29}$$

The functions f^δ determine f by the Peter-Weyl theorem for vector-valued functions ([GGA], Ch. V, Cor. 3.4) which states

$$f(x) = \sum_{\delta\in\widehat{K}} d(\delta) \int_K \chi_\delta(k^{-1})f(k \cdot x)dk, \tag{30}$$

where χ_δ is the character of δ. Because of (26) (and $f^\delta = 0$ if $V_\delta^M = 0$) we can write this

$$f = \sum_{\delta\in\widehat{K}_M} \mathrm{Trace}(f^\delta). \tag{31}$$

It is obvious that the linear transformations $f^\delta(a \cdot o)$, $\widetilde{f^\delta}(\lambda)$, $\Phi_{\lambda,\delta}(a \cdot o)$ map the space V_δ into V_δ^M so the orthogonal projection $E_\delta : V_\delta \longrightarrow V_\delta^M$ satisfies $f^\delta(a \cdot o) = E_\delta f^\delta(a \cdot o)$. Thus (31) implies the following result.

Lemma 5.6. *Vertical bar denoting restriction we have for each $f \in \mathcal{E}(X)$,*

$$f(a \cdot o) = \sum_{\delta\in\widehat{K}_M} \mathrm{Trace}(f^\delta(a \cdot o)|V_\delta^M).$$

Lemma 5.7. *Let $f \in \mathcal{D}(X)$ and $\delta \in \widehat{K}_M$. Then the matrix-valued Fourier transform $\widetilde{f^\delta}(\lambda)$ satisfies the functional equations*

$$\widetilde{f^\delta}(s\lambda)v = \frac{C_{s^{-1}}(s\overline{\lambda})^*}{c(-\lambda)}\widetilde{f^\delta}(\lambda)v, \qquad v \in V_\delta^M,$$

for each s in the Weyl group W.

This is a direct consequence of Corollary 5.5 combined with the functional equation in Theorem 2.10.

4. Completion of the Proof of the Paley-Wiener Theorem. The Range $\mathcal{E}'(X)^{\sim}$.

We have now collected the tools needed for the proof of Theorem 5.1. We shall use the expansion in Theorem 2.7, the estimates in Lemma 5.2, the identity in Theorem 5.3 and the functional equations in Lemma 5.7.

Let $f \in \mathcal{D}(X)$ have support in $C\ell(B^R(o))$. Because of Lemma 1.2 the Fourier transform satisfies the identities (3). We also know from [DS], Ex. B 2 (ii) in Ch. VI that the distance d on X satisfies

$$d(o, a \cdot o) = d(o, ka \cdot \xi_o)$$

so by (25) and the classical Paley-Wiener theorem in $\boldsymbol{R}^n$, $\widetilde{f}(\lambda, b)$ has uniform exponential type and satisfies (2).

Conversely, suppose ψ satisfies (2) and (3). We shall find $f \in \mathcal{D}(X)$, with support in $C\ell(B^R(o))$ such that $\widetilde{f} = \psi$. Because of the inversion formula in Theorem 1.3 we define a function f on X by

$$f(x) = w^{-1} \int\limits_{\mathfrak{a}^* \times B} \psi(\lambda, b) e^{(i\lambda+\rho)(A(x,b))} |\boldsymbol{c}(\lambda)|^{-2} d\lambda db. \tag{32}$$

If $\delta \in \widehat{K}_M$ we derive from (32)

$$f^\delta(x) = w^{-1} \int\limits_{\mathfrak{a}^*} \Phi_{\lambda,\delta}(x) \psi^\delta(\lambda) |\boldsymbol{c}(\lambda)|^{-2} d\lambda, \tag{33}$$

where

$$\psi^\delta(\lambda, b) = d(\delta) \int\limits_K \psi(\lambda, k \cdot b) \delta(k^{-1}) dk, \quad \psi^\delta(\lambda) = \psi^\delta(\lambda, eK).$$

The relations (3) have important implications for $\psi^\delta(\lambda)$. In (3) replace x by $k \cdot x$, use $A(k \cdot x, b) = A(x, k^{-1} \cdot b)$, make substitution $b \longrightarrow k \cdot b$ multiply by $\delta(k^{-1})$ and integrate over K. The result is

$$\int\limits_B e^{(is\lambda+\rho)(A(x,b))} \psi^\delta(s\lambda, b) db = \int\limits_B e^{(i\lambda+\rho)(A(x,b))} \psi^\delta(\lambda, b) db.$$

Using here $\psi^\delta(s\lambda, kM) = \delta(k)\psi^\delta(s\lambda)$ we deduce

$$\Phi_{s\lambda,\delta}(x) \psi^\delta(s\lambda) = \Phi_{\lambda,\delta}(x) \psi^\delta(\lambda)$$

which by Theorems 2.10 and 5.3 implies

$$\psi^\delta(s\lambda) = \frac{\boldsymbol{C}_{s^{-1}}(s\bar{\lambda})^*}{\boldsymbol{c}(-\lambda)}\psi^\delta(\lambda). \tag{34}$$

First we want to prove $f(x) = 0$ for $d(o,x) > R$. For this it suffices to prove $f(a \cdot o) = 0$ for $a \in A^+$, $d(o, a \cdot o) > R$ because we can then apply this conclusion to the rotated function $x \longrightarrow f(k \cdot x)$, the assumptions on $\psi(\lambda, b)$ being rotation-invariant in b. But then Lemma 5.6 shows that it suffices to show for each $\delta \in \widehat{K}_M$ that

$$f^\delta(a \cdot o)v = 0 \quad \text{for} \quad a \in A^+, \quad |\log a| > R, \quad v \in V_\delta^M. \tag{35}$$

In order to prove (35) we substitute the expansion from Theorem 2.7 into (33). Then if $v \in V_\delta$

$$e^{\rho(H)} f^\delta(\exp H \cdot o)v =$$

$$w^{-1} \sum_{s \in W} \int_{\mathfrak{a}^*} \sum_{\mu \in \Lambda} e^{-\mu(H)} \boldsymbol{\Gamma}_\mu(s\lambda) \boldsymbol{C}_s(\lambda) e^{is\lambda(H)} \psi^\delta(\lambda) |\boldsymbol{c}(\lambda)|^{-2} d\lambda\, v \tag{36}$$

for $H \in \mathfrak{a}^+$. Recalling the Remark following Theorem 2.7 the meromorphic functions appearing are holomorphic at all points $\lambda \in \mathfrak{a}^*$. Now $|\boldsymbol{c}(\lambda)|^{-1}$ has at most polynomial growth on $\mathfrak{a}^*$ ([GGA], Ch. IV, Prop. 7.2) so considering Exercise **F1** and the polynomial decay of $\psi^\delta(\lambda)$ on $\mathfrak{a}^*$ we see that the function $\boldsymbol{C}_s(\lambda)\psi^\delta(\lambda)|\boldsymbol{c}(\lambda)|^{-2}$ is integrable on $\mathfrak{a}^*$. Combining this with Lemma 5.2 for $H/2$ in place of H we see that in (36) the integral and Σ_μ can be interchanged. Thus it remains to prove the following result.

Lemma 5.8. *For each* $\mu \in \Lambda$,

$$\int_{\mathfrak{a}^*} \boldsymbol{\Gamma}_\mu(s\lambda) \boldsymbol{C}_s(\lambda) e^{is\lambda(H)} \psi^\delta(\lambda) |\boldsymbol{c}(\lambda)|^{-2} d\lambda = 0$$

if $H \in \mathfrak{a}^+, |H| > R$.

Proof. For $\lambda \in \mathfrak{a}^*$ we have $|\boldsymbol{c}(\lambda)|^2 = \boldsymbol{c}(\lambda)\boldsymbol{c}(-\lambda) = \boldsymbol{c}(s\lambda)\boldsymbol{c}(-s\lambda)$ so the integral can be rewritten as

$$\int_{\mathfrak{a}^*} e^{-i\lambda(H)} \boldsymbol{\Gamma}_\mu(-\lambda) \boldsymbol{c}(\lambda)^{-1} \boldsymbol{C}_s(-s^{-1}\lambda) \boldsymbol{c}(-\lambda)^{-1} \psi^\delta(-s^{-1}\lambda) d\lambda\, v.$$

In this integral appears now a most fortunate cancellation. Because of (34) the integral equals

$$\int_{\mathfrak{a}^*} e^{-i\lambda(H)}\boldsymbol{\Gamma}_\mu(-\lambda)\boldsymbol{c}(\lambda)^{-1}\left\{\frac{\boldsymbol{C}_s(-s^{-1}\lambda)}{\boldsymbol{c}(-\lambda)}\frac{\boldsymbol{C}_s(-s^{-1}\overline{\lambda})^*}{\boldsymbol{c}(\lambda)}\right\}\psi^\delta(-\lambda)d\lambda\, v$$

and by Theorem 5.3 this reduces to

$$\int_{\mathfrak{a}^*} e^{-i\xi(H)}\boldsymbol{\Gamma}_\mu(-\xi)\boldsymbol{c}(\xi)^{-1}\psi^\delta(-\xi)d\xi. \tag{37}$$

Although we are dealing with vector-valued functions $\boldsymbol{\Gamma}_\mu$ and ψ^δ the proof of Theorem 7.3, Ch. IV in [GGA] now works without change and shows that the integral (37) vanishes for $H \in \mathfrak{a}^+$, $|H| > R$. This then proves that the C^∞ function f defined by (32) has support contained in $C\ell(B^R(o))$.

It remains to prove that $\widetilde{f} = \psi$. By the inversion formula for f we have for $\varphi = \widetilde{f} - \psi$

$$\int_{\mathfrak{a}^*}\Big(\int_B \varphi(\lambda, b)e^{(i\lambda+\rho)(A(x,b))}|\boldsymbol{c}(\lambda)|^{-2}db\Big)d\lambda = 0. \tag{38}$$

Since both $\widetilde{f}$ and ψ satisfy the compatibility conditions (3) the inner integral in (38) is invariant under each transformation $\lambda \longrightarrow s\lambda$ ($s \in W$). Thus we can in (38) replace $\mathfrak{a}^*$ by $\mathfrak{a}^*_+$. We integrate the resulting relation against an arbitrary $F \in \mathcal{D}(X)$. Then we have (after the substitution $\lambda \longrightarrow -\lambda$),

$$\int_{-\mathfrak{a}^*_+\times B} \varphi(-\lambda, b)\widetilde{F}(\lambda, b)|\boldsymbol{c}(\lambda)|^{-2}db\, d\lambda = 0. \tag{39}$$

Now by Theorem 1.5 the functions $\widetilde{F}(\lambda, b)$ span a dense subspace of $L^2(-\mathfrak{a}^*_+ \times B)$ so (39) implies $\varphi(\lambda, b) = 0$ for almost all $(\lambda, b) \in \mathfrak{a}^*_+ \times B$, hence for all $(\lambda, b) \in \mathfrak{a}^*_+ \times B$ by continuity. Thus $\varphi^\delta(\lambda) = 0$ for $\lambda \in \mathfrak{a}^*_+$ whence by (34), $\varphi^\delta(\lambda) = 0$ for all $\lambda \in \mathfrak{a}^*$. Since the $\varphi^\delta(\lambda)$ are the Fourier coefficients of the function $b \longrightarrow \varphi(\lambda, b)$ we deduce $\varphi \equiv 0$ and Theorem 5.1 is proved.

From this theorem it is easy to deduce a similar characterization of the Fourier transform of the space $\mathcal{E}'(X)$ of distributions of compact support. Let $\mathcal{E}'_\natural(X)$ denote the set of $T \in \mathcal{E}'(X)$ invariant under K.

Definition. A C^∞ function ψ on $\mathfrak{a}_{\mathbf{c}}^* \times B$ holomorphic in λ is called a *holomorphic function of uniform exponential type and slow growth* if there exist constants $R \geq 0$ $N \in \mathbf{Z}^+$ such that

$$|\psi(\lambda, b)| \leq C(1 + |\lambda|)^N e^{R|Im\ \lambda|}, \qquad \lambda \in \mathfrak{a}_{\mathbf{c}}^*, b \in B, \tag{40}$$

where C is a constant.

Given $R \geq 0$ let $\mathcal{K}^R(\mathfrak{a}^* \times B)$ denote the space of these ψ satisfying (40) (for some N and C) and $\mathcal{K}(\mathfrak{a}^* \times B)$ their union over all $R \geq 0$. Let $\mathcal{K}(\mathfrak{a}^* \times B)_W$ denote the space of functions in $\mathcal{K}(\mathfrak{a}^* \times B)$ satisfying (3) and $\mathcal{K}_W(\mathfrak{a}_{\mathbf{c}}^*)$ the space of functions in $\mathcal{K}(\mathfrak{a}^* \times B)$ constant in $b \in B$, W-invariant in $\lambda \in \mathfrak{a}_{\mathbf{c}}^*$.

Corollary 5.9. *The Fourier transform $T \longrightarrow \widetilde{T}$ where*

$$\widetilde{T}(\lambda, b) = \int_X e^{(-i\lambda+\rho)(A(x,b))} dT(x)$$

is a bijection of $\mathcal{E}'(X)$ onto the space $\mathcal{K}(\mathfrak{a}^* \times B)_W$. Moreover, $\widetilde{T}$ satisfies (40) if and only if $\text{supp}(T) \subset C\ell(B^R(o))$.

In particular, $T \longrightarrow \widetilde{T}$ is a bijection of $\mathcal{E}'_{\natural}(X)$ onto $\mathcal{K}_W(\mathfrak{a}_{\mathbf{c}}^*)$.

Proof. Let $\eta_\epsilon \in \mathcal{D}(X)$ be K-invariant, positive, with $\int \eta_\epsilon dx = 1$, $\text{supp}(\eta_\epsilon) \subset C\ell(B_\epsilon(o))$. Then

$$\widetilde{\eta}_\epsilon(\lambda, b) = \widetilde{\eta}_\epsilon(\lambda) = \int_{B^\epsilon(o)} \eta_\epsilon(x)(\varphi_{-\lambda}(x) - 1)dx + 1.$$

Given a compact set $C \subset \mathfrak{a}_{\mathbf{c}}^*$ and $\delta > 0$ there exists $\epsilon > 0$ such that $|\varphi_{-\lambda}(x) - 1| < \delta$ for $d(o, x) < \epsilon$, $\lambda \in C$. Thus $\widetilde{\eta}_\epsilon \longrightarrow 1$ uniformly on compact sets. If $T \in \mathcal{E}'(X)$ then by Lemma 1.4, $(T \times \eta_\epsilon)^\sim(\lambda, b) = \widetilde{T}(\lambda, b)\widetilde{\eta}_\epsilon(\lambda)$. Since $T \times \eta_\epsilon \in \mathcal{D}(X)$ we deduce from Theorem 5.1 that $T \longrightarrow \widetilde{T}$ is injective on $\mathcal{E}'(X)$, and that $\widetilde{T}$ satisfies (3).

To show that the holomorphic function $\lambda \longrightarrow T(\lambda, b)$ satisfies (40) if $\text{supp}(T) \subset C\ell(B_R(o))$ consider the diffeomorphism $(Z, H) \longrightarrow \exp Z \exp H{\cdot}o$ of $\mathfrak{n} \oplus \mathfrak{a}$ onto X. With this identification each differential operator on X can be written as a locally finite linear combination

$$\Sigma f_{ij} \tau(D_i)\tau(E_j),$$

where $f_{ij} \in C^\infty(X)$, $D_i \in \boldsymbol{D}(N)$, $E_j \in \boldsymbol{D}(A)$ and if $X_1, \ldots, X_p$ are members of $\mathfrak{g}$,

$$(\tau(X_1 \ldots X_p)f)(na \cdot o) = \left\{ \frac{\partial^p}{\partial t_1 \ldots \partial t_p} f(\exp(-t_p X_p) \ldots \exp(-t_1 X_1) na \cdot o) \right\}_{t=0}.$$

Then there exist some $D_i \in \boldsymbol{D}(N), E_i \in \boldsymbol{D}(A)$ such that for $\zeta \in \mathcal{E}(X)$,

$$|T(\zeta)| \leq \sup_{x \in B^R(o)} \sum_i |\tau(D_i)\tau(E_i)\zeta|(x). \tag{41}$$

Since $\tau(k)$ $(k \in K)$ leaves $B_R(o)$ invariant (41) implies

$$|T^{\tau(k)}(\zeta)| \leq \sup_{x \in B^R(o)} \sum_i |((\tau(D_i)\tau(E_i))^{\tau(k)}\zeta|(x) \tag{42}$$

and

$$(\tau(D_i)\tau(E_i))^{\tau(k)} = \sum_{p,q} f_{pq}(k, x)\tau(D_p)\tau(E_q),$$

where $f_{pq} \in C^\infty(K \times X)$, $D_p \in \boldsymbol{D}(N), E_q \in \boldsymbol{D}(A)$. In (42) we take $\zeta(x) = e^{(-i\lambda+\rho)(A(x,eM)}$ so $\zeta(na \cdot o) = e^{(-i\lambda+\rho)(\log a)}$. This ζ is an eigenfunction of each $\tau(E_q)$ with eigenvalue a polynomial in λ, ζ is annihilated by each $\tau(D_p)$ without constant term and each f_{pq} is bounded on $B^R(o)$. Hence (42) implies (with k^{-1} replacing k),

$$|\widetilde{T}(\lambda, kM)| \leq C(1 + |\lambda|)^N e^{R|\text{Im } \lambda|}, \qquad \lambda \in \mathfrak{a}_c^*.$$

Conversely, suppose ψ satisfies (3) and (40). Then the assignment

$$\varphi \longrightarrow T(\varphi) = \int_{\mathfrak{a}^* \times B} \psi(\lambda, b)\widetilde{\varphi}(-\lambda, b)|\boldsymbol{c}(\lambda)|^{-2} d\lambda\, db$$

is a well-defined linear functional on $\mathcal{D}(X)$ because of Theorem 5.1. Let T_ϵ be the functional

$$T_\epsilon(\varphi) = \int_{\mathfrak{a}^* \times B} \psi(\lambda, b)\widetilde{\eta}_\epsilon(\lambda)\widetilde{\varphi}(-\lambda, b)|\boldsymbol{c}(\lambda)|^{-2} d\lambda db.$$

By (40) and $\text{supp}(\eta_\epsilon) \subset C\ell(B^\epsilon(o))$ we have for each $N \in \boldsymbol{Z}^+$ a constant C_N such that

$$|\psi(\lambda, b)\widetilde{\eta}_\epsilon(\lambda)| \leq C_N(1 + |\lambda|)^{-N} e^{(R+\epsilon)|Im\ \lambda|}. \tag{43}$$

By Theorem 5.1, $\psi\widetilde{\eta}_\epsilon$ is the Fourier transform of a function $f_\epsilon \in \mathcal{D}(X)$ but by the Plancherel formula

$$\int_X f_\epsilon(x)\varphi(x)dx = \int_{\mathfrak{a}^*\times B} \widetilde{f}_\epsilon(\lambda, b)\widetilde{\varphi}(-\lambda, b)|\mathbf{c}(\lambda)|^{-2}d\lambda\, db$$

we have $T_\epsilon = f_\epsilon$. By (43), $\operatorname{supp}(T_\epsilon) \subset C\ell(B^\epsilon(o))$ and since $\widetilde{\eta}_\epsilon \longrightarrow 1$ uniformly on compacts, $T_\epsilon(\varphi) \longrightarrow T(\varphi)$ $(\varphi \in \mathcal{D}(X))$ so $\operatorname{supp}(T) \subset C\ell(B^R(o))$. Since $T_\epsilon(\varphi) \longrightarrow T(\varphi)$ for all $\varphi \in \mathcal{E}(X)$, we have

$$\widetilde{T}(\lambda, b) = \int_X e^{-i\lambda+\rho)(A(x,b))}dT(x) = \lim_{\epsilon\to 0}\int_X e^{(-i\lambda+\rho)(A(x,b))}dT_\epsilon(x)$$
$$= \lim_{\epsilon\to 0}\int_X e^{(-i\lambda+\rho)(A(x,b))}f_\epsilon(x)dx = \lim_{\epsilon\to 0}\psi(\lambda, b)\widetilde{\eta}_\epsilon(\lambda)$$

so $\widetilde{T} = \psi$ as desired.

5. A Topological Paley-Wiener Theorem for the K-types.

In Theorem 5.1 we have described the range $\mathcal{D}(X)^\sim$ without describing the topology of this space in intrinsic terms. By specializing and reformulating the Fourier transform for functions of a fixed K-type we shall obtain a more specific topological Paley-Wiener theorem.

With $\delta \in \widehat{K}$ acting on V_δ consider the space

$$\mathcal{D}(X, \operatorname{Hom}(V_\delta, V_\delta)) \tag{44}$$

of C^∞ functions on X of compact support, having values in $\operatorname{Hom}(V_\delta, V_\delta)$. We define

$$\mathcal{D}^\delta(X) = \{F \in \mathcal{D}(X, \operatorname{Hom}(V_\delta, V_\delta)) : \ F(k\cdot x) \equiv \delta(k)F(x)\}\,. \tag{45}$$

The space (44) carries a natural topology in which it is an LF-space, in fact the inductive limit (cf. e.g. [GGA], p. 5) of the Fréchet spaces

$$\mathcal{D}_R(X, \operatorname{Hom}(V_\delta, V_\delta)) \quad (R = 0, 1, 2, \dots)$$

of functions in (44) having support in $C\ell(B^R(o))$. Let $\mathcal{D}_\delta(X)$ denote the space of K-finite functions in $\mathcal{D}(X)$ of type δ. The spaces $\mathcal{D}^\delta(X)$ and $\mathcal{D}_\delta(X)$

are given the topologies induced from the space (44) and the space $\mathcal{D}(X)$, respectively. For $\varphi \in C(K), f \in C(X)$ we write

$$(\varphi * f)(x) = \int_K \varphi(k) f(k^{-1} \cdot x) dk, \qquad x \in X. \tag{46}$$

Proposition 5.10.

(i) *The map*

$$Q: \ F(x) \longrightarrow \mathrm{Trace}(F(x))$$

is a homeomorphism of $\mathcal{D}^\delta(X)$ *onto* $\mathcal{D}_{\check{\delta}}(X)$ *and its inverse is given by*

$$f \longrightarrow f^\delta.$$

(ii) *The maps*

$$p: \ f \in \mathcal{D}(X) \longrightarrow d(\delta)\chi_\delta * f \in \mathcal{D}_{\check{\delta}}(X)$$

$$q: \ f \in \mathcal{D}(X) \longrightarrow f^\delta \in \mathcal{D}^\delta(X)$$

are continuous open surjections and the images $\mathcal{D}_{\check{\delta}}(X)$ *and* $\mathcal{D}^\delta(X)$ *are LF-spaces, closed in* $\mathcal{D}(X)$ *and* $\mathcal{D}(X, \mathrm{Hom}(V_\delta, V_\delta))$, *respectively.*

The proof is analogous to that of a similar result about the space of continuous functions on X (cf. [GGA], Ch. IV, §1). In connection with the last statement of the proposition it should be recalled that a closed subspace of an LF-space is not necessarily an LF-space.

If $f \in \mathcal{D}_{\check{\delta}}(X)$ then $f = d(\delta)\chi_\delta * f$ so

$$\widetilde{f}(\lambda, kM) = d(\delta) \int_X f(x) \left(\int_K e^{(-i\lambda+\rho)(A(x,vK))} \chi_\delta(kv^{-1}) dv \right) dx.$$

This leads to the following definition of the *δ-spherical transform* on $\mathcal{D}_{\check{\delta}}(X)$, $\delta \in \widehat{K}_M$.

Definition. For $f \in \mathcal{D}_{\check{\delta}}(X)$, the δ-spherical transform $\widetilde{f}$ is defined by

$$\widetilde{f}(\lambda) = d(\delta) \int_X f(x) \Phi_{\overline{\lambda},\delta}(x)^* dx. \tag{47}$$

If δ is the trivial representation, (29) shows that $f \longrightarrow \widetilde{f}$ is just the spherical transform of K-invariant functions on X. In general $\delta(m)\widetilde{f}(\lambda) = \widetilde{f}(\lambda)$ so $\widetilde{f}$ is a smooth function from $\mathfrak{a}_{\mathbf{c}}^*$ to $\mathrm{Hom}(V_\delta, V_\delta^M)$. In addition (29) and the last formula for $\widetilde{f}(\lambda, kM)$ show that

$$(48) \qquad \widetilde{f}(\lambda) = d(\delta) \int_K \widetilde{f}(\lambda, kM)\delta(k^{-1})dk, \quad \widetilde{f}(\lambda, kM) = \mathrm{Tr}(\delta(k)\widetilde{f}(\lambda))$$

so we conclude from (25) that $\widetilde{f}$ belongs to the space

$$(49) \qquad \mathcal{H}(\mathfrak{a}_{\mathbf{c}}^*, \mathrm{Hom}(V_\delta, V_\delta^M))$$

of holomorphic functions on $\mathfrak{a}_{\mathbf{c}}^*$ of exponential type with values in $\mathrm{Hom}(V_\delta, V_\delta^M)$.

Recall now that the Fourier transform in $\boldsymbol{R}^n$ is a bijection of $\mathcal{D}(\boldsymbol{R}^n)$ onto the space $\mathcal{H}(\boldsymbol{C}^n)$ of holomorphic functions of exponential type. The space $\mathcal{H}(\boldsymbol{C}^n)$ has an intrinsic topology with the property that the Fourier transform is a homeomorphism of $\mathcal{D}(\boldsymbol{R}^n)$ onto $\mathcal{H}(\boldsymbol{C}^n)$ and the multiplication by a fixed nonzero polynomial P is a homeomorphism of $\mathcal{H}(\boldsymbol{C}^n)$ onto the subspace $P\mathcal{H}(\boldsymbol{C}^n)$ (see Theorem 5.21 below). This induces a topology on the space (49) of matrix-valued functions such that Euclidean Fourier transform is a homeomorphism of the space $\mathcal{D}(A, \mathrm{Hom}(V_\delta, V_\delta^M))$ (of C^∞ functions on A of compact support with values in $\mathrm{Hom}(V_\delta, V_\delta^M)$) onto the space (49). In particular, the space (49) is an LF-space.

We recall now the linear transformation $Q^\delta(\lambda) : E_\delta \longrightarrow V_{\check{\delta}}^M$ from (22) §2 and consider the subspace

$$(50) \quad \mathcal{H}^\delta(\mathfrak{a}^*) = \left\{ F \in \mathcal{H}(\mathfrak{a}_{\mathbf{c}}^*, \mathrm{Hom}(V_\delta, V_\delta^M)) : \ (Q^{\check{\delta}})^{-1}F \qquad W\text{-invariant} \right\}$$

and give it the topology induced by that of the space (49). We can now state the topological Paley-Wiener theorem.

Theorem 5.11. *The δ-spherical transform $f \longrightarrow \widetilde{f}$ where*

$$\widetilde{f}(\lambda) = d(\delta) \int_X f(x) \Phi_{\overline{\lambda},\delta}(x)^* dx$$

is a homeomorphism of $\mathcal{D}_{\check{\delta}}(X)$ onto $\mathcal{H}^\delta(\mathfrak{a}^*)$.

Proof. We know from Theorem 2.10 that on V_δ^M,

$$\Phi_{s\lambda,\delta}(x) = \Phi_{\lambda,\delta}(x)\Gamma_{s,\lambda}$$

so the δ-spherical transform satisfies the functional equations

$$\widetilde{f}(s\lambda) = (\Gamma_{s,\overline{\lambda}})^* \widetilde{f}(\lambda), \qquad s \in W. \tag{51}$$

However, by (54), (55) in §2 we have

$$(\Gamma_{s,\overline{\lambda}})^* = B(-\lambda, s) \qquad \text{on } V_\delta^M \tag{52}$$

so by virtue of Theorem 3.5, the identities (51) imply

$$Q^{\overset{\vee}{\delta}}(\lambda)^{-1}\widetilde{f}(\lambda) \qquad \text{is } W\text{-invariant}, \tag{53}$$

whence $\widetilde{f} \in \mathcal{H}^\delta(\mathfrak{a}^*)$.

For the surjectivity let $\psi \in \mathcal{H}^\delta(\mathfrak{a}^*)$. Then the function

$$\Psi(\lambda, kM) = \operatorname{Tr}(\delta(k)\psi(\lambda)) \tag{54}$$

on $\mathfrak{a}_{\mathbf{c}}^* \times B$ is clearly a holomorphic function of uniform exponential type. To see that it satisfies the compatibility relations (3) it would suffice to verify

$$\Phi_{s\lambda,\delta}(x)\psi(s\lambda) = \Phi_{\lambda,\delta}(x)\psi(\lambda). \tag{55}$$

By assumption,

$$Q^{\overset{\vee}{\delta}}(s\lambda)^{-1}\psi(s\lambda) = Q^{\overset{\vee}{\delta}}(\lambda)^{-1}\psi(\lambda)$$

so by Theorem 3.5

$$\psi(s\lambda) = B(-\lambda, s)\psi(\lambda).$$

Combining this with (52) and (23) we conclude

$$\Gamma_{s,\lambda}\psi(s\lambda) = \psi(\lambda)$$

so left-multiplying by $\Phi_{\lambda,\delta}(x)$ we get (55). The assumptions of Theorem 5.1 being verified there exists a unique $F \in \mathcal{D}(X)$ such that $\widetilde{F}(\lambda, b) = \Psi(\lambda, b)$. Then by Proposition 5.10 the function $f = d(\delta)\chi_\delta * F$ belongs to $\mathcal{D}_{\overset{\vee}{\delta}}(X)$ and

$$\begin{aligned}\widetilde{f}(\lambda) &= d(\delta)^2 \int_K \int_K \widetilde{F}(\lambda, u^{-1}kM)\chi_\delta(u)\delta(k^{-1})du\, dk \\ &= d(\delta)^2 \int_K \int_K \operatorname{Tr}(\delta(k)\psi(\lambda))\chi_\delta(u)\delta(k^{-1}u^{-1})dk\, du \\ &= \psi(\lambda)\end{aligned}$$

by virtue of the orthogonality relations

$$d(\delta)\int_K \operatorname{conj}(\delta_{ij}(k))\delta(k)dk = E_{ij}. \tag{56}$$

The injectivity of the transform $f \longrightarrow \widetilde{f}$ is clear from (48) and Theorem 1.3.

It remains to prove the topological statement of the theorem. Now it will be most convenient to think of $\overset{\vee}{Q}{}^{\delta}$ and $\mathcal{H}^{\delta}(\mathfrak{a}^*)$ in terms of matrices. We fix an orthonormal basis $v_1, \ldots, v_{d(\delta)}$ of V_δ such that $v_1, \ldots, v_{\ell(\delta)}$ span V_δ^M. Then the members of $\mathcal{H}^{\delta}(\mathfrak{a}^*)$ become matrix-valued holomorphic functions on $\mathfrak{a}_{\mathbf{c}}^*$ and $\overset{\vee}{Q}{}^{\delta}(\lambda)$ is an $\ell(\delta) \times \ell(\delta)$ matrix whose entries are polynomial functions on $\mathfrak{a}_{\mathbf{c}}^*$ (cf. (22) in §2 and the subsequent remark). Let $\mathcal{J}^{\delta}(\mathfrak{a}^*)$ denote the set of Weyl group invariants in $\mathcal{H}(\mathfrak{a}_{\mathbf{c}}^*, \operatorname{Hom}(V_\delta, V_\delta^M))$, with the relative topology. We now need the following lemma.

Lemma 5.12. *The mapping*

$$\psi(\lambda) \longrightarrow \overset{\vee}{Q}{}^{\delta}(\lambda)\psi(\lambda) \tag{57}$$

is a homeomorphism of $\mathcal{J}^{\delta}(\mathfrak{a}^*)$ onto $\mathcal{H}^{\delta}(\mathfrak{a}^*)$.

Proof. Clearly (57) maps $\mathcal{J}^{\delta}(\mathfrak{a}^*)$ into $\mathcal{H}^{\delta}(\mathfrak{a}^*)$. Next let $F \in \mathcal{H}^{\delta}(\mathfrak{a}^*)$. Then we know that

$$\overset{\vee}{Q}{}^{\delta}(s\lambda)^{-1}F(s\lambda) = \overset{\vee}{Q}{}^{\delta}(\lambda)^{-1}F(\lambda). \tag{58}$$

Now

$$\overset{\vee}{Q}{}^{\delta}(\lambda)^{-1} = Q_c(\lambda)(\det \overset{\vee}{Q}{}^{\delta}(\lambda))^{-1},$$

where $Q_c(\lambda)$ is a matrix whose entries are cofactors of $\overset{\vee}{Q}{}^{\delta}(\lambda)$; in particular the entries in $Q_c(\lambda)$ are polynomial functions on $\mathfrak{a}_c^*$, i. e. polynomials in the coordinates of λ. Now by Lemma 2.11 and Proposition 4.1,

$$\det Q^{\delta}(\lambda) \neq 0 \text{ for } \lambda \in \mathfrak{a}^* + iC\ell(\mathfrak{a}_+^*).$$

But then since the function

$$\psi(\lambda) = \overset{\vee}{Q}{}^{\delta}(\lambda)^{-1}F(\lambda) = (\det \overset{\vee}{Q}{}^{\delta}(\lambda))^{-1}Q_c(\lambda)F(\lambda) \tag{59}$$

is W-invariant it must be holomorphic on all of $\mathfrak{a}_c^*$. Since each entry in the matrix $Q_c(\lambda)F(\lambda)$ is an entire function of exponential type (the Euclidean

Fourier transform of a function in $\mathcal{D}(A)$) Lemma 5.13 below shows that ψ is of exponential type, so $\psi \in \mathcal{J}^\delta(\mathfrak{a}^*)$. Thus the map (57) is surjective.

The mapping $F(\lambda) \longrightarrow \Sigma_{s\in W} F(s\lambda)$ is a continuous open mapping of the LF-space $\mathcal{H}(\mathfrak{a}_c^*, \mathrm{Hom}(V_\delta, V_\delta^M))$ onto $\mathcal{J}^\delta(\mathfrak{a}^*)$ and by [GGA], Ch. IV, Lemma 1.13, the latter space is also an LF-space. The map (57) is obviously continuous. The inverse is a composite $I_1 \circ I_2$ of the maps

$$I_2 : \; F \in \mathcal{H}^\delta(\mathfrak{a}^*) \longrightarrow Q_c F \in Q_c \mathcal{H}^\delta(\mathfrak{a}^*)$$

$$I_1 : \; G \in Q_c\mathcal{H}^\delta(\mathfrak{a}^*) \longrightarrow (\det \check{Q}^\delta)^{-1} G \in \mathcal{J}^\delta(\mathfrak{a}^*)$$

Here I_2 obviously continuous; I_1 is continuous because by (59) each $G \in Q_c\mathcal{H}^\delta(\mathfrak{a}^*)$ is a multiple of $\det(\check{Q}^\delta)$ so I_1^{-1} is just multiplication by $\det(\check{Q}^\delta)$ on a subspace of $\mathcal{H}(\mathfrak{a}_c^*, \mathrm{Hom}(V_\delta, V_\delta^M))$. To conclude the proof of Lemma 5.12 we just need the following result.

Lemma 5.13. *Suppose $H(\zeta)$ is an entire function of exponential type in $\boldsymbol{C}^n$ and suppose $P(\zeta)$ is a polynomial such that $F(\zeta) = H(\zeta)/P(\zeta)$ is holomorphic in $\boldsymbol{C}^n$. Then $F(\zeta)$ is also entire of exponential type.*

Proof. This is essentially a special case of a result of Malgrange [1955] p. 306 (where both H and P are entire of exponential type) but we sketch a proof for the present case. The assumption about $H(\zeta)$ means that

$$|H(\zeta)| \leq C_N(1+|\zeta|)^{-N} e^{A|Im\ \zeta|}, \tag{60}$$

$A > 0$ being fixed, $N \in \boldsymbol{Z}^+$ being arbitrary.

Let $\alpha = (\alpha_1, \dots, \alpha_n)$, where $\alpha_i \in \boldsymbol{Z}^+$ and put

$$P^{(\alpha)}(\zeta) = (D^\alpha P)(\zeta), \quad \zeta^\alpha = \zeta_1^{\alpha_1} \cdots \zeta_n^{\alpha_n}$$

$$|\zeta| = (|\zeta_1|^2 + \cdots + |\zeta_n|^2)^{\frac{1}{2}}, \quad m = \deg(P).$$

Then we have the following inequality which can be derived from the Cauchy integral formula in one variable,

$$|F(0)P^{(\alpha)}(0)| \leq C_{m,\alpha} \int\limits_{|\zeta|\leq 1} |F(\zeta)P(\zeta)| d\zeta, \tag{61}$$

where $d\zeta$ is the Lebesgue measure in $\boldsymbol{C}^n$ and the constant $C_{m,\alpha}$ depends only on m and α (cf. Hörmander [1963], Lemma 3.1.4). Replacing in (61), $F(\zeta)$ by $F(\lambda+\zeta), P(\zeta)$ by $P(\lambda+\zeta)$ we obtain with another constant C,

$$|F(\lambda)P^{(\alpha)}(\lambda)| \leq C \sup_{|\zeta|\leq 1} |H(\lambda+\zeta)|, \qquad \lambda \in \boldsymbol{C}^n. \tag{62}$$

Choosing α such that $P^{(\alpha)}$ is a constant $\neq 0$ and using (60) for H we obtain the desired inequality for F. This proves Lemma 5.13 and hence also Lemma 5.12.

We can now conclude the proof of Theorem 5.11. Because of Lemma 5.12, $\mathcal{H}^\delta(\mathfrak{a}^*)$ is an LF-space (like $\mathcal{J}^\delta(\mathfrak{a}^*)$) so the δ-spherical transform $f \longrightarrow \widetilde{f}$ is a continuous bijection of the LF-space onto the LF-space $\mathcal{H}^\delta(\mathfrak{a}^*)$, hence a homeomorphism.

Corollary 5.14. *Representing* $\mathrm{Hom}(V_\delta, V_\delta^M)$ *and* $\check{Q}^\delta(\lambda)$ *by an orthonormal basis* (v_i) *of* V_δ *(cf. (11) and (12) §2) the map* $f \longrightarrow \widetilde{f}$ *is a homeomorphism of* $\mathcal{D}_{\check{\delta}}(X)$ *onto the space* $\check{Q}^\delta \mathcal{J}^\delta(\mathfrak{a}^*)$.

Using (52), (23) and Theorem 3.5 we can restate the functional equation in Theorem 2.10 in a more explicit form.

Theorem 5.15. *The restriction* $\Phi_{\lambda,\delta}|A{\cdot}o$ *belongs to* $\mathcal{E}(A, \mathrm{Hom}(V_\delta, V_\delta^M))$ *and satisfies the following functional equations*

$$\int_K e^{(i\lambda+\rho)(A(a\cdot o,kM))}\delta(k)dk\ \check{Q}^\delta(\lambda) = \int_K e^{(is\lambda+\rho)(A(a\cdot o,kM))}\delta(k)dk\ \check{Q}^\delta(s\lambda)$$

for $s \in W$ *and*

$$\check{Q}^\delta(\lambda)^{-1}\int_K e^{(-i\lambda+\rho)(A(a\cdot o,kM))}\delta(k^{-1})dk =$$
$$\check{Q}^\delta(s\lambda)^{-1}\int_K e^{(-is\lambda+\rho)(A(a\cdot o,kM))}\delta(k^{-1})dk.$$

Both sides are holomorphic on all of $\mathfrak{a}_{\mathbf{c}}^*$, *in other words,* $\Phi_{\overline{\lambda},\delta}(a{\cdot}o)^*$ *is divisible by* $\check{Q}^\delta(\lambda)$.

Proof. By Theorem 2.10

$$\Phi_{s\lambda,\delta}(a\cdot o) = \Phi_{\lambda,\delta}(a\cdot o)\Gamma_{s,\lambda},$$

which by (52) equals $\Phi_{\lambda,\delta}(a\cdot o)B(-\overline{\lambda}, s)^*$. Multiplying by $B(-\lambda, s)$ on the right and using (23) and Theorem 3.5 we derive the first formula. For the second we take adjoint above, getting

$$(\Gamma_{s,\overline{\lambda}})^*(\Phi_{\overline{\lambda},\delta}(a\cdot o))^* = (\Phi_{s\overline{\lambda},\delta}(a\cdot o))^*.$$

Plugging in (52) and (29) the second formula follows from Theorem 3.5. For the last statement we observe as in (59) that the holomorphy is clear on $\mathfrak{a}^* + i(C\ell(\mathfrak{a}^*_+))$, hence by the Weyl group invariance on all of $\mathfrak{a}^*_c$.

6. The Inversion — and the Plancherel Formula for the δ-spherical Transform.

In principle the inversion formula and the Plancherel formula for the δ-spherical transform are contained in the corresponding formulas for the full Fourier transform $f(x) \longrightarrow \widetilde{f}(\lambda, b)$. For convenience we write them down explicitly.

Theorem 5.16. *The δ-spherical transform (47) is inverted by*

$$f(x) = w^{-1}\operatorname{Tr}\Big[\int_{\mathfrak{a}^*} \Phi_{\lambda,\delta}(x)\widetilde{f}(\lambda)|c(\lambda)|^{-2}d\lambda\Big], \quad f \in \mathcal{D}_{\check{\delta}}(X). \tag{63}$$

Moreover,

$$\int_X |f(x)|^2 dx = w^{-1}d(\delta)^{-1}\int_{\mathfrak{a}^*} \operatorname{Tr}(\widetilde{f}(\lambda)\widetilde{f}(\lambda)^*)|c(\lambda)|^{-2}d\lambda. \tag{64}$$

Here Tr *and* $*$ *denote trace and adjoint on V_δ, respectively, and w is the order of the Weyl group.*

Proof. If $f \in \mathcal{D}_{\check{\delta}}(X)$ we have by (48), $\widetilde{f}(\lambda, kM) = \operatorname{Tr}(\delta(k)\widetilde{f}(\lambda))$ so (63) follows from Theorem 1.3. On the other hand (48), together with the orthogonality relations, show that

$$d(\delta)\int |\widetilde{f}(\lambda, kM)|^2 dk = \operatorname{Tr}(\widetilde{f}(\lambda)\widetilde{f}(\lambda)^*)$$

so (64) follows from Theorem 1.5.

We shall now combine this result with Corollary 5.14. As in Lemma 3.6 consider the dual orthonormal bases

$$(v_i) \text{ of } V_\delta, \qquad (v_i') \text{ of } V_{\check{\delta}}$$

and the bases (ϵ_j) of E_δ, (ϵ_j') of $E_{\check{\delta}}$. The following result is a restatement of Theorem 2.6.

Corollary 5.17. *The matrix $\Phi_{\lambda,\delta}(x)Q^{\check{\delta}}(\lambda)$ can be written for $1 \leq p \leq d(\delta), 1 \leq q \leq \ell(\delta)$,*

$$\left((\Phi_{\lambda,\delta}(gK))Q^{\check{\delta}}(\lambda)\right)_{pq} = (\tau(\epsilon'_q(v'_p))\varphi_\lambda)(g). \tag{65}$$

Thus if $D^{\check{\delta}}$ is the differential operator matrix

$$D^{\check{\delta}} = \begin{pmatrix} \tau(\epsilon'_1(v'_1)), \ldots, \tau(\epsilon'_{\ell(\delta)}(v'_1)) \\ \vdots \\ \tau(\epsilon'_1(v'_{d(\delta)})), \ldots, \tau(\epsilon'_{\ell(\delta)}(v'_{d(\delta)})) \end{pmatrix}$$

then

$$\Phi_{\lambda,\delta}(gK)|V_\delta^M = (D^{\check{\delta}}\varphi_\lambda)(g)Q^{\check{\delta}}(\lambda)^{-1}. \tag{66}$$

In fact replace in Theorem 2.6 δ by $\check{\delta}$, v by v'_p, v_i by v'_i. Since

$$\langle v'_p, \check{\delta}(k)v'_i\rangle = \langle v'_p, \sum_q \overline{\delta(k)_{qi}v'_q}\rangle = \delta(k)_{pi}$$

(65) is a restatement of Theorem 2.6.

Corollary 5.18. *In the notation of Theorem 5.16 let*

$$\Phi(\lambda) = Q^{\check{\delta}}(\lambda)^{-1}\widetilde{f}(\lambda) \qquad (\ell(\delta) \times d(\delta)\text{-}matrix)$$

and let φ be the inverse spherical transform

$$\varphi(x) = w^{-1}\int_{\mathfrak{a}^*} \varphi_\lambda(x)\Phi(\lambda)|c(\lambda)|^{-2}d\lambda.$$

Then (63) can be written

$$f(gK) = \mathrm{Tr}(D^{\check{\delta}}\varphi)(g). \tag{67}$$

Corollary 5.19. *Each K-finite function in $\mathcal{D}(X)$ is a finite linear combination of derivatives of functions in $\mathcal{D}^\natural(X)$ by members of $\tau(\mathbf{D}(G))$.*

We conclude this section with a justification of the properties used about the topology of $\mathcal{H}(\boldsymbol{C}^n)$. Although the result is only needed here for $X = \boldsymbol{R}^n$ we shall prove it for an open convex subset $X \subset \boldsymbol{R}^n$.

Suppose $K_1 \subset K_2 \subset \cdots$ are compact, convex subsets of X such that $K_i \subset \overset{\circ}{K}_{i+1}$ for each i and $X = \bigcup_i K_i$. The topology of $\mathcal{D}(X)$ is the inductive limit topology of the Fréchet spaces $\mathcal{D}_{K_i}(X)$. Let H_K denote the supporting function of a set $K \subset \boldsymbol{R}^n$

$$H_K(\xi) = \sup_{x \in K} \langle x, \xi \rangle \qquad \xi \in \boldsymbol{R}^n.$$

If $K = B_r(0)$ then $H_K(\xi) = r|\xi|$. The following result (Hörmander [1983], Theorem 15.4.2) characterizes the neighborhoods of 0 in $\mathcal{D}(X)$ in terms of Fourier transforms $\widetilde{u}$.

Theorem 5.20. *With $X = \bigcup K_j$ as above the following conditions on a convex subset $V \subset \mathcal{D}(X)$ are equivalent:*

(i) V is a neighborhood of 0.

(ii) There exist δ_j, $M_j > 0$ such that V contains all $u \in \mathcal{D}(X)$ such that

$$|\widetilde{u}(\lambda)| \leq \sum_j \delta_j (1 + |\lambda|)^{-M_j} \exp H_{K_j}(\operatorname{Im} \lambda). \tag{68}$$

This result implies an important property of constant-coefficient differential operators.

Theorem 5.21. *Let $X \subset \boldsymbol{R}^n$ be an open convex subset and D a differential operator on $\boldsymbol{R}^n$ with constant coefficients. Then the mapping $u \longrightarrow Du$ is a homeomorphism of $\mathcal{D}(X)$ onto the subspace $D\mathcal{D}(X)$ (with the relative topology).*

Proof. The continuity is obvious and the injectivity follows taking Fourier transforms. To prove the continuity of the inverse we must show that if V is a convex neighborhood of 0 in $\mathcal{D}(X)$ then there exists a convex neighborhood W of 0 in $\mathcal{D}(X)$ such that

$$u \in \mathcal{D}(X), \quad Du \in W \Longrightarrow u \in V. \tag{69}$$

Let V be defined by (68). We shall show that if W is defined as the set of $f \in \mathcal{D}(X)$ such that

$$|\widetilde{f}(\lambda)| \leq \sum_j \epsilon_j (1 + |\lambda|)^{-M_j} \exp H_{K_j}(\operatorname{Im} \lambda) \tag{70}$$

then (69) is valid if the ϵ_j are small enough. For the proof we write $f = Du$ which implies $\widetilde{f}(\lambda) = P(\lambda)\widetilde{u}(\lambda)$ where P is a polynomial. As a consequence of (62) we have for a constant C,

$$|\widetilde{u}(\lambda)| \leq C \sup_{|\zeta|\leq 1} |\widetilde{f}(\zeta + \lambda)|.$$

With $h_j = \sup_{|\eta|=1} H_{K_j}(\eta)$ we have for $|\zeta| \leq 1$

$$(1+|\lambda+\zeta|)^{-M_j} \leq 2^{M_j}(1+|\lambda|)^{-M_j}, \qquad H_{K_j}(\operatorname{Im}(\lambda+\zeta)) \leq H_{K_j}(\operatorname{Im}\lambda) + h_j.$$

Thus (70) implies (68) so (69) is valid provided

$$C\ 2^{M_j} e^{h_j} \epsilon_j \leq \delta_j.$$

This concludes the proof.

Remark. I am indebted to Hörmander for communicating this proof of Theorem 5.21 to me. For the case $X = \boldsymbol{R}^n$ the theorem appears in Ehrenpreis [1956], p. 692.

Corollary 5.22. *If $X \subset \boldsymbol{R}^n$ is an open convex subset and D a differential operator on $\boldsymbol{R}^n$ with constant coefficients then*

$$D\mathcal{D}'(X) = \mathcal{D}'(X).$$

In fact let $T \in \mathcal{D}'(X)$ and consider, by the theorem, the continuous linear functional

$$D^*u \longrightarrow T(u)$$

on $D^*\mathcal{D}(X)$. By the Hahn-Banach theorem this extends to a distribution $S \in \mathcal{D}'(X)$. Thus $S(D^*u) = T(u)$ so $DS = T$ as claimed.

§6. EIGENFUNCTIONS AND EIGENSPACE REPRESENTATIONS.

On the basis of the topological Paley-Wiener theorem (Theorem 5.11) we shall now determine the K-finite joint eigenfunctions of $\boldsymbol{D}(G/K)$ and settle the irreducibility question for the eigenspace representations T_λ (Ch. II, §2, No. 3) for $X = G/K$.

1. THE K-FINITE EIGENFUNCTIONS OF $\boldsymbol{D}(X)$.

Theorem 6.1. *The K-finite joint eigenfunctions of all the operators $D \in \boldsymbol{D}(X)$ are precisely*

(1) $$f(x) = \int_B e^{(i\lambda+\rho)(A(x,b))} F(b)db,$$

where $\lambda \in \mathfrak{a}_c^*$ and F is a K-finite function on B.

Proof. Let $\lambda \in \mathfrak{a}_{\mathbf{c}}^*, D \in \boldsymbol{D}(X)$. As in Ch. II, Prop. 3.14 let $\Gamma(D)$ be the W-invariant polynomial function on $\mathfrak{a}_{\mathbf{c}}^*$ such that

$$D\varphi_\lambda = \Gamma(D)(i\lambda)\varphi_\lambda, \qquad D \in \boldsymbol{D}(X),$$

and as usual let $\mathcal{E}_\lambda(X)$ denote the joint eigenspace in Prop. 2.6, Ch. II. For $\delta \in \widehat{K}_M$, let $\mathcal{E}_{\lambda,\delta}(X) \subset \mathcal{E}_\lambda(X)$ and $\mathcal{E}_\delta(B) \subset \mathcal{E}(B)$ denote the subspaces of K-finite elements of type δ. Each joint eigenfunction of $\boldsymbol{D}(X)$ lies in $\mathcal{E}_\lambda(X)$ for some $\lambda \in \mathfrak{a}_{\mathbf{c}}^*$ and $\mathcal{E}_{s\lambda}(X) = \mathcal{E}_\lambda(X)$, $\mathcal{E}_{s\lambda,\delta}(X) = \mathcal{E}_{\lambda,\delta}(X)$ for $s \in W$. Selecting, by Theorem 4.4, an element $s \in W$ such that $s\lambda$ is simple we obtain by the injectivity of the Poisson transform $\mathcal{P}_{s\lambda} : F \longrightarrow f$ the relation

(2) $$\dim \mathcal{E}_{\lambda,\delta}(X) = \dim \mathcal{E}_{s\lambda,\delta}(X) \geq \dim \mathcal{E}_\delta(B) = \dim V_\delta \dim V_\delta^M.$$

On the other hand, let $h \in \mathcal{E}_{\lambda,\delta}(X)$. Viewing h as a distribution we define its Fourier transform $\widetilde{h}$ as a linear functional on $\mathcal{H}^\delta(\mathfrak{a}^*)$:

(3) $$\widetilde{h}(\widetilde{f}) = h(f) = \int_X h(x)f(x)dx, \quad f \in \mathcal{D}_{\check{\delta}}(X).$$

Because of Theorem 5.11, $\widetilde{h}$ is a continuous linear functional on $\mathcal{H}^\delta(\mathfrak{a}^*)$ and by Lemma 5.12 the mapping

(4) $$\psi \in \mathcal{J}^\delta(\mathfrak{a}^*) \longrightarrow \widetilde{h}(\check{Q}^\delta \psi)$$

is a continuous linear functional on $\mathcal{J}^\delta(\mathfrak{a}^*)$. By Corollary 5.14 (for δ the identity representation) the mapping (4) is the spherical transform of a $\operatorname{Hom}(V_\delta, V_\delta^M)$-valued K-invariant distribution j on X, that is

(5) $$\widetilde{j}(\widetilde{\varphi}) = \widetilde{h}(\check{Q}^\delta \widetilde{\varphi}) = j(\varphi)$$

for all K-invariant φ in $\mathcal{D}(X, \operatorname{Hom}(V_\delta, V_\delta^M))$.

Now $Dh = \Gamma(D)(i\lambda)h$ and by (47) and (29) §5,

$$(6) \qquad (Df)^{\sim}(\mu) = \Gamma(D^*)(-i\mu)\widetilde{f}(\mu) \qquad \mu \in \mathfrak{a}_{\mathbf{c}}^*.$$

Let $p_D(\mu) = \Gamma(D)(-i\mu)$ for $\mu \in \mathfrak{a}_{\mathbf{c}}^*$, $D \in \boldsymbol{D}(X)$. Then

$$(Dh)(f) = h(D^*f) = \widetilde{h}((D^*f)^{\sim}) = \widetilde{h}(p_D\widetilde{f}) = (p_D\widetilde{h})(\widetilde{f})$$

so by (3) and $Dh = p_D(-\lambda)h$

$$(7) \qquad p_D\widetilde{h} = p_D(-\lambda)\widetilde{h}.$$

But then we deduce from (5)–(7),

$$\begin{aligned}(p_D\widetilde{j})(\widetilde{\varphi}) &= \widetilde{j}(p_D\widetilde{\varphi}) = \widetilde{j}((D^*\varphi)^{\sim}) = \widetilde{h}(\overset{\vee}{Q^\delta}p_D\widetilde{\varphi}) \\ &= (p_D\widetilde{h})(\overset{\vee}{Q^\delta}\widetilde{\varphi}) = p_D(-\lambda)\widetilde{h}(\overset{\vee}{Q^\delta}\widetilde{\varphi}) = p_D(-\lambda)\widetilde{j}(\widetilde{\varphi}),\end{aligned}$$

so $\widetilde{j}$ satisfies (7) too. Thus j is an eigendistribution of all $D \in \boldsymbol{D}(X)$. Being K-invariant and $\mathrm{Hom}(V_\delta, V_\delta^M)$-valued, j equals $\varphi_\lambda A$ where A is a fixed element in $\mathrm{Hom}(V_\delta, V_\delta^M)$. Thus the map $h \longrightarrow j$ gives an injection of $\mathcal{E}_{\lambda,\delta}(X)$ into $\mathrm{Hom}(V_\delta, V_\delta^M)$ so

$$(8) \qquad \dim \mathcal{E}_{\lambda,\delta}(X) \leq \dim V_\delta \dim V_\delta^M.$$

By comparing with (2) we have equality sign. This proves the theorem.

Remark. The proof shows that in (1) we can always take λ simple. In that case,

$$\mathcal{P}_\lambda(\mathcal{E}_\delta(B)) = \mathcal{E}_{\lambda,\delta}(X).$$

2. The Irreducibility Criterion for the Eigenspace Representations on G/K.

Preserving the notation from No. 1 we now define the Gamma function for X by

$$(9) \quad \Gamma_X(\lambda) = \prod_{\alpha\in\Sigma_o} \Gamma(\tfrac{1}{2}(\tfrac{1}{2}m_\alpha + 1 + \langle i\lambda, \alpha_o\rangle))\Gamma(\tfrac{1}{2}(\tfrac{1}{2}m_\alpha + m_{2\alpha} + \langle i\lambda, \alpha_o\rangle)),$$

where $\alpha_o = \alpha/\langle\alpha,\alpha\rangle$. In terms of the $\boldsymbol{e}_s$ function (Ch. II, §5 (24)) we have

$$(10) \qquad \Gamma_X(\lambda) = (\boldsymbol{e}_{s^*}(\lambda)\boldsymbol{e}_{s^*}(-\lambda))^{-1},$$

where $s^* \in W$ is as usual the element interchanging $\mathfrak{a}^+$ and $-\mathfrak{a}^+$. Given $\lambda \in \mathfrak{a}_c^*$ we consider again the joint eigenspace $\mathcal{E}_\lambda(X)$ and as in Ch. II, §2, No. 3 we let T_λ denote the natural representation of G on $\mathcal{E}_\lambda(X)$, i.e.

$$(T_\lambda(g)f)(x) = f(g^{-1} \cdot x), \qquad x \in X, g \in G, f \in \mathcal{E}_\lambda(X).$$

Theorem 6.2. *The eigenspace representation T_λ is irreducible if and only if*

$$1/\Gamma_X(\lambda) \neq 0. \tag{11}$$

Proof. Suppose first (11) holds and let $V \subset \mathcal{E}_\lambda(X)$ be a closed invariant subspace. Then $\varphi_\lambda \in V$. Since $-\lambda$ is simple, Lemma 2.3 implies that V contains the space $\mathcal{H}_\lambda$ of functions

$$f(x) = \int_B e^{(i\lambda+\rho)(A(x,b))} F(b)db, \qquad F \in L^2(B). \tag{12}$$

Secondly, since λ is simple, Theorem 6.1 and the subsequent remark shows that $\mathcal{H}_\lambda$ contains the space $\mathcal{E}_{\lambda,\delta}(X)$ for each $\delta \in \widehat{K}_M$. The expansion (30) §5 which also holds in $\mathcal{E}(X)$ then shows that $\mathcal{H}_\lambda$ is dense in $\mathcal{E}_\lambda(X)$ so $V = \mathcal{E}_\lambda(X)$ and T_λ is irreducible.

Conversely, suppose T_λ is irreducible. Since $\mathcal{E}_\lambda(X) = \mathcal{E}_{s\lambda}(X)$ $(s \in W)$, since condition (11) is W-invariant and since at least one of the elements $s\lambda$ is simple we may assume λ simple. Then we can make the space $\mathcal{H}_\lambda$ of functions f in (12) into a Hilbert space by giving f the L^2-norm of F on B. Let $0 \neq E \subset \mathcal{H}_\lambda$ be a closed invariant subspace. Since T_λ is irreducible, E is dense in $\mathcal{E}_\lambda(X)$. Hence $\overline{\chi}_\delta * E$ is dense in $\overline{\chi}_\delta * \mathcal{E}_\lambda(X) = \mathcal{E}_{\lambda,\delta}(X)$ so by finite-dimensionality, $\overline{\chi} * E = \mathcal{E}_{\lambda,\delta}(X)$. Now the space $E^* = \{F \in L^2(B) : \ f \in E\}$ is a K invariant closed subspace of $L^2(B)$. Since $\overline{\chi}_\delta * E^* \subset E^*$ (convolution on K/M) we deduce $\overline{\chi}_\delta * E \subset E$ so $\mathcal{E}_{\lambda,\delta}(X) \subset E$ for each $\delta \in \widehat{K}_M$. Hence $E = \mathcal{H}_\lambda$ so G acts irreducibly on $\mathcal{H}_\lambda$. In particular, the translates $\varphi_\lambda^{\tau(g)}$ $(g \in G)$ generate a dense subspace of $\mathcal{H}_\lambda$. Because of the symmetry of φ_λ (Theorem 1.1) this means that the functions $kM \longrightarrow e^{(-i\lambda+\rho)(A(gK,kM))}$ on K/M span a dense subspace of $L^2(B)$ (as g runs through G), in other words, $-\lambda$ is simple. Now condition (11) follows by virtue of Theorem 4.4.

§7. TANGENT SPACE ANALYSIS.

1. Discussion.

Let X_o denote the tangent space to the symmetric space $X = G/K$ at the origin o and let G_o denote the group of affine transformations of X_o generated by all the translations and the natural action $d\tau(k)_o$ ($k \in K$) of K on X_o. Under the natural map $\pi : G \longrightarrow G/K$, $d\pi$ is an isomorphism of $\mathfrak{p}$ onto X_o intertwining $Ad_G(K)|\mathfrak{p}$ and $d\tau(K)$. Thus $X_o = G_o/d\tau(K)$ which we write $X_o = G_o/K$ for simplicity.

In this section we shall extend to X_o the principal results for X developed so far in this chapter. While the results for $X_o = G_o/K$ show considerable analogies to those for $X = G/K$ the proofs are quite different. For the spherical transforms on X and X_o we have already seen this in [GGA], Ch. IV. The proofs of the Paley-Wiener theorems there (Theorems 7.1 and 9.2) are indeed entirely different. One reason is that Harish-Chandra's convergent expansion for the zonal spherical function φ_λ which is a basic tool in analysis on $X = G/K$, does not in general have a good analog for the zonal spherical function ψ_λ (cf. (24) below) on the flat space X_o. In fact, if $\beta_1, \ldots, \beta_p$ are the indivisible restricted roots for G/K the expansion of φ_λ is a convergent series in the exponentials $e^{-n_1\beta_1-\cdots-n_p\beta_p}$ with recursively determined coefficients; for X_o the natural analog would be an expansion of ψ_λ in terms of the monomials $\beta_1^{-n_1} \ldots \beta_p^{-n_p}$. If G/K has rank one, ψ_λ does indeed have such an asymptotic (but not convergent) expansion for large arguments: Being a radial eigenfunction of the Laplacian with eigenvalue $-\lambda^2, \psi_\lambda$ is a constant multiple of the function

$$(1) \qquad (\lambda r)^{-\frac{n-2}{2}} J_{\frac{n-2}{2}}(\lambda r) \qquad n = \dim X_o$$

and $J_m(z)$ has an asymptotic expansion

$$J_m(z) \sim (2/\pi z)^{\frac{1}{2}} \left[\cos(z - \frac{m\pi}{2} - \frac{\pi}{4})U_m(z) - \sin\left(z - \frac{m\pi}{2} - \frac{\pi}{4}\right) V_m(z)\right],$$

where $U_m(z)$ and $V_m(z)$ are power series in z^{-1} (see e.g. Whittaker-Watson [1927], p. 368). For G complex, Harish-Chandra's integral formula for ψ_λ (cf. [GGA], Ch. II, Theorem 5.35) gives such a (finite) expansion. For general G one has the serious difficulty that the monomials above may not be linearly independent. For example, $\Sigma_{i \neq j}\langle\beta_i, \beta_j\rangle\beta_i^{-1}\beta_j^{-1} = 0$ for $G = \boldsymbol{SL}(3, \boldsymbol{R})$. Nevertheless, Barlet and Clerč have shown in [1986] the existence of an asymptotic expansion for ψ_λ in terms of the monomials above. However, since the expansion is not convergent and the coefficients non-unique (because of the above remark) it seems unlikely that it can play a role analogous to the expansion for φ_λ.

The analogy between X and X_o is nevertheless quite helpful in formulating conjectures. But in some respects, X_o is more subtle. An example is the divisibility of the Eisenstein integral in Theorem 5.15 and 7.5. For X the divisibility by $\overset{\vee}{Q}{}^{\delta}(\lambda)$ is a consequence of the Weyl group invariance. This divisibility argument does not extend to X_o and $\overset{\vee}{J}{}^{\delta}(\lambda)$ in Theorem 7.5 in spite of the fact that the Weyl group invariance is valid and can be proved directly. Instead we use the result for X, combined with a suitable limit argument, to obtain the desired result for X_o.

2. The J-Polynomials

In the Cartan decomposition $\mathfrak{g} = \mathfrak{k} + \mathfrak{p}$ we introduce the orthogonal decomposition

$$\mathfrak{p} = \mathfrak{a} + \mathfrak{q}. \tag{2}$$

The mapping $X \longrightarrow X - \theta X$ is then a bijection of $\mathfrak{n}$ onto $\mathfrak{q}$. The (complex) symmetric algebra $S(\mathfrak{g})$ which decomposes $S(\mathfrak{g}) = S(\mathfrak{p}) \oplus S(\mathfrak{g})\mathfrak{k}$ then decomposes further as

$$S(\mathfrak{g}) = S(\mathfrak{a}) \oplus (S(\mathfrak{g})\mathfrak{k} + \mathfrak{q}S(\mathfrak{g})). \tag{3}$$

Let $p \longrightarrow j^p$ denote the corresponding projection of $S(\mathfrak{g})$ onto $S(\mathfrak{a})$.

As in §2 we identify $\mathfrak{p}$ and its dual $\mathfrak{p}^*$ via the Killing form whereby $S(\mathfrak{p})$ is identified with $S(\mathfrak{p}^*)$. Let $I(\mathfrak{p}) = I(\mathfrak{p}^*)$ denote the subalgebra of K-invariants and $H(\mathfrak{p}) = H(\mathfrak{p}^*)$ the corresponding space of harmonic polynomials.

Again consider a $\delta \in \widehat{K}_M$ acting unitarily on the vector space V_δ with inner product $\langle , \rangle$ and $\delta(M)$-fixed subspace V_δ^M. Let $v_1, \dots, v_{d(\delta)}$ be an orthonormal basis of V_δ with $v_1, \dots, v_{\ell(\delta)}$ a basis of V_δ^M.

Under the adjoint action of K on $\mathfrak{p}$ let $S_\delta(\mathfrak{p}) \subset S(\mathfrak{p})$, $H_\delta(\mathfrak{p}) \subset H(\mathfrak{p})$ be the subspaces of type δ. Given $p \in S(\mathfrak{p})$ let $\partial(p)$ denote the corresponding constant coefficient differential operator on $\mathfrak{p}$. Note that $\partial(Ad(k)p) = \partial(p)^{Ad(k)}$. Observe also that by (13) §2 and Prop. 2.5 the $Ad(K)$ action of $H_\delta(\mathfrak{p})$ decomposes into $\ell(\delta)$ copies of δ. Thus we write

$$H_\delta(\mathfrak{p}) = \bigoplus_{j=1}^{\ell(\delta)} H_{\delta,j}, \tag{4}$$

where the action of K on $H_{\delta,j}$ is equivalent to δ. By decomposing $H_\delta(\mathfrak{p})$ first into homogeneous components we can assume the $H_{\delta,j}$ to consist of homogeneous polynomials of degree, say $d_j(\delta)$. By (4) the vector space $F_\delta = \mathrm{Hom}_K(V_\delta, H_\delta(\mathfrak{p}))$ of linear maps η of V_δ into $H_\delta(\mathfrak{p})$ satisfying

$$\eta(\delta(k)v) = Ad(k)(\eta(v)) \qquad k \in K, v \in V_\delta \tag{5}$$

has dimension $\ell(\delta)$. For $\lambda \in \mathfrak{a}_{\mathbf{c}}^*$ consider the linear map

$$J^\delta(\lambda):\ F_\delta \longrightarrow V_{\check{\delta}}^M$$

given by

$$(J^\delta(\lambda)(\eta))(v) = j^{\eta(v)}(-i\lambda) \qquad v \in V_\delta.$$

The right hand side is indeed invariant under $v \longrightarrow \delta(m)v$ by virtue of (3). Expressing $J^\delta(\lambda)$ in any basis $\eta_1, \ldots, \eta_{\ell(\delta)}$ of F_δ and the basis (v_i') of $V_{\check{\delta}}^M$ we have

$$J^\delta(\lambda)_{ij} = j^{\eta_j(v_i)}(-i\lambda). \tag{6}$$

We call the J^δ the *J-polynomials.* If we change the basis (η_j) to another one (ζ_k) where $\zeta_k = \sum\limits_j a_{jk}\eta_j$ then the matrix $J^\delta(\lambda)$ changes to $J^\delta(\lambda)A$ where $A = (a_{jk})$. If the basis (η_j) had been chosen such that, by (4), $\eta_j(V_\delta) = H_{\delta,j}$ $(1 \leq j \leq \ell(\delta))$ we see that the matrix entry $J^\delta(\lambda)_{ij}$ is a homogeneous polynomial of degree $d_j(\delta)$. This proves the following result.

Lemma 7.1. *For any orthonormal basis (v_i) of $V_{\check{\delta}}^M$ and any basis (η_j) of F_δ, the function $\lambda \longrightarrow \det(J^\delta(\lambda))$ is a homogeneous polynomial of degree $\sum\limits_1^{\ell(\delta)} d_j(\delta)$.*

As before let $p \longrightarrow p^*$ be the symmetrization mapping of $S(\mathfrak{g})$ onto $U(\mathfrak{g})$. We shall now compare the matrix $J^\delta(\lambda)$ with $Q^\delta(\lambda)$ in §2. From a basis (η_j) of F_δ we obtain a basis (ϵ_j) of E_δ by $\epsilon_j(v) = \eta_j(v)^*$ $(v \in V_\delta)$. Then $(22)'$ in §2 becomes

$$Q^\delta(\lambda)_{ij} = q^{\eta_j(v_i)^*}(\rho - i\lambda), \qquad (1 \leq i, j \leq \ell(\delta)). \tag{7}$$

Lemma 7.2. *Let the basis (η_j) be chosen such that*

$$\eta_j(V_\delta) = H_{\delta,j} \subset S_{d_j(\delta)}, \qquad 1 \leq j \leq \ell(\delta).$$

Then

$$\lim_{r \longrightarrow \infty} r^{-d_j(\delta)} Q^\delta(r\lambda)_{ij} = J^\delta(\lambda)_{ij}$$

uniformly as λ varies in any compact set.

Proof. Let $S^d(\mathfrak{g})$ denote the space of homogeneous elements $p \in S(\mathfrak{g})$ of degree d. Fix $p \in S^d(\mathfrak{g})$. Because of the decomposition

$$S(\mathfrak{g}) = \mathfrak{k}S(\mathfrak{g}) + \mathfrak{n}S(\mathfrak{g}) + S(\mathfrak{a}) \tag{8}$$

there exist $T_i \in \mathfrak{k}$, $Z_j \in \mathfrak{n}, p_i, r_j \in S^{d-1}(\mathfrak{g}), h \in S^d(\mathfrak{a})$ such that

$$p = \sum_i p_i T_i + \sum_j Z_j r_j + h. \tag{9}$$

Using the symmetrization,

$$p^* = \sum_i p_i^* T_i^* + \sum_j Z_j^* r_j^* + h^* + \text{ terms of degree } < d.$$

Following this by the projection $D \longrightarrow q^D$ of $\boldsymbol{D}(G)$ onto $\boldsymbol{D}(A)$ according to (18) §2 we deduce

$$q^{p^*} = h^* + \text{ terms of degree } < d. \tag{10}$$

On the other hand, (9) implies

$$p = \sum_i p_i T_i + \sum_j \tfrac{1}{2}(Z_j + \theta Z_j) r_j + \sum_j \tfrac{1}{2}(Z_j - \theta Z_j) r_j + h$$

so, since $\frac{1}{2}(Z_j - \theta Z_j) \in \mathfrak{q}$ we deduce

$$j^p = h. \tag{11}$$

Since $\mathfrak{a}$ is abelian, $S(\mathfrak{a}) = S(\mathfrak{a})^*$ so (10) and (11) imply

$$q^{\eta_j(v_i)^*}(-i\lambda) = j^{\eta_j(v_i)}(-i\lambda) + s(\lambda), \qquad \lambda \in \mathfrak{a}^*, \tag{12}$$

where degree $(s) < d_j(\delta)$. Since the first term on the right has degree $d_j(\delta)$ the left hand side does too. Also it differs from (7) by a lower degree polynomial. Consequently,

$$q^{\eta_j(v_i)^*}(\rho - i\lambda) = j^{\eta_j(v_i)}(-i\lambda) + t(\lambda), \quad \lambda \in \mathfrak{a}^*, \tag{13}$$

where degree $(t) < d_j(\delta)$, and now Lemma 7.2 follows.

For $g \in G, k \in K$ let $A(g), A(gK, kM) \in \mathfrak{a}$ be as in (3) §1. Thus $A(gK, kM) = A(k^{-1}g)$. Geometrically, $\exp A(g)$ is the composite distance from o to the horocycle through $g \cdot o$ parallel to $N \cdot o$ (Ch. II, §1). For $H \in \mathfrak{a}$ the plane $H + \mathfrak{q}$ in $\mathfrak{p}$ is the tangent space analog to the horocycle $\xi_H = (\exp H)N \cdot o$. (Actually the diffeomorphism $\operatorname{Exp} : \mathfrak{p} \longrightarrow X$ induces a bijection between the tangent spaces:

$$d\operatorname{Exp}_H(\mathfrak{q}) = (\xi_H)_{\operatorname{Exp} H}$$

as is readily seen from the formula for $d\operatorname{Exp}_H$ in [DS], Ch. IV, Theorem 4.1. The following lemma is therefore geometrically plausible. Let $A_o : \mathfrak{p} \longrightarrow \mathfrak{a}$ denote the orthogonal projection (corresponding to the decomposition $\mathfrak{p} = \mathfrak{a} \oplus \mathfrak{q}$).

Lemma 7.3. *Let $H \in \mathfrak{a}$. Then*

$$\left\{\frac{d}{dt} A(k \exp tH)\right\}_{t=0} = A_o(Ad(k)H)$$

uniformly for $k \in K$.

Proof. We have $Ad(k)H = A_o(Ad(k)H) + X$ where $X \in \mathfrak{q}$ so

$$\begin{aligned} A(k \exp tH) &= A(k \exp tHk^{-1}) = A(\exp(Ad(k)tH)) \\ &= A(\exp t\{A_o(Ad(k)H) + X\}). \end{aligned}$$

Now $X = Z - \theta Z$ where $Z \in \mathfrak{n}$ so $X = 2Z - (Z + \theta Z)$. However, since $A(ngk) \equiv A(g)$, and since $\exp tU \exp tV = \exp(t(U + V) + O(t^2))$ we have

$$\begin{aligned} tA_o(Ad(k)H) &= A(\exp\{tA_o(Ad(k)H)\}) \\ &= A(\exp t(2Z) \exp\{tA_o(Ad(k)H)\} \exp t(-Z - \theta Z)) \\ &= A(\exp[t(2Z + A_o(Ad(k)H) - (Z + \theta Z)) + O(t^2)]) \end{aligned}$$

where $O(t^2)/t^2$ is bounded, uniformly in k, for small t. Thus

$$tA_o(Ad(k)H) = A(\exp[tAd(k)H + O(t^2)]),$$

proving the lemma.

For $\lambda \in \mathfrak{a}_c^*$ let as usual $A_\lambda \in \mathfrak{a}^c$ be determined by $B(H, A_\lambda) = \lambda(H)$, $(H \in \mathfrak{a})$ B being the Killing form. In analogy with the generalized spherical function $\Phi_{\lambda,\delta}(x)$ (§2, No. 1) we consider now the *generalized Bessel function*

$$\Psi_{\lambda,\delta}(X) = \int_K e^{iB(k \cdot A_\lambda, X)} \delta(k)\, dk. \tag{14}$$

Here $\delta \in \widehat{K}_M$ and $k \cdot A_\lambda$ stands for $Ad(k)A_\lambda$. Since these functions can be written

$$\Phi_{\lambda,\delta}(gK) = \int_K e^{(i\lambda+\rho)(A(k^{-1}g))}\delta(k)dk \tag{15}$$

$$\Psi_{\lambda,\delta}(X) = \int_K e^{i\lambda(A_o(k^{-1}\cdot X))}\delta(k)dk \tag{15a}$$

they are indeed analogous. We shall now exploit this analogy further using Lemmas 7.2 and 7.3. First we need an expression for the determinant of $J^\delta(\lambda)$. Let $\Sigma_o^+ \subset \Sigma^+$ denote the set of indivisible positive roots and for $\beta \in \Sigma_o^+$ consider the rank-one space G_β/K_β as in §3, No. 3. Consider the K_β-irreducible representations $\delta(i,\beta)$ $(1 \leq i \leq \ell(\delta))$ in (50) §3 and let $D(\delta(i,\beta))$ denote the degree of the K_β-harmonic polynomials in the space $H_{\delta(i,\beta)}(\mathfrak{p}_\beta)$. Let

$$D(\beta) = \sum_{i=1}^{\ell(\delta)} D(\delta(i,\beta)). \tag{16}$$

Proposition 7.4. *For a certain constant $c \neq 0$,*

$$\det J^\delta(\lambda) = c \prod_{\beta\in\Sigma_o^+} \langle\lambda,\beta\rangle^{D(\beta)}.$$

Proof. For G/K of rank one this is clear from Lemma 7.1. For the general case we use the product formula (50) §3 (with P_δ constant) combined with Lemma 7.2.

Our next result states that the generalized Bessel function is divisible by the J-polynomial.

Theorem 7.5. *For $H \in \mathfrak{a}$ we have the identities*

$$\check{J}^\delta(\lambda)^{-1}\int_K e^{-iB(kA_\lambda,H)}\delta(k^{-1})dk = \check{J}^\delta(s\lambda)^{-1}\int_K e^{-iB(kA_{s\lambda},H)}\delta(k^{-1})dk$$

for $s \in W$, $H \in \mathfrak{a}$ and both sides are holomorphic in λ on all of $\mathfrak{a}_c^*$.

Proof. Let $s \in W$ and let m_s be a representative of s in K. Viewing elements of $S(\mathfrak{a})$ as functions on $\mathfrak{a}^*$ we have for $s \in \boldsymbol{GL}(\mathfrak{a})$,

$$(17) \qquad s \cdot h = h \circ {}^t s, \qquad h \in S(\mathfrak{a}).$$

Let $s \in W$ and let m_s be a representative of s in K. Then $s\mu = {}^t Ad(m_s^{-1})\mu$ for $\mu \in \mathfrak{a}^*$ so if $p \in S(\mathfrak{p})$ then

$$(18) \qquad j^p(s\mu) = (j^p)^{{}^t Ad(m_s)}(\mu) = (Ad(m_s^{-1})(j^p))(\mu) = (j^{Ad(m_s^{-1})p})(\mu).$$

Now if $\eta \in F_\delta$ we have by (5) and the definition of J^δ,

$$\begin{aligned} &(J^\delta(s^{-1}\lambda)(\eta))(v) = j^{\eta(v)}(-is^{-1}\lambda) = j^{Ad(m_s)(\eta(v))}(-i\lambda) \\ &= j^{\eta(\delta(m_s)v)}(-i\lambda) = J^\delta(\lambda)(\eta)(\delta(m_s)v) \\ &= (\check{\delta}(m_s^{-1})J^\delta(\lambda)(\eta))(v). \end{aligned}$$

Replacing δ by $\check{\delta}$ and s by s^{-1} we obtain

$$(19) \qquad J^{\check{\delta}}(s\lambda) = \delta(m_s)J^{\check{\delta}}(\lambda),$$

and now the Weyl group invariance in Theorem 7.5 follows.

To conclude the proof of the theorem we now establish the holomorphy statement. We write

$$(20) \qquad J^{\check{\delta}}(\lambda)^{-1} = (\det J^{\check{\delta}}(\lambda))^{-1} J_c(\lambda),$$

where $J_c(\lambda)$ is a matrix whose entries are the cofactors of $J^{\check{\delta}}(\lambda)$; in particular the entries in $J_c(\lambda)$ are polynomials in the coordinates of λ. Consider now the third expression in Theorem 5.15; the ij^{th} matrix element is (with $H = \log a$ and $Q_c(\lambda)$ as in (59) §5)

$$(21) \qquad \det Q^{\check{\delta}}(\lambda)^{-1} \left[\sum_{p=1}^{\ell(\delta)} Q_c(\lambda)_{ip} \int_K e^{(-i\lambda+\rho)(A(k\exp H))} \delta_{pj}(k)\,dk \right].$$

Let $e_1, \ldots e_\ell$ be the basis of $\mathfrak{a}$, dual to the basis of simple roots (α_i). For $\alpha \in \Sigma^+$ let $m_i(\alpha) \in \boldsymbol{Z}^+$ be determined by $\alpha = \sum_1^\ell m_i(\alpha)\alpha_i$. Let $\mathfrak{a}_c' = \{\lambda \in \mathfrak{a}_c^* : \det J^{\check{\delta}}(\lambda) \neq 0\}$. Writing $A_\lambda = \Sigma_1^\ell \lambda_i e_i$, $\mathfrak{a}_c'$ is the complement in $\mathfrak{a}_c^*$

with the hyperplanes $\sum_1^\ell m_i(\alpha)\lambda_i = 0$ removed ($\alpha \in \Sigma^+$). Now we have by Lemma 7.2, since $Q_c(\lambda)_{ip}$ has degree $\Sigma_{k\neq i}\, d_k(\check{\delta})$ and $\det Q^{\check{\delta}}(\lambda)$ has degree $\Sigma_k\, d_k(\check{\delta})$,

$$\lim_{r\longrightarrow\infty} r^{d_i(\check{\delta})}(\det Q^{\check{\delta}}(r\lambda))^{-1}Q_c(r\lambda)_{ip} = (\det J^{\check{\delta}}(\lambda))^{-1}J_c(\lambda)_{ip} \tag{22}$$

uniformly for λ varying in any compact subset of $\mathfrak{a}'_c$. Now we replace in (21) λ by $r\lambda$, H by H/r and multiply by $r^{d_i(\check{\delta})}$. In the resulting expression we let $r \longrightarrow \infty$. Using (22) and Lemma 7.3 we find that the limit is

$$\left(\det J^{\check{\delta}}(\lambda)\right)^{-1}\left[\sum_{p=1}^{\ell(\delta)} J_c(\lambda)_{ip}\int_K e^{-i\lambda(A_o(k\cdot H))}\delta_{pj}(k)dk\right] \tag{23}$$

and convergence is uniform for λ in each compact subset of $\mathfrak{a}'_{\mathbf{c}}$. Since Σ^+ is a finite set we can select numbers $d_1 > 0\ldots, d_\ell > 0$ such that the set

$$\mathfrak{a}_d = \{\lambda \in \mathfrak{a}^*_{\mathbf{c}} :\ |\lambda_1| = d_1,\ldots,|\lambda_\ell| = d_\ell\}$$

belongs to $\mathfrak{a}'_{\mathbf{c}}$. (For example take $d_i > m(d_{i+1}+\cdots+d_\ell)$ for $1\leq i<\ell$ where $m = Max_{i,\alpha}(m_i(\alpha))$). Cauchy's formula for the holomorphic function (21) in the polydisk $|\lambda_1|\leq d_1,\ldots,|\lambda_\ell|\leq d_\ell$ then implies that the above convergence to (23) is uniform in each polydisk $|\lambda_i|\leq d'_i < d_i$ $(1\leq i\leq \ell)$. Since $\mathfrak{a}_{td}\subset\mathfrak{a}'_c$ for each $t>0$ we can conclude that (23) is holomorphic on all of $\mathfrak{a}^*_{\mathbf{c}}$. This concludes the proof of Theorem 7.5.

3. Generalized Bessel Functions and Zonal Spherical Functions.

We shall now show that the generalized Bessel function $\Psi_{\lambda,\delta}$ can be written as a certain derivative of the zonal spherical function ψ_λ given by

$$\psi_\lambda(X) = \int_K e^{iB(kA_\lambda,X)}dk. \tag{24}$$

This is analogous to Theorem 2.6 (alias Cor. 5.17) but in the present flat case the result can be proved rather directly. Let $p \longrightarrow p^t$ denote the complex-linear automorphism of $S(\mathfrak{p})$ induced by the endomorphism $X \longrightarrow -X$ of $\mathfrak{p}$. Consider the orthonormal basis (v_i) of V_δ such that $v_1,\ldots,v_{\ell(\delta)} \in V_\delta^M$ and define (v'_i) by $\langle w, v_i\rangle = v'_i(w)$ $(w \in V_\delta)$. Let (ζ_j) be any basis of $F_{\check{\delta}}$.

Note that if $\partial(p)$ is the constant-coefficient differential operator determined by $p \in S(\mathfrak{p})$ then the adjoint $\partial(p)^*$ equals $\partial(p^t)$.

Consider now the matrix $L^{\check{\delta}}$ with $d(\delta)$ rows, $\ell(\delta)$ columns defined by

$$(25) \qquad (L^{\check{\delta}} f)_{ij}(X) = (\partial(\zeta_j(v_i'))^* f)(X), \qquad f \in C^\infty(\mathfrak{p}).$$

Thus

$$L^{\check{\delta}} = \begin{pmatrix} \partial(\zeta_1(v_1'))^* & \dots & \partial(\zeta_{\ell(\delta)}(v_1'))^* \\ \vdots & & \vdots \\ \partial(\zeta_1(v_{d(\delta)}'))^* & \dots & \partial(\zeta_{\ell(\delta)}(v_{d(\delta)}'))^* \end{pmatrix}.$$

Theorem 7.6. *The generalized Bessel function can be written*

$$\Psi_{\lambda,\delta}(X)|V_\delta^M = (L^{\check{\delta}}\psi_\lambda)(X)J^{\check{\delta}}(\lambda)^{-1}.$$

Proof. We may assume the basis (ζ_j) chosen such that $\zeta_j(V_{\check{\delta}}) = H_{\check{\delta},j} \subset S_{d_j(\check{\delta})}$. In fact if we go to a new basis $(B\zeta_j)$ then $L^{\check{\delta}} f$ and $J^{\check{\delta}}(\lambda)$ are replaced by $(L^{\check{\delta}} f)B$ and $J^{\check{\delta}}(\lambda)B$ respectively. Now by (5)

$$(26) \qquad k^{-1}\zeta_j(v_i') = \zeta_j(\check{\delta}(k^{-1})v_i'), \qquad \check{\delta}(k^{-1})v_i' = \sum_q \langle \delta(k)v_q, v_i\rangle v_q'$$

and for each $p \in S(\mathfrak{p})$, $k \in K$, $\lambda \in \mathfrak{a}_c^*$ the function

$$f_k(X) = e^{iB(kA_\lambda, X)} \qquad (X \in \mathfrak{p})$$

satisfies

$$(27) \qquad \partial(p)f_k = j^{(k^{-1}\cdot p)}(i\lambda)f_k.$$

By (26)–(27)

$$\partial(\zeta_j(v_i'))\psi_\lambda(X) = \int_K j^{\zeta_j(\check{\delta}(k^{-1})v_i')}(i\lambda)e^{iB(kA_\lambda,X)}dk$$

$$= \sum_{q=1}^{d(\delta)} j^{\zeta_j(v_q')}(i\lambda) \int_K \langle \delta(k)v_q, v_i\rangle e^{iB(kA_\lambda,X)}dk.$$

Replacing k by km and integrating over M we see that the integral vanishes if $q > \ell(\delta)$. Noting now that if $p \in S^k(\mathfrak{p})$ then

$$j^{p^t}(i\lambda) = j^p(-i\lambda), \qquad \partial(p)^* = (-1)^k\partial(p)$$

we obtain

$$(\partial(\zeta_j(v_i'))^*\psi_\lambda)(X) = \sum_{q=1}^{\ell(\delta)} \Psi_{\lambda,\delta}(X)_{iq}\overset{\vee}{J^\delta}(\lambda)_{qj}$$

for $1 \leq i \leq d(\delta)$, $1 \leq j \leq \ell(\delta)$. Multiplying by $(\overset{\vee}{J^\delta}(\lambda)^{-1})_{jk}$ and summing on j we obtain the result.

4. The Fourier Transform of K-finite Functions.

We shall now study the flat analog of the δ-spherical transform from §5, No. 5. If f is a K-finite function on $\mathfrak{p}$ of type $\overset{\vee}{\delta}$ we define its *generalized Bessel transform* by

(28)
$$\widetilde{f}(\lambda) = d(\delta)\int_{\mathfrak{p}} f(X)\Psi_{\overline{\lambda},\delta}(X)^*dX = d(\delta)\int_{\mathfrak{p}} f(X)\int_K e^{-iB(kA_\lambda,X)}\delta(k^{-1})dk\,dX$$

for all $\lambda \in \mathfrak{a}_c^*$ for which the integral converges. Here $*$ denotes the adjoint on $\mathrm{Hom}(V_\delta, V_\delta)$. Note that $\delta(m)\widetilde{f}(\lambda) = \widetilde{f}(\lambda)$ for $m \in M$.

Consider now the space $\mathcal{H} = \mathcal{H}(\mathfrak{a}_c^*, \mathrm{Hom}(V_\delta, V_\delta^M))$ in (49) §5 with the topology described in §5, No. 5. We denote by $\mathcal{K}^\delta(\mathfrak{a}^*)$ the topological subspace of $\mathcal{H}$ given by

$$(29) \qquad \mathcal{K}^\delta(\mathfrak{a}^*) = \{F \in \mathcal{H} : \ (\overset{\vee}{J^\delta})^{-1}F \qquad W\text{-invariant and holomorphic}\}.$$

This is analogous to the space $\mathcal{H}^\delta(\mathfrak{a}^*)$ defined in (50) §5. Note however that the Weyl group invariance of $(\overset{\vee}{Q^\delta})^{-1}F$ in (50) §5 implies automatically that it is holomorphic whereas the analogous statement is false for $\mathcal{K}^\delta(\mathfrak{a}^*)$. For example if $\mathrm{rank}(G/K) = 1$ then $\dim V_\delta^M = 1$ and $J^\delta(\lambda)$ and $J^\delta(-\lambda)$ differ by a constant so the holomorphy of $(\overset{\vee}{J^\delta})^{-1}F$ is not a consequence of it being even. Our definition (29) takes this into account.

Consider now the "polar coordinate" representation of $\mathfrak{p}$

$$(30) \qquad (kM, H) \longrightarrow Ad(k)H \qquad k \in K, H \in C\ell(\mathfrak{a}^+)$$

and the corresponding integral formula

$$(31) \qquad \int_{\mathfrak{p}} F(X)dX = \int_{\mathfrak{a}^+} \left(\int_K F(Ad(k)H)dk \right) \omega(H)dH, \qquad F \in C_c(\mathfrak{p}).$$

Here dX and dH are the Euclidean measures and the density function ω is given by

$$(32) \qquad \omega(H) = c \prod_{\alpha \in \Sigma^+} \alpha(H)^{m_\alpha}, \qquad H \in \mathfrak{a},$$

where c is a constant.

We now state the "Paley-Wiener type theorem" for the generalized Bessel transform.

Theorem 7.7. *The mapping $f \longrightarrow \widetilde{f}$ where*

$$\widetilde{f}(\lambda) = d(\delta) \int_{\mathfrak{p}} f(X) \Psi_{\overline{\lambda},\delta}(X)^* dX$$

is a homeomorphism of $\mathcal{D}_{\check{\delta}}(\mathfrak{p})$ onto $\mathcal{K}^\delta(\mathfrak{a}^*)$. The inverse is given by

$$(33) \qquad f(X) = c \ \mathrm{Tr} \left[\int_{\mathfrak{a}^*} \Psi_{\lambda,\delta}(X) \widetilde{f}(\lambda) |\omega(\lambda)| d\lambda \right],$$

where c is a constant, independent of f, Tr *means Trace on V_δ and $\omega(\lambda) = \omega(A_\lambda)$.*

Proof. First we note that writing $X = Ad(\ell)H$

$$\check{J}^\delta(\lambda)^{-1} \Psi_{\overline{\lambda},\delta}(X)^* = \check{J}^\delta(\lambda)^{-1} \Psi_{\overline{\lambda},\delta}(H)^* \delta(\ell^{-1})$$

so by Theorem 7.5, $\widetilde{f} \in \mathcal{K}^\delta(\mathfrak{a}^*)$ for each $f \in \mathcal{D}_{\check{\delta}}(\mathfrak{p})$. For the surjectivity, let $F \in \mathcal{K}^\delta(\mathfrak{a}^*)$ and define f by

$$(34) \qquad f(X) = \mathrm{Tr} \left[\int_{\mathfrak{a}^*} \Psi_{\lambda,\delta}(X) F(\lambda) |\omega(\lambda)| d\lambda \right].$$

Since $F(\lambda)$ maps V_δ into $V_{\check{\delta}}^M$ we can substitute in (34) the formula for $\Psi_{\lambda,\delta}$ in Theorem 7.6. We can also write $F = J^{\check{\delta}}\Phi$, where Φ is W-invariant and holomorphic. Writing (cf.(20))

$$\Phi(\lambda) = \det J^{\check{\delta}}(\lambda)^{-1} J_c(\lambda) F(\lambda) \tag{35}$$

we see from Lemma 5.13 that Φ is of exponential type. Thus Φ belongs to the space $\mathcal{J}^\delta(\mathfrak{a}^*)$ of W-invariants in $\mathcal{H}(\mathfrak{a}_c^*, \mathrm{Hom}(V_\delta, V_{\check{\delta}}^M))$. With the zonal spherical function ψ_λ consider the inverse spherical transform (cf. [GGA], Ch. IV, §9) of the vector-valued function Φ

$$\varphi(X) = \int_{\mathfrak{a}^*} \psi_\lambda(X)\Phi(\lambda)|\omega(\lambda)|d\lambda. \tag{36}$$

Using Theorem 7.6 and (34), (36) we obtain

$$f = \mathrm{Tr}(L^{\check{\delta}}\varphi). \tag{37}$$

Here $L^{\check{\delta}}$ is an $d(\delta) \times \ell(\delta)$ matrix, φ an $\ell(\delta) \times d(\delta)$ matrix so $L^{\check{\delta}}\varphi$ is a square matrix of order $d(\delta)$. By [GGA], Ch. IV, Theorem 9.2, φ has compact support so by (34) (and Prop. 5.10) $f \in \mathcal{D}_{\check{\delta}}(\mathfrak{p})$. To compute the generalized Bessel transform we write in the orthonormal basis (v_i) of V_δ (with $v_1, \dots, v_{\ell(\delta)} \in V_{\check{\delta}}^M$)

$$\varphi(X)v_i = \sum_{j=1}^{\ell(\delta)} \varphi_{ji}(X)v_j, \qquad \Phi(\lambda)v_i = \sum_{j=1}^{\ell(\delta)} \Phi_{ji}(\lambda)v_j, \qquad (1 \le i \le d(\delta)).$$

Then

$$\widetilde{f}(\lambda) = d(\delta) \int_K \delta(k)dk \sum_{i,j} \int_{\mathfrak{p}} (\partial(\zeta_j(v_i'))^* \varphi_{ji})(X) e^{-iB(k^{-1}A_\lambda, X)} dX.$$

The integral over $\mathfrak{p}$ is by (27) and (5) equal to

$$\int_{\mathfrak{p}} \varphi_{ji}(X) j^{\zeta_j(\check{\delta}(k)v_i')}(-i\lambda) e^{-iB(k^{-1}A_\lambda, X)} dX$$

so, since

$$\check{\delta}(k)v_i' = \sum_{q=1}^{d(\delta)} \overline{\delta(k)_{qi}} v_q' = \sum_{q=1}^{d(\delta)} \langle v_q, \delta(k)v_i\rangle v_q'$$

we obtain, by the K-invariance of φ_{ji},

$$(38)\qquad \widetilde{f}(\lambda) = \sum_{1\leq i,q\leq d(\delta),1\leq j\leq \ell(\delta)} A_{ijq}(\lambda)\int_K \delta(k)\langle v_q, \delta(k)v_i\rangle dk,$$

where

$$A_{ijq}(\lambda) = j^{\zeta_j(v_q')}(-i\lambda)\widetilde{\varphi}_{ji}(\lambda)\, d(\delta),$$

$\widetilde{\varphi}_{ji}$ denoting the spherical transform of φ_{ji}. By the definition (28), $\delta(m)\widetilde{f}(\lambda) = \widetilde{f}(\lambda)$ for $m \in M$. Hence if $m \in M$,

$$\begin{aligned}\widetilde{f}(\lambda) &= \sum_{i,j,q} A_{ijq}(\lambda)\int_K \delta(mk)\langle v_q, \delta(k)v_i\rangle dk\\ &= \sum_{i,j,q} A_{ijq}(\lambda)\int_K \delta(\ell)\langle \delta(m)v_q, \delta(\ell)v_i\rangle d\ell.\end{aligned}$$

Integrating this over M we see that the terms for $q > \ell(\delta)$ drop out so the formula becomes

$$(39)\qquad \widetilde{f}(\lambda) = \sum_{1\leq j,q\leq \ell(\delta),1\leq i\leq d(\delta)} A_{ijq}(\lambda)\int_K \delta(k)\langle v_q, \delta(k)v_i\rangle dk.$$

However the orthogonality relations can be interpreted as

$$\int_K \langle v_r, \delta(k)v_s\rangle\delta(k)dk = d(\delta)^{-1}E_{rs} \qquad 1 \leq r, s \leq d(\delta)$$

so (39) becomes

$$(40)\qquad \widetilde{f}(\lambda) = \sum_{\substack{1\leq j,q\leq \ell(\delta)\\ 1\leq i\leq d(\delta)}} \check{J}^{\delta}(\lambda)_{qj}\widetilde{\varphi}_{ji}(\lambda)E_{qi}.$$

By the spherical inversion formula ([GGA], Ch. IV, Theorem 9.1) we have

$$\widetilde{\varphi}_{ji}(\lambda) = c\Phi_{ji}(\lambda),$$

where c is a constant. Hence (40) becomes

$$(41)\qquad \widetilde{f}(\lambda) = c\sum_{\substack{1\leq q\leq \ell(\delta)\\ 1\leq i\leq d(\delta)}} \left(\check{J}^{\delta}(\lambda)\Phi(\lambda)\right)_{qi} E_{qi} = c\check{J}^{\delta}(\lambda)\Phi(\lambda) = cF(\lambda).$$

This proves the surjectivity as well as (33).

For the injectivity note that if $f \in \mathcal{D}_{\check{\delta}}(\mathfrak{p})$ then

$$f = d(\delta)\chi_\delta * f, \tag{42}$$

where we put for $\varphi \in C(K)$, $h \in C(\mathfrak{p})$,

$$(\varphi * h)(X) = \int_K \varphi(k)h(Ad(k^{-1})X)dk.$$

Hence

$$\widetilde{f}(\lambda, kM) = d(\delta) \int_{\mathfrak{p}} \int_K \mathrm{Tr}(\delta(v)f(v^{-1}X))dv\, e^{-iB(kA_\lambda, X)}dX,$$

which implies

$$\widetilde{f}(\lambda, kM) = \mathrm{Tr}(\delta(k)\widetilde{f}(\lambda)) \qquad \text{for} \qquad f \in \mathcal{D}_{\check{\delta}}(\mathfrak{p}). \tag{43}$$

Now the injectivity is obvious.

Finally, we must prove the topological statement of the theorem. The space $\mathcal{J}^\delta(\mathfrak{a}^*)$ is an LF-space as remarked in §5 and so is the space $\mathcal{D}_{\check{\delta}}(\mathfrak{p})$ (compare Proposition 5.10). We claim now that $\mathcal{K}^\delta(\mathfrak{a}^*)$ is an LF-space. The mapping $\Phi \longrightarrow \check{J}^\delta \Phi$ of the LF-space $\mathcal{J}^\delta(\mathfrak{a}^*)$ onto $\mathcal{K}^\delta(\mathfrak{a}^*)$ is clearly continuous; its inverse is the composite $I_1^{-1} \circ I_2$ where the maps I_1 and I_2 are given by

$$I_2 : \ F \in \mathcal{K}^\delta(\mathfrak{a}^*) \longrightarrow J_c F \in J_c \mathcal{K}^\delta(\mathfrak{a}^*)$$

$$I_1 : \ \Phi \in \mathcal{J}^\delta(\mathfrak{a}^*) \longrightarrow \det(\check{J}^\delta)\Phi \in \det(\check{J}^\delta)\mathcal{J}^\delta(\mathfrak{a}^*).$$

Now I_2 is continuous and I_1 is a homeomorphism so $I_1^{-1} \circ I_2$ is indeed continuous. The above map $\Phi \longrightarrow \check{J}^\delta \Phi$ of $\mathcal{J}^\delta(\mathfrak{a}^*)$ onto $\mathcal{K}^\delta(\mathfrak{a}^*)$ is thus a linear homeomorphism; hence $\mathcal{K}^\delta(\mathfrak{a}^*)$ is an LF-space too. Now the mapping $f \longrightarrow \widetilde{f}$ in the theorem is a continuous mapping from one LF space onto another, hence it is a homeomorphism as stated. This concludes the proof of Theorem 7.7.

Corollary 7.8. *Each K-finite function $f \in \mathcal{D}(\mathfrak{p})$ is a linear combination of constant coefficient derivatives of K-invariant functions in $\mathcal{D}(\mathfrak{p})$.*

This is an immediate consequence of (37) because (33) shows that each $f \in \mathcal{D}_{\check{\delta}}(\mathfrak{p})$ appears in the form (37).

We conclude this section with the analog of Theorem 7.7 for the space $\mathcal{S}_{\check{\delta}}(\mathfrak{p})$ of rapidly decreasing functions on $\mathfrak{p}$ of type $\check{\delta}$. Let

$$\mathcal{S}^{\delta}(\mathfrak{a}^*) = \{F \in \mathcal{S}(\mathfrak{a}^*, \mathrm{Hom}(V_\delta, V_\delta^M)) : (J^{\check{\delta}})^{-1}F \quad W\text{-invariant and } C^\infty\}.$$

Now if $x \longrightarrow h(x)$ is in $\mathcal{S}(\boldsymbol{R})$ and $h(x)/x$ smooth then $h(x)/x$ belongs to $\mathcal{S}(\boldsymbol{R})$. Writing for $F \in \mathcal{S}^{\delta}(\mathfrak{a}^*)$

$$J^{\check{\delta}}(\lambda)^{-1}F(\lambda) = (\det J^{\check{\delta}}(\lambda))^{-1}J_c(\lambda)F(\lambda) \tag{44}$$

and using Proposition 7.4 the function (44) is rapidly decreasing on $\mathfrak{a}^*$.

Theorem 7.9. *The mapping $f \longrightarrow \widetilde{f}$, where*

$$\widetilde{f}(\lambda) = d(\delta)\int_{\mathfrak{p}} f(x)\Psi_{\overline{\lambda},\delta)}(X)^* dX,$$

is a homeomorphism of $\mathcal{S}_{\check{\delta}}(\mathfrak{p})$ onto $\mathcal{S}^{\delta}(\mathfrak{a}^*)$.

Proof. That $\widetilde{f}$ is rapidly decreasing is clear from (28) and the statement $\widetilde{f} \in \mathcal{S}^{\delta}(\mathfrak{a}^*)$ then follows from (44) above together with Theorem 7.5. For the surjectivity let $F \in \mathcal{S}^{\delta}(\mathfrak{a}^*)$ and define f by (34). Then $F = J^{\check{\delta}}\Phi$ where Φ is W-invariant and rapidly decreasing. The function φ in (36) is then rapidly decreasing K-invariant function on $\mathfrak{p}$ ([GGA), Ch. IV, Theorem 9.1) and $f \in \mathcal{S}_{\check{\delta}}(\mathfrak{p})$ by (37). That $\widetilde{f} = cF$ follows just as in the proof of (41). The injectivity follows from (43) as before. The topological statement in the theorem is now trivial since the spaces involved are closed subspaces of Fréchet spaces, hence Fréchet.

Again (37) for $f \in \mathcal{S}_{\check{\delta}}(\mathfrak{p})$ implies the following result.

Corollary 7.10. *Each K-finite $f \in \mathcal{S}(\mathfrak{p})$ is a linear combination of constant coefficient derivatives of K-invariant functions in $\mathcal{S}(\mathfrak{p})$.*

5. The Range $\mathcal{D}(\mathfrak{p})^{\sim}$ inside $\mathcal{H}(\mathfrak{a}^* \times K/M)$.

We shall now use Theorem 7.7 to determine the range $\mathcal{D}(\mathfrak{p})^{\sim}$ as a function space on $K/M \times \mathfrak{a}^*$. Let $\mathcal{H}(\mathfrak{a}^* \times K/M)$ denote the space of holomorphic functions of uniform exponential type in the sense of (2) §5. We recall from [GGA], Introduction, §2 that as f runs through $\mathcal{D}(\boldsymbol{R}^n)$ the function $\widetilde{f}(\lambda\omega) = \varphi(\lambda, \omega)$ runs through the space of functions $\varphi \in \mathcal{H}(\boldsymbol{R} \times \boldsymbol{S}^{n-1})$ satisfying the condition:

(45) *For each $k \in \boldsymbol{Z}^+$ and each isotropic vector $a \in \boldsymbol{C}^n$ the function*

$$\lambda \longrightarrow \lambda^{-k} \int_{\boldsymbol{S}^{n-1}} \varphi(\lambda, \omega)(a, \omega)^k d\omega$$

is even and holomorphic on $\boldsymbol{C}$.

We shall now generalize this to the Fourier transform

$$\widetilde{f}(Y) = \int_{\mathfrak{p}} f(X) e^{-iB(Y,X)} dX \qquad f \in L^1(\mathfrak{p})$$

written in the "polar coordinate" form

$$\widetilde{f}(kA_\lambda) = \int_{\mathfrak{p}} f(X) e^{-iB(kA_\lambda, X)} dX, \qquad k \in K, \tag{46}$$

with $\lambda \in \mathfrak{a}^*$ and $A_\lambda \in \mathfrak{a}$ as before. We ask, which functions $\varphi(\lambda, kM)$ on $\mathfrak{a}^* \times K/M$ have the form $\varphi(\lambda, kM) = \widetilde{f}(kA_\lambda)$ for $f \in \mathcal{D}(\mathfrak{p})$? The answer will be useful in the theory of the horocycle plane Radon transform developed in the next chapter.

Theorem 7.11. *The range $\mathcal{D}(\mathfrak{p})^{\sim}$ of $\mathcal{D}(\mathfrak{p})$ under the Fourier transform (46) is the set of functions $\varphi(\lambda, kM)$ on $\mathfrak{a}^* \times K/M$ satisfying*

(i) $\varphi \in \mathcal{H}(\mathfrak{a}^* \times K/M)$.

(ii) For each $\delta \in \widehat{K}_M$

$$J^{\check{\delta}}(\lambda)^{-1} \int_K \varphi(\lambda, kM)\delta(k^{-1}) dk$$

is W-invariant and holomorphic on $\mathfrak{a}^*_{\boldsymbol{c}}$.

Proof. Let $f \in \mathcal{D}(\mathfrak{p})$. That the function $(\lambda, kM) \longrightarrow \widetilde{f}(kA_\lambda)$ belongs to $\mathcal{H}(\mathfrak{a}^* \times K/M)$ is immediate from Lemma 5.2 in the next chapter. In order to verify condition (ii) for $\varphi(\lambda, kM) = \widetilde{f}(kA_\lambda)$ we use (31) and (46). This gives

$$\int_K \widetilde{f}(k \cdot A_\lambda)\delta(k^{-1})dk = \int_{\mathfrak{p}} f(X)\Psi_{\overline{\lambda},\delta}(X)^* dX$$

$$= \int_{\mathfrak{a}^+} \left(\int_K f(k \cdot H)\Psi_{\overline{\lambda},\delta}(H)^*\delta(k^{-1})dk \right) \omega(H)dH$$

and now (ii) follows from Theorem 7.5.

On the other hand suppose φ satisfies (i) and (ii). We define f by

$$f(X) = \int_{K/M} dk_M \int_{\mathfrak{a}^*_+} \varphi(\lambda, kM)e^{iB(k\cdot A_\lambda, X)}\omega(\lambda)d\lambda, \tag{47}$$

where $\omega(\lambda) = \omega(A_\lambda)$. Here wc decompose by the Peter-Weyl theorem ([GGA], Ch. V, §3 (13))

$$\varphi(\lambda, kM) = \sum_{\delta \in \widehat{K}_M} d(\delta) \operatorname{Tr}(\varphi_\delta(\lambda)\delta(k)), \tag{48}$$

where

$$\varphi_\delta(\lambda) = \int_K \varphi(\lambda, kM)\delta(k^{-1})dk.$$

We substitute (48) into (47) and observe that

$$\int_K \operatorname{Tr}\left(\varphi_\delta(\lambda)e^{iB(kA_\lambda, X)}\delta(k)\right) dk = \operatorname{Tr}(\varphi_\delta(\lambda)\Psi_{\lambda,\delta}(X))$$

so

$$f(X) = \sum_{\delta \in \widehat{K}_M} d(\delta) \int_{\mathfrak{a}^*_+} \operatorname{Tr}(\varphi_\delta(\lambda)\Psi_{\lambda,\delta}(X))\omega(\lambda)d\lambda. \tag{49}$$

Now observe from (19) and condition (ii) that for each $s \in W$

$$\varphi(s^{-1}\lambda, km_s M) = \varphi(\lambda, kM) \tag{50}$$

so

$$\varphi_\delta(s\lambda) = \delta(m_s)\varphi_\delta(\lambda). \tag{51}$$

Since $\Psi_{s\lambda,\delta}(X) = \Psi_{\lambda,\delta}(X)\delta(m_s^{-1})$ we can rewrite (49) in the form

$$f(X) = |W|^{-1} \sum_{\delta\in\widehat{K}_M} d(\delta) \int_{\mathfrak{a}^*} \operatorname{Tr}(\varphi_\delta(\lambda)\Psi_{\lambda,\delta}(X))|\omega(\lambda)|d\lambda. \tag{52}$$

By (ii), $\overset{\vee}{J^\delta}(\lambda)^{-1}\varphi_\delta(\lambda)$ is W-invariant and holomorphic of exponential type R (the exponential type of φ) so by Theorem 7.7 $\varphi_\delta(\lambda)$ equals $\widetilde{F}_\delta(\lambda)$ for the function

$$F_\delta(X) = c\operatorname{Tr}\left[\int_{\mathfrak{a}^*} \Psi_{\lambda,\delta}(X)\varphi_\delta(\lambda)|\omega(\lambda)|d\lambda\right], \tag{53}$$

which has support in the ball $|X| \leq R$. Hence by (52) $f \in \mathcal{D}(\mathfrak{p})$ and now the inversion formula and (47) show that $\varphi(\lambda, kM)$ is a constant multiple of $\widetilde{f}(k \cdot A_\lambda)$. This proves the theorem.

§8. EIGENFUNCTIONS AND EIGENSPACE REPRESENTATIONS ON X_o.

1. Simplicity.

In analogy with the joint eigenspace $\mathcal{E}_\lambda(X)$ (cf. Ch. II, §2, (21)) we consider for each $\lambda \in \mathfrak{a}_c^*$ the joint eigenspace

$$\mathcal{E}_\lambda(\mathfrak{p}) = \{f \in \mathcal{E}(\mathfrak{p}) : \ \partial(p)f = p(i\lambda)f \text{ for } p \in I(\mathfrak{p})\} \tag{1}$$

with the topology induced by that of $\mathcal{E}(\mathfrak{p})$. Then we have the following analog of Proposition 2.6 in Ch. II which is proved in the same way.

Proposition 8.1. *The joint eigenspaces of the operators $\partial(I(\mathfrak{p}))$ are just the spaces $\mathcal{E}_\lambda(\mathfrak{p})$ for a suitable $\lambda \in \mathfrak{a}_c^*$. Also $\mathcal{E}_{s\lambda}(\mathfrak{p}) = \mathcal{E}_\lambda(\mathfrak{p})$ for $s \in W$. The members of $\mathcal{E}_\lambda(\mathfrak{p})$ are characterized by*

$$\int_K f(X + k \cdot Y)dk = f(X)\psi_\lambda(Y), \qquad X, Y \in \mathfrak{p}. \tag{2}$$

In analogy with (42) in Ch. II, §3 we define for each $\lambda \in \mathfrak{a}_c^*$ the *Poisson Transform* P_λ by

$$(P_\lambda F)(X) = \int_{K/M} e^{iB(k\cdot A_\lambda, X)}F(kM)dk_M \tag{3}$$

for $F \in C(K/M)$, dk_M being the normalized K-invariant measure on K/M. From (1) it is clear that P_λ maps $C(K/M)$ into $\mathcal{E}_\lambda(\mathfrak{p})$.

Definition. The point $\lambda \in \mathfrak{a}_c^*$ is called *simple* if P_λ is injective.

We recall that $\lambda \in \mathfrak{a}_c^*$ is said to be *regular* if $\alpha(A_\lambda) \neq 0$ for all $\alpha \in \Sigma$, otherwise λ is said to be *singular.*

Theorem 8.2. *A point $\lambda \in \mathfrak{a}_c^*$ is simple if and only if it is regular.*

We first establish a simple description of regularity. Let K_λ denote the stabilizer of λ in K.

Lemma 8.3. *A point $\lambda \in \mathfrak{a}_c^*$ is regular if and only if $K_\lambda = M$.*

Proof. If λ is singular then K_λ has a Lie algebra larger than $\mathfrak{m}$ so $K_\lambda \neq M$. If λ is regular then K_λ and M have the same Lie algebra. If $k \in K_\lambda$, $\lambda = \zeta + i\eta$ $(\xi, \eta \in \mathfrak{a}^*)$ then $k \cdot \xi = \xi$, $k \cdot \eta = \eta$. For each $\alpha \in \Sigma$ either $\alpha(A_\xi)$ or $\alpha(A_\eta)$ is $\neq 0$. Thus the centralizers $\mathfrak{z}_\xi$ and $\mathfrak{z}_\eta$ of ξ and η, respectively, in $\mathfrak{p}$ have intersection $\mathfrak{a}$. But k leaves each of these centralizers invariant so $k \in M'$, the normalizer of $\mathfrak{a}$ in K. Thus the Weyl group element $s = Ad(k)|\mathfrak{a}$ fixes the plane $\mathfrak{a}_o = \boldsymbol{R}A_\xi + \boldsymbol{R}A_\eta$ pointwise. But each point in $\mathfrak{a}_o$ outside the sets $\{H \in \mathfrak{a}_o : \alpha(H) = 0\}$ where $\alpha \in \Sigma$ is regular. Thus s fixed a regular element so $s = e$ and $k \in M$. This proves the lemma.

For the proof of Theorem 8.2 we write (3) in the form (cf. [GGA], Ch. I, Prop. 1.13),

$$\int_{K/M} e^{iB(k\cdot A_\lambda, X)} F(kM)\,dk_M = \int_{K/K_\lambda} e^{iB(kA_\lambda, X)}\,dk_{K_\lambda} \int_{K_{\lambda/M}} F(k\ell M)\,d\ell_M, \tag{4}$$

where dk_{K_λ} is the normalized invariant measure on K/K_λ. If λ is singular then by the lemma, $K_\lambda \neq M$, and we shall find an $F \not\equiv 0$ such that the last integral (over K_λ/M) is 0 for all k. Select a continuous function $g \not\equiv 0$ on K_λ, bi-invariant under M, such that

$$\int_{K_\lambda} g(\ell)\,d\ell = 0. \tag{5}$$

If $\varphi \in C(K/M)$ is arbitrary put

$$(6) \qquad F(kM) = \int_{K_\lambda} \widetilde{\varphi}(k\ell)g(\ell)d\ell, \qquad k \in K,$$

where $\widetilde{\varphi}$ is the "lift" of φ to K. Here F is well-defined by the M-invariance of g and

$$\int_{K_\lambda/M} F(kuM)du_M = \int_{K_\lambda} g(\ell)d\ell \int_{K_\lambda} \widetilde{\varphi}(ku\ell)du.$$

This vanishes by (5) for all k because ℓ can, by the right invariance of du, be removed in the last integral. On the other hand we can choose φ such that on the subset $K_\lambda/M \subset K/M$ it coincides with $\operatorname{conj}(g_o)$ where $g_o(\ell M) = g(\ell)$. Then (6) implies $F(eM) \neq 0$ so (4) shows that we have a function $F \not\equiv 0$ with $P_\lambda F \equiv 0$. Thus each singular λ is not simple, i.e. λ simple $\Longrightarrow$ λ regular. Because of Proposition 7.4 this conclusion could also be drawn from the more technical Theorem 7.5.

Conversely, suppose λ regular and assume $F \in C(K/M)$ such that $P_\lambda F \equiv 0$. By analyticity this is equivalent to all derivatives of $P_\lambda F$ vanishing at the origin which in turn amounts to

$$(7) \qquad \int_{K/M} p(ik \cdot A_\lambda)F(kM)dk_M = 0 \qquad \text{for } p \in S(\mathfrak{p}).$$

Since λ is regular, Lemma 8.3 shows that the algebra of functions $kM \longrightarrow p(ikA_\lambda)$ on K/M separates points but the lack of self-adjointness makes the Stone-Weierstrass theorem inapplicable. We shall therefore use another method to conclude $F \equiv 0$ on the basis of (7).

For the orthonormal basis (v_i) of V_δ with $v_1, \ldots, v_{\ell(\delta)} \in V_\delta^M$ and any basis (η_j) of F_δ we put $p = \eta_j(v_i)$ in (7). Then

$$\int_K (\eta_j(v_i))^{k^{-1}}(i\lambda)\widetilde{F}(k)dk = 0,$$

where $\widetilde{F}$ is the lift of F to K. But

$$\begin{aligned}(\eta_j(v_i))^{k^{-1}} &= Ad(k^{-1})(\eta_j(v_i)) = \eta_j(\delta(k^{-1})v_i) \\ &= \sum_{p=1}^{d(\delta)} \langle v_i, \delta(k)v_p\rangle \eta_j(v_p)\end{aligned}$$

so

$$\sum_{p=1}^{d(\delta)} \int_K \langle v_i, \delta(k)v_p\rangle j^{\eta_j(v_p)}(i\lambda)\widetilde{F}(k)dk = 0.$$

Using $\int_K = \int_{K/M}\int_M$ this becomes

$$\sum_{p=1}^{\ell(\delta)} \int_{K/M} \langle v_i, \delta(k)v_p\rangle j^{\eta_j(v_p)}(i\lambda)F(kM)dk_M = 0$$

with summation to $\ell(\delta)$ only because if $p > \ell(\delta)$ then $\int_M \delta(m)v_p dm = 0$. Hence we have for $1 \leq i \leq d(\delta)$, $1 \leq j \leq \ell(\delta)$

$$\sum_{p=1}^{\ell(\delta)} J^\delta(-\lambda)_{pj} \int_{K/M} \langle v_i, \delta(k)v_p\rangle F(kM)dk_M = 0. \tag{8}$$

If F were $\not\equiv 0$ this integral would be $\neq 0$ for some δ, i and p. Fix such δ and i and denote the integral by A_p. Then the relation

$$\sum_{p=1}^{\ell(\delta)} J^\delta(-\lambda)_{pj}A_p = 0 \qquad 1 \leq j \leq \ell(\delta)$$

would imply $\det J^\delta(-\lambda) = 0$ which by Proposition 7.4 contradicts the regularity of λ. Thus λ is simple and Theorem 8.2 is proved.

Remark. Theorem 8.2 shows a considerable contrast with the curved case. For the flat case the result shows that the set of non-simple λ is Weyl group invariant whereas in the curved case each Weyl group orbit in $\mathfrak{a}_c^*$ contains a simple λ (Proposition 4.1).

2. The K-finite joint eigenfunctions of $\boldsymbol{D}(G_o/K)$.

In this subsection we shall describe the K-finite joint eigenfunctions of the algebra $\boldsymbol{D}(G_o/K)$ of G_o-invariant differential operators on $\mathfrak{p}$. Recall that each joint eigenfunction lies in one of the spaces $\mathcal{E}_\lambda(\mathfrak{p})$.

Theorem 8.4. *Assume $\lambda \in \mathfrak{a}_c^*$ regular. Then the K-finite elements in $\mathcal{E}_\lambda(\mathfrak{p})$ are the functions*

$$f(X) = \int_{K/M} e^{iB(k\cdot A_\lambda, X)}F(kM)dk_M,$$

where F is a K-finite continuous function on K/M.

Proof. The proof is quite similar to that of Theorem 6.1. Let $\mathcal{E}_{\lambda,\delta}(\mathfrak{p})$ denote the space of functions in $\mathcal{E}_\lambda(\mathfrak{p})$ of type δ. Since λ is simple we then have

$$\dim \mathcal{E}_{\lambda,\delta}(\mathfrak{p}) \geq \dim V_\delta \dim V_\delta^M \tag{9}$$

in analogy with (2) §6.

We shall prove the converse inequality

$$\dim \mathcal{E}_{\lambda,\delta}(\mathfrak{p}) \leq \dim V_\delta \dim V_\delta^M \tag{10}$$

for *arbitrary* $\lambda \in \mathfrak{a}_{\mathbf{c}}^*$. Let $h \in \mathcal{E}_{\lambda,\delta}(\mathfrak{p})$. Viewing h as a distribution we define its Fourier transform $\widetilde{h}$ as a linear functional on $\mathcal{K}^\delta(\mathfrak{a}^*)$:

$$\widetilde{h}(\widetilde{f}) = h(f) = \int_{\mathfrak{p}} h(X)f(X)dX, \qquad f \in \mathcal{D}_{\check{\delta}}(\mathfrak{p}). \tag{11}$$

Because of Theorem 7.7, $\widetilde{h}$ is a continuous linear functional on $\mathcal{K}^\delta(\mathfrak{a}^*)$. As noted in the proof of that theorem the mapping $\psi \longrightarrow \overset{\vee}{J^\delta}\psi$ is a homeomorphism of $\mathcal{J}^\delta(\mathfrak{a}^*)$ onto $\mathcal{K}^\delta(\mathfrak{a}^*)$ so the map

$$\psi \in \mathcal{J}^\delta(\mathfrak{a}^*) \longrightarrow \widetilde{h}(\overset{\vee}{J^\delta}\psi) \tag{12}$$

is a continuous linear functional on $\mathcal{J}^\delta(\mathfrak{a}^*)$. By the Paley-Wiener type theorem for the spherical transform on $\mathfrak{p}$ ([GGA], Ch. IV, Theorem 9.2) the mapping (12) is the spherical transform of a $\mathrm{Hom}(V_\delta, V_\delta^M)$-valued K-invariant distribution ℓ on $\mathfrak{p}$, that is

$$\widetilde{\ell}(\widetilde{\varphi}) = \widetilde{h}(\overset{\vee}{J^\delta}\widetilde{\varphi}) = \ell(\varphi)$$

for all K-invariant $\varphi \in \mathcal{D}(\mathfrak{p}, \mathrm{Hom}(V_\delta, V_\delta^M))$.

Now $\partial(p)h = p(i\lambda)h$ and of course by (28) §7,

$$(\partial(p)f)^\sim(\mu) = (\partial(p)^*(-i\mu))\widetilde{f}(\mu) = (\partial(p)(i\mu))\widetilde{f}(\mu), \qquad \mu \in \mathfrak{a}_{\mathbf{c}}^*. \tag{13}$$

Let $P(\mu) = \partial(p)(i\mu)$ ($\mu \in \mathfrak{a}_{\mathbf{c}}^*, p \in I(\mathfrak{p})$). By (11) and (13) we have

$$(P\widetilde{h})(\widetilde{f}) = h(\partial(p)f) = (\partial(p)^*h)(f) = p(-i\lambda)h(f)$$

so

(14) $$P\widetilde{h} = P(-\lambda)\widetilde{h}.$$

But then

$$(P\widetilde{\ell})(\widetilde{\varphi}) = \widetilde{\ell}(P\widetilde{\varphi}) = \widetilde{\ell}((\partial(p)\varphi)^{\sim}) = \widetilde{h}(\overset{\vee}{J^{\delta}} P\widetilde{\varphi})$$
$$= (P\widetilde{h})(\overset{\vee}{J^{\delta}}\widetilde{\varphi}) = P(-\lambda)\widetilde{h}(\overset{\vee}{J^{\delta}}\widetilde{\varphi}) = P(-\lambda)\widetilde{\ell}(\widetilde{\varphi})$$

so $\widetilde{\ell}$ satisfies (14) too. Thus ℓ is an eigendistribution of all $D \in \boldsymbol{D}(X)$, in particular ℓ is a function. Being K-invariant and $\mathrm{Hom}(V_\delta, V_\delta^M)$-valued, ℓ equals $\psi_\lambda A$ where A is a fixed element in $\mathrm{Hom}(V_\delta, V_\delta^M)$. Thus the map $h \longrightarrow \ell$ gives an injection of $\mathcal{E}_{\lambda,\delta}(\mathfrak{p})$ into $\mathrm{Hom}(V_\delta, V_\delta^M)$ so (10) follows.

If λ is regular both (9) and (10) hold so the theorem is proved.

In spite of the analogies in the proofs, Theorems 6.1 and 8.4 show significant differences. For G/K any $\lambda \in \mathfrak{a}_c^*$ has at least one of its Weyl group transforms $s\lambda$ simple so each K-finite $f \in \mathcal{E}_\lambda(X)$ is the Poisson transform $\mathcal{P}_{s\lambda}F$ for some K-finite $F \in \mathcal{E}(K/M)$. The analog fails for G_o/K; for example if $\lambda = 0$ only the constants in $\mathcal{E}_o(\mathfrak{p})$ are Poisson transforms. We shall now prove a variation of Theorem 8.4 which deals with singular $\lambda \in \mathfrak{a}_c^*$ as well. As usual let $\delta \in \widehat{K}_M$ and let $v_1, \ldots, v_{d(\delta)}$ be a basis of V_δ such that $v_1, \ldots, v_{\ell(\delta)}$ span V_δ^M.

Theorem 8.5. *Let $\lambda \in \mathfrak{a}_c^*$ and $\delta \in \widehat{K}_M$ be arbitrary. Then the following statements hold.*

(i) $\dim \mathcal{E}_{\lambda,\delta}(\mathfrak{p}) = \dim V_\delta \dim V_\delta^M$.

(ii) Let $\mu \in \mathfrak{a}_c^*$ be regular. Then each $f \in \mathcal{E}_{\lambda,\delta}(\mathfrak{p})$ is a linear combination of functions of the form

(15) $$\int_{K/M} e^{iB(kA_\lambda, X)} Q_{ij}(B(k \cdot A_\mu, X)) \langle v_j, \delta(k)v_i\rangle dk_M,$$

where $1 \le i \le \ell(\delta), 1 \le j \le d(\delta)$ and Q_{ij} is a polynomial in one variable. If λ is regular, the Q_{ij} can be taken to be constants.

Proof. We already know that $\dim \mathcal{E}_{\lambda,\delta}(\mathfrak{p})$ is bounded by $\dim V_\delta \dim V_\delta^M$ and that if λ is regular the functions

(16) $$f_{ij}^\lambda(X) = \int_{K/M} e^{iB(kA_\lambda, X)} \langle v_j, \delta(k)v_i\rangle dk_M$$

$(1 \leq j \leq d(\delta), 1 \leq i \leq \ell(\delta))$ form a basis of $\mathcal{E}_{\lambda,\delta}(\mathfrak{p})$. If λ is singular, linear dependencies will occur among the functions f_{ij}^λ (since λ is not simple). However one can produce additional functions in $\mathcal{E}_{\lambda,\delta}(\mathfrak{p})$ by applying to f_{ij}^λ derivatives in the variable λ. We shall now verify that all the functions in $\mathcal{E}_{\lambda,\delta}(\mathfrak{p})$ are obtained by means of such a process. For this we need the following proposition.

Let D denote the open unit disk in $\boldsymbol{C}$, V a Banach space, and $\mathcal{H}(D, V)$ the space of holomorphic maps $\varphi : D \longrightarrow V$.

Proposition 8.6. *Let $\varphi_1, \dots, \varphi_k \in \mathcal{H}(D, V)$ and suppose that for each $z \in D - \{0\}$ the vectors $\varphi_1(z), \dots, \varphi_k(z)$ in V are linearly independent.*

Then there exist $\psi_1, \dots, \psi_k \in \mathcal{H}(D, V)$ such that

(a) For each $z \in D$ the vectors $\psi_i(z)$ $(1 \leq i \leq k)$ are linearly independent.

(b) $$\psi_j(z) = z^{n_j} \sum_{i=1}^{k} c_j^i(z)\varphi_i(z) \qquad (z \neq 0),$$

where $n_j \in \boldsymbol{Z}$ and each c_j^i is a holomorphic function in D.

The proof is based on the following lemma.

Lemma 8.7. *Let $f, \varphi_1, \dots, \varphi_m \in \mathcal{H}(D, V)$ such that*

(i) For each $z \in D - \{0\}$,

$$f(z) \notin \operatorname{span}(\varphi_1(z), \dots, \varphi_m(z)).$$

Then there exist $n \in \boldsymbol{Z}^+$, and complex polynomials $p_1, \dots, p_m$ of degree $< n$ such that the function

$$g(z) = z^{-n} \left(f(z) - \sum_1^m p_i(z)\varphi_i(z) \right) \qquad z \in D - \{0\}$$

extends to $g \in \mathcal{H}(D, V)$ and such that

$$g(0) \notin \operatorname{span}(\varphi_1(0), \dots, \varphi_m(0)).$$

Proof. We write each function as a power series

$$f(z) = \sum_{n=0}^{\infty} z^n u_n, \quad \varphi_i(z) = \sum_{n=0}^{\infty} z^n v_{i,n}$$

with $u_n, v_{i,n} \in V$. Then there exists a $\rho \geq 1$ such that

$$\|u_n\| \leq \rho^{n+1} \qquad \|v_{i,n}\| \leq \rho^{n+1}, \qquad n \geq 0.$$

Put

$$\Phi_n(\alpha) = \sum_{i=1}^{m} \alpha_i v_{i,n} \qquad \alpha = (\alpha_1, \dots, \alpha_m) \in \boldsymbol{C}^m.$$

Then

$$\|\Phi_n(\alpha)\| \leq \rho^{n+1}|\alpha|, \qquad |\alpha| = \sum_1^m |\alpha_i|. \tag{17}$$

We may assume that $\varphi_1(0), \dots, \varphi_m(0)$ are linearly independent (otherwise replace the set $\varphi_i (1 \leq i \leq m)$ by a suitable subfamily). Since $\Phi_o(\alpha) = \sum_i \alpha_i \varphi_i(0)$ there exists a constant $C > 0$ such that

$$|\alpha| \leq C\|\Phi_o(\alpha)\| \qquad \alpha \in \boldsymbol{C}^m. \tag{18}$$

Scaling the φ_i we may assume $C = 1$. Put

$$V_o = \Phi_o(\boldsymbol{C}^m) = \mathrm{Span}(\varphi_1(0), \dots, \varphi_m(0)).$$

If $u_o \notin V_o$ we can take $g = f$.

If $u_o = \Phi_o(\alpha_o)$, $u_1 - \Phi_1(\alpha_o) \notin V_o$ we can take

$$g(z) = z^{-1}\left(f(z) - \sum_{i=1}^{m} \alpha_{o,i}\varphi_i(z)\right) = \sum_{n=1}^{\infty} z^{n-1}\left(u_n - \Phi_n(\alpha_o)\right).$$

In fact, by (18) and (17)

$$|\alpha_o| \leq \|u_o\| \leq \rho, \qquad \|\Phi_n(\alpha_o)\| \leq \rho^{n+2}$$

so g is holomorphic at $z = 0$, hence $g \in \mathcal{H}(D, V)$.

If $u_o = \Phi_o(\alpha_o), u_1 - \Phi_1(\alpha_o) \in V_o$ then

$$u_1 - \Phi_1(\alpha_o) = \Phi_o(\alpha_1).$$

Continuing in this fashion, we get in n steps the m tuples $\alpha_o, \alpha_1, \dots, \alpha_{n-1} \in \boldsymbol{C}^m$ such that

$$u_k = \sum_{j=0}^{k} \Phi_{k-j}(\alpha_j) \qquad \text{for } 0 \leq k \leq n-1.$$

Then if

$$u_n - \sum_0^{n-1} \Phi_{n-j}(\alpha_j) \notin V_o \tag{19}$$

we can take

$$g(z) = z^{-n}\big(f(z) - \sum_{i=1}^{m}\sum_{j=0}^{n-1} \alpha_{j,i} z^j \varphi_j(z)\big)$$
$$= \sum_{k=n}^{\infty} z^{k-n}\big(u_k - \sum_{j=0}^{n-1} \Phi_{k-j}(\alpha_j)\big).$$

If (19) fails we have a $\alpha_n \in \boldsymbol{C}^n$ such that

$$u_n = \sum_{j=0}^{n} \Phi_{n-j}(\alpha_j).$$

Using this and (17)–(18) we have by induction on n,

$$|\alpha_n| \leq 2^n \rho^{2n+1}.$$

In fact

$$|\alpha_n| \leq \|\Phi_o(\alpha_n)\| = \|u_n - \sum_{j=0}^{n-1} \Phi_{n-j}(\alpha_j)\|$$
$$\leq \rho^{n+1} + \sum_{j=0}^{n-1} \rho^{n-j+1}|\alpha_j| \leq \rho^{n+1} + \sum_{j=0}^{n-1} \rho^{n-j+1} 2^j \rho^{2j+1}$$
$$\leq \rho^{2n+1}\left(1 + \sum_{j=0}^{n-1} 2^j\right) = 2^n \rho^{2n+1}.$$

However this choice of the α_j has to stop with some $j = n$ because otherwise we would get

$$f(z) = \sum_0^{\infty} z^n u_n = \sum_0^{\infty} z^n \sum_{j=0}^{n} \Phi_{n-j}(\alpha_j) = \sum_{j=0}^{\infty}\sum_{k=0}^{\infty} z^{j+k} \Phi_k(\alpha_j)$$
$$= \sum_{i=1}^{m}\left(\sum_{j=0}^{\infty} z^j \alpha_{j,i}\right)\left(\sum_{k=0}^{\infty} z^k v_{i,k}\right) = \sum_{i=1}^{m}\left(\sum_{j=0}^{\infty} z^j \alpha_{j,i}\right)\varphi_i(z)$$

contradicting (i). (The series manipulation is permissible because of the estimates for $|\alpha_n|$.) If α_{n-1} is the last α_j then (19) holds so $g(0) \notin V_o$ as stated.

For Proposition 8.6 we use induction on k. Assuming $\varphi_1(0), \cdots, \varphi_{k-1}(0)$ linearly independent we take $m = k-1$, $f = \varphi_k$ in the lemma.

Remark. Prop. 8.6 is proved in Oshima and Sekiguchi [1980] even for the more subtle case when V is replaced by the space of distributions on a manifold. The version above, whose proof rests on the same idea but is more elementary because of the existence of a norm on V, was kindly communicated to the author by Schlichtkrull.

We now come back to the proof of Theorem 8.5. Let $\lambda \in \mathfrak{a}_c^*$ be arbitrary. For each $\alpha \in \Sigma$, $\alpha(\lambda + z\mu) \neq 0$ for all complex z in some punctured disk $0 < |z| < \epsilon_\alpha$. Hence $\lambda + z\mu$ is regular for z in some punctured disk $0 < |z| < \epsilon$. For such z the functions $(f_{ij})^{\lambda+z\mu}$ $(1 \leq i \leq \ell(\delta), 1 \leq j \leq d(\delta))$ form a basis of $\mathcal{E}_{\lambda+z\mu,\delta}(\mathfrak{p})$. From (16) it is clear that each $f \in \mathcal{E}_{\lambda+z\mu,\delta}(\mathfrak{p})$ has an exponential bound uniform for $|z| < \epsilon$,

$$|f(X)| \leq Ce^{D|X|} \qquad X \in \mathfrak{p}$$

where C and D are constants. Thus

$$\mathcal{E}_{\lambda+z\mu}(\mathfrak{p}) \subset V \qquad |z| < \epsilon$$

where V is the Banach space

$$V = \{f \in C(\mathfrak{p}) : \sup_{X \in \mathfrak{p}} e^{-D|X|}|f(X)| < \infty\}.$$

Because of Proposition 8.6 there exist functions $C_{ij,rs}(z)$ holomorphic in $|z| < \epsilon$ and integers n_{rs} such that the functions

$$g_{rs}^z(X) = z^{-n_{rs}} \sum_{i,j} C_{ij,rs}(z) f_{ij}^{\lambda+z\mu}(X) \tag{20}$$

$1 \leq r \leq d(\delta)$, $1 \leq s \leq \ell(\delta)$ are for each z linearly independent. Combining this with (10) we deduce that the functions g_{rs}° form a basis of $\mathcal{E}_{\lambda,\delta}(X)$. Writing

$$e^{iB(k\cdot A_{\lambda+z\mu},X)} = e^{iB(kA_\lambda,X)}\{1 + z(iB(kA_\mu, X)) + \dots\}$$

we see that g_{rs}° is a linear combination of functions (15). This concludes the proof.

3. The Irreducibility Criterion for the Eigenspace Representations of G_o/K.

We can now settle the irreducibility question for the eigenspace representations of $D(G_o/K)$. We recall that each joint eigenspace of these operators has the form $\mathcal{E}_\lambda(\mathfrak{p})$ for some $\lambda \in \mathfrak{a}_{\mathfrak{c}}^*$.

Theorem 8.8. *Let $\lambda \in \mathfrak{a}_{\mathfrak{c}}^*$ and let T_λ denote the natural representation of G_o on the joint eigenspace $\mathcal{E}_\lambda(\mathfrak{p})$. Then T_λ is irreducible if and and only if λ is regular.*

Proof. Suppose λ is regular. Let $0 \neq V \subset \mathcal{E}_\lambda(\mathfrak{p})$ be a closed invariant subspace. Then V contains a function ψ with $\psi(0) \neq 0$ and by averaging over K, ψ can be taken K-invariant. Thus the zonal spherical function ψ_λ belongs to $\mathcal{E}_\lambda(\mathfrak{p})$. Since λ is simple we can turn the space

$$\mathcal{H}_\lambda = \{f(X) = \int_{K/M} e^{iB(kA_\lambda,X)} F(kM)dk_M \; : \; F \in L^2(K/M)\} \tag{21}$$

into a Hilbert space by giving f the L^2 norm of F. Since $-\lambda$ is also simple the functions

$$e_Y : \; kM \longrightarrow e^{-iB(k\cdot A_\lambda,Y)}$$

as Y runs through $\mathfrak{p}$ span a dense subspace of $L^2(K/M)$. Hence the Poisson transforms of the functions e_Y span a dense subspace of $\mathcal{H}_\lambda$. However, these Poisson transforms are just the translates $X \longrightarrow \psi_\lambda(X-Y)$ of ψ_λ which belong to V. The embedding $\mathcal{H}_\lambda \longrightarrow \mathcal{E}_\lambda(\mathfrak{p})$ being continuous (by Schwartz' inequality) we deduce $\mathcal{H}_\lambda \subset V$. Therefore, by Theorem 8.4, V contains all $\mathcal{E}_{\lambda,\delta}(\mathfrak{p})$, $\delta \in \widehat{K}_M$, so, by Fourier expansion along $K, V = \mathcal{E}_\lambda(\mathfrak{p})$. This proves the irreducibility of T_λ.

For the converse we assume λ singular and consider the subspace $\mathcal{E}_\lambda^o = P_\lambda(\mathcal{E}(K/M))$ of $\mathcal{E}_\lambda(\mathfrak{p})$ consisting of the Poisson transforms of functions $F \in \mathcal{E}(K/M)$. Let W be the closure of $\mathcal{E}_\lambda^o$ in $\mathcal{E}_\lambda(\mathfrak{p})$. Since $P_\lambda \neq 0$ we have $W \neq 0$ and W is G_o-invariant. We shall prove $W \neq \mathcal{E}_\lambda(\mathfrak{p})$.

For $\delta \in \widehat{K}_M$ let

$$\mathcal{E}_{\lambda,\delta}^o \subset W_\delta \subset \mathcal{E}_{\lambda,\delta}(\mathfrak{p}), \qquad \mathcal{E}_\delta(K/M),$$

be the respective subspaces of K-finite elements of type δ. Then

$$\mathcal{E}_{\lambda,\delta}^o = P_\lambda(\mathcal{E}_\delta(K/M)). \tag{22}$$

Since λ is not simple there exists an $F \in \mathcal{E}(K/M)$ ($F \neq 0$) in the kernel of P_λ. By Fourier decomposition along K/M we may assume that this F is in $\mathcal{E}_\delta(K/M)$ for a certain δ. Then by (22) and Theorem 8.5,

$$\dim \mathcal{E}^o_{\lambda,\delta} < \dim \mathcal{E}_\delta(K/M) = \dim V_\delta \dim V_\delta^M = \dim \mathcal{E}_{\lambda,\delta}(\mathfrak{p}). \tag{23}$$

Denoting by χ_δ the character of δ the convolution

$$f(X) \longrightarrow \int_K \overline{\chi}_\delta(k) f(k^{-1} \cdot X) dk \tag{24}$$

is a continuous surjection of the spaces

$$\mathcal{E}^o_\lambda \subset W \subset \mathcal{E}_\lambda(\mathfrak{p})$$

onto

$$\mathcal{E}^o_{\lambda,\delta} \subset W_\delta \subset \mathcal{E}_{\lambda,\delta}(\mathfrak{p}).$$

Since $\mathcal{E}^o_\lambda$ is dense in W the continuity of (24) implies that the space $\mathcal{E}^o_{\lambda,\delta}$ is dense in the space W_δ, so by the finite-dimensionality they are equal. Then by (23) $W_\delta \neq \mathcal{E}_{\lambda,\delta}(\mathfrak{p})$ so $W \neq \mathcal{E}_\lambda(\mathfrak{p})$ proving the non-irreducibility of T_λ.

§9. THE COMPACT CASE.

1. Motivation.

If U is a compact Lie group the Peter-Weyl theorem on U can be written

$$F = \sum_{\delta \in \widehat{U}} d(\delta) F * \chi_\delta, \qquad F \in \mathcal{E}(U), \tag{1}$$

(Weil [1940], §22, [GGA], Ch. IV, Cor. 1.8). Here $\widehat{U}$ is the set of equivalence classes δ of irreducible representations of U, $d(\delta)$ and χ_δ denote the dimension and character of δ, respectively, and $*$ denotes convolution on U. Let V_δ be a representation space of δ with inner product $\langle , \rangle$. Then the Peter-Weyl theorem can also be expressed

$$L^2(U) = \bigoplus_{\delta \in \widehat{U}} H_\delta(U), \tag{2}$$

where

$$H_\delta(U) = \{F \in C(U) : \; F(u) = \mathrm{Tr}(\delta(u)C), \; C \in \mathrm{Hom}(V_\delta, V_\delta)\} \tag{3}$$

and the space H_δ is irreducible under the representation τ_δ of $U \times U$ which to $(u, v) \in U \times U$ assigns the endomorphism

$$F \longrightarrow F^{L(u)R(v^{-1})}$$

of H_δ (loc. cit. Theorem 1.6). Note that $F * \chi_\delta = \chi_\delta * F$ for each $F \in C(U)$.

Let $K \subset U$ be a closed subgroup and V_δ^K the space of vectors $v \in V_\delta$ fixed under $\delta(K)$. Let $\widehat{U}_K = \{\delta \in \widehat{U} : V_\delta^K \neq 0\}$. For the homogeneous space U/K, (2) takes the form

$$L^2(U/K) = \bigoplus_{\delta \in \widehat{U}_K} H_\delta(U/K), \tag{4}$$

where

$$H_\delta(U/K) = \{f \in C(U/K) : f(uK) = \operatorname{Tr}(\delta(u)C),\ C \in \operatorname{Hom}(V_\delta, V_\delta^K)\}, \tag{5}$$

(cf. [GGA], Ch. V, Theorem 3.5). If $\pi : U \longrightarrow U/K$ is the natural mapping, convolution $\times$ on U/K is as usual defined by

$$(f_1 \times f_2) \circ \pi = (f_1 \circ \pi) * (f_2 \circ \pi). \tag{6}$$

In the case when U/K is symmetric, V_δ^K is spanned by a single unit vector $\boldsymbol{e}$, so the space $H_\delta(U/K)$ is given by

$$H_\delta(U/K) = \{f_v(uK) = \langle \delta(u)\boldsymbol{e}, v\rangle : v \in V_\delta\}. \tag{7}$$

It contains a unique K-invariant function Φ_δ satisfying $\Phi_\delta(e) = 1$, namely the spherical function

$$\Phi_\delta(uK) = \langle \delta(u)\boldsymbol{e}, \boldsymbol{e}\rangle. \tag{8}$$

Proposition 9.1. *For the symmetric space U/K we have the convergent expansion*

$$f = \sum_{\delta \in \widehat{U}_K} f \times \Phi_\delta d(\delta), \qquad f \in \mathcal{E}(U/K). \tag{9}$$

Proof. Using (1) on the function $F = f \circ \pi$ we have for $k \in K$

$$(F * \chi_\delta)(u) \equiv (F * \chi_\delta)(uk) = \int_U \int_K F(v)\chi_\delta(v^{-1}uk)dvdk$$

and

$$\int_K \chi_\delta(uk)dk \equiv \begin{cases} 0 & \text{if} \quad \delta \notin \widehat{U}_K \\ \Phi_\delta(u) & \text{if} \quad \delta \in \widehat{U}_K \end{cases}$$

([GGA], Ch. IV, Theorem 4.2) so (9) follows.

The analog of (9) for a symmetric space $X = G/K$ of the noncompact type is

$$(10) \qquad f(x) = c \int_{\mathfrak{a}^*} (f \times \varphi_\lambda)(x)|\mathbf{c}(\lambda)|^{-2} d\lambda, \qquad f \in \mathcal{D}(X),$$

c being a fixed constant. In fact, the function

$$F(h) = \int_K f(gkhK)dk$$

is K-bi-invariant so by Harish-Chandra's inversion formula for K-bi-invariant functions in $\mathcal{D}(G)$, ([GGA], Ch. IV, Theorem 7.5)

$$(11) \qquad F(h) = c \int_{\mathfrak{a}^*} \widetilde{F}(\lambda)\varphi_\lambda(h)|\mathbf{c}(\lambda)|^{-2} d\lambda.$$

Then (§1, No. 1) $\widetilde{F}(\lambda) = (f \times \varphi_\lambda)(gK)$ so (10) follows by putting $h = e$ in (11).

The Fourier inversion formula

$$(12) \qquad f(x) = w^{-1} \int_{\mathfrak{a}^*} \int_B e^{(i\lambda+\rho)(A(x,b))} \widetilde{f}(\lambda, b)|\mathbf{c}(\lambda)|^{-2} d\lambda db$$

from Theorem 1.3 is a refinement of (10). It involves a genuine Fourier transform $f(x) \longrightarrow \widetilde{f}(\lambda, b)$ whereby range questions like $\mathcal{D}(X)^\sim =?$, $L^2(X)^\sim =?$ become meaningful problems (cf. Theorems 1.5 and 5.1, which in turn have significant applications to differential equations, developed in Ch. V).

The question now arises: Is there a Fourier transform on the symmetric space U/K which refines Proposition 9.1 in the same way as (12) refines (10)? We shall now see that this is so — with some limitation.

2. Compact Symmetric Spaces.

Consider again the compact symmetric space $S = U/K$; now we assume U simply connected and K connected. Along with S we consider the dual symmetric space $X = G/K$ with both G and U analytic subgroups of the simply connected group G^c whose Lie algebra is the complexification $\mathfrak{g}^c$ of the Lie algebra $\mathfrak{g}$ of G. We write as usual (§7 (2))

$$\mathfrak{g} = \mathfrak{k} + \mathfrak{p} = \mathfrak{k} + \mathfrak{a} + \mathfrak{q}$$
$$\mathfrak{u} = \mathfrak{k} + i\mathfrak{p} = \mathfrak{k} + i\mathfrak{a} + i\mathfrak{q}.$$

Let Σ denote the set of roots of $\mathfrak{g}$ with respect to $\mathfrak{a}$ and Σ^+ the set of positive roots corresponding to the Weyl chamber $\mathfrak{a}^+$ and the Lie algebra $\mathfrak{n} = \Sigma_{\alpha\in\Sigma^+}\mathfrak{g}^\alpha$. As usual we extend $\mathfrak{a}$ to a Cartan subalgebra $\mathfrak{h}$ of $\mathfrak{g}$. Then $\mathfrak{h}^C$ is a Cartan subalgebra of $\mathfrak{g}^C$. Let s^* denote the Weyl group element which interchanges $\mathfrak{a}^+$ and $-\mathfrak{a}^+$ and consider the set from Ch. II, Theorem 4.12 and (27),

$$\Lambda = \left\{\mu \in \mathfrak{a}^* : \frac{\langle \mu, \alpha\rangle}{\langle \alpha, \alpha\rangle} \in \boldsymbol{Z}^+ \text{ for } \alpha \in \Sigma^+\right\}. \tag{13}$$

Given $\mu \in \Lambda$ let π_μ denote the representation of G^C whose restriction to G is spherical (and conical) and has highest restricted weight μ. Let $V_\mu = V_{\pi_\mu}$ be the corresponding representation space with $\pi_\mu(U)$-invariant inner product $\langle , \rangle$. Viewing μ as a linear form on $\mathfrak{h}^C$, 0 on $\mathfrak{h} \cap \mathfrak{k}$, let e_μ belong to the weight μ and let $v_\mu \in V_\mu$ be a unit vector fixed under $\pi_\mu(K)$. We put $d(\mu) = d(\pi_\mu)$.

If $\mu \in \Lambda$ then $\mu^* = -s^*\mu$ belongs to Λ and π_{μ^*} is the representation of G^C on V_{μ^*} contragredient to π_μ. The mapping $v \longrightarrow v'$ where $v'(w) = \langle w, v\rangle$ is a conjugate linear mapping of V_μ onto V_{μ^*}. Then $(v_\mu)' = v_{\mu^*}$ is a unit vector in V_{μ^*} fixed under $\pi_{\mu^*}(K)$. Consider now the spherical function

$$\psi_\mu(g) = \{\pi_\mu(g^{-1})v_\mu, v_{\mu^*}\}, \qquad g \in G^C, \tag{14}$$

where $\{,\}$ is the canonical bilinear form on $V_\mu \times V_{\mu^*}$. Putting

$$f_w(g) = \{\pi_\mu(g^{-1})w, v_{\mu^*}\}, \qquad w \in V_\mu, \tag{15}$$

we have

$$f_w(h^{-1}g) = f_{\pi_\mu(h)w}(g)$$

so the map $w \longrightarrow f_w$ is an equivalence between π_μ and the natural representation of G^C on the space of translates of ψ_μ all of which are holomorphic functions on G^C. Let us calculate the function corresponding to the vector e_μ.

We have as usual

$$G = NAK, \qquad g = n \exp A(g)k, \qquad A(g) \in \mathfrak{a},$$
$$\mathfrak{g} = \mathfrak{n} + \mathfrak{a} + \mathfrak{k}, \qquad \mathfrak{g}^C = \mathfrak{n}^C + \mathfrak{a}^C + \mathfrak{k}^C.$$

The mapping

$$(16)\qquad (X,H,T)\longrightarrow \exp X\exp H\exp T\qquad (X\in\mathfrak{n}^{\boldsymbol{C}},H\in\mathfrak{a}^{\boldsymbol{C}},T\in\mathfrak{k}^{\boldsymbol{C}})$$

is a holomorphic diffeomorphism of a neighborhood of 0 in $\mathfrak{u}^{\boldsymbol{C}}=\mathfrak{g}^{\boldsymbol{C}}$ onto a neighborhood $U_o^{\boldsymbol{C}}$ of e in $G^{\boldsymbol{C}}$. We can take $U_o^{\boldsymbol{C}}$ as the diffeomorphic image (under exp) of a ball $B(0)$ relative to an $\mathrm{Ad}(U)$-invariant quadratic form on $\mathfrak{g}^{\boldsymbol{C}}$. In particular, the neighborhood $U_o=U_o^{\boldsymbol{C}}\cap U$ is invariant under the maps $u\longrightarrow kuk^{-1}$ $(k\in K)$. Since (16) is a diffeomorphism we can consider the map

$$(17)\qquad A:\ \exp X\exp H\exp T\longrightarrow H$$

of $U_o^{\boldsymbol{C}}$ into $\mathfrak{a}^{\boldsymbol{C}}$. The notation reflects the fact that on $U_o^{\boldsymbol{C}}\cap G$ the map A agrees with the earlier one. Now e_μ (or more precisely the function f_{e_μ}) is a holomorphic function on $G^{\boldsymbol{C}}$. Since it is N-invariant on the left, K-invariant on the right we have in the notation of (16)

$$e_\mu(\exp X\exp H\exp T)=e_\mu(\exp H)=e^{-\mu(H)}e_\mu(e)$$

so normalizing e_μ by $e_\mu(e)=1$ we have

$$(18)\qquad e_\mu(u)=e^{-\mu(A(u))}\qquad u\in U_o^{\boldsymbol{C}}.$$

We put also

$$(19)\qquad f_\mu(u)=e^{(\mu+2\rho)(A(u))}\qquad u\in U_o^{\boldsymbol{C}}$$

and set

$$S_o=\{uK:\ u\in U_o\},$$
$$e_\mu(uK)=e_\mu(u),\quad f_\mu(uK)=f_\mu(u)\quad \text{for } u\in U_o.$$

Lemma 9.2. *Let $s,t\in S_o$ and $\mu\in\Lambda$. Then, writing $t=uK$,*

$$(20)\qquad \int_K e_\mu(k\cdot s)f_\mu(k\cdot t)dk=\psi_\mu(u^{-1}\cdot s).$$

Proof. This follows from the symmetry of the spherical function ((7) §1) by holomorphic continuation. Let $\lambda=-i(\mu+\rho)$ so $\mu=i\lambda-\rho$. Since

$$\psi_\mu(u)=\int_K e_\mu(ku)dk=\int_K e^{(-i\lambda+\rho)(A(ku))}dk\qquad u\in U_o^{\boldsymbol{C}}$$

we see that on $U_o^{\boldsymbol{C}} \cap G$, ψ_μ equals the spherical function $\varphi_{-\lambda}$. Now we know from Theorem 1.1, if $g, h \in G$,

(21)

$$\int_K e^{(-i\lambda+\rho)(A(kgk^{-1}))} e^{(i\lambda+\rho)(A(khk^{-1}))} dk = \varphi_\lambda(g^{-1}h) = \varphi_{-\lambda}(h^{-1}g) = \psi_\mu(h^{-1}g).$$

The expressions on the extreme sides of (21) are holomorphic for $g, h \in U_o^{\boldsymbol{C}}$ and agree for $g, h \in U_o^{\boldsymbol{C}} \cap G$. By holomorphic continuation both sides agree on $U_o^{\boldsymbol{C}}$. In particular (20) holds and the lemma is proved.

Definition. For $f \in \mathcal{D}(S_o)$ its *Fourier transform* $\widetilde{f}$ is defined by

$$\widetilde{f}(\mu, kM) = \int_S f(s) f_\mu(k^{-1} \cdot s) ds, \quad \mu \in \Lambda, \quad kM \in K/M, \tag{22}$$

where ds is the normalized volume element on S.

Theorem 9.3. *The Fourier transform (22) has the following inversion formula*

$$f(s) = \sum_{\mu \in \Lambda} d(\mu) \int_{K/M} \widetilde{f}(\mu, kM) e_\mu(k^{-1} \cdot s)\, dk_M, \qquad s \in S_o, \tag{23}$$

valid for $f \in \mathcal{D}(S_o)$.

The proof is a direct consequence of Lemma 9.2. The integral on the right hand side of (23) equals

$$\int_{K/M} \left(\int_U f(uK) f_\mu(k^{-1}uK) du \right) e_\mu(k^{-1} \cdot s)\, dk_M$$
$$= \int_U f(uK) \psi_\mu(u^{-1}s) du = (f \times \psi_\mu)(s)$$

so the desired formula reduces to (9).

Example. In thc case of the unit sphere $\boldsymbol{S}^d \subset \boldsymbol{R}^{d+1}$ Theorem 9.3 can be put into more explicit form ([GGA], Introduction, Problem C.1). We take $a \in \boldsymbol{S}^d$ as the North Pole $a = (0, \ldots, 0, 1)$, and B the Equator

$= \{s \in \boldsymbol{S}^d : (a,s) = 0\}$. Here the functions e_μ and f_μ become (with $n \in \boldsymbol{Z}^+$)

$$e_{b,n}(s) = (a+ib,s)^n, \qquad s \in \boldsymbol{S}^d$$
$$f_{b,n}(s) = \{sgn(s,a)\}^{d-1}(a+ib,s)^{-n-d-1}, \quad s \in \boldsymbol{S}^d - B$$

and (23) takes the form

$$f(s) = \sum_{n=0}^{\infty} d(n) \int_B \widetilde{f}(n,b) e_{b,n}(s)\,db,$$

where $d(n) = \dim H_n(\boldsymbol{R}^{d+1})$, the dimension of the space of homogeneous n-degree harmonic polynomials on $\boldsymbol{R}^{d+1}$.

Theorem 9.3, found by Sherman [1977], can be improved in the case when U/K has rank one. The condition that $\mathrm{supp}(f) \subset S_o$ can then be dropped by defining the integrals (21), (22) through delicate regularization. For the details we refer to Sherman [1990].

3. Analogies.

For the noncompact space G/K and the compact space U/K we have proved the following inversion formulas

$$f(x) = w^{-1} \int_{\mathfrak{a}^*} \int_B \widetilde{f}(\lambda,b) e^{(i\lambda+\rho)(A(x,b))} |\boldsymbol{c}(\lambda)|^{-2} d\lambda\, db$$
$$f(s) = \sum_{\mu \in \Lambda} d(\mu) \int_{K/M} \widetilde{f}(\mu, kM) e_\mu(k^{-1}\cdot s)\, dk_M.$$

To what extent are they analytic continuations of each other as the kernels

$$e^{(i\lambda+\rho)(A(x,b))}, \qquad e_\mu(k^{-1}\cdot s)$$

indeed are? What is still missing is a relationship of $d(\mu)$ to the $\boldsymbol{c}$-function. We shall later in this section establish the formula

$$d(\mu) = \left\{ \frac{\boldsymbol{c}(\lambda+i\mu)\boldsymbol{c}(-\lambda-i\mu)}{\boldsymbol{c}(\lambda)\boldsymbol{c}(-\lambda)} \right\}_{\lambda=-i(\mu+\rho)}. \tag{24}$$

Through this formula, the Weyl product formula for $d(\mu)$ ([GGA], Ch. V, Theorem 1.8) corresponds to the product formula of Gindikin-Karpelevič for $\boldsymbol{c}(\lambda)$ (loc. cit. Ch. IV, §6, No. 1).

4. The Product Decomposition.

It is known from the theory of infinite-dimensional representations that the tensor product of two unitary irreducible representations decomposes into a direct integral of irreducible unitary representations. Since spherical functions are representation coefficients this would give a decomposition of a product of two spherical functions into a "continuous sum" of spherical functions. We shall obtain a fairly explicit decomposition of this type in case one of the spherical functions comes from a finite-dimensional representation.

Let Σ_* denote the set of unmultipliable roots in Σ and let $\beta_1, \ldots, \beta_\ell$ be a basis of this root system Σ_*. If $\mu_1, \ldots, \mu_\ell \in \mathfrak{a}^*$ are determined by

$$\frac{\langle \mu_i, \beta_j \rangle}{\langle \beta_j, \beta_j \rangle} = \delta_{ij}, \qquad 1 \leq i, j \leq \ell, \tag{25}$$

we know from Prop. 4.23, Ch. II that

$$\Lambda = \left\{ \sum_1^\ell n_i \mu_i \, : \, n_i \in \boldsymbol{Z}^+ \right\}. \tag{26}$$

Proposition 9.4. *Let $\mu \in \Lambda$ and let ψ_μ be the spherical function (14). Then*

$$\psi_\mu(\exp H) = \sum_{i=0}^q c_i e^{-\nu_i(H)}, \qquad H \in \mathfrak{h}^{\boldsymbol{C}}, \tag{27}$$

where $c_i \geq 0$ and $\nu_i \in \boldsymbol{Z}\mu_1 + \cdots + \boldsymbol{Z}\mu_\ell$.

Proof. As usual we extend each $\nu \in \mathfrak{a}^*$ to a complex linear form on $\mathfrak{h}^{\boldsymbol{C}}$ making it 0 on $\mathfrak{h} \cap \mathfrak{k}$. Let $e_o, \ldots, e_q$ be an orthonormal basis of V_μ such that e_i belongs to the weight λ_i (where $\lambda_o = \mu$ and several λ_i may coincide). Then

$$v_\mu = \sum_0^q a_i e_i, \quad v_{\mu^*} = (v_\mu)' = \sum_0^q \overline{a}_i e_i', \quad a_i = \langle v_\mu, e_i \rangle.$$

Hence

$$\begin{aligned} \psi_\mu(\exp H) &= \{\pi_\mu(\exp(-H))v_\mu, v_{\mu^*}\} \\ &= \left\{ \sum_0^q a_i \pi_\mu(\exp(-H))e_i, \sum_0^q \overline{a}_j e_j' \right\} \end{aligned}$$

so

$$\psi_\mu(\exp H) = \sum_0^q |a_i|^2 e^{-\lambda_i(H)}, \qquad H \in \mathfrak{h}^{\boldsymbol{C}}. \tag{28}$$

Taking $H \in \mathfrak{h} \cap \mathfrak{k}$ we get

$$1 = \sum_0^q |a_i|^2 e^{-\lambda_i(H)},$$

which together with the relation $1 = \sum_0^q |a_i|^2$ and the fact that the restriction $\lambda_i|\mathfrak{h} \cap \mathfrak{k}$ is purely imaginary, implies

$$\lambda_i|\mathfrak{h} \cap \mathfrak{k} = 0 \qquad \text{if} \quad a_i \neq 0. \tag{29}$$

Now recall from ([DS], Ch. VII, §8) that the lattice

$$\mathfrak{a}_K = \{H \in i\mathfrak{a} : \ \exp H \in K\}$$

is spanned by the vectors

$$\frac{2\pi i}{\langle \beta, \beta \rangle} A_\beta, \qquad (\beta \in \Sigma). \tag{30}$$

By general weight theory, each λ_i is purely imaginary on $i\mathfrak{a}$. Let $H \in \mathfrak{a}_K$ and $a_i \neq 0$. Then $\psi_\mu(\exp H) = 1$ so by (28), $\lambda_i(H) \in 2\pi i \boldsymbol{Z}$. Hence by (30)

$$\frac{\langle \lambda_i, \beta \rangle}{\langle \beta, \beta \rangle} \in \boldsymbol{Z} \qquad (\beta \in \Sigma)$$

so $\lambda_i \in \sum_1^\ell \boldsymbol{Z}\mu_j$. Taking $\nu_i = \lambda_i$ if $a_i \neq 0$ and $\nu_i = 0$ otherwise, formula (27) follows.

Remark. Relation (29) does not state that $\lambda_i|\mathfrak{h} \cap \mathfrak{k} = 0$ except for those i for which $a_i \neq 0$.

Let $\widetilde{\Lambda} = \boldsymbol{Z}\mu_1 + \cdots + \boldsymbol{Z}\mu_\ell$ and consider the following *partial ordering* in $\widetilde{\Lambda}$:

$$\nu_1 \prec \nu_2 \qquad \text{if } \langle \nu_1, \mu_i \rangle \leq \langle \nu_2, \mu_i \rangle \text{ for all } i.$$

Proposition 9.5. *Fix $\mu \in \Lambda$ and put $\mu^* = -s^*\mu$. Then there exists constants $c_\nu(\nu_o)$ such that*

$$\psi_\mu \psi_{\nu_o} = \sum_{-\mu^* \prec \nu \prec \mu, \nu + \nu_o \in \Lambda} c_\nu(\nu_o) \psi_{\nu + \nu_o} \tag{31}$$

for $\nu_o \in \Lambda$.

Proof. The tensor product $\pi_\mu \otimes \pi_{\nu_o}$ is a spherical representation of U and its irreducible components are also spherical. Since the highest weight of $\pi_\mu \otimes \pi_{\nu_o}$ is the sum of the highest weights of π_μ and π_{ν_o} we have for certain constants $c_\nu(\mu, \nu_o)$

$$\psi_\mu \psi_{\nu_o} = \sum_{\nu \prec \mu + \nu_o} c_\nu(\mu, \nu_o) \psi_\nu.$$

For different $\zeta \in \Lambda$ the restrictions $\psi_\zeta | U$ are orthogonal with respect to the standard inner product $(\, , \,)$ on $L^2(U)$. Thus

$$(32) \qquad (\psi_\mu \psi_{\nu_o}, \psi_\nu) \neq 0 \Longrightarrow \nu \prec \mu + \nu_o.$$

However, $\mathrm{conj}(\psi_\mu) = \psi_{-s^*\mu} = \psi_{\mu^*}$ so $(\psi_\mu \psi_{\nu_o}, \psi_\nu) = \mathrm{conj}(\psi_{\mu^*} \psi_\nu, \psi_{\nu_o})$ so by (32)

$$(\psi_\mu \psi_{\nu_o}, \psi_\nu) \neq 0 \Longrightarrow \nu_o \prec \mu^* + \nu$$

so $-\mu^* + \nu_o \prec \nu$. Replacing ν in the sum above by $\nu + \nu_o$ the result follows.

With μ fixed let $\nu_1, \ldots, \nu_p$ be the elements $\nu \in \widetilde{\Lambda}$ such that $-\mu^* \prec \nu \prec \mu$ with $c_\nu(\nu_o) \neq 0$ for some $\nu_o \in \Lambda$. We recall also from (21) that

$$(33) \qquad \psi_{\nu_o} = \varphi_{i(\nu_o + \rho)} \qquad \text{on } G.$$

The Laplacian L on X satisfies

$$(34) \qquad L\varphi_\lambda = -(\langle \lambda, \lambda \rangle + \langle \rho, \rho \rangle) \varphi_\lambda$$

([GGA], Ch. IV (7)) so

$$L\psi_{\nu_o} = \chi(\nu_o) \psi_{\nu_o}, \qquad \chi(\nu) = \langle \nu, \nu + 2\rho \rangle.$$

We now write (31) in the form

$$(35) \qquad \psi_\mu \psi_{\nu_o} = \sum_{i=1}^{p} c_{\nu_i}(\nu_o) \psi_{\nu_i + \nu_o}.$$

For $i \neq j$ let σ_{ij} denote the affine hyperplane

$$\{\nu \in \mathfrak{a}^* : \ \chi(\nu_i + \nu) = \chi(\nu_j + \nu)\}$$

and Λ' the complement

$$\Lambda' = \{\nu \in \Lambda : \ \nu \notin \sigma_{ij} \text{ for all } i \neq j\}.$$

Proposition 9.6. *There exist rational functions C_{ν_i} on $\mathfrak{a}_{\mathbf{c}}^*$ $(1 \leq i \leq p)$ such that $C_{\nu_i}(\nu_o) = c_{\nu_i}(\nu_o)$ for $\nu_o \in \Lambda'$, $1 \leq i \leq p$.*

Proof. We apply the operators $L^o = I, L^1, \ldots, L^{p-1}$ to (35) and evaluate at e. Then we get p equations for the $c_{\nu_i}(\nu_o)$ with the Vandermonde determinant

$$\prod_{1 \leq k < \ell \leq p} (\chi(\nu_o + \nu_\ell) - \chi(\nu_o + \nu_k)). \tag{36}$$

On the other hand, it is clear from (21), (25) in Ch. III, §2 that $L^j(\psi_\mu \psi_{\nu_o})(e)$ is a polynomial in μ and ν_o. This implies the proposition.

We now restate (35) on the group G by invoking (33). Then since $\varphi_{-\lambda}(g) = \varphi_\lambda(g^{-1})$,

$$\varphi_{-i(\mu+\rho)}\varphi_{-i(\nu_o+\rho)} = \sum_{j=1}^{p} c_{\nu_j}(\nu_o)\varphi_{-i(\nu_j+\nu_o+\rho)}, \qquad \nu_o \in \Lambda'. \tag{37}$$

Here we would like to undertake holomorphic extension in the variable ν_o. For this we consider for a fixed $a \in A^+$ the meromorphic function on $\mathfrak{a}_{\mathbf{c}}^*$,

$$f_1(\lambda) = \varphi_{-i(\mu+\rho)}(a)\varphi_{-i(\lambda+\rho)}(a) - \sum_{j=1}^{p} C_{\nu_j}(\lambda)\varphi_{-i(\nu_j+\lambda+\rho)}(a).$$

For a suitable polynomial $A(\lambda)$ vanishing on each σ_{ij} the function

$$f(\lambda) = e^{-\lambda(\log a)}A(\lambda)f_1(\lambda)$$

is holomorphic on $\mathfrak{a}_{\mathbf{c}}^*$. By (37) and the choice of $A(\lambda)$, we have

$$f(\lambda) = 0 \qquad \text{for } \lambda \in \Lambda. \tag{38}$$

Lemma 9.7. *There exists a polynomial P on $\mathfrak{a}_{\mathbf{c}}^*$ such that*

$$|f(\lambda)| \leq |P(\lambda)| \qquad \textit{for } Re\, \lambda \in \mathfrak{a}_+^*.$$

Proof. We have

$$\varphi_{-i(\nu_j+\lambda+\rho)}(a) = \int_K e^{(\nu_j+\lambda)(H(ak))}dk$$

and (cf. [GGA], Ch. IV, Lemma 6.5)

$$\zeta(\log a) \geq \zeta(H(ak)) \qquad\qquad k \in K, a \in A^+$$

if $\zeta \in \mathfrak{a}_+^*$. Thus the integral is bounded by $\exp((\nu_j + Re\ \lambda)(\log a))$ so the lemma is proved, a being fixed.

In order to conclude $f \equiv 0$ we use the following result of Carlson (Titchmarsh [1939], Ch. V).

Theorem 9.8. *Let $F(z)$ be holomorphic in $\boldsymbol{C}$ such that for some a $(0 < a < \pi)$,*

$$F(z) = \begin{cases} O(e^{a|z|}) & \text{for } Re\ z \geq 0 \\ 0 & \text{for } z \in \boldsymbol{Z}^+ \end{cases}$$

Then $F \equiv 0$.

Taking $F(z) = f(z\mu_1 + \mu)$ the assumptions of Theorem 9.8 are satisfied (Lemma 9.7 and (38)) so $F \equiv 0$. Similarly $f(z_1\mu_1 + z_2\mu_2 + \mu) = 0$ for $z_1, z_2 \in \boldsymbol{C}$, and after ℓ steps we conclude $f \equiv 0$. We have thus proved the following result.

Theorem 9.9. *Fix $\mu \in \Lambda$ and put $\mu^* = -s^*\mu$. Then*

$$(39) \qquad \varphi_{-i\lambda}\varphi_{-i(\mu+\rho)} = a(\lambda)\varphi_{-i(\lambda-\mu^*)} + b(\lambda)\varphi_{-i(\lambda+\mu)} + \sum_{\nu \in I} e_\nu(\lambda)\varphi_{-i(\lambda+\nu)}$$

where $a(\lambda), b(\lambda)$ and $e_\nu(\lambda)$ are rational functions of λ and I is the finite set

$$I = \{\nu \in \mathfrak{a}^* - \{0\} : \ \nu + \mu^*, \mu - \nu \in \Lambda - (0)\}.$$

We shall now use this expansion to establish the relation (24) between the degree of π_μ and the Harish-Chandra $\boldsymbol{c}$-function. In (39) we put $\lambda = \rho + \mu^*$ so the left hand side becomes $\varphi_{-i(\mu^*+\rho)}\varphi_{-i(\mu+\rho)}$ which integrated over U gives $d(\mu)^{-1}$. Here the formula (39) corresponds to the decomposition of $\pi_\mu \otimes \pi_{\mu^*}$ into irreducibles. Integrating the right hand side over U we obtain since $\varphi_{-i\rho} \equiv 1$ and since the $\varphi_{-i(\rho+\mu^*+\mu)}, \varphi_{-i(\rho+\mu^*+\nu)}$ are nonconstant spherical functions,

$$(40) \qquad\qquad d(\mu)^{-1} = a(\rho + \mu^*).$$

On the other hand, we know ([GGA], Ch. IV, Theorem 6.14) for the asymptotic behavior of the spherical function on $\mathfrak{a}^+$

$$\varphi_\lambda \sim \mathbf{c}(\lambda)e^{i\lambda-\rho} \qquad \text{if } Re(\langle i\lambda, \alpha\rangle) > 0 \quad \text{for } \alpha \in \Sigma^+.$$

We can now compare the asymptotic behavior on $\mathfrak{a}^+$ of both sides of (39). Assuming $Re(\langle\lambda,\alpha\rangle) > 0$ ($\alpha \in \Sigma^+$) we obtain (the condition $Re(\langle\mu+\rho,\alpha\rangle) > 0$ ($\alpha \in \Sigma^+$) being automatic)

$$\mathbf{c}(-i\lambda)\mathbf{c}(-i(\mu+\rho)) = b(\lambda)\mathbf{c}(-i(\lambda+\mu)). \tag{41}$$

Replacing in (39) λ by $s^*\lambda$ we get from the W-invariance of φ_λ in λ, $a(s^*\lambda) = b(\lambda)$. Now the desired formula (24) follows from (40) and (41).

Thus we have proved the following result.

Theorem 9.10. *The Plancherel measure for the simply connected symmetric space U/K is given by $d(\mu)dk_M$ (Theorem 9.3) where $d(\mu)$, the dimension of the representation with highest restricted weight μ, is given by*

$$d(\mu) = \left\{\frac{\mathbf{c}(\lambda+i\mu)\mathbf{c}(-\lambda-i\mu)}{\mathbf{c}(\lambda)\mathbf{c}(-\lambda)}\right\}_{\lambda=-i(\mu+\rho)}.$$

Example. Let us work out (24) in the case $G = \mathbf{SL}(2,\mathbf{R})$. Let

$$t = \frac{\langle i\lambda,\alpha\rangle}{\langle\alpha,\alpha\rangle}, \qquad n = \frac{\langle\mu,\alpha\rangle}{\langle\alpha,\alpha\rangle}, \qquad \Sigma^+ = (\alpha).$$

Since $i\lambda = \mu+\rho$ we evaluate the expression (24) for $t = n+\frac{1}{2}$. We have by [GGA], Ch. IV, (12) and Theorem 6.14

$$\mathbf{c}(\lambda) = \frac{1}{\pi^{\frac{1}{2}}}\frac{\Gamma(t)}{\Gamma(t+\frac{1}{2})}$$

so

$$\frac{\mathbf{c}(\lambda+i\mu)\mathbf{c}(-\lambda-i\mu)}{\mathbf{c}(\lambda)\mathbf{c}(-\lambda)} = \frac{\Gamma(t-n)\Gamma(n-t)\Gamma(\frac{1}{2}+t)\Gamma(\frac{1}{2}-t)}{\Gamma(\frac{1}{2}+t-n)\Gamma(\frac{1}{2}-t+n)\Gamma(t)\Gamma(-t)},$$

which as $t \longrightarrow n+\frac{1}{2}$ has limit

$$\begin{aligned} &\frac{\Gamma(\frac{1}{2})\Gamma(-\frac{1}{2})\Gamma(n+1)}{\Gamma(n+\frac{1}{2})\Gamma(-n-\frac{1}{2})\Gamma(1)} \lim_{t\to n+\frac{1}{2}} \frac{\Gamma(\frac{1}{2}-t)}{\Gamma(n-t+\frac{1}{2})} \\ &= \frac{-2\pi\; n!}{-\pi(n+\frac{1}{2})^{-1}\cos(n\pi)} \lim_{z\longrightarrow -n} \frac{\Gamma(z)}{\Gamma(z+n)} = 2n+1. \end{aligned}$$

This number $2n+1$ is just the dimension of the space H_n of spherical harmonics on $\boldsymbol{S}^2$ of degree n ([GGA], Introduction, Exercise A5) and the natural representation of H_n has highest weight $n\alpha$ ([GGA], Ch. V, Exercise A11).

§10. ELEMENTS OF $D(G/K)$ AS FRACTIONS.

We shall now use the Fourier transform on the noncompact symmetric space $X = G/K$ to relate the center $\boldsymbol{Z}(G)$ of the universal enveloping algebra $\boldsymbol{D}(G)$ to the algebra $\boldsymbol{D}(G/K)$ of G-invariant differential operators on X. With $\boldsymbol{D}_K(G)$ denoting the algebra of operators $D \in \boldsymbol{D}(G)$ which are right invariant under K we have considered the surjective homomorphism μ : $\boldsymbol{D}_K(G) \longrightarrow \boldsymbol{D}(G/K)$ given by $(\mu(D)f)\circ\pi = D(f\circ\pi)$, π denoting the natural map of G onto G/K and $f \in \mathcal{E}(X)$. At the same time we have a homomorphism $\nu : \boldsymbol{D}(G) \longrightarrow \boldsymbol{E}(X)$ into the algebra $\boldsymbol{E}(X)$ of all differential operators on X. This ν is just the differential of the natural representation of G on $\mathcal{E}(X)$ (cf. §3, No. 1). We know from [GGA], Ch. II, Cor. 3.1, that

$$(1) \qquad \mu(D) = \nu(D^*) \qquad D \in \boldsymbol{Z}(G),$$

D^* denoting the adjoint of D. Let $\boldsymbol{Z}(G/K)$ denote the image of $\boldsymbol{Z}(G)$ under μ (and ν). As we shall see later, $\boldsymbol{Z}(G/K) \neq \boldsymbol{D}(G/K)$ in general. However we shall prove the following general result.

Theorem 10.1. *Given $D \in \boldsymbol{D}(G/K)$ there exist $Z_1 \neq 0, Z_2$ in $\boldsymbol{Z}(G/K)$ such that*

$$DZ_1 = Z_2.$$

As usual let $\mathfrak{a} \subset \mathfrak{p}$ be a maximal abelian subspace, $\mathfrak{h}$ a Cartan subalgebra of $\mathfrak{g}$ containing $\mathfrak{a}$, and $\mathfrak{a}^{\boldsymbol{C}}$, $\mathfrak{h}^{\boldsymbol{C}}$, $\mathfrak{g}^{\boldsymbol{C}}$ the corresponding complexifications. Consider the root systems $\Sigma = \Sigma(\mathfrak{g},\mathfrak{a})$, $\Delta = \Delta(\mathfrak{g}^{\boldsymbol{C}},\mathfrak{h}^{\boldsymbol{C}})$ and let W and $\widetilde{W}$ denote the corresponding Weyl groups. The space $\mathfrak{h}_{\boldsymbol{R}}$ of $H \in \mathfrak{h}^{\boldsymbol{C}}$ on which all $\alpha \in \Delta$ are real contains $\mathfrak{a}$. Using the orthogonal decomposition $\mathfrak{a} + i(\mathfrak{h} \cap \mathfrak{k}) = \mathfrak{h}_{\boldsymbol{R}}$, the dual $\mathfrak{a}^*$ can be viewed as a subset of $\mathfrak{h}^*_{\boldsymbol{R}}$. We select orderings in $\mathfrak{a}^*$ and $\mathfrak{h}^*_{\boldsymbol{R}}$ which are compatible in the sense that $\lambda \in \mathfrak{h}^*_{\boldsymbol{R}}$ is > 0 whenever $\overline{\lambda} > 0$, the bar denoting restriction to $\mathfrak{a}$. The set Δ^+ of positive roots is then divided into two classes

$$(2) \qquad P = \{\alpha \in \Delta^+ : \overline{\alpha} \neq 0\}, \; P_0 = \{\alpha \in \Delta^+ : \overline{\alpha} = 0\}$$

and then $\Sigma^+ = \Sigma^+(\mathfrak{g},\mathfrak{a})$ equals the set $\overline{P}$ of restrictions to $\mathfrak{a}$. Let

$$(3) \qquad \rho = \tfrac{1}{2}\sum_{\alpha \in P} \alpha, \quad \rho_o = \tfrac{1}{2}\sum_{\alpha \in P_o} \alpha, \quad \rho^* = \tfrac{1}{2}\sum_{\alpha \in \Delta^+} \alpha = \rho_o + \rho.$$

Since $\alpha \longrightarrow -\theta\alpha$ is a permutation of P we see that

$$\rho|(\mathfrak{h} \cap \mathfrak{k}) \equiv 0, \qquad \rho_o|\mathfrak{a} \equiv 0. \tag{4}$$

The group W acts on $\mathfrak{a}, \mathfrak{a}^{\boldsymbol{C}}$, the duals $\mathfrak{a}^*, (\mathfrak{a}^{\boldsymbol{C}})^*$ and the symmetric algebras $S(\mathfrak{a}), \ldots, S((\mathfrak{a}^{\boldsymbol{C}})^*)$ resulting in corresponding algebras $I(\mathfrak{a}), \ldots, I((\mathfrak{a}^{\boldsymbol{C}})^*)$ of W-invariants. Similar notions pertain to $\widetilde{W}$ acting on $\mathfrak{h}_{\boldsymbol{R}}$ and $\mathfrak{h}^{\boldsymbol{C}}$. We recall the following relationships between W and $\widetilde{W}$ (cf. [DS], Ch. VII, §8):

(i) W is the set of linear transformations of $\mathfrak{a}$ induced by the subgroup of $\widetilde{W}$ leaving $\mathfrak{a}$ invariant.

(ii) If two elements $H_1, H_2 \in \mathfrak{a}^{\boldsymbol{C}}$ are conjugate under $\widetilde{W}$ they are conjugate under W.

We also have to consider the translation of $\widetilde{W}$ by ρ_o, that is the group $\widetilde{W}_o$ of transformations

$$\mu \longrightarrow s\mu + s\rho_o - \rho_o \qquad \mu \in (\mathfrak{h}^{\boldsymbol{C}})^* \tag{5}$$

as $s \in \widetilde{W}$. Then the corresponding algebra $I_o(\mathfrak{h}^{\boldsymbol{C}})$ of invariants consists of the functions $\mu \longrightarrow P(\rho_o + \mu)$ on $(\mathfrak{h}^{\boldsymbol{C}})^*$ as P runs through $I(\mathfrak{h}^{\boldsymbol{C}})$.

Lemma 10.2. *The following properties of G/K are equivalent:*

(i) $\boldsymbol{Z}(G/K) = \boldsymbol{D}(G/K)$

(ii) The restriction from $\mathfrak{h}^*_{\boldsymbol{R}}$ to the subspace $\mathfrak{a}^*$ maps $I_o(\mathfrak{h}^{\boldsymbol{C}})$ onto $I(\mathfrak{a}^{\boldsymbol{C}})$.

Proof. Let $\widetilde{\mathfrak{n}}_+$ and $\widetilde{\mathfrak{n}}_-$ denote the sums of the positive and negative root spaces, respectively,

$$\widetilde{\mathfrak{n}}_+ = \sum_{\alpha\in\Delta_+} (\mathfrak{g}^{\boldsymbol{C}})^\alpha, \qquad \widetilde{\mathfrak{n}}_- = \sum_{-\alpha\in\Delta+} (\mathfrak{g}^{\boldsymbol{C}})^\alpha. \tag{6}$$

Choosing $X_\alpha \neq 0$ in each $(\mathfrak{g}^{\boldsymbol{C}})^\alpha$ and a basis $H_1, \ldots, H_r$ of $\mathfrak{h}^{\boldsymbol{C}}$ each $D \in \boldsymbol{D}(G)$ can be uniquely written as a linear combination of monomials

$$X_{-\alpha_1}^{m_1} \ldots X_{-\alpha_p}^{m_p} H_1^{n_1} \ldots H_r^{n_r} X_{\alpha_1}^{q_1} \ldots X_{\alpha_p}^{q_p} \tag{7}$$

([DS], Ch. II, Cor. 1.10). If $[D, H] = 0$ for each $H \in \mathfrak{h}^{\boldsymbol{C}}$ then each term (7) appearing satisfies $\sum_1^p m_i\alpha_i = \sum_1^p q_i\alpha_i$. This means that there exists a unique $\gamma'(D) \in S(\mathfrak{h})^{\boldsymbol{C}}$ such that

$$D - \gamma'(D) \in \widetilde{\mathfrak{n}}_-\boldsymbol{D}(G) + \boldsymbol{D}(G)\widetilde{\mathfrak{n}}_+ \tag{8}$$

and that if $D = Z \in \boldsymbol{Z}(G)$ then

$$Z - \gamma'(Z) \in \sum_{\alpha \in \Delta^+} \boldsymbol{D}(G)(\mathfrak{g}^{\boldsymbol{C}})^{\alpha}. \tag{9}$$

If $P \longrightarrow P^o$ is the automorphism of $S(\mathfrak{h}^{\boldsymbol{C}})$ given by $H^o = H - \rho^*(H)$ and we put $\gamma(Z) = (\gamma'(Z))^o$ then, by a theorem of Harish-Chandra, $Z \longrightarrow \gamma(Z)$ is an isomorphism of $\boldsymbol{Z}(G)$ onto $I(\mathfrak{h}^{\boldsymbol{C}})$ (cf. Harish-Chandra [1951]; for a shorter proof see Wallach [1988], Theorem 3.2.3; compare also [GGA], Ch. V, Theorem 1.9 where γ is defined differently although it can be shown to be the same as above (cf. Exercise **A3**)).

As usual, if $g \in G$ let $H(g) \in \mathfrak{a}$ be determined by $g \in k \exp H(g) N$ and consider the function

$$f(g) = e^{(i\lambda - \rho)(H(g))} \qquad (g \in G) \tag{10}$$

$\lambda \in \mathfrak{a}_{\boldsymbol{c}}^*$ being fixed. Then f is right invariant under MN so since $\tilde{\mathfrak{n}}_+ \subset \mathfrak{m}^{\boldsymbol{C}} + \mathfrak{n}^{\boldsymbol{C}}$, (9) implies

$$Zf = \gamma'(Z)f. \tag{11}$$

Next we note that $f(g \exp H) = e^{(i\lambda-\rho)(H_1)} f(g)$ where H_1 is the $\mathfrak{a}$-component of H. We have by (4)

$$(i\lambda - \rho)(H_1) = (i\lambda - \rho)(H) = (i\lambda - \rho^* + \rho_o)(H)$$

so

$$(Hf)(g) = (i\lambda - \rho^* + \rho_o)(H) f(g). \tag{12}$$

Viewing $P \in S(\mathfrak{h}^{\boldsymbol{C}})$ as a polynomial function on $(\mathfrak{h}^{\boldsymbol{C}})^*$ we have $P^o = P \circ T(-\rho^*)$ where $T(\mu)$ denotes translation by μ. Hence by (11) and (12)

$$(Zf)(g) = \gamma'(Z)(i\lambda - \rho^* + \rho_o) f(g) = \gamma(Z)(i\lambda + \rho_o) f(g). \tag{13}$$

Consider now the spherical function

$$\varphi_\lambda(g) = \varphi_\lambda(gK) = \int_K e^{(i\lambda-\rho)(H(gk))} dk. \tag{14}$$

Since Z commutes with right translations (13) implies

$$Z\varphi_\lambda = \gamma(Z)(i\lambda + \rho_o)\varphi_\lambda. \tag{15}$$

On the other hand the isomorphism Γ of $\boldsymbol{D}(G/K)$ onto $I(\mathfrak{a}^{\boldsymbol{C}})$ satisfies

(16) $$D\varphi_\lambda = \Gamma(D)(i\lambda)\varphi_\lambda.$$

From (15) and (16) we deduce that if $\boldsymbol{Z}(G/K) = \boldsymbol{D}(G/K)$ then $I_o(\mathfrak{h}^{\boldsymbol{C}})^- = I(\mathfrak{a}^{\boldsymbol{C}})$ the bar denoting restriction.

Conversely, we assume the surjectivity in (ii) and let $D \in \boldsymbol{D}(G/K)$. Then there exists a $Q \in I_o(\mathfrak{h}^{\boldsymbol{C}})$ such that $Q(i\lambda) = \Gamma(D)(i\lambda)$ for all $\lambda \in \mathfrak{a}^*$, in other words $P(\rho_o + i\lambda) = \Gamma(D)(i\lambda)$ for some $P \in I(\mathfrak{h}^{\boldsymbol{C}})$. By the surjectivity of γ, there exists a $\widetilde{Z} \in \boldsymbol{Z}(G)$ such that $\gamma(\widetilde{Z}) = P$ so by (15) $Z = \mu(\widetilde{Z})$ satisfies

(17) $$Z\varphi_\lambda = P(\rho_o + i\lambda)\varphi_\lambda = \Gamma(D)(i\lambda)\varphi_\lambda = D\varphi_\lambda.$$

But then $\boldsymbol{Z}$ and D have the same eigenvalue for the function $x \longrightarrow e^{(i\lambda+\rho)(A(x,b))}$ so by the inversion formula (Theorem 1.3) $\mu(\widetilde{Z})$ and D coincide.

Passing to the proof of the theorem consider the automorphism $\tau : Q \longrightarrow Q \circ T(-\rho_o)$ of $S(\mathfrak{h}^{\boldsymbol{C}})$ which maps $I_o(\mathfrak{h}^{\boldsymbol{C}})$ onto $I(\mathfrak{h}^{\boldsymbol{C}})$. Consider the commutative diagram

$$\begin{array}{ccc} I_0(\mathfrak{h}^{\boldsymbol{C}}) & \xrightarrow{\ \tau\ } & I(\mathfrak{h}^{\boldsymbol{C}}) \\ {\scriptstyle\sigma}\big\downarrow & & \big\downarrow{\scriptstyle\zeta} \\ I(\mathfrak{a}^{\boldsymbol{C}}) & \xrightarrow{\ I\ } & I(\mathfrak{a}^{\boldsymbol{C}}) \end{array}$$

where $\sigma(Q) = Q|\mathfrak{a}^*$, $\zeta(P) = (P \circ T(\rho_o))|\mathfrak{a}^*$, and I is the identity operator. The surjectivity of $\gamma : \boldsymbol{Z}(G) \longrightarrow I(\mathfrak{h}^{\boldsymbol{C}})$ and (15) show that σ and ζ do map the respective spaces $I_o(\mathfrak{h}^{\boldsymbol{C}})$, $I(\mathfrak{h}^{\boldsymbol{C}})$ into $I(\mathfrak{a}^{\boldsymbol{C}})$. Again using the Fourier transform and (15)–(16) we see that the operators DZ_1 and Z_2 ($D \in \boldsymbol{D}(G/K)$, $Z_1, Z_2 \in \boldsymbol{Z}(G/K)$) coincide if and only if

$$\Gamma(D)(i\lambda)\gamma(Z_1)(\rho_o + i\lambda) = \gamma(Z_2)(\rho_o + i\lambda),$$

i.e.

(18) $$\Gamma(D)\zeta(\gamma(Z_1)) = \zeta(\gamma(Z_2)).$$

Thus the problem is to prove that the image $\zeta(I(\mathfrak{h}^{\boldsymbol{C}}))$ has quotient field $\boldsymbol{C}(\zeta(I(\mathfrak{h}^{\boldsymbol{C}})))$ equal to $\boldsymbol{C}(I(\mathfrak{a}^{\boldsymbol{C}}))$. By the commutativity of the diagram above this amounts to proving that $J_o = \sigma(I_o(\mathfrak{h}^{\boldsymbol{C}}))$ has quotient field $\boldsymbol{C}(J_o)$ equal to $\boldsymbol{C}(I(\mathfrak{a}^{\boldsymbol{C}}))$.

The ring $S(\mathfrak{h}^{\boldsymbol{C}})$ is integral over $I(\mathfrak{h}^{\boldsymbol{C}})$ ([GGA], Ch. III, §3 (1)) so since τ^{-1} is an automorphism, $S(\mathfrak{h}^{\boldsymbol{C}})$ is also integral over $I_o(\mathfrak{h}^{\boldsymbol{C}})$. Applying the homomorphism $Q \longrightarrow Q|\mathfrak{a}^*$ we see that $S(\mathfrak{a}^{\boldsymbol{C}})$ is integral over J_o.

Suppose now $\boldsymbol{C}(J_o)$ were a proper subset of $\boldsymbol{C}(I(\mathfrak{a}^{\boldsymbol{C}}))$. Then we select $R \in I(\mathfrak{a}^{\boldsymbol{C}})$, $R \notin \boldsymbol{C}(J_o)$. By the above, R is algebraic over $\boldsymbol{C}(J_o)$. Because of the homomorphism $\sigma \circ \tau^{-1} : I(\mathfrak{h}^{\boldsymbol{C}}) \longrightarrow J_o$, the ring J_o has finitely many generators, say $\xi_1, \ldots, \xi_r$. Let

$$p_R(x) = x^n + f_1(\xi)x^{n-1} + \cdots + f_n(\xi)$$

be a polynomial with coefficients in $\boldsymbol{C}(J_o) = \boldsymbol{C}(\xi_1, \ldots, \xi_r)$ of lowest degree and leading coefficient 1 having R as a root. By the assumption, $n \geq 2$. Let $f(\xi)$ be the product of the denominators of the $f_i(\xi)$ and let $q(\xi)$ denote the product of $f(\xi)$ with the discriminant of the polynomial $x \longrightarrow f(\xi)p_R(x)$. Then $q(\xi)$ is a polynomial function on $(\mathfrak{a}^{\boldsymbol{C}})^*$ which is $\not\equiv 0$.

Select $\mu^* \in \mathfrak{a}^*$ such that $\lambda^* = i\mu^*$ satisfies

$$q(\xi)(\lambda^*) \neq 0, \tag{19}$$

and consider the homomorphism

$$\chi : j \longrightarrow j(\lambda^*) \qquad \text{of } J_o \quad \text{into } \boldsymbol{C}. \tag{20}$$

The image of $p_R(x)$ under χ is a polynomial with coefficients in $\boldsymbol{C}$ and discriminant $\neq 0$ having n distinct roots $\alpha_1, \ldots, \alpha_n \in \boldsymbol{C}$. For each i ($1 \leq i \leq n$) we wish to extend χ to a homomorphism $\chi_i : \boldsymbol{C}[\xi_1, \ldots, \xi_r, R] \longrightarrow \boldsymbol{C}$ by putting $\chi_i(R) = \alpha_i$. To see that this is possible, suppose $p(\xi, x)$ is a polynomial in x with coefficients in $J_o = \boldsymbol{C}[\xi_1, \ldots, \xi_r]$ having R as a root. Then $p_R(x)$ divides $p(\xi, x) : p(\xi, x) = p_R(x)q(\xi, x)$. The coefficients of $q(\xi, x)$ will be rational expressions in $\xi_1, \ldots, \xi_r$ with denominators dividing $f(\xi)$. By (19), $\chi(f(\xi)) \neq 0$ so the image of $p(\xi, x)$ under χ has $x = \alpha_i$ as a zero. Thus χ_i can indeed be defined by

$$\chi_i(t(\xi_1, \ldots, \xi_r, R)) = t(\chi(\xi_1), \ldots, \chi(\xi_r), \alpha_i)$$

t being an arbitrary polynomial in $r+1$ variables. Since $S(\mathfrak{a}^{\boldsymbol{C}})$ is integral over J_o and hence over $\boldsymbol{C}[\xi_1, \ldots, \xi_r, R]$ there exists a homomorphism $\Lambda_i : S(\mathfrak{a}^{\boldsymbol{C}}) \longrightarrow \boldsymbol{C}$ extending χ_i. This is given by evaluation at a point $\lambda_i \in (\mathfrak{a}^{\boldsymbol{C}})^*$, i.e. $\Lambda_i(P) = P(\lambda_i)$ for $P \in S(\mathfrak{a}^{\boldsymbol{C}})$. In particular, if $Q \in I_o(\mathfrak{h}^{\boldsymbol{C}})$ then by (20), $Q(\lambda^*) = \chi(\sigma(Q))$ and this must equal both $\sigma(Q)(\lambda_1) = Q(\lambda_1)$ and $\sigma(Q)(\lambda_2) = Q(\lambda_2)$. Hence

$$P(\rho_o + i\mu^*) = P(\rho_o + \lambda_1) = P(\rho_o + \lambda_2), \qquad P \in I(\mathfrak{h}^{\boldsymbol{C}}).$$

This implies that the elements $\rho_o + i\mu^*$, $\rho_o + \lambda_1$, and $\rho_o + \lambda_2$ are conjugate under $\widetilde{W}$. Writing $\lambda_j = \nu_j + i\mu_j$ ($\mu_j, \nu_j \in \mathfrak{a}^*, j = 1, 2$) we obtain for some $s_1, s_2 \in \widetilde{W}$,

$$s_1(\rho_o + i\mu^*) = \rho_o + \nu_1 + i\mu_1, \qquad s_2(\rho_o + i\mu^*) = \rho_o + \nu_2 + i\mu_2. \tag{21}$$

Since $\widetilde{W}$ preserves $(\mathfrak{h}_{\boldsymbol{R}})^*$ (21) implies $s_j\rho_o = \rho_o + \nu_j$, $s_j\mu^* = \mu_j$ so, since $\nu_j \perp \rho_o$, $\nu_1 = \nu_2 = 0$ and $s_2 s_1^{-1}\lambda_1 = \lambda_2$. By the property (ii) above relating $\widetilde{W}$ and W we have that λ_1 and λ_2 are conjugate under W. Thus

$$\alpha_1 = \chi_1(R) = R(\lambda_1) = R(\lambda_2) = \chi_2(R) = \alpha_2,$$

contradicting $\alpha_1 \neq \alpha_2$. This proves the theorem.

Concerning the question of the equality $\boldsymbol{Z}(G/K) = \boldsymbol{D}(G/K)$ we have the following result from Helgason [1992a]. Using classification and a lemma of D. Vogan one proves that $I_o(\mathfrak{h}^{\boldsymbol{C}})^- = I(\mathfrak{a}^{\boldsymbol{C}})$ if and only if $I(\mathfrak{h}^{\boldsymbol{C}})^- = I(\mathfrak{a}^{\boldsymbol{C}})$. Then one verifies that this last relation fails only for the four cases (22) so Lemma 10.2 yields the result.

Theorem 10.3. *Assume G/K irreducible. Then $\boldsymbol{Z}(G/K) = \boldsymbol{D}(G/K)$ if and only if $(\mathfrak{g}, \mathfrak{k})$ is not one of the pairs*

(22) $(\mathfrak{e}_6, \mathfrak{so}(10) + \boldsymbol{R}), \qquad (\mathfrak{e}_6, \mathfrak{f}_4), \quad (\mathfrak{e}_7, \mathfrak{e}_6 + \boldsymbol{R}), \quad (\mathfrak{e}_8, \mathfrak{e}_7 + \mathfrak{su}(2)).$

§11. THE RANK ONE CASE.

In this section we assume that our noncompact symmetric space G/K has rank one. We can then work out our harmonic analysis theory from §§1,5 in more explicit and concrete form.

1. An Explicit Formula for the Eisenstein Integral.

Let $\delta \in \widehat{K}_M$ and let $v_1, \ldots, v_{d(\delta)}$ be an orthonormal basis of V_δ such that $v = v_1$ spans V_δ^M (which as we have seen is one-dimensional).

Lemma 11.1. *For $\lambda \in \mathfrak{a}_c^*$ let*

$$f_{\delta,j}(x) = \int_K e^{(i\lambda+\rho)(A(x,kM))} \langle \delta(k)v, v_j \rangle dk, \qquad 1 \leq j \leq d(\delta),$$

and

$$\varphi_{\lambda,\delta}(x) = \int_K e^{(i\lambda+\rho)(A(x,kM))} \langle \delta(k)v, v \rangle \, dk. \tag{1}$$

Then

$$f_{\delta,j}(ka\cdot o) = \langle \delta(k)v, v_j\rangle\, \varphi_{\lambda,\delta}(a\cdot o), \qquad k\in K,\, a\in A. \tag{2}$$

Proof. Define $F:\ X \longrightarrow V_\delta$ by

$$F(x) = \int_K e^{(i\lambda+\rho)(A(x,kM))}\delta(k)v\,dk = \Phi_{\lambda,\delta}(x)\,v.$$

Then

$$f_{\delta,j}(x) = \langle F(x),\, v_j\rangle$$

and since $F(k\cdot x) = \delta(k)F(x)$ we have $F(a\cdot o)\in V_\delta^M$. Since $\dim V_\delta^M = 1$ we deduce

$$F(a\cdot o) = \varphi_{\lambda,\delta}(a\cdot o)\,v.$$

Calculating $\langle F(ka\cdot o),\, v_j\rangle$ we derive (2).

We shall now determine $\varphi_{\lambda,\delta}$ quite explicitly in terms of the hypergeometric function $F(a,b;c;x)$.

Let α be the single root in Σ_o^+, determine $H\in\mathfrak{a}^+$ by $\alpha(H)=1$ and put $\ell = (i\lambda-\rho)(H)$, $a_t = \exp tH$. Consider the elements $\omega_\alpha, \omega_{2\alpha}\in \boldsymbol{D}(K)$ defined by (28) §2.

Theorem 11.2. *To each $\delta\in\widehat{K}_M$ we associate unique numbers $r\geq 0$, $s\geq 0$ such that*

$$\delta(\omega_\alpha)|V_\delta^M = \{r(r+m_{2\alpha}-1) - s(s+m_\alpha+m_{2\alpha}-1)\}\,[2(m_\alpha+4m_{2\alpha})]^{-1}, \tag{3}$$

$$\delta(\omega_{2\alpha})|V_\delta^M = -2r(r+m_{2\alpha}-1)(m_\alpha+4m_{2\alpha})^{-1} \tag{4}$$

(with the understanding that $r=0$ if $m_{2\alpha}=0$). Then r and s are integers, $r\leq s$, and the function $\varphi_{\lambda,\delta}$ is given by

$$\varphi_{\lambda,\delta}(a_t\cdot o) = c_{\lambda,\delta}\tanh^s t\cosh^\ell t\times \tag{5}$$
$$F(\tfrac12(s+r-\ell), \tfrac12(s-r-\ell+1-m_{2\alpha});\, s+\tfrac12(m_\alpha+m_{2\alpha}+1); \tanh^2 t),$$

where the constant $c_{\lambda,\delta}$ is

$$c_{\lambda,\delta} = p_{r,s}(\lambda)q_{r,s}(\lambda)c_s \tag{6}$$

and $p_{r,s}$ and $q_{r,s}$ are given by

$$(7)\qquad p_{r,s}(\lambda)=\frac{\Gamma(\frac{1}{2}(\langle i\lambda+\rho,\alpha_o\rangle+s+r))}{\Gamma(\frac{1}{2}(\langle i\lambda+\rho,\alpha_o\rangle))},\qquad \alpha_o=\alpha/\langle\alpha,\alpha\rangle,$$

$$(8)\qquad q_{r,s}(\lambda)=\frac{\Gamma(\frac{1}{2}(\langle i\lambda+\rho,\alpha_o\rangle+1-m_{2\alpha}+s-r))}{\Gamma(\frac{1}{2}(\langle i\lambda+\rho,\alpha_o\rangle+1-m_{2\alpha}))},$$

and

$$c_s=\frac{\Gamma(\frac{1}{2}(m_\alpha+m_{2\alpha}+1))}{\Gamma(s+\frac{1}{2}(m_\alpha+m_{2\alpha}+1))}.$$

If $m_{2\alpha}>0$ then $p_{r,s}$ and $q_{r,s}$ are polynomials.

If $m_{2\alpha}=0$ then $c_{\lambda,\delta}$ is the polynomial

$$(9)\qquad c_{\lambda,\delta}=\frac{\Gamma(\langle i\lambda+\rho,\alpha_o\rangle+s)}{\Gamma(\langle i\lambda+\rho,\alpha_o\rangle)}\frac{c_s}{2^s}.$$

Proof. We have

$$B(H,H)=2\sum_{\beta\in\Sigma^+}m_\beta\beta(H)^2=2(m_\alpha+4m_{2\alpha})$$

so, writing $A_\lambda=\langle\lambda,\lambda\rangle H/\lambda(H)$, we deduce

$$\lambda(H)^2=2(m_\alpha+4m_{2\alpha})\langle\lambda,\lambda\rangle,\qquad A_\alpha=\tfrac{1}{2}(m_\alpha+4m_{2\alpha})^{-1}H,$$

whence

$$\begin{aligned}\langle\lambda,\lambda\rangle+\langle\rho,\rho\rangle&=\tfrac{1}{2}(m_\alpha+4m_{2\alpha})^{-1}(\lambda(H)^2+\rho(H)^2)\\&=-\tfrac{1}{2}(m_\alpha+4m_{2\alpha})^{-1}\ell(\ell+m_\alpha+2m_{2\alpha}).\end{aligned}$$

Now if $F\in\mathcal{E}(A)$

$$(HF)(\exp tH)=\frac{d}{dt}F(\exp tH)$$

so in the coordinate system $\exp tH\longrightarrow t$, H is identified with d/dt. Thus the Laplacian L_A satisfies

$$2(m_\alpha+4m_{2\alpha})L_A=\frac{d^2}{dt^2},$$

and

$$2(m_\alpha + 4m_{2\alpha})A_\alpha = \frac{d}{dt}.$$

The differential equation for the restriction $\overline{\Phi}_{\lambda,\delta} = \Phi_{\lambda,\delta}|A \cdot o$ in §2 thus reduces to the following equation for the function $\varphi(t) = \varphi_{\lambda,\delta}(\exp tH \cdot o)$:

$$\begin{aligned}&\varphi''(t) + (m_\alpha \coth t + 2m_{2\alpha} \coth 2t)\varphi'(t)\\ &+ [d_\alpha \operatorname{sh}^{-2} t + d_{2\alpha} \operatorname{sh}^{-2}(2t) - \ell(\ell + m_\alpha + 2m_{2\alpha})]\varphi = 0.\end{aligned}$$

Here the constants d_α and $d_{2\alpha}$ are determined by

$$\begin{aligned}\delta(\omega_\alpha)|V_\delta^M &= d_\alpha(2(m_\alpha + 4m_{2\alpha}))^{-1},\\ \delta(\omega_{2\alpha})|V_\delta^M &= d_{2\alpha}(2(m_\alpha + 4m_{2\alpha}))^{-1}.\end{aligned}$$

The equations for r and s are

$$\begin{aligned}r(r + m_{2\alpha} - 1) &= -\tfrac{1}{4}d_{2\alpha}\\ s(s + m_\alpha + m_{2\alpha} - 1) &= -d_\alpha - \tfrac{1}{4}d_{2\alpha}\end{aligned}$$

and clearly d_α and $d_{2\alpha}$ are ≤ 0. Under the stipulation $r = 0$ if $m_{2\alpha} = 0$, r and s are uniquely determined under the condition $r \geq 0$, $s \geq 0$. Putting temporarily $p = m_{2\alpha}$, $q = m_\alpha + m_{2\alpha}$ the equation for φ can be written

$$\text{(10)}\qquad \begin{aligned}&\operatorname{ch}^{-p} t \operatorname{sh}^{-q} t \tfrac{d}{dt}\left(\operatorname{ch}^p t \operatorname{sh}^q t \tfrac{d\varphi}{dt}\right)\\ &+ \left[\frac{r(r+p-1)}{\operatorname{ch}^2 t} - \frac{s(s+q-1)}{\operatorname{sh}^2 t} - \ell(\ell + p + q)\right]\varphi = 0.\end{aligned}$$

According to Vilenkin [1968], p. 543 the following functions are solutions to the equation

$$\begin{aligned}u_1 = {} & \operatorname{th}^s t \operatorname{ch}^\ell t\, F(\tfrac{1}{2}(s - \ell + r),\ \tfrac{1}{2}(s - r - \ell + 1 - p);\ s + \tfrac{1}{2}(q+1);\ \operatorname{th}^2 t),\\ u_2 = {} & \operatorname{cth}^{q+s-1} t \operatorname{ch}^\ell t \times\\ & F(\tfrac{1}{2}(r - \ell - s - q + 1),\ 1 - \tfrac{1}{2}(p + s + q + \ell - r);\ -s - \tfrac{1}{2}(q - 3);\ \operatorname{th}^2 t).\end{aligned}$$

If $s+q-1 > 0$ the second solution is singular at $t = 0$ so we conclude formula (5) where $c_{\lambda,\delta}$ is a constant. By Theorem 3.16 in Chapter II we have for $\operatorname{Re}(\langle i\lambda, \alpha\rangle) > 0$,

$$\text{(11)}\qquad e^{(-i\lambda+\rho)(tH)}\varphi_{\lambda,\delta}(a_t \cdot o) \longrightarrow \mathbf{c}(\lambda) \qquad \text{as} \qquad t \longrightarrow +\infty.$$

Since we know $\boldsymbol{c}(\lambda)$ and since the value of $F(a,b;c;x)$ at $x=1$ is given by

$$(12)\qquad F(a,b;c;1)=\frac{\Gamma(c)\Gamma(c-a-b)}{\Gamma(c-a)\Gamma(c-b)},\qquad -c\notin \boldsymbol{Z}^+,\ \mathrm{Re}(c)>\mathrm{Re}(a+b)$$

it is easy to derive formulas (6) and (9) for $c_{\lambda,\delta}$. So far they are proved for $Re(\langle i\lambda,\alpha\rangle)>0$.

It remains to consider the case $s+q-1=0$. Then $s=0$, $m_\alpha=1$, $m_{2\alpha}=0$ and $r=0$ and δ is the identity representation. The right hand side of (5) reduces to

$$\mathrm{ch}^\ell t F(-\tfrac{1}{2}\ell,\tfrac{1}{2}(1-\ell);1;\ \mathrm{th}^2 t),$$

which by the transformation formula

$$F(a,b;c;z)=(1-z)^{-a}F(c-b,a;c;z/(z-1))$$

coincides with

$$F(\tfrac{1}{2}(\ell+1),-\tfrac{1}{2}\ell;1,-\mathrm{sh}^2 t).$$

However, this is just the zonal spherical function for the hyperbolic plane ([GGA], Ch. IV, Exercise B8). Thus (5) holds in general.

Note that $\frac{1}{2}(1-m_{2\alpha})$ in (6) is an integer except in the case when $m_{2\alpha}=0$ and then $r=0$. In this case we can simplify (6) by using the duplication formula

$$\Gamma(2z)=2^{2z-1}\pi^{-\frac{1}{2}}\Gamma(z)\Gamma(z+\tfrac{1}{2}).$$

The result is formula (9).

We see from (5) since the function $t\longrightarrow\varphi_{\lambda,\delta}(a_t\cdot o)$ is smooth that $s\in\boldsymbol{Z}^+$. Applying d^s/dt^s at $t=0$ it is clear from (1) that $c_{\lambda,\delta}$ is a polynomial in λ. Thus (6) and (9) hold for all λ by analytic continuation.

For the conclusion about r in case $m_{2\alpha}>0$ we recall again that $\frac{1}{2}(m_{2\alpha}-1)\in\boldsymbol{Z}^+$. Unless both $\frac{1}{2}(s+r)$ and $\frac{1}{2}(s-r)$ are integers the denominator in (6) would have infinitely many poles which are not cancelled by the poles of the numerator. Thus $r,s\in\boldsymbol{Z}^+$. To see that $s\geq r$, suppose to the contrary that $r>s$ and put $\mu=\frac{1}{2}\langle i\lambda+\rho,\alpha_o\rangle$ in (6). Then the expression in (6) is

$$\frac{(\mu+\frac{1}{2}(s+r)-1)\cdots(\mu+1)\mu\cdot c_s}{[\mu+\frac{1}{2}(1-m_{2\alpha}+s-r)+\frac{1}{2}(r-s)-1]\cdots[\mu+\frac{1}{2}(1-m_{2\alpha}+s-r)]},$$

but clearly the factor $\mu+\frac{1}{2}(1-m_{2\alpha}+s-r)$ can not divide the numerator. This contradiction with $c_{\lambda,\delta}$ being a polynomial proves $r\leq s$, concluding the proof of the theorem.

We now have from Theorem 5.15,

$$\varphi_{\lambda,\delta}(a_t \cdot o)Q^{\check{\delta}}(\lambda) = \varphi_{-\lambda,\delta}(a_t \cdot o)Q^{\check{\delta}}(-\lambda). \tag{13}$$

Comparing this with (5) we deduce that

$$\frac{Q^{\check{\delta}}(\lambda)}{Q^{\check{\delta}}(-\lambda)} = \frac{p_{r,s}(-\lambda)q_{r,s}(-\lambda)}{p_{r,s}(\lambda)q_{r,s}(\lambda)}. \tag{14}$$

Now if $m_{2\alpha} > 0$,

$$\begin{aligned} p_{r,s}(\lambda) &= 0 \qquad \text{if} \quad i\lambda(H) = -\rho(H), -\rho(H) - 2, \cdots \\ q_{r,s}(\lambda) &= 0 \qquad \text{if} \quad i\lambda(H) = -\tfrac{1}{2}(m_\alpha + 1),\ -\tfrac{1}{2}(m_\alpha + 1) - 2, \cdots \end{aligned}$$

so $p_{r,s}(-\lambda)q_{r,s}(-\lambda)$ and $p_{r,s}(\lambda)q_{r,s}(\lambda)$ are relatively prime. Because of (9) this holds also for $m_{2\alpha} = 0$. We have already remarked (since either λ or $-\lambda$ is simple) that $Q^{\check{\delta}}(\lambda)$ and $Q^{\check{\delta}}(-\lambda)$ are relatively prime. Thus for a suitable normalization in $E_{\check{\delta}}$,

$$Q^{\check{\delta}}(\lambda) = p_{r,s}(-\lambda)q_{r,s}(-\lambda). \tag{15}$$

For the contragredient representation $\check{\delta}$ the representation coefficient $\langle \check{\delta}(k)v', v' \rangle$ is the conjugate of $\langle \delta(k)v, v \rangle$ ([GGA],(14) p. 393). Consequently,

$$\varphi_{\lambda,\check{\delta}}(x) = (\varphi_{-\bar{\lambda},\delta}(x))^{-}. \tag{16}$$

Since the product $p_{r,s}q_{r,s}$ is a polynomial in $i\lambda$ with real coefficients we thus deduce from (13), (15), and (16),

$$Q^{\check{\delta}}(\lambda) = Q^{\delta}(\lambda). \tag{17}$$

Corollary 11.3. *For a suitable normalization the polynomial $Q^{\delta}(\lambda)$ is given by*

$$Q^{\delta}(\lambda) = p_{r,s}(-\lambda)q_{r,s}(-\lambda),$$

where $p_{r,s}$ and $q_{r,s}$ are given by (7) and (8).

2. Harmonic Analysis of K-Finite Functions.

We shall now reformulate the Paley-Wiener theorem (Theorem 5.11) in the rank-one case. With the notation of No. 1 let $\delta \in \widehat{K}_M$ and let $\mathcal{D}^\delta(A)$ denote the space of functions $\Phi \in \mathcal{D}(A, \mathrm{Hom}(V_\delta, V_\delta))$ satisfying the conditions:

(i) $\Phi(kak^{-1}) = \delta(k)\Phi(a)$ if $a \in A$ and $kak^{-1} \in A$ for some $k \in K$.

(ii) $\Phi(a)/\rho(\log a)^{s(\delta)}$ is smooth and even on A.

Here $s(\delta)$ is the number s from Theorem 11.2.

It is clear from (i) that if $\Phi \in \mathcal{D}^\delta(A)$ then $\Phi(a)$ maps V_δ into V_δ^M. For simplicity in notation we identify A with the submanifold $A \cdot o$ of X. We consider now the space $\mathcal{D}^\delta(X)$ from (45) in §5.

Proposition 11.4. *The restriction $F \longrightarrow F|A$ is a bijection of $\mathcal{D}^\delta(X)$ onto $\mathcal{D}^\delta(A)$.*

Proof. If $F \in \mathcal{D}^\delta(X)$ then $F(ka \cdot o) = \delta(k)F(a)$ so (i) is satisfied and clearly the restriction map is one-to-one. For (ii) put $f(x) = Tr(F(x))$. Then by Prop. 5.10, $f \in \mathcal{D}_{\check{\delta}}$, $F = f^\delta$ and by Corollary 5.5,

$$F(a) = w^{-1} \int_{\mathfrak{a}^*} \Phi_{\lambda,\delta}(a \cdot o)\widetilde{f^\delta}(\lambda)|c(\lambda)|^{-2} d\lambda. \tag{18}$$

Fix a unit vector $v \in V_\delta^M$. Since $\widetilde{f^\delta}(\lambda)$ maps V_δ into the one-dimensional space V_δ^M we have

$$\widetilde{f^\delta}(\lambda)u = \langle \widetilde{f^\delta}(\lambda)u, v\rangle v$$

and it follows from Theorem 11.2 that the function

$$a \longrightarrow \langle \Phi_{\lambda,\delta}(a \cdot o)v, v\rangle \rho(\log a)^{-s(\delta)}$$

is smooth and even on A. Then (18) implies that condition (ii) holds.

For the surjectivity let $\Phi \in \mathcal{D}^\delta(A)$ and define F on X by

$$F(ka \cdot o) = \delta(k)\Phi(a). \tag{19}$$

Since by (i) $\Phi(e) = \delta(k)\Phi(e)$ for all k, $F(o)$ is well defined. Moreover, if $a \neq e$ and $ka \cdot o = k_1 a_1 \cdot o$ then $ka = k_1 a_1 k_2$. Applying the Cartan involution θ and eliminating k_2 we get $ka^2k^{-1} = k_1 a_1^2 k_1^{-1}$ so $k_1^{-1}k = m$ normalizes A and $a_1 = mam^{-1}$. Then by (i),

$$\delta(k)\Phi(a) = \delta(k_1)\delta(m)\Phi(a) = \delta(k_1)\Phi(mam^{-1}) = \delta(k_1)\Phi(a_1)$$

so F is well defined on X.

It remains to prove that F is smooth. Let $S \subset \mathfrak{p}$ be the unit sphere and fix $H_o \in \mathfrak{a} \cap S$. Under the map $kM \longrightarrow \mathrm{Ad}(k)H_o$, S becomes the homogeneous space K/M, the action of K on S being the adjoint action. Under this action let $\mathcal{E}_\delta(S)$ denote the set of elements in $\mathcal{E}(S)$ which are K-finite of type δ. Then

$$\dim \mathcal{E}_\delta(S) = \dim \mathcal{E}_\delta(K/M) = \dim V_\delta \dim V_\delta^M = d(\delta)$$

so $\mathcal{E}_\delta(S)$ is irreducible under K. If H denotes the space of harmonic polynomial functions on $\mathfrak{p}$ then the restriction map $p \longrightarrow p|S$ is an injection of H into $\mathcal{E}(S)$. Also if H_δ is the set of $h \in H$ of type δ, K acts irreducibly on H_δ and it consists of homogeneous polynomials of degree equal to $\deg(Q^\delta(\lambda)) = s(\delta)$. Thus we take V_δ as the space H_δ.

The maps

$$w \in V_\delta \longrightarrow (w|S) \in \mathcal{E}_\delta(S)$$
$$w \in V_\delta \longrightarrow h \in \mathcal{E}_\delta(S) \qquad \text{where } h(\mathrm{Ad}(k)H_o) = \langle w, \delta(k)v\rangle$$

both commute with the action of K so by Schur's lemma they are proportional; thus

$$w(\mathrm{Ad}(k)H_o) = c_\delta\langle w, \delta(k)v\rangle, \qquad w \in V_\delta,\ k \in K,$$

where c_δ is a constant. If $w' \in V_\delta$ then

$$\begin{aligned}\langle w',\, F(ka\cdot o)w\rangle &= \langle w',\, \delta(k)\Phi(a)w\rangle = \langle w',\, \delta(k)\langle \Phi(a)w, v\rangle v\rangle \\ &= \langle v,\, \Phi(a)w\rangle\langle w',\, \delta(k)v\rangle = c_\delta^{-1}\langle v,\, \Phi(a)w\rangle w'(\mathrm{Ad}(k)H_o) \\ &= c\langle v, \rho(\log a)^{-s(\delta)}\Phi(a)w\rangle \quad w'(\mathrm{Ad}(k)\log a),\end{aligned}$$

where c is another constant. The scalar function $a \longrightarrow \langle v, \rho(\log a)^{-s(\delta)}\Phi(a)w\rangle$ on A is by (ii) smooth and W-invariant so extends uniquely to a smooth K-invariant function ψ on X (cf. [GGA], II, Theorem 5.8). But then if $\mathrm{Exp} : \mathfrak{p} \longrightarrow X$ is the exponential mapping, the formulas above imply

$$\langle w',\, F(\mathrm{Exp}\, Y)w\rangle = c\,\psi(\mathrm{Exp}\, Y)\, w'(Y).$$

Thus F is smooth and the lemma is proved.

Because of Proposition 5.10 the functions $f^\delta (f \in \mathcal{D}_{\check{\delta}}(X))$ fill up the space $\mathcal{D}^\delta(X)$ so, taking Proposition 11.4 into account, we can restate Corollary 5.5. Observe that $f^\delta(a\cdot o)$ maps V_δ into V_δ^M and that $\Phi_{\bar{\lambda},\delta}(a\cdot o)^*$ on V_δ^M is just multiplication by the scalar

$$\langle \Phi_{\bar{\lambda},\delta}(a\cdot o)^* v, v\rangle = \varphi_{-\lambda,\check{\delta}}(a\cdot o)$$

in the notation of (1). Thus we have the following result (with Δ as in (10) §5).

Corollary 11.5. *The mapping* $F \longrightarrow \widetilde{F}$ *given by*

$$\widetilde{F}(\lambda) = w^{-1} \int_A \varphi_{-\lambda,\check{\delta}}(a) F(a) \Delta(a)\, da \tag{20}$$

is a bijection of $\mathcal{D}^\delta(A)$ *onto* $\mathcal{H}^\delta(\mathfrak{a}^*)$ *with inverse*

$$F(a) = w^{-1} \int_{\mathfrak{a}^*} \varphi_{\lambda,\delta}(a) \widetilde{F}(\lambda) |\mathbf{c}(\lambda)|^{-2}\, d\lambda. \tag{21}$$

Taking traces on both sides in these formulas we obtain an integral transform for scalar-valued functions with a hypergeometric function as kernel. Let

$$K_\delta(\lambda, a) = \tfrac{1}{2}\varphi_{-\lambda,\check{\delta}}(a \cdot o) \rho(\log a)^{-s(\delta)} Q^\delta(\lambda)^{-1}$$
$$\varphi(a) = \operatorname{Tr}(F(a))\rho(\log a)^{-s(\delta)}, \qquad \widetilde{\varphi}(\lambda) = \operatorname{Tr}(\widetilde{F}(\lambda)) Q^\delta(\lambda)^{-1}$$
$$\Delta^\delta(a) = \rho(\log a)^{2s(\delta)} \Delta(a), \qquad \mathbf{c}^\delta(\lambda) = \mathbf{c}(\lambda) Q^\delta(-\lambda)^{-1}.$$

Because of (5) and Corollary 11.3 the kernel K_δ is smooth in both variables and is in fact a hypergeometric function. Corollary 11.5 can now be restated as a scalar-valued integral transform in one variable.

Corollary 11.6. *The mapping* $\varphi(a) \longrightarrow \widetilde{\varphi}(\lambda)$ *where*

$$\widetilde{\varphi}(\lambda) = \int_A K_\delta(\lambda, a) \varphi(a) \Delta^\delta(a)\, da \tag{22}$$

is a bijection of the space of even functions in $\mathcal{D}(A)$ *onto the space of even functions on* $\mathfrak{a}^*$ *which extend to entire functions of exponential type on* $\mathfrak{a}^*_{\mathbf{c}}$. *The inverse is given by*

$$\varphi(a) = \int_{\mathfrak{a}^*} \overline{K_\delta(\lambda, a)}\, \widetilde{\varphi}(\lambda) |\mathbf{c}^\delta(\lambda)|^{-2}\, d\lambda. \tag{23}$$

In the case when δ is the identity representation this is just the spherical transform.

§12. THE SPHERICAL TRANSFORM REVISTED.

In [GGA], Ch. IV the *spherical transform* $f \longrightarrow \widetilde{f}$ was studied with main emphasis on the space $\mathcal{D}^{\natural}(G)$ of K-bi-invariant functions in $\mathcal{D}(G)$. The principal results were the Paley-Wiener theorem (characterization of $\mathcal{D}^{\natural}(G)^{\sim}$), the inversion formula for the transform $f \longrightarrow \widetilde{f}$ on $\mathcal{D}^{\natural}(G)$, the Plancherel formula (with an explicit dual measure) and a determination of the maximal ideal space of the Banach algebra $L^{\natural}(G)$ of K-bi-invariant integrable functions on G, that is the set of $\lambda \in \mathfrak{a}_{\mathbf{c}}^*$ for which φ_λ is bounded.

We shall now supplement this with a functionalanalytic treatment which applies to more general G and K and which leads to further results also in the special case above.

1. Positive Definite Functions.

We recall that a continuous complex-valued function φ on a locally compact group G is said to be *positive definite* if

$$(1) \qquad \sum_{i,j} \varphi(x_i^{-1}x_j)\alpha_i\overline{\alpha}_j \geq 0$$

for all finite sets $x_1, \cdots x_n$ of elements in G and any complex numbers $\alpha_1, \cdots \alpha_n$.

Such a function φ satisfies

$$(2) \qquad \varphi(e) \geq 0, \quad \varphi(x^{-1}) = \overline{\varphi(x)}, \quad |\varphi(x)| \leq \varphi(e).$$

Theorem 12.1.

(i) Let π be a unitary representation of a locally compact group G on a Hilbert space $\mathfrak{H}$. For each vector $\mathbf{e} \in \mathfrak{H}$ the function $x \longrightarrow \langle \mathbf{e}, \pi(x)\mathbf{e}\rangle$ on G is positive definite.

(ii) If $\varphi \not\equiv 0$ is a positive definite function on G then there exists a unitary representation π of G on a Hilbert space $\mathfrak{H}$ such that $\varphi(x) = \langle \mathbf{e}, \pi(x)\mathbf{e}\rangle$ for a suitable vector $\mathbf{e}$.

(iii) The space $\mathfrak{H}$, the representation π and the vector $\mathbf{e}$ in (ii) can be chosen such that $\mathbf{e}$ is a cyclic vector (the vector space spanned by $\pi(G)\mathbf{e}$ is dense in $\mathfrak{H}$).

If $(\pi', \mathfrak{H}', \mathbf{e}')$ is another such triple such that

$$(3) \qquad \varphi(x) = \langle \mathbf{e}, \pi(x)\mathbf{e}\rangle = \langle \mathbf{e}', \pi'(x)\mathbf{e}'\rangle,$$

then there exists an isomorphism A of $\mathfrak{H}$ onto $\mathfrak{H}'$ such that

$$Ae = e', \quad \pi'(x) = A\pi(x)A^{-1} \qquad x \in G.$$

Proof. (i) and (ii) are proved in [GGA] Ch. IV, Theorem 1.5, where it is also proved that e in (ii) can be chosen cyclic. For the last statement of (iii) note that by (3),

$$\langle \pi(x)e, \pi(y)e\rangle = \langle \pi(y^{-1}x)e, e\rangle =$$
$$\langle \pi'(y^{-1}x)e', e'\rangle = \langle \pi'(x)e', \pi'(y)e'\rangle.$$

Since e and e' are cyclic vectors this implies that $\pi(x)e = e$ if and only if $\pi'(x)e' = e'$. Thus the mapping $A: \ \pi(x)e \longrightarrow \pi'(x)e'$ extends to an isomorphism of $\mathfrak{H}$ onto $\mathfrak{H}'$ which clearly has the stated properties.

Let $P(G)$ denote the set of positive definite functions on G. It is contained in the space $BC(G)$ of continuous bounded functions on G which is a Banach space in the uniform norm $\| \ \|_\infty$.

If $\varphi \in P(G)$ the triple $(\mathfrak{H}, \pi, e)$ can be chosen in the following concrete fashion (cf. [GGA], Ch. IV, Theorem 1.5). Let V_φ denote the set of complex linear combinations of left translates $\varphi^{L(x)}$ $(x \in G)$ of φ. Define a Hermitian form on V_φ by the formula

$$\langle f, g\rangle = \sum_{i,j} \alpha_i \overline{\beta_j} \varphi(x_i^{-1}y_j) \tag{4}$$

if $f = \Sigma\alpha_i\varphi^{L(x_i)}$, $g = \Sigma\beta_j\varphi^{L(y_j)}$. Since

$$\langle f, g\rangle = \Sigma\alpha_i\overline{g(x_i)} = \Sigma\overline{\beta_j}f(y_j)$$

it is clear that (4) depends only on f and g but not on their special expression in terms of φ. Since $\langle f, f\rangle \geq 0$ we have the Schwartz inequality

$$|\langle f, g\rangle|^2 \leq \langle f, f\rangle\langle g, g\rangle. \tag{5}$$

Using this on $g = g^{L(x)}$ and observing that

$$f(x) = \langle f, \varphi^{L(x)}\rangle \tag{6}$$

we see that

$$\langle f, f\rangle = 0 \quad \text{if and only if} \quad f = 0. \tag{7}$$

From (5) and (6) we deduce

$$\|f\|_\infty \leq \langle f, f\rangle^{\frac{1}{2}}\,\varphi(e)^{\frac{1}{2}} \tag{8}$$

so a Cauchy sequence in V_φ is also a Cauchy sequence in the norm $\|\ \|_\infty$. The completion $\mathfrak{H}_\varphi$ of V_φ in the norm $\|f\| = \langle f, f\rangle^{\frac{1}{2}}$ can therefore be viewed as a subspace of $BC(G)$:

$$\mathfrak{H}_\varphi \subset BC(G). \tag{9}$$

Because of (8) formula (6) extends to $\mathfrak{H}_\varphi$, i.e.,

$$f(x) = \langle f, \varphi^{L(x)}\rangle \qquad \text{for} \quad f \in \mathfrak{H}_\varphi. \tag{10}$$

Each $x \in G$ gives rise to an endomorphism $f \longrightarrow f^{L(x)}$ of V_φ; this endomorphism preserves the inner product (4) so extends uniquely to a unitary operator $\pi_\varphi(x)$ of $\mathfrak{H}_\varphi$. By (4) we also find

$$\begin{aligned}&\langle f^{L(x)} - f, f^{L(x)} - f\rangle\\ &= 2\sum_{i,j}\alpha_i\bar{\alpha}_j\varphi(x_i^{-1}x_j) - \sum_{i,j}\alpha_i\bar{\alpha}_j\varphi(x_i^{-1}x^{-1}x_j) - \sum_{i,j}\alpha_i\bar{\alpha}_j\varphi(x_i^{-1}xx_j),\end{aligned}$$

which tends to 0 for $x \longrightarrow e$. Thus for a dense set of vectors $\boldsymbol{a} \in \mathfrak{H}_\varphi$ the mapping $x \longrightarrow \pi_\varphi(x)\boldsymbol{a}$ is continuous at $x = e$, hence at all $x \in G$. From the inequality

$$\|\pi_\varphi(x)\boldsymbol{b} - \boldsymbol{b}\| \leq \|\pi_\varphi(x)(\boldsymbol{a} - \boldsymbol{b})\| + \|\pi_\varphi(x)\boldsymbol{a} - \boldsymbol{a}\| + \|\boldsymbol{a} - \boldsymbol{b}\|$$

it then follows that for each $\boldsymbol{b} \in \mathfrak{H}_\varphi$ the mapping $x \longrightarrow \pi_\varphi(x)\boldsymbol{b}$ is continuous. Thus π_φ is a unitary representation of G on $\mathfrak{H}_\varphi$ and φ is a cyclic vector which by (6) satisfies

$$\varphi(x) = \langle \varphi, \pi_\varphi(x)\varphi\rangle.$$

The pair $(\pi_\varphi, \mathfrak{H}_\varphi)$ is called the unitary representation *associated to* φ.

Theorem 12.2. *Let* $\varphi = \varphi_1 + \varphi_2$ $(\varphi_1, \varphi_2 \in P(G))$. *Then* $\varphi_1, \varphi_2 \in \mathfrak{H}_\varphi$ *and moreover,*

$$\mathfrak{H}_\varphi = \mathfrak{H}_{\varphi_1} + \mathfrak{H}_{\varphi_2},$$

as an (algebraic) sum of vector spaces.

Proof. Consider the Hilbert space product $\mathfrak{H}_{\varphi_1} \times \mathfrak{H}_{\varphi_2}$ on which G acts via $\pi_{\varphi_1} \oplus \pi_{\varphi_2}$. Let $\mathfrak{H} \subset \mathfrak{H}_{\varphi_1} \times \mathfrak{H}_{\varphi_2}$ be the closed invariant subspace generated by the vector $(\varphi_1, \varphi_2) \in \mathfrak{H}_{\varphi_1} \times \mathfrak{H}_{\varphi_2}$. Then

$$\begin{aligned}\langle(\varphi_1,\varphi_2),(\pi_{\varphi_1} \oplus \pi_{\varphi_2})(x)(\varphi_1,\varphi_2)\rangle &= \langle\varphi_1, \pi_{\varphi_1}(x)\varphi_1\rangle + \langle\varphi_2, \pi_{\varphi_2}(x)\varphi_2\rangle \\ &= \varphi_1(x) + \varphi_2(x) = \varphi(x).\end{aligned}$$

Thus by Theorem 12.1 (iii) there is an isomorphism $A : \mathfrak{H}_\varphi \longrightarrow \mathfrak{H}$ such that

$$A\varphi = (\varphi_1, \varphi_2), \qquad (\pi_{\varphi_1} \oplus \pi_{\varphi_2})(x) = A\pi_\varphi(x)A^{-1}.$$

Let $f \in \mathfrak{H}_\varphi$, $(f_1, f_2) = Af$. Then by (10), since A preserves the scalar products,

$$\begin{aligned}f(x) = \langle f, \varphi^{L(x)}\rangle &= \langle(f_1, f_2), (\pi_{\varphi_1} \oplus \pi_{\varphi_2})(x)(\varphi_1, \varphi_2)\rangle \\ &= \langle f_1, \varphi_1^{L(x)}\rangle + \langle f_2, \varphi_2^{L(x)}\rangle = f_1(x) + f_2(x).\end{aligned}$$

Thus

$$\mathfrak{H}_\varphi \subset \mathfrak{H}_{\varphi_1} + \mathfrak{H}_{\varphi_2}.$$

On the other hand let $f_1 \in \mathfrak{H}_{\varphi_1}, f_2 \in \mathfrak{H}_{\varphi_2}$. Then

$$f_1(x) = \langle f_1, \varphi_1^{L(x)}\rangle, \quad f_2(x) = \langle f_2, \varphi_2^{L(x)}\rangle.$$

Let (g_1, g_2) be the orthogonal projection of (f_1, f_2) on the subspace $\mathfrak{H} \subset \mathfrak{H}_{\varphi_1} \times \mathfrak{H}_{\varphi_2}$. Then if $(g_1, g_2) = Ag$ we have, since A preserves scalar products,

$$\begin{aligned}&f_1(x) + f_2(x) = \langle(f_1, f_2), (\pi_{\varphi_1} \oplus \pi_{\varphi_2})(x)(\varphi_1, \varphi_2)\rangle = \\ &\langle(g_1, g_2), (\pi_{\varphi_1} \oplus \pi_{\varphi_2})(x)(\varphi_1, \varphi_2)\rangle = \langle g, \pi_\varphi(x)\varphi\rangle = g(x).\end{aligned}$$

Thus $\mathfrak{H}_{\varphi_1} + \mathfrak{H}_{\varphi_2} \subset \mathfrak{H}_\varphi$ so the theorem is proved.

Definition. Let V be a locally convex topological vector space over $\boldsymbol{R}$ and $S \subset V$ a convex subset. A point $v \in S$ is an *extreme point* if it is not contained in any open segment contained in S.

The subset $P_1(G) = \{\varphi \in P(G) : \ \varphi(e) = 1\}$ is clearly convex.

Corollary 12.3. *Let $\varphi \neq 0$ in $P_1(G)$. Then φ is an extreme point if and only if the representation π_φ on $\mathfrak{H}_\varphi$ is irreducible.*

Proof. Suppose $\varphi \in P_1(G)$ is not extreme. Then $\varphi = \lambda\varphi_1 + (1-\lambda)\varphi_2$ where $0 < \lambda < 1$, $\varphi_1 \neq \varphi_2$ in $P_1(G)$. Then $\psi = \lambda\varphi_1$ is not proportional to φ and $\varphi - \psi \in P(G)$. By Theorem 12.2, $\psi \in \mathfrak{H}_\varphi$ so

$$(11) \qquad \langle \psi, \varphi^{L(x)} \rangle = \psi(x) = \overline{\psi(x^{-1})} = \langle \varphi^{L(x^{-1})}, \psi \rangle = \langle \varphi, \psi^{L(x)} \rangle.$$

Consider the mapping

$$A: \ \xi = \Sigma\alpha_i\varphi^{L(x_i)} \longrightarrow \eta = \Sigma\alpha_i\psi^{L(x_i)},$$

which is well defined because (11) implies

$$\langle \xi, \psi^{L(x)} \rangle = \langle \eta, \varphi^{L(x)} \rangle$$

so $\eta = 0$ if $\xi = 0$. Also, since $\varphi - \psi \in P(G)$ we have

$$\langle \Sigma\alpha_i\psi^{L(x_i)}, \Sigma\alpha_j\psi^{L(x_j)} \rangle = \Sigma\alpha_i\overline{\alpha}_j\psi(x_i^{-1}x_j) \leq \Sigma\alpha_i\overline{\alpha}_j\varphi(x_i^{-1}x_j)$$

so

$$\langle A\xi, A\xi \rangle \leq \langle \xi, \xi \rangle.$$

Thus A extends to a continuous linear map of $\mathfrak{H}_\varphi$ into $\mathfrak{H}_\varphi$ commuting with each $\pi_\varphi(x)$. Since $A\varphi \notin \boldsymbol{C}\varphi$, π_φ is not irreducible by Schur's lemma.

On the other, suppose π_φ is not irreducible. Then $\mathfrak{H}_\varphi = \mathfrak{H}_1 \oplus \mathfrak{H}_2$ where $\mathfrak{H}_1$ and $\mathfrak{H}_2$ are nonzero closed invariant orthogonal subspaces. Let $P_i : \mathfrak{H}_\varphi \longrightarrow \mathfrak{H}_i$ be the two orthogonal projections and consider the functions

$$\varphi_i(x) = \langle P_i\varphi, \pi_\varphi(x)P_i\varphi \rangle = \langle \varphi, \pi_\varphi(x)P_i\varphi \rangle = (P_i\varphi)(x)$$

for $i = 1, 2$. Then $\varphi = \varphi_1 + \varphi_2$ and being a cyclic vector, $\varphi \notin \mathfrak{H}_1 \cup \mathfrak{H}_2$. Thus $\varphi_1 \neq 0, \varphi_2 \neq 0$ and since they are orthogonal they are not proportional. The relation

$$\varphi = \varphi_1(o)\psi_1 + \varphi_2(o)\psi_2 \qquad \psi_i = \varphi_i/\varphi_i(o) \in P_1(G)$$

shows that φ is not an extreme point.

Proposition 12.4. *If* $\varphi_1, \varphi_2 \in P(G)$ *then* $\varphi_1\varphi_2 \in P(G)$.

Proof. Let $\pi_i = \pi_{\varphi_i}$ $(i = 1, 2)$ and consider the tensor product representation $\pi_1 \otimes \pi_2$ of G on the Hilbert space $\mathcal{H}_{\varphi_1} \otimes \mathcal{H}_{\varphi_2}$. If

$$\varphi_i(x) = \langle \varphi_i, \pi_i(x)\varphi_i \rangle \qquad (i = 1, 2)$$

then

$$\langle \varphi_1 \otimes \varphi_2, (\pi_1 \otimes \pi_2)(x)\varphi_1 \otimes \varphi_2 \rangle = \langle \varphi_1 \otimes \varphi_2, \pi_1(x)\varphi_1 \otimes \pi_2(x)\varphi_2 \rangle$$
$$= \langle \varphi_1, \pi_1(x)\varphi_1 \rangle \langle \varphi_2, \pi_2(x)\varphi_2 \rangle = \varphi_1(x)\varphi_2(x).$$

2. The Spherical Transform For Gelfand Pairs.

Let A be a commutative Banach algebra over $\boldsymbol{C}$, $\mathfrak{M}$ the set of continuous homomorphisms of A onto $\boldsymbol{C}$. If $x \in A$ the *Gelfand transform* $\widetilde{x}$ of x is the function on $\mathfrak{M}$ given by

$$\widetilde{x}(h) = h(x) \qquad h \in \mathfrak{M}. \tag{12}$$

The *Banach algebra topology* of $\mathfrak{M}$ is the weakest topology for which all the functions $\widetilde{x}$ are continuous. Then(cf. e.g. Loomis [1953]) $\mathfrak{M}$ is locally compact, it is compact if A has an identity, and if $\mathfrak{M}$ is noncompact all the functions $\widetilde{x}$ vanish at infinity on $\mathfrak{M}$. This topology of $\mathfrak{M}$ coincides with the relative topology of the weak* topology $\sigma(L^\infty(G), L^1(G))$.

Let G be a locally compact unimodular group, K a compact subgroup. Let $P^\natural(G)$, $C_c^\natural(G)$, $C^\natural(G)$, $L^p(G)^\natural$ denote the subspace of K-bi-invariant functions in $P(G)$, $C_c(G)$, $C(G)$ and $L^p(G)$, respectively. We often write $L^\natural(G)$ or even $L^\natural$ for $L^1(G)^\natural$ and $P^\natural$ for $P^\natural(G)$. As usual if f is a function on G we put $f^*(x) = \overline{f(x^{-1})}$. If $S \subset G$ is compact we put

$$C_S(G) = \{f \in C_c(G) : \text{supp}(f) \subset S\}.$$

The pair (G, K) is called a *Gelfand pair* if the convolution algebra $C_c^\natural(G)$ is commutative. This will now be assumed.

A *spherical function* on G is a continuous function $\varphi \not\equiv 0$ satisfying the functional equation

$$\int_K \varphi(xky)dk = \varphi(x)\varphi(y), \qquad x, y \in G, \tag{13}$$

dk being the normalized Haar measure on K.

Theorem 12.5.

(i) If $C_c(G)$ is taken with the inductive limit topology of the spaces $C_S(G)$ (S compact) the continuous homomorphisms of $C_c^\natural(G)$ onto $\boldsymbol{C}$ are given by

$$f \longrightarrow \int_G f(x)\varphi(x^{-1})dx, \tag{14}$$

where φ is a spherical function on G.

(ii) The Banach algebra $L^\natural(G)$ is semisimple and the continuous homomorphisms of $L^\natural(G)$ onto $\mathbf{C}$ are given by (14) where φ is a bounded spherical function on G.

For a proof see [GGA], pp. 409, 486 (the Lie group assumption made there did not enter into the proofs).

Let $\mathfrak{P} \subset P^\natural(G)$ denote the set of spherical functions on G which are positive definite. Also let

$$P_1^\natural(G) = \{\varphi \in P^\natural(G) : \varphi(e) = 1\}.$$

Theorem 12.6.

(i) The mapping $\varphi \longrightarrow \pi_\varphi$ (cf. Cor. 12.3) is a bijection of $\mathfrak{P}$ onto the set of equivalence classes of unitary irreducible spherical representations of G.

(ii) Let $\varphi \in P_1^\natural(G)$. The following properties are equivalent:

a) φ is a spherical function.

b) The representation π_φ is irreducible.

c) φ is an extreme point of $P_1^\natural(G)$.

Part (i) is proved in [GGA], Ch. IV, §3 and (ii) follows by combining (i) with Cor. 12.3.

Let $\mathfrak{B}$ denote the set of bounded spherical functions on G. If $f \in L^\natural(G)$ its *spherical transform* $\widetilde{f}$ is given by

$$\widetilde{f}(\varphi) = \int_G f(x)\varphi(x^{-1})dx, \qquad \varphi \in \mathfrak{B}. \tag{15}$$

Then (13) implies easily that

$$f * \varphi = \widetilde{f}(\varphi)\varphi. \tag{16}$$

Proposition 12.7. *The Banach algebra topology of $\mathfrak{B}$ (cf. Theorem 12.5 (ii)) coincides with the topology on $\mathfrak{B}$ given by uniform convergence on compact subsets of G.*

For a proof see e.g. Dieudonné [1978], Vol. VI, Ch. XXII, (6.9) (or Loomis [1953], 34C for an analogous result).

From remarks above about $\mathfrak{M}$, the space $\mathfrak{B}$ is locally compact, and if it is noncompact the functions $\widetilde{f}$ vanish at ∞ on $\mathfrak{B}$. Similar statement holds for the closed subspace $\mathfrak{P}$ (given the relative topology of $\mathfrak{B}$).

Proposition 12.8. *The space of restrictions*

$$\{\widetilde{f}|\mathfrak{P} : \ f \in L^{\natural}(G)\} \tag{17}$$

is dense in $C_o(\mathfrak{P})$. *(If* $\mathfrak{P}$ *is compact* $C_o(\mathfrak{P})$ *should here be replaced by* $C(\mathfrak{P})$*).*

Proof. Since

$$(f_1 * f_2)^{\sim} = \widetilde{f}_1\widetilde{f}_2 \qquad\qquad f_1, f_2 \in L^{\natural}(G) \tag{18}$$

the space (17) is an algebra. Since $\varphi^* = \varphi$ ($\varphi \in \mathfrak{P}$) we have $(f^*)^{\sim} = (\widetilde{f})^-$ on $\mathfrak{P}$. Since the family (17) separates points on $\mathfrak{P}$ the proposition follows from the Stone-Weierstrass theorem.

Theorem 12.9. *For each* $f \in P^{\natural}(G)$ *there exists a unique positive bounded measure* μ_f *on* $\mathfrak{P}$ *such that*

$$f(x) = \int_{\mathfrak{P}} \varphi(x)d\mu_f(\varphi). \tag{19}$$

Proof. Note that by Prop. 12.7 the function $\varphi \longrightarrow \varphi(x)$ is continuous on $\mathfrak{P}$ so the integral is well-defined. For the existence consider the set $P_1^{\natural}(G)$ which is a convex subset of the unit ball in $L^{\infty}(G)$, closed in the weak* topology $\sigma(L^{\infty}(G), L^1(G))$, hence compact in the induced topology. Also $\mathfrak{P}$ is a topological subspace of $P_1^{\natural}(G)$. Let $\mathfrak{M}(\mathfrak{P})$ denote the space of bounded measures on $\mathfrak{P}$ and

$$\mathfrak{M}_1^+(\mathfrak{P}) = \{\mu \in \mathfrak{M}(\mathfrak{P}) : \ \mu \geq 0, \int d\mu = 1\}.$$

Then $\mathfrak{M}_1^+(\mathfrak{P})$ is compact in the weak* topology $\sigma(\mathfrak{M}(\mathfrak{P}), BC(\mathfrak{P}))$. Consider the map $A : \ \mathfrak{M}_1^+(\mathfrak{P}) \longrightarrow P_1^{\natural}(G)$ given by

$$(A\mu)(x) = \int_{\mathfrak{P}} \varphi(x)d\mu(\varphi). \tag{20}$$

Then

$$\int_G f(x)(A\mu)(x^{-1})dx = \int_{\mathfrak{P}} \widetilde{f}(\varphi)d\mu(\varphi), \qquad f \in L^\natural(G),$$

so A is continuous for the topologies considered and the image is compact and convex. Considering point measures μ in (20) this image contains the extreme points of $P_1^\natural(G)$ (by Theorem 12.6). By the Krein-Milman theorem this image is all of $P_1^\natural(G)$. This proves the existence of μ_f.

For the uniqueness let $g \in L^\natural(G)$. Then by (16) and (19),

$$\int_G f(x)g(x^{-1})dx = \int_{\mathfrak{P}} \widetilde{g}(\varphi)d\mu_f(\varphi)$$

so Prop. 12.8 proves the uniqueness of μ_f.

Theorem 12.10, *There exists a unique measure μ on $\mathfrak{P}$, such that*

$$\mu_f = \widetilde{f}\mu \qquad \textit{for all} \quad f \in L^\natural(G) \cap P^\natural(G). \tag{21}$$

Moreover, if $f \in L^\natural(G) \cap P^\natural G$ then

$$\widetilde{f} \in L^1(\mathfrak{P}, \mu), \quad \widetilde{f} \geq 0 \textit{ on } \operatorname{supp}(\mu) \tag{22}$$

and

$$f(x) = \int_{\mathfrak{P}} \widetilde{f}(\varphi)\varphi(x)d\mu(\varphi). \tag{23}$$

The measure μ is called *the Plancherel measure.*

Proof. Let $f, g \in L^\natural(G) \cap P^\natural(G)$. Convolving (19) with g and then interchanging f and g we obtain

$$(g * f)(x) = \int_{\mathfrak{P}} \widetilde{g}(\varphi)\varphi(x)d\mu_f(\varphi) = \int_{\mathfrak{P}} \widetilde{f}(\varphi)\varphi(x)d\mu_g(\varphi).$$

Integrating against $h \in L^\natural(G)$ we conclude

$$\int_{\mathfrak{P}} \widetilde{h}\widetilde{g}d\mu_f = \int_{\mathfrak{P}} \widetilde{h}\widetilde{f}d\mu_g$$

whence by Prop. 12.8,

$$\widetilde{f}\mu_g = \widetilde{g}\mu_f. \tag{24}$$

On the open set $\mathfrak{P}_f = \{\varphi \in \mathfrak{P} : \widetilde{f}(\varphi) \neq 0\}$ we put $m_f = (1/\widetilde{f})\mu_f$. Then by (24), m_f and m_g coincide on $\mathfrak{P}_f \cap \mathfrak{P}_g$. Since the $\mathfrak{P}_f$ ($f \in L^\natural \cap P^\natural$) cover $\mathfrak{P}$ (consider $f = F * F^*$ for $F \in C_c^\natural(G)$) there exists a unique measure μ on $\mathfrak{P}$ such that $m_f = \mu$ on $\mathfrak{P}_f$ for each $f \in L^\natural \cap P^\natural$ (Bourbaki [1952], Ch. III). Then $\mu_f = \widetilde{f}\mu$ on $\mathfrak{P}_f$. To see that this holds on all of $\mathfrak{P}$ note that any function $h \in C_c(\mathfrak{P}_g)$ can be written $h = \widetilde{g}F$ where $F \in C_c(\mathfrak{P}_g)$. Hence

$$\begin{aligned}\mu_f(h) &= \mu_f(\widetilde{g}F) = (\widetilde{g}\mu_f)(F) = (\widetilde{f}\mu_g)(F) = \mu_g(\widetilde{f}F)\\ &= (\widetilde{g}\mu)(\widetilde{f}F) = (\widetilde{f}\mu)(\widetilde{g}F) = (\widetilde{f}\mu)(h)\end{aligned}$$

so

$$\mu_f = \widetilde{f}\mu \quad \text{on } \mathfrak{P}_g \qquad (g \in L^\natural \cap P^\natural). \tag{25}$$

Given $F \in C_c(\mathfrak{P})$ finitely many $\mathfrak{P}_{g_i}$ cover $\operatorname{supp}(F)$ so by partition of unity $F = \Sigma F_i$ with $\operatorname{supp}(F_i) \subset \mathfrak{P}_{g_i}$. Thus (25) implies (21) and the inversion formula (23) follows from Theorem 12.8. Now let

$$\mathfrak{N}_f = \{\varphi \in \mathfrak{P} : \widetilde{f}(\varphi) < 0\}.$$

If $\varphi_o \in \mathfrak{N}_f \cap \operatorname{supp}(\mu)$ then taking $F \in C_c(\mathfrak{N}_f)$, $F \geq 0$, $F(\varphi_o) > 0$, we have the contradiction

$$0 \leq \mu_f(F) = \int_{\mathfrak{P}} \widetilde{f}(\varphi)F(\varphi)d\mu < 0.$$

Thus $\mathfrak{N}_f \cap \operatorname{supp}(\mu) = \emptyset$, i.e., $\widetilde{f} \geq 0$ on $\operatorname{supp}(\mu)$ and $\mu_f(1) = \int \widetilde{f}\, d\mu < \infty$.

Theorem 12.11. *If $f \in L^1(G)^\natural \cap L^2(G)^\natural$ then $\widetilde{f} \in L^2(\mathfrak{P}, \mu)$ and the mapping $f \longrightarrow \widetilde{f}$ extends to an isometry of $L^2(G)^\natural$ onto $L^2(\mathfrak{P}, \mu)$.*

Proof. If $f \in L^1(G)^\natural \cap L^2(G)^\natural$ then (22) – (23) applied to the function $f * f^*$ gives $\widetilde{f} \in L^2(\mathfrak{P}, \mu)$ and the identity

$$\int_G |f(x)|^2\, dx = \int_{\mathfrak{P}} |\widetilde{f}(\varphi)|^2 d\mu(\varphi). \tag{26}$$

For the surjectivity it suffices to show that $(L^1(G)^\natural \cap L^2(G)^\natural)^\sim$ is dense in $L^2(\mathfrak{P}, \mu)$. Let $F \in C_c(\mathfrak{P})$ and $\epsilon > 0$. Then by Prop. 12.8 there exist

a) A function $g \in C_c^\natural(G)$ such that $\widetilde{g} > 0$ on $\operatorname{supp}(F)$ and a function $F_1 \in C_c(\mathfrak{P})$ such that $F = \widetilde{g}F_1$ on $\mathfrak{P}$.

b) A function $h \in C_c^\natural(G)$ such that $\|F_1 - \widetilde{h}\|_\infty < \epsilon/\|\widetilde{g}\|_2$ where $\|\widetilde{g}\|_2$ is the L^2-norm of $\widetilde{g}$.

Then

$$|F(w) - (g * h)^\sim(w)| \leq |\widetilde{g}(w)|\,|F_1(w) - \widetilde{h}(w)|$$

so

$$\int |F(w) - (g * h)^\sim(w)|^2 d\mu \leq \epsilon^2,$$

proving the theorem.

These last theorems can now be complemented by the following analogs of some results in classical harmonic analysis.

Theorem 12.12. *Suppose G separable.*

(a) If $f \in L^\natural(G)$ and if $\widetilde{f} \in L^1(\mathfrak{P}, \mu)$ then the inversion formula (23) holds.

(b) Suppose $f \in L^1(G)^\natural \cap L^2(G)^\natural$ is continuous at e and that $\widetilde{f} \geq 0$ on $\operatorname{supp}(\mu)$. *Then $\widetilde{f} \in L^1(\mathfrak{P}, \mu)$ and the inversion formula holds.*

(c) If $f, g \in L^\natural \cap P^\natural$ then $f * g \in L^\natural \cap P^\natural$.

(d) If $f \in L^p(G)^\natural \cap BC(G)$ for some p $(1 \leq p \leq 2)$ then $\widetilde{f} \geq 0$ almost everywhere on $\operatorname{supp}(\mu)$ *if and only if $f \in P^\natural(G)$. In this case*

$$f(x) = \int_{\mathfrak{P}} \varphi(x)\widetilde{f}(\varphi)\, d\mu(\varphi) \qquad \text{for all } x \in G.$$

In particular, $\widetilde{f} \in L^1(\mathfrak{P}, \mu)$.

Proof. a) Let $h_n \in C_c^\natural(G)$ be an approximate δ-function, $h_n(x) = h_n(x^{-1})$, $h_n \geq 0$, $\int h_n = 1$, $\operatorname{supp}(h_n) \longrightarrow \{e\}$ and (replacing h_n by $h_n * h_n$) $\widetilde{h}_n \geq 0$. Then if $f \in L^\natural(G)$,

$$\widetilde{f}(\varphi)\widetilde{h}_n(\varphi) = \int_G (f * h_n)(x)\varphi(x^{-1})dx \qquad \varphi \in \mathfrak{P}, \tag{27}$$

so $\widetilde{f}\widetilde{h}_n \longrightarrow \widetilde{f}$ uniformly on $\mathfrak{P}$. By Prop. 12.8 this holds with $\widetilde{f}$ replaced by any $F \in C_c(\mathfrak{P})$ so $\widetilde{h}_n \longrightarrow 1$ uniformly on compacts. Now the function $f * h_n$

is in the span of $L^\natural \cap P^\natural$ so the inversion formula holds for it so a) follows by dominated convergence.

b) We have by Theorem 12.11 and the continuity of f at e,

$$\lim_n \int_{\mathfrak{P}} \widetilde{f}(\varphi)\widetilde{h}_n(\varphi)d\mu(\varphi) = \lim_n (f * h_n)(e) = f(e).$$

If $C \subset \mathfrak{P}$ is a compact subset it follows that

$$\int_C \widetilde{f}\, d\mu = \lim_n \int_C \widetilde{f}\widetilde{h}_n\, d\mu \leq \lim_n \int_{\mathfrak{P}} \widetilde{f}\widetilde{h}_n\, d\mu = f(e)$$

so $\widetilde{f} \in L^1(\mathfrak{P}, \mu)$.

c) By (22), $\widetilde{f}$ and $\widetilde{g}$ are ≥ 0 on $\operatorname{supp}(\mu)$. Since f and g are also in $L^2(G)^\natural$ we know from b) that $\widetilde{f}, \widetilde{g}$ and also $\widetilde{f}\widetilde{g}$ belong to $L^1(\mathfrak{P}, \mu)$. Hence (23) holds for $f * g$ and $f * g \in P^\natural(G)$ follows.

d) Since $|f|^2 = |f|^{2-p}|f|^p$, we have $f \in L^2(G)^\natural$ and $\widetilde{f}$ is defined according to Theorem 12.11 as follows: Let $(f_n) \subset C_c^\natural(G)$ converge to f, $\|f_n - f\|_2 \longrightarrow 0$; then $\|\widetilde{f}_n - \widetilde{f}\|_2 \longrightarrow 0$. Passing to a subsequence we can assume that $f_n \longrightarrow f$ almost everywhere on G and that $\widetilde{f}_n \longrightarrow \widetilde{f}$ almost everywhere on $\mathfrak{P}$. If $g \in L^\natural \cap P^\natural$ we have by Theorem 12.10

$$g(x) = \int_{\mathfrak{P}} \varphi(x)d\mu_g(\varphi) = \int_{\mathfrak{P}} \varphi(x)\widetilde{g}(\varphi)d\mu(\varphi).$$

Convolving with f_n we get

(28) $$(g * f_n)(x) = \int_{\mathfrak{P}} \varphi(x)\widetilde{f}_n(\varphi)d\mu_g(\varphi).$$

Since $g \in L^2(G)^\natural$ we get $(g*f_n)(x) \longrightarrow (g*f)(x)$ uniformly on G by Schwarz' inequality.

Using $\mu_g = \widetilde{g}\mu$ and the inequality

$$\left|\int_{\mathfrak{P}} \varphi(x)\widetilde{g}(\varphi)(\widetilde{f}_n(\varphi) - \widetilde{f}(\varphi))\, d\mu\right| \leq \|g\|_2\|\widetilde{f}_n - \widetilde{f}\|_2$$

we deduce

(29) $$(g * f)(x) = \int_{\mathfrak{P}} \varphi(x)\widetilde{f}(\varphi)\widetilde{g}(\varphi)d\mu.$$

If $\widetilde{f} \geq 0$ almost everywhere on supp(μ), (29) shows that $g * f \in P^\natural(G)$. Letting g run through the sequence h_n in a) we conclude $f \in P^\natural(G)$.

On the other hand, if $f \in P^\natural(G)$ we have

$$f(x) = \int\limits_{\mathfrak{P}} \varphi(x) d\mu_f(\varphi)$$

so convolving with g,

$$(30) \qquad (g * f)(x) = \int\limits_{\mathfrak{P}} \varphi(x)\widetilde{g}(\varphi) d\mu_f(\varphi).$$

Integrating (29) and (30) against $h(x^{-1})$ $(h \in C_c^\natural(G))$ we deduce

$$\int \widetilde{h}\widetilde{g}\, d\mu_f = \int \widetilde{h}\widetilde{f}\widetilde{g}\, d\mu.$$

Taking g of the form $h_1 * h_1^*$ $(h_1 \in C_c^\natural(G))$ we deduce that

$$(31) \qquad \int \widetilde{k}\, d\mu_f = \int \widetilde{k}\widetilde{f}\, d\mu$$

for all $\widetilde{k}$ of the form $\widetilde{h}_1\widetilde{h}_2\widetilde{h}_3$ $(h_i \in C_c^\natural(G))$. These span an algebra closed under complex conjugation, separating points in $\mathfrak{P}$. Hence the $\widetilde{k}$ span a dense subspace of $C_o(\mathfrak{P})$ so (31) implies $\mu_f = \widetilde{f}\mu$ and d) is proved.

3. The Case of a Symmetric Space G/K.

Now we assume that the pair (G, K) consists of a connected semisimple Lie group G with finite center and a maximal compact subgroup K. Then we can compare the abstract results in No. 2 with the spherical transform theory for the symmetric space G/K ([GGA], Ch. IV). In the notation of §1 of this chapter the set of spherical functions is parametrized by $\mathfrak{a}_{\mathbf{c}}^*/W$ so we write for the spherical transform

$$(32) \qquad \widetilde{f}(\varphi_\lambda) = \widetilde{f}(\lambda) = \int\limits_G f(x)\varphi_\lambda(x^{-1})dx = \int\limits_G f(x)\varphi_{-\lambda}(x)dx.$$

We also know from *loc. cit,* Theorem 8.1 that if $C(\rho)$ denotes the convex hull of the orbit $W \cdot \rho$ then the points $\lambda \in \mathfrak{a}^* + iC(\rho)$ are exactly those for which the spherical function φ_λ is bounded. In view of Theorem 12.5 (ii) we thus have the identification

$$(33) \qquad \mathfrak{M} \sim (\mathfrak{a}^* + iC(\rho))/W$$

Proposition 12.13. *The following two topologies on $\mathfrak{M}$ coincide:*

(i) The Banach algebra topology.

(ii) The Euclidean topology on $\mathfrak{M}$ given by (33).

Proof. The functions $\widetilde{f}(f \in L^{\natural}(G))$ vanish at ∞ on $\mathfrak{M}$ (in the Euclidean topology). In fact, if $\lambda = \xi + i\eta$ $(\xi, \eta \in \mathfrak{a}^*)$, $f \in L^{\natural}(G)$, then

$$\lim_{\xi \to \infty} \widetilde{f}(\xi + i\eta) = 0 \qquad \text{uniformly for } \eta \in C(\rho). \tag{34}$$

If $f \in \mathcal{D}^{\natural}(G)$ this is obvious from the Paley-Wiener theorem for the spherical transform. In general it follows by approximating $f \in L^{\natural}(G)$ in the L^1 norm by a sequence $(f_n) \subset \mathcal{D}^{\natural}(G)$ because then $\widetilde{f}_n \longrightarrow \widetilde{f}$ uniformly on $\mathfrak{a}^* + iC(\rho)$. Thus

$$L^{\natural}(G)^{\sim} \subset C_o(\mathfrak{M})$$

and since the functions $\widetilde{f}$ also separate points on $\mathfrak{M}$ it follows by general principles (e.g. Loomis [1955], Theorem 5G) that the topologies (i) and (ii) coincide.

Remark. In this context see also Exercise **F2**.

By Harish-Chandra's formula for the spherical transform (cf. [GGA], Ch. IV, Theorem 7.5 and p. 588 for normalization),

$$\int_G |f(x)|^2 \, dx = \int_{\mathfrak{a}^*/W} |\widetilde{f}(\lambda)|^2 |\mathbf{c}(\lambda)|^{-2} d\lambda, \tag{35}$$

for $f \in L^2(G)^{\natural}$. Viewing $\mathfrak{P}$ as a subset of $(\mathfrak{a}^* + iC(\rho))/W$ and transporting the Plancherel measure $d\mu$ accordingly, we have by Theorem 12.11,

$$\int_{\mathfrak{P}} F(\lambda) d\mu = \int_{\mathfrak{a}^*/W} F(\lambda) |\mathbf{c}(\lambda)|^{-2} d\lambda \tag{36}$$

for all $F(\lambda)$ of the form $|\widetilde{f}(\lambda)|^2$ $(f \in L^2(G)^{\natural})$. If $\mu(\mathfrak{P} - \mathfrak{a}^*/W) > 0$ let $S \subset \mathfrak{P} - \mathfrak{a}^*/W$ be a measurable set such that $0 < \mu(S) < \infty$, and $f \in L^2(G)^{\natural}$ such that $\widetilde{f}$ is the characteristic function of S. Then (36) gives a contradiction so $\mu(\mathfrak{P} - \mathfrak{a}^*/W) = 0$ and $d\mu = |\mathbf{c}(\lambda)|^{-2} d\lambda$ on $\mathfrak{a}^*/W$.

Proposition 12.14. *The function $K(\lambda, \mu, \nu)$ on $\mathfrak{a}^* \times \mathfrak{a}^* \times \mathfrak{a}^*$ given by*

$$K(\lambda, \mu, \nu) = \int_G \varphi_\mu(x) \varphi_\nu(x) \varphi_{-\lambda}(x) dx \tag{37}$$

is positive, analytic and W-invariant in all three variables. Moreover,

(38)
$$\varphi_\mu(x)\varphi_\nu(x) = w^{-1} \int_{\mathfrak{a}^*} \varphi_\lambda(x) K(\lambda, \mu, \nu) |\mathbf{c}(\lambda)|^{-2} d\lambda.$$

Proof. Consider Harish-Chandra's formula

(39)
$$\varphi_\lambda(g) = \int_K e^{(i\lambda - \rho)(H(gk))} dk, \qquad \lambda \in \mathfrak{a}^*_{\mathbf{c}}.$$

If $\lambda \in \mathfrak{a}^*$ then $|\varphi_\lambda(x)| \leq \varphi_o(x)$ and (cf. [GGA], p. 483)

(40)
$$|\varphi_o(a)| \leq c_o(1 + |\log a|)^d e^{-\rho(\log a)}, \quad a \in A^+,$$

where $d = \text{Card}(\Sigma_o^+)$ and c_o is a constant. In terms of the decomposition $G/K \simeq (K/M) \times \overline{A^+}$, dx behaves like $e^{2\rho(\log a)} da\, dk_M$. Let $c > 0$ be such that $\rho(\log a) \geq c|\log a|$ for $a \in A^+$ and note that

(41)
$$\int_{A^+} e^{-\delta|\log a|}(1 + |\log a|)^\ell da < \infty$$

for each $\delta > 0, \ell \in \boldsymbol{Z}^+$. This means that φ_o^2 and therefore $\varphi_\mu\varphi_\nu$ (for $\mu, \nu \in \mathfrak{a}^*$) belongs to $L^p(G)^\natural$ for each $p > 1$. Also $\varphi_\mu\varphi_\nu\varphi_{-\lambda}$ ($\mu, \nu, \lambda \in \mathfrak{a}^*$) belongs to $L^\natural(G)$ so $K(\lambda, \mu, \nu)$ is well defined and since $\varphi_\mu\varphi_\nu$ is positive definite (Prop. 12.4) K is positive and (38) holds (Theorem 12.12 (d)).

For the analyticity statement let $\epsilon < c/3$, let $B_\epsilon(0)$ be the ϵ-ball in $\mathfrak{a}^*$ and take $\mu, \nu, \lambda \in \mathfrak{a}^* + iB_\epsilon(0)$. By [GGA], (14), p. 476,

(42)
$$|H(ak)| \leq |\log a| \qquad a \in A,\ k \in K,$$

we have by (39)–(40), for $a \in A^+$,

$$|\varphi_\mu(a)\varphi_\nu(a)\varphi_{-\lambda}(a)| \leq c_o^3(1 + |\log a|)^{3d} e^{(3\epsilon - c)|\log a|} e^{-2\rho(\log a)}$$

so (37) gives a holomorphic function for $\mu, \nu, \lambda \in \mathfrak{a}^* + iB_\epsilon(0)$. This proves the proposition.

If $(\pi_\mu, \mathfrak{H}_\mu)$ and $(\pi_\nu, \mathfrak{H}_\nu)$ are the unitary representations associated to the positive definite functions φ_μ and φ_ν, respectively, formula (38) can be viewed as the decomposition of the tensor product $\pi_\mu \otimes \pi_\nu$ into irreducible representations. The result is a continuous analog of Theorem 9.9.

The kernel $K(\lambda, \mu, \nu)$ can now be used to express the spherical transform of the product of two functions in $\mathcal{D}^\natural(G)$.

Theorem 12.15. *Let $f, g \in \mathcal{D}^\natural(G)$ and $\widetilde{f}, \widetilde{g}$ their spherical transforms. Then the spherical transform of fg is given by*

$$(fg)^\sim(\lambda) = w^{-2} \int_{\mathfrak{a}^*} \int_{\mathfrak{a}^*} \widetilde{f}(\mu)\widetilde{g}(\nu)K(\lambda, \mu, \nu)|\mathbf{c}(\mu)|^{-2}|\mathbf{c}(\nu)|^{-2}d\mu d\nu.$$

Proof. We have

$$f(x)g(x) = w^{-2} \int_{\mathfrak{a}^*} \varphi_\mu(x)\widetilde{f}(\mu)|\mathbf{c}(\mu)|^{-2}d\mu \int_{\mathfrak{a}^*} \varphi_\nu(x)\widetilde{g}(\nu)|\mathbf{c}(\nu)|^{-2}d\nu.$$

Here we plug in (38) and compare the result with

$$f(x)g(x) = w^{-1} \int_{\mathfrak{a}^*} (fg)^\sim(\lambda)\varphi_\lambda(x)|\mathbf{c}(\lambda)|^{-2}d\lambda.$$

This gives the desired formula.

Remark. Writing (38) in the form

$$\varphi_\mu(a)\varphi_\nu(a) = \int_{\mathfrak{a}^*} \varphi_\lambda(a)\, dm_{\mu,\nu}(\lambda) \qquad a \in A \tag{43}$$

where $m_{\mu,\nu}$ is a positive W-invariant measure on $\mathfrak{a}^*$ we can compare it with the formula (cf. [GGA], Ch. IV, §10, (32))

$$\varphi_\lambda(b)\varphi_\lambda(c) = \int_{A^+} \varphi_\lambda(a)d\mu_{b,c}(a), \qquad b, c \in A^+, \tag{44}$$

where $\mu_{b,c}$ is a positive W-invariant measure on A with compact support. While the first formula is related to the product structure on $\mathcal{D}^\natural(G)$, the second is related to its convolution structure. It is obviously an interesting problem to relate $m_{\mu,\nu}$ and $\mu_{b,c}$ to the structure of G.

Examples. (i) *The groups $G = \boldsymbol{SL}(2, \boldsymbol{R})$, $\boldsymbol{SL}(2, \boldsymbol{C})$.*

Here we list without proofs and without our verification the explicit formulas for the measures $dm_{\mu,\nu}$ and $d\mu_{b,c}$ following Mizony [1976] and Koornwinder [1975], respectively. The latter paper has $d\mu_{b,c}$ for all G of rank one, but for the case $\boldsymbol{SL}(2,\boldsymbol{R})$ we state the result as it appears in the former paper.

A. The measure $dm_{\mu,\nu}$.

In both cases $\Sigma_0^+ = \{\alpha\}$ and we put (with $\langle , \rangle$ the Killing form)

$$r = \frac{\langle \lambda, \alpha \rangle}{\langle \alpha, \alpha \rangle}, \qquad s = \frac{\langle \mu, \alpha \rangle}{\langle \alpha, \alpha \rangle}, \qquad t = \frac{\langle \nu, \alpha \rangle}{\langle \alpha, \alpha \rangle}.$$

The case $G = \boldsymbol{SL}(2,\boldsymbol{R})$. Here

$$(45) \quad \varphi_\lambda(a) = F(\tfrac{1}{4} + \tfrac{ir}{2}, \tfrac{1}{4} - \tfrac{ir}{2}, 1, -\mathrm{sh}^2\alpha(\log a)), \qquad \mathbf{c}(\lambda) = \pi^{-\frac{1}{2}} \frac{\Gamma(ir)}{\Gamma(\frac{1}{2}+ir)},$$

and for a suitable normalization of dx,

$$(46) \qquad K(\lambda,\mu,\nu)/\mathrm{ch}(\pi r)\mathrm{ch}(\pi s)\mathrm{ch}(\pi t) =$$
$$|\Gamma(\tfrac{1}{4}+\tfrac{i}{2}(r+s+t))\Gamma(\tfrac{1}{4}+\tfrac{i}{2}(r+s-t))\Gamma(\tfrac{1}{4} + \tfrac{i}{2}(-r+s+t))\Gamma(\tfrac{1}{4}+\tfrac{i}{2}(r-s+t))|^2.$$

The case $G = \boldsymbol{SL}(2,\boldsymbol{C})$. Here

$$(47) \qquad \varphi_\lambda(a) = r^{-1}\frac{\sin(\lambda(\log a))}{\mathrm{sh}(\alpha(\log a))}, \quad \mathbf{c}(\lambda) = \frac{1}{2ir}$$

and the computation of (37) is fairly elementary. The result is, with dx suitably normalized,

$$(48) \qquad K(\lambda,\mu,\nu)\; rst =$$
$$\mathrm{th}(\tfrac{\pi}{2}(-r+s+t)) + \mathrm{th}(\tfrac{\pi}{2}(r-s+t) + \mathrm{th}(\tfrac{\pi}{2}(r+s-t)) + \mathrm{th}(\tfrac{\pi}{2}(r+s+t)).$$

B. The measure $d\mu_{b,c}$.

We put $x = \alpha(\log b), y = \alpha(\log c), z = \alpha(\log a)$.

The case $G = \boldsymbol{SL}(2,\boldsymbol{R})$. Here

$$d\mu_{b,c}(a) = L(z,x,y)\;\mathrm{sh}\, z\, dz,$$

where

$$L(z,x,y)=\begin{cases}0 & \text{if} \quad z\notin[|x-y|,x+y]\\ [\mathrm{sh}(\frac{x+y+z}{2})\mathrm{sh}(\frac{x+y-z}{2})\mathrm{sh}(\frac{x-y+z}{2})\mathrm{sh}(\frac{-x+y+z}{2})]^{-1} & \text{else.}\end{cases}$$

The case $G=\boldsymbol{SL}(2,\boldsymbol{C})$. Here

$$d\mu_{b.c}(a)=L(z,x,y)\,\mathrm{sh}^2 z\,dz,$$

where

$$L(z,x,y)=\begin{cases}(\mathrm{sh}\ x\ \mathrm{sh} y\ \mathrm{sh} z)^{-1} & \text{for } |x-y|\le z<x+y\\ 0 & \text{else.}\end{cases}$$

(ii) Since the groups $\boldsymbol{SL}(2,\boldsymbol{R})$ and $\boldsymbol{SL}(2,\boldsymbol{C})$ appear as motion groups for the hyperbolic spaces $\boldsymbol{H}^2$ and $\boldsymbol{H}^3$, respectively, we shall work out the analogs of $dm_{\mu,\nu}$ and $d\mu_{b,c}$ for the motion groups $\boldsymbol{M}(2)$ and $\boldsymbol{M}(3)$ of $\boldsymbol{R}^2$ and $\boldsymbol{R}^3$.

As in §7, let ψ_λ ($\lambda\in\boldsymbol{C}$) denote the radial eigenfunction of the Laplacian on $\boldsymbol{R}^n$ with eigenvalue $-\lambda^2$. If $\boldsymbol{e}_n=(0,\cdots,0,1)$ this is given by

$$\psi_\lambda(X)=\int\limits_{\boldsymbol{O}(n)} e^{i\lambda\langle \boldsymbol{e}_n,k\cdot X\rangle}dk=\Omega_n^{-1}\int\limits_{\boldsymbol{S}^{n-1}} e^{i\lambda\langle w,X\rangle}dw$$

which by [GGA], Introduction, Lemma 3.6 equals

$$\psi_\lambda(X)=\Omega_n^{-1}(2\pi)^{\frac{n}{2}}\frac{J_{\frac{n}{2}-1}(\lambda r)}{(\lambda r)^{\frac{n}{2}-1}},\qquad r=|X|. \tag{49}$$

For $\lambda\in\boldsymbol{R}$ the representation of $\boldsymbol{M}(n)$ asociated to ψ_λ is unitary and since $\psi_\lambda=\psi_{-\lambda}$ we can take $\lambda\ge 0$. Then we can also write

$$\psi_\lambda(X)=(M^{\lambda|X|}j)(0),\qquad j(Y)=e^{i\langle \boldsymbol{e}_n,Y\rangle}, \tag{50}$$

where $(M^r f)(Y)$ is the average of f on $S_r(Y)$. Being a power series in the Laplacian (Ch. II, §2, No. 4) M^r has j as an eigenfunction so the iteration formula in Exercise **E3** gives a formula for $\psi_\mu(X)\psi_\nu(X)$ which we now write down.

The case $\boldsymbol{M}(2)$. Here

$$\psi_\lambda(X)=J_o(\lambda|X|),\qquad d\mu(\lambda)=\lambda\,d\lambda$$

with $d\mu$ as in Theorem 12.11. From Exercise **E3** we deduce the following counterpart to (43) and (46):

$$\psi_\mu(X)\psi_\nu(X) = \frac{2}{\pi}\int_{\mathbf{R}^+} K(\lambda,\mu,\nu)\psi_\lambda(X)\lambda\,d\lambda, \tag{51}$$

where

$$K(\lambda,\mu,\nu) = \begin{cases} 0 \text{ for } \lambda \notin [|\mu-\nu|,\mu+\nu] \\ [(\lambda+\mu+\nu)(\lambda+\mu-\nu)(\lambda-\mu+\nu)(-\lambda+\mu+\nu)]^{-\frac{1}{2}} & \text{else.}\end{cases}$$

The case $\boldsymbol{M}(3)$. Here (49) reduces to

$$\psi_\lambda(X) = \sin(\lambda|X|)/(\lambda|X|), \qquad d\mu = \lambda^2 d\lambda$$

and again Exercise **E3** gives the counterpart to (43) and (48):

$$\psi_\mu(X)\psi_\nu(X) = \frac{1}{2}\int_{\mathbf{R}^+} K(\lambda,\mu,\nu)\psi_\lambda(X)\lambda^2 d\lambda, \tag{52}$$

where

$$K(\lambda,\mu,\nu) = \begin{cases} (\lambda\mu\nu)^{-1} & \text{for } |\mu-\nu| < \lambda < \mu+\nu \\ 0 & \text{else.}\end{cases}$$

Formulas (51)–(52) give the flat analogs of the measure $dm_{\mu,\nu}$. Since the formula (49) is symmetric in r and λ the measures $dm_{\mu,\nu}$ and $d\mu_{b,c}$ have the same flat analogs.

EXERCISES AND FURTHER RESULTS.

A. Differential Operators.

1. As in §3 let

$$\eta_\lambda(g) = e^{(i\lambda-\rho)(H(g))}.$$

Show that for each $D \in \boldsymbol{D}(G)$ and each $f \in \mathcal{E}(G)$ right invariant under K,

$$\int_G (D\eta_\lambda)(g)f(g)dg = (D\eta_\lambda)(e)\int_G \eta_\lambda(g)f(g)dg.$$

2. In the notation of §10 show that

$$\gamma(\Omega)(\rho_0 + i\lambda) = L_A(i\lambda) - \langle\rho,\rho\rangle,$$

where $\Omega \in \boldsymbol{Z}(G)$ is the Casimir operator.

3.* In §10 we have an isomorphism $\gamma : \boldsymbol{Z}(G) \longrightarrow I(\mathfrak{h}^{\boldsymbol{C}})$ and in [GGA], Ch. V, Theorem 1.9 we obtained an isomorphism $\gamma : D \longrightarrow \delta^{\frac{1}{2}}\Delta(D) \circ \delta^{-\frac{1}{2}}$ of $\boldsymbol{Z}(G)$ onto $I(\mathfrak{h}^{\boldsymbol{C}})$ obtained by viewing D as a differential operator on a compact real form U of $G^{\boldsymbol{C}}$, $\Delta(D)$ being its radial part on a maximal torus of U. These isomorphisms coincide (Harish-Chandra [1956b], Theorem 2).

B. Rank one Results.

1. (G/K of rank one). For $\delta \in \widehat{K}_M$ let as in §11, No. 1,

$$\varphi_{\lambda,\delta}(x) = \int_K e^{(i\lambda+\rho)(A(x,kM))}\langle\delta(k)v, v\rangle dk$$

and prove that

$$e^{(i\lambda+\rho)(A(x,eM))} = \sum_{\delta\in\widehat{K}_M} d(\delta)\varphi_{\lambda,\delta}(x).$$

2.* The rank one theory from §11, No. 1 can be extended to spaces G/H where $G = \boldsymbol{O}(p,q), \boldsymbol{U}(p,q)$ or $\boldsymbol{Sp}(p,q)$, H the stabilizer of a line $\boldsymbol{F}(1,0,\cdots 0)$ in $\boldsymbol{F}^{p+q}(\boldsymbol{F} = \boldsymbol{R}, \boldsymbol{C}$ or $\boldsymbol{H})$. (See Faraut [1979] and Schlichtkrull [1987] and references given there.)

3.* Suppose $X = G/K$ has rank one and let 1_r denote the characteristic function of the ball $B_r(o)$. Then the function $\lambda \longrightarrow \widetilde{1}_r(\lambda)(= \widetilde{1}_r(\lambda, b))$ has all its zeros on the real space $\mathfrak{a}^*$ (Berenstein and Shahshahani [1983] and Shahshahani and Sitaram [1987]).

C. Jacobi Functions and Jacobi Transforms.

In [GGA], Ch. IV, Exercise B8 it is pointed out (following Harish-Chandra [1958a]) how the spherical function $\varphi_\lambda(\exp H)$ for G/K of rank one is a hypergeometric function of $\operatorname{sh}^2(\alpha(H))$. For the compact case U/K there is an analogous formula for the spherical function ([GGA], Ch. V, Theorem 4.5) but then the hypergeometric function reduces to a (Jacobi) polynomial.

More generally, the *Jacobi function*

$$\varphi_\mu^{(\alpha,\beta)}(t) \qquad (\alpha, \beta, \mu \in \boldsymbol{C},\ \alpha \neq -1, -2, \cdots)$$

is defined as the even smooth function on $\boldsymbol{R}$ which has value 1 at $t = 0$ and which satisfies the differential equation

$$\psi''(t) + [(2\alpha+1)\coth t + (2\beta+1)\operatorname{th} t]\psi'(t) + [\mu^2 + (\alpha+\beta+1)^2]\psi(t) = 0.$$

Then

$$\varphi_\mu(t) = \varphi_\mu^{(\alpha,\beta)}(t) = F(\tfrac{1}{2}(\alpha+\beta+1-i\mu),\ \tfrac{1}{2}(\alpha+\beta+1+i\mu);\ \alpha+1;\ -\mathrm{sh}^2 t)$$

and the *Jacobi transform* $f \longrightarrow \widetilde{f}$ is given by

$$\widetilde{f}(\mu) = \int_{-\infty}^{\infty} f(t)\varphi_\mu(t)\Delta(t)dt, \qquad (f \text{ even}),$$

where

$$\Delta(t) = \Delta_{\alpha,\beta}(t) = (2\,\mathrm{sh}\ t)^{2\alpha+1}(2\,\mathrm{ch}\ t)^{2\beta+1}.$$

1.* The mapping $f \longrightarrow \widetilde{f}$ is a bijection of the space of even functions in $\mathcal{D}(\boldsymbol{R})$ onto the space of even functions in $\boldsymbol{C}$ of exponential type (cf. Flensted-Jensen [1972] and Koornwinder [1975]).

2. Using the transformation formula

$$F(a,b;c;z) = (1-z)^{-a}F(a,c-b;c;z/(z-1))$$

the Jacobi function can be written

$$\varphi_\mu^{(\alpha,\beta)}(t) = (\mathrm{ch}\ t)^{i\mu-\alpha-\beta-1}F(\tfrac{1}{2}(\alpha+\beta+1-i\mu), \tfrac{1}{2}(\alpha-\beta+1-i\mu); \alpha+1; \mathrm{th}^2\, t).$$

Deduce that the function $\varphi_{\lambda,\delta}(a_t \cdot o)$ from Theorem 11.2 can be written

$$\varphi_{\lambda,\delta}(a_t \cdot o) = c_{\lambda,\delta}\,\mathrm{sh}^s\, t\,\mathrm{ch}^r\, t\,\varphi_{\lambda(H)}^{(\alpha,\beta)}(t),$$

with

$$\alpha = s + \tfrac{1}{2}(m_\alpha + m_{2\alpha} - 1), \qquad \beta = \tfrac{1}{2}(m_{2\alpha} - 1) + r$$

(Koornwinder [1984], p. 29). For this choice of α, β the theorem in **C1** resembles, at least formally, the range statement in our Corollary 11.6.

D. The Compact Case.

1. Let $U = \boldsymbol{SU}(2)$ and prove the Fourier inversion formula

$$f(e) = \sum_{\pi\in\widehat{U}} \chi_\pi(f)d_\pi \qquad f \in \mathcal{E}(U),$$

(where χ_π and d_π, respectively, are the character and degree of π) through the following steps:

(i) When $\boldsymbol{SU}(2)$ is identified with $\boldsymbol{S}^3$ by means of the mapping

$$x = \begin{pmatrix} x_1 + ix_2 & x_3 + ix_4 \\ -x_3 + ix_4 & x_1 - ix_2 \end{pmatrix} \longrightarrow (x_1, x_2, x_3, x_4)$$

the conjugacy classes become sections of $\boldsymbol{S}^3$ with hyperplanes in $\boldsymbol{R}^4$ perpendicular to the x_1-axis.

(ii) Let $T = \{t_\theta : \ 0 \leq \theta \leq 2\pi\}$ where

$$t_\theta = \begin{pmatrix} e^{i\theta} & 0 \\ 0 & e^{-i\theta} \end{pmatrix}.$$

Prove the special case of Weyl's integral formula (with du normalized Haar measure on U)

$$\int_U f(u)du = \tfrac{1}{4\pi} \int_0^{2\pi} |e^{i\theta} - e^{-i\theta}|^2 \int_U f(ut_\theta u^{-1}) du \, d\theta$$

by polar coordinate integration on $\boldsymbol{S}^3$.

(iii) Prove that the orbital integral

$$F_f(\theta) = (e^{i\theta} - e^{-i\theta}) \int_U f(ut_\theta u^{-1}) du$$

satisfies

$$\left(\frac{dF_f}{d\theta}\right)_{\theta=0} = 2if(e).$$

(iv) For $\ell \geq 0$ let π_ℓ be the natural representation of U on the space V_ℓ of polynomials $\Sigma_{p+q=\ell} a_{p,q} z^p w^q$. Show that the corresponding character χ_ℓ is given by

$$\chi_\ell(t_\theta) = e^{i\ell\theta} + e^{i(\ell-2)\theta} + \cdots + e^{-i\ell\theta}$$

and

$$\chi_\ell(f) = \frac{1}{2\pi} \int_0^{2\pi} F_f(\theta) e^{-i(\ell+1)\theta} d\theta$$

(v) Show that π_ℓ is irreducible.

(vi) Writing $(dF_f/d\theta)_{\theta=0}$ as the sum of Fourier coefficients show that

$$f(e) = \sum_{\ell=0}^{\infty} (\ell+1)\chi_\ell(f)$$

(vii) Show that $\widehat{U} = \{\pi_\ell : \ell \in \boldsymbol{Z}^+\}$.

2. Let φ be a spherical function on a coset space U/K (K compact) and let $f \in C_c(U/K)$. Lifting φ and f to functions $\widetilde{\varphi}$ and $\widetilde{f}$ on U prove that

$$(\widetilde{\varphi} * \widetilde{f})(u) = \widetilde{\varphi}(u) \int_U \widetilde{f}(v)\widetilde{\varphi}(v^{-1})dv.$$

More generally, each joint eigenfunction of $\boldsymbol{D}(U/K)$ is an eigenfunction of the convolution operator $F \longrightarrow F \times f$.

3. Using Harish-Chandra's formula

$$\boldsymbol{c}(\lambda) = \prod_{\alpha>0} \frac{\langle \rho, \alpha\rangle}{\langle i\lambda, \alpha\rangle}$$

for the $\boldsymbol{c}$-function for G complex and Vretare's formula (24) §9 for $d(\mu)$ deduce Weyl's product formula for $d(\mu)$.

E. The Flat Case.

1.* For $\lambda \in \mathfrak{a}_c^*$ rewrite (4) §8 in the form

$$(P_\lambda F)(X) = (Q_\lambda \dot{F})(X) \qquad X \in \mathfrak{p}^c$$

where

$$(Q_\lambda f)(X) = \int_{K/K_\lambda} e^{iB(k\cdot A_\lambda, X)} f(\dot{k})d\dot{k}, \qquad \dot{F}(\dot{k}) = \int_{K_\lambda/M} F(k\ell M)d\ell_M.$$

Then if $f \in C(K/K_\lambda)$,

$$\lim_{t\longrightarrow\infty} \left(\frac{Q_\lambda f}{\psi_\lambda}\right)(-itk\cdot \overline{A}_\lambda) = f(kA_\lambda),$$

the bar denoting conjugation of $\mathfrak{a}^c$ with respect to $\mathfrak{a}$ (Korányi [1982a]).

For λ regular, $P_\lambda = Q_\lambda$ so this extends the "if" part of Theorem 8.2; see also Champetier and Delorme [1981] for another proof of the injectivity of Q_λ and Stetkaer [1991] for a further generalization.

2. Let $\lambda \in \mathfrak{a}_c^*$ and K_λ the stabilizer of λ in K. Let h be a K_λ-harmonic polynomial on $\mathfrak{p}$. Then the function

$$X \longrightarrow h(X)e^{iB(A_\lambda, X)} \qquad X \in \mathfrak{p}$$

belongs to the joint eigenspace $\mathcal{E}_\lambda(\mathfrak{p})$ and

$$\int_K h(k\cdot X)e^{iB(A_\lambda,k\cdot X)}dk = h(0)\psi_\lambda(X).$$

3. For $x \in \boldsymbol{R}^n$ let $M^x = M^{|x|}$ be the spherical mean value operator

$$(M^x f)(z) = \int_K f(z+k\cdot x)dk, \qquad K = \boldsymbol{O}(n).$$

Prove that if $y \in \boldsymbol{R}^n$, $|x| = r$, $|y| = s$,

$$(M^y M^x f)(z) = \int_K (M^{x+k\cdot y} f)(z)\, dk =$$

$$\frac{\Omega_{n-1}}{\Omega_n}\int_0^\pi (M^{(r^2+s^2-2rs\cos\theta)^{\frac{1}{2}}} f)(z)\sin^{n-2}\theta\, d\theta,$$

where Ω_k is the area of $\boldsymbol{S}^{k-1}$. Deduce

$$(M^s M^r f)(z) =$$

$$\frac{2\Omega_{n-1}}{\Omega_n(2rs)^{n-2}}\int_{|r-s|}^{r+s} [(t+r-s)(t+r+s)(t-r+s)(-t+r+s)]^{\frac{n-3}{2}} t(M^t f)(z)\, dt$$

F. The Noncompact Case.

1. Deduce from Theorem 5.3 that for the Hilbert-Schmidt norm $\|A\| = \mathrm{Tr}(AA^*)^{\frac{1}{2}}$

$$\|\boldsymbol{C}_s(\lambda)\|^2 = |\boldsymbol{c}(\lambda)|^2 \dim V_\delta^M \qquad \text{for } \lambda \in \mathfrak{a}^*.$$

2. Show that the topology of $\mathfrak{a}_c^*/W$ coincides with the compact open topology of the family φ_λ $(\lambda \in \mathfrak{a}_c^*)$.

3.* For the symmetric space $X = G/K$ the following conditions are equivalent:

(i) Each closed G-invariant subspace of $L^2(X)$ is invariant under complex conjugation.

(ii) Each spherical function $\varphi_\lambda(\lambda \in \mathfrak{a}^*)$ is real-valued.

(iii) The Weyl group W contains $-I$ (where I is the identity operator of $\mathfrak{a}$).

If these conditions are satisfied then each closed G-invarinat subalgebra of $C_o(X)$ containing a nonzero function in $L^1(X)$ coincides with $C_o(X)$. (Here $C_o(X)$ denotes the Banach space of continuous functions on X vanishing at ∞) (Wolf [1964], Agranovskii [1985]).

4. Let $X = G/K$ be a symmetric space of the noncompact type. Let $f \neq 0$ in $L^1(X) \cap L^2(X)$ and V_f the closed G-invariant subspace of $L^2(X)$ generated by f. Then $V_f = L^2(X)$ (Sitaram [1988]).

(Hint: First assume f K-invariant and show by spherical transform theory that the functions $\varphi \times f$ ($\varphi \in \mathcal{D}^\natural(X)$) are dense in the space $L^2(X)^\natural$ of K-invariants in $L^2(X)$.)

NOTES

§1. The Fourier transform (4) on X was introduced in the author's paper [1965b], including the inversion and the Plancherel formula. The symmetry formula (7) was obtained in the course of the proofs ([1970a], p. 116). This Fourier transform on X has the merit of being scalar-valued in contrast to the standard operator-valued Fourier transform on G given by (1) below.

§2. The generalized spherical function $\Phi_{\lambda,\delta}$ is a special case of Eisenstein integrals considered by Harish-Chandra in [1959], [1960] where (and in Warner [1972], Ch. 9) an expansion more general than that of Theorem 2.7 is proved. The proof takes 50 pages. In [1976] the author proved Theorem 2.1 showing that the generalized spherical function is a suitable derivative of the zonal spherical function. This leads, in particular, to a shorter proof of the expansion in Theorem 2.7. The functional equation in Theorem 2.10 is proved by the author in [1970a], p. 95 (also [1973a]). It plays an important role in the topological Paley-Wiener theorem (Theorem 5.11) in [1976]. The proof of Lemma 2.11 is related to that of Lemma 1.5.12 in Kostant [1975a].

§3. Theorem 2.4 on harmonic polynomials (Kostant [1963], Kostant and Rallis [1971], and Wallach [1992], 11.2.4) leads to the Q^δ-matrices of Parthasarathy, Rao and Varadarajan [1967] (for G complex) and Kostant [1975a] (for G real). These authors use the Q^δ in an algebraic study of the spherical principal series employing the rank-one reduction. The relationship with the operator $B(\lambda, s)$ (the δ-component of the intertwining operator $I_{\lambda,s}$) in Theorem 3.5 is from Johnson and Wallach [1977] (see also Kostant [1991]). The product formula for $\det(Q^\delta)$ in Theorem 4.2 is proved by Kostant in [1975a] in an algebraic fashion; our method is analytic and uses the method of Johnson and the author [1969] for determining for which λ the spherical function φ_λ is bounded. Through the rank-one reduction the appearance of the subcase $X = \boldsymbol{H}^2$ causes us some complications in the proof because as explained before Lemma 3.9 the spherical functions on the circle can be either $e^{i\ell\theta}$ or $\cos \ell\theta$.

§4. The simplicity criterion (Theorem 4.4) was proved by the author in [1970a] and [1976].

§5. The original Paley-Wiener theorem [1934] described $L^2(B_r(o))^\sim$ in $\boldsymbol{R}^n$. With distribution theory, the interest shifted to $\mathcal{D}(\boldsymbol{R}^n)^\sim$ and $\mathcal{E}'(\boldsymbol{R}^n)^\sim$ (Schwartz [1966], Hörmander [1963]). The Paley-Wiener theorem for the symmetric space X (Theorem 5.1) was proved by the author [1973a]. Torasso in [1977] gave another proof of Theorem 5.1. This proof also yielded the new result Corollary 5.19 as a by-product. The proof of it in the text is rather different. Corollary 5.17 is from Dadok [1976].

The material in No. 5–6 on the K-types, in particular the topological Paley-Wiener theorem (Theorem 5.11) and the inversion in Theorem 5.16 are from the author's paper [1976]. These results extend earlier work on the spherical transform documented in [GGA], p. 493.

For the group G the Paley-Wiener problem is to describe the image of $\mathcal{D}(G)$ under the operator Fourier transform

$$f \longrightarrow \pi(f) = \int_G f(x)\pi(x)dx, \qquad \pi \in \widehat{G}. \tag{1}$$

The problem is usually restricted to f $K \times K$-finite under a given double representation of K. For $G = \boldsymbol{SL}(2, \boldsymbol{R})$ this was carried out by Ehrenpreis and Mautner [1957], by Campoli [1977], Kawazoe [1979] and Johnson [1979a] for G of real rank one, by Zhelobenko [1974] for G complex and by Delorme [1982] for G with all Cartan subgroups conjugate. General results of varying completeness were obtained by Johnson [1976b], [1979], Kawazoe [1979], [1980], Clozel and Delorme [1984] and Arthur [1983]. For the case of non-Riemannian symmetric spaces, see Delorme and Flensted-Jensen [1991] and forthcoming work of Ban and Schlichtkrull.

The extension to $\mathcal{E}'(X)$ in Corollary 5.9 was proved by the author in connection with Theorem 8.5 in [1973a] and by Eguchi, Hashizume and Okamoto [1973]. The simple proof in the text is from Dadok [1976].

For the sake of brevity we have not discussed the Schwartz spaces $\mathcal{J}^p(G) \subset \mathcal{E}^\natural(G)$ (see [GGA] p. 489 for the definition). These and the analogous spaces in $\mathcal{E}(G)$ play a prominent role in Harish-Chandra's work. These spaces are studied in detail in the book Gangolli-Varadarajan [1988]. The image $\mathcal{J}^2(G)^\sim$ was determined by Harish-Chandra [1958b] and $\mathcal{J}^p(G)^\sim$ $(1 \leq p \leq 2)$ by Trombi and Varadarajan [1971] (see also Clerč [1980], where the case of general p is reduced to the case $p = 2$ and Helgason [1970a], Ch. II, §2 where the case $p = 1$ is done for G either complex or of real rank one). A strikingly simple proof of these results was given by Anker [1991b] on the basis of the spherical Paley-Wiener theorem; he simply proves that the inverse spherical transform $\widetilde{f} \longrightarrow f$ from $\mathcal{H}_W(\mathfrak{a}_\mathfrak{c}^*)$ onto $\mathcal{D}^\natural(G)$ is continuous in the Schwartz space topologies. Eguchi in [1979] has given a similar characterization of the *Fourier transforms* of the Schwartz spaces $\mathcal{S}^p(X)$. So far, Anker's simplification has not been extended to this case.

§6. The eigenfunction representation (Theorem 6.1) and the eigenspace irreducibility criterion (Theorem 6.2) are from the author's papers [1970a] and [1976]. Another proof of Theorem 6.1 comes from the expansion theory in Ban and Schlichtkrull [1987] (Theorem 14.1).

§§7–8. The results of these sections which deal with K-invariant Paley-Wiener theorems on the flat space X_o, joint eigenfunctions and eigenspace representations, are from the author's papers [1980a], [1992b]. Of many papers dealing with the representation theory of G we mention Champetier and Delorme [1981] (which includes a generalization of Theorem 8.2) and Rader [1988]. In this context we call particular attention to the papers of Rouviére ([1986], [1990], [1991a], and [1993]) which establish profound connections between the analysis on G/K (particularly the convolution structure) and on G_o/K, also for the non-Riemannian case.

§9. Theorem 9.3 is due to Sherman [1977], and for U/K of rank one he gave a stronger (global) version in [1990]. The mapping (17) was also used by Clerč [1976] and Stanton [1976] in a different context. The product decomposition in Theorem 9.9 and the ensuing relation of $d(\mu)$ with the $\mathbf{c}$-function (Theorem 9.10) is due to Vretare [1976], [1977]. In the proof of Theorem 9.10 I have used a method kindly communicated to me by T. Oshima. Another and more general method has been given by Heckman (cf. Heckman and Schlichtkrull [1994]).

§10. Theorems 10.1 and 10.3 relating the center $\boldsymbol{Z}(G)$ to $\boldsymbol{D}(G/K)$ are from the author's paper [1992a]. This has a certain bearing on Flensted-Jensen's method [1978] for spherical functions as explained in [GGA], p. 491.

§11. The explicit formula for the Eisenstein integral (Theorem 11.2) is from Helgason [1976] and so are the Paley-Wiener theorem refinements in Corollaries 11.5–11.6. The formula for Q^δ in Corollary 11.3 (in a different parametrization of $\widehat{K}_M$) was determined in an algebraic fashion by Kostant [1975a], Ch. II, §8. Our analytic derivation which appeared in [1976] depends in a crucial fashion on properties of the Gamma function.

§**12.** The results on positive definite functions come from Gelfand and Raikov [1943] and Godement [1948] (in particular Theorem 12.2). Theorem 12.6 is from Gelfand and Naimark [1952] and Godement [1957]. Theorem 12.9 goes back to Herglotz for $G = \boldsymbol{T}$, to Bochner [1932] for $G = \boldsymbol{R}^n$, to Weil [1940] for G abelian and Godement [1957] for the result in the text. An extension to positive definite distributions was given by Barker [1975]. Theorem 12.11 is also from Godement [1957] (a less specific version is in Mautner [1951]). Our proof uses simplifications from van Dijk [1969]. Proposition 12.14 is from Flensted-Jensen and Koornwinder [1979a,b] and Mayer-Lindeberg [1981]. In §12 we have also used Faraut [1982a], Koornwinder [1984] and, Gangolli-Varadarajan [1988].

CHAPTER IV

THE RADON TRANSFORM ON X AND ON X_o. RANGE QUESTIONS.

One of the principal problems concerning the Radon transform mentioned in Chapter I, §1, No. 2, is the range problem. In this chapter we turn to this problem for the Radon transform $f \longrightarrow \widehat{f}$ and its dual $\varphi \longrightarrow \overset{\vee}{\varphi}$ for the symmetric space $X = G/K$ and its horocycle space $\Xi = G/MN$.

This is carried out in §§1–3, using tools developed in Chapter III. The characterization of the range $\mathcal{D}(X)^\wedge$, while less transparent than that for $\mathcal{D}(\boldsymbol{R}^n)^\wedge$, shows that this range is closed. This in turn shows that $\mathcal{D}(X)^\wedge = \mathcal{N}^\perp$ where $\mathcal{N}$ is the kernel of the dual transform. Invoking several further tools we prove a similar characterization of $\mathcal{E}'(X)^\wedge$ which in turn implies the surjectivity of $\varphi \longrightarrow \overset{\vee}{\varphi}$.

Restricting to fixed K-types yields more explicit descriptions in §3 of the range and the kernel. The case of K-invariant functions treated in §4 gives particularly precise results.

Finally in §5 we discuss the principal problems for the Radon transform (inversion-, range- and support problems) for the horocycle planes in the tangent space X_o. Again the principal tools come from the tangent space Fourier analysis developed in Chapter III.

§1. THE SUPPORT THEOREM.

We shall follow the notation of §5 in the last chapter where $\widetilde{f}$ and $\widehat{f}$, respectively, denote the Fourier transform and the Radon transform of a function f on X. We shall now prove a result, *the support theorem* for $f \longrightarrow \widehat{f}$, which will be applied later to differential equations on X.

Theorem 1.1. *Let $f \in \mathcal{D}(X)$ and V a closed ball in X. Assume*

$$\widehat{f}(\xi) = 0 \qquad \text{if } \xi \cap V = \emptyset. \tag{1}$$

Then

$$f(x) = 0 \qquad \text{for } x \notin V. \tag{2}$$

This is the symmetric space analog of the support theorems for $\boldsymbol{R}^n$ and the hyperbolic space $\boldsymbol{H}^n$ ([GGA], Ch. I, Theorems 2.6 and 4.2) but Theorem 1.1 is more difficult and requires an entirely different proof. For this consider the connections from (26) §5,

$$\widetilde{f}(\lambda, kM) = \int_A \widehat{f}(ka \cdot \xi_o)e^{(-i\lambda+\rho)(\log a)}da, \tag{3}$$

$$\widehat{f}(ka \cdot \xi)e^{\rho(\log a)} = \int_{\mathfrak{a}^*} \widetilde{f}(\lambda, kM)e^{i\lambda(\log a)}d\lambda. \tag{4}$$

For Theorem 1.1 we may assume V is centered at the origin, so $V = C\ell(B_R(o))$ for some $R > 0$. Since the point $a \cdot o$ minimizes the distance from o to $a \cdot \xi_o$ ([DS], Ch. VI, Ex. B2) it is clear that $ka \cdot \xi_o \cap V = \emptyset$ if $|\log a| > R$. Thus (1) implies that $\widehat{f}(ka \cdot \xi_o) = 0$ for $|\log a| > R$ and all $k \in K$. For each $D \in \boldsymbol{D}(A)$ the derivative $D_a(\widehat{f}(ka \cdot \xi_o))$ has a bound on A uniformly in k. Thus (3) and the Euclidean Paley-Wiener theorem on A imply that $\widetilde{f}$ has uniform exponential type R (Ch. III §5 (2)) so by Theorem 5.1, Ch. III, $\operatorname{supp}(f) \subset C\ell(B_R(o))$ proving Theorem 1.1.

It is easy to extend Theorem 1.1 to distributions. For $R > 0$ let

$$\beta_R = \{\xi \in \Xi : d(o, \xi) < R\}, \qquad \sigma_R = \{\xi \in \Xi : d(o, \xi) = R\} \tag{5}$$

Corollary 1.2. *Let $T \in \mathcal{E}'(X)$ satisfy the condition*

$$\operatorname{supp}(\widehat{T}) \subset \overline{\beta_R}.$$

Then

$$\operatorname{supp}(T) \subset \overline{B_R(o)}.$$

Proof. For $F \in \mathcal{D}(G)$, $\varphi \in \mathcal{D}(\Xi)$ we consider the mixed convolution

$$(F \times \varphi)(\xi) = \int_G F(g)\varphi(g^{-1} \cdot \xi)dg$$

for which

$$(F \times \varphi)^\vee(x) = \int_G F(g)\check{\varphi}(g^{-1} \cdot x)dg.$$

The support assumption on $\widehat{T}$ amounts to

$$T(\check{\varphi}) = 0 \qquad \text{if} \qquad \operatorname{supp}(\varphi) \cap \overline{\beta_R} = \emptyset.$$

Let $\epsilon > 0$ and let $\varphi \in \mathcal{D}(\Xi)$ satisfy $\operatorname{supp}(\varphi) \cap \overline{\beta_{R+\epsilon}} = \emptyset$. Let N_ϵ be a symmetric neighborhood of e in G such that $g\beta_R \subset \beta_{R+\epsilon}$ for $g \in N_\epsilon$. Then

$$g \cdot \operatorname{supp}(\varphi) \cap \beta_R = \emptyset \qquad \text{for} \qquad g \in N_\epsilon.$$

If F is symmetric and $\operatorname{supp}(F) \subset N_\epsilon$ we have

$$\operatorname{supp}(F \times \varphi) \cap \overline{\beta_R} = \emptyset.$$

Thus the distribution $F \times T$ defined by

$$(F \times T)(f) = \int_G F(g) T(f^g)\, dg$$

satisfies

$$(F \times T)(\overset{\vee}{\varphi}) = \int_G F(g) T((\overset{\vee}{\varphi})^g)\, dg$$

$$= \int_G F(g) \left(\int_X \overset{\vee}{\varphi}(g^{-1} \cdot x)\, dT(x) \right) dg = T((F \times \varphi)^\vee) = 0.$$

Hence

$$(F \times T)^\wedge(\varphi) = (F \times T)(\overset{\vee}{\varphi}) = 0$$

so $(F \times T)^\wedge$ has support in $\overline{\beta_{R+\epsilon}}$. On the other hand, $F \times T$ is a function (as we see by lifting T to G) so by Theorem 1.1, $\operatorname{supp}(F \times T) \subset \overline{B_{R+\epsilon}(o)}$. Letting $\epsilon \longrightarrow 0$ we obtain the corollary.

Remark. Let us call a function $f \in \mathcal{E}(X)$ rapidly decreasing if for each $g \in G$ the function $h \longrightarrow \int f(gkh \cdot o)\, dk$ belongs to the Schwartz space $\mathcal{J}^2(G)$ [GGA], p. 489). Since the inversion formula (Theorem 1.3 in Ch. III) holds for such f we can replace the assumption $f \in \mathcal{D}(X)$ in the support theorem by the assumption "f is rapidly decreasing".

We state here another special support result which has applications to the wave equation on X. Accordingly the proof is postponed until Chapter V, §5. We use the notation in (5) above.

Proposition 1.3. *Suppose G has all its Cartan subgroups conjugate. Then for $T \in \mathcal{E}'(X)$*

$$\operatorname{supp}(\widehat{T}) \subset \sigma_R \Longrightarrow \operatorname{supp}(T) \subset S_R(o).$$

§2. THE RANGES $\mathcal{D}(X)^\wedge$, $\mathcal{E}'(X)^\wedge$ AND $\mathcal{E}(\Xi)^\vee$.

The range $\mathcal{D}(X)^\wedge$ can be described by combining Theorem 1.1 and Theorem 5.1, Ch. III. If $D \in \boldsymbol{D}(A)$ then the function

$$(1) \qquad P_D(b,\lambda) = \left\{D_a\left(e^{(i\lambda+\rho)(A(a\cdot o,b))}\right)\right\}_{a=e} \qquad b \in B$$

is a polynomial in $\lambda \in \mathfrak{a}^*_C$ with coefficients in $\mathcal{E}(B)$. Let $L(b,D)$ denote the differential operator on A which under the Fourier transform on $\mathfrak{a}^*$ corresponds to multiplication by $P_D(b,\lambda)$, i.e.

$$(L(b,D)F)(a) = \int_{\mathfrak{a}^*} P_D(b,\lambda)F^*(\lambda)e^{i\lambda(\log a)}d\lambda$$

if

$$F(a) = \int_{\mathfrak{a}^*} F^*(\lambda)e^{i\lambda(\log a)}d\lambda.$$

We can view $L(\cdot,D)$ as an element in the tensor product $\mathcal{E}(B) \otimes \boldsymbol{D}(A)$.

Theorem 2.1. *The Radon transform $f \longrightarrow \widehat{f}$ is a bijection of $\mathcal{D}(X)$ onto the set of functions $\varphi \in \mathcal{D}(\Xi)$ satisfying the condition:*

For each $u \in K$ and each $D \in \boldsymbol{D}(A)$ the function

$$(2) \qquad a \longrightarrow \int_K L(ukM,D)_a(e^{\rho(\log a)}\varphi(ka\cdot\xi_o))dk, \qquad a \in A,$$

is W-invariant.

Proof. Because of (4) §1 we must first characterize the Euclidean Fourier transform of the functions $\lambda \longrightarrow \psi(\lambda,kM)$ which satisfy (3) Ch. III §5 and are holomorphic of uniform exponential type. Writing in the quoted formula, $x = ka\cdot o$, the condition reads

$$(3) \qquad \int_B e^{(i\lambda+\rho)(A(a\cdot o,b))}\psi(\lambda,k\cdot b)db \quad \text{is } W\text{-invariant in } \lambda$$

for each $k \in K$. As a function of a the function (3) is analytic and thus determined by its derivatives at $a = e$. Thus (3) is equivalent to the condition that for each $D \in \boldsymbol{D}(A)$ and each $k \in K$

$$(4) \qquad \int_B P_D(b,\lambda)\psi(\lambda,k\cdot b)db \quad \text{is } W\text{-invariant in } \lambda.$$

Because of (4) §1 the theorem follows by taking Fourier transform of the expression in (4).

Consider now the bilinear form

$$(\varphi, \psi) \longrightarrow \int_{\Xi} \varphi(\xi)\psi(\xi)d\xi \quad \text{on } \mathcal{D}(\Xi) \times \mathcal{E}(\Xi).$$

This being nonsingular there is induced a topology $\sigma(\mathcal{D}(\Xi), \mathcal{E}(\Xi))$ (Ch. I, §2, No. 5) on $\mathcal{D}(\Xi)$ which is weaker than the customary one.

Corollary 2.2. *The range $\mathcal{D}(X)^\wedge$ is a closed subspace of $\mathcal{D}(\Xi)$ in the topology $\sigma(\mathcal{D}(\Xi), \mathcal{E}(\Xi))$ (and thus also in the usual topology of $\mathcal{D}(\Xi)$).*

For this we first observe that a function $f \in \mathcal{D}(A)$ is W-invariant if and only if for each root $\alpha \in \Sigma$

$$\int_A f(a)\psi(a)da = 0 \tag{5}$$

for all $\psi \in \mathcal{E}(A)$ satisfying $\psi^{s_\alpha} = -\psi$ (s_α being the reflection corresponding to α). In fact for each $h \in \mathcal{E}(A)$, $\psi = h - h^{s_\alpha}$ satisfies this condition so $f - f^{s_\alpha}$ is orthogonal to h for each α, hence is 0.

For $F \in \mathcal{E}(K)$, $D \in \mathbf{D}(A)$, $\alpha \in \Sigma$ put $\mathcal{D}_{F,D,\alpha}$ for the intersection over all $\psi \in \mathcal{E}(A)$ satisfying $\psi^{s_\alpha} = -\psi$ of the spaces

$$\Big\{\varphi \in \mathcal{D}(\Xi) : \int_{K\times K\times A} F(u)L(ukM, D)_a(e^{\rho(\log a)}\varphi(ka\cdot\xi_o)\psi(a))da\,du\,dk = 0\Big\}.$$

Then by Theorem 2.1,

$$\mathcal{D}(X)^\wedge = \bigcap_{F,D,\alpha} \mathcal{D}_{F,D,\alpha}$$

and is thus an intersection of subspaces of $\mathcal{D}(\Xi)$ which are closed in the topology $\sigma(\mathcal{D}(\Xi), \mathcal{E}(\Xi))$. This proves Corollary 2.2.

Definition. Let $\mathcal{N}$ denote the kernel of the dual transform $\varphi \longrightarrow \check{\varphi}$ of $\mathcal{E}(\Xi)$ into $\mathcal{E}(X)$.

Because of (8), Ch. I, §3 we have that

$$\mathcal{N} = (\mathcal{D}(X)^\wedge)^\perp, \tag{6}$$

the annihilator of $\mathcal{D}(X)^\wedge$ in $\mathcal{E}(\Xi)$. The double annihilator $((E)^\perp)^\perp$ of a subspace E being the weak closure, Cor. 2.2 and (6) imply the following result.

Corollary 2.3.

$$\mathcal{D}(X)^\wedge = \{\varphi \in \mathcal{D}(\Xi) : \int_\Xi \varphi(\xi)\psi(\xi)d\xi = 0 \textit{ for } \psi \in \mathcal{N}\}$$

in short,

$$\mathcal{D}(X)^\wedge = \mathcal{N}^\perp. \tag{7}$$

Next we shall determine the range $\mathcal{E}'(X)^\wedge$ and shall prove the following result.

Theorem 2.4. *The image of $\mathcal{E}'(X)$ under the Radon transform is given by*

$$\mathcal{E}'(X)^\wedge = \{\Sigma \in \mathcal{E}'(\Xi) : \Sigma(\mathcal{N}) = 0\}. \tag{8}$$

The proof will depend on the inversion formula and the Plancherel formula (Theorems 3.9 and 3.13, Ch. II) and the support theorem and Cor. 2.3. The proof is quite similar to that of Theorem 2.6, Ch. I so we just summarize the main steps.

The Plancherel formula mentioned is

$$\int_X f(x)g(x)dx = \int_\Xi (\Omega\widehat{f})(\xi)\widehat{g}(\xi)d\xi, \tag{9}$$

where Ω is the operator $w^{-1}\overline{\Lambda}\Lambda$ in the notation of Theorem 3.13, Ch. II.

The inclusion $\subset$ in (8) being obvious let $\Sigma \in \mathcal{E}'(\Xi)$ satisfy $\Sigma(\mathcal{N}) = 0$. Motivated by (9) we define the distribution S on X by

$$S(f) = \Sigma(\Omega\widehat{f}) \tag{10}$$

The problem is to show that S has compact support and that $\widehat{S} = \Sigma$. This is done in the following steps:

(i) For $F \in \mathcal{D}(G)$ define $\sigma \in \mathcal{D}'(\Xi)$ by

$$\sigma(\varphi) = \int_G F(g)\Sigma^{g^{-1}}(\varphi)dg. \tag{11}$$

Then σ has compact support but it also equals the smooth function

$$gMN \longrightarrow (F' * \widetilde{\Sigma})(g), \tag{12}$$

where $F'(g) = F(g^{-1})$ and $\Psi \in \mathcal{D}'(\Xi) \longrightarrow \widetilde{\Psi} \in \mathcal{D}'(G)$ is the adjoint of the mapping $F \in \mathcal{D}(G) \longrightarrow \dot{F} \in \mathcal{D}(\Xi)$ where

$$\dot{F}(gMN) = \int_{MN} F(gmn)dmdn. \tag{13}$$

Thus $\sigma \in \mathcal{D}(\Xi)$.

(ii) Since $\mathcal{N}$ is G-invariant, $\sigma(\mathcal{N}) = 0$ so by Cor. 2.3, $\sigma = \widehat{s}$ where $s \in \mathcal{D}(X)$.

(iii) Using $S^g(f) = \Sigma^g(\Omega\widehat{f})$ and the inversion formula (Theorem 3.13, Ch. II) we get

$$\int_X s(x)f(x)dx = \int_X s(x)(\Omega\widehat{f})^\vee(x)dx = \sigma(\Omega\widehat{f}) = \int_G F(g)S^{g^{-1}}(f)dg,$$

which implies

$$s(gK) = (F' * \widetilde{S})(g). \tag{14}$$

(iv) Since $F \in \mathcal{D}(G)$ was arbitrary (14) shows

$$\mathcal{D}(G) * \widetilde{S} \subset \mathcal{D}(G),$$

more precisely, given a compact neighborhood C of e there exists another such, say C', such that $C_1 \subset C_2 \Longrightarrow C_1' \subset C_2'$ and

$$\mathcal{D}_C(G) * \widetilde{S} \subset \mathcal{D}_{C'}(G).$$

Letting C shring to $\{e\}$ we deduce that S has compact support.

(v) Since $s(\overset{\vee}{\varphi}) = \sigma(\varphi)$ we deduce quickly that

$$\int_G F(g)S^{g^{-1}}(\overset{\vee}{\varphi})dg = (F' * \widetilde{\Sigma})(\Phi)$$

if $\dot{\Phi} = \varphi$. Letting F approximate the delta function we deduce

$$S(\overset{\vee}{\varphi}) = \widetilde{\Sigma}(\Phi) = \Sigma(\varphi)$$

so $\widehat{S} = \Sigma$ as desired.

Corollary 2.5. *The dual transform $\varphi \longrightarrow \overset{\vee}{\varphi}$ on $\mathcal{E}(\Xi)$ is surjective, i.e.*

$$\mathcal{E}(\Xi)^\vee = \mathcal{E}(X).$$

For this we combine Theorem 2.4 (which assures that $\mathcal{E}'(X)^\wedge$ is closed) with Theorem 3.7 in Ch. I. It remains to check that $S \longrightarrow \widehat{S}$ is injective on $\mathcal{E}'(X)$. However the inversion formula $f = (\Omega\widehat{f})^\vee$ extends to $\mathcal{E}'(X)$ as follows. Defining for $\Sigma \in \mathcal{E}'(\Xi)$, $\Omega^*\Sigma$ by

$$(\Omega^*\Sigma)(\varphi) = \Sigma(\Omega\varphi), \qquad \varphi \in \mathcal{D}(\Xi)$$

we have for $S \in \mathcal{E}'(X)$, $f \in \mathcal{D}(X)$,

$$\begin{aligned}(\Omega^*\widehat{S})^\vee(f) &= (\Omega^*\widehat{S})(\widehat{f}) = \widehat{S}(\Omega\widehat{f})\\ &= S((\Omega\widehat{f})^\vee) = S(f),\end{aligned}$$

proving the injectivity needed. In the process we have obtained the following extension of the inversion formula.

Corollary 2.6. *The Radon transform on X is extended to $\mathcal{E}'(X)$ by the definition*

$$\widehat{S}(\varphi) = S(\overset{\vee}{\varphi}) \qquad \varphi \in \mathcal{E}(\Xi).$$

Then we have the inversion formula

(15) $$S = (\Omega^*\widehat{S})^\vee, \qquad S \in \mathcal{E}'(X).$$

§3. THE RANGE AND KERNEL FOR THE K-TYPES.

1. The general case.

Theorem 2.1 characterizes $\mathcal{D}(X)^\wedge$ by means of differential equations and leads to general conclusions like Corollaries 2.3 and 2.5. We shall now obtain more transparent results by specializing the range characterization to individual K-types. This will be based on related results for the Fourier transform in Ch. III, §5.

For $\delta \in \widehat{K}$ let V_δ be a corresponding representation space, $d(\delta)$ its dimension and $\overset{\vee}{\delta}$ the contragredient representation. For $f \in \mathcal{E}(X)$, $x \in X$, $\varphi \in \mathcal{E}(\Xi)$, $\xi \in \Xi$, we put

(1) $$f^\delta(x) = d(\delta)\int_K f(k\cdot x)\delta(k^{-1})dk \in \mathrm{Hom}(V_\delta, V_\delta)$$

$$\varphi^\delta(\xi) = d(\delta) \int_K \varphi(k \cdot \xi)\delta(k^{-1})dk \in \operatorname{Hom}(V_\delta, V_\delta). \tag{2}$$

As before, let

$$V_\delta^M = \{v \in V_\delta : \ \delta(m)v = v \text{ for } m \in M\}, \quad \ell(\delta) = \dim V_\delta^M \tag{3}$$

and

$$\widehat{K}_M = \{\delta \in \widehat{K} : \ V_\delta^M \neq 0\}. \tag{4}$$

For f of sufficient decrease at ∞ we define the Radon transform

$$(f^\delta)^\wedge(\xi) = \int_\xi f^\delta(x)ds(x), \quad \xi \in \Xi. \tag{5}$$

Since $f^\delta(k \cdot x) = \delta(k)f^\delta(x)$ we have

$$(f^\delta)^\wedge(ka \cdot \xi_o) = \delta(k)(f^\delta)^\wedge(a), \text{ where } (f^\delta)^\wedge(a) = (f^\delta)^\wedge(a \cdot \xi_o). \tag{6}$$

Then we have (cf. Ch. III, §5),

$$f(x) = \sum_{\delta \in \widehat{K}_M} \operatorname{Trace} f^\delta(x). \tag{7}$$

Also $\varphi^\delta(k \cdot \xi) = \delta(k)\varphi^\delta(\xi)$ and $\delta(m)\varphi^\delta(a \cdot \xi_o) = \varphi^\delta(a \cdot \xi_o)$ so $\varphi^\delta = 0$ unless $\delta \in \widehat{K}_M$. Hence ([GGA], Ch. V, Cor. 3.4)

$$\varphi(\xi) = \sum_{\delta \in \widehat{K}_M} \operatorname{Trace}(\varphi^\delta(\xi)). \tag{8}$$

Since $(f^\delta)^\wedge = (\widehat{f})^\delta$ this implies

$$\widehat{f}(\xi) = \sum_{\delta \in \widehat{K}_M} \operatorname{Trace}((f^\delta)^\wedge(\xi)). \tag{9}$$

The range $\mathcal{D}(X)^\wedge$ is thus determined by the functions $(f^\delta)^\wedge$ which by (6) are determined by their restrictions to $A \cdot \xi_o$ which we identify with A. Because of (6) $(f^\delta)^\wedge$ maps A into $\operatorname{Hom}(V_\delta, V_\delta^M)$.

Consider the matrix $Q^\delta(\lambda)$ defined in (22)′ Ch. III, whose entries are certain polynomials in λ. Let $Q^\delta(D)$ be the corresponding matrix of differential operators with constant coefficients. This means

$$(Q^\delta(D)_{ij}F)(a) = \int_{\mathfrak{a}^*} Q^\delta(\lambda)_{ij}F^*(\lambda)e^{i\lambda(\log a)}da, \tag{10}$$

where $F \in \mathcal{D}(A)$ and F^* its Fourier transform

$$F^*(\lambda) = \int_A F(a)e^{-i\lambda(\log a)}da, \qquad F \in \mathcal{D}(A).$$

Then if $H \in \mathcal{D}(A, \mathrm{Hom}(V_\delta, V_\delta^M))$ the function $\check{Q}^\delta(D)H$ belongs to the space $\mathcal{D}(A, \mathrm{Hom}(V_\delta, V_\delta^M))$ too. By (10) we have for $1 \leq i \leq \ell(\delta), 1 \leq j \leq d(\delta)$,

$$(\check{Q}^\delta(D)H)_{ij}(a) = \sum_{k=1}^{\ell(\delta)}(\check{Q}^\delta(D)_{ik}H_{kj})(a),$$

$$= \sum_{k=1}^{\ell(\delta)}\int_{\mathfrak{a}^*} \check{Q}^\delta(\lambda)_{ik}H^*_{kj}(\lambda)e^{i\lambda(\log a)}d\lambda,$$

which we write in the form

$$(\check{Q}^\delta(D)H)(a) = \int_{\mathfrak{a}^*} \check{Q}^\delta(\lambda)H^*(\lambda)e^{i\lambda(\log a)}d\lambda. \tag{11}$$

Theorem 3.1. *Fix $\delta \in \widehat{K}_M$. As f runs through $\mathcal{D}(X)$ the function $(f^\delta)^\wedge$ on A runs through the set of functions ψ of the form*

$$\psi(a) = e^{-\rho(\log a)}(\check{Q}^\delta(D)\Psi)(a), \qquad a \in A,$$

where $\Psi \in \mathcal{D}(A, \mathrm{Hom}(V_\delta, V_\delta^M))$ is W-invariant.

Proof. We have from Ch. III, §5 (25) if $g \in \mathcal{D}(X)$,

$$\widehat{g}(ka \cdot \xi_o)e^{\rho(\log a)} = \int_A \widetilde{g}(\lambda, kM)e^{i\lambda(\log a)}da.$$

Now if $f \in \mathcal{D}_{\check{\delta}}(X)$, we know (Ch. III, §5, (47)) that

$$\begin{aligned}
\widetilde{f}(\lambda) &= d(\delta)\int_X f(x)\Phi_{\overline{\lambda},\delta}(x)^* dx \\
&= d(\delta)\int_X f(x)\Big(\int_K e^{(-i\lambda+\rho)(A(x,kM))}\delta(k^{-1})dk\Big)dx \\
&= d(\delta)\int_K \widetilde{f}(\lambda, kM)\delta(k^{-1})dk \\
&= d(\delta)\int_K\Big(\int_A \widehat{f}(ka\cdot\xi_o)e^{(-i\lambda+\rho)(\log a)}da\Big)\delta(k^{-1})dk \\
&= \int_A (f^\delta)^\wedge(a\cdot\xi_o)e^{(-i\lambda+\rho)(\log a)}da.
\end{aligned}$$

Thus

$$(f^\delta)^\wedge(a) = e^{-\rho(\log a)}\int_{\mathfrak{a}^*}\widetilde{f}(\lambda)e^{i\lambda(\log a)}d\lambda. \tag{12}$$

By Theorem 5.11, Ch. III the functions $\widetilde{f}(\lambda)$ run through the function $\check{Q}^\delta(\lambda)\Psi^*(\lambda)$, where $\Psi \in \mathcal{D}(A, \mathrm{Hom}(V_\delta, V_\delta^M))$ is W-invariant so the theorem follows from (11) and (12).

We recall that the map $f \longrightarrow \mathrm{Tr}(f)$ is a bijection of $\mathcal{D}^\delta(X)$ onto $\mathcal{D}_{\check{\delta}}(X)$ (Ch. III, Prop. 5.10). Similar statement holds for Ξ. Thus we can restate Theorem 3.1 as follows.

Corollary 3.2. *Let $\delta \in \widehat{K}_M$. The range $\mathcal{D}_{\check{\delta}}(X)^\wedge$ consists of the functions*

$$\varphi(ka\cdot\xi_o) = e^{-\rho(\log a)}\,\mathrm{Tr}\Big(\delta(k)(\check{Q}^\delta(D)\Psi)(a))\Big) \tag{13}$$

with $\Psi \in \mathcal{D}(A, \mathrm{Hom}(V_\delta, V_\delta^M))$ W-invariant.

We shall now give a description of the kernel $\mathcal{N}$ of the dual transform $\varphi \in \mathcal{E}(\Xi) \longrightarrow \check{\varphi} \in \mathcal{E}(X)$ in the spirit of Theorem 3.1. We have the decomposition

$$\mathcal{N} = \sum_{\delta\in\widehat{K}_M}\mathcal{N}_\delta, \tag{14}$$

where $\mathcal{N}_\delta$ is the space of $\varphi \in \mathcal{N}$ of type δ. Recall that $\mathrm{Tr}(\varphi^\delta)$ in (8) belongs to $\mathcal{N}_{\check{\delta}}$.

Definition. A (vector-valued) function ζ on A will be called *W-odd* if $\sum_{s \in W} \zeta^s = 0$.

For X of rank one, W-odd means the same thing as skew. In general we have

$$\text{(15)} \qquad \zeta \text{ is } W\text{-odd} \iff \int_A \zeta(a)\psi(a)da = 0 \text{ for } \psi \quad W\text{-invariant.}$$

Theorem 3.3. *Let $\delta \in \widehat{K}_M$. The space $\mathcal{N}_\delta$ consists of the functions*

$$\varphi(ka \cdot \xi_o) = \mathrm{Tr}(\check{\delta}(k)Z(a)),$$

where $Z \in \mathcal{E}(A, \mathrm{Hom}(V_{\check{\delta}}, V_{\check{\delta}}^M))$ and the function

$$\zeta(a) = ({}^tQ^{\check{\delta}}(D)^*(e^\rho Z))(a)$$

is W-odd.

Here ${}^tA$ denotes the transpose of the matrix A and if the entries of A are differential operators, $(A^*)_{ij} = (A_{ij})^*$.

Proof. Since $d\xi = e^{2\rho(\log a)}dk_M da$ (Ch. II, Lemma 3.1) we see from the orthogonality relations on K that $\mathcal{D}_{\check{\delta}}(X)^\wedge$ is orthogonal to $\mathcal{E}_\pi(\Xi)$ for $\pi \in \widehat{K}_M$ not equivalent to δ. Thus

$$\text{(16)} \qquad \mathcal{N}_\delta = \left\{\zeta \in \mathcal{E}_\delta(\Xi) : \int_\Xi \zeta(\xi)\varphi(\xi)d\xi = 0 \quad \text{for} \quad \varphi \in \mathcal{D}_{\check{\delta}}(X)^\wedge\right\}.$$

Writing (13) out in matrix form we have

$$\varphi(ka \cdot \xi_o) = \sum_{1 \le i,j \le d(\delta)} \delta_{ij}(k)\psi_{ji}(a) = \sum_{\substack{1 \le i \le d(\delta) \\ 1 \le j \le \ell(\delta)}} \delta_{ij}(k)\psi_{ji}(a)$$

where we have taken into account the fact that $\Psi(a)$ maps V_δ into V_δ^M. Writing $\zeta \in \mathcal{N}_\delta$ in the form

$$\text{(17)} \qquad \zeta(ka \cdot \xi_o) = \mathrm{Tr}(\check{\delta}(k)Z(a))$$

we have $\mathrm{Tr}(\check{\delta}(k)\check{\delta}(m)Z(a)) = \mathrm{Tr}(\check{\delta}(k)Z(a)) \quad (m \in M)$ so, by the irreducibility of $\check{\delta}$, we have $\check{\delta}(m)Z(a) = Z(a)$, in other words, $Z \in \mathcal{E}(A, \mathrm{Hom}(V_{\check{\delta}}, V_{\check{\delta}}^M))$. Since $\check{\delta}(k)_{ij} = (\delta(k)_{ij})^-$ expression (17) equals

$$\zeta(ka \cdot \xi_o) = \sum_{1 \le p \le d(\delta), 1 \le q \le \ell(\delta)} \delta_{pq}(k)^- Z_{qp}(a).$$

Hence by the orthogonality relations and (16) we have (with $1 \le i \le d(\delta)$, $1 \le j \le \ell(\delta)$)

$$\sum_{i,j} \int_A \sum_{k=1}^{\ell(\delta)} \left(Q_{jk}^{\check{\delta}}(D)\Psi_{ki} \right)(a) Z_{ji}(a) e^{\rho(\log a)} da = 0,$$

where Ψ_{ki} is the matrix expression of Ψ. This amounts to

$$\sum_{i,k} \int_A \Psi_{ki}(a) \left({}^t Q^{\check{\delta}}(D)^* (Z e^{\rho}) \right)_{ki} (a) da = 0.$$

The entries Ψ_{ki} are arbitrary W-invariants in $\mathcal{D}(A)$ so by the W-odd characterization (15) the theorem follows.

2. Examples: $\boldsymbol{H}^2$ and $\boldsymbol{R}^2$.

Consider the hyperbolic plane $\boldsymbol{H}^2$, that is the unit disk $|z| < 1$ with the Riemannian structure

$$ds^2 = \frac{dx^2 + dy^2}{(1 - x^2 - y^2)^2}. \tag{18}$$

This is a constant multiple of the Riemannian structure which $\boldsymbol{H}^2$ acquires from its representation

$$\boldsymbol{H}^2 = \boldsymbol{SU}(1,1)/\boldsymbol{SO}(2) \tag{19}$$

and the Killing form of the Lie algebra of $\boldsymbol{SU}(1,1)$. We shall now write out Theorems 3.1 and 3.3 for this case. The horocycles are the circles tangential to the boundary $|z| = 1$. Let $\xi_{t,\theta}$ denote the horocycle through $e^{i\theta}$ having (Riemannian) distance t (with sign) from 0. A function $\psi \in \mathcal{E}(\Xi)$ can be expanded in a Fourier series

$$\psi(\xi_{t,\theta}) = \sum_{n \in \boldsymbol{Z}} \psi_n(t) e^{in\theta}, \qquad \psi_n \in \mathcal{E}(\boldsymbol{R}). \tag{20}$$

Theorem 3.4. *Let* $X = \boldsymbol{H}^2$.

(i) The range $\mathcal{D}(X)^\wedge$ *consists of the functions (20) such that for each* $n \in \boldsymbol{Z}$,

$$\psi_n(t) = e^{-t}\left(\frac{d}{dt} - 1\right)\left(\frac{d}{dt} - 3\right)\dots\left(\frac{d}{dt} - 2|n| + 1\right)\varphi_n(t), \tag{21}$$

where $\varphi_n \in \mathcal{D}(\boldsymbol{R})$ *is even.*

(ii) The kernel $\mathcal{N}$ *consists of the functions (20) such that for each* $n \in \boldsymbol{Z}$,

$$\left(\frac{d}{dt} + 1\right)\left(\frac{d}{dt} + 3\right)\dots\left(\frac{d}{dt} + 2|n| - 1\right)(e^t\psi_n) \ \ \textit{is odd.} \tag{22}$$

Proof. Part (i) is already proved in [GGA], pp. 76, 555. Note that the proof used the elementary fact that if $f(z)$ is holomorphic of exponential type, $p(z)$ a polynomial such that $f(z)/p(z)$ is holomorphic, then f/p is of exponential type.

Now we have the general fact that in the parametrization $\xi = ka\cdot\xi_o$, $d\xi$ equals $e^{2\rho(\log a)}dk_M da$. Comparing with (21) above and (13) §3 we deduce that

$$d\xi = e^{2t}d\theta dt \qquad \text{if} \qquad \xi = \xi_{t,\theta} \tag{23}$$

a fact which is easy to prove directly. Thus we see from (6) §2 that $\psi \in \mathcal{N}$ if and only if for each $n \in \boldsymbol{Z}$

$$\int_{\boldsymbol{R}} \psi_{-n}(t)\left(e^{-t}(\frac{d}{dt} - 1)\dots(\frac{d}{dt} - 2|n| + 1)\varphi_n(t)\right)e^{2t}dt = 0 \tag{24}$$

whenever $\varphi_n \in \mathcal{D}(\boldsymbol{R})$ is even. This translates to the desired condition (22).

It is of interest to write down the analog of Theorem 3.4 for $X = \boldsymbol{R}^2$ and the space $\Xi = \boldsymbol{P}^2$ of lines in $\boldsymbol{R}^2$. Let $\xi_{p,\theta}$ denote the line $\cos\theta\, x + \sin\theta\, y = p$. A function ψ on Ξ can be expanded

$$\psi(\xi_{p,\theta}) = \sum_n \psi_n(p)e^{in\theta}, \qquad \psi_n \in \mathcal{E}(\boldsymbol{R}). \tag{25}$$

Theorem 3.5.

(i) The range $\mathcal{D}(\boldsymbol{R}^2)^\wedge$ consists of the functions (25) such that for each $n \in \boldsymbol{Z}$,

$$\psi_n(p) = \frac{d^{|n|}}{dp^{|n|}}\varphi_n(p) \qquad \varphi_n \in \mathcal{D}(\boldsymbol{R}) \ \textit{even}\,. \tag{26}$$

(ii) The kernel $\mathcal{N}$ of the map $\psi \longrightarrow \overset{\vee}{\psi}$ consists of the functions (25) such that for each $n \in \boldsymbol{Z}$,

$$\frac{d^{|n|}}{dp^{|n|}}\psi_n(p) \qquad \textit{is odd.} \tag{27}$$

This result would be conjectured from Theorem 3.4 by rewriting it for $\boldsymbol{H}^2$ of curvature ϵ and then letting $\epsilon \longrightarrow 0$. However it is easier to prove directly. For part (i) we note that the characterization in Ch. I, Theorem 2.4, (ii) is equivalent to

$$\int_{\boldsymbol{R}} \psi_n(p)p^k dp = 0 \qquad \text{for } |n| > k \qquad k \geq 0 \tag{28}$$

which is equivalent to (26). In fact (26) implies (28) and (28) implies (26) for

$$\varphi_n(p) = \int_{-\infty}^{p} dt_1 \int_{-\infty}^{t_1} \cdots \int_{-\infty}^{t_{n-1}} \psi_n(t_n)dt_n.$$

The evenness of φ_n comes from the fact that $\xi_{-p,\theta+\pi} = \xi_{p,\theta}$ so

$$\psi_n(-p) = (-1)^n\psi_n(p). \tag{29}$$

For (ii) we note that (29) implies that $\psi_n^{(|n|)}$ is even so (27) in fact amounts to

$$\frac{d^{|n|}}{dp^{|n|}}\psi_n(p) = 0.$$

Combining this with (29) we see that $\psi_n(p)$ is a linear combination of the powers $p^{|n|-2k}$ $(2 \leq 2k \leq |n|)$ and this again is equivalent to Theorem 2.5(i), Ch. I.

§4. THE RADON TRANSFORM AND ITS DUAL FOR K-INVARIANTS.

In the case of K-invariant functions the transforms $f \longrightarrow \widehat{f}$, $\varphi \longrightarrow \check{\varphi}$ can be written out in a more precise form. As usual, let the space of K-invariants in $\mathcal{D}(X)$, $\mathcal{E}(X)$, $\mathcal{D}'(X)$ and $\mathcal{E}'(X)$ be denoted by $\mathcal{D}^\natural(X)$, $\mathcal{E}^\natural(X)$, $\mathcal{D}'_\natural(X)$, and $\mathcal{E}'_\natural(X)$, respectively. Also denote the subspace of Weyl group invariants in $\mathcal{D}(A)$, $\mathcal{E}(A)$, $\mathcal{D}'(A)$ and $\mathcal{E}'(A)$ by $\mathcal{D}_W(A)$, $\mathcal{E}_W(A)$, $\mathcal{D}'_W(A)$ and $\mathcal{E}'_W(A)$, respectively.

The mapping

$$\mathcal{A} : f \longrightarrow F_f \tag{1}$$

given by

$$F_f(a) = e^{\rho(\log a)} \int_N f(an \cdot o)dn, \qquad f \in \mathcal{D}^\natural(X) \tag{2}$$

is often called the *Abel transform* because if $X = \boldsymbol{H}^2$ then $\mathcal{A}$ can be expressed in terms of the Abel integral equation ([GGA], p. 41). This transform has the following properties ([GGA], Ch. II, §5 (37), (39), Ch. IV, Cor. 7.4 and this book Ch. II, Theorem 2.1).

Theorem 4.1.

(i) The mapping $\mathcal{A}$ is a homeomorphism of $\mathcal{D}^\natural(X)$ onto $\mathcal{D}_W(A)$ and has the transmutation property

$$\mathcal{A}Df = \Gamma(D)\mathcal{A}f, \qquad f \in \mathcal{D}^\natural(X), \tag{3}$$

for each $D \in \boldsymbol{D}(G/K)$.

(ii) $\mathcal{A}$ connects the Fourier transform on A and the spherical transform on X, i.e.,

$$\int_X f(x)\varphi_{-\lambda}(x)dx = \int_A (\mathcal{A}f)(a)e^{-i\lambda(\log a)}da. \tag{4}$$

Moreover,

$$\mathcal{A}(f_1 \times f_2) = \mathcal{A}f_1 * \mathcal{A}f_2, \tag{5}$$

where $\times$ is the convolution on X and $*$ the convolution on A.

For the transpose operator $\mathcal{A}^*$ we shall now prove the following result.

Theorem 4.2.

(i) The transpose operator $\mathcal{A}^* : \mathcal{D}'_W(A) \longrightarrow \mathcal{D}'_\natural(X)$ *is a bijection and it satisfies*

$$DA^* = \mathcal{A}^*\Gamma(D) \qquad \textit{for} \qquad D \in \mathcal{D}(G/K). \tag{6}$$

(ii) The restriction $\mathcal{A}^*|\mathcal{E}_W(A)$ *is given by*

$$(\mathcal{A}^*\varphi)(gK) = \int_K \varphi(\exp H(gk))e^{-\rho(H(gk))}dk \tag{7}$$

and it is a homeomorphism of $\mathcal{E}_W(A)$ *onto* $\mathcal{E}^\natural(X)$.

Proof. Taking transpose in (3) we obtain $D^*\mathcal{A}^* = \mathcal{A}^*\Gamma(D)^*$ on $\mathcal{D}'_W(A)$ and $\mathcal{A}^*$ is injective. However, by [GGA], Ch. II (Cor. 5.3 and Lemma 5.21) we have $\Gamma(D^*) = \Gamma(D)^*$ so (6) follows.

Recalling the identification $\Xi = K/M \times A$ a function φ on A extends to the function $1 \otimes \varphi : (kM, a) \longrightarrow \varphi(a)$ on Ξ which is of course K-invariant. The surjectivity in Corollary 2.5 implies a similar one for the K-invariants, so the map

$$1 \otimes \varphi \longrightarrow (1 \otimes \varphi)^\vee \tag{8}$$

is a surjection of $1 \otimes \mathcal{E}(A)$ onto $\mathcal{E}^\natural(X)$. For $\varphi \in \mathcal{E}(A)$ we put

$$\varphi^\natural(a) = w^{-1} \sum_{s\in W} \varphi(s \cdot a).$$

Then if $f \in \mathcal{D}^\natural(X)$, $\quad \varphi \in \mathcal{E}(A)$,

$$\int_X f(x)(1 \otimes e^{-\rho}\varphi)^\vee(x)dx = \int_\Xi \widehat{f}(\xi)(1 \otimes e^{-\rho}\varphi)(\xi)d\xi$$

$$= \int_A e^{-\rho(\log a)}(\mathcal{A}f)(a)e^{-\rho(\log a)}\varphi(a)e^{2\rho(\log a)}da$$

$$= \int_A (\mathcal{A}f)(a)\varphi(a)da = \int_A (\mathcal{A}f)(a)\varphi^\natural(a)da$$

$$= \varphi^\natural(\mathcal{A}f) = (\mathcal{A}^*\varphi^\natural)(f)$$

so

$$(1 \otimes e^{-\rho}\varphi)^{\vee} = \mathcal{A}^*\varphi^{\natural}. \tag{9}$$

Coupled with the surjectivity of the map (8) this shows that the map

$$\mathcal{A}^* : \mathcal{E}_W(A) \longrightarrow \mathcal{E}^{\natural}(X) \tag{10}$$

is surjective. We already know that it is injective and by (9) continuous. The spaces being Fréchet spaces, the map (10) is even a homeomorphism (Ch. I, Theorem 2.16). Writing out (9) we get

$$\begin{aligned}(\mathcal{A}^*\varphi^{\natural})(gK) &= \int_K (1 \otimes e^{-\rho}\varphi)(gk \cdot \xi_o)dk \\ &= \int_K \varphi(\exp H(gk))e^{-\rho(H(gk))}dk.\end{aligned}$$

Taking $\varphi = \varphi^{\natural}$ this proves (7) so Theorem 4.2 is proved.

In the process we obtained the following result.

Corollary 4.3. *For each $s \in W$, $\varphi \in \mathcal{E}(A)$*

$$\int_K \varphi(\exp H(gk))e^{-\rho(H(gk))}dk = \int_K \varphi(\exp s \cdot H(gk))e^{-\rho(H(gk))}dk.$$

For $\varphi = e^{i\lambda}$ this is just the invariance $\varphi_{s\lambda} = \varphi_\lambda$.

Now let $\mathcal{F}$ denote the spherical transform

$$f \in \mathcal{D}^{\natural}(X) \longrightarrow \widetilde{f} \in \mathcal{H}_W(\mathfrak{a}_{\mathbf{c}}^*)$$

and $\mathcal{F}_o$ the Euclidean Fourier transform

$$(\mathcal{F}_o\varphi)(\lambda) = \int_A \varphi(a)e^{-i\lambda(\log a)}da.$$

Let $\mathcal{F}'$ denote the inverse spherical transform

$$(\mathcal{F}'\psi)(x) = \int_{\mathfrak{a}^*} \psi(\lambda)\varphi_\lambda(x)|\mathbf{c}(\lambda)|^{-2}d\lambda$$

and $\mathcal{F}_o'$ the inverse Euclidean Fourier transform

$$(\mathcal{F}_o'\psi)(a) = \int_{\mathfrak{a}^*} \psi(\lambda) e^{i\lambda(\log a)} d\lambda.$$

Then (4), (7), Cor. 4.3, and the Fourier inversion formulas imply the following result.

Theorem 4.4. *We have*

$$(11) \qquad \mathcal{F}f = \mathcal{F}_o\mathcal{A}f \qquad f \in D^\natural(X),$$

$$(12) \qquad \mathcal{F}'\psi = \mathcal{A}^*\mathcal{F}_o'(\psi|\mathbf{c}|^{-2}), \qquad \psi \in \mathcal{H}_W(\mathfrak{a}_c^*).$$

Moreover,

$$\mathcal{F}_o\mathcal{F}_o' = I \quad on \quad \mathcal{H}(\mathfrak{a}_c^*), \quad \mathcal{F}_o'\mathcal{F}_o = I \quad on \quad \mathcal{D}(A),$$
$$\mathcal{F}\mathcal{F}' = wI \quad on \quad \mathcal{H}_W(\mathfrak{a}_c^*), \quad \mathcal{F}'\mathcal{F} = wI \quad on \quad \mathcal{D}^\natural(X).$$

Consider now the Schwartz space $\mathcal{S}^\natural(X)$ of L^2 K-invariant rapidly decreasing functions on X. This space is denoted by $\mathcal{J}^2(G)$ in [GGA], Ch. IV, Exercise C6. As proved for the most part by Harish-Chandra [1958b] (see Anker [1991b] for a much shorter proof) the spherical transform $f \longrightarrow \widetilde{f}$ is a bijection of $\mathcal{S}^\natural(X)$ onto the space $\mathcal{S}_W(\mathfrak{a}^*)$ of W-invariant rapidly decreasing functions on $\mathfrak{a}^*$. The relation (4) remaining valid for $f \in \mathcal{S}^\natural(X)$ we can thus add the following statements to Theorems 4.1–4.2

(13) $\qquad \mathcal{A}$ *is a bijection of* $\mathcal{S}^\natural(X)$ *onto* $\mathcal{S}_W(A)$

(14) $\qquad \mathcal{A}^*$ *is a bijection of* $\mathcal{S}_W'(A)$ *onto* $\mathcal{S}_\natural'(X)$,

the notation $\mathcal{S}'$ referring to the space of tempered distributions.

§5. THE RADON TRANSFORM ON X_o.

1. Preliminaries.

As in Ch. III, §7, let X_o denote the tangent space to the symmetric space G/K at the origin. Again we view X_o as the coset space G_o/K. We shall now define a Radon transform on X_o which is a kind of a flat analog to the horocycle Radon transform investigated in Chapter II.

Consider the decompositions

$$\mathfrak{g} = \mathfrak{k} + \mathfrak{p}, \qquad \mathfrak{p} = \mathfrak{a} + \mathfrak{q}, \tag{1}$$

where $\mathfrak{q}$ is the orthogonal complement in $\mathfrak{p}$ to the maximal abelian subspace $\mathfrak{a}$. Under the natural isomorphisms

$$X_o = (G/K)_o = (G \cdot o)_o \approx \mathfrak{g}/\mathfrak{k} \approx \mathfrak{p}$$

we have by restriction

$$(A \cdot o)_o \approx \mathfrak{a}, \qquad (N \cdot o)_o \approx \mathfrak{q} \tag{2}$$

because (Ch. II, Lemma 1.2) the manifolds $A \cdot o$ and $N \cdot o$ are orthogonal in o. This motivates the following definition (in view of the fact that each horocycle in X is an image of $\xi_o = N \cdot o$ under some $g \in G$).

Definition. A *horocycle plane* in $\mathfrak{p}$ is the image $g \cdot \mathfrak{q}$ of $\mathfrak{q}$ under some element $g \in G_o$.

Let Ξ_o denote the set of all horocycle planes. Each $g \in G_o$ has the form $g \cdot Y = Ad(k_o)(X_o + Y)$ for some fixed $k_o \in K$, $X_o \in \mathfrak{p}$. Because of (1) each $\xi \in \Xi_o$ can thus be written

$$\xi = Ad(k)(H + \mathfrak{q}), \qquad k \in K, \ H \in \mathfrak{a}. \tag{3}$$

As usual, let M' denote the normalizer of A in K. let $M'\mathfrak{q}$ denote the group of affine transformations $X \longrightarrow Ad(m')(Y + X)$ of $\mathfrak{p}$, with $m' \in M'$ and $Y \in \mathfrak{q}$. Since $Ad(m')\mathfrak{q} \subset \mathfrak{q}$, $M'\mathfrak{q}$ is indeed a group.

Let $(K/M \times \mathfrak{a})_W$ denote the product $K/M \times \mathfrak{a}$ with the equivalence relation

$$(k_1M, H_1) \sim (k_2M, H_2)$$

if $H_2 = s^{-1}H_1$, $k_2 = k_1 m_s$ where $s \in W$ and $m_s \in M'$ is a representative of s. We shall often write $k \cdot X = Ad(k)(X)$ for short.

Lemma 5.1.

(i) $M'\mathfrak{q}$ is the subgroup of G_o leaving $\mathfrak{q}$ invariant so

$$\Xi_o = G_o/M'\mathfrak{q}.$$

(ii) The mapping $(kM, H) \longrightarrow Ad(k)(H + \mathfrak{q})$ induces a bijection of $(K/M \times \mathfrak{a})_W$ onto Ξ_o.

Proof. (i) If $g \cdot \mathfrak{q} = \mathfrak{q}$ then $g \cdot 0 = Y_o \in \mathfrak{q}$ so the transformation $X \longrightarrow -Y_o + g \cdot X$ $(X \in \mathfrak{p})$ is an element $k \in K$ and $k\mathfrak{q} = \mathfrak{q}$. Hence $k\mathfrak{a} = \mathfrak{a}$ so $k \in M'$ whence $g \in M'\mathfrak{q}$.

(ii) By (3) it is clear that $(kM, H) \longrightarrow Ad(k)(H + \mathfrak{q})$ induces a well-defined mapping of $(K/M \times \mathfrak{a})_W$ onto Ξ_o. For injectivity suppose $Ad(k_1)(H_1 + \mathfrak{q}) = Ad(k_2)(H_2 + \mathfrak{q})$. Since H_1 is the point in $H_1 + \mathfrak{q}$ closest to the origin we deduce $Ad(k_1)H_1 = Ad(k_2)H_2$ so $Ad(k_1)\mathfrak{q} = Ad(k_2)\mathfrak{q}$ whence $k_2 = k_1 m_s$ for some $s \in W$. This proves (ii).

By (i) we have the double fibration

(4)
$$\begin{array}{ccc} & G_o/M' & \\ & \swarrow \quad \searrow & \\ X_o = G_o/K & & \Xi_o = G_o/M'\mathfrak{q} \end{array}$$

which is the infinitesimal (flat) analog of the double fibration

(5)
$$\begin{array}{ccc} & G/M & \\ & \swarrow \quad \searrow & \\ X = G/K & & \Xi = G/MN \end{array}$$

from Chapter II. The difference between $\Xi_o = G_o/M'\mathfrak{q} = \big(K/M \times \mathfrak{a}\big)_W$ and $\Xi = G/MN = \big(K/M\big) \times \mathfrak{a}$ is of considerable interest. While $Ad(m')$ $(m' \in M')$ leaves the flat space $\mathfrak{q}$ invariant it will permute the various horocycles tangent to $\mathfrak{q}$ at o and there are exactly $|W|$ of those (Ch. II, Prop. 1.7).

Definition. The Radon transform on X_o is defined by

(6)
$$\widehat{f}(\xi) = \int_{\xi} f(Y)\,dm(Y), \qquad \xi \in \Xi_o,$$

dm being the Euclidean measure given by the Killing form B on $\mathfrak{p}$.

Along with (6) we consider also the usual Fourier transform on $\mathfrak{p}$ given by

(7)
$$\widetilde{f}(Z) = \int_{\mathfrak{p}} f(Y) e^{-iB(Y,Z)}\,dY.$$

For $\lambda \in \mathfrak{a}_{\mathbf{c}}^*$ let $A_\lambda \in \mathfrak{a}^{\mathbf{c}}$ as usual be given by $\lambda(H) = B(A_\lambda, H)$ for $H \in \mathfrak{a}$. As for $\boldsymbol{R}^n$ and $X = G/K$ there is a simple relation between these transforms.

The measures dH on $\mathfrak{a}$ and $d\lambda$ on $\mathfrak{a}^*$ will be normalized as in Ch. II, §3, No. 1.

Lemma 5.2. *Let $f \in L^1(\mathfrak{p})$ and $k \in K$. Then for each $\lambda \in \mathfrak{a}^*$*
$$\widetilde{f}(k \cdot A_\lambda) = c_o \int_{\mathfrak{a}} \widehat{f}(k \cdot (H + \mathfrak{q}))e^{-i\lambda(H)}dH \qquad c_o = \textit{const.}$$

Proof. We have

$$\widetilde{f}(k \cdot A_\lambda) = \int_{\mathfrak{p}} f(X)e^{-iB(k\cdot A_\lambda,X)}dX = \int_{\mathfrak{p}} f(k \cdot Z)e^{-iB(A_\lambda,Z)}dZ$$
$$= c_o \int_{\mathfrak{a}}\int_{\mathfrak{q}} f(k(H+Q))e^{-iB(A_\lambda,H)}dHdQ,$$

proving the lemma.

This implies immediately that the Radon transform $f \longrightarrow \widehat{f}$ is injective. We shall now use the classical theory of the Fourier transform to give an explicit inversion of $f \longrightarrow \widehat{f}$.

Let $f \in \mathcal{S}(\mathfrak{p})$. Then by Fourier transform theory

$$f(0) = c_1 \int_{\mathfrak{p}} \widetilde{f}(Y)dY, \tag{8}$$

where c_1 is the constant $(2\pi)^{-\dim \mathfrak{p}}$. Writing dY according to the polar coordinate representation $\mathfrak{p} \approx K/M \times \overline{\mathfrak{a}^+}$ the measure dY is given by $dY = c_2\, \omega(\lambda)d\lambda dk_M$ where

$$\omega(\lambda) = \prod_{\alpha\in\Sigma^+} \langle\alpha, \lambda\rangle^{m_\alpha}. \tag{9}$$

Then by (8)

$$f(0) = c_3 \int_{K/M} dk_M \int_{\mathfrak{a}^*_+} \widetilde{f}(kA_\lambda)\omega(\lambda)d\lambda. \tag{10}$$

For the Euclidean Fourier inversion we would like to integrate over all of $\mathfrak{a}^* = \bigcup_{s\in W} sC\ell(\mathfrak{a}^*_+)$. Let $m_s \in M'$ be a representative of s. Since m_s normalizes M the map $kM \longrightarrow km_sM$ is a well-defined diffeomorphism of K/M which clearly preserves the measure dk_M. With w denoting the order of W, relation (10) becomes

$$wf(0) = c_3 \int_{K/M} dk_M \int_{\mathfrak{a}^*} \widetilde{f}(k \cdot A_\lambda)|\omega(\lambda)|d\lambda. \tag{11}$$

Because of the identification above $(K/M \times \mathfrak{a})_W = \Xi_o$ it will be convenient to view functions φ on Ξ_o as functions on $K/M \times \mathfrak{a}$ satisfying $\varphi(s^{-1}H, km_sK) \equiv \varphi(H, kM)$ $(s \in W)$. The function spaces $\mathcal{D}(\Xi_o)$, $\mathcal{S}(\Xi_o)$ will be viewed in this way. The functions

$$(H, kM) \longrightarrow \widehat{f}(k \cdot (H + q)), \quad (\lambda, kM) \longrightarrow \widetilde{f}(k \cdot A_\lambda) \tag{12}$$

satisfy these invariance conditions.

We consider now the *signum* function for Σ^+ defined by

$$\operatorname{sgn}_\Sigma(\lambda) = \omega(\lambda)/|\omega(\lambda)|, \qquad \lambda \in \mathfrak{a}^*, \tag{13}$$

where, by definition, the value is 1 if $\omega(\lambda) = 0$. Then by (9)

$$\operatorname{sgn}_\Sigma(\lambda) = \prod_{\substack{\alpha \in \Sigma_o^+ \\ m_\alpha + m_{2\alpha} \text{odd}}} \epsilon_\alpha(\lambda), \tag{14}$$

where, as usual, Σ_o^+ denotes the set of indivisible (short) roots in Σ^+ and

$$\epsilon_\alpha(\lambda) = \begin{cases} 1 & \langle \alpha, \lambda \rangle \geq 0 \\ -1 & \langle \alpha, \lambda \rangle < 0. \end{cases}$$

Note that $\operatorname{sgn}_\Sigma$ depends not only on Σ but also on the multiplicity function on Σ. For example, in the case $G = \boldsymbol{SO}_o(n, 1)$ we have $\operatorname{sgn}_\Sigma \equiv 1$ if and only if n is odd. In any case $\operatorname{sgn}_\Sigma$ is a tempered distribution on $\mathfrak{a}^*$. Corresponding to [GGA], Ch. I, §2, ((40) and (69)) we define the Hilbert transform $\mathcal{H}_\Sigma$ as the convolution operator with the Fourier transform of $\operatorname{sgn}_\Sigma$ on $\mathfrak{a}$, i.e.

$$(\mathcal{H}_\Sigma F)^\sim(\lambda) = \operatorname{sgn}_\Sigma(\lambda)\widetilde{F}(\lambda). \tag{15}$$

Viewing A_α $(\alpha \in \Sigma^+)$ as a differential operator on $\mathfrak{a}$ we define $\Lambda_o\varphi$ for $\varphi \in \mathcal{S}(\Xi_o)$ by

$$(\Lambda_o\varphi)(H, kM) = \left(\mathcal{H}_\Sigma \prod_{\alpha \in \Sigma^+} A_\alpha^{m_\alpha}\right)_H (\varphi(H, kM)), \tag{16}$$

where the subscript H indicates action on the first argument. Since the Fourier transform $F \longrightarrow F^*$ on $\mathfrak{a}$ satisfies

$$(17) \qquad \big(\Lambda_o\varphi\big)^*(\lambda, kM) = |\omega(\lambda)|\varphi^*(\lambda, kM)(i)^{\Sigma m_\alpha}$$

and since $\psi(s^{-1}H, km_sM) \equiv \psi(H, kM)$ implies $\psi^*(s^{-1}\lambda, km_sM) = \psi^*(\lambda, kM)$ we see that

$$(18) \qquad (\Lambda_o\varphi)(s^{-1}H, km_sM) = (\Lambda_o\varphi)(H, kM)$$

so $\Lambda_o\varphi$ is function on Ξ_o.

Combining Lemma 5.2 and (11) we get

$$f(0) = c_4 \int\limits_{K/M} dk_M \int\limits_{\mathfrak{a}} dH \int\limits_{\mathfrak{a}^*} \widehat{f}(k \cdot (H + \mathfrak{q}))e^{-i\lambda(H)} \operatorname{sgn}_\Sigma(\lambda)\, \omega(\lambda)\, d\lambda.$$

Replacing F in (15) by $\left(\prod\limits_{\alpha\in\Sigma+} A_\alpha^{m_\alpha}\right)(F)$ the last formula can be rewritten,

$$f(0) = c \int\limits_{K/M} dk_M \int\limits_{\mathfrak{a}} dH \int\limits_{\mathfrak{a}^*} \big(\Lambda_o\widehat{f}\big)(H, kM)e^{-i\lambda(H)}d\lambda,$$

which by the Fourier inversion formula on $\mathfrak{a}$ equals

$$f(0) = c \int\limits_{K/M} \big(\Lambda_o\widehat{f}\big)(0, kM)dk_M.$$

The dual transform $\varphi \longrightarrow \overset{\vee}{\varphi}$ for the double fibration (4) is given by

$$\overset{\vee}{\varphi}(X) = \int\limits_{K/M'} \varphi(X + k \cdot \mathfrak{q})dk_{M'},$$

which of course can be expressed by a similar integral over K/M. Thus

$$f(0) = c\big(\big(\Lambda_o\widehat{f}\big)^\vee\big)(0).$$

By the invariance of the operators this proves the following result.

Theorem 5.3. *The Radon transform $f \longrightarrow \widehat{f}$ on $\mathfrak{p}$ is inverted by the formula*

$$(19) \qquad f = c\big(\Lambda_o\widehat{f}\big)^\vee, \qquad\qquad f \in \mathcal{S}(\mathfrak{p}),$$

where Λ_o is defined by (15)–(16) and c is a constant independent of f.

Note that if each m_α is even then $\mathcal{H}_\Sigma$ in (15) is $\equiv 1$. This happens if G is complex or more generally if $\mathfrak{g}$ has all its Cartan subgroups conjugate ([DS], Ch. X, Exercise F6). In these cases Λ_o is a differential operator.

2. The Support Theorem.

In this subsection we shall prove the following support theorem for the Radon transform $f \longrightarrow \widehat{f}$ on $X_o = \mathfrak{p}$. Let $|X|^2 = B(X,X)$ and $d(X,Y) = |X - Y|$.

Theorem 5.4. *Let $f \in \mathcal{S}(\mathfrak{p})$ and assume*

$$(20) \qquad \widehat{f}(\xi) = 0 \qquad \text{for } \xi \in \Xi_o \qquad \text{with } d(o,\xi) > R.$$

Then

$$f(X) = 0 \qquad \text{for} \qquad |X| > R.$$

Remark. It is of interest to compare this with the support theorem for the d-plane transform on $\boldsymbol{R}^n$ ([GGA], Ch. I, Cor. 2.26). Then the requirement that $\widehat{f}(\xi) = 0$ for $d(0,\xi) > R$ is a condition on $(n-d)(d+1)$-dimensional family of d-planes. The condition (20) is much weaker because it only involves an n-dimensional family of planes (where $n = \dim \Xi_o = \dim \mathfrak{p}$).

The proof of Theorem 5.4 relies on the Fourier transform theorem of Ch. III, §7. If $\delta \in \widehat{K}_M$ we put for $f \in \mathcal{E}(\mathfrak{p})$, $\varphi \in \mathcal{E}(\Xi_o)$,

$$(21) \quad f^\delta(X) = d(\delta) \int_K f(k \cdot X)\delta(k^{-1})dk, \quad \varphi^\delta(\xi) = d(\delta) \int_K \varphi(k \cdot \xi)\delta(k^{-1})dk$$

and

$$(22) \qquad (f^\delta)^\wedge(\xi) = \int_\xi f^\delta(X)dm(X), \qquad f \in \mathcal{S}(\mathfrak{p}).$$

Substituting from (21) we see that

$$(23) \qquad (f^\delta)^\wedge = (\widehat{f})^\delta, \qquad f \in \mathcal{S}(\mathfrak{p}).$$

Lemma 5.5. *Let $f \in \mathcal{S}_{\check{\delta}}(\mathfrak{p})$ and let $\widetilde{f}(\lambda)$ denote its generalized Bessel transform (Ch. III, §7, No. 4). Then (with c_o as in Lemma 5.2),*

$$\widetilde{f}(\lambda) = c_o \int_{\mathfrak{a}} \widehat{f^\delta}(H + \mathfrak{q})e^{-i\lambda(H)}dH.$$

Proof. Immediate consequence of Lemma 5.2.

Passing to the proof of Theorem 5.4, let $f \in \mathcal{S}(\mathfrak{p})$ satisfy (20). Note that if $\xi = k \cdot (H + \mathfrak{q})$ then $d(0, \xi) = |H|$. As usual we have

$$f = \sum_{\delta \in \widehat{K}_M} \mathrm{Tr}(f^\delta). \tag{24}$$

The function $F_\delta = \mathrm{Tr}(f^\delta)$ belongs to $\mathcal{S}_{\check{\delta}}(\mathfrak{p})$ and $\left(F_\delta\right)^\delta = f^\delta$ ([GGA], Ch. IV, Prop. 1.11). Also by (23)

$$\left(f^\delta\right)^\wedge(\xi) = d(\delta) \int_K \widehat{f}(k \cdot \xi)\delta(k^{-1})dk = 0 \qquad \text{for } d(0, \xi) > R,$$

and using Lemma 5.5 on $f = F_\delta$ we get

$$\widetilde{F}_\delta(\lambda) = c_o \int_{\mathfrak{a}} \left(f^\delta\right)^\wedge(H + \mathfrak{q})e^{-i\lambda(H)}dH.$$

Thus

$$\widetilde{F}_\delta \in \mathcal{H}\big(\mathfrak{a}^*, \mathrm{Hom}\big(V_\delta, V_\delta^M\big)\big) \tag{25}$$

and has exponential type R. By Ch. III, Theorem 7.9, $\widetilde{F}_\delta \in \mathcal{S}^\delta(\mathfrak{a}^*)$ so in particular, the function $\lambda \longrightarrow \check{J}^\delta(\lambda)^{-1}\widetilde{F}_\delta(\lambda)$ is smooth on $\mathfrak{a}^*$. Thus, using Ch. III, §7, (20), the function

$$\lambda \longrightarrow \det\big(\check{J}^\delta(\lambda))^{-1}\big)J_c(\lambda)\widetilde{F}_\delta(\lambda) \tag{26}$$

is smooth on $\mathfrak{a}^*$ and by (25) $\lambda \longrightarrow J_c(\lambda)\widetilde{F}_\delta(\lambda)$ is holomorphic of exponential type R on all of $\mathfrak{a}_{\mathfrak{c}}^*$. The same is the case with the function (26) if we recall that $\det \check{J}^\delta(\lambda)$ is a product of linear factors and in addition observe the following elementary fact. If

(i) $F(z)$ is holomorphic in $\boldsymbol{C}$;

(ii) $x^{-1}F(x)$ smooth on $\boldsymbol{R}$,

then $z^{-1}F(z)$ is holomorphic in $\boldsymbol{C}$.

We have thus proved that $\widetilde{F}_\delta \in \mathcal{K}^\delta(\mathfrak{a}^*)$ (Ch. III, §7, (29)) and it has exponential type R. By Theorem 7.7 Ch. III, the function $F_\delta = \mathrm{Tr}(f^\delta)$ has compact support and examining the proof (particularly relations (36)–(37)) we see that the function has support in the closed ball $|X| \leq R$. By (24) the same holds for f so Theorem 5.4 is proved.

3. The Range and Kernel for the K-types.

On the basis of Theorem 7.7 in Ch. III and Lemma 5.5 above it is a simple matter to describe the range $\mathcal{D}_{\check{\delta}}(\mathfrak{p})^\wedge$ under the Radon transform and the K-finite elements in the kernel of the dual transform. The proofs are similar to those of §3 and will therefore only be sketched.

Given $\delta \in \widehat{K}_M$ we define the differential operator matrix $J^\delta(D)$ by

$$(27) \qquad \left(J^\delta(D)_{ij}F\right)(H) = \int_{\mathfrak{a}^*} J^\delta(\lambda)_{ij} F^*(\lambda) e^{i\lambda(H)}\, d\lambda, \qquad F \in \mathcal{S}(\mathfrak{a}),$$

and then if $\Phi \in \mathcal{D}(\mathfrak{a}, \mathrm{Hom}(V_\delta, V_\delta^M))$ we have the analog of (11) §3,

$$(28) \qquad \left(J^{\check{\delta}}(D)\Phi\right)(H) = \int_{\mathfrak{a}^*} J^{\check{\delta}}(\lambda)\Phi^*(\lambda) e^{i\lambda(H)}\, d\lambda.$$

Then we prove in the same way as Theorems 3.1–3.3, the following result. Here $\mathcal{N} \subset \mathcal{E}(\Xi_o)$ is the kernel of $\varphi \longrightarrow \check{\varphi}$ and $\mathcal{N}_\delta$ the set of $\varphi \in \mathcal{N}$ of type δ.

Theorem 5.6. *Fix $\delta \in \widehat{K}_M$.*

(i) As f runs through $\mathcal{D}(\mathfrak{p})$ the function $(f^\delta)^\wedge$ on $\mathfrak{a}$ runs through the set of functions of the form

$$J^{\check{\delta}}(D)\Psi$$

where $\Psi \in \mathcal{D}(\mathfrak{a}, \mathrm{Hom}(V_\delta, V_\delta^M))$ is W-invariant.

(ii) The range $\mathcal{D}_{\check{\delta}}(\mathfrak{p})^\wedge$ consists of the functions

$$\psi\left(k \cdot (H + \mathfrak{q})\right) = \mathrm{Tr}\left(\delta(k)\left(J^{\check{\delta}}(D)\Psi\right)(H)\right)$$

with Ψ as in (i).

(iii) The space $\mathcal{N}_\delta$ consists of the functions

$$\varphi(k \cdot (H + \mathfrak{q})) = \mathrm{Tr}\left(\check{\delta}(k)\Phi(H)\right)$$

where $\Phi \in \mathcal{E}(\mathfrak{a}, \operatorname{Hom}(V_{\check{\delta}}, V_{\check{\delta}}^M)$ *and the function*

$$ {}^tJ^{\check{\delta}}(D)^*\Phi \qquad \text{on } \mathfrak{a} $$

is W*-odd.*

In (iii) ${}^tA$ denotes the transpose of the matrix A and if A has differential operator entries, $(A^*)_{ij} = (A_{ij})^*$, star denoting adjoint.

4. The Ranges $\mathcal{E}'(X_o)^\wedge$ and $\mathcal{E}(\Xi_o)^\vee$.

We preserve the notation from the earlier subsections and put

$$ \langle \varphi, \psi \rangle = \int_{\Xi_o} \varphi(\xi)\psi(\xi)d\xi_o, \tag{29} $$

where $d\xi_o$ is the G_o-invariant measure on Ξ_o. Under the bijection $(H, kM) \longrightarrow k \cdot (H + \mathfrak{q})$ of $(\mathfrak{a} \times K/M)_W$ onto Ξ_o we can take

$$ d\xi_o = dH\, dk_M. \tag{30} $$

We shall now prove the flat analog of Corollary 2.3; however the proof is quite different since we do not have the flat analog of Theorem 2.1. Instead we shall use Theorem 5.6 and Ch. III, Theorem 7.11.

Theorem 5.7. *In terms of the bilinear form* $\langle , \rangle$ *on* $\mathcal{D}(\Xi_o) \times \mathcal{E}(\Xi_o)$,

$$ \mathcal{D}(\mathfrak{p})^\wedge = \mathcal{N}^\perp. \tag{31} $$

In particular, $\mathcal{D}(\mathfrak{p})^\wedge$ *is a closed subspace of* $\mathcal{D}(\Xi_o)$.

Proof. Consider the bilinear form $\langle \, , \rangle$ on $\mathcal{D}(\Xi_o) \times \mathcal{E}(\Xi_o)$. If $\delta, \pi \in \widehat{K}_M$ it is clear from the orthogonality relations that $\mathcal{D}_\delta(\Xi_o)$ and $\mathcal{E}_\pi(\Xi_o)$ are orthogonal unless $\pi \sim \check{\delta}$. From the general formula

$$ \int_{\mathfrak{p}} f(Y)\check{\varphi}(Y)dY = c \int_{\Xi_o} \widehat{f}(\xi_o)\varphi(\xi_o)d\xi_o \tag{32} $$

(Ch. I, (8) §3) where c is a constant it is clear that

$$ \mathcal{N} = (\mathcal{D}(\mathfrak{p})^\wedge)^\perp. \tag{33} $$

Thus by the remark above

$$(34)\qquad \mathcal{N}_\delta = \left(\mathcal{D}_{\check{\delta}}(\mathfrak{p})^\wedge\right)^\perp \qquad \text{(annihilator in } \mathcal{E}_{\check{\delta}}(\Xi_o)).$$

If L is a constant-coefficient differential operator on $\boldsymbol{R}^n$, then the adjoint ${}^tL$ maps $\mathcal{E}(\boldsymbol{R}^n)$ onto itself so, since L is injective on $\mathcal{E}'(\boldsymbol{R}^n)$ the image $L\mathcal{E}'(\boldsymbol{R}^n)$ is closed in $\mathcal{E}'(\boldsymbol{R}^n)$ (Theorem 2.16, Ch. I).) Hence, using a fundamental solution of L, we see that the subspace $L\mathcal{D}(\boldsymbol{R}^n)$ of $\mathcal{D}(\boldsymbol{R}^n)$ is closed. Hence by Theorem 5.6 (ii), $\mathcal{D}_{\check{\delta}}(\mathfrak{p})^\wedge$ is a closed subspace of $\mathcal{D}_{\check{\delta}}(\Xi_o.)$ Therefore (34) implies

$$(35)\qquad \mathcal{D}_{\check{\delta}}(\mathfrak{p})^\wedge = (\mathcal{N}_\delta)^\perp \qquad \text{(annihilator in } \mathcal{D}_{\check{\delta}}(\Xi_o)).$$

The inclusion $\subset$ in (31) is obvious from (32). For the converse, let $\varphi \in \mathcal{N}^\perp$. Consider the decomposition ([GGA], Ch. V, Cor. 3.4)

$$(36)\qquad \varphi = \sum_{\delta \in \widehat{K}_M} d(\delta)(\chi_\delta * \varphi),$$

where

$$d(\delta)(\chi_\delta * \varphi)(H, kM) = d(\delta)\int_K \varphi(H, ukM)\chi_\delta(u^{-1})du$$

$$= \operatorname{Tr}\left(d(\delta)\int_K \varphi(H, ukM)\delta(u^{-1})du\right) = \operatorname{Tr}\left(\varphi^\delta(H, kM)\right)$$

in the notation of (21). The function

$$(37)\qquad \varphi_{\check{\delta}} = d(\delta)\chi_\delta * \varphi$$

belongs to $(\mathcal{N}_\delta)^\perp$ which equals $\mathcal{D}_{\check{\delta}}(\mathfrak{p})^\wedge$ by (35). Hence, by Theorem 5.6

$$(38)\qquad \varphi_{\check{\delta}}(H, kM) = \operatorname{Tr}\left(\delta(k)\left(J^{\check{\delta}}(D)\Psi\right)(H)\right),$$

where $\Psi \in \mathcal{D}(\mathfrak{a}, \operatorname{Hom}(V_\delta, V_\delta^M))$ is W-invariant. This means that

$$\operatorname{Tr}\left(\delta(k)d(\delta)\int_K \varphi(H, uM)\delta(u^{-1})du\right) = \operatorname{Tr}\left(\delta(k)J^{\check{\delta}}(D)\Psi\right)$$

so $\varphi^\delta(H, eM) = \big(\overset{\vee}{J^\delta}(D)\Psi\big)(H)$ and

$$\varphi^\delta(H, kM) = \delta(k)\big(\overset{\vee}{J^\delta}(D)\Psi\big)(H). \tag{39}$$

We consider now the Euclidean Fourier transform on $\mathfrak{a}$,

$$\varphi^*(\lambda, kM) = \int_{\mathfrak{a}} \varphi(H, kM)e^{-i\lambda(H)}dH. \tag{40}$$

Then by (39),

$$\begin{aligned} &d(\delta)\int_K \varphi^*(\lambda, kM)\delta(k^{-1})dk \\ &= d(\delta)\int_{\mathfrak{a}}\left(\int_K \varphi(H, kM)\delta(k^{-1})dk\right)e^{-i\lambda(H)}dH \\ &= \int_{\mathfrak{a}} \varphi^\delta(H, eM)e^{-i\lambda(H)}dH = \int_{\mathfrak{a}} \big(\overset{\vee}{J^\delta}(D)\Psi\big)(H)e^{-i\lambda(H)}dH \end{aligned}$$

so by (28)

$$d(\delta)\int_K \varphi^*(\lambda, kM)\delta(k^{-1})dk = \overset{\vee}{J^\delta}(\lambda)\widetilde{\Psi}(\lambda). \tag{41}$$

Now Theorem 7.11 in Ch. III shows that

$$\varphi^*(\lambda, kM) = \widetilde{f}(kA_\lambda)$$

for some $f \in \mathcal{D}(\mathfrak{p})$ whence by Lemma 5.2 in the present chapter and (40),

$$\widehat{f} = \varphi.$$

This proves Theorem 5.7.

While this proof was completely different from the proof of the curved analog, Corollary 2.3, there is no difficulty in extending the proofs of Theorem 2.4 and Corollary 2.5 to the present situation. The tools needed are the inversion formula (Theorem 5.3) the support theorem (Theorem 5.4) and the range theorem (Theorem 5.7). We can therefore state the following result.

Corollary 5.8. *With the explained notation,*

$$\mathcal{E}'(X_o)^\wedge = \{\Sigma \in \mathcal{E}'(\Xi_o) : \ \Sigma(\mathcal{N}) = 0\}. \tag{42}$$

$$\mathcal{E}(\Xi_o)^\vee = \mathcal{E}(X_o). \tag{43}$$

It is of some interest to compare (43) with Theorem 3.22 in Ch. I. In the latter case the dual transform involves *all* d-planes through a given point; in (43) only very special d-planes are used, forming a lower-dimensional family. Yet the range is the same in both cases.

EXERCISES AND FURTHER RESULTS.

1. For $\varphi_1, \varphi_2 \in \mathcal{D}(\Xi)$ define the convolution $\varphi_1 * \varphi_2$ by

$$(\varphi_1 * \varphi_2)(kM, a) = \int_A \varphi_1(kM, ab^{-1})\varphi_2(kM, b)db.$$

Show that if $f_1, f_2 \in \mathcal{D}(X)$ then

$$(f_1 \times f_2)^\wedge = \widehat{f_1} * \widehat{f_2},$$

provided f_2 is K-invariant. This restriction on f_2 cannot be dropped (in contrast to the Euclidean case, cf. Ch. I, §2, (7)).

2. Let $f \in \mathcal{D}(X)$, $\varphi \in \mathcal{D}(\Xi)$ and assume φ K-invariant. Then

$$f \times \overset{\vee}{\varphi} = (\widehat{f} * \varphi)^\vee$$

with $*$ as in **1.** The invariance restriction on φ cannot be dropped (in contrast to the Euclidean case, see Ch. I Exercise **A6**) (See Bray and Solmon [1990]).

3. Define $\mathcal{A}$ on $\mathcal{E}'_{\natural}(X)$ as the transpose of $\mathcal{A}^*|\mathcal{E}_W(A)$. Let $h \in A$ and μ_h the invariant measure $f \longrightarrow \int_K f(kh \cdot o)dk$ on the orbit $K \cdot h \cdot o$. Then $\mathcal{A}\mu_h$ is a measure ν_h such that $\operatorname{supp}(\nu_h) = C(h)$ the convex hull of $\{s \cdot h : s \in W\}$ in A.

In particular if X has rank one and μ is the characteristic function of a sphere $S_r(o)$ then $\operatorname{supp}(\mathcal{A}\mu)$ is an interval.

4.* Let f be a continuous exponentially decreasing function on $X = G/K$ ([GGA], Ch. I, §4, No. 4). Fix r and $R > 0$ and assume

$$\int_{B_r(y)} f(x)dx = 0 \qquad \text{for} \qquad y \notin B_R(o).$$

Then

$$\operatorname{supp}(f) \subset C\ell(B_{r+R}(o))$$

(Shahshahani and Sitaram [1987]).

5. For $X = \boldsymbol{H}^2$ prove the following characterization of the range $\mathcal{E}'(X)^\wedge$ extending Theorem 3.4. For $\sigma \in \mathcal{D}'(X)$ define for each $n \in \boldsymbol{Z}$ the distribution $\sigma_n \in \mathcal{D}'(\boldsymbol{R})$ (the Fourier coefficients) by

$$\sigma_n(\varphi) = \sigma(e^{-2t}\varphi(t) \otimes e^{-in\theta}).$$

Then in the notation of §3, (20),

(i) $\sigma(\psi) = \sum_n (e^{2t}\sigma_n)(\psi_{-n})$.

(ii) $\mathcal{E}'(X)^\wedge$ consists of the distributions $\sigma \in \mathcal{E}'(\Xi)$ for which the Fourier coefficients σ_n have the form

$$\sigma_n = e^{-t}\left(\frac{d}{dt} - 1\right)\cdots\left(\frac{d}{dt} - 2|n| + 1\right)\tau_n,$$

where $\tau_n \in \mathcal{E}'(\boldsymbol{R})$ is even.

NOTES

The results of this chapter are from the author's papers [1973a,b], [1983a,b,c], [1987] and [1992b,c]. For a generalization of Theorem 1.1 to non-invariant measures see Quinto [1993] and Gonzalez and Quinto [1994]. The inversion in Theorem 5.3 in a different form was obtained by Gindikin [1967]. In the case of X of rank one Radon transforms for X and for X_o with more general measures have been considered by Quinto [1983] and Orloff [1990a,b], obtaining support theorems, in particular, as well as Lemma 5.1.

CHAPTER V

DIFFERENTIAL EQUATIONS ON SYMMETRIC SPACES.

In [GGA], Ch. II, §4 we discussed briefly what we consider the principal problems for invariant differential operators on a homogeneous space G/H :

- **A. Description of the Algebra $\boldsymbol{D}(G/H)$.**
- **B. Solvability.**
- **C. Determination of the Joint Eigenfunctions.**
- **D. Description of the Eigenspace Representations.**
- **E. Global Properties of Solutions.**

Having in the last three chapters developed tools for analysis on the symmetric space $X = G/K$ we turn in this chapter (and the next) to the applications of these tools to differential equations with Problems $\boldsymbol{A}$–$\boldsymbol{E}$ in mind.

In §1 we use Radon transform results to establish range results and existence of fundamental solutions as well as a Fourier transform criterion for solvability. §2 deals with generalizations of classical mean value theorems. In §3 we discuss harmonic functions on X, effect of a boundedness assumption, and prove analogs of the classical theorems of Poisson, Schwarz, and Fatou. The passage to higher rank of X involves new features which have no analogs in the classical case of the unit disk. In §4 we specialize to the case of bounded symmetric domains D. A harmonic function on D which is continuous on $\overline{D}$ has a Poisson integral representation involving the Shilov boundary $S(D)$, and satisfies additional differential equations (the "Hua Equations"). In §5 we discuss the wave equation on X (shifted by $|\rho|^2$), write down its solution and derive the implications about the appearance of Huygens' principle. The polynomial nature of $1/\boldsymbol{c}(\lambda)$ has implications for Huygens' principle and equidistribution of energy related to the consequences for the Radon transform in Ch. II. We conclude the chapter with a description of the joint eigenfunctions of $\boldsymbol{D}(X)$ in terms of the Poisson transform.

§1. SOLVABILITY.

We shall now use range characterizations from earlier chapters to settle solvability questions for an arbitrary invariant differential operator D on the noncompact symmetric space $X = G/K$ where G is a connected semisimple Lie group with finite center and K is a maximal compact subgroup. Let o denote the origin $\{K\}$ in X, and d the distance function on X.

1. Fundamental Solution of D.

As usual a distribution J on X is called a *fundamental solution* for D if $DJ = \delta$, δ denoting the delta distribution $f \longrightarrow f(o)$ on X.

Theorem 1.1. *Each G-invariant differential operator $D \neq 0$ on X has a fundamental solution.*

Proof. Consider the usual Iwasawa decomposition $G = KAN$ and the Haar measure dn on N normalized as in Ch. II, §3, No. 1. As usual, let $\mathcal{D}^\natural(X)$ and $\mathcal{D}'_\natural(X)$ denote the spaces of K-invariant elements in $\mathcal{D}(X)$ and $\mathcal{D}'(X)$, respectively, and $\mathcal{D}_W(A)$ and $\mathcal{D}'_W(A)$ the space of Weyl group invariants in $\mathcal{D}(A)$ and $\mathcal{D}'(A)$.

As before, let $\boldsymbol{D}(G/K)$ denote the algebra of differential operators on G/K which are invariant under the action of G and let $\boldsymbol{D}_W(A)$ denote the algebra of constant coefficient differential operators on A which are invariant under the action of the Weyl group W. Under the action of N on G/K with transversal manifold $A \cdot o \sim A$ let $\Delta_N(D)$ denote the radial part of $D \in \boldsymbol{D}(G/K)$; then (cf. Theorem 2.1, Ch. II) the map

$$\Gamma : \; D \longrightarrow e^{-\rho}\Delta_N(D) \circ e^{\rho} \tag{1}$$

is an isomorphism of $\boldsymbol{D}(G/K)$ onto $\boldsymbol{D}_W(A)$. Now Theorem 1.1 rests on the following result (Part (i) of Theorem 4.1 in Ch. IV).

Theorem 1.2. *For $f \in \mathcal{D}^\natural(X)$ put*

$$F_f(a) = e^{\rho(\log a)} \int_N f(an \cdot o)dn.$$

Then $F_f \in \mathcal{D}_W(A)$ and the mapping

$$\mathcal{A} : \; f \longrightarrow F_f$$

is a homeomorphism of $D^\natural(X)$ onto $\mathcal{D}_W(A)$ satisfying

$$\mathcal{A}Df = \Gamma(D)\mathcal{A}f \qquad f \in \mathcal{D}^\natural(X) \tag{2}$$

for each $D \in \boldsymbol{D}(G/K)$.

Relation (2) means that $\mathcal{A}$ is a simultaneous "transmutation operator" for all $D \in \boldsymbol{D}(G/K)$, that is transforms them into constant coefficient operators at least when acting on K-invariant functions.

We now come to the proof of Theorem 1.1. The dual spaces of $\mathcal{D}_W(A)$ and $\mathcal{D}^\natural(X)$, are, by the finiteness of W and compactness of K, naturally identified with $\mathcal{D}'_W(A)$ and $\mathcal{D}'_\natural(X)$, respectively. Moreover, by Theorem 1.2 the adjoint

$$\mathcal{A}^* : \mathcal{D}'_W(A) \longrightarrow \mathcal{D}'_\natural(X) \tag{3}$$

is a bijection and

$$D^*\mathcal{A}^* = \mathcal{A}^*\Gamma(D)^*. \tag{4}$$

Because of (3) there exists a $S_o \in \mathcal{D}'_W(A)$ such that $\mathcal{A}^*S_o = \delta$. Since a constant coefficient differential operator on $\mathbf{R}^n$ maps $\mathcal{D}'(\mathbf{R}^n)$, onto itself (Ch. III, Cor. 5.22) there exists an $S \in \mathcal{D}'(A)$ such that $\Gamma(D^*)^*S = S_o$. Since S_o and $\Gamma(D^*)^*$ are W-invariant we can take $S \in \mathcal{D}'_W(A)$. But then (4) implies

$$D\mathcal{A}^*S = \delta$$

so $J = \mathcal{A}^*S$ is the desired fundamental solution.

Corollary 1.3. *Let $D \neq 0$ in $\mathbf{D}(G/K)$. Then the map $D : \mathcal{E}'(X) \longrightarrow \mathcal{E}'(X)$ is injective.*

Let J be a fundamental solution of D. If $\times$ denotes the convolution product on X induced by the convolution $*$ on G (Ch. II, §3 (32)) we have for $S \in \mathcal{E}'(X), T \in \mathcal{D}'(X)$

$$D(S \times T) = DS \times T = S \times DT \tag{5}$$

([GGA], Ch. II, Theorem 5.5). In particular, if $T = J$ we get (since $S \times \delta = S$),

$$DS \times J = S, \tag{6}$$

whence the corollary.

Remark. In the case G complex, Theorem 1.1 can be proved in a more direct fashion. See Exercise 4.

2. Solvability in $\mathcal{E}(X)$.

Equation (5) shows that if $f \in \mathcal{D}(X)$ then the function $u = f \times J$ is a solution to the equation $Du = f$. We shall now prove a stronger result.

Theorem 1.4. *Let $D \neq 0$ in $\mathbf{D}(G/K)$. Then*

$$D\mathcal{E}(X) = \mathcal{E}(X), \tag{7}$$

that is, the equation $Du = f$ *has for each* $f \in \mathcal{E}(X)$ *a solution* $u \in \mathcal{E}(X)$.

The principal tool in the proof is the support theorem for the Radon transform on X which in turn depends on results about the Fourier transform on X (Ch. IV, Theorem 1.1). First we prove a sufficient condition for the surjectivity (7).

Lemma 1.5. *Let* $D \neq 0$ *in* $\mathbf{D}(G/K)$. *Assume that for each closed ball* $V \subset X$,

(8) $$f \in \mathcal{D}(X), \quad \operatorname{supp}(Df) \subset V \textit{ implies } \operatorname{supp}(f) \subset V.$$

Then

(9) $$D\mathcal{E}(X) = \mathcal{E}(X).$$

We begin by showing how (8) implies a more general support property.

Lemma 1.6. *If the operator* D *satisfies (8) then*

(10) $$T \in \mathcal{E}'(X), \quad \operatorname{supp}(D^*T) \subset V \Longrightarrow \operatorname{supp}(T) \subset V.$$

Proof Let $V_r(x)$ denote the closed ball in X of radius r and center x. Let R denote the radius of V. By the invariance of D we may assume $V = V_R(o)$. For $0 < \epsilon < 1$ let ψ_ϵ be a nonnegative radial function in $\mathcal{D}(X)$ with support in $V_\epsilon(o)$ and total integral 1.

Now suppose $T \in \mathcal{E}'(X)$, $\operatorname{supp}(D^*T) \subset V$. Denoting by s the geodesic symmetry of X with respect to o we recall that $D^* = D^s$ ([GGA], Ch. II, Cor. 5.3). Thus, by (5) the function $T^s \times \psi_\epsilon$ satisfies

$$D(T^s \times \psi_\epsilon) = (D^*T)^s \times \psi_\epsilon.$$

Thus if $\pi : G \longrightarrow G/K$ is the natural map and we put $L_r = \pi^{-1}(V_r(o))$ we have by $V = V_R(o)$

$$\operatorname{supp}(D(T^s \times \psi_\epsilon)) \subset \pi(L_R L_\epsilon).$$

Hence by (8),

(11) $$\operatorname{supp}(T \times \psi_\epsilon) \subset V_{R(\epsilon)}(o)$$

if $R(\epsilon) = \sup(d(o, q))$ for q varying in $\pi(L_R L_\epsilon)$. Fix η $(0 < \eta < 1)$. Then by (11)

(12) $$\operatorname{supp}(T \times \psi_\epsilon) \subset V_{R(\eta)}(o), \qquad 0 < \epsilon \leq \eta.$$

On the other hand let $F \in \mathcal{D}(X)$, let $\widetilde{F} = F \circ \pi$ and $g' = g^{-1}$ $(g \in G)$. Since $(\widetilde{\psi}_\epsilon)' = \widetilde{\psi}_\epsilon$ we have

$$(T \times \psi_\epsilon)(F) = T(F \times \psi_\epsilon) \tag{13}$$

and of course

$$\lim_{\epsilon \to 0} F \times \psi_\epsilon = F$$

in $\mathcal{D}(X)$. Thus by (12) and (13),

$$\operatorname{supp}(T) \subset V_{R(\eta)}(o)$$

so, taking intersection over η, $\operatorname{supp}(T) \subset V$ as desired.

We can now prove Lemma 1.5, utilizing some results on Fréchet spaces summarized in Ch. I, §2, No. 6. Since $D: \mathcal{E}(X) \longrightarrow \mathcal{E}(X)$ is a continuous linear mapping of a Fréchet space into itself, its surjectivity is by Ch. I, Theorem 2.16 equivalent to the adjoint D^* being one-to-one on $\mathcal{E}'(X)$ and $D^*\mathcal{E}'$ being closed in $\mathcal{E}'$ (for the topology $\sigma(\mathcal{E}', \mathcal{E})$). Thus by Corollary 1.3 it remains to prove that $D^*\mathcal{E}'$ is closed in $\mathcal{E}'$, or equivalently, by Ch. I, Theorem 2.17, that whenever $B' \subset \mathcal{E}'$ is weak* bounded and closed, $(D^*\mathcal{E}') \cap B'$ is weak* closed. But by Ch. I, Prop. 2.18, $B' \subset \mathcal{E}'_C(X)$ where $C \subset X$ is a compact subset. Since $(D^*\mathcal{E}') \cap B' \subset (D^*\mathcal{E}') \cap \mathcal{E}'_C$ is a closed subset in the topology of $(D^*\mathcal{E}') \cap \mathcal{E}'_C$ induced by $\sigma(\mathcal{E}', \mathcal{E})$ it suffices to prove

$$(D^*\mathcal{E}') \cap \mathcal{E}'_C \text{ is closed in } \mathcal{E}'. \tag{14}$$

By Ch. I, Prop. 2.18 it suffices to prove this for the strong topology of $\mathcal{E}'(X)$. Let $T_j = D^*S_j$ $(S_j \in \mathcal{E}')$ be a net in $(D^*\mathcal{E}') \cap \mathcal{E}'_C$ converging to $T \in \mathcal{E}'$. Since $\mathcal{E}'_C$ is closed, $T \in \mathcal{E}'_C$. Select r such that $C \subset V_r(o)$. Then $\operatorname{supp}(T_j) \subset V_r(o)$ so by Lemma 1.6, $\operatorname{supp}(S_j) \subset V_r(o)$. If J^* is a K-invariant fundamental solution for D^* we have by (6),

$$S_j = T_j \times J^* \tag{15}$$

which converges (in $\sigma(\mathcal{D}', \mathcal{D})$) to $T \times J^*$. In particular, $\operatorname{supp}(T \times J^*) \subset V_r(o)$. By (15),

$$T_j = D^*(T_j \times J^*)$$

so taking limits $T = D^*(T \times J^*) \in D^*\mathcal{E}'$. This proves (14) and therefore Lemma 1.5.

We shall now need the following result which actually is a special case of the convexity theorem of Lions and Titchmarsh (Lions [1952]).

Lemma 1.7. *Let $L \neq 0$ be a differential operator in $\boldsymbol{R}^n$ with constant coefficients. Then for each $f \in \mathcal{D}(\boldsymbol{R}^n)$, the supports of the functions f and Lf have the same convex hull.*

Proof. Viewing a convex set as an intersection of half spaces it suffices to prove that if Lf has support in a closed half space H then so does f. It suffices to consider the case when H is the half space $x_n \leq 0$. Consider the Fourier transform in the variable $x' = (x_1, \ldots, x_{n-1})$,

$$(\mathcal{F}_n f)(\xi', x_n) = \int_{\boldsymbol{R}^{n-1}} f(x', x_n) e^{-i\langle x', \xi' \rangle} dx'. \tag{16}$$

The Fourier transform $\mathcal{F}_n Lf$ is then given by

$$(\mathcal{F}_n Lf)(\xi', x_n) = L_n(\xi')(\mathcal{F}_n f)(\xi', x_n) \tag{17}$$

where $L_n(\xi')$ is a differential operator in the x_n variable whose coefficients are certain polynomials in ξ'. Since $f \in \mathcal{D}(\boldsymbol{R}^n)$, (16) shows that $(\mathcal{F}_n f)(\xi', x_n) = 0$ for x_n sufficiently large. On the other hand $(\mathcal{F}_n Lf)(\xi', x_n) = 0$ for $x_n > 0$. Applying (17) for a fixed ξ' we deduce from the uniqueness theorem for ordinary differential equations in x_n that $(\mathcal{F}_n f)(\xi', x_n) = 0$ for $x_n > 0$ unless the operator $L_n(\xi')$ has all its coefficients 0. This can only happen when ξ' belongs to a certain algebraic set in $\boldsymbol{R}^{n-1}$ so by continuity of $\mathcal{F}_n f$ we deduce $(\mathcal{F}_n f)(\xi', x_n) = 0$ for all ξ' and all $x_n > 0$. Hence by (16) $f(x', x_n) = 0$ for all x' and $x_n \geq 0$, proving the lemma.

Because of Lemma 1.5, Theorem 1.4 can be proved by verifying the support property (8). This we now do by means of the support theorem for the Radon transform on X.

Because of the invariance of D we may assume V in (8) to be a ball $V_R(o)$ centered at the origin. Let $f \in \mathcal{D}(X)$ satisfy $\text{supp}(Df) \subset V$. Then we have for the Radon transform,

$$(Df)^\wedge(\xi) = 0 \qquad \text{if } d(o, \xi) > R, \tag{18}$$

d denoting distance. Now by Ch. II, Theorem 3.3, $(Df)^\wedge = \widehat{D}\widehat{f}$ where $\widehat{D}$ is a G-invariant differential operator on the horocycle space $\Xi = G/MN$. But under the diffeomorphism $(kM, a) \longrightarrow ka \cdot \xi_o$ of $(K/M) \times A$ onto Ξ the operator $\widehat{D}$ is given by a constant coefficient operator Δ on the Euclidean space A (Ch. II, Theorem 2.2). In addition, the distance d satisfies

$$d(o, ka \cdot \xi_o) = |\log a| \qquad k \in K, \ a \in A$$

([DS], Ch. VI, Exercise B2). Combining these facts, (18) can be restated,

$$\Delta_a(\widehat{f}(ka\cdot\xi_o)) = 0, \qquad k\in K, |\log a| > R.$$

From this and Lemma 1.7 we deduce $\widehat{f}(ka\cdot\xi_o) = 0$ for $|\log a| > R,\ k\in K$ or equivalently $\widehat{f}(\xi) = 0$ for $d(o,\xi) > R$. Hence by the support theorem (Ch. IV, Theorem 1.1) $f(x) = 0$ for $x\notin V_R(o)$. This verifies condition (8) whereby Theorem 1.4 is proved.

3. Solvability in $\mathcal{E}'(X)$.

In general if D is an invariant differential operator on G/K the differential equation $DS = T$ can only be solved in $\mathcal{E}'(X)$ under certain necessary conditions.

Let $\widetilde{T}$ denote the Fourier transform

$$\widetilde{T}(\lambda,b) = \int_X e^{(-i\lambda+\rho)(A(x,b))}dT(x) \qquad T\in\mathcal{E}'(X) \tag{19}$$

as defined in Ch. III, §5, No. 4. Also let $\Gamma:\ \boldsymbol{D}(G/K)\longrightarrow\boldsymbol{D}_W(A)$ be the isomorphism in Ch. II, Theorem 2.1. We shall view $\boldsymbol{D}_W(A)$ as the set of W-invariants in the symmetric algebra $S(\mathfrak{a})$ which we can also view as the algebra of W-invariant polynomial functions on $\mathfrak{a}^*_C$.

Theorem 1.8. *If $D\in\boldsymbol{D}(G/K)$ and $T\in\mathcal{E}'(X)$ the differential equation*

$$DS = T \tag{20}$$

has a solution $S\in\mathcal{E}'(X)$ if and only if for each $b\in B$ the function $\lambda\longrightarrow\widetilde{T}(\lambda,b)/\Gamma(D)(i\lambda)$ is an entire function on $\mathfrak{a}^*_C$.

Proof. Assuming (20) let us take its Fourier transform. We have (Ch. II, Prop. 3.14),

$$\begin{aligned}(DS)^{\sim}(\lambda,b) &= \int_X D^*_x(e^{(-i\lambda+\rho)(A(x,b))})dS(x)\\ &= \Gamma(D^*)(-i\lambda)\widetilde{S}(\lambda,b).\end{aligned}$$

By [GGA], Ch. II, Cor. 5.3 and Lemma 5.21 we have $\Gamma(D^*)(-i\lambda) = \Gamma(D)(i\lambda)$ so the necessity in the theorem follows.

On the other hand, suppose $\lambda \longrightarrow \widetilde{T}(\lambda, b)/\Gamma(D)(i\lambda)$ is an entire function on $\mathfrak{a}_C^*$. We know from Ch. III, Cor. 5.9 that $\widetilde{T}(\lambda, b)$ is holomorphic of uniform exponential type and slow growth. By a trivial modification of Lemma 5.13 in Ch. III, the function $\widetilde{T}(\lambda, b)/\Gamma(D)(i\lambda)$ has these properties, being assumed holomorphic. Using Cor. 5.9 in Ch. III once more the existence of S in (20) follows, proving the theorem.

§2. MEAN VALUE THEOREMS.

1. The Mean Value Operators.

Let G be a connected unimodular Lie group and K a compact subgroup. We do not yet assume the space $X = G/K$ to be symmetric. As usual we let $\pi: G \longrightarrow G/K$ denote the natural map and put $o = \pi(e), \widetilde{f} = f \circ \pi$ for $f \in C(X)$. For $x \in G$ we have in Ch. II §2 defined the mean value operator

$$(1) \qquad (M^x f)(gK) = \int_K f(gkx \cdot o)dk, \qquad f \in C(X),$$

and have proved in Theorem 2.7 that for x sufficiently close to e, M^x is formally a "noncommutative power series" in generators $\Delta_1, \ldots, \Delta_\ell$ of $\boldsymbol{D}(G/K)$ without constant term. The theorem quoted implies the following characterization of the joint eigenfunctions of $\boldsymbol{D}(G/K)$.

Theorem 2.1. *Let V be an open subset of G/K and let $f \in C^\infty(V)$ be a joint eigenfunction of $\boldsymbol{D}(G/K)$, i.e.*

$$Df = c_D f \qquad D \in \boldsymbol{D}(G/K)$$

where $c_D \in \boldsymbol{C}$. Then for each $q \in V$

$$(2) \qquad (M^x f)(q) = c_x f(q) \qquad c_x \in \boldsymbol{C}$$

for all $x \in G$ sufficiently close to e. Conversely, a continuous function satisfying (2) is a joint eigenfunction of $\boldsymbol{D}(G/K)$.

Proof. Since the Laplace-Beltrami operator is a member of $\boldsymbol{D}(G/K)$, f is necessarily analytic. Thus (2) is an immediate consequence of Theorem 2.7 in Chapter II. On the other hand suppose $f \in C(V)$ satisfies (2). Let $q \in V$ and select $g_o \in G$ such that $\pi(g_o) = q$. Then

$$(3) \qquad \int_K \widetilde{f}(gkx)dk = c_x \widetilde{f}(g)$$

for all g in a neighborhood G_{g_o} of g_o in G and all x in a neighborhood N_{g_o} of e in G. Equation (3) implies c_x continuous in x. Let $\rho \in C_c^\infty(G)$ have support inside N_{g_o} such that $\int \rho(x)c_x dx = 1$. Then (3) implies

$$\widetilde{f}(g) = \int_G \rho(x) \int_K \widetilde{f}(gkx)dkdx = \int_G \widetilde{f}(y) \int_K \rho(yk^{-1}g^{-1})dkdy,$$

which shows that $\widetilde{f}$ is differentiable near g_o so $f \in C^\infty(V)$. Thus by (3) $c_x = c(x)$ depends differentiably on x. Let $D \in \boldsymbol{D}_K(G)$ and apply it to (3) in the variable x. Then

$$\int_K (D\widetilde{f})(gkx)dk = (Dc)(x)\widetilde{f}(g) \tag{4}$$

so, taking $x = e$,

$$(D\widetilde{f})(g) = (Dc)(e)\widetilde{f}(g), \tag{5}$$

so f is a joint eigenfunction of $\boldsymbol{D}(G/K)$.

Corollary 2.2. (Godement) *Let V be an open subset of G/K. A function u on V satisfying the equations*

$$\Delta_1 u = \cdots = \Delta_\ell u = 0$$

is characterized by the mean value property

$$(M^x u)(q) = u(q) \tag{6}$$

for each $q \in V$, x being sufficiently close to e in G.

In fact, (4) shows that $(Dc)(e) = 0$ for each $D \in \boldsymbol{D}_K(G)$. Since c is bi-invariant under K this implies $(Dc)(e) = 0$ for each $D \in \boldsymbol{D}(G)$ annihilating the constants so by analyticity, c is a constant. Since $c(e) = 1$ (6) follows.

Remark. Consider the special case when $X = G/K$ is a symmetric space of rank one and G the identity component of the group of isometries of X. In this case ℓ above is 1 and $\boldsymbol{D}(G/K)$ is generated by the Laplace-Beltrami operator L. Here Cor. 2.2 states that the solutions to $Lu = 0$ are characterized by the mean-value of u over spheres equals the value at the center. For $X = \boldsymbol{R}^n$ this is Gauss' mean value theorem for harmonic functions. One can also take G as the semidirect product of the torus $K = \boldsymbol{T}^k$ acting on $\boldsymbol{C}^k$ and $\boldsymbol{C}^k$. Then harmonic means harmonic in each variable.

Corollary 2.3. *The joint eigenfunctions of $\boldsymbol{D}(G/K)$ are characterized by the functional equation*

$$\int_K f(gkx\cdot o)dk = f(g\cdot o)\varphi(x\cdot o) \qquad x,g\in G \tag{7}$$

φ being some spherical function on G/K.

In fact (7) is a restatement of (3) which by analyticity of $\widetilde{f}$ extends to $x, g \in G$. The result was already noted in Ch. II, §2 for X symmetric.

Remark. Relation (7) implies

$$f \times F = c_F f \qquad \text{for} \quad F \in C_c(X) \tag{8}$$

where the constant c_F equals $\int \widetilde{\varphi}(x^{-1})\widetilde{F}(x)dx$.

This is clear since

$$\int_G \left(\int_K \widetilde{f}(gkx^{-1})dk\right)\widetilde{F}(x)dx = \int_G \widetilde{f}(gy)\widetilde{F}(y^{-1})dy = (f\times F)(g\cdot o).$$

We now specialize to the case when (G/K) is a symmetric pair, that is there is an involutive automorphism θ of G such that $(K_\theta)_o \subset K \subset K_\theta$ where K_θ is the fixed point set of θ and $(K_\theta)_o$ its identity component.

Theorem 2.4. *Let (G/K) be a symmetric pair, K compact. Let $f \in \mathcal{E}(X)$ and define F on $X \times X$ by*

$$F(gK, hK) = \int_K f(gkh\cdot o)dk. \tag{9}$$

Then if $D \in \boldsymbol{D}(G/K)$,

$$D_1 F = D_2 F \tag{10}$$

where the subscripts indicate action on the first and second variable, respectively. Moreover

$$M^x D = DM^x \qquad \textit{for each } \; x \in G. \tag{11}$$

Proof. The first part was proved as Theorem 3.12 in Ch. II although there it was assumed that $X = G/K$ was a symmetric space of the noncompact type. However the proof used only the relation

$$D(f \times \varphi) = Df \times \varphi = f \times D\varphi \qquad f \in \mathcal{E}(X),\ \varphi \in \mathcal{D}(X)$$

and this holds in the present more general situation ([GGA], Ch. II, Theorem 5.5). For (11) note that

$$F(gK, hK) = (M^h f)(gK)$$

so by (10)

$$D_{gK}\left((M^h f)(gK)\right) = D_{hK}\left((M^h f)(gK)\right)$$
$$= D_{hK}\left(\int_K f(gkh \cdot o)dk\right) = \int_K (Df)(gkh \cdot o)dk = \left(M^h Df\right)(gK)$$

and this proves (11).

2. Approximations by Analytic Functions.

Since Theorem 2.7, Ch. II is limited to analytic functions we must, in order to apply it to C^∞ functions, approximate these by analytic functions. For this problem we recall some facts concerning analytic vectors of representations of Lie groups.

Let M be an analytic manifold and $\mathfrak{H}$ a (complex) Banach space. A mapping $f : M \longrightarrow \mathfrak{H}$ is said to be *analytic* at a point $p \in M$ if there exists a coordinate system $\{x_1, \cdots, x_m\}$ on a neighborhood U of p such that $x_1(p) = \cdots = x_m(p) = 0$ and

$$f(q) = \Sigma a_{n_1 \cdots n_m} x_1(q)^{n_1} \cdots x_m(q)^{n_m} \qquad (q \in U),$$

where the coefficients $a_{n_1 \ldots n_m}$ belong to $\mathfrak{H}$ and the series converges absolutely. (A series Σa_n where $a_n \in \mathfrak{H}$ is said to converge absolutely if $\Sigma\|a_n\|$ converges, $\|\ \|$ denoting the norm in $\mathfrak{H}$.)

The following two lemmas are obvious.

Lemma 2.5. *Consider two mappings $\varphi : M \longrightarrow M$ and $f : M \longrightarrow \mathfrak{H}$. If φ is analytic at $p \in M$ and f is analytic at $\varphi(p)$, then the composite mapping $f \circ \varphi : M \longrightarrow \mathfrak{H}$ is analytic at p.*

Lemma 2.6. *Let the mapping $f : M \longrightarrow \mathfrak{H}$ be analytic at p and let F be a continuous linear mapping of $\mathfrak{H}$ into C. Then the function $F \circ f$ is analytic at p.*

Definition. Let G be a Lie group and let π be a representation of G on a Banach space $\mathfrak{H}$. A vector $\boldsymbol{e} \in \mathfrak{H}$ is said to be *analytic* (under π) if the mapping $x \longrightarrow \pi(x)\boldsymbol{e}$ is an analytic mapping of G into $\mathfrak{H}$.

Let $\mathcal{A}_\pi$ denote the set of analytic vectors under π. It is obvious that $\mathcal{A}_\pi$ is a linear subspace of $\mathfrak{H}$. If $\boldsymbol{e} \in \mathcal{A}_\pi$ and $y \in G$, then the mapping $x \longrightarrow \pi(x)\pi(y)\boldsymbol{e}$ is composed of the analytic mappings $x \longrightarrow xy$ and $x \longrightarrow \pi(x)\boldsymbol{e}$. Using Lemma 2.5 it follows that $\pi(y)\boldsymbol{e} \in \mathcal{A}_\pi$.

Lemma 2.7. *The space $\mathcal{A}_\pi$ is invariant under each $\pi(y), y \in G$.*

If $\mathfrak{H}$ is finite-dimensional, then each vector $e \in \mathfrak{H}$ is analytic. This is a special case of the theorem that a continuous homomorphism of one Lie group into another is analytic. For Banach spaces $\mathfrak{H}$ of infinite dimension the basic fact is:

Theorem 2.8. *The space $\mathcal{A}_\pi$ of analytic vector is dense in $\mathfrak{H}$.*

This theorem was proved by Harish-Chandra [1953], p. 220, for semisimple Lie groups and generalized by E. Nelson to arbitrary Lie groups [1959]. The quickest proof is due to L. Gårding [1960] who, like Nelson, uses the heat equation on the group. We shall use this theorem in order to prove the following lemma on approximation by analytic functions.

Lemma 2.9. *Let $F_1, \dots F_n$ be a finite set of bounded, continuous functions on G, let C be a compact subset of G and ϵ a number > 0. Then there exists an analytic, integrable function φ on G such that the convolution $\varphi * F_i$ is analytic and such that*

$$|(\varphi * F_i)(x) - F_i(x)| < \epsilon \qquad (1 \leq i \leq n) \tag{12}$$

for all $x \in C$. If each F_i has compact support, φ can be chosen such that (12) holds for all $x \in G$.

Proof. Let $L^1(G)$ denote the Banach space of integrable functions on G with respect to a left invariant measure dg on G, the norm on $L^1(G)$ being

$$\|f\| = \int_G |f(g)| dg.$$

The space $L^1(G)$ is the completion of the group algebra in this norm. For each $x \in G$ let $\pi(x)$ denote the endomorphism of $L^1(G)$ given by $[\pi(x)f](y) = f(x^{-1}y)$, $f \in L^1(G)$, $y \in G$. By the left invariance of dg, $\|\pi(x)f\| = \|f\|$. If $f \in C_c(G)$ and $\delta > 0$ there exists a neighborhood V of e in G such that $|f(x^{-1}y) - f(y)| < \delta$ for all $x \in V$ and all $y \in G$. This implies that the mapping $x \longrightarrow \pi(x)f$ of G into $L^1(G)$ is continuous. Using the fact that the group algebra is dense in $L^1(G)$ this continuity follows for all $f \in L^1(G)$. Hence the mapping $x \longrightarrow \pi(x)$ is a representation of G on $L^1(G)$. Let $h \in L^1(G)$ be an analytic vector under π. If $x \in G$, the vector $\pi(x)h$ is also analytic (Lemma 2.7). If F is a bounded continuous function on G then by Lemma 2.6 the function

$$x \longrightarrow \int_G F(y^{-1})[\pi(x^{-1})h](y)dy,$$

which is just the function $h*F$, is analytic. Let U be a compact neighborhood of e in G and $\gamma_U \in C(G), \gamma_U \geq 0$, $\operatorname{supp}(\gamma_U) \subset U$ and $\int \gamma_U(x)dx = 1$. Then if U is sufficiently small we have

$$\text{(13)} \qquad \left|(\gamma_U * \gamma_U * F_i)(x) - F_i(x)\right| < \frac{\epsilon}{2} \qquad 1 \leq i \leq n$$

for all $x \in C$. According to Theorem 2.8 there exists a sequence (φ_n) of analytic vectors converging to γ_U. Since $\|\varphi_n - \gamma_U\| \longrightarrow 0$ as $n \longrightarrow \infty$ it is clear that if F is continuous and bounded on G then $\varphi_n * F \longrightarrow \gamma_U * F$ uniformly on G. In particular, if N is sufficiently large, φ_N satisfies

$$\text{(14)} \quad |(\varphi_N * \gamma_U * F_i)(x) - (\gamma_U * \gamma_U * F_i)(x)| \leq \frac{\epsilon}{2} \qquad (x \in G) \qquad (1 \leq i \leq n).$$

The first statement of Lemma 2.9 (with $\varphi = \varphi_N * \gamma_U$) follows from (13) and (14). Also, if each F_i has compact support, U can be chosen such that (13) holds for all $x \in G$. This proves the lemma.

3. Ásgeirsson's Mean Value Theorem extended to Homogeneous Spaces.

We shall now consider functions on the product $G/K \times G/K$ and if E is an operator on functions on G/K, the operators E_1 and E_2 are defined on functions on $G/K \times G/K$ by letting E act on the first and second variable respectively. A function $u \in \mathcal{E}(G)$ will be said to be of *bounded growth* if for each $D \in \boldsymbol{D}(G)$, Du is bounded. A function $u \in \mathcal{E}(X)$ is said to be of bounded growth if $\widetilde{u}$ is.

Theorem 2.10. *Let $V \subset G/K$ be a connected open and K-invariant neighborhood of o and let $u \in \mathcal{E}(V \times V)$ satisfy the system*

$$D_1 u = D_2 u \qquad \textit{for} \quad D \in \boldsymbol{D}(G/K). \tag{15}$$

Then

$$\int_K u(k\cdot x, o)\,dk = \int_K u(o, k\cdot x)\,dk \qquad x \in V \tag{16}$$

provided either

(i) *u is analytic on $V \times V$*

or

(ii) *$V = X$ and u is of bounded growth.*

Proof. If $z \in G$ is such that $z \cdot o = x$ then (16) amounts to

$$(M_1^z u)\,(o, o) = (M_2^z u)\,(o, o). \tag{17}$$

Thus Theorem 2.7 in Ch. II implies that (16) holds for $x \in V$ near the origin so in case (i) the analyticity of u implies that (16) holds for all x in the connected set V.

Next we consider case (ii). Let $E \in \boldsymbol{D}_K(G)$. Then $E\widetilde{f} = (Df)^{\sim}$ where $D \in \boldsymbol{D}(G/K)$. Thus by (15),

$$E_1\widetilde{u} = E_2\widetilde{u} \qquad \text{for } E \in \boldsymbol{D}_K(G).$$

If $\varphi \in \mathcal{E}(G \times G) \cap L^1(G \times G)$ the convolution

$$(\varphi \times u)(x_1, x_2) = \int_{G\times G} \varphi(y_1, y_2)\,u(y_1^{-1}\cdot x_1, y_2^{-1}\cdot x_2)\,dy_1\,dy_2 \tag{18}$$

exists and since $\widetilde{u}$ is of bounded growth,

$$E_1(\varphi * \widetilde{u}) = \varphi * E_1\widetilde{u}, \quad E_2(\varphi * \widetilde{u}) = \varphi * E_2\widetilde{u},$$

whence

$$D_1(\varphi \times u) = D_2(\varphi \times u).$$

Choosing φ in accordance with Lemma 2.9, $\varphi \times u$ is analytic so by (i) relation (16) holds for $\varphi \times u$ and all x. For a suitable sequence (φ_i) of such φ, $\varphi_i \times u \longrightarrow u$ uniformly on compacts so (16) follows in general.

In the case when G/K is a noncompact symmetric space of rank one the theorem can be sharpened a bit by means of a different proof. Here $\boldsymbol{D}(G/K)$ consists of the polynomials in the Laplacian L.

Theorem 2.11. *Let X be a noncompact symmetric space of rank one. Let $(x_o, y_o) \in X \times Y$ and $0 < R \leq \infty$. Suppose $u \in C^2\big(B_R(x_o) \times B_R(y_o)\big)$ satisfies the differential equation*

$$L_x\big(u(x,y)\big) = L_y\big(u(x,y)\big). \tag{19}$$

Then if dw is the surface element on $S_r(x_o)$,

$$\int\limits_{S_r(x_o)} u(x, y_o)dw(x) = \int\limits_{S_r(y_o)} u(x_o, y)dw(y) \qquad (r < R). \tag{20}$$

Proof. A geometric proof for the case $R = \infty$ is given in [GGA], Ch. II, §5. This proof is also valid for the present case $R < \infty$. Here we shall give a different proof, using the Fourier transform on X. By group invariance we may assume $x_o = y_o = o$. Taking convolution $\varphi \times u$ as in (18) with a smooth function on $G \times G$ with support in a neighborhood of e we may assume that u is smooth. We consider the distribution $T \in \mathcal{E}'(X \times X)$ given by

$$T(f) = \int\limits_{S_r(o)} f(x, o)dw(x) - \int\limits_{S_r(o)} f(o, y)dw(y). \tag{21}$$

We shall compute the Fourier transform $\widetilde{T}$. Since the Iwasawa decomposition of $G \times G$ is given by

$$(G \times G) = (N \times N)(A \times A)(K \times K)$$

we see that

$$A\big((g_1, g_2)(K \times K), (k_1, k_2)(M \times M)\big) = A(g_1K, k_1M) + A(g_2K, k_2M).$$

Thus, if $(\lambda_1, \lambda_2) \in \mathfrak{a}_c^* \times \mathfrak{a}_c^*$, $b_1 = k_1M$, $b_2 = k_2M$,

$$\widetilde{T}\big((\lambda_1, \lambda_2), (b_1, b_2)\big) = \int\limits_{X \times X} e^{(-i\lambda_1+\rho)(A(x_1,b_1))}e^{(-i\lambda_2+\rho)(A(x_2,b_2))}dT(x_1, x_2)\big)$$

so by (21),

$$\widetilde{T}\big((\lambda_1, \lambda_2), (b_1, b_2)\big) = \big(\varphi_{-\lambda_1}(a_r) - \varphi_{-\lambda_2}(a_r)\big)A(r), \tag{22}$$

where $a_r \in S_r(o) \cap A\cdot o$ and $A(r)$ is the area of $S_r(o)$. Now $\lambda \longrightarrow \varphi_\lambda(a_r)$ is an even holomorphic function on $\mathfrak{a}_C^*$ so has the form $F(\langle\lambda,\lambda\rangle)$ where F is holomorphic on $\boldsymbol{C}$. Thus by (22),

$$\widetilde{T}\big((\lambda_1,\lambda_2),(b_1,b_2)\big) = A(r)\big(F(\langle\lambda_1,\lambda_1\rangle) - F(\langle\lambda_2,\lambda_2\rangle)\big). \tag{23}$$

Since

$$F(z_1) - F(z_2) = (z_1 - z_2)\int_o^1 F'\big(z_2 + t(z_1 - z_2)\big)dt$$

we see from (23) that

$$\widetilde{T}\big((\lambda_1,\lambda_2),(b_1,b_2)\big)\Big/\big(\langle\lambda_1,\lambda_1\rangle - \langle\lambda_2,\lambda_2\rangle\big) \tag{24}$$

is an entire function in $\langle\lambda_1,\lambda_1\rangle$, $\langle\lambda_2,\lambda_2\rangle$ and can thus be viewed as a holomorphic function on $\mathfrak{a}_C^* \times \mathfrak{a}_C^*$. Also, by Ch. III, Cor. 5.9, $\widetilde{T}$ has uniform exponential type and slow growth. Because of Lemma 5.13 in Ch. III the same is the case for the quotient (24) so again by Cor. 5.9 there is a distribution $S \in \mathcal{E}'(X \times X)$ with support contained in $\mathrm{Cl}(B_r(o) \times B_r(o))$ such that

$$\widetilde{T}\big((\lambda_1,\lambda_2)\big) = \big(\langle\lambda_1,\lambda_1\rangle - \langle\lambda_2,\lambda_2\rangle\big)\widetilde{S}\big((\lambda_1,\lambda_2)\big), \tag{25}$$

where we have, by (23), suppressed the variable (b_1, b_2). Now

$$\big((-L_1 + L_2)S\big)^{\sim} = \big\{\langle\lambda_1,\lambda_1\rangle + \langle\rho,\rho\rangle - \big(\langle\lambda_2,\lambda_2\rangle + \langle\rho,\rho\rangle\big)\big\}\widetilde{S}$$

so by (25)

$$(-L_1 + L_2)S = T.$$

Now select $\epsilon > 0$ such that $r + 3\epsilon < R$, $\varphi_o \in \mathcal{E}(X)$ such that $\varphi_o \equiv 1$ on $B_{r+2\epsilon}(o)$, $\varphi_o \equiv 0$ outside $B_{r+3\epsilon}(o)$ and put $\varphi(x,y) = \varphi_o(x)\varphi_o(y)$. Then $u\varphi$ belongs to $\mathcal{E}(X \times X)$ and satisfies (19) on $B_{r+\epsilon}(o) \times B_{r+\epsilon}(o)$. It follows that

$$T(u\varphi) = S((-L_1 + L_2)(u\varphi)) = 0,$$

proving (20).

§3. HARMONIC FUNCTIONS ON SYMMETRIC SPACES.

1. Generalities.

Let G be a Lie group, $K \subset G$ a compact subgroup. Let $V \subset G/K$ be a connected open subset. We recall ([GGA] Ch. IV, §2) that a function $f \in \mathcal{E}(V)$ is said to be *harmonic* if $Df = 0$ for each $D \in \boldsymbol{D}(G/K)$ which annihilates the constants. Such functions are characterized by Godement's mean value theorem (Cor. 2.2 above)

$$\int_K f(gkx \cdot o)dk = f(g \cdot o), \tag{1}$$

which for $G/K = \boldsymbol{M}(n)/\boldsymbol{O}(n) = \boldsymbol{R}^n$ reduces to the Gauss' mean value theorem for harmonic functions on $\boldsymbol{R}^n$.

We shall now investigate harmonic functions in greater detail in the case of a symmetric space.

2. Bounded Harmonic Functions.

Suppose now $X = G/K$ is a symmetric space of the noncompact type and as usual let L denote the Laplace-Beltrami operator on X.

Theorem 3.1. *Suppose f is a bounded C^2-function on X satisfying*

$$Lf = 0.$$

Then f is harmonic.

The proof starts with the following lemma where S denotes a fixed sphere $S_r(o)$ in X with center $o = \{K\}$ and radius $r > 0$.

Lemma 3.2. *There exists a positive K-invariant measure ν on G/K such that*

$$\textit{support } (\nu) = S, \qquad \nu(S) = 1$$

and such that any function f satisfying $Lf = 0$ has the form

$$f(g \cdot o) = \int_S f(g \cdot s)d\nu(s).$$

Proof. Consider the open ball $B = B_r(o)$ with boundary S and let $C_{\mathbf{R}}(S)$ denote the space of real-valued continuous functions on S. By the theory of the classical Dirichlet problem for the elliptic operator L, given $\varphi \in C_{\mathbf{R}}(S)$ there exists a unique function $\Phi \in C^\infty(B)$ such that $L\Phi = 0$ and such that the function

$$\begin{aligned} U(x) &= \Phi(x) \qquad && x \in B \\ U(s) &= \varphi(s) \qquad && s \in S \end{aligned}$$

is continuous on $\overline{B} = B \cup S$. Also, by the maximum principle, if U is not constant U cannot take its maximum in B. Clearly

$$|\Phi(x)| \leq \|\varphi\|_\infty \qquad x \in B \tag{2}$$

for the uniform norm. For a fixed $x \in B$ the functional $\varphi \longrightarrow \Phi(x)$ is by (2) given by a unique positive measure ν_x on S for which

$$\Phi(x) = \int_S \varphi(s)d\nu_x(s) = \nu_x(\varphi), \qquad \varphi \in C_{\mathbf{R}}(S). \tag{3}$$

In particular, $\Phi(o) = \nu_o(\varphi)$. We claim that $\operatorname{supp}(\nu_o) = S$. Otherwise, choose $\varphi \not\equiv 0$, $\varphi \geq 0$ with $\operatorname{supp}(\varphi) \subset S - \operatorname{supp}(\nu_o)$. Then by (3), $\Phi \geq 0$ and nonconstant so by the maximum principle for $-\Phi$ stated above, $\Phi(o) = 0$ is impossible, yet $\nu_o(\varphi) = 0$.

If $k \in K$ the mapping $\varphi \longrightarrow \Phi$ commutes with the action of k so

$$\nu_o^{\tau(k)}(\varphi) = \nu_o\big(\varphi^{\tau(k^{-1})}\big) = \Phi^{\tau(k^{-1})}(o) = \Phi(o) = \nu_o(\varphi).$$

Thus by the uniqueness, ν_o is K-invariant. Define the measure ν on X by $\nu(A) = \nu_o(A \cap S)$ if A is a Borel set. Then if $Lf = 0$ we have

$$f(o) = \nu_o\big(f|S\big) = \nu(f).$$

Using this on the function $f^{\tau(g^{-1})}$ the relation of the lemma follows.

The idea is now to modify this relation in such a way that the integration only takes place over a K-orbit on S. For this we use an abstract theorem about commuting contractions on a Banach space E. An operator P on E is called a *contraction* if its norm $\|P\|$ is ≤ 1.

Lemma 3.3. *Let P and Q be two commuting contractions on a Banach space E. Let $0 < \alpha < 1$. Then each vector fixed by the operator $\alpha P + (1-\alpha)Q$ is also fixed by P and Q.*

Proof. If A is a bounded operator on E, e^A is a well-defined operator on E given by the power series expansion. We shall prove that

$$\lim_{n\to+\infty} \|(I-P)e^{-n(I-P)}\| = 0, \tag{4}$$

where I is the identity operator. We have

$$\begin{aligned} e^{-n(I-P)} &= \sum_{k=0}^{\infty} \frac{n^k}{k!}(P-I)^k = \sum_{k=0}^{\infty} \frac{n^k}{k!} \sum_{j=0}^{k} \binom{k}{j} P^j (-1)^{k-j} \\ &= \sum_{j=0}^{\infty} \frac{P^j n^j}{j!} \sum_{k=j}^{\infty} \frac{n^{k-j}}{(k-j)!}(-1)^{k-j} = \sum_{j=0}^{\infty} \alpha_{j,n} P^j, \end{aligned}$$

where

$$\alpha_{j,n} = n^j e^{-n} (j!)^{-1}.$$

Observe that

$$\frac{\alpha_{j+1,n}}{\alpha_{j,n}} = \frac{n}{j+1} = \begin{cases} <1 & \text{for } j \geq n \\ \geq 1 & \text{for } j \leq n-1. \end{cases}$$

Putting $\alpha_{-1,n} = 0$ we thus have

$$\begin{aligned} \|(I-P)e^{-n(I-P)}\| &\leq \sum_{j=0}^{\infty} |\alpha_{j,n} - \alpha_{j-1,n}| \\ = \sum_{j=0}^{n} (\alpha_{j,n} - \alpha_{j-1,n}) &+ \sum_{j=n+1}^{\infty} (\alpha_{j-1,n} - \alpha_{j,n}) = 2\alpha_{n,n}. \end{aligned}$$

By Stirling's formula $n! \sim n^n e^{-n} (2\pi n)^{\frac{1}{2}}$ so $\alpha_{n,n} \longrightarrow 0$ and (4) follows.

Now we put $S = \alpha P + (1-\alpha)Q$. By the commutativity,

$$\begin{aligned} e^{-\alpha^{-1}} e^{\alpha^{-1}S} &= e^{-(I-P)} e^{(1-\alpha^{-1})} e^{(\alpha^{-1}-1)Q} \\ &= e^{-(I-P)} e^{(1-\alpha^{-1})(I-Q)} \end{aligned}$$

so

$$e^{-n\alpha^{-1}} e^{n\alpha^{-1}S} = e^{-n(I-P)} U^n, \tag{5}$$

where

$$U = e^{(1-\alpha^{-1})(I-Q)} = e^{-\beta I} e^{\beta Q} \qquad (\beta > 0)$$

Since

$$\|U\| \leq e^{-\beta} \sum_{n=0}^{\infty} \frac{(\beta\|Q\|)^n}{n!} \leq 1,$$

U is a contraction. Now if $x \in E$ is fixed by S then (5) implies

$$x = e^{-n(I-P)} U^n x$$

so

$$(I-P)x = (I-P)e^{-n(I-P)} U^n x.$$

Letting $n \longrightarrow \infty$ relation (4) implies $Px = x$ proving the lemma.

With the natural map $\pi : G \longrightarrow G/K$ we put $F = f \circ \pi$ and lift ν to a measure μ on G by $\mu(\varphi) = \nu(\dot{\varphi})$ where $\dot{\varphi}(gK) = \int \varphi(gk)dk$. Then μ is bi-invariant under K and by Lemma 3.2,

$$F(g) = \int_G F(gh)d\mu(h) = (F * \check{\mu})(g), \tag{6}$$

where $\check{\mu}(A) = \mu(A^{-1})$. If μ_1, μ_2 are two K-bi-invariant measures of compact support on G then (cf. [GGA], II, Cor. 5.2) the symmetry of G/K implies,

$$\mu_1 * \mu_2 = \mu_2 * \mu_1. \tag{7}$$

Coming back to Theorem 3.1 we shall now prove that each bounded function f on G/K satisfying $Lf = 0$ also satisfies (1) so by Cor. 2.2, f is harmonic.

Let $\ell = \dim A$. If $\ell = 1$, $\boldsymbol{D}(G/K)$ is generated by L so there is nothing to prove. Hence we assume $\ell > 1$.

Let $A^+ \subset A$ be the open positive Weyl chamber with closure $\overline{A^+}$. Let $\sigma^+ = S \cap (A^+ \cdot o)$, the set of elements in $A^+ \cdot o$ at distance r from o. Since $X = K \cdot \overline{A^+} \cdot o$ we have $S = K \cdot \overline{\sigma^+}$. Fix $a_o \in \sigma^+$, let N_o' be a compact neighborhood of a_o in σ^+ with interior N_o and put

$$V = K \cdot N_o.$$

Each $s \in S$ has the form $s = ka^+$ where $a^+ \in \overline{\sigma^+}$ is unique. Since N_o' is a proper subset of $\overline{\sigma^+}$ it follows that $\overline{V}$ is a proper subset of S.

Let ν_V and ν_V', respectively, be the restrictions of ν to V and to the complement $S - V$, normalized such that $\nu_V(V) = \nu_V'(S - V) = 1$. This is possible since $\text{supp}(\nu) = S$. Also

$$\nu = \alpha\nu_V + (1-\alpha)\nu_V' \qquad (0 < \alpha < 1).$$

Let μ_V and μ'_V be the lifts of ν_V and ν'_V to G as defined above.

We shall now prove the plausible statement

$$\lim_{N_o \longrightarrow \{a_o\}} \int_S \psi(s) d\nu_V(s) = \int_K \psi(k \cdot a_o) dk \tag{8}$$

for $\psi \in C(S)$. For this we put

$$\psi^\natural(s) = \int_K \psi(k \cdot s) dk, \qquad \overline{\psi}(n_o) = \psi(n_o K), \qquad n_o \in N_o,$$

and if $\varphi \in C(N_o)$ we can define $\widetilde{\varphi} \in C(V)$ by $\widetilde{\varphi}(kaK) = \varphi(a)$. Then the map

$$\varphi \longrightarrow \int_V \widetilde{\varphi}(s) d\nu_V(s) \tag{9}$$

is a probability measure m on N_o and by the K-invariance

$$\int_S \psi(s) d\nu_V(s) = \int_V \psi(s) d\nu_V(s)$$

$$= \int_V \psi^\natural(s) d\nu_V(s) = \int_{N_o} (\psi^\natural)^-(n_o) dm(n_o),$$

which by the continuity of $\psi^\natural$ tends to $\psi^\natural(a_o)$ as $N_o \longrightarrow \{a_o\}$. This proves (8).

Consider now the space E of bounded continuous functions φ on G satisfying

$$\varphi = \varphi * \check{\mu}. \tag{10}$$

The measure μ having compact support, E is a Banach space under the uniform norm. Since both μ_V and μ'_V are bi-invariant under K, (7) shows that right convolutions with them are operators satisfying the hypothesis of Lemma 3.3. Consequently, (6) implies

$$F = F * \check{\mu}_V,$$

in other words,

$$f(g \cdot o) = \int_S f(g \cdot s) d\nu_V.$$

Using (8) on the function $\psi(s) = f(g \cdot s)$ we thus deduce

$$f(g \cdot o) = \int_K f(gk \cdot a_o) dk. \tag{11}$$

By continuity, this extends to all $a_o \in \overline{\sigma^+}$ and then, by invariance of dk, to all $a_o \in S$. Since the radius of S was arbitrary, f satisfies the mean value theorem and thus is harmonic.

3. The Poisson Integral Formula for X.

We recall the classical fact that a solution of Laplace's equation

$$\left(\frac{\partial^2}{\partial x^2} + \frac{\partial^2}{\partial y^2} \right) u = 0 \tag{12}$$

in the disk $D: \ x^2 + y^2 < 1$ is bounded if and only if it is the Poisson integral

$$u(re^{i\theta}) = \frac{1}{2\pi} \int_o^{2\pi} \frac{1 - r^2}{1 - 2r\cos(\theta - \varphi) + r^2} U(e^{i\varphi}) d\varphi \tag{13}$$

of a function $U \in L^\infty(S^1)$.

In the non-Euclidean metric

$$ds^2 = \frac{dx^2 + dy^2}{(1 - x^2 - y^2)^2}$$

the Laplace Beltrami operator L is given by

$$L = (1 - x^2 - y^2)^2 \left(\frac{\partial^2}{\partial x^2} + \frac{\partial^2}{\partial y^2} \right)$$

and the Poisson kernel by

$$\frac{1 - r^2}{1 - 2r\cos(\theta - \varphi) + r^2} = \frac{1 - |z|^2}{|z - b|^2} = e^{2\langle z, b \rangle},$$

where $z = re^{i\theta}$, $b = e^{i\varphi}$ and $\langle z, b \rangle$ is the (signed) distance from o to the horocycle through z, tangential to the unit circle at b (cf. [GGA], Introd. §4, No. 1). Thus (12) can be rewritten: The bounded solutions of the equation

$$Lu = 0 \qquad \text{in } D \tag{14}$$

are the integrals

$$(15) \qquad u(z) = \int_{S^1} e^{2\langle z,b\rangle} U(b)\,db, \qquad z \in D,$$

where $U \in L^\infty(S^1)$. We shall now generalize this to the symmetric space $X = G/K$. Roughly speaking, the idea is to let $a_o \longrightarrow \infty$ in (11) and interpret the result properly.

We begin by establishing a classical fixed point property of solvable groups. A group S is said to act on a topological vector space V if there is given a homomorphism $\pi : S \longrightarrow Aut(V)$.

Definition. A group S has the *fixed point property* if whenever it acts on a locally convex topological vector space by continuous linear transformations leaving invariant a compact convex set $Q \neq \emptyset$ it has a fixed point in Q.

Proposition 3.4. *Connected solvable Lie groups have the fixed point property.*

Proof. Let V be a locally convex topological space and S an abelian family of continuous linear transformations of V. Let $s \in S$ and put

$$s_n = \frac{1}{n}\left(I + \cdots + s^{n-1}\right).$$

Let $\widetilde{S}$ be the set of all products

$$s_{n_1} \ldots s_{n_k}, \quad s \in S, \quad n_i \in \mathbf{Z}^+.$$

All the elements in $\widetilde{S}$ commute. Let $Q \subset V$ be compact, convex, $\neq \emptyset$ and invariant under each $s \in S$. Then by the convexity, $hQ \subset Q$ for $h \in \widetilde{S}$. Let $h_1, \ldots, h_r \in \widetilde{S}$ $(1 \leq i \leq r)$. Then by commutativity,

$$h_1 \ldots h_r Q = h_i h_1 \ldots \widehat{h_i} \ldots h_r Q \subset h_i Q \neq \emptyset$$

so

$$h_1 \ldots h_r Q \subset \bigcap_{i=1}^{r} h_i Q \neq \emptyset.$$

Hence, by compactness of Q,

$$(16) \qquad Q^* = \bigcap_{h \in \widetilde{S}} hQ \neq \emptyset.$$

Let $x \in Q^*$. If $s \in S$ and $n \in \mathbf{Z}^+$ then $x \in s_n Q$ so for some $y \in Q$,

$$x = \frac{1}{n}(y + sy + \cdots + s^{n-1}y),$$

whence

$$s \cdot x - x = \frac{1}{n}(s^n y - y) \subset \frac{1}{n}(Q + (-Q)).$$

Since n is arbitrary, $s \cdot x = x$.

Next let S be a connected solvable Lie group acting on V. Let the action be given by a homomorphism $\pi : S \longrightarrow \mathrm{Aut}(V)$. Since the Lie algebra $\mathfrak{s}$ of S is solvable we have a composition series

$$\mathfrak{s} = \mathfrak{s}_0 \supset \mathfrak{s}_1 \supset \cdots \supset \mathfrak{s}_m = \{0\}, \quad \mathfrak{s}_{m-1} \neq 0$$

where $\mathfrak{s}_i$ is an *ideal* in $\mathfrak{s}_{i-1}$ $(1 \leq i \leq m)$ and $\mathfrak{s}_i/\mathfrak{s}_{i-1}$ abelian. Let

$$S = S_o \supset S_1 \supset \cdots \supset S_m = \{e\} \tag{17}$$

be the corresponding sequence of analytic subgroups; m is called the *length* of the composition series.

We shall now prove Proposition 3.4 by induction on m. If $m = 1$, S is abelian so we are done. Assume now the proposition holds for all S that have a composition series of length $< m$. Suppose S in (17) acts on V via a homomorphism $\pi : S \longrightarrow \mathrm{Aut}(V)$ leaving a convex compact set $Q \neq \emptyset$ invariant. Let Q_1 be the set of points in Q fixed under $\pi(S_1)$. By induction assumption $Q_1 \neq \emptyset$; also Q_1 is compact and convex.

Let $\sigma \in \pi(S)$. If $s \in S_1$ then $\sigma\pi(s)\sigma^{-1} \in \pi(S_1)$ so if $x \in Q_1$ we have $\sigma\pi(s)\sigma^{-1} \cdot x = x$ so $\pi(s)\sigma^{-1} \cdot x = \sigma^{-1} \cdot x$. Thus $\sigma^{-1} \cdot x$ is fixed by all members of $\pi(S_1)$ and since it is in Q we have $\sigma^{-1} \cdot x \in Q_1$. Thus $\sigma^{-1} \cdot Q_1 \subset Q_1$. The closed subspace V_1 of V generated by Q_1 is locally convex, $\pi(S)$-invariant, and $\pi(S_1)$ acts trivially on it. The subgroup $S_1 \subset S$ being normal, the restrictions $\pi(S)|V_1$ form an abelian group so by the first part of the proof there exists a $x_1 \in V_1$ fixed by all $\pi(S)$. This concludes the proof.

Now we consider our symmetric space $X = G/K$ and let $G = KAN$ be the usual Iwasawa decomposition of the semisimple Lie group G. Let M be the centralizer of A in K, put $B = K/M$ and let db denote the normalized K-invariant measure on B.

Corollary 3.5. *The group MAN has the fixed point property.*

In fact if MAN acts on V leaving Q as above invariant, Prop. 3.4 applies to the connected solvable Lie group AN. If $x \in Q$ is a fixed point for it then by compactness and convexity of Q the point

$$x^* = \int_M m{\cdot}x \, dm,$$

defined as a limit of approximating sums, belongs to Q and since M normalizes AN $sx^* = x^*$ for $s \in AN$.

Note that the group G acts both on $X = G/K$ and on $B = K/M = G/MAN$. It is convenient to denote these two actions by $g{\cdot}x$ and $g(b)$ respectively.

We have seen in [GGA], Ch. I, Lemma 5.19 that the Jacobian of the action on B is given by

$$\frac{d(g^{-1}(b))}{db} = e^{2\rho(A(g\cdot o,b))}, \tag{18}$$

which is the *Poisson kernel* $P(gK, b)$ from Ch. II, §3.

Theorem 3.6. *Let u be a bounded harmonic function on X. Then there exists a unique function $U \in L^\infty(B)$ such that*

$$u(g{\cdot}o) = \int_B U(g(b))db, \qquad g \in G. \tag{19}$$

The mapping $u \longrightarrow U$ is a bijection of the space of bounded harmonic functions on X onto $L^\infty(B)$.

Formula (19) can also be written as the Poisson integral formula

$$u(x) = \int_B e^{2\rho(A(x,b))}U(b)db, \qquad x \in X. \tag{20}$$

Proof. We may assume u real-valued. Consider the Banach space $BC(G)$ of real-valued bounded continuous functions on G with the uniform norm $\| \ \|_\infty$, $BC^*(G)$ its dual with the weak* topology $\sigma(BC^*(G), BC(G))$. Then $BC^*(G)$ consists of the bounded measures on G. Let Q be the set of positive $\mu \in BC^*(G)$ satisfying $\mu(1) = 1$. Decomposing a $f \in BC(G)$, $f = f^+ + f^-$ ($f^+ = \max(f,0)$, $f^- = \min(f,0)$) we see that $|\mu(f)| \leq 2\|f\|_\infty$ so $\frac{1}{2}Q$ is a nonempty convex, closed subset of the unit ball in $BC^*(G)$, hence compact. Since the left translations on G induce continuous linear transformations of $BC^*(G)$ there exists, by Cor. 3.5, an element $\mu \in Q$ invariant under left translations from MAN.

Suppose now u is bounded and harmonic on X and put $f = u \circ \pi$. Define $\widehat{f}$ on G by

$$\widehat{f}(x) = \mu(f^{L(x^{-1})}) = \int_G f(xg)\,d\mu(g) = (f * \check{\mu})(x). \tag{21}$$

Then $\widehat{f}$ is right invariant under MAN and is bounded. Viewing f as the pointwise limit of a sequence (f_n) in $C_c(G)$ the function $\widehat{f}$ is the pointwise limit of the sequence $(f_n * \check{\mu})$ in $C(G)$ so $\widehat{f}$ is measurable. Moreover, if $h \in G$,

$$\int_K \widehat{f}(hk)\,dk = \int_K \int_G f(hkg)\,d\mu(g)\,dk = f(h) \tag{22}$$

by the mean value theorem. Defining $U(gMAN) = \widehat{f}(g)$ this gives the desired formula (19).

Conversely, let U be a bounded measurable function on B and define $\widehat{f}$ on G by $\widehat{f}(g) = U(gMAN)$. Define f by (22). Then

$$f(gkh) = \int_K \widehat{f}(gkhk')\,dk'.$$

Writing $hk' = k_1a_1n_1$ we have

$$\begin{aligned} \int_K f(gkh)\,dk &= \int_K dk \int_K \widehat{f}(gkhk')\,dk' = \int_K \widehat{f}(gkk_1a_1n_1)\,dk \\ &= \int_K dk' \int_K \widehat{f}(gkk_1)\,dk = \int_K dk' \int_K \widehat{f}(gk)\,dk = f(g) \end{aligned}$$

so f satisfies the mean value theorem. Integrating the equation $f(g) = \int f(gkh)\,dk$ against $\rho(h)$, $\rho \in C_c(G)$, we see that f is continuous. Thus $gK \longrightarrow f(g)$ is harmonic by Theorem 2.1.

For the uniqueness we observe first that if u in (19)–(20) is $\equiv 0$ and U continuous then $U \equiv 0$ by Ch. II, Theorem 3.16.

For the general case suppose $U \in L^\infty(B)$ and u in (19) identically 0. Let $\varphi \in C_c(G)$ and consider

$$U_1(b) = \int_G \varphi(g)U\big(g(b)\big)\,dg. \tag{23}$$

Then U_1 is continuous and if

$$u_1(g\cdot o) = \int_B U_1\big(g(b)\big)db$$

then by (23) and (19)

$$u_1(g\cdot o) = \int_B \left(\int_G \varphi(g)U\big(gh(b)\big)dg\right)dh = \int_G \varphi(g)u(gh\cdot o)dg.$$

Since $u \equiv 0$, we have $u_1 \equiv 0$ so by continuity of U_1, $U_1 \equiv 0$. Since φ is arbitrary, $U \equiv 0$. This proves Theorem 3.6.

4. The Fatou Theorem.

Preserving the notation from No. 3 let $\mathfrak{g} = \mathfrak{k} + \mathfrak{p}$ be the Cartan decomposition of the Lie algebra $\mathfrak{g}$ of G corresponding to G/K. By a *Weyl chamber* in $\mathfrak{p}$ we mean a Weyl chamber in a maximal abelian subspace of $\mathfrak{p}$. Since K acting on $\mathfrak{p}$ permutes the Weyl chambers transitively and since M is the subgroup of K mapping $\mathfrak{a}^+$ into itself the space $B = K/M$ is identified with the set of Weyl chambers. As usual we put $\overline{N} = \theta N$ where θ is the Cartan involution of G with fixed point set K.

Proposition 3.7. *Let $U \in L^\infty(B)$ and u the Poisson integral*

$$u(x) = \int_B e^{2\rho(A(x,b))}U(b)db. \tag{24}$$

Suppose U is continuous at a point $b_o \in B$. Then

$$\lim_{t\to+\infty} u(\exp t Z\cdot o) = U(b_o) \tag{25}$$

for each Z in the Weyl chamber b_o.

Although this is contained in Theorem 3.16, Ch. II we give a direct proof here because this special case is much simpler and besides, the proof motivates the proof of the more subtle Fatou theorem.

Without restriction of generality we can take $b_o = eM$ and $Z = H \in \mathfrak{a}^+$. Putting $a_t = \exp tH$ we must by (19) prove

$$\int_{K/M} U\big(a_t(kM)\big)dk_M \longrightarrow U(eM) \qquad \text{as } t \longrightarrow +\infty. \tag{26}$$

The action of a_t on K/M is best understood by using the mapping $\overline{n} \longleftrightarrow k(\overline{n})M$ where $k(\overline{n})$ is the K-component of $\overline{n} \in \overline{N}$ in the decomposition $G = KAN$. This mapping is a diffeomorphism of $\overline{N}$ onto an open submanifold of K/M whose complement is a null set for the invariant measure dk_M. The mapping has the useful feature that if $a \in A$ then

$$a(k(\overline{n})M) = ak(\overline{n})MAN = k(a\overline{n})MAN = k(a\overline{n}a^{-1})MAN$$

so

$$a(k(\overline{n})M) = k(\overline{n}^a)M. \tag{27}$$

Thus the action of a on K/M corresponds to conjugation by a on $\overline{N}$ which is much easier to handle.

Because of the properties of the mapping $\overline{n} \longrightarrow k(\overline{n})M$ stated above we have an integral formula

$$\int_{K/M} F(kM)dk_M = \int_{\overline{N}} F(k(\overline{n})M)\psi(\overline{n})d\overline{n}, \qquad F \in C(K/M) \tag{28}$$

where ψ is a function in $L^1(\overline{N})$. Actually

$$\psi(\overline{n}) = e^{-2\rho(H(\overline{n}))}, \tag{29}$$

(cf. [GGA], Ch. I, §5) but we do not need this fact for Prop. 3.7. Using (27) and (28) the left hand side of (26) equals

$$\int_{\overline{N}} U\big(a_t(k(\overline{n})M)\big)\psi(\overline{n})d\overline{n} = \int_{\overline{N}} U\big(k(\overline{n}^{a_t})M\big)\psi(\overline{n})d\overline{n}. \tag{30}$$

Now corresponding to the decomposition $\overline{\mathfrak{n}} = \sum_{\alpha<0} \mathfrak{g}_\alpha$ of the Lie algebra of $\overline{\mathfrak{n}}$ into restricted root spaces, we have

$$\overline{n} = \exp\left(\sum_{\alpha<0} X_\alpha\right),$$

where $X_\alpha \in \mathfrak{g}_\alpha$. Moreover

$$\begin{aligned}\overline{n}^{\exp H} &= \exp H \exp\left(\sum_\alpha X_\alpha\right)\exp(-H) = \exp\left\{\mathrm{Ad}(\exp H)\left(\sum_\alpha X_\alpha\right)\right\} \\ &= \exp\left(\sum_\alpha e^{\alpha(H)}X_\alpha\right).\end{aligned}$$

Since $\alpha(H) < 0$ for all α above we have

$$\overline{n}^{\exp(tH)} \longrightarrow e$$

for each $\overline{n} \in \overline{N}$ as $t \longrightarrow +\infty$. By the dominated convergence theorem the right hand side of (30) converges to $U(eM)$ as $t \longrightarrow +\infty$.

Remarks. **(i)** For the classical case $X = \boldsymbol{H}^2$ this theorem is the classical theorem of Schwarz in potential theory. The proof above then reduces to an elementary one which seems somewhat simpler than the classical one (cf. [GGA], Introduction, Theorem 4.25).

(ii) It is of particular interest to note that the limit (25) is the same for all Z in the Weyl chamber $\mathfrak{b}_o$.

The classical Fatou theorem states that a bounded harmonic function u in the unit disk has boundary values almost everywhere, i.e. for almost all θ,

$$\lim_{r \to 1} u\left(re^{i\theta}\right) \qquad \text{exists.} \tag{31}$$

The customary proof uses the explicit formula for the Poisson kernel and its derivative combined with integration by parts. In contrast, the group-theoretic proof of (31) (cf. [GGA], Introd., Theorem 4.26) relies mostly on the Lebesgue differentiation theorem

$$\lim_{h \to 0} \frac{1}{2h} \int_{-h}^{h} |\varphi(y+x) - \varphi(y)| dx = 0 \qquad \text{for almost all } y. \tag{32}$$

This is the proof which we shall generalize to the symmetric space $X = G/K$.

Theorem 3.8. (The Fatou theorem) *Let u be a bounded harmonic function on $X = G/K$. Then for almost all unit vectors $Z \in \mathfrak{p}$, the limit*

$$\lim_{t \to +\infty} u(\exp tZ \cdot o) \tag{33}$$

exists.

For the proof we shall need an analog of (32) for the group $\overline{N}$ on which the group A acts by conjugation. Again (27) will play a role in the proof. Consider the root space decomposition

$$\overline{\mathfrak{n}} = \sum_{\alpha \in \Sigma^+} \mathfrak{g}_{-\alpha} = \sum_{\alpha \in \Sigma_o^+} \overline{\mathfrak{n}}_\alpha, \tag{34}$$

where Σ_o^+ is the set of indivisible roots. As already used in Ch. II, §5, No. 2 the mapping

$$(Z_1, \ldots, Z_p) \longrightarrow \exp Z_1 \ldots \exp Z_p$$

is a diffeomorphism of $\overline{\mathfrak{n}}_{\alpha_1} \times \cdots \times \overline{\mathfrak{n}}_{\alpha_p}$ onto $\overline{N}_{\alpha_1} \ldots \overline{N}_{\alpha_p}$. In addition, $\overline{\mathfrak{n}}_\alpha = \mathfrak{g}_{-\alpha} + \mathfrak{g}_{-2\alpha}$ and $\exp(X+Y) = \exp X \exp Y$ $(X \in \mathfrak{g}_{-\alpha}, Y \in \mathfrak{g}_{-2\alpha})$. Putting these facts together we observe the following:

There exist one-parameter subgroups $N_1, \ldots, N_r$ of $\overline{N}$ such that

(i) each N_j is invariant under A;
(ii) $\overline{N} = N_1 \ldots N_r$ (diffeomorphically);
(iii) for each j, $N_j N_{j+1} \ldots N_r$ is a normal subgroup of N.

We now need the classical maximal theorem of Hardy-Littwood (see e.g. Stein [1970], Ch. I). If f is a measurable function on $\boldsymbol{R}$ put

$$f^*(x) = \sup_r \frac{1}{2r} \int_{-r}^{r} |f(x+y)| dm(y), \tag{35}$$

m denoting the Lebesgue measure. For $1 \leq p \leq \infty$ let $\| \ \|_p$ denote the L^p norm.

Theorem 3.9.

(i) *If $f \in L^1(\boldsymbol{R})$ then for each $\alpha > 0$*

$$m(\{x : \ f^*(x) > \alpha\}) \leq \frac{C}{\alpha} \|f\|_1,$$

where C is a constant independent of f and α.

(ii) *If $f \in L^p(\boldsymbol{R})$ $(1 < p \leq \infty)$ then $f^* \in L^p(\boldsymbol{R})$ and*

$$\|f^*\|_p \leq C_p \|f\|_p, \tag{36}$$

where C_p is independent of f.

We now extend this theorem to the group $\overline{N}$. Put $\overline{n}^a = a\overline{n}a^{-1}$ let m denote the Haar measure on the various subgroups of $\overline{N}$ that enter.

Proposition 3.10. *Let $V \subset \overline{N}$ be a bounded measurable subset with $m(V) > 0$. For $1 < p \leq \infty$ and $f \in L^p(\overline{N})$ put*

$$f^*(\overline{n}) = \sup_{a \in A} \frac{1}{m(V^a)} \int_{V^a} |f(\overline{n}\overline{n}')| d\overline{n}'. \tag{37}$$

Then

$$\|f^*\|_p \leq C_p \|f\|_p, \tag{38}$$

where the constant C_p is independent of f.

Proof. The case $p = \infty$ being trivial suppose $1 < p < \infty$. In the product representation (ii) of $\overline{N}$ suppose first $V = I_1 \ldots I_r$ where I_j is a symmetric interval around the identity in N_j. We prove the result by induction on $r - j$ for the subgroups $N_j \ldots N_r (1 \leq j \leq r)$. If $r - j = 0$ the result reduces to Theorem 3.9. Now assume the proposition proved for the group $T = N_j \ldots N_r$, and put $S = N_{j-1}$. We shall prove (38) for the group $Q = ST$ which by (iii) equals TS. As already used in Ch. II, §5, No. 2 we have for $f \in L^1(Q)$

$$\int_Q f(q)dq = \int_{S\times T} f(st)ds\, dt = \int_{S\times T} f(ts)ds\, dt, \tag{39}$$

with appropriate normalization of the Haar measures. Put $I_S = I_{j-1}$, $I_T = I_j \ldots I_r$ and define for $f \in L^p(Q)$, $q \in Q$,

$$f_1(q) = \sup_{a\in A} \frac{1}{m(I_T^a)} \int_{I_T^a} |f(qt)|dt,$$

$$f^{**}(q) = \sup_{a\in A} \frac{1}{m(I_S^a)} \int_{I_S^a} f_1(qs)ds.$$

Using (39) and

$$\int_{I_T^a} |f(qst)|dt \leq m(I_T^a) f_1(qs)$$

we deduce

$$f^*(q) \leq f^{**}(q), \qquad q \in Q. \tag{40}$$

By (39) and the Fubini theorem, for almost all $s \in S$ the function $f^{L(s^{-1})}$: $t \longrightarrow f(st)$ is in $L^p(T)$. For such s the inductive hypothesis implies

$$\int_T |f_1(st)|^p dt \leq C_p^1 \int_T |f(st)|^p dt, \tag{41}$$

which by integration gives

$$\|f_1\|_p \leq C_p^2 \|f\|_p. \tag{42}$$

This and the Fubini theorem shows that for almost all t the function $f_1^{L(t^{-1})}$: $s \longrightarrow f_1(ts)$ belongs to $L^p(S)$. For such t (36) implies

$$\int_S \left| \left(f_1^{L(t^{-1})}\right)^* (x)\right|^p ds \leq C_p^3 \int_S \left|f_1^{L(t^{-1})}(s)\right|^p ds. \tag{43}$$

But $\left(f_1^{L(t^{-1})}\right)^* (s) = f^{**}(ts)$ so integrating (43) over t,

$$\left\|f^{**}\right\|_p \leq C_p^4 \left\|f_1\right\|_p. \tag{44}$$

Now (40), (42), and (44) give the inequality (38) for the group Q, completing the induction.

Finally, if V is arbitrary we have $V \subset I = I_1 \dots I_r$ with the I_j as above. Then

$$\frac{1}{m(V^a)} \int_{V^a} \left|f(\overline{n}\overline{n}')\right| d\overline{n}' \leq \frac{m(I^a)}{m(V^a)} \frac{1}{m(I^a)} \int_{I^a} \left|f(\overline{n}\overline{n}')\right| d\overline{n}'$$

and $m(I^a)/m(V^a) = m(I)/m(V)$ and this proves (38) in general.

We can now derive the desired differentiation result. As before, we let $a \longrightarrow +\infty$ for $a \in A$ mean $\alpha(\log a) \longrightarrow \infty$ for each $\alpha \in \Sigma^+$.

Corollary 3.11. *For any $f \in L^p(\overline{N})$ $(1 < p \leq \infty)$ and any bounded $V \subset \overline{N}$ of positive measure we have*

$$\lim_{a \to +\infty} \frac{1}{m(V^a)} \int_{V^a} f(\overline{n}\overline{n}') d\overline{n}' = f(\overline{n}) \tag{45}$$

for almost all $\overline{n} \in \overline{N}$.

Proof. The statement is obvious if f is continuous because (cf. Prop. 3.7) V^a is contained in an arbitrary neighborhood of e in $\overline{N}$ provided a is large enough.

Since the general statement (45) is local it suffices to prove it for $1 < p < \infty$. Let $f \in L^p(\overline{N})$. If $\epsilon > 0$ we shall show that there is a subset $\overline{N}_\epsilon \subset \overline{N}$ such that $m(\overline{N}_\epsilon) \leq \epsilon$ and

$$\limsup_{a \to +\infty} \left|\left(M_a f\right)(\overline{n}) - f(\overline{n})\right| > \epsilon \ \text{ implies } \overline{n} \in \overline{N}_\epsilon. \tag{46}$$

Here

$$\left(M_a f\right)(\overline{n}) = \frac{1}{m(V^a)} \int_{V^a} f(\overline{n}\overline{n}') d\overline{n}'.$$

For this we decompose f into a sum $f = f_1 + f_2$ where f_1 is continuous and

$$\|f_2\|_p^p \leq \frac{\epsilon^{p+1}}{(C_p+1)^p}.$$

Then

$$\left|(M_a f)(\overline{n}) - f(\overline{n})\right| \leq \left|(M_a f_1)(\overline{n}) - f_1(\overline{n})\right| + \left|(M_a f_2)(\overline{n}) - f_2(\overline{n})\right|.$$

Taking $\limsup_{a\to+\infty}$ the first term on the right gives contribution 0 so

(47) $$\limsup_{a\to+\infty}\left|(M_a f)(\overline{n}) - f(\overline{n})\right| \leq f_2^*(\overline{n}) + \left|f_2(\overline{n})\right|.$$

The right hand side has L^p-norm bounded by $(C_p+1)\|f_2\|_p$. Thus the set of points $\overline{n} \in \overline{N}$ where

$$h(\overline{n}) = \limsup_{a\to+\infty}\left|(M_a f)(\overline{n}) - f(a)\right| > \epsilon$$

has measure at most

$$\epsilon^p\,(C_p+1)^p\,(\|f_2\|_p)^p \leq \epsilon.$$

Writing

$$\{\overline{n} \in \overline{N} :\ h(\overline{n}) > 0\} = \bigcup_{k=1}^{\infty}\left\{\overline{n} \in \overline{N} :\ h(\overline{n}) > \frac{\epsilon}{2^k}\right\}$$

we have, since $\epsilon > 0$ above was arbitrary,

$$m\left(\{\overline{n} \in \overline{N} :\ h(\overline{n}) > 0\}\right) \leq \sum_{1}^{\infty} \epsilon/2^k = \epsilon.$$

Again, ϵ being arbitrary, this shows that $h(\overline{n}) = 0$ for almost all $\overline{n}$ so Corollary 3.11 is proved.

Corollary 3.12. *With the assumptions of Cor. 3.11 we have*

(48) $$\lim_{a\to+\infty}\frac{1}{m(V^a)}\int_{V^a}\left|f(\overline{n}_o\overline{n}) - f(\overline{n}_o)\right|d\overline{n} = 0$$

for almost all $\overline{n}_o \in \overline{N}$.

The proof is the same as for $\boldsymbol{R}$: let r be a rational number and apply Corollary 3.11 to $|f(\overline{n}) - r|$. Then

$$\lim_{a \to +\infty} \frac{1}{m(V^a)} \int_{V^a} |f(\overline{n}\overline{n}') - r| d\overline{n}' = |f(\overline{n}) - r| \tag{49}$$

for all $\overline{n}$ outside a fixed null set E independent of r. Now let $\overline{n} \notin E$ and $\alpha \in \boldsymbol{R}$. Let $\epsilon > 0$ and $|r - \alpha| < \epsilon$. Since

$$\big||f(\overline{n}) - \alpha| - |f(\overline{n}) - r|\big| \leq |\alpha - r|$$

we have

$$(M_a|f - \alpha|)(\overline{n}) - (M_a|f - r|)(\overline{n}) \leq |\alpha - r|. \tag{50}$$

Hence

$$\begin{aligned} \big|(M_a|f-\alpha|)(\overline{n}) - |f(\overline{n}) - \alpha|\big| &\leq \big|(M_a|f-\alpha|)(\overline{n}) - (M_a|f-r|)(\overline{n})\big| \\ &+ \big|(M_a|f-r|)(\overline{n}) - |f(\overline{n}) - r|\big| + \big||f(\overline{n}) - r| - |f(\overline{n}) - \alpha|\big|. \end{aligned}$$

Because of (49) and (50) this is $\leq 3\epsilon$ for a sufficiently large. Taking $\alpha = f(\overline{n})$ the corollary follows.

We can now finish the proof of Theorem 3.8. We can write the bounded harmonic function u in the form

$$u(g \cdot o) = \int_B U(g(b)) db, \tag{51}$$

where $U \in L^\infty(B)$. Let $f(\overline{n}) = U(\overline{n}MAN)$ for $\overline{n} \in \overline{N}$. We shall prove that if $\overline{n}_o \in \overline{N}$ is such that (48) holds then for each $\epsilon > 0$ and each compact neighborhood C of e in $\overline{N}$ there exists $L > 0$ such that

$$|u(\overline{n}_0 a \overline{n}_1 \cdot o) - f(\overline{n}_o)| < \epsilon \tag{52}$$

for $\overline{n}_1 \in C$, $\alpha(\log a) > L$ $(\alpha \in \Sigma^+)$. Using (51) and writing

$$\overline{n}_o a \overline{n}_1 MAN = \overline{n}_o \overline{n}_1^a MAN$$

and

$$\int_B U(g(b)) dg = \int_{\overline{N}} U(g\overline{n}MAN) e^{-2\rho(H(\overline{n}))} d\overline{n},$$

we have

$$u(\overline{n}_o a\overline{n}_1 \cdot o) - f(\overline{n}_o) = \int_{\overline{N}} [U(\overline{n}_o a\overline{n}_1\overline{n}MAN) - f(\overline{n}_o)]\, e^{-2\rho(H(\overline{n}))} d\overline{n}$$

$$= \int_{\overline{N}} [f(\overline{n}_o(\overline{n}_1\overline{n})^a) - f(\overline{n}_o)]\, e^{-2\rho(H(\overline{n}))} d\overline{n}$$

so

$$(53)\quad \left|u(\overline{n}_o a\overline{n}_1 o) - f(\overline{n}_o)\right| \leq \Big(\int_{\overline{N}-C_1} + \int_{C_1}\Big) \left|f(\overline{n}_o(\overline{n}_1\overline{n})^a) - f(\overline{n}_o)\right| e^{-2\rho(H(\overline{n}))} d\overline{n}.$$

Choose C_1 a compact neighborhood of e such that

$$\int_{\overline{N}-C_1} e^{-2\rho(H(\overline{n}))} d\overline{n} < \epsilon/4\|f\|_\infty.$$

Then the first integral on the right in (53) is bounded by $\epsilon/2$. Also $\rho(H(\overline{n})) \geq 0$ so the second integral is bounded by

$$(54)\qquad e^{2\rho(\log a)} \int_{(\overline{n}_1 C_1)^a} \left|f(\overline{n}_o\overline{n}) - f(\overline{n}_o)\right| d\overline{n}$$

(after a change of variable). Now $\overline{n}_1 \in C$ so putting $V = CC_1$ and noting that $m(V^a) = m(V)e^{-2\rho(\log a)}$ the expression (54) is bounded by

$$m(V)\frac{1}{m(V^a)} \int_{V^a} \left|f(\overline{n}_o\overline{n}) - f(\overline{n}_o)\right| d\overline{n},$$

which by Cor. 3.12 is $< \epsilon/2$ if L is large enough. This proves (52).

Now take $H \in \mathfrak{a}^+$, put $a_t = \exp tH$ and write by the Iwasawa decomposition

$$(55)\qquad \overline{n}_o = k(\overline{n}_o)(a_1 n_1)^{-1}.$$

Then

$$k(\overline{n}_o)a_t \cdot o = \overline{n}_o a_1 n_1 a_t \cdot o = \overline{n}_o a_1 a_t (n_1)^{a-t} \cdot o.$$

By $G = A\overline{N}K$,

$$n_1^{a-t} = a(t)\overline{n}(t)k(t),$$

where each factor tends to e as $t \longrightarrow +\infty$. Thus by (52),

$$\lim_{t\to+\infty} u(k(\overline{n}_o)a_t \cdot o) = U(k(\overline{n}_o)M) \tag{56}$$

proving Theorem 3.8.

Formula (56) implies the analog to Remark (ii) following Proposition 3.7.

Corollary 3.13. *For almost all $b \in B$ the limit*

$$\lim_{t\to+\infty} u(\exp tZ \cdot o)$$

is the same for all Z in the Weyl chamber b.

Estimate (52) gives a further refinement of Theorem 3.8 in that the convergence does not have to be "radial".

Corollary 3.14. *There exists a null set $\mathcal{N} \subset K$ (depending on u) such that if $k \notin \mathcal{N}$, $C_o \subset X$ compact then*

$$\lim_{a\to+\infty} u(ka\cdot c) = U(kM) \tag{57}$$

uniformly for $c \in C_o$.

For this let $\overline{N}_o$ be the set of $\overline{n}_o$ for which (48) holds and let $m \in M$. Then if $\overline{n}_o \in \overline{N}_o$, $c \in C_o$, $a \in A^+$ we have by (55)

$$k(\overline{n}_o)m\ a\cdot c = \overline{n}_o a_1 n_1\ am\cdot c = \overline{n}_o a a_1 n_1^{a^{-1}}\ m\cdot c = \overline{n}_o a a_2 \overline{n}_2\cdot o,$$

where a_2 and $\overline{n}_2$ belong to compact subsets of A and $\overline{N}$, respectively. Thus by (52)

$$|u(k(\overline{n}_o)m\ a\cdot c) - U(k(\overline{n}_o)M)| < \epsilon$$

for $a \in A^+$ sufficiently large. Since $k(\overline{N}_o)M$ fills up K up to a null set this proves (57).

Remark. Suppose G/K is the non-Euclidean disk. Then the orbits of A are the circular arcs joining -1 and 1 ([GGA], pp. 76, 555). If C_o is a disk $|z| \leq R$ then $kA\cdot C_o$ contains a sector with vertex kM so (57) amounts to non-tangential convergence.

5. The Furstenberg Compactification.

Formula (26) can be interpreted in terms of the space $\mathfrak{M}(B)$ of measures on B. In fact it states that if $(db)^g$ denotes the image of db under the map $b \longrightarrow g(b)$ of B onto itself and if $\delta = \delta_{eM}$ then

$$(db)^{a_t} \longrightarrow \delta \qquad \text{as } t \longrightarrow +\infty \tag{58}$$

in the *weak sense,* i.e. $(db)^{a_t}(U) \longrightarrow U(eM)$ for $t \longrightarrow +\infty$, $U \in C(B)$ being arbitrary.

We shall accordingly consider the mapping $\varphi: \; X \longrightarrow \mathfrak{M}(B)$ given by

$$\varphi(gK) = (db)^g. \tag{59}$$

Since

$$(db)^g(F) = (db)(F^{g^{-1}}) = \int_B F(g(b))db$$

$$= \int_B F(b)\frac{d(g^{-1}(b))}{db}db = \int_B F(b)P(gK,b)db$$

we can also write

$$\varphi(gK) = P(gK,b)db = e^{2\rho(A(gK,b))}db. \tag{60}$$

As usual $\mathfrak{M}(B)$ is taken with the weak*-topology, cf. Ch. I, §2, No. 6, i.e. the topology $\sigma\big(\mathfrak{M}(B), C(B)\big)$. Let $\overline{X}$ denote the closure of $\varphi(X)$ in $\mathfrak{M}(B)$.

Theorem 3.15.

(i) *φ is a homeomorphism of X onto $\varphi(X)$ (with the relative topology of $\mathfrak{M}(B)$).*

(ii) *$\overline{X}$ is compact.*

(iii) *Let u be a bounded harmonic function on X with continuous boundary values U on B. Then $u \circ \varphi^{-1}$ extends to a continuous function on $\overline{X}$.*

Remark. Identifying X with its image under φ, $\overline{X}$ can be viewed as a compactification of X, the *Furstenberg compactification.*

Proof. (ii) The unit ball in the dual of a Banach space is weak*-compact. Since $\int P(gK,b)db = 1$, $\varphi(X)$ is contained in this unit ball so

(ii) follows. In (i) the continuity of φ is obvious. For the injectivity let us rewrite (47) in Ch. II, §3 in the form

$$A\big(ghK, g(b)\big) = A(hK, b) + A\big(gK, g(b)\big). \tag{61}$$

Now if

$$\varphi(ghK) = \varphi(gK)$$

we deduce from (60) and (61) that

$$e^{2\rho(A(hK,b))} = 1 \qquad \text{for all } b \in B$$

so

$$2\rho\big(H(h^{-1}k)\big) = 0 \qquad \text{for all } k \in K.$$

Writing $h^{-1} = k_1 a k_2$, $a \in \overline{A^+}$, this implies $\rho(\log a) = 0$ so $a = e$ and $h \in K$. Thus φ is injective. For continuity of φ^{-1} on $\varphi(X)$ let $(g_\gamma K)$ be a net in $X = G/K$ such that $(db)^{g_\gamma}$ converges to $(db)^{g_o}$ in the topology of $\mathfrak{M}(B)$. If the points $g_\gamma K$ all lie in a compact subset $S \subset X$ then $g_\gamma K \longrightarrow g_o K$ since $\varphi|S$ is a homeomorphism. If on the other hand the set of points $g_\gamma K$ is not relatively compact we choose a subsequence $(k_n \exp H_n)K$ such that $(k_n) \subset K$ is convergent and $H_n \in \overline{\mathfrak{a}^+}$ and $|H_n| \longrightarrow \infty$. Put $a_n = \exp H_n$. If all the H_n belong to a fixed closed subchamber of $\mathfrak{a}^+$ then $(db)^{a_n} \longrightarrow \delta$ (by (58)) which would contradict $(db)^{g_n} \longrightarrow (db)^{g_o}$. In general if $\alpha_1, \ldots, \alpha_\ell$ are the simple roots we can, by going to a subsequence of (H_n), assume that for each i

$$\big\{\alpha_i(H_n)\big\}_{n=1,2\ldots} \qquad \text{is convergent}$$

(the limit $+\infty$ allowed). Hence we divide Σ^+ into roots $\beta_1, \ldots, \beta_p$ and roots $\gamma_1 \ldots, \gamma_q$ such that

$$b_i = \lim_{n\to\infty} \beta_i(H_n) < \infty, \qquad \lim_{n\to\infty} \gamma_j(H_n) = \infty.$$

Hence if

$$\overline{n} = \exp\Big(\sum_i c_i X_{-\beta_i} + \sum_j d_j X_{-\gamma_j}\Big)$$

$$\overline{n}^{a_n} \longrightarrow \exp\Big(\sum_i c_i e^{-b_i} X_{-\beta_i}\Big).$$

Thus replacing in (30) a_t by a_n we deduce the convergence

$$(db)^{a_n}(U) \longrightarrow \mu(U),$$

where μ is a measure with support in a subset of B of lower dimension. This would contradict the assumption $(db)^{g_\gamma} \longrightarrow (db)^{g_o}$. This proves (i).

Finally, for (iii) we have

$$u(gK) = \int_B U(b)d(\varphi(gK))(b), \qquad U \in C(B).$$

Defining

$$\widetilde{u}(\mu) = \int_B U(b)d\mu(b), \qquad \mu \in \mathfrak{M}(B)$$

$\widetilde{u}$ is continuous on $\mathfrak{M}(B)$ and $(\widetilde{u} \circ \varphi)(gK) = u(gK)$ as claimed.

Remark. Because of (58) we can view B as a subset of $\overline{X}$.

Example. Let $X = G/K$ denote the bidisk $\boldsymbol{H}^2 \times \boldsymbol{H}^2 = (G'/K') \times (G''/K'')$ in $\boldsymbol{C}^2$ where $G' = G'' = \boldsymbol{SU}(1,1)$, $K' = K'' = \boldsymbol{SO}(2)$. Here

(i) $\overline{X}$ is naturally homeomorphic to the closure of $\boldsymbol{H}^2 \times \boldsymbol{H}^2$ in $\boldsymbol{C}^2$

(ii) $B = (K'/M') \times (K''/M'') = \boldsymbol{S}^1 \times \boldsymbol{S}^1$.

While (ii) is obvious let us verify (i). Let $(g_\gamma K) = (g'_\gamma K', g''_\gamma K'')$ be a net in X such that $\varphi(g_\gamma K)$ has a limit point $\mu \in \mathfrak{M}(B)$. If all $g_\gamma K$ belong to a compact subset of X then $\mu \in \varphi(X)$. Otherwise we can choose a subsequence $g_n K = (k_n \exp H_n)K$, where $k_n \longrightarrow (k', k'') \in K' \times K''$, $H_n = (H'_n, H''_n) \in \overline{\mathfrak{a}^+}$ such that (H'_n) and (H''_n) have limits, say H' and H'', at least one of them infinite, and such that $\varphi(g_n K)$ has μ as a limit point. If both H', H'' are infinite then μ is the point measure $\delta_{(k'M', k''M'')}$. If H' is finite and $g' = (k' \exp H')K$ then

$$(db)^{g_n} = (db' \times db'')^{g_n} \longrightarrow (db')^{g'} \times \delta_{k''M''},$$

so μ is identified with the pair $(g'K', k''M'')$ in $\boldsymbol{H}^2 \times K''/M''$. In this way $\overline{X}$ is identified with

$$X \cup (\boldsymbol{H}^2 \times K''/M'') \cup (K'/M' \times \boldsymbol{H}^2) \cup (K'/M' \times K''/M'')$$

proving (i).

§4. HARMONIC FUNCTIONS ON BOUNDED SYMMETRIC DOMAINS.

1. The Bounded Realization of a Hermitian Symmetric Space.

In this subsection we review the Cartan-Harish-Chandra realization of an irreducible Hermitian symmetric space X of the noncompact type as a bounded domain in a complex vector space. The proofs are given in [DS], Ch. VIII, §7 and we shall follow the notation there.

By the irreducibility of X the identity component $I_o(X)$ of the isometry group is simple. Let $\mathfrak{g}_o$ denote its Lie algebra and $\mathfrak{g}_o = \mathfrak{k}_o + \mathfrak{p}_o$ a Cartan decomposition.

Let $\mathfrak{c}_o$ denote the (one-dimensional) center of $\mathfrak{k}_o$ and $\mathfrak{h}_o \subset \mathfrak{k}_o$ any maximal abelian subalgebra. Then $\mathfrak{c}_o \subset \mathfrak{h}_o \subset \mathfrak{k}_o$ and $\mathfrak{h}_o$ is maximal abelian in $\mathfrak{g}_o$. We pass to the complexifications $\mathfrak{c} \subset \mathfrak{h} \subset \mathfrak{k}$ and consider the compact real form $\mathfrak{u} = \mathfrak{k}_o + i\mathfrak{p}_o$ of $\mathfrak{g}$ as well as the conjugation σ and τ of $\mathfrak{g}$ with respect to $\mathfrak{g}_o$ and $\mathfrak{u}$, respectively. Then $\mathfrak{h}$ is a Cartan subalgebra of $\mathfrak{g}$ and we denote by Δ the corresponding set of roots. For each $\alpha \in \Delta$ the root space $\mathfrak{g}^\alpha$ lies either in $\mathfrak{k}$ or in $\mathfrak{p}$; in the two respective cases α is called *compact* or *noncompact*. Let Q denote the set of noncompact roots. Then

$$(1) \qquad \mathfrak{k} = \mathfrak{h} + \sum_{\alpha \in \Delta - Q} \mathfrak{g}^\alpha \qquad \mathfrak{p} = \sum_{\beta \in Q} \mathfrak{g}^\beta$$

and recall that a root α is compact if and only if it vanishes identically on $\mathfrak{c}$.

We introduce compatible orderings in the duals of the real vector spaces $i\mathfrak{c}_o$ and $i\mathfrak{h}_o$. Let Δ^+ denote the corresponding set of positive roots and let $Q^+ = \Delta^+ \cap Q$. Then the spaces

$$(2) \qquad \mathfrak{p}_+ = \sum_{\beta \in Q^+} \mathfrak{g}^\beta, \qquad \mathfrak{p}_- = \sum_{-\beta \in Q^+} \mathfrak{g}^\beta$$

are abelian and

$$(3) \qquad [\mathfrak{k}, \mathfrak{p}_+] \subset \mathfrak{p}_+, \quad [\mathfrak{k}, \mathfrak{p}_-] \subset \mathfrak{p}_- \qquad \mathfrak{p} = \mathfrak{p}_- + \mathfrak{p}_+.$$

As usual, $H_\alpha \in \mathfrak{h}$ is determined by $B(H, H_\alpha) = \alpha(H)$ $(H \in \mathfrak{h})$. We then select ([DS, VI, Lemma 3.1) a vector $X_\alpha \in \mathfrak{g}^\alpha$ $(\alpha \in \Delta)$ such that

$$(4) \qquad X_\alpha - X_{-\alpha}, \; i(X_\alpha + X_{-\alpha}) \in \mathfrak{u},$$

$$(5) \qquad [X_\alpha, X_{-\alpha}] = (2/\alpha(H_\alpha))H_\alpha.$$

By the quoted result, $H_\alpha \in i\mathfrak{k}_o$ $(= i\mathfrak{u} \cap \mathfrak{k})$ for each $\alpha \in \Delta$. Two roots $\alpha, \beta \in \Delta$ are said to be *strongly orthogonal* if $\alpha \pm \beta \notin \Delta$. By root theory, "strongly orthogonal" implies "orthogonal". We use these to construct a special maximal abelian subspace $\mathfrak{a}_o \subset \mathfrak{p}_o$ as follows: For $Q \subset Q^+$ put

$$\mathfrak{p}_Q = \sum_{\gamma \in Q} (\mathfrak{g}^\gamma + \mathfrak{g}^{-\gamma}).$$

Let β be the lowest root in Q and put

$$Q(\beta) = \{\gamma \in Q : \ \gamma \neq \beta, \qquad \gamma - \beta, \gamma + \beta \notin \Delta\}.$$

Then

$$\text{centralizer of } \mathfrak{g}^\beta + \mathfrak{g}^{-\beta} \text{ in } \mathfrak{p}_Q \text{ equals } \mathfrak{p}_{Q(\beta)}; \tag{6}$$

$$\text{centralizer of } X_\beta + X_{-\beta} \text{ in } \mathfrak{p}_Q \text{ equals } \boldsymbol{C}(X_\beta + X_{-\beta}) + \mathfrak{p}_{Q(\beta)}. \tag{7}$$

Let $Q_1 = Q^+$ and γ_1 the lowest root in Q_1, γ_2 the lowest root in $Q_2 = Q_1(\gamma_1)$ etc. With $\ell = \operatorname{rank} X$ this gives

$$\gamma_1, \ldots, \gamma_\ell \quad \text{strongly orthogonal}$$

$$\mathfrak{a}_o = \sum_{i=1}^{\ell} \boldsymbol{R}(X_{\gamma_i} + X_{-\gamma_i}) \text{ maximal abelian in } \mathfrak{p}_o, \tag{8}$$

and

$$\mathfrak{p}_{Q_1} \supset \mathfrak{p}_{Q_2} \supset \cdots \supset \mathfrak{p}_{Q_\ell} \supset \mathfrak{p}_{Q_{\ell+1}} = 0. \tag{9}$$

Let G denote the simply connected Lie group with Lie algebra $\mathfrak{g}^{\boldsymbol{R}}$. Let U, K, P_-, P_+, G_o, K_o and A_o denote the analytic subgroups of G with Lie algebras $\mathfrak{u}, \mathfrak{k}, \mathfrak{p}_-, \mathfrak{p}_+, \mathfrak{g}_o, \mathfrak{k}_o$ and $\mathfrak{a}_o$, respectively. Then we have the diagrams (loc. cit.)

$$\begin{array}{ccccc}
G_oKP_+/KP_+ & \xrightarrow{I_1} & P_-KP_+/KP_+ & \xrightarrow{I_2} & G/KP_+ \\
\psi_1 \big\downarrow & & \psi_2 \big\downarrow & & \\
G_o/K_o & \xrightarrow{\psi_o} & P_- & \xrightarrow{\log} & \mathfrak{p}_-
\end{array}$$

where I_1, I_2 are inclusions, and

$$\psi_1 : \; gkpKP_+ \longrightarrow gK_o \qquad (g \in G_o, k \in K, p \in P_+)$$
$$\psi_2 : \; qkpKP_+ \longrightarrow q \qquad (q \in P_-, k \in K, p \in P_+)$$

$\psi_o = \psi_2 \circ I_1 \circ \psi_1^{-1}$, ψ_1, ψ_2 are diffeomorphisms. Then the mapping $\psi = \log \circ \psi_o$ is a holomorphic diffeomorphism (the Harish-Chandra imbedding) of G_o/K_o onto a bounded open domain D in $\mathfrak{p}_-$. Its differential $d\psi : \; \mathfrak{p}_o \longrightarrow \mathfrak{p}_-$ is given by

$$d\psi(X) = \frac{1}{2}(X - iJ_o(X)), \tag{10}$$

where J_o is the complex structure of G_o/K_o at the origin. Moreover

$$J_o = ad_{\mathfrak{g}_o}(H_o), \tag{11}$$

where $H_o \in \mathfrak{c}_o$ is given by

$$\alpha(H_o) = -i \qquad \text{for } \alpha \in Q^+. \tag{12}$$

In the diagram above G/KP_+ is via the diffeomorphism

$$f : \; uK_o \longleftrightarrow uKP_+ \tag{13}$$

identified with the Hermitian symmetric space $X^* = U/K_o$, the dual to X. The map $f \circ I_2 \circ I_1 \circ \psi_1^{-1}$ thus imbeds X as an open subset of X^* (Borel imbedding).

2. The Geodesics in a Bounded Symmetric Domain.

Let o denote the origin in G_o/K_o, let $H \in \mathfrak{a}_o$ and γ_H the geodesic

$$\gamma_H(t) = \exp tH \cdot o. \tag{14}$$

Denoting by Γ the set $\{\gamma_1, \dots \gamma_\ell\}$ of strongly orthogonal roots we have

$$H = \sum_{\gamma \in \Gamma} t_\gamma (X_\gamma + X_{-\gamma})$$

and (cf. [DS], VIII, Lemma 7.11) the diffeomorphism

$$\psi : \; G_o/K_o \longrightarrow D \subset \mathfrak{p}_- \tag{15}$$

satisfies

(16) $$\psi(\gamma_H(t)) = \sum_{\gamma\in\Gamma} \tanh(tt_\gamma)X_{-\gamma}.$$

This shows that

(17) $$D = Ad_{G_o}(K_o)(\Box),$$

where $\Box$ denotes the "cube"

(18) $$\Box = \left\{ \sum_{\gamma\in\Gamma} x_\gamma X_{-\gamma} : \ |x_\gamma| < 1 \right\}.$$

Also (16) implies that if $H \in \mathfrak{a}_o$ belongs to the octant $t_\gamma > 0$ $(\gamma \in \Gamma)$ then the geodesics $\psi(\gamma_H(t))$ all converge to the same point, in fact,

(19) $$\lim_{t\to+\infty} \psi(\gamma_H(t)) = \sum_{\gamma\in\Gamma} X_{-\gamma}.$$

3. The Restricted Root Systems for Bounded Symmetric Domains.

We consider now the automorphism ν of $\mathfrak{g}$ given by

(20) $$\nu = \exp \frac{\pi}{4} \operatorname{ad} \left(\sum_{\gamma\in\Gamma} (X_\gamma - X_{-\gamma}) \right).$$

Then by [DS] p. 528, 581,

(21) $$\nu(X_\gamma + X_{-\gamma}) = \frac{2}{\langle\gamma,\gamma\rangle} H_\gamma \quad \text{so} \quad \nu(\mathfrak{a}_o) = \sum_{\gamma\in\Gamma} \boldsymbol{R} H_\gamma.$$

Thus $\nu(\mathfrak{a}) \subset \mathfrak{h}$ and we have an orthogonal decomposition

(22) $$\mathfrak{a}_{\mathfrak{k}} + \nu(\mathfrak{a}) = \mathfrak{h}$$

relative to the positive definite Hermitian form $B_\tau(X,Y) = -B(X,\tau Y)$ on $\mathfrak{g}$. By this orthogonality, $\gamma(H) = 0$ for $\gamma \in \Gamma$, $H \in \mathfrak{a}_{\mathfrak{k}}$ so ν fixes $\mathfrak{a}_{\mathfrak{k}}$ elementwise. Thus, if

$$\widetilde{\mathfrak{a}} = \mathfrak{a}_{\mathfrak{k}} + \mathfrak{a},$$

we have $\nu(\widetilde{\mathfrak{a}}) = \mathfrak{h}$ so $\widetilde{\mathfrak{a}}$ is a Cartan subalgebra of $\mathfrak{g}$.

In order to determine the restricted root system $\Sigma = \Sigma(\mathfrak{g}_o, \mathfrak{a}_o)$ it suffices via the automorphism ν to study the restrictions of the roots in $\Delta = \Delta(\mathfrak{g}, \mathfrak{h})$ to $\nu(\mathfrak{a})$ and for this we can take advantage of the division into compact and noncompact roots.

Lemma 4.1. *For the ordering of $\Delta = \Delta(\mathfrak{g}, \mathfrak{h})$ specified above let $\alpha_1, \ldots, \alpha_r$ denote the simple roots and α_1 the highest one. Then α_1 is noncompact, $\alpha_2, \ldots, \alpha_r$ compact and each $\beta \in Q^+$ has the form*

$$\beta = \alpha_1 + \sum_2^r m_i \alpha_i, \qquad m_i \in \boldsymbol{Z}. \tag{23}$$

Proof. If $\alpha = \Sigma_i \alpha_i \in \Delta^+$ and if at least one α_i occurring is noncompact then the restriction $\alpha | i\mathfrak{c}_o$ is strictly positive so $\alpha \in \boldsymbol{Q}^+$. Thus if α is compact then $\alpha = \Sigma m_i \alpha_i$ where all α_i are compact. The subspace of $\mathfrak{h}$ where all compact roots vanish is by (1) the center $\mathfrak{c}$. Thus the compact roots span an $(r-1)$-dimensional space so only one of the α_i is noncompact. By the compatibility of the orderings, α_1 is the noncompact one.

On the other hand, K acts irreducibly on $\mathfrak{p}_+$ so $\mathfrak{p}_+ = U(\mathfrak{k}) X_{\alpha_1}$ where $U(\mathfrak{k})$ is the universal enveloping algebra of $\mathfrak{k}$. This implies (23).

Lemma 4.2. *Let $\gamma, \delta \in \boldsymbol{Q}^+$. Then $\gamma(H_\delta) \geq 0$.*

Proof. If $\gamma(H_\delta) < 0$ then $\gamma + \delta$ would be a root contradicting Lemma 4.1.

Lemma 4.3. *Let $\alpha \in \Delta$ and $H_\alpha \in \mathfrak{a}_{\mathfrak{k}}$. Then α is compact and $\gamma_i \pm \alpha$ is never a root $(1 \leq i \leq \ell)$.*

Proof. We may assume $\alpha > 0$. If α were noncompact then $\alpha + \gamma_i \notin \Delta$ by Lemma 4.1 and since by $H_\alpha \in \mathfrak{a}_{\mathfrak{k}}$ and (22) $\alpha(H_{\gamma_i}) = 0$, we have $\alpha - \gamma_i \notin \Delta$ by general root theory. Thus X_α commutes with $X_{\gamma_i}, X_{-\gamma_i}$ which contradicts $\mathfrak{a}$ being maximal abelian in $\mathfrak{p}$. Thus α is compact.

Now we define the sequence

$$\mathfrak{g} = \mathfrak{g}_1 \supset \mathfrak{g}_2 \supset \cdots \supset \mathfrak{g}_\ell \supset \mathfrak{g}_{\ell+1} \tag{24}$$

in the same way as (9), that is $\mathfrak{g}_{i+1}$ is the centralizer of $\mathfrak{g}^{\gamma_i} + \mathfrak{g}^{-\gamma_i}$ in $\mathfrak{g}_i$ $(1 \leq i \leq \ell)$. Then by (9),

$$\mathfrak{p}_{Q_i} = \mathfrak{p} \cap \mathfrak{g}_i, \qquad \mathfrak{g}_{\ell+1} \subset \mathfrak{k}.$$

We claim that $X_\alpha \in \mathfrak{g}_{i+1}$ for each i. Otherwise choose the smallest $r \geq 0$ such that $X_\alpha \notin \mathfrak{g}_{r+1}$. Then either $\gamma_r + \alpha$ or $\gamma_r - \alpha$ is a root. By minimality of r, $X_\alpha \in \mathfrak{g}_r$ and hence $X_{-\alpha} \in \mathfrak{g}_r$ so $[X_{\gamma_r}, X_{-\alpha}] \in \mathfrak{g}_r \cap \mathfrak{p} = \mathfrak{p}_{Q_r}$ contradicting the minimality of γ_r in Q_r. Thus $\gamma_r - \alpha$ is not a root, so $\gamma_r + \alpha$ is a root. Considering the α-string through γ_r we see that $\gamma_r(H_\alpha) < 0$ contradicting the orthogonality in (22). This contradiction shows $X_\alpha \in \mathfrak{g}_{i+1}$ for each i $(0 \leq i \leq \ell)$ so $\gamma_i \pm \alpha \notin \Delta$ $(1 \leq i \leq \ell)$. This proves the lemma.

Lemma 4.4. *Let α be any root. Then for any i $(1 \leq i \leq \ell)$ $\gamma_i + \alpha$ and $\gamma_i - \alpha$ cannot both be roots.*

Proof. We may take $\alpha > 0$. If α is noncompact, $\gamma_i + \alpha$ is not a root by Lemma 4.1. Hence we can assume α compact. Suppose i is such that $\gamma_i + \alpha$ and $\gamma_i - \alpha$ are both roots. Then both are in Q^+ so by Lemma 4.2

$$(\gamma_i \pm \alpha)(H_\delta) \geq 0, \qquad \delta \in Q^+. \tag{25}$$

But $\gamma_i(H_{\gamma_j}) = 0$ $(i \neq j)$ by the strong orthogonality so (25) implies

$$\alpha(H_{\gamma_j}) = 0 \qquad \text{for } j \neq i. \tag{26}$$

The α-series containing γ_i, that is the set of roots $\gamma_i + n\alpha$ $(p \leq n \leq q)$ contains $\gamma_i - \alpha, \gamma_i, \gamma_i + \alpha$. Let $\gamma_o = \gamma_i + q\alpha$ be the top of the series and let $\gamma_o + m\alpha$ $(p_o \leq m \leq 0)$ be the α-series through γ_o Then by root theory (cf. [DS], III, Theorem 4.3)

$$-p_o = 2\frac{\gamma_o(H_\alpha)}{\alpha(H_\alpha)}. \tag{27}$$

We claim this integer is ≤ 2. In fact suppose it is ≥ 3. We know $\alpha + \gamma_o \notin \Delta$ so since $\alpha \in \Delta$ we have $\gamma_o - \alpha \in \Delta$ and hence also $-\gamma_o + \alpha \in \Delta$. The γ_o-series through $\alpha, \ldots, -\gamma_o + \alpha, \alpha$, shows

$$1 \leq 2\frac{\alpha(H_{\gamma_o})}{\gamma_o(H_{\gamma_o})}. \tag{28}$$

Consider the root

$$\gamma' = s_{\gamma_o} s_\alpha \gamma_o = (c-1)\gamma_o - d\alpha,$$

where c is the product of the right hand sides of (27) and (28) and d a scalar. Since $c - 1 \geq 2$ this would contradict Lemma 4.1. Thus $p_o \geq -2$ so the α-series through γ_o contains at most $\gamma_o - 2\alpha, \gamma_o - \alpha$, γ_o. Hence the α series through γ_i contains just $\gamma_i - \alpha, \gamma_i, \gamma_i + \alpha$. Hence (loc. cit.) $\alpha(H_{\gamma_i}) = 0$ which together with (26) implies $H_\alpha \in \mathfrak{a}_\mathfrak{k}$ and now Lemma 4.3 gives the desired contradiction.

Let λ and μ be two linear forms on $\mathfrak{h}$. We write $\lambda \sim \mu$ if $\lambda - \mu$ vanishes identically on $\nu(\mathfrak{a})$.

Lemma 4.5. *Let $\alpha \in \Delta^+$ be compact. Then there are only the following three mutually exclusive possibilities:*

(i) $H_\alpha \in \mathfrak{a}_\mathfrak{k}$ so $\alpha \sim 0$ and $\gamma_i \pm \alpha$ $(1 \leq i \leq \ell)$ is never a root.

(ii) For a unique index i $(1 \leq i \leq \ell)$, $\alpha + \frac{1}{2}\gamma_i \sim 0$.

(iii) There exist two unique indices i, j $(1 \leq i < j \leq \ell)$ such that $\alpha \sim \frac{1}{2}(\gamma_j - \gamma_i)$.

Proof. Case (i) is covered by Lemma 4.3 so we assume $H_\alpha \notin \mathfrak{a}_\mathfrak{k}$. Then $\alpha(H_{\gamma_r}) \neq 0$ for some r so $X_\alpha \notin \mathfrak{g}_{r+1}$. Let i be the smallest index such that $X_\alpha \notin \mathfrak{g}_{i+1}$. Since $\alpha > 0$ we have $\gamma_i + \alpha \in \Delta$, $\gamma_i - \alpha \notin \Delta$ by the proof of Lemma 4.3. Suppose $\gamma_j + \epsilon\alpha \in \Delta$ for some $j(1 \leq j \leq \ell, \epsilon = \pm 1)$. If $j \neq i$ we claim $\epsilon = -1$. Otherwise suppose $\gamma_j + \alpha \in \Delta$. Then by Lemma 4.2, $0 \leq (\gamma_j + \alpha)(H_{\gamma_i}) = \alpha(H_{\gamma_i})$. On the other hand, $\gamma_i + \alpha \in \delta$, $\gamma_i - \alpha \notin \Delta$ so $\alpha(H_{\gamma_i}) < 0$. This contradiction shows $\epsilon = -1$. This leaves us with two cases. Either

(1) $\gamma_j \pm \alpha \notin \Delta$ for all $j \neq i$

(2) $\gamma_j - \alpha \in \Delta$ for some $j \neq i$.

In case (1) $\alpha(H_{\gamma_j}) = 0$ $(j \neq i)$. Also $\alpha, \alpha + \gamma_i \in \Delta$, $\alpha - \gamma_i \notin \Delta$ and by Lemma 4.1, $\alpha + 2\gamma_i \notin \Delta$. The γ_i-series through α being $\alpha, \alpha + \gamma_i$ we deduce $2\alpha(H_{\gamma_i}) = -\gamma_i(H_{\gamma_i})$. Consequently $(\alpha + \frac{1}{2}\gamma_i)$ $(H_{\gamma_j}) = 0$ for $1 \leq j \leq \ell$ so $\alpha + \frac{1}{2}\gamma_i \sim 0$.

In case (2) let j be the smallest index such that $\gamma_j - \alpha \in \Delta$. Then $j \neq i$. Also $\gamma_j - \alpha \in \Delta$ implies $X_\alpha \notin \mathfrak{g}_{j+1}$ so by the definition of i, $i < j$. We now claim that if $k \neq i, j$ then $\gamma_k \pm \alpha \notin \Delta$. We have already seen this for $\gamma_k + \alpha$. Suppose then $\gamma_k - \alpha \in \Delta$. By Lemma 4.2, $(\gamma_k - \alpha)(H_{\gamma_j}) \geq 0$ so $-\alpha(H_{\gamma_j}) \geq 0$. Since $\alpha - \gamma_j, \alpha \in \Delta, \alpha + \gamma_j \notin \Delta$ we deduce $\alpha(H_{\gamma_j}) > 0$ which is a contradiction. This proves

$$\gamma_k \pm \alpha \notin \Delta \qquad k \neq i, j. \tag{29}$$

We have also

$$\alpha, \alpha + \gamma_i \in \Delta, \;\; \alpha, \alpha - \gamma_i, \alpha + 2\gamma_i \notin \Delta \;\; \text{(Lemma 4.1)}$$

so $2\alpha(H_{\gamma_i}) = -\gamma_i(H_{\gamma_i})$. Similarly,

$$\alpha, \alpha - \gamma_j \in \Delta, \quad \alpha + \gamma_j, \; \alpha + 2\gamma_j \notin \Delta$$

so $2\alpha(H_{\gamma_j}) = \gamma_j(H_{\gamma_j})$. Also by (29) $\alpha(H_{\gamma_k}) = 0$ for $k \neq i, j$. These relations imply

$$\alpha(H_{\gamma_k}) = \frac{1}{2}\gamma_j(H_{\gamma_k}) - \frac{1}{2}\gamma_i(H_{\gamma_k}) \qquad (1 \leq k \leq \ell)$$

so $\alpha \sim \frac{1}{2}(\gamma_j - \gamma_i)$. Finally note that since the γ_k vanishes identically on $\mathfrak{a}_{\mathfrak{k}}$ their restrictions to $\nu(\mathfrak{a})$ are linearly independent. The uniqueness in (ii) and (iii) and the mutual exclusiveness of the three possibilities therefore follows.

For each i $(1 \leq i \leq \ell)$ let C_i denote the set of compact roots α such that $\alpha \sim -\frac{1}{2}\gamma_i$ and Q_i the set of noncompact roots β for which $\beta \sim \frac{1}{2}\gamma_i$. If $\alpha \in C_i$ then $-\alpha \notin \Delta^+$ by Lemma 4.5 so C_i and Q_i are contained in Δ^+.

Lemma 4.6 The map $\alpha \longrightarrow \gamma_i + \alpha$ is a bijection of C_i onto Q_i.

Proof. If $\alpha \in C_i$ then $\alpha \sim -\frac{1}{2}\gamma_i$ so $\alpha(H_{\gamma_i}) = -\frac{1}{2}\gamma_i(H_{\gamma_i})$. Thus $s_{\gamma_i}\alpha = \alpha + \gamma_i$ which belongs to Q_i. On the other hand, if $\beta \in Q_i$ then $\beta \sim \frac{1}{2}\gamma_i$ so $\beta(H_{\gamma_i}) = \frac{1}{2}\gamma_i(H_{\gamma_i})$. Thus $s_{\gamma_i}\beta = \beta - \gamma_i \in \Delta$. By Lemma 4.1, $\alpha = \beta - \gamma_i$ is compact and $\alpha + \frac{1}{2}\gamma_i \sim 0$ so $\alpha \in C_i$. This proves the lemma.

For any pair i, j $(1 \leq i < j \leq \ell)$ let C_{ij} denote the set of all compact roots α such that $\alpha \sim \frac{1}{2}(\gamma_j - \gamma_i)$. Also let Q_{ij} denote the set of noncompact roots β such that $\beta \sim \frac{1}{2}(\gamma_j + \gamma_i)$. Again Lemma 4.5 implies $C_{ij} \subset \Delta^+$ and $Q_{ij} \subset \Delta^+$ is obvious.

Lemma 4.7. *The map $\alpha \longrightarrow \gamma_i + \alpha$ is a bijection of C_{ij} onto Q_{ij}.*

Proof. If $\alpha \in C_{ij}$ then $\alpha \sim \frac{1}{2}(\gamma_j - \gamma_i)$ and $\alpha(H_{\gamma_i}) = -\frac{1}{2}\gamma_i(H_{\gamma_i})$. Thus $s_{\gamma_i}\alpha = \alpha + \gamma_i$ and clearly $\gamma_i + \alpha \in Q_{ij}$. Conversely, if $\beta \in Q_{ij}$ then $\beta(H_{\gamma_i}) = \frac{1}{2}\gamma_i(H_{\gamma_i})$ so $s_{\gamma_i}\beta = \beta - \gamma_i$ which clearly is an element $\alpha \in C_{ij}$.

Let C_o denote the set of $\alpha \in \Delta$ such that $\alpha \sim 0$. Recall that Γ denotes the set $\gamma_1, \ldots, \gamma_\ell$.

Proposition 4.8. *Δ^+ is the disjoint union of the sets $C_o, C_i, C_{ij}, \Gamma, Q_i$, Q_{ij} $(1 \leq i < j \leq \ell)$.*

The linear independence of the γ_i on $\nu(\mathfrak{a})$ shows that these sets are disjoint. Let P denote their union. We must show that each $\gamma \in \Delta^+$ lies in P. If γ is compact this is assured by Lemma 4.5. Hence assume $\gamma \in Q^+$. Choose an index i such that $X_\gamma \in \mathfrak{p}_{Q_i}$, $X_\gamma \notin \mathfrak{p}_{Q_{i+1}}$. Since $\gamma_i \in \Gamma \subset P$ we can assume $\gamma \neq \gamma_i$. Then the definition of γ_i implies $\gamma > \gamma_i$. By Lemma 4.1, $\gamma + \gamma_i \notin \Delta$. Hence, since $X_\gamma \notin \mathfrak{g}_{i+1}$, $\alpha = \gamma - \gamma_i \in \Delta$ which then is compact and positive. Since $\gamma = \alpha + \gamma_i \in \Delta$, Lemma 4.5 implies that $\alpha \notin C_o$ and that either $\alpha \sim -\frac{1}{2}\gamma_j$ or $\alpha \sim \frac{1}{2}(\gamma_k - \gamma_j)$ for some j or (k, j) $(1 \leq j < k \leq \ell)$. In the first case $\gamma \sim \gamma_i - \frac{1}{2}\gamma_j$ but since Lemma 4.2 implies $\gamma(H_{\gamma_j}) \geq 0$ we must have $i = j$ whence $\gamma \in Q_i$. In the second case $\gamma \sim \gamma_i + \frac{1}{2}(\gamma_k - \gamma_j)$ and again $\gamma(H_{\gamma_j}) \geq 0$ implies $i = j$ so $\gamma \in Q_{jk} \subset P$ and the proposition is proved.

We now proceed with more explicit determination of the set $\Sigma_\nu = \Delta|\nu(\mathfrak{a})$ of restricted roots. All $\gamma_1, \ldots, \gamma_\ell$ belong there.

Now consider the set of integers,

$$\{j \mid 2 \leq j \leq \ell,\ \tfrac{1}{2}(\pm\gamma_1 \pm \gamma_j) \in \Sigma_\nu\}.$$

Disregarding temporarily our previous numbering of the γ_i we renumber them such that for a certain k,

$$\tfrac{1}{2}(\pm\gamma_1 \pm \gamma_j) \in \Sigma_\nu \ (2 \leq j \leq k),\ \tfrac{1}{2}(\pm\gamma_1 \pm \gamma_i) \notin \Sigma_\nu \ (i > k). \tag{30}$$

Lemma 4.9.

(a) $$\tfrac{1}{2}(\pm\gamma_i \pm \gamma_j) \in \Sigma_\nu \qquad \textit{for } 1 \leq i < j \leq k$$

(b) $$\tfrac{1}{2}(\pm\gamma_i \pm \gamma_j) \notin \Sigma_\nu \qquad \textit{for } 1 \leq i \leq k,\ j > k.$$

Proof. (a) We have

$$\tfrac{1}{2}(\gamma_1 - \gamma_j),\ \tfrac{1}{2}(-\gamma_1 + \gamma_i), \in \Sigma_\nu,\quad 1 < i, j \leq k\ (i \neq j).$$

Their inner product is < 0 so the sum $\frac{1}{2}(\gamma_i - \gamma_j)$ is a root. So is $\frac{1}{2}(\gamma_i + \gamma_j)$ by Lemma 4.7.

(b) Suppose for example $\frac{1}{2}(\gamma_i - \gamma_j) \in \Sigma_\nu$. Since we know $\frac{1}{2}(\gamma_1 - \gamma_i) \in \Sigma_\nu$ and since these two have a negative inner product the sum $\frac{1}{2}(\gamma_1 - \gamma_j)$ would be a root which is a contradiction.

We can now repeat the construction (30) for the roots $\gamma_{k+1}, \ldots, \gamma_\ell$. The result is a splitting of Σ_ν into mutually orthogonal root systems. If $k < \ell$ this would mean Σ_ν not irreducible (cf. [DS], X, Prop. 3.7) which would contradict the irreducibility of G_o/K_o (cf. loc. cit. VII, Lemma 11.7). Consequently $k = \ell$, and we can take the numbering of the γ_i as in (8) (9).

Theorem 4.10. *Let $\lambda \longrightarrow \overline{\lambda}$ denote the restriction from $\mathfrak{h}$ to the subspace $\sum_{i=1}^{\ell} CH_{\gamma_i}$ and write γ_i for $\overline{\gamma}_i$. Then there are two possibilities for $\Delta(\mathfrak{g}, \mathfrak{h})^-$:*

(i) $$\Delta(\mathfrak{g}, \mathfrak{h})^- = \{\tfrac{1}{2}(\pm\gamma_i \pm \gamma_j)\ (1 \leq i < j \leq \ell),\ \pm\gamma_i\ (1 \leq i \leq \ell),\ 0\}$$

(ii) $\Delta(\mathfrak{g},\mathfrak{h})^- = \{\frac{1}{2}(\pm\gamma_i \pm \gamma_j)(1 \leq i < j \leq \ell), \pm\frac{1}{2}\gamma_i, \pm\gamma_i(1 \leq i \leq \ell), 0\}$.

Moreover, all γ_i have the same length.

Proof. We already know from Lemma 4.9 that (i) occurs. Also

$$s_{\frac{1}{2}(\gamma_i-\gamma_j)}\gamma_j = \gamma_j + c(\gamma_i - \gamma_j) \qquad c = 1 \text{ or } \tfrac{1}{2}$$
$$s_{\frac{1}{2}(\gamma_i-\gamma_j)}\gamma_i = \gamma_i + d(\gamma_j - \gamma_i) \qquad d = 1 \text{ or } \tfrac{1}{2}.$$

Since $s_\alpha\alpha = -\alpha$ we deduce by subtracting, $c + d = 2$ so $c = d = 1$. Thus $s_{\frac{1}{2}(\gamma_i-\gamma_j)}$ exchanges γ_i and γ_j. Hence all γ_i have the same length and if $\frac{1}{2}\gamma_i \in \Sigma_\nu$ for one i then $\frac{1}{2}\gamma_j \in \Sigma_\nu$ for all j. This concludes the proof.

Remark. For completeness we remark that not only (i) but also (ii) occurs. The simplest example is $G_o = \boldsymbol{SU}(2,1)$ in which case $\Sigma_\nu = \{\pm\gamma_1, \pm\frac{1}{2}\gamma_1\}$.

Corollary 4.11. *With the Hermitian form B_τ on $\mathfrak{g}$ let $\| \ \|$ denote the corresponding operator norm. With the holomorphic diffeomorphism*

$$\psi : \ G_o/K_o \longrightarrow D$$

with differential

$$d\psi : \ \mathfrak{p}_o \longrightarrow \mathfrak{p}_-$$

let $\mathcal{D} = d\psi^{-1}(D)$. Then

$$\mathcal{D} = \{X \in \mathfrak{p}_o : \ \|adX\| < 2\}.$$

This follows quickly from Theorem 4.10 (cf. [DS], p. 581).

Now we know that the automorphism ν in (20) maps $\tilde{\mathfrak{a}} = \mathfrak{a}_{\mathfrak{k}} + \mathfrak{a}$ onto $\mathfrak{h}$. The transpose maps Δ onto $\Delta(\mathfrak{g}, \tilde{\mathfrak{a}})$ and Σ_ν onto the system $\Sigma(\mathfrak{g}_o, \mathfrak{a}_o)$ of restricted roots. Putting $\lambda_i = {}^t\nu(\gamma_i)$, we have an orthogonal basis $\lambda_1, \ldots, \lambda_\ell$ of $\mathfrak{a}_o^*$ and $\Sigma = \Sigma(\mathfrak{g}_o, \mathfrak{a}_o)$ is given by

(i) $$\Sigma = \left\{\tfrac{1}{2}(\pm\lambda_i \pm \lambda_j)\ (1 \leq i < j \leq \ell);\ \pm\lambda_i\ (1 \leq i \leq \ell)\right\}$$

(ii) $$\Sigma = \left\{\tfrac{1}{2}(\pm\lambda_i \pm \lambda_j)\ (1 \leq i < j \leq \ell),\ \pm\tfrac{1}{2}\lambda_i, \pm\lambda_i\ (1 \leq i \leq \ell)\right\}.$$

Carrying the ordering of the γ_i over from (9) we have $\lambda_1 < \lambda_2 < \cdots < \lambda_\ell$. Putting $\epsilon = 1, \frac{1}{2}$ in the respective cases (i) and (ii) the simple roots are

(31) $$\epsilon\lambda_1,\ \tfrac{1}{2}(\lambda_2 - \lambda_1), \ldots,\ \tfrac{1}{2}(\lambda_\ell - \lambda_{\ell-1})$$

so the positive Weyl chamber is given by

$$\mathfrak{a}_o^+ = \{H \in \mathfrak{a}_o : \ 0 < \lambda_1(H) < \lambda_2(H) < \cdots < \lambda_\ell(H)\}. \tag{32}$$

Since $\nu(X_\gamma + X_{-\gamma}) = 2H_\gamma/\langle\gamma,\gamma\rangle$ and $\lambda_i = {}^t\nu(\gamma_i)$ we see that

$$\lambda_i\big(\tfrac{1}{2}(X_{\gamma_j} + X_{-\gamma_j})\big) = \delta_{ij}. \tag{33}$$

Defining as usual for $\mu \in \mathfrak{a}_o^*$ the vector $A_\mu \in \mathfrak{a}$ by $\mu(H) = B(H, A_\mu)$ we see from (33) that

$$A_{\lambda_i} = \tfrac{1}{2}\langle\gamma_i,\gamma_i\rangle\big(X_{\gamma_i} + X_{-\gamma_i}\big). \tag{34}$$

The Weyl group W of $\Sigma = \Sigma(\mathfrak{g}_o, \mathfrak{a}_o)$ consists of all the signed permutations of the λ_i since we saw above the similar statement for the γ_i.

4. The Action of G_o on D and the Polydisk in D.

Now we need to clarify the action of G_o on D. Let o and o_c denote the identity cosets in G_o/K_o and G/KP_+, respectively. Consider the mapping

$$\xi : Y \longrightarrow (\exp Y)\cdot o_c \qquad Y \in \mathfrak{p}_-. \tag{35}$$

Then we have the diagram

$$\begin{array}{c} \mathfrak{p}_- \xrightarrow{\ \xi\ } G/KP_+ \xrightarrow{\ f^{-1}\ } U/K_o \\ D \xrightarrow{\ \xi\ } G_o\cdot o_c = G_o/K_o = X \end{array} \tag{36}$$

so the action of G_o on D, $(g, Y) \longrightarrow g\cdot Y$ is given by

$$\xi(g\cdot Y) = g\xi(Y), \qquad Y \in D,\ g \in G_o. \tag{37}$$

Let

$$\mathfrak{a}_- = \sum_{\gamma\in\Gamma} \boldsymbol{R}X_{-\gamma}$$

and put

$$Z = \sum_{\gamma\in\Gamma} x_\gamma X_{-\gamma}, \qquad a = \exp\sum_{\gamma\in\Gamma} y_\gamma(X_\gamma + X_{-\gamma}) \in A_o.$$

Then, as proved by Korányi-Wolf (cf. [DS], p. 398),

$$a\cdot Z = \sum_{\gamma\in\Gamma} \frac{x_\gamma\cosh y_\gamma + \sinh y_\gamma}{x_\gamma\sinh y_\gamma + \cosh y_\gamma} X_{-\gamma} \tag{38}$$

and this extends the action of $a \in A_o$ to a map of $\mathfrak{a}_-$ into itself. It also shows that

$$A_o \cdot 0 = \Box \tag{39}$$

with the notation of (18). We also know that the differential $d\psi : \mathfrak{p}_o \longrightarrow \mathfrak{p}_-$ of the map $\psi : G_o/K_o \longrightarrow D$ satisfies

$$d\psi(X_\gamma + X_{-\gamma}) = X_{-\gamma} \qquad \gamma \in \Gamma \tag{40}$$

and thus maps $\mathfrak{a}_o$ onto $\mathfrak{a}_-$. Furthermore $d\psi$ commutes with the action of $Ad(K_o)$ on $\mathfrak{p}_o$ and $\mathfrak{p}_-$, respectively, and by (34)

$$d\psi(2A_{\lambda_i}/\langle\gamma_i, \gamma_i\rangle) = X_{-\gamma_i} \qquad (1 \leq i \leq \ell). \tag{41}$$

The Weyl group W carried over on $\mathfrak{a}_-$ by $d\psi$ consists of all the signed permutations of the $X_{-\gamma}$ ($\gamma \in \Gamma$).

The imbedding $G_o \cdot o_c \subset G/KP_+$ in (36) shows that each $g \in G_o$ acts on the compact space G/KP_+ extending the action of g on D in (37). In particular, the action of g on D extends continuously to the closure $\overline{D}$ in $\mathfrak{p}_-$.

As a general property of the Weyl group ([DS], VII, Prop. 2.2) we have

$$\big(Ad(K_o) \cdot H\big) \cap \mathfrak{a}_o = W \cdot H, \qquad H \in \mathfrak{a}_o. \tag{42}$$

Since W carried over on $\mathfrak{a}_-$ permutes the $\pm X_{-\gamma}$ we thus deduce from (17) and (18),

$$D \cap \mathfrak{a}_- = \Box. \tag{43}$$

In particular, the points $\sum_{\gamma \in \Gamma} \epsilon_\gamma X_{-\gamma}$ ($\epsilon_\gamma = 0, \pm 1$) are all boundary points of D when not all ϵ_γ are 0.

Now recall from (4) that if $\gamma \in \Gamma$ then

$$X_\gamma - X_{-\gamma} \subset \mathfrak{u} \cap \mathfrak{p} = i\mathfrak{p}_o$$

so $i(X_\gamma - X_{-\gamma}) \in \mathfrak{p}_o$. For $\gamma \in \Gamma$, let $K_\gamma \subset G_\gamma \subset G_o$ denote the analytic subgroups with Lie algebras

$$\mathfrak{k}(\gamma) = \boldsymbol{R}(iH_\gamma) \quad \mathfrak{g}(\gamma) = \boldsymbol{R}(iH_\gamma) + \boldsymbol{R}(X_\gamma + X_{-\gamma}) + \boldsymbol{R}(i(X_\gamma - X_{-\gamma})) \tag{44}$$

and let $K^* \subset G^* \subset G_o$ be the analytic subgroups with Lie algebras $\sum_{\gamma \in \Gamma} \mathfrak{k}(\gamma) \subset \sum_{\gamma \in \Gamma} \mathfrak{g}(\gamma)$. Because of the strong orthogonality, the different Lie algebras $\mathfrak{g}(\gamma)$

commute elementwise. The group K_γ is a circle group ([DS], Ch. VII, Lemma 7.6) so the group $K_{\gamma_1} \cdots K_{\gamma_\ell}$ being closed, hence analytic, must coincide with K^*.

Proposition 4.12. *The orbit $G^* \cdot 0 = D^*$ is a totally geodesic complex submanifold of D and is a polydisk.*

Proof. Since G^* is invariant under the Cartan involution θ, D^* is a totally geodesic submanifold of D. The tangent space

$$\boldsymbol{R}(X_\gamma + X_{-\gamma}) + \boldsymbol{R}(i(X_\gamma - X_{-\gamma})) = (G^*/K^*)_o \subset (G_o/K_o)_o = \mathfrak{p}_o$$

is under the differential $d\psi : \mathfrak{p}_o \longrightarrow \mathfrak{p}_-$ sent (by (10) – (12)) to the complex subspace $\boldsymbol{R}X_{-\gamma} + \boldsymbol{R}(iX_{-\gamma})$ of $\mathfrak{p}_-$ so D^* is a complex submanifold of D.

By the mentioned commutation, we have if $I_\gamma = \{x_\gamma X_{-\gamma} : \ |x_\gamma| < 1\}$

$$G^* \cdot 0 = K^* A_o \cdot 0 = K_{\gamma_1} \ldots K_{\gamma_\ell} \cdot \square = \prod_{\gamma \in \Gamma} K_\gamma \cdot I_\gamma = D^*. \tag{45}$$

5. The Shilov Boundary of a Bounded Symmetric Domain.

Let $D \subset \boldsymbol{C}^n$ be a bounded domain. A subset $S \subset \overline{D}$ is called a *Shilov boundary* of D if it is minimal for the following property (P) :

S is closed and

$$\max_S |f| = \max_{\overline{D}} |f| \tag{46}$$

for every function f which is holomorphic on D and continuous on $\overline{D}$.

It is known that a Shilov boundary exists and is unique. We shall prove this in the case of a bounded symmetric domain D and shall describe this set $S(D)$ geometrically in terms of the restricted roots.

By the maximum modulus theorem for the unit disk Δ, (46) holds with S the unit circle $\partial\Delta$. By symmetry this S is minimal so $S(\Delta) = \partial\Delta$. For a polydisk $G^*/K^* = D^*$ the Shilov boundary is the product B^* of the (circle) boundaries. In fact, if $b \in B^*$, a repeated application of the mean value theorem shows $|f(0)| \leq \mathrm{Max}_{K^* \cdot b} |f|$. Applying this to the composite $f \circ g$ ($g \in G^*$) we see that $|f|$ takes its maximum on B^*. Applying this to a product of functions each holomorphic on a single factor of D^*, constant on the other factors, we deduce $S(D^*) = B^*$. Hence for the D^* in Proposition 4.12 we have

$$S(D^*) = K^* \cdot b_\Gamma, \qquad b_\Gamma = \sum_{\gamma \in \Gamma} X_{-\gamma}. \tag{47}$$

Theorem 4.13. *Let* $b_\Gamma = \sum_{\gamma\in\Gamma} X_{-\gamma}$. *Then*

(i) *The* K_o*-orbit* $K_o \cdot b_\Gamma$ *is the only* K_o*-orbit on* $\overline{D}$ *which is also a* G_o*-orbit.*

(ii) *This orbit is the unique Shilov boundary* $S(D)$ *of* D.

Proof. (i) Let $b = \sum_{\gamma\in\Gamma} x_\gamma X_{-\gamma}$, $|x_\gamma| \leq 1$ and consider the intersections

$$(G_o \cdot b) \cap \mathfrak{a}_-, \qquad (K_o \cdot b) \cap \mathfrak{a}_-. \tag{48}$$

The second set is finite by (42). The first is by (38) infinite (and thus $G_o \cdot b \neq K_o \cdot b$) except possibly if $x_\gamma = \pm 1$ for all $\gamma \in \Gamma$.

Assuming now $x_\gamma = \pm 1$ for all $\gamma \in \Gamma$ we have by the remark about W acting on $\mathfrak{a}_-$,

$$(K_o \cdot b) \cap \mathfrak{a}_- = \Big\{\sum_{\gamma\in\Gamma} x_\gamma X_{-\gamma} : \ x_\gamma = \pm 1 \ (\gamma \in \Gamma)\Big\}. \tag{49}$$

In particular

$$(K_o \cdot b_\Gamma) \cap \mathfrak{a}_- = \text{ set of vertices of } \square. \tag{50}$$

We shall now prove that $K_o \cdot b_\Gamma$ is a Shilov boundary for D. Let f be holomorphic on D and continuous on $\overline{D}$. Let $z \in D$ and fix $k_o \in K_o$ such that (cf. (17)) $z = k_o \cdot X$ ($X \in \square$). Then the composite function $f \circ k_o$ is holomorphic on D^* so by (47) since $X \in D^*$,

$$|f(z)| = |f(k_o \cdot X)| = |(f \circ k_o)(X)| \leq \sup_{K^* \cdot b_\Gamma} |f \circ k_o| \leq \sup_{K_o \cdot b_\Gamma} |f \circ k_o|$$

so

$$|f(z)| \leq \sup_{K_o \cdot b_\Gamma} |f|. \tag{51}$$

On the other hand let $w \in K_o \cdot b_\Gamma$ and let S be the sphere in $\mathfrak{p}_-$ (for the B_τ metric) with center 0, passing through w. Then tangent plane S_w is given by a real linear function $r(z)$ with the property that $r(z) \leq 0$ for all $z \in \overline{D}$ with equality only for $z = w$ (since by (17), $\overline{D} \subset \overline{B_{|w|}(0)}$). Now $r(z) = Re(\ell(z))$ with a complex-linear function $\ell(z)$ ($ax + by - c = Re(az - ibz - c)$ for $a, b, c \in \boldsymbol{R}$) and the holomorphic function $f(z) = e^{\ell(z)}$ satisfies

$$|f(w)| = 1, \ |f(z)| < 1, \qquad z \in \overline{D} - \{w\}. \tag{52}$$

This shows that w is contained in any Shilov boundary of D. Combining with (51) we deduce that $K_o \cdot b_\Gamma$ is the unique Shilov boundary of D.

It remains to prove that $G_o \cdot b_\Gamma = K_o \cdot b_\Gamma$. Let $g \in G_o$. Considering the composite function $f \circ g$ we deduce that $gK_o \cdot b_\Gamma$ is a Shilov boundary so by uniqueness it equals $K_o \cdot b_\Gamma$. This concludes the proof.

Since K_o acts on D as the linear group $Ad_{G_o}(K_o)$ we can by Theorem 4.13 write

$$S(D) = K_o/L_o, \tag{53}$$

where

$$L_o = \{k \in K_o : \; Ad(k_o)b_\Gamma = b_\Gamma\}.$$

As in §3, No. 4 we view K_o/M_o as the set of Weyl chambers in $\mathfrak{p}_o$. Under the mapping

$$d\psi : \; \mathfrak{p}_o \longrightarrow \mathfrak{p}_-$$

these Weyl chambers in $\mathfrak{p}_o$ go into Weyl chambers in $\mathfrak{p}_-$. In particular, we put

$$(\mathfrak{a}_-)^+ = d\psi(\mathfrak{a}_o^+). \tag{54}$$

The line $\boldsymbol{R}b_\Gamma$ is the image of the line $\boldsymbol{R}\left(\sum_{i=1}^{\ell} A_{\lambda_i}\right)$ under $d\psi$ which is annihilated by all the simple roots (31) except one, namely $\epsilon\lambda_1$. Thus L_o is the isotropy group of an edge of the positive Weyl chamber $(\mathfrak{a}_-)^+$. If we define

$$H_\Gamma = \sum_{\gamma \in \Gamma} H_\gamma$$

then since $\nu(A_\lambda) = H_\gamma$ for $\lambda = {}^t\nu(\gamma)$ we have

$$d\psi(\nu^{-1}(H_\Gamma)) = b_\Gamma.$$

The edge $\boldsymbol{R}b_\Gamma$ is not defined by the choice of set Q_+ of positive noncompact roots alone but by the choice of a maximal set $\Gamma \subset Q^+$ of strongly orthogonal roots.

Denoting by P_Γ the isotropy subgroup

$$P_\Gamma = \{g \in G_o : \; g \cdot b_\Gamma = b_\Gamma\}$$

we have

$$S(D) = K_o/L_o = G_o/P_\Gamma. \tag{55}$$

Also, if $P_o = M_o A_o N_o$ we have

$$B = K_o/M_o = G_o/P_o. \tag{56}$$

By analogy with the Poisson kernel on $X \times B$ ((18), §3) we define the Poisson kernel P_o on $X \times S(D)$ as the Jacobian

$$P_o(gK_o, b) = \frac{d(g^{-1} \cdot b)}{db}, \quad g \in G_o,\ b \in S(D). \tag{57}$$

In other words, if $F \in C(B)$, $b_1 = g \cdot b$, and db the normalized K_o-invariant volume element on $S(D)$,

$$\int_{S(D)} F(g \cdot b)db = \int_{S(D)} F(b_1)\frac{db}{db_1}db_1 = \int_{S(D)} F(b_1)\frac{d(g^{-1} \cdot b_1)}{db_1}db_1$$
$$= \int_{S(D)} F(b)P_o(gK_o, b)db.$$

In particular,

$$\int_{S(D)} P_o(gK_o, b)db \equiv 1. \tag{58}$$

Proposition 4.14. *For each fixed $b \in S(D)$ the function*

$$x \longrightarrow P_o(x, b) \qquad \text{on } X \tag{59}$$

is harmonic.

Proof. In analogy with the proof of Theorem 3.6 we prove that the function (59) satisfies the mean value theorem (1) in §3. In fact,

$$P_o(ghK_o, b) = \frac{d(h^{-1}g^{-1} \cdot b)}{d(g^{-1} \cdot b)}\frac{d(g^{-1} \cdot b)}{db} = P_o(hK_o, g^{-1} \cdot b)P_o(gK_o, b). \tag{60}$$

Here we replace g by gk and integrate over K_o. Using (58) we get

$$\int_{K_o} P_o(gkhK_o, b)dk = P_o(gK, b)$$

proving the proposition.

Consider now the Furstenberg compactification $\overline{X}$ of X which includes both X and the boundary B.

$$\begin{array}{ccccc} & & \mathfrak{M}(B) & & \\ & & | & & \\ & & \overline{X} & & \\ & \nearrow & & \nwarrow & \\ X & & & & B \end{array} \tag{61}$$

Theorem 4.15. *The bijection $\psi: X \longrightarrow D$ extends to a continuous map $\widetilde{\psi}$ of $\overline{X}$ onto $\overline{D}$ which commutes with the action of G_o and maps B onto the Shilov boundary $S(D)$.*

Moreover, if f is harmonic on D and continuous on $\overline{D}$ we have

$$f(g \cdot 0) = \int_{S(D)} f(g(\beta))d\beta = \int_{S(D)} P_o(g \cdot 0, \beta) f(\beta) d\beta, \tag{62}$$

$d\beta$ being the normalized K_o-invariant measure on $S(D)$.

We begin with a simple lemma.

Lemma 4.16. *Let u be a bounded harmonic function on X such that the family of functions $(x \in X)$*

$$k \longrightarrow v_x(k) = u(k \cdot x)$$

is an equicontinuous family on K_o. Then the boundary function U (§3, (19)) on B is continuous.

Proof. Let $b_n \longrightarrow b$ and write $b_n = k_n \cdot b$ ($k_n \in K_o$ and $k_n \longrightarrow e$). The function $v_n(x) = u(k_n \cdot x) - u(x)$ is harmonic on D, with boundary function $V_n(b) = U(k_n \cdot b) - U(b)$. From (20) and (55) in §3 we have

$$\|v_n\|_\infty = \|V_n\|_\infty.$$

By the assumption, $\|v_n\|_\infty \longrightarrow 0$ so $U(k_n \cdot b) \longrightarrow U(b)$; hence U is continuous.

Lemma 4.17. *If f is a harmonic function on D, continuous on $\overline{D}$ then $f \circ \psi$ is a harmonic function on X which extends to a continuous function on the compactification $\overline{X}$.*

Proof. If $k_n \longrightarrow k$, then $f(k_n \cdot z) \longrightarrow f(z)$ uniformly on $\overline{D}$ so $u = f \circ \psi$ satisfies the assumptions of Lemma 4.16. Thus U is continuous on B and by Theorem 3.15, u extends continuously to $\overline{X}$.

To prove Theorem 4.15 recall first that the kernel function metric Q on D and the metric B_o derived by means of ψ from the Killing form on $\mathfrak{g}_o$ agree up to a constant factor: $B_o = 4Q$ ([DS], Ch. VIII, Exercise B1 and (2) §7; the factor 4 is missing). The manifold D is a Kähler manifold so if $g_{ij^*} = B_o(\partial/\partial z_i,\ \partial/\partial \overline{z}_j)$ the Laplace-Beltrami operator reduces to

$$L_D = \sum_{i,j} g^{ij^*} \frac{\partial^2}{\partial z_i \partial \overline{z}_j} \tag{63}$$

([DS], Ch. VIII, (4) and (6) in §2; [GGA], Ch. II, Prop. 2.6). The coordinate functions being also bounded are therefore harmonic (Theorem 3.1). By Lemma 4.17, $z_i \circ \psi$ extends continuously so the surjective extension $\widetilde{\psi}$: $\overline{X} \longrightarrow \overline{D}$ is established. Since ψ commutes with the actions of G_o and since these actions extend continuously to $\overline{X}$ and to $\overline{D}$, $\widetilde{\psi}$ commutes with G_o. We put $\widetilde{B} = \widetilde{\psi}(B)$.

Now let f be any real-valued harmonic function on D which is continuous on $\overline{D}$. Then $f \circ \widetilde{\psi}$ is harmonic on X, continuous on $\overline{X}$, and the restriction to B is the boundary function F of $f \circ \psi$, i.e. $F(b) = f(\widetilde{\psi}(b))$ for $b \in B$. Thus the Poisson formula for $f \circ \psi$

$$(f \circ \psi)(g \cdot o) = \int_B F(g(b))db = \int_B f(\widetilde{\psi}(g(b)))db \tag{64}$$

becomes a representation formula

$$f(g \cdot 0) = \int_{\widetilde{B}} f(g(\beta))d\beta, \tag{65}$$

where $d\beta = \widetilde{\psi}(db)$ is the normalized K_o-invariant measure on $\widetilde{B}$. This formula implies that the function f on $\overline{D}$ reaches its maximum on $\widetilde{B}$.

Now suppose $F(z)$ is a holomorphic function on D, continuous on $\overline{D}$. Then $|F|$ has a maximum M at a boundary point d_1. Writing $F(d_1) = e^{i\theta}|F|(d_1)$ the function $F_1(z) = e^{-i\theta}F(z) = U_1(z) + iV_1(z)$ has the property that $|U_1| \leq |F(d_1)|$ and $U_1(z) \leq U_1(d_1) = M$. Taking $f = U_1$ in the argument above we see that $U_1(d_o) = U_1(d_1)$ for some $d_o \in \widetilde{B}$. But then

$$M \leq \left(U_1(d_o)^2 + V_1(d_o)^2\right)^{\frac{1}{2}} = |F_1(d_o)| = |F(d_o)| \leq M$$

so $|F|$ takes its maximum on $\widetilde{B}$. By definition of the Shilov boundary, $S(D) \subset \widetilde{B}$. Since both of these sets are K_o-orbits they must coincide. This proves Theorem 4.15.

Remark. See Exercise 14 for an alternative proof of the harmonicity of any holomorphic function on D.

Proposition 4.18.

(i) *In the identifications* (cf. (53))

$$B = K_o/M_o = G_o/P_o$$
$$S(D) = K_o/L_o = G_o/P_\Gamma$$

we have $P_o \subset P_\Gamma$, $P_\Gamma = L_oA_oN_o$ and

$$\widetilde{\psi}(kM_o) = kL_o, \qquad \widetilde{\psi}(gP_o) = gP_\Gamma.$$

(ii) $\int_{L_o} P(gK_o, k\ell M_o)d\ell = P_o(g \cdot 0, kL_o)$.

(iii) *The ray* $r_\Gamma = \boldsymbol{R}^+(\sum_1^\ell A_{\lambda_i})$ *is an edge of the Weyl chamber* $\mathfrak{a}_o^+$. *If* $C \subset \mathfrak{p}_o$ *is any Weyl chamber having* r_Γ *as an edge then* $\mathfrak{a}_o^+$ *and* C *are conjugate under* L_o *and for each* $H \in C$,

$$\lim_{t\to+\infty} \psi(\exp tH \cdot o) = b_\Gamma.$$

Proof. (i) We use (64) on the coordinate functions f and take $g = ka_t$. Then by (25) §3 and (19) §4,

$$\widetilde{\psi}(kM_o) = \lim_{t\to+\infty} \int_B \widetilde{\psi}(ka_t(b))db = \lim_{t\to+\infty} \psi(ka_t \cdot o) = k \cdot b_\Gamma = kL_o.$$

Also, since $\widetilde{\psi}$ commutes with the action of G_o,

$$\widetilde{\psi}(gP_o) = g\widetilde{\psi}(P_o) = g \cdot b_\Gamma.$$

Thus $\widetilde{\psi}(\{P_o\}) = b_\Gamma$ and $P_o \subset P_\Gamma$ so using the Iwasawa decomposition, $P_\Gamma = L_oA_oN_o$.

For (ii) we note that by (64) and (65),

$$\int_B P(g \cdot o, b)(f \circ \widetilde{\psi})(b)db = \int_{S(D)} P_o(g \cdot 0, \beta)f(\beta)d\beta. \tag{66}$$

On the left hand side we consider the fibration $K_o/M_o \longrightarrow K_o/L_o$ and use the integral formula

$$\int_{K_o/M_o} F(kM_o)dk_{M_o} = \int_{K_o/L_o} dk_{L_o} \int_{L_o/M_o} F(k\ell M_o)d\ell_{M_o} \tag{67}$$

([GGA], Ch. I, Prop. 1.13). By (i), $(f \circ \widetilde{\psi})(kM_o) = f(kL_o)$ so (ii) follows from (66).

For (iii) we have already observed that r_Γ is annihilated by all the simple restricted roots except one so it is indeed an edge of $\mathfrak{a}_o^+$. Suppose first $C \subset \mathfrak{a}_o$ and let W_L be the subgroup of W induced by those $\ell \in L_o$ which leave $\mathfrak{a}_o$ invariant. Let $s \in W$ be the element which maps $\mathfrak{a}_o^+$ onto C. If $sr_\Gamma \neq r_\Gamma$ we would have elements $H_1 \neq sH_1$ both in $\overline{C}$. This is impossible since each orbit of W intersects $\overline{C}$ in just one point. Thus $s \in W_L$ so $\mathfrak{a}_o^+$ and C are conjugate under L. Finally, let $C \subset \mathfrak{p}_o$ be arbitrary having r_Γ as edge. Consider the centralizer G_Γ of r_Γ in G_o. Then $G_\Gamma \cap K = L_o$ and G_Γ/L_o is symmetric. Its tangent space $\mathfrak{p}_\Gamma \subset \mathfrak{p}_o$ contains both $\mathfrak{a}_o$ and the space $\widetilde{C}$ spanned by C as maximal abelian subspaces. Thus $\mathfrak{a}_o$ and $\widetilde{C}$ are conjugate under some $\ell \in L_o$ so we are reduced to the previous case. The limit relation in (iii) is now obvious from (19).

Remark. Part (i) shows that when acting on $\overline{D}$ the group A_oN_o fixes the point b_Γ. For A_o this is of course obvious from (38).

6. The Dirichlet Problem for the Shilov Boundary.

We can now solve the Dirichlet problem of constructing a harmonic function on D with preassigned boundary values on $S(D)$.

Let F be a continuous function on $S(D) = K_o/L_o$ and $\widetilde{F} = F \circ \pi$ where $\pi : K_o/M_o \longrightarrow K_o/L_o$ is the natural map. Let u denote the Poisson integral

$$u(gK_o) = \int_B P(gK_o, b)\widetilde{F}(b)db = \int_{S(D)} P_o(g \cdot 0, \beta)F(\beta)d\beta. \tag{68}$$

Lemma 4.19. *The limit*

$$\lim_{t \to +\infty} u(\exp tZ \cdot o)$$

exists and is the same for all vectors in the Weyl chamber union

$$\bigcup_{\ell \in L_o} Ad(\ell)\mathfrak{a}_0^+.$$

Proof. We know from Prop. 3.7 that if $H \in \mathfrak{a}_o^+$ then

$$\lim_{t \to +\infty} u(k \exp tH \cdot o) = \widetilde{F}(kM) \tag{69}$$

and this proves the lemma since $\widetilde{F}(\ell M) = F(\{L_o\})$.

Corollary 4.20. *Given $F \in C(S(D))$ the Poisson integral*

$$f(g \cdot 0) = \int_{S(D)} P_o(g \cdot 0, \beta) F(\beta) d\beta$$

is harmonic on D and has boundary values F, i.e.

$$\lim_{t \to +\infty} f(k \exp tH \cdot 0) = F(kL_o).$$

Proof. We have $u = f \circ \psi$ so the limit relation is just a restatement of (69).

7. The Hua Equations.

Since K_o/L_o is the base of the fibration $\pi : K_o/M_o \longrightarrow K_o/L_o$ of K_o/M_o and since the Poisson integral §3 (24) gives a bijection of $L^\infty(K_o/M_o)$ onto the space of bounded harmonic functions on X it is clear that the Poisson integral (66) on $D \times S(D)$

$$u(z) = \int_{S(D)} P_o(z, \beta) U(\beta) d\beta = (P_o U)(z) \tag{70}$$

maps $L^\infty(S(D))$ onto a proper subspace of the space of harmonic functions on D. A natural question, raised by E. Stein, is how can this subspace be characterized? Here we shall survey the results which solve this problem. Instead of giving detailed proofs we shall refer to the literature for individual results. The first case of this problem, namely with $X = \boldsymbol{Sp}(2, \boldsymbol{R})/\boldsymbol{U}(2)$ was settled by Korányi and Malliavin [1975]. A new proof with an extension to higher dimensions was given by Johnson [1977], [1978].

Example. (Hua [1963], Johnson-Korányi [1980]). Consider for D the bounded domain A III ($p = q = n$) consisting of the $n \times n$ complex matrices Z such that $I_n - {}^tZ\overline{Z} > 0$ (positive definite). This space is given by

$$X = \boldsymbol{SU}(n, n)/\boldsymbol{S}(\boldsymbol{U}_n \times \boldsymbol{U}_n)$$

and the action by G_o on D by

$$Z \longrightarrow (AZ+B)(CZ+D)^{-1} \qquad Z=(z_{jk})$$

the $n \times n$ matrices A, B, C, D satisfying

$$(71) \quad AA^*-BB^* = I_n, \; AC^* = BD^*, \; CC^*-DD^* = -I_n, \; \det\begin{pmatrix} AB \\ CD \end{pmatrix} = 1,$$

$*$ denoting conjugate transpose. (Our double use of D hopefully does not cause confusion.)

Here the Shilov boundary $S(D)$ and the Poisson kernel P_o on $D \times S(D)$ are given by

$$(72) \qquad S(D) = \{Z: \; I_n = Z^*Z\} = \boldsymbol{U}(n)$$

$$(73) \qquad P_o(Z,U) = c\frac{\det(I_n - Z^*Z)^n}{|\det(I_n - U^*Z)|^{2n}},$$

where c is a constant. Consider the matrices $\partial = (\partial/\partial z_{jk})$, $\overline{\partial} = (\partial/\partial \overline{z}_{jk})$. Then we have the following solution to the question raised above.

Theorem 4.21. *The Poisson integrals u in (70) with $U \in L^\infty(S(D))$ are precisely the bounded solutions of the system of differential equations*

$$(74) \qquad \{(I_n - ZZ^*)\overline{\partial}(I_n - Z^*Z)^t\partial\}(u) = 0$$

$$(74') \qquad \{{}^t\partial(I_n - ZZ^*)\overline{\partial}(I_n - Z^*Z)\}(u) = 0.$$

(Here it is understood that $\overline{\partial}$ and ${}^t\partial$ do not differentiate the entries of $(I_n - Z^*Z)$ and $I_n - ZZ^*$.)

Hua showed [1963], p. 116–120 that the Poisson integrals u satisfy (74′); the sufficiency of (74)–(74′), which is far more difficult was proved by Johnson and Korányi [1980].

In this latter paper, a similar theorem is proved for all D for which the restricted root system $\Sigma = \Sigma(\mathfrak{g}_o, \mathfrak{a}_o)$ is of type C_ℓ, i.e. corresponding to the first possibility in Theorem 4.10. As shown in work by Koszul and by Korányi and Wolf [1965] Theorem 4.9 these irreducible Hermitian symmetric spaces X are those which can be realized as tube domains over a self-dual cone C. In these cases the Shilov boundary is a compact symmetric space (and dual to the symmetric space C). From Table VI p. 532 in [DS] we thus get the following description:

Tube Type:

$$\begin{aligned} \boldsymbol{A}\ \mathit{III}\ (p=q=n), &\quad X=\boldsymbol{SU}(n,n)/\boldsymbol{S}(\boldsymbol{U}_n\times\boldsymbol{U}_n)\\ \boldsymbol{D}\ \mathit{III}\ n \text{ even}, &\quad X=\boldsymbol{SO}^*(4m)/\boldsymbol{U}(2m)\\ \boldsymbol{C}\ \mathit{I} &\quad X=\boldsymbol{Sp}(n,\boldsymbol{R})/\boldsymbol{U}(n)\\ \boldsymbol{BD}\ \mathit{I}\ (q=2) &\quad X=\boldsymbol{SO}_o(p,2)/\boldsymbol{SO}(p)\times\boldsymbol{SO}(2)\\ \boldsymbol{E}\ \mathit{VII} &\quad X=\boldsymbol{E}_{7(-25)}/\boldsymbol{E}_6\times\boldsymbol{T} \end{aligned}$$

Non Tube Type:

$$\begin{aligned} \boldsymbol{A}\ \mathit{III}\ (p\neq q) &\quad X=\boldsymbol{SU}(p,q)/\boldsymbol{S}(\boldsymbol{U}_p\times\boldsymbol{U}_q)\\ \boldsymbol{D}\ \mathit{III}\ n \text{ odd}, &\quad X=\boldsymbol{SO}^*(4m-2)/\boldsymbol{U}(2m-1)\\ \boldsymbol{E}\ \mathit{III} &\quad X=\boldsymbol{E}_{6(-14)}/\boldsymbol{SO}(10)\times\boldsymbol{T} \end{aligned}$$

We shall now describe the extension by Johnson and Korányi [1980] of Theorem 4.21 to D of tube type and also the extension by Berline and Vergne [1981] to the non tube type. We follow the presentation in this latter paper. A further study of the Hua equations in the tube type case within the framework of Jordan algebras is given by Faraut and Korányi [1993]. See also Lasalle [1984], described below.

Let δ be a finite-dimensional representation of K_o on a vector space V over $\boldsymbol{C}$. Consider (as in Ch. VI, §1, No. 3) the *associated vector bundle* $G_o\times_{K_o}V$ over G_o/K_o which is defined as the quotient of $G_o\times V$ by the equivalence relation $(g,v)\approx(gk,\delta(k^{-1})v)$. The C^∞ sections of this bundle are identified with the C^∞ maps $F:\ G_o\longrightarrow V$ satisfying $F(gk)=\delta(k^{-1})F(g)$. In fact, given such an F the mapping $s(g)=(g,F(g))$ satisfies $s(gk)\approx s(g)$. Thus the map

$$gK_o\longrightarrow \text{ equivalence class of } (g,F(g))$$

is a section and vice-versa. This space $\Gamma\big(G_o\times_{K_o}V\big)$ of C^∞ sections is acted on by G_o on the left.

Let V' be the dual of V with the contragredient representation $\check{\delta}$ of K_o acting. Let $\boldsymbol{D}(G_o)$ denote the algebra of left invariant differential operators on G_o. Given a linear map

$$d: V'\longrightarrow \boldsymbol{D}(G_o) \tag{75}$$

commuting with the action of K_o (i.e. $d\check{\delta}(k) = \mathrm{Ad}(k)d$) we can define a differential operator

$$D: \mathcal{E}(G_o/K_o) \longrightarrow \Gamma(G_o \times_{K_o} V) \tag{76}$$

commuting with the actions of G_o, by the formula

$$\langle Df(g), v'\rangle = (d(v')\tilde{f})(g) \qquad f \in \mathcal{E}(G_o/K_o), g \in G_o,\ v' \in V', \tag{77}$$

where $\tilde{f}$ is the lift of f to G_o. In fact,

$$\begin{aligned}\langle Df(gk), v'\rangle &= (d(v')\tilde{f})(gk) = (d(v')\tilde{f})^{R(k^{-1})}(g)\\ = (\mathrm{Ad}(k)(d(v'))\tilde{f})(g) &= (d(\check{\delta}(k)v')\tilde{f})(g) = \langle \delta(k^{-1})Df(g), v'\rangle\end{aligned}$$

so indeed $Df \in \Gamma(G_o \times_{K_o} V)$. The invariance under G_o is obvious.

If (e_i) is a basis of V, (e_i') the dual basis of V' then

$$(Df)(g) = \sum_i (d(e_i')\tilde{f})(g)e_i. \tag{78}$$

More generally, let V_1, V_2 and W be finite-dimensional vector spaces on which K_o acts by representations $\delta_1, \delta_2, \delta_o$. Let $C: V_1 \otimes V_2 \longrightarrow W$ be a linear map commuting with K_o and

$$d_1: V_1' \longrightarrow \boldsymbol{D}(G_o),\ d_2: V_2' \longrightarrow \boldsymbol{D}(G_o)$$

linear maps commuting with the action of K_o. These give rise to D_1, D_2 as above. In addition we have a linear map

$$V_1' \otimes V_2' \longrightarrow \boldsymbol{D}(G_o), \tag{79}$$

commuting with the action of K_o given by the bilinear map

$$(v_1', v_2') \longrightarrow d_1(v_1')d_2(v_2').$$

Composing this map (79) with the transpose ${}^tC: W' \longrightarrow V_1' \otimes V_2'$ we obtain a map of $W' \longrightarrow \boldsymbol{D}(G_o)$ commuting with the action of K_o. The corresponding differential operator of

$$\mathcal{E}(G_o/K_o) \longrightarrow \Gamma(G_o \times_{K_o} W)$$

is denoted $C(D_1, D_2)$. If (e_i) is a basis of V_1, (f_j) a basis of V_2, (e_i') and (f_j') the corresponding dual bases, we have

$$(C(D_1, D_2)f)(g) = \sum_{i,j} \big(d_1(e_i')d_2(f_j')\widetilde{f}\big)(g)C\big(e_i \otimes f_j\big). \tag{80}$$

This construction can be extended to several V_i.

We now apply this construction to K_o acting on $\mathfrak{p}_-$ in (2). Its dual $(\mathfrak{p}_-)'$ can by means of the Killing form B be identified with $\mathfrak{p}_+$. Viewing $Z \in \mathfrak{p}_+$ as a left invariant vector field on G_o we have $\mathfrak{p}_+ \subset \boldsymbol{D}(G_o)$. Thus we have a map $(\mathfrak{p}_-)' \longrightarrow \boldsymbol{D}(G_o)$ commuting with the action of K_o. The corresponding operator of $\mathcal{E}\big(G_o/K_0\big) \longrightarrow \Gamma\big(G_o \times_{K_o} \mathfrak{p}_-\big)$ is denoted by $\overline{\partial}$. If (X_i) is a basis of $\mathfrak{p}_-$ and (X_i') the dual basis of $\mathfrak{p}_+$ (i.e. $B(X_i, X_j') = \delta_{ij}$) then by (78)

$$(\overline{\partial}f)(g) = \sum_i \big(X_i'\widetilde{f}\big)(g)X_i, \qquad f \in \mathcal{E}\big(G_o/K_o\big).$$

Interchanging $\mathfrak{p}_-$ and $\mathfrak{p}_+$ we get the operator

$$\partial : \ \mathcal{E}\big(G_o/K_o\big) \longrightarrow \Gamma\big(G_o \times_{K_o} \mathfrak{p}_+\big)$$

given by

$$(\partial f)(g) = \sum_i \big(X_i\widetilde{f}\big)(g)X_i', \qquad f \in \mathcal{E}\big(G_o/K_o\big).$$

The bracket $\big(Y, Z\big) \longrightarrow \big[Y, Z\big]$ gives a linear map of $\mathfrak{p}_- \otimes \mathfrak{p}_+$ into $\mathfrak{k}$ commuting with the action of K_o. The operator

$$C\big(\overline{\partial}, \partial\big) : \ \mathcal{E}\big(G_o/K_o\big) \longrightarrow \Gamma\big(G_o \times_{K_o} \mathfrak{k}\big)$$

defined above has by (80) the explicit form

$$\big(C\big(\overline{\partial}, \partial\big)f\big)(g) = \sum_{i,j} \big(\widetilde{X}_i'\widetilde{X}_j\widetilde{f}\big)(g)[X_i, X_j']. \tag{81}$$

The generalization indicated above applies to the linear map

$$\big(X, Y, Z\big) \longrightarrow \big[X, \big[Y, Z\big]\big] \text{ of } \mathfrak{p}_+ \times \mathfrak{p}_- \times \mathfrak{p}_+ \text{ into } \mathfrak{p}_+$$

giving rise to an operator

$$C\big(\partial, \overline{\partial}, \partial\big) : \ \mathcal{E}\big(G_o/K_o\big) \longrightarrow \Gamma\big(G_o \times_{K_o} \mathfrak{p}_+\big),$$

where

$$(C(\partial,\overline{\partial},\partial)f)(g) = \sum_{i,j,k} (\widetilde{X}_i \widetilde{X}'_j \widetilde{X}_k \widetilde{f})(g)[X'_i,[X_j,X'_k]]. \tag{82}$$

The original problem has very satisfactory answers in terms of these operators.

Theorem 4.22. (Johnson and Korányi [1980]). *Assume G_o/K_o of tube type. The Poisson integrals u of $U \in L^\infty(S(D))$ are precisely the bounded solutions of the system of differential equations*

$$C(\overline{\partial},\partial)u = 0. \tag{83}$$

Theorem 4.23. (Berline and Vergne [1981]). *Let G_o/K_o be an arbitrary irreducible Hermitian symmetric space. The Poisson integrals u of $U \in L^\infty(S(D))$ are precisely the bounded harmonic solutions of the system*

$$C(\partial,\overline{\partial},\partial)u = 0. \tag{84}$$

Lasalle has shown in [1984] that the system (83) can be replaced by an equivalent system containing fewer equations. Recall that in the tube case the Shilov boundary K_o/L_o is a symmetric space. Let

$$\mathfrak{k}_o = \mathfrak{l}_o \oplus \mathfrak{q}_o \tag{85}$$

be the decomposition of $\mathfrak{k}_o$ into the eigenspaces of the corresponding involution of $\mathfrak{k}_o$. Let $\mathfrak{q} = \mathfrak{q}_o^C$ and let $C_\mathfrak{q}$ be the element in $\boldsymbol{D}(G_o) \otimes \mathfrak{q}$ which is obtained from $C(\overline{\partial},\partial)$ by means of the projection of $\mathfrak{k}$ onto $\mathfrak{q}$. The refinement of Theorem 4.22 is the following.

Theorem 4.24. *Assume G_o/K_o of tube type. The range $P_o(L^\infty(S(D)))$ consists of the bounded solutions u to the system*

$$C_\mathfrak{q} u = 0. \tag{86}$$

While (83) amounts to dim K_o equations (86) consists of only dim$\mathfrak{q}_o$ equations.

Remark. The papers quoted in connection with Theorems 4.22 and 4.24 prove the characterization of the Poisson integrals without the boundedness condition by showing that the equations (83), (86) imply that u is harmonic. Taking into account Theorem 6.6 below and denoting by $\mathcal{A}'(S(D))$ the space of hyperfunctions on $S(D)$ these theorems can be stated as follows.

Theorem 4.25.

(i) *If G_o/K_o is of tube type*

$$\begin{aligned} P_o\big(\mathcal{A}'(S(D)\big) &= \{u \in \mathcal{E}\big(G_o/K_o\big) : \quad C\big(\overline{\partial},\partial\big)u = 0\} \\ &= \{u \in \mathcal{E}\big(G_o/K_o\big) : \quad C_{\mathfrak{q}}u = 0\} \end{aligned}$$

(ii) *For G_o/K_o arbitrary,*

$$P_o\big(\mathcal{A}'\big(S(D)\big)\big) = \{u \in \mathcal{E}\big(G_o/K_o\big) : u \text{ harmonic and } C\big(\partial,\overline{\partial},\partial\big)u = 0\}$$

An extension of this result to Poisson transforms from G/P to G/K (P a parabolic subgroup of a semisimple Lie group G) was given by Johnson [1984].

8. Integral Geometry Interpretation.

Considering the double fibration

$$\begin{array}{ccc} & G_o/L_o & \\ & \swarrow \qquad \searrow & \\ S(D) = G_o/P_\Gamma & & D = G_o/K_o \end{array} \tag{87}$$

and viewing the Poisson integral

$$u(z) = \int_{S(D)} P_o(z,\beta)U(\beta)d\beta \tag{88}$$

in the form

$$u(gK_o) = \int_{K_o/L_o} U\big(gkP_\Gamma\big)dk_{L_o} \tag{89}$$

the Poisson integral becomes a special case of the Radon transform $u = \widehat{U}$ defined in Ch. I, §3, No. 1. Since dim $G_o/P_\Gamma < \dim G_o/K_o$ it becomes particularly relevant to compare (89) with the d-plane transform (Ch. I, §3, (79)

$$\widehat{f}(gH) = \int_{H/H\cap K} f(ghK)dh \qquad f \in \mathcal{S}(G/K), \tag{90}$$

where

$$(91)\quad X = G/K = \boldsymbol{M}(n)/\boldsymbol{O}(n), \qquad \Xi = G/H = \boldsymbol{M}(n)/\boldsymbol{M}(d) \times \boldsymbol{O}(n-d).$$

If $d \leq n-2$ we have again $\dim X < \dim \Xi$. In these two cases the Radon transform goes from functions on a space to functions on a higher dimensional space. It is natural that the image functions will be "special" and indeed in the two cases above they are characterized by a system of differential equations. While Richter's result (Ch. I, Theorem 3.18) could be thus compared to Theorems 4.22–4.23 above, Gonzalez' refinement (Ch. I, Theorem 3.19) is analogous to Theorem 4.24. We recall also that for the homomorphisms from Ch. I, §3, (86)

$$(92)\qquad \lambda : \boldsymbol{D}(G) \longrightarrow \boldsymbol{E}(G/K)$$

$$(93)\qquad \Lambda : \boldsymbol{D}(G) \longrightarrow \boldsymbol{E}(G/H)$$

the differential equations $\Lambda(D)\widehat{f} = 0$ characterizing the range $\mathcal{S}(G/K)^\wedge$ came from the D in the kernel of λ. By analogy if we consider the homomorphisms

$$(94)\qquad \nu : \boldsymbol{D}(G_o) \longrightarrow \boldsymbol{E}(G_o/P_\Gamma)$$

$$(95)\qquad \widehat{\nu} : \boldsymbol{D}(G_o) \longrightarrow \boldsymbol{E}(G_o/K_o)$$

as in Proposition 3.4, Ch. I then by (89)

$$\widehat{\nu}(D)u = (\nu(D)U)^\wedge$$

so

$$(96)\qquad \widehat{\nu}(D)u = 0 \qquad \text{for } D \in \nu^{-1}(0).$$

The Hua equations imply (96). The converse question and a more explicit description of the kernel $\nu^{-1}(0)$ are of considerable interest. Compare Ch. I, Theorem 3.23.

The dual transform to (89) would be

$$\check{\varphi}(gP_\Gamma) = \int_{P_\Gamma/L_o} \varphi(gqK_o)\,dq_{L_o} \qquad \varphi \in \mathcal{D}(G_o/K_o).$$

The left invariant measure on $L_oA_oN_o$ is just $d\ell_o da dn$ and up to a constant

$$\int_X \varphi(x)dx = \int_{A_oN_o} \varphi(an \cdot o)dadn.$$

Hence the dual transform just maps φ into the constant function

$$\check{\varphi}(gP_\Gamma) = \int_X \varphi(x)dx.$$

Since

$$(\widehat{\nu}(D)\varphi)^\vee = \nu(D)\check{\varphi}$$

we see that if D is in the kernel of ν then $\widehat{\nu}(D)\mathcal{D}(G_o/K_o)$ is in the kernel of the dual transform $\varphi \longrightarrow \check{\varphi}$. Here one could inquire about analogs of and perhaps sharpening of Ch. I, Theorem 3.24. A result in this spirit is the following result which is included among the exercises in this chapter:

If $X = G/K$ is a symmetric space of rank one then

$$L_X(\mathcal{D}(X)) = \{f \in \mathcal{D}(X) : \int_X f(x)dx = 0, \qquad \int_X f(x)P(x,b)dx \equiv 0\},$$

P being the Poisson kernel on $X \times B$.

§5. THE WAVE EQUATION ON SYMMETRIC SPACES.

1. Introduction. Huygens' Principle.

It is a familiar phenomenon from daily life that waves propagate quite differently in 2 and 3 dimensions. When a pebble falls in water at a certain point P, circular ripples around P are formed. A given point Q near P will be hit by an initial ripple and later by residual waves.

In three dimensions the situation is quite different. A flash of light at a point P has an effect on the surface of a sphere around P after a certain time interval but then no more. There are no residual waves as those present on the water surface. The same is the case with sound waves; one has pure propagation without residual waves; thus music can exist in $\boldsymbol{R}^3$ but not in $\boldsymbol{R}^2$.

There is a mathematical explanation of these phenomena. The propagation of waves in $\boldsymbol{R}^n$ is governed by the wave equation

$$(1) \qquad \frac{\partial^2 u}{\partial t^2} = \frac{\partial^2 u}{\partial x_1^2} + \cdots + \frac{\partial^2 u}{\partial x_n^2}, \qquad u(x,0) = f_o(x), \; u_t(x,0) = f_1(x).$$

Here $u_t = \partial u/\partial t$. The problem of solving (1) in terms of the initial data $f_o, f_1 \in \mathcal{E}(\boldsymbol{R}^n)$ is a special case of the so called Cauchy problem for hyperbolic equations. Equation (1) can be solved by various methods; here we restate the solution obtained in [GGA], Ch. II, Exercise F1 by use of the Ásgeirsson mean value theorem. The result has different forms depending on the parity of n.

Case I n odd. With $(M^r f)(x)$ denoting as usual the mean value of a function f on the sphere $S_r(x)$ we put

$$(I_r f)(x) = \left(\frac{\partial}{\partial(r^2)}\right)^{\frac{n-3}{2}} \left(r^{n-2}(M^r f)(x)\right). \tag{2}$$

Then the solution to (1) is given by

$$u(x,t) = c_n \left[\frac{\partial}{\partial t}(I_t f_o(x)) + (I_t f_1)(x)\right], \tag{3}$$

where c_n is the constant

$$c_n = \frac{\frac{1}{2}\Omega_n}{\frac{1}{2}(n-3)!\Omega_{n-1}}.$$

Case II n even. Here we put

$$(J_r f)(x) = \left(\frac{\partial}{\partial(r^2)}\right)^{\frac{n-2}{2}} \left(r^{n-2}(M^r f)(x)\right). \tag{4}$$

Then the solution to (1) is given by

$$u(x,t) = C_n \left[\frac{\partial}{\partial t}\int_o^t r\left(t^2-r^2\right)^{-\frac{1}{2}}(J_r f_o)(x)dr + \int_o^t r\left(t^2-r^2\right)^{-\frac{1}{2}}(J_r f_1)(x)\right] \tag{5}$$

where C_n is the constant

$$C_n = \frac{1}{2}\left(\frac{1}{2}(n-2)!\right)^{-1}.$$

In Case II formula (5) shows that $u(x,t)$ is determined by the initial data f_o, f_1 in a ball $B_{t+\epsilon}(x)$ (for $\epsilon > 0$ arbitrarily small). In Case I a stronger statement holds:

(6) $u(x,t)$ is determined by the values of the initial data in an arbitrarily thin shell around $S_t(x)$.

To relate this better to light waves and sound waves suppose f_o and f_1 have support in a small ball $N_o = B_\epsilon(x_o)$ and consider the conical shell (see figure)

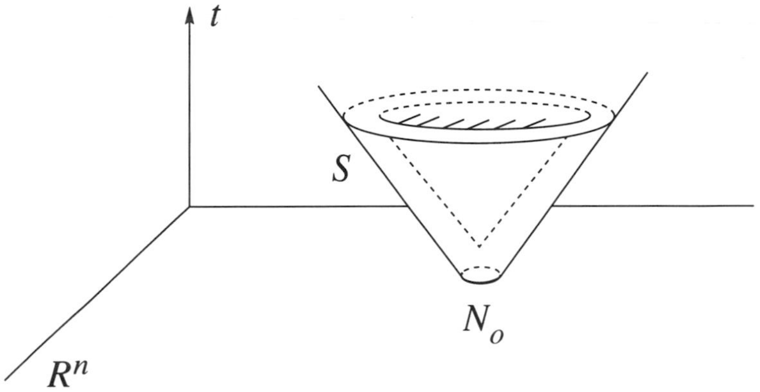

$S = \bigcup_{x \in N_o} C_x$ where C_x denotes the surface of the forward light cone with vertex x. Then Property (6) can also be stated in the equivalent form:

(6′) For all $\epsilon > 0$ the support of the function $(x,t) \longrightarrow u(x,t) \quad (t > 0)$ is contained in the conical shell
$$S = \{(x,t) \| \, t - \epsilon < |x_o - x| < t + \epsilon\},$$
whenever f_o and $f_1$have support in $B_\epsilon(x_o)$.

If (6′) holds then for a fixed $x \notin \overline{N}_o$ the function $t \longrightarrow u(x,t)$ is supported in a short time interval. This means pure propagation without residual effects.

For $n = 2$ formula (5) only implies $\text{supp}(u) \subset \widetilde{S} = \bigcup_{x \in N_o} D_x$ $(t > 0$ where D_x is the solid forward light cone with vertex x. Thus if $x \notin \overline{N}_o$ is fixed, the function $t \longrightarrow u(x,t)$ will have support in an interval $t \geq |x - x_o| - \epsilon$ as illustrated by the water waves.

Property (6) is an instance of the so called *Huygens' principle* (Hadamard's "Minor Premise"), which for general Riemannian manifolds X can be formulated as follows. Let L_X denote the Laplace operator on X and consider the Cauchy problem,

$$(7) \qquad \frac{\partial^2 u}{\partial t^2} = (L_X + c)u, \qquad u(x,0) = f_o(x), \quad u_t(x,0) = f_1(x),$$

c being some constant. *Huygens' principle* is said to hold for (7) if the solution u satisfies (6). Hadamard raised the question of finding all X for which (7) satisfies Huygens' principle. He proved that $\dim X =$ odd is a necessary condition. More generally, the problem was raised with the operator $\partial^2/\partial t^2 - L_X$ replaced with the Laplacian on a Lorentzian manifold and the surface $t = 0$ by a spacelike hypersurface ([GGA], Ch. I, §6).

2. Huygens' Principle for Compact Groups and Symmetric Spaces $X = G/K$ (G complex).

To begin with let $X = G/K$ be an arbitrary symmetric space of the noncompact type and Exp:$\mathfrak{p} \longrightarrow X$ the usual exponential mapping. As in [GGA], Ch. II, §3 let J denote the function on $\mathfrak{p}$ defined by

$$\int_{G/K} f(x)dx = \int_{\mathfrak{p}} f(\operatorname{Exp} X)J(X)dX \tag{8}$$

and put

$$\Omega(\operatorname{Exp} Y) = J^{-\frac{1}{2}}(Y)L_{\mathfrak{p}}(J^{\frac{1}{2}})(Y). \tag{9}$$

Proposition 5.1. *Let E be a K-invariant distribution on $\mathfrak{p}$ and define $\mathcal{E}_E \in \mathcal{D}'(X)$ by*

$$\mathcal{E}_E(f) = E\big(J^{\frac{1}{2}}(f \circ \operatorname{Exp})\big), \qquad f \in \mathcal{D}(X). \tag{10}$$

Then

$$(L_X + \Omega)\mathcal{E}_E = \mathcal{E}_{L_{\mathfrak{p}}E}. \tag{11}$$

Proof. We know form [GGA], Ch. II, §3 that

$$L_X^{\operatorname{Exp}^{-1}} F = \big(L_{\mathfrak{p}} + \text{ grad } (\log J)\big)F \tag{12}$$

if F is a K-invariant function on $\mathfrak{p}$. Since

$$L_{\mathfrak{p}}(uv) = uL_{\mathfrak{p}}(v) + 2 \text{ grad } u(v) + vL_{\mathfrak{p}}(u)$$

(12) can be written,

$$(L_X + \Omega)(F^{\operatorname{Exp}}) = \Big(J^{-\frac{1}{2}}L_{\mathfrak{p}}(J^{\frac{1}{2}}F)\Big)^{\operatorname{Exp}}.$$

Hence

$$((L_X+\Omega)(\mathcal{E}_E))\left((J^{-\frac{1}{2}}F)^{\mathrm{Exp}}\right) = \mathcal{E}_E\left\{(L_X+\Omega)\left((J^{-\frac{1}{2}}F)^{\mathrm{Exp}}\right)\right\}$$
$$= \mathcal{E}_E\left\{\left(J^{-\frac{1}{2}}L_{\mathfrak{p}}(F)\right)^{\mathrm{Exp}}\right\}$$
$$= E\left\{J^{\frac{1}{2}}\left(J^{-\frac{1}{2}}L_{\mathfrak{p}}(F)\right)^{\mathrm{Exp}} \circ \mathrm{Exp}\right\} = E(L_{\mathfrak{p}}(F)) =$$
$$(L_{\mathfrak{p}}E)(F) = \left(J^{\frac{1}{2}}(L_{\mathfrak{p}}E)\right)(J^{-\frac{1}{2}}F) = \left(J^{\frac{1}{2}}L_{\mathfrak{p}}E\right)^{\mathrm{Exp}}\left(J^{-\frac{1}{2}}F\right)^{\mathrm{Exp}}.$$

Since both $(L_X+\Omega)(\mathcal{E}_E)$ and $(J^{\frac{1}{2}}L_{\mathfrak{p}}E)^{\mathrm{Exp}}$ are K-invariant the calculation above implies that they coincide. But if $f \in \mathcal{D}(X)$ then

$$\left(J^{\frac{1}{2}}L_{\mathfrak{p}}E\right)^{\mathrm{Exp}}(f) = (L_{\mathfrak{p}}E)\left(J^{\frac{1}{2}}(f\circ \mathrm{Exp})\right) = \mathcal{E}_{L_{\mathfrak{p}}E}(f)$$

proving the proposition.

Now suppose G is complex. Then the K-radial part of $L_{\mathfrak{p}}$ ([GGA], Ch. II, §3) is given by

$$\Delta(L_{\mathfrak{p}}) = \pi^{-1}L_{\mathfrak{a}}\circ\pi, \tag{13}$$

where $\pi = \prod_{\alpha\in\Sigma^+}\alpha$. Also

$$J(H) = \prod_{\alpha\in\Sigma^+}(\sinh\alpha(H))^2\Big/\pi(H)^2. \tag{14}$$

Since

$$\prod_{\alpha\in\Sigma^+} 2\sinh\alpha(H) = \sum_{s\in W}\epsilon(s)e^{s\rho(H)} \tag{15}$$

we thus deduce from (9), (13)–(14) that

$$\Omega(\exp H) = \langle\rho,\rho\rangle \qquad\qquad H\in\mathfrak{a}$$

so by K-invariance, $\Omega \equiv \langle\rho,\rho\rangle$. Because of Prop. 5.1 we thus take $c = \langle\rho,\rho\rangle$ in (7) and take the shifted wave equation

$$\frac{\partial^2}{\partial t^2} = \left(L_X + |\rho|^2\right)u, \qquad u(x,0) = f_o(x), \qquad u_t(x,0) = f_1(x) \tag{16}$$

as the natural one on X. (The so-called Klein-Gordon equation has opposite sign on the constant term).

Proposition 5.2. *Define the distribution E_t as follows:*

$$E_t(\varphi) = \begin{cases} c_n(I_t\varphi)(0), & n \text{ odd} \\ C_n \int\limits_o^t r(t^2 - r^2)^{-\frac{1}{2}}(J_r\varphi)(0)dr, & n \text{ even} \end{cases} \tag{17}$$

where $\varphi \in \mathcal{D}(\mathfrak{p})$ and $n = \dim X$. Then the solution to

$$\frac{\partial^2 u}{\partial t^2} = L_{\mathfrak{p}} u, \qquad u(x, 0) = f_o(x), \quad u_t(x, 0) = f_1(x) \tag{18}$$

is given by

$$u(x, t) = \big(f_o * E_t'\big)(x) + \big(f_1 * E_t\big)(x). \tag{19}$$

Here $*$ denotes convolution on $\mathfrak{p}$ and $'$ denotes d/dt.

Proof. This is just a restatement of (3) and (5).

In accordance with Prop. 5.1 we put

$$\mathcal{E}_t(f) = E_t\big(J^{\frac{1}{2}}(f \circ \operatorname{Exp})\big) \qquad f \in \mathcal{D}(X). \tag{20}$$

Theorem 5.3. *Let $X = G/K$ be a symmetric space of the noncompact type, G complex. The solution to the shifted wave-equation (16) with $f_o, f_1 \in \mathcal{E}(X)$ is given by*

$$u(x, t) = (f_o \times \mathcal{E}_t')(x) + (f_1 \times \mathcal{E}_t)(x), \tag{21}$$

where $\times$ denotes the convolution on X.

If $\dim X$ *is odd then Huygens' principle holds for (16).*

Proof. Apply $L_{\mathfrak{p}}$ as well as $\partial^2/\partial t^2$ to (19) and equate the results. Since f_o and f_1 are arbitrary, we deduce that

$$L_{\mathfrak{p}} E_t = \frac{\partial^2 E_t}{\partial t^2}, \tag{22}$$

Thus, by Prop. 5.1,

$$\frac{\partial^2 \mathcal{E}_t}{\partial t^2} = (L_X + |\rho|^2)\mathcal{E}_t. \tag{23}$$

Applying L_X to (21) we get by [GGA], Ch. II, Theorem 5.5,

$$(L_X)_x\,(u(x,t)) = (f_o \times L_X\mathcal{E}_t')\,(x) + (f_1 \times L_X\mathcal{E}_t)\,(x)$$

so by (23)

$$\left(L_X + |\rho|^2\right)_x (u(x,t)) = f_o \times \frac{\partial^2 \mathcal{E}_t'}{\partial t^2} + f_1 \times \frac{\partial^2 \mathcal{E}_t}{\partial t^2} = \frac{\partial^2 u}{\partial t^2}.$$

Secondly

$$\lim_{t \to 0} \mathcal{E}_t'(f_o) = \lim_{t \to 0} E_t'\left(J^{\frac{1}{2}}(f_0 \circ \operatorname{Exp})\right) = f_0(o) \tag{24}$$

and since E_t' is K-invariant and invariant under the symmetry $Y \longrightarrow -Y$, the lift $\widetilde{\mathcal{E}}_t'$ of $\mathcal{E}_t'$ to G is invariant under the inversion $g \longrightarrow g^{-1}$. Hence

$$(f_0 \times \mathcal{E}_t')(g \cdot o) = \int f_0(g \cdot x) d\mathcal{E}_t'(x), \tag{25}$$

which by (24) has limit $f_0(g \cdot o)$ as $t \longrightarrow 0$. Similarly

$$f_0 \times \mathcal{E}_t'' \longrightarrow 0 \text{ and } f_1 \times \mathcal{E}_t \longrightarrow 0 \tag{26}$$

as $t \longrightarrow 0$. This proves that (21) does indeed satisfy the shifted wave equation.

Finally, if $\dim X$ is odd, E_t and E_t' have support contained in $S_t(0)$ and by (20), $\mathcal{E}_t$ and $\mathcal{E}_t'$ have support in $S_t(o)$. Since

$$u(g \cdot o, t) = \int_X f_0(g \cdot x) d\mathcal{E}_t'(x) + \int_X f_1(g \cdot x) d\mathcal{E}_t(x), \tag{27}$$

the value $u(g \cdot o, t)$ is determined by the values of f_0 and f_1 in an arbitrary shell around $g \cdot S_t(o) = S_t(g \cdot o)$. This concludes the proof.

Remark. The constant $\langle \rho, \rho \rangle$ in (16) is given by

$$\langle \rho, \rho \rangle = \frac{1}{12} \dim X. \tag{28}$$

To see this we extend $\mathfrak{a}$ to a Cartan subalgebra $\mathfrak{h}^C$ of the complexification $\mathfrak{g}^C$. Let $\Delta(\mathfrak{g}^C, \mathfrak{h}^C)$ be the corresponding set of nonzero roots. We order the roots in such a way that if β is a positive root then the restriction $\overline{\beta} = \beta|\mathfrak{a}$ belongs to Σ^+ (cf. [DS], Ch. VI, Lemma 6.1). Let Δ_+ be the set of positive elements in $\Delta(\mathfrak{g}^C, \mathfrak{h}^C)$ and let $\widetilde{\rho} = \frac{1}{2} \sum_{\beta \in \Delta_+} \beta$. Extending the Cartan involution θ of $\mathfrak{g}$ to a complex automorphism of $\mathfrak{g}^C$ we know that θ leaves

$\mathfrak{h}^C$ invariant ([DS], Ch. VI, lemma 3.2) and by the compatibility of the orderings, $\beta \longrightarrow -\theta\beta$ is a permutation of Δ_+. Hence $\widetilde{\rho} = -\theta\widetilde{\rho}$ so $\widetilde{\rho}$ vanishes on $\mathfrak{h}^C \cap \mathfrak{k}$ whence $\widetilde{\rho} = \rho$. The Killing forms B and B^C of $\mathfrak{g}^C$ considered as a real (respectively complex) Lie algebra are related by $B = 2ReB^C$. On the other hand, if G is simple we have by the "strange formula" of Freudenthal-deVries (cf. [GGA], p. 544) that

$$B^C(\widetilde{\rho},\widetilde{\rho}) = \frac{1}{24}\dim_C G = \frac{1}{24}\dim X \tag{29}$$

so

$$B(\rho,\rho) = 2B^C(\widetilde{\rho},\widetilde{\rho}) = \frac{1}{12}\dim X$$

if G is simple. In the general semisimple case, X is a product and both sides of this equation are additive. This proves (28) in general.

Next we shall prove an analog of Theorem 5.3 for compact semisimple Lie groups K. Let $R > 0$ be such that the mapping $\exp : \mathfrak{k} \longrightarrow K$ is a diffeomorphism of $B_R(0)$ onto $B_R(e)$. The group K is furnished with an invariant Riemannian metric given by the negative of the Killing form Q of $\mathfrak{k}$.

Again let J be the volume element ratio given by

$$\int_K f(k)dk = \int_{B_R(0)} f(\exp Y)J(Y)dY$$

and if E is a distribution on $B_R(0) \subset \mathfrak{k}$ invariant under $Ad(K)$ we put (as in (10))

$$\mathcal{E}_E(f) = E\big(J^{\frac{1}{2}}(f \circ \exp)\big) \qquad f \in \mathcal{D}(B_R(e)).$$

In analogy with (20) we put

$$\mathcal{E}_t(f) = E_t\big(J^{\frac{1}{2}}(f \circ \exp)\big) \quad \text{for} \quad f \in \mathcal{D}(B_R(e)), \tag{30}$$

where E_t is defined in (17) and $|t| < R$.

Theorem 5.4. *Let K be a simply connected compact semisimple Lie group with the bi-invariant metric given by the negative of the Killing form. Then the solution to the shifted wave equation on K*

$$\frac{\partial^2 u}{\partial t^2} = \big(L_K - \frac{\dim K}{24}\big)u, \qquad u(k,0) = f_0(k), \quad u_t(k,0) = f_1(k) \tag{31}$$

where $f_0, f_1 \in \mathcal{D}(B_R(e))$ has the solution

$$u(k,t) = \big(f_o * \mathcal{E}'_t\big)(k) + \big(f_1 * \mathcal{E}_t\big)(k), \tag{32}$$

where $*$ denotes convolution on K.

If $\dim K$ *is odd, then Huygens' principle holds for (31).*

Proof. Let $\mathfrak{t} \subset \mathfrak{k}$ be a maximal abelian subalgebra, $\mathfrak{t}^C \subset \mathfrak{k}^C$ the corresponding complexifications, $\Delta(\mathfrak{k}^C, \mathfrak{t}^C)$ the corresponding set of roots and 2ρ the sum of the corresponding set of positive roots relative to a specific ordering. Under the action of K on itself by conjugation $T = \exp \mathfrak{t}$ is a transversal submanifold in the sense of [GGA], Ch. II, Lemma 3.3. As a consequence of *loc. cit.*, Prop. 3.12, the corresponding radial part of the Laplacian L_K is given by

$$\Delta(L_K) = \delta^{-\frac{1}{2}} L_T \circ \delta^{\frac{1}{2}} + |\rho|^2, \tag{33}$$

where

$$\delta^{\frac{1}{2}}(\exp H) = \sum_{s \in W} (\det s) e^{s\rho(H)} \qquad (H \in \mathfrak{t}), \tag{34}$$

W being the Weyl group $W(\mathfrak{k}, \mathfrak{t})$. On the other hand, under the adjoint action of K of $\mathfrak{k}$ with $\mathfrak{t}$ as transversal submanifold the radial part $\Delta(L_{\mathfrak{k}})$ given by Harish-Chandra's formula

$$\delta(L_{\mathfrak{k}}) = \pi^{-1} L_{\mathfrak{t}} \circ \pi, \tag{35}$$

where π is the product of the positive roots. On the positive Weyl chamber in $\mathfrak{t}$ J equals the ratio $(\delta \circ \exp)/\pi^2$. We can now prove the following result.

Lemma 5.5. *Let $F \in C_c^\infty(B_R(0))$ be invariant under* $\operatorname{Ad}(K)$. *Then*

$$L_K^{\exp^{-1}} F = \left(J^{-\frac{1}{2}} L_{\mathfrak{k}} \circ J^{\frac{1}{2}}\right) F + |\rho|^2 F. \tag{36}$$

Also, if $E \in \mathcal{D}'(B_R(0))$ is $\operatorname{Ad}(\mathfrak{k})$*-invariant, and*

$$\mathcal{E}_E(f) = E\left(J^{\frac{1}{2}}(f \circ \exp)\right) \qquad f \in \mathcal{D}(B_R(e)) \tag{37}$$

then

$$\left(L_K - |\rho|^2\right) \mathcal{E}_E = \mathcal{E}_{L_{\mathfrak{k}} E} \qquad \text{on } B_R(e). \tag{38}$$

Proof. For (36) it suffices to verify that both sides agree on $\mathfrak{t}$. If $H \in \mathfrak{t}$ we obtain from (33) and (35) if $f = F^{\exp}$

$$\begin{aligned}
&\left(L_K^{\exp^{-1}} F\right)(H) = \left(L_K f\right)(\exp H) \\
&= \delta^{-\frac{1}{2}}(\exp H) L_T(\delta^{\frac{1}{2}} f)(\exp H) + |\rho|^2 f(\exp H). \\
&\left(J^{-\frac{1}{2}} L_{\mathfrak{k}}(J^{\frac{1}{2}} F)\right)(H) = J^{-\frac{1}{2}}(H) \pi^{-1}(H) \left(L_{\mathfrak{t}}(\pi J^{\frac{1}{2}} F)\right)(H) \\
&= \delta^{-\frac{1}{2}}(\exp H) L_{\mathfrak{t}} \left((\delta^{\frac{1}{2}} \circ \exp)(f \circ \exp)\right)(H),
\end{aligned}$$

so (36) follows since $L_{\mathfrak{k}}(g \circ \exp) = L_T(g) \circ \exp$ for $g \in \mathcal{D}(\mathfrak{t} \cap B_R(0))$.

For (38) let $f \in \mathcal{D}(B_R(e))$ be invariant under K-conjugation. Then using (36),

$$\begin{aligned} ((L_K - |\rho|^2)\mathcal{E}_E)(f) &= E\left(J^{\frac{1}{2}}(L_K - |\rho|^2)f \circ \exp\right) \\ &= E\left(L_{\mathfrak{k}}(J^{\frac{1}{2}}(f \circ \exp))\right) = (\mathcal{E}_{L_{\mathfrak{k}}E})(f). \end{aligned}$$

Next we can apply (29) to the simple components of the complexification of K and deduce

$$|\rho|^2 = \frac{1}{24} \dim K.$$

To finish the proof of Theorem 5.4, we note that by (24) and (38)

$$\begin{aligned} \frac{\partial^2 \mathcal{E}_t(f)}{\partial t^2} &= \frac{\partial^2 E_t(J^{\frac{1}{2}}(f \circ \exp))}{\partial t^2} = (L_{\mathfrak{k}}E_t)(J^{\frac{1}{2}}(f \circ \exp)) \\ &= \mathcal{E}_{L_{\mathfrak{k}}E_t}(f) = ((L_K - |\rho|^2)\mathcal{E}_t)(f). \end{aligned}$$

Thus $u(x,t)$ as defined by (32) is a solution to the shifted wave equation. The initial conditions in (31) are verified in the same way as in Theorem 5.3. Again $\mathcal{E}_t$ and $\mathcal{E}_t'$ are invariant under the inversion $k \longrightarrow k^{-1}$ so

$$u(k,t) = \int_K f_o(kh)\,d\mathcal{E}_t'(h) + \int_K f_1(kh)\,d\mathcal{E}_t(h).$$

If $\dim K$ is odd, $\mathcal{E}_t'$ and $\mathcal{E}_t$ are supported in $S_t(e)$ $(t < R)$. Thus if k is such that $kS_t(e) \subset B_R(e)$, $u(k,t)$ is determined by the restrictions of f_0 and f_1 to an arbitrarily thin shell around $S_t(e)$. With these restrictions on t and k, the Huygens' principle (6) is verified and Theorem 5.4 proved.

3. Huygens' Principle and Cartan Subgroups.

In this subsection we shall study the solution of the shifted wave equation

$$(39) \qquad \frac{\partial^2 u}{\partial t^2} = \left(L_X + |\rho|^2\right)u, \qquad u(x,0) = f_0(x), \quad u_t(x,0) = f_1(x)$$

for arbitrary symmetric spaces $X = G/K$ of the noncompact type, particularly as regards Huygens' principle. While Theorem 5.3 resulted from the special structure of L_X for G complex here we shall use the Fourier transform on X. We shall temporarily assume $f_0, f_1 \in \mathcal{D}(X)$.

Since the function

$$e_{\lambda,b}(x) = e^{(i\lambda+\rho)(A(x,b))} \qquad x \in X$$

satisfies

$$L_X e_{\lambda,b} = -\left(\langle\lambda,\lambda\rangle + \langle\rho,\rho\rangle\right) e_{\lambda,b}$$

we see that if $f \in \mathcal{D}(X)$,

$$(L_X f)^{\sim}(\lambda, b) = -\left(\langle\lambda,\lambda\rangle + \langle\rho,\rho\rangle\right) \widetilde{f}(\lambda, b).$$

Thus taking formally the Fourier transform of (39) we would get, assuming the Fourier transform

$$\widetilde{u}(\lambda, b, t) = \int_X u(x,t) e^{(-i\lambda+\rho)(A(x,b))}\, dx \tag{40}$$

defined for u as well as for the derivatives of u,

$$\widetilde{u}_{tt}(\lambda, b, t) + \langle\lambda,\lambda\rangle \widetilde{u}(\lambda, b, t) = 0. \tag{41}$$

For $\lambda \in \mathfrak{a}^*$ this would give

$$\widetilde{u}(\lambda, b, t) = \widetilde{f}_0(\lambda, b) \cos|\lambda|t + \widetilde{f}_1(\lambda, b) \frac{\sin|\lambda|t}{|\lambda|}. \tag{42}$$

Lemma 5.6. *The function $\lambda \longrightarrow |\lambda|^{-1}\sin(|\lambda|t)$ on $\boldsymbol{R}^n$ is the Fourier transform*

$$\frac{\sin(|\lambda|t)}{|\lambda|} = \int_{\boldsymbol{R}^n} e^{-i(x,\lambda)}\, dT_t(x), \tag{43}$$

where $T_t \in \mathcal{E}'(\boldsymbol{R}^n)$ has support in the closed ball $|x| \leq t$. If n is odd and > 1 then $\operatorname{supp}(T_t) \subset S_t(0)$.

Proof. While this could be proved directly (see Remark below) the lemma is here essentially contained in Prop. 5.2. In fact, take $f_0 \equiv 0$ in (18) and take Fourier transform in the x variable. Then

$$\widetilde{u}(\lambda, t) = \widetilde{f}_1(\lambda) \frac{\sin|\lambda|t}{|\lambda|}$$

so comparing with (19) we get $E_t = T_t$ whence the lemma.

Remark. The Fourier transform of the distribution $\varphi \longrightarrow (M^t\varphi)(0)$ is expressible in terms of the Bessel function $J_{(n-2)/2}(t|\lambda|)$ ([GGA], Introduction, Lemma 3.6). If n is odd, say $n = 2m+3$ then

$$\left(\frac{d}{zdz}\right)^m \left[z^{m+\frac{1}{2}} J_{m+\frac{1}{2}}(z)\right] = z^{\frac{1}{2}} J_{\frac{1}{2}}(z) = \left(\frac{2}{\pi}\right)^{\frac{1}{2}} \sin z.$$

This readily implies (43) for n odd and > 1, even with T_t supported in the *sphere* $S_t(0)$. To prove (43) for n replaced by $n-1$ we use the "method of descent" and simply restrict λ to the subspace $\boldsymbol{R}^{n-1} \times 0$. If T_t^* denotes the distribution

$$f \longrightarrow \int_{\boldsymbol{R}^n} f(x_1, \ldots, x_{n-1}) dT_t(x), \qquad f \in \mathcal{E}(\boldsymbol{R}^{n-1}),$$

then $T_t^*(f) = 0$ for supp(f) disjoint from the *ball* $|x| \leq t$ in $\boldsymbol{R}^{n-1}$. But then if $\mu \in \boldsymbol{R}^{n-1}$,

$$\frac{\sin|\mu|t}{|\mu|} = \int_{\boldsymbol{R}^{n-1}} e^{-i(x,\mu)} dT_t^*(x), \tag{44}$$

proving (43) in the even dimensional case too.

Because of Lemma 5.6 we deduce from Euclidean Fourier transform theory that the function

$$\lambda \longrightarrow |\lambda|^{-1} \sin(|\lambda| t)$$

on $\mathfrak{a}^*$ extends to a holomorphic function on $\mathfrak{a}_C^*$ which is of exponential type and slow growth in the sense of (40) Ch. III, §5. As a result of Cor. 5.9 in Ch. III there exists a distribution $\mathcal{E}_t \in \mathcal{E}'(X)$ with support in $\overline{B_t(o)}$ such that for $\lambda \in \mathfrak{a}_C^*, b \in B$,

$$\frac{\sin(|\lambda|t)}{|\lambda|} = \int_X e^{(-i\lambda+\rho)(A(x,b))} d\mathcal{E}_t(x). \tag{45}$$

While we arived at the "propagator" $\mathcal{E}_t$ by assuming f_o and f_1 of compact support we can now solve (39) without this assumption.

Theorem 5.7. *With $f_0, f_1 \in \mathcal{E}(X)$ the function*

$$u(x,t) = (f_0 \times \mathcal{E}_t')(x) + (f_1 \times \mathcal{E}_t)(x) \tag{46}$$

is a solution of the shifted wave equation (39).

If $\dim X$ *is odd and if* G *has all its Cartan subgroups conjugate, then Huygens' principle holds.*

Proof. We have

$$(L_X u)(x,t) = (f_o \times L_X \mathcal{E}_t')(x) + (f_1 \times L_X \mathcal{E}_t)(x)$$

and by (45), $\mathcal{E}_0 = 0$, $\mathcal{E}_0' = \delta$. For (39) it thus remains to prove

$$(L_X + |\rho|^2)\mathcal{E}_t = \frac{\partial^2}{\partial t^2}\mathcal{E}_t.$$

Both sides give the same result when applied to the function $e_{\lambda,b}$ (Prop. 3.14, Ch. II) so they must coincide.

For the last statement of the theorem we first rewrite (43) for A :

$$\frac{\sin|\lambda|t}{|\lambda|} = \int_A e^{-i\lambda(\log a)} dT_t(a), \tag{47}$$

where $\operatorname{supp}(T_t) \subset \overline{B_t(e)}$ and $\operatorname{supp}(T_t) \subset S_t(e)$ if $\dim A$ is odd and > 1.

Lemma 5.8. *In terms of the product decomposition* $\Xi = \big(K/M\big) \times A$ *we have for the Radon transform* $\widehat{\mathcal{E}}_t$,

$$\widehat{\mathcal{E}}_t = 1 \otimes e^{\rho} T_t.$$

Proof. Because of the uniqueness in (45), $\mathcal{E}_t$ is invariant under K. Integrating (45) over b we get

$$\frac{\sin|\lambda|t}{|\lambda|} = \int_X \varphi_{-\lambda}(x) d\mathcal{E}_t(x), \tag{48}$$

where $\varphi_{-\lambda}$ is the spherical function. If $g = kan$ relative to the decomposition $G = KAN$ we put as usual, $a = \exp H(g)$. Let $\alpha \in \mathcal{E}(K/M)$, $\beta(a) = e^{-(i\lambda+\rho)(\log a)}$ and $\alpha^\natural$ the integral of α over K/M. Then by the definition of $\widehat{\mathcal{E}}_t$ (Ch. I, §3, (12)) and the K-invariance we have with $(\alpha \otimes \beta)(kM, a) = \alpha(kM)\beta(a)$,

$$\widehat{\mathcal{E}}_t(\alpha \otimes \beta) = \alpha^\natural \widehat{\mathcal{E}}_t(1 \otimes \beta) = \alpha^\natural \mathcal{E}_t\big((1 \otimes \beta)^\vee\big),$$

$$(1 \otimes \beta)^\vee(g \cdot o) = \int_K (1 \otimes \beta)(gk \cdot \xi_o) dk = \int_K \beta(\exp H(gk)) dk = \varphi_{-\lambda}(g)$$

so by (47)–(48),

$$\widehat{\mathcal{E}}_t(\alpha \otimes \beta) = \alpha^\natural \mathcal{E}_t(\varphi_{-\lambda}) = \alpha^\natural T_t(e^{-i\lambda}) = (1 \otimes e^\rho T_t)(\alpha \otimes \beta).$$

Since the functions $\alpha \otimes \beta$ span a dense subspace of $\mathcal{E}(\Xi)$ the lemma follows.

Let as in Ch. IV, §1,

$$\beta_R = \{\xi \in \Xi : \ d(o,\xi) < R\}\,, \quad \sigma_R = \{\xi \in \Xi : \ d(o,\xi) = R\}\,.$$

Lemma 5.9. *Suppose G has all its Cartan subgroups conjugate. Then if $T \in \mathcal{E}'(X)$,*

$$\operatorname{supp}(\widehat{T}) \subset \sigma_R \Longrightarrow \operatorname{supp}(T) \subset S_R(o).$$

Proof. From Ch. II, Theorem 3.13 we have the inversion formula for the Radon transform which we write in the form

$$f = (L\widehat{f})^\vee \tag{49}$$

and under our assumption, L is a differential operator on Ξ.

Let $\epsilon > 0$ and suppose f satisfies $\operatorname{supp}(f) \subset B_{R-\epsilon}(o)$. Then $\operatorname{supp}(\widehat{f}) \subset \overline{\beta}_{R-\epsilon}$ and since L in (49) is a differential operator, $\operatorname{supp}(L\widehat{f}) \subset \overline{\beta}_{R-\epsilon}$. Hence

$$T(f) = T\left((L\widehat{f})^\vee\right) = \widehat{T}(L\widehat{f}) = 0$$

so $\operatorname{supp}(T) \cap B_{R-\epsilon}(o) = \emptyset$. Since ϵ is arbitrary, $\operatorname{supp}(T) \cap B_R(o) = \emptyset$. On the other hand, by Ch. IV, Cor. 1.2, $\operatorname{supp}(T) \subset \overline{B_R(o)}$ so $\operatorname{supp}(T) \subset S_R(o)$ as desired.

The assumption on G implies that all restricted root subspaces are even-dimensional so since $\dim X$ is odd, $\dim A$ is odd. Assuming also $\dim A > 1$, T_t in (47) has support in $S_t(e)$ so by Lemma 5.8–5.9, $\operatorname{supp}(\mathcal{E}_t) \subset S_t(o)$. Since $\varphi_{-\lambda}(g \cdot o) = \varphi_\lambda(g^{-1} \cdot o)$ formula (48) implies that the lift of $\mathcal{E}_t$ to G is invariant under $g \longrightarrow g^{-1}$. Thus the solution formula (46) can be written

$$u(g \cdot o, t) = \int_X f_o(g \cdot x)\, d\mathcal{E}_t'(x) + \int_X f_1(g \cdot x)\, d\mathcal{E}_t(x)$$

and now Huygens' principle is obvious.

Finally, if X is of odd dimension and of rank one, X is an odd-dimensional hyperbolic space. In this case we have for the Euclidean Fourier transform

$$2\frac{\sin|\lambda|t}{|\lambda|} = \left(\Phi_{B_{t(e)}}\right)^{\sim}(\lambda),\ 2\cos|\lambda|t = \left(\Phi_{S_{t(e)}}\right)^{\sim}(\lambda),$$

Φ_E denoting the characteristic function of a subset $E \subset A$, so the argument above gives Huygens' principle for (39) in the case when $f_1 \equiv 0$. However, it holds also in the general case because of the solution formula

$$u(x,t) = c_n\left[\frac{\partial}{\partial t}(J_t f_o)(x) + (J_t f_1)(x)\right] \tag{50}$$

with c_n as in (3) and

$$(J_t f)(x) = \left(\frac{\partial}{\partial(2\operatorname{ch} t)}\right)^{\frac{n-3}{2}} \left(\operatorname{sh}^{n-2} t(M^t f)(x)\right). \tag{51}$$

Here $\boldsymbol{H}^n$ is taken with metric g with sectional curvature -1 (cf. [GGA], pp. 343, 577), and the shifted wave equation considered is

$$\frac{\partial^2 v}{\partial s^2} = \left(L + \left(\tfrac{n-1}{2}\right)^2\right)v, \qquad v(x,0) = v_o(x),\ v_s(x,0) = v_1(x). \tag{52}$$

For completeness this should be related to (39). We know from [DS], p. 566 that $g = B/2(n-1)$, where B is the Killing form. From [GGA], p. 582, we have for the single restricted root α, $|\alpha|^2 = \frac{1}{2}m_\alpha^{-1} = (2(n-1))^{-1}$ so

$$|\rho|^2 = |\tfrac{1}{2}(n-1)\alpha|^2 = \tfrac{1}{8}(n-1).$$

Thus

$$L_X + |\rho|^2 = \frac{1}{2(n-1)}L + \frac{1}{8}(n-1) = \frac{1}{2(n-1)}\left(L + \left(\tfrac{n-1}{2}\right)^2\right).$$

Thus the solutions to (39) and (52) correspond with

$$v(x,s) = u\left(x, (2(n-1))^{\frac{1}{2}}s\right).$$

4. Orbital Integrals and Huygens' Principle.

Let X be an n-dimensional Lorentzian manifold, that is a pseudo-Riemannian manifold whose pseudo-Riemannian structure has signature

$(1, n-1)$. We assume X has constant sectional curvature $\varkappa = 0, -1$ or $+1$ and therefore locally isometric to the respective spaces

$$\boldsymbol{O}_o(1,n-1)\boldsymbol{R}^n/\boldsymbol{O}_o(1,n-1), \qquad \boldsymbol{O}_o(1,n)/\boldsymbol{O}_o(1,n-1), \qquad \boldsymbol{O}_o(2,n-1)/\boldsymbol{O}_o(1,n-1), \tag{53}$$

the subscript o denoting identity component (cf. [GGA], Ch. I, §6).

Let $X = G/H$ be one of the spaces (53). For $y \in X$ let $\boldsymbol{C}_y \subset X$ be the *light cone* with vertex y and $\boldsymbol{D}_y \subset \boldsymbol{C}_y$ the *retrograde cone* with vertex y as defined in [GGA], I, §6. On the figure we consider the case $\boldsymbol{O}_o(1,2)/\boldsymbol{O}_o(1,1)$ which is identified with the hyperboloid of one sheet in $\boldsymbol{R}^3$.

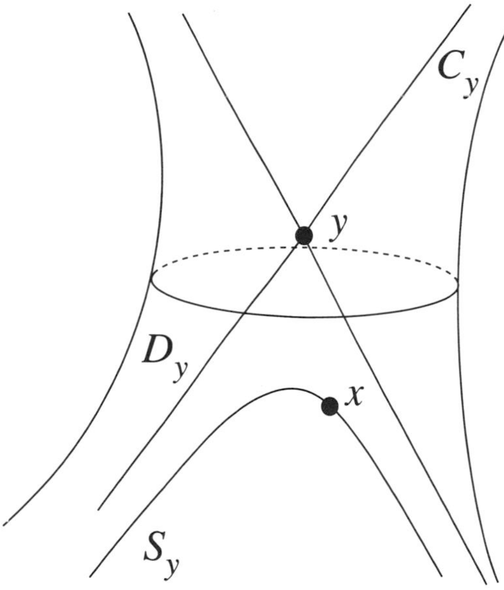

The group $G = \boldsymbol{O}_o(1,2)$ acts transitively on the hyperboloid. Here the light cone $\boldsymbol{C}_y$ consists of two lines through y and $\boldsymbol{D}_y$ of two half lines.

If $n > 2$ the isotropy subgroup of G at y acts transitively on the connected set $\boldsymbol{D}_y - \{y\}$. Let o denote the origin eH in X, $Y_1, \cdots, Y_n$ a basis of the tangent space $(G/H)_o$ in which the Lorentzian structure g_o has the form

$$g_o(Y,Y) = y_1^2 - \cdots - y_n^2 \quad \text{if} \quad Y = \Sigma y_i Y_i. \tag{54}$$

Then if $\mathrm{Exp} : (G/H)_o \longrightarrow G/H$ is the usual exponential mapping the retrograde cone $\boldsymbol{D}_o$ is given by $\boldsymbol{D}_o = \mathrm{Exp}(D_o)$ where D_o is the retrograde cone in $(G/H)_o$:

$$y_1^2 - y_2^2 - \cdots - y_n^2 = 0, \qquad y_1 \leq 0.$$

The component of the hyperboloid $g_o(Y, Y) = r^2$ which lies in D_o is denoted $S_r(o)$; if $y \in X$ and if $g \in G$ is chosen such that $g \cdot o = y$ we put $S_r(y) = g \cdot S_r(o)$. This is a valid definition because $h \cdot D_o \subset D_o$ for $h \in H$ by the connectedness of H. Finally, we put

$$\boldsymbol{S}_r(y) = \text{Exp } S_r(y) \tag{55}$$

for $r > 0$; for the last case in (53) we limit r to the interval (o, π) in order for Exp to be injective (see [GGA], I, §6).

Let dh denote the bi-invariant measure on the unimodular group H. Let $y \in X, r > 0$ and select $g \in G$ such that $g \cdot o = y$ and let $x \in \boldsymbol{S}_r(o)$. We can then define the operator M^r (*the orbital integral*) by

$$(M^r u)(y) = \int_H u(gh \cdot x)dh, \qquad u \in C_c(X). \tag{56}$$

As shown *loc. cit.*, Lemma 6.11, the measure dh can be normalized such that

$$(M^r u)(y) = \frac{1}{A(r)} \int_{\boldsymbol{S}_r(y)} u(z)d\boldsymbol{w}_r(z) \tag{57}$$

where $d\boldsymbol{w}_r$ is the Riemannian measure on $\boldsymbol{S}_r(y)$ and

$$A(r) = r^{n-1}, \qquad (\sinh r)^{n-1}, \qquad (\sin r)^{n-1} \tag{58}$$

in the three cases in (53). We then have the following result (*loc. cit.* §6).

Theorem 5.10. *Let X be one of the Lorentzian manifolds (53) of constant curvature $\varkappa = 0, -1, +1$ and dimension $n > 2$.*

(i) *If $v \in C_c(X)$ then*

$$a = \lim_{r \to o} r^{n-2}(M^r v)(y) \text{ exists and is } \not\equiv 0. \tag{59}$$

(ii) $M^r(Lu) = L_r(M^r u) \qquad u \in \mathcal{D}(X)$

where $L_r = d^2/dr^2 + A'(r)/A(r)d/dr$ is the radial part of the Laplacian L.

(iii) *Suppose n is even, $n = 2m$, and put*

$$Q(L) = (L - \varkappa(n-3)2)(L - \varkappa(n-5)4)\cdots(L - \varkappa(1)(n-2)).$$

Then if $u \in \mathcal{D}(X)$,

$$u(y) = c \lim_{r \to o} r^{n-2} Q(L_r)((M^r u)(y)) \qquad (n > 2) \tag{60}$$

where the constant c is given by

$$c^{-1} = (4\pi)^{m-1}(m-2)!.$$

Also if f is such that $f \circ \mathrm{Exp}$ has support in a ball $B_A : \Sigma y_i^2 < A^2 < 1$ in X_o then (*loc. cit.* §6, No. 4) we have

$$|r^{n-2}(M^r f)(o)| \leq CA \sup |f|, \tag{61}$$

where C is a constant independent of r. If φ is the "smoothed-out" characteristic function of B_A we write with $u \in \mathcal{D}(X)$, $u_1 = u\varphi$, $u_2 = u(1-\varphi)$

$$r^{n-2}(M^r u)(o) = r^{n-2}(M^r u_1)(o) + r^{n-2}(M^r u_2)(o). \tag{62}$$

Take $y = o$ in (59) and $v \in C_c(X - \{o\})$. The limit (59) is a positive, linear and H-invariant functional on $C_c(X - \{o\})$ with support in the H-orbit $\boldsymbol{D}_o - \{o\}$; hence it is "the" H-invariant measure on $\boldsymbol{D}_o - \{o\}$ and we denote it by μ. Then we have

$$|r^{n-2}(M^r u)(o) - \int_{\boldsymbol{D}_o} u(z)\,d\mu(z)|$$

$$\leq |r^{n-2}(M^r u_1)(o) - \int_{\boldsymbol{D}_o} u_1(z)\,d\mu(z)| + |r^{n-2}(M^r u_2)(o) - \int_{\boldsymbol{D}_o} u_2(z)\,d\mu(z)|.$$

By (61) the first term on the right is $O(A)$ uniformly in r and the second tends to 0 as $r \to 0$. Since A is arbitrary we conclude

$$\lim_{r \to o} r^{n-2}(M^r u)(o) = \int_{\boldsymbol{D}_o} u(z)\,d\mu(z). \tag{63}$$

Corollary 5.11. *Let $n > 2$ be even and δ the delta distribution at o. Then*

$$\delta = c\,Q(L)\mu \tag{64}$$

In fact, using (ii) and (63) equation (60) becomes

$$u(o) = c\lim_{r\to o} r^{n-2}(M^r(Q(L)u))(o) = c\int_{D_o}(Q(L)u)(z)d\mu(z),$$

which is (64).

Note that by (57), formula (63) can also be written

$$\lim_{r\to o}\frac{1}{r}\boldsymbol{w}_r = \mu. \tag{63'}$$

Remark. Formula (64) shows that each factor

$$L_m = L - \varkappa(n-m)(m-1) \qquad (m = 3, 5, \cdots, n-1)$$

in $Q(L)$ has a fundamental solution supported on the retrograde cone $\boldsymbol{D}_o$. It is known that this is equivalent to Huygens' principle for the Cauchy problem for the equation $L_m u = 0$ with initial data given on some spacelike hypersurface (see Günther [1991] and [1988], Ch. IV, Cor. 1.13).

For the operator $L-\frac{\varkappa}{4}n(n-2)$ which is L_m for $m = 1+2\left[\frac{n}{4}\right]$ this is *a priori* clear by the conformal invariance of Huygens' principle (Ørsted [1981]). Our space X is in fact conformally equivalent to $\boldsymbol{R}^n$ so by the classical transformation formula for the Laplacian under conformal diffeomorphisms (see e.g. [GGA], p. 332) the null spaces of L_m and $L_{\boldsymbol{R}^n}$ correspond. Huygens' principle for $L_{\boldsymbol{R}^n}$ explained in No. 1 of this section thus implies the same for L_m on X.

5. Energy Equipartition.

Let X be an oriented manifold with Riemannian structure $\langle,\rangle$ and let L_X denote the Laplace-Beltrami operator on X. For a vector field Y on X let ω_Y be the 1-form given by $\omega_Y(Z) = \langle Y, Z\rangle$ and extend $\langle,\rangle$ to 1-forms by $\langle\omega_Y, \omega_Z\rangle = \langle Y, Z\rangle$. The star operator $*$ on the Grassmann algebra $\mathfrak{A}(X)$ maps p-forms to $(n-p)$-forms (if $n = \dim X$), it operates pointwise, i.e. $*(f\omega) = f(*\omega)$ if $f \in \mathcal{E}(X)$ and it has the property that $*(\omega_{i_1} \wedge \cdots \wedge \omega_{i_p}) = \pm\omega_{j_i} \wedge \cdots \wedge \omega_{j_{n-p}}$ if (ω_i) is an orthonormal basis of 1-forms and $\{i_1, \ldots, i_p, j_1, \ldots j_{n-p}\}$ is a permutation of $\{1, \ldots, n\}$, the sign being $+$ or $-$ depending on whether the permutation is even or odd (cf. [DS], p. 142 and [GGA], p. 330). The operator δ is the linear operator on $\mathfrak{A}(X)$ given by

$$\delta\omega = (-1)^{np+n+1} * d * \omega, \qquad \omega \in \mathfrak{A}_p(X),$$

d denoting as usual the exterior derivative. Then we know ([GGA], Ch. II, Ex. A.1) that for $f \in \mathcal{E}(X)$, Y a vector field

$$(65) \qquad \delta(\omega_Y) = -\operatorname{div} Y, \qquad \omega_{\operatorname{grad} f} = df$$

and on $\mathcal{E}(X)$, $L_X = -\delta d$. The familiar formula

$$\operatorname{div}(fY) = f \operatorname{div} Y + Yf$$

thus becomes

$$(66) \qquad \delta(f\omega) = f\delta\omega - \langle df, \omega\rangle, \qquad \omega \text{ 1-form}, \ f \in \mathcal{E}(X).$$

Consider now the shifted wave equation (7) with real valued initial data f_0 and f_1. From general theory it is known that if f_0 and f_1 have compact support then so does the function $x \longrightarrow u(x,t)$. More precisely, if $\operatorname{supp}(f_i) \subset B_R(x_o)$ then for each $t_o > 0$ the solution $x \longrightarrow u(x, t_o)$ has support in $B_{R+t_o}(x_o)$. Thus if $f_0, f_1 \in \mathcal{D}(X)$ we can consider the *energy*

$$(67) \qquad (\mathcal{E}_c u)(t) = \int_X (E_c u)(x,t)\,dx,$$

where

$$(68) \qquad (E_c u)(x,t) = \frac{1}{2}\left(u_t^2 + \langle du, du\rangle - cu^2\right).$$

Lemma 5.12. *(Conservation of energy).*

$$\big(\mathcal{E}_c u\big)(t) \text{ is independent of } t,$$

if u is a solution to (7).

Proof. From (68) we get

$$\begin{aligned}
\big(E_c u\big)_t &= u_t u_{tt} + \langle du, du_t\rangle - cuu_t \\
&= u_t\big(u_{tt} - cu\big) + \langle du, du_t\rangle = u_t\big(-\delta du\big) + \langle du, du_t\rangle \\
&= -\delta\big(u_t du\big)
\end{aligned}$$

by (66). Since by (65) this last expression is a divergence $\operatorname{div} Z$ and since $E_c u$ has compact support in the x-variable we have

$$\partial_t \int_X (E_c u)(x,t)\,dx = \int_X \big(E_c u\big)_t(x,t)\,dx = \int_X \operatorname{div} Z = 0,$$

proving the lemma.

We now specialize X to the spaces considered in Theorem 5.7 and take $c = |\rho|^2$ so our shifted wave equation is

$$\frac{\partial^2 u}{\partial t^2} = (L_X + |\rho|^2)u, \qquad u(x,0) = f_0(x), \quad u_t(x,0) = f_1(x) \tag{69}$$

with f_0, f_1, u real-valued and $f_0, f_1 \in \mathcal{D}(X)$. We split the energy $\mathcal{E} = \mathcal{E}_{|\rho|^2}$ in two parts, the *kinetic energy* and the *potential energy,*

$$(\mathcal{K}u)(t) = \tfrac{1}{2} \int\limits_X u_t^2 \, dx, \tag{70}$$

$$\mathcal{P}u(t) = \tfrac{1}{2} \int\limits_X (\langle du, du\rangle - |\rho|^2 u^2) \, dx. \tag{71}$$

Since

$$\langle du, du\rangle = u\,\delta du - \delta(u\,du)$$

and since by (65), $\int\limits_X \delta(u\,du)dx = 0$ (71) can also be written

$$(\mathcal{P}u)(t) = -\frac{1}{2} \int\limits_X u(L_X + |\rho|^2)u \, dx. \tag{72}$$

We can now state the energy equipartition theorem.

Theorem 5.13. *Let X be a symmetric space G/K of the noncompact type. We assume $\dim X$ odd and that all Cartan subgroups of G are conjugate. Then if $\operatorname{supp} f_0$, $\operatorname{supp} f_1 \subset B_R(o)$ we have*

$$\mathcal{K}u(t) = \mathcal{P}u(t) \qquad \textit{for } |t| \geq R.$$

Proof. Again we use the Fourier transform on X which, by (42), gives for the solution,

$$\widetilde{u}(\lambda, b, t) = \Phi(\lambda, b) \cos pt + \Psi(\lambda, b) p^{-1} \sin pt, \tag{73}$$

where $\Phi = \widetilde{f_0}$, $\Psi = \widetilde{f_1}$, $p = |\lambda|$. Hence

$$\widetilde{u}_t(\lambda, b, t) = -p\,\Phi(\lambda, b) \sin pt + \Psi(\lambda, b) \cos pt$$

so by the Plancherel formula,

$$2w(\mathcal{K}u)(t) = \int\limits_{\mathfrak{a}^*\times B} \cos^2 pt|\Psi(\lambda,b)|^2|\mathbf{c}(\lambda)|^{-2}d\lambda\, db$$
$$+ \int\limits_{\mathfrak{a}^*\times B} \sin^2 pt\; p^2|\Phi(\lambda,b)|^2|\mathbf{c}(\lambda)|^{2}d\lambda\, db$$
$$- \int\limits_{\mathfrak{a}^*\times B} \sin pt \cos pt\; p\big(\Phi\overline{\Psi}+\Psi\overline{\Phi}\big)(\lambda,b)|\mathbf{c}(\lambda)|^{-2}d\lambda\, db.$$

Hence, by the double-angle identities,

$$\begin{aligned} 2w(\mathcal{K}u)(t) &= \tfrac{1}{2}\|\Psi\|^2 + \tfrac{1}{2}\|p\Phi\|^2 \\ &+ \tfrac{1}{2}\int\limits_{\mathfrak{a}^*\times B} \cos 2pt\big(|\Psi|^2 - |p\Phi|^2\big)|\mathbf{c}(\lambda)|^{-2}d\lambda\, db \\ &- \tfrac{1}{2}\int\limits_{\mathfrak{a}^*\times B} \sin 2pt\; p\big(\Phi\overline{\Psi}+\Psi\overline{\Phi}\big)|\mathbf{c}(\lambda)|^{-2}d\lambda\, db, \end{aligned} \tag{74}$$

the norms being taken in $L^2\big(\mathfrak{a}^*\times B, |\mathbf{c}(\lambda)|^{-2}db\, d\lambda\big)$. On the other hand, by (70), (72),

$$2\big(\mathcal{E}u\big)(t) = \int\limits_X u_t^2 dx - \int\limits_X u\big(L_X + |\rho|^2\big)u\, dx,$$

which by Lemma 5.12 and the Plancherel formula equals

$$2\big(\mathcal{E}u\big)(0) = \big(\|\Psi\|^2 + \|p\Phi\|^2\big)w^{-1}.$$

Since $\mathcal{E}u(t) = (\mathcal{K}u)(t) + (\mathcal{P}u)(t)$ it will suffice to prove that

$$\mathcal{I}(t) = 0 \qquad \text{for } |t| > R,$$

where $\mathcal{I}(t)$ is the sum of the two last terms in (74). Actually each of the terms will be shown to vanish for $|t| > R$. Now we have by Ch. III, (25) in §5,

$$\begin{aligned} \Phi(\lambda, kM) &= \int\limits_X f_0(x)e^{(\rho - i\lambda)(A(x,kM))}dx \\ &= \int\limits_A \widehat{f_0}\big(ka\cdot\xi_o\big)e^{(\rho-i\lambda)(\log a)}da, \quad \lambda\in\mathfrak{a}^*. \end{aligned}$$

Here we put $q = \lambda/p$ for the angular spherical coordinate in $\mathfrak{a}^*$ and write

$$\Phi(p;q,kM) = \Phi(\lambda,kM) = \int_X f_0(x)e^{(\rho-ipq)(A(x,kM))}dx$$
$$= \int_A \widehat{f_0}(ka\cdot\xi_o)e^{(\rho-ipq)(\log a)}da.$$

For (q,kM) fixed this defines an extension of $p \longrightarrow \Phi(p;q,kM)$ from $\boldsymbol{R}^+$ to a holomorphic function $\Phi^*(p,q,kM)$ of a single complex variable $p=\sigma+i\tau$. The last formula shows, since $\widehat{f_0}(ka\cdot\xi_o)=0$ for $|\log a|>R$ and $|q|\le 1$, that $p \longrightarrow \Phi(p;q,kM)$ is an entire function on $\boldsymbol{C}$ of exponential type R, uniformly in (q,kM). The complex conjugate $\overline{\Phi}(\bullet,q,kM)$ has a similar extension to an entire function $p \longrightarrow \overline{\Phi}^*(p;q,kM)$ of exponential type R. Since f_0 is real-valued we actually have

$$\overline{\Phi}^*(p;q,b) = \Phi^*(p;-q,b).$$

Similar construction is carried out for Ψ. Recall also ([GGA], Ch. IV, Cor. 6.15) that under our assumption on G, $\boldsymbol{c}(\lambda)^{-1}$ is a polynomial $\gamma(\lambda)$ in $i\lambda$ with real coefficients. We put $\gamma(p;q)=\gamma(pq)$ for $p\in\boldsymbol{R}^+$, $|q|=1$. Now

$$|\boldsymbol{c}(\lambda)|^{-2} = \boldsymbol{c}(\lambda)^{-1}\boldsymbol{c}(-\lambda)^{-1}, \qquad \lambda\in\mathfrak{a}^*$$

so for a fixed $q\in\mathfrak{a}^*$, $|q|=1$, $(|\boldsymbol{c}(\lambda)|^{-2})|\mathfrak{a}^*$ has the holomorphic extension $\gamma(p;q)\gamma(p;-q)$ to $p\in\boldsymbol{C}$. We now switch to polar coordiantes (p,q) in the integrals in (74) so $d\lambda = p^{\ell-1}dp\,dq$ where dq is the (suitably normalized) measure on the unit sphere S_1 in $\mathfrak{a}^*$. Then

$$\mathcal{I}(t) = \int_o^\infty \Big(p^{\ell-1}\cos 2pt \int_{S_1\times B} Y_o(p,q,b)db\,dq\Big)dp$$
$$+ \int_o^\infty \Big(p^{\ell-1}\sin 2pt \int_{S_1\times B} Y_1(p,q,b)db\,dq\Big)dp,$$

where

$$2Y_o(p;q,b) =$$
$$\big[\Psi^*(p;q,b)\Psi^*(p,-q,b) - p^2\Phi^*(p;q,b)\,\Phi^*(p;-q,b)\big]\,\gamma(p;q)\gamma(p;-q)$$
$$2Y_1(p;q,b) =$$
$$-\,p\,[\Phi^*(p;q,b)\Psi^*(p;-q,b)+\Phi^*(p;-q,b)\Psi^*(p;q,b)]\,\gamma(p;q)\gamma(p;-q).$$

These functions Y_0, Y_1 are for fixed (q, b) entire functions of p of exponential type $2R$. By the uniformity mentioned, the same is the case for the functions

$$Z_j(p) = p^{\ell-1} \int_{S_1 \times B} Y_j(p; q, b) db\, dq, \qquad (j = 0, 1)$$

and the expression for $\mathcal{I}(t)$ becomes

$$\mathcal{I}(t) = \int_0^\infty Z_o(p) \cos 2pt\, dp + \int_o^\infty Z_1(p) \sin 2pt dp.$$

On the other hand, $\Phi^*(-p; q, b) = \Phi^*(p; -q, b)$ and similarly for Ψ^*. The assumptions on X and G imply that $\dim A$ is odd so

$$Z_o(-p) = Z_0(p), \qquad Z_1(-p) = -Z_1(p),$$

whence

$$\mathcal{I}(t) = \frac{1}{2} \int_{-\infty}^{\infty} (Z_0(p) + iZ_1(p)) e^{-2ipt} dp.$$

Here Z_0 and Z_1 are entire functions of exponential type $2R$ so by the classical Paley-Wiener theorem $\mathcal{I}(t) = 0$ for $|2t| > 2R$. This proves Theorem 5.13.

§6. EIGENFUNCTIONS AND HYPERFUNCTIONS

1. Arbitrary Eigenfunctions.

As in the beginning of this chapter let $X = G/K$ be a symmetric space of the noncompact type and $\boldsymbol{D}(X)$ the algebra of G-invariant differential operators on X. As established in Ch. II, Prop. 2.6, the joint eigenfunctions f of $\boldsymbol{D}(X)$ are characterized by the functional equation

$$\text{(1)} \qquad \int_K f(gk \cdot x) dk = \varphi_\lambda(x) f(g \cdot o),$$

λ being some element of $\mathfrak{a}_{\mathbf{c}}^*$. By the Fourier inversion formula (Ch. III, Theorem 1.3) it is natural to view the exponentials $e_{\lambda,b} : x \longrightarrow e^{(i\lambda+\rho)(A(x,b))}$ as building blocks for "arbitrary" functions. This is further borne out by the following result from Ch. III, §6.

Theorem 6.1. *The K-finite joint eigenfunctions of $\mathbf{D}(X)$ are the functions*

(2) $$f(x) = \int_B e^{(i\lambda+\rho)(A(x,b))} F(b)\,db,$$

where F is a K-finite continuous function on B and $\lambda \in \mathfrak{a}_{\mathbf{c}}^*$.

Given $\delta \in \widehat{K}$ let $\check{\delta}$ denote its contravariant and let V_δ be a representation space of δ. Put $d(\delta) = \dim V_\delta$. As usual, let $\mathcal{E}_\delta(X)$ (resp. $C_\delta(B)$) denote the space of functions $f \in \mathcal{E}(X)$ (resp. $F \in C(B)$) which are K-finite of type δ. Under the map $f \longrightarrow f^\delta$ where

(3) $$f^\delta(x) = d(\delta) \int_K f(k^{-1} \cdot x)\delta(k)\,dk$$

the space $\mathcal{E}_{\check{\delta}}(X)$ corresponds bijectively to the space

$$\mathcal{E}^\delta(X) = \{\varphi \in \mathcal{E}(X, \operatorname{Hom}(V_\delta, V_\delta) : \ \varphi(k \cdot x) \equiv \delta(k)\varphi(x)\}$$

([GGA], Prop. 1.11, Ch. IV). Thus $\mathcal{E}_{\check{\delta}}(X) = 0$ unless $\delta(M)$ has a nonzero fixed vector, in other words, $\delta \in \widehat{K}_M$ (equivalently, $\check{\delta} \in \widehat{K}_M$). Also $C_\delta(B) = 0$ unless $\delta \in \widehat{K}_M$. We put

(4) $$\mathcal{E}_\infty(X) = \bigoplus_{\delta \in \widehat{K}_M} \mathcal{E}_\delta(X), \qquad C_\infty(B) = \bigoplus_{\delta \in \widehat{K}_M} C_\delta(B).$$

As usual, let $\mathcal{E}_\lambda(X)$ denote the joint eigenspace of $\mathbf{D}(X)$ containing φ_λ and put

(5) $$\mathcal{E}_{\lambda,\delta}(X) = \mathcal{E}_\lambda(X) \cap \mathcal{E}_\delta(X), \qquad \lambda \in \mathfrak{a}_{\mathbf{c}}^*, \qquad \delta \in \widehat{K}_M.$$

By a *functional* on B we shall mean a linear form T on a subspace of $C(B)$ containing $C_\infty(B)$. It is convenient to write $\int_B F(b)\,dT(b)$ instead of $T(F)$ for a function F on B. The following result is a simple consequence of Theorem 6.1.

Theorem 6.2. *Let f be an arbitrary joint eigenfunction of $\mathbf{D}(X)$. Then there exists a $\lambda \in \mathfrak{a}_{\mathbf{c}}^*$ and a functional T on B such that:*

(i) *f is given by the Poisson transform*

(6) $$f(x) = \int_B e^{(i\lambda+\rho)(A(x,b))}\,dT(b).$$

(ii) *T has the Fourier series*

(7) $$T \sim \sum_{\delta \in \widehat{K}_M} d(\delta)\,\mathrm{Tr}(A_\delta \delta(k)),$$

in the sense that

(8) $$A_\delta = \int_K \delta(k^{-1}) d\widetilde{T}(k),$$

$\widetilde{T}$ *being the lift of* T *to* K.

(iii) $f(x)$ *has a convergent expansion in* $\mathcal{E}(X)$,

(9) $$f(x) = \sum_{\delta \in \widehat{K}_M} d(\delta)\,\mathrm{Tr}(A_\delta \Phi_{\lambda,\delta}(x)),$$

where $\Phi_{\lambda,\delta}$ *is the Eisenstein integral (Ch. III, §2) and* A_δ *given by (8).*

Proof. First we recall that $f \in \mathcal{E}_\lambda(X)$ for some $\lambda \in \mathfrak{a}_{\mathfrak{c}}^*$. Since $\mathcal{E}_\lambda(X) = \mathcal{E}_{s\lambda}(X)$ for $s \in W$ and since $s\lambda$ is simple for some s we can take λ simple. Using Fourier decomposition along K ([GGA], Ch. V, Theorem 3.1) we see that f has an expansion, convergent in the topology of $\mathcal{E}(X)$,

(10) $$f(x) = \sum_{\delta \in \widehat{K}_M} d(\delta) f_{\check{\delta}}(x),$$

where $f_{\check{\delta}} \in \mathcal{E}_{\lambda,\check{\delta}}(X)$. Now by Theorem 6.1,

(11) $$\mathcal{E}_{\lambda,\check{\delta}}(X) = \mathcal{P}_\lambda(\mathcal{E}_{\check{\delta}}(B)),$$

where $\mathcal{P}_\lambda$ is the Poisson transform. If $v_i (1 \le i \le d(\delta))$ is a basis of V_δ such that $v_j (1 \le j \le \ell(\delta))$ spans V_δ^M then the functions $\langle \delta(k)v_j, v_i\rangle$ form a basis of $\mathcal{E}_{\check{\delta}}(B)$. Hence $f_{\check{\delta}}$ can be written

(12) $$f_{\check{\delta}}(x) = \mathrm{Tr}(A_\delta \Phi_{\lambda,\delta}),$$

where $A_\delta \in \mathrm{Hom}(V_\delta, V_\delta^M)$. This proves (9). Now we define the functional T on B by

(13) $$T(F) = \sum_{\delta \in \widehat{K}_M} d(\delta) \int_K \mathrm{Tr}(A_\delta \delta(k)) \widetilde{F}(k) dk$$

for all F in $\mathcal{E}(B)$ for which the sum over δ converges, $\widetilde{F}$ denoting the lift of F to K. By the orthogonality relations, (8) holds. Putting $F(b) = e^{(i\lambda+\rho)(A(x,b))}$ $(b \in B)$ in (13) we obtain (6).

The natural question is now: As f runs through $\mathcal{E}_\lambda(X)$ (λ fixed, simple) what are the required functionals T in (6), or equivalently, how are the Fourier coefficients A_δ in (7) characterized?

The following result was proved by the author in 1970 ([GGA], Introduction, Theorem 4.3). Here the Laplace-Beltrami operator L generates $\boldsymbol{D}(X)$.

Theorem 6.3. *Let $X = \boldsymbol{SU}(1,1)/\boldsymbol{SO}(2)$ be the non-Euclidean disk. Then the eigenfunctions f of L are precisely*

$$f(x) = \int_B e^{(i\lambda+\rho)(A(x,b))}dT(b), \tag{14}$$

where T is an analytic functional on B and $\lambda \in \mathfrak{a}_{\mathbf{c}}^*$.

Remarks.

1. *A priori* one would expect that the class of functionals T needed in (14) would depend on λ, that is depend on the eigenvalue, but surprisingly enough, this class is the same for all λ.

2. While analytic functionals predate distributions and enter in work of Carleman [1944], Köthe [1952], they existed in 1970 as rather isolated objects outside the mainstream of analysis. Their natural appearance in hyperbolic geometry via Theorem 6.3 was thus rather unexpected.

Theorem 6.3 was proved by showing that the functionals T in Proposition 6.2 with Fourier series

$$T \sim \sum_n a_n e^{in\theta} \tag{15}$$

are characterized by the condition

$$\sum_{n\in\boldsymbol{Z}} |a_n| r^{|n|} < \infty, \qquad 0 \leq r < 1, \tag{16}$$

which in fact characterizes the Fourier coefficients of analytic functionals. The proof relied on estimates for hypergeometric functions which actually describe the K-finite eigenfunctions of L in this case.

More generally, let B be a compact analytic Riemannian manifold with Laplace-Beltrami operator L. Let $\{\varphi_n\}_{n\in \boldsymbol{Z}^+}$ be the complete orthonormal basis of $L^2(B)$ formed by the analytic eigenfunctions of L and let λ_n be the corresponding eigenvalues of $-L$. We assume these enumerated such that

$$0 \leq \lambda_o \leq \lambda_1 \leq \cdots \tag{17}$$

Let $\mathcal{A}(B)$ denote the vector space of analytic functions on B. For $T > 0$ put

$$|F|_T = \sup_{k\in \boldsymbol{Z}^+} \left(\frac{1}{(2k!)} T^k \|L^k T\| \right), \tag{18}$$

where $\| \quad \|$ is the L^2 norm on B, and

$$\mathcal{A}_T(B) = \{F \in \mathcal{E}(B) : \, |F|_T < \infty\}. \tag{19}$$

Then $\mathcal{A}_T(B)$ is a Banach space, $\mathcal{A}(B)$ is the union of the spaces $\mathcal{A}_T(B)$ and is accordingly given the inductive limit topology. The *analytic functionals (hyperfunctions)* are the functionals in the dual space $\mathcal{A}'(B)$ of $\mathcal{A}(B)$ (Lions-Magenes [1963]). Generalizing (15) we associate to each $T \in \mathcal{A}'(B)$ the expansion

$$T \sim \sum_o^\infty a_n \varphi_n, \qquad a_n = T(\overline{\varphi_n}). \tag{20}$$

Then (16) generalizes as follows (Hashizume, Minemura and Okamoto [1973]).

Theorem 6.4. *The mapping $T \longrightarrow (a_n)$ in (20) is a bijection of $\mathcal{A}'(B)$ onto the set of sequences $(a_n) \subset \boldsymbol{C}$ satisfying*

$$\sum_o^\infty |a_n| e^{-t\sqrt{\lambda_n}} < \infty \qquad \textit{for each } t > 0. \tag{21}$$

Using such a characterization of $\mathcal{A}'(B)$ the following result was proved by the author in [1974].

Theorem 6.5. *Let $X = G/K$ have rank one. Let $\lambda \in \mathfrak{a}^*$. Then the mapping $T \longrightarrow f$ given by (6) is a bijection of $\mathcal{A}'(B)$ onto $\mathcal{E}_\lambda(X)$.*

The idea is that since the series (9) is convergent and since the entries in $A_\delta \Phi_{\lambda,\delta}(x)$ are here expressible in terms of hypergeometric functions (Ch. III, Lemma 11.1) known asymtotic properties of the hypergeometric functions will imply that the entries in A_δ satisfy (21).

Theorems 6.3 and 6.5 suggested the conjecture that the functionals T in (6) are just the elements in $\mathcal{A}'(B)$. This conjecture was proved in the six-author paper, Kashiwara, Kowata, Minemura, Okamoto, Oshima and Tanaka [1978].

Theorem 6.6. *($X = G/K$ of any rank). The joint eigenfunctions of $\boldsymbol{D}(X)$ are the functions*

$$f(x) = \int_B e^{(i\lambda+\rho)(A(x,b))}\,dT(b),$$

where $\lambda \in \mathfrak{a}_{\mathbf{c}}^*$ and $T \in \mathcal{A}'(B)$.

The proof does not use the Fourier decomposition along K, but instead relies on general boundary-value theory of Kashiwara and Oshima [1977]. For a fine exposition of much of the proof, with some simplifications, see Schlichtkrull [1984].

According to Schmid [1992], the theorem can also be proved by Fourier decomposition along K, using Theorem 6.1; this requires verifying (21) on the basis of the convergence of the series (9), combined with an estimate of the entries of $\Phi_{\lambda,\delta}(x)$ from below.

2. Exponentially Bounded Eigenfunctions.

The theorems in the previous subsection suggest the problem of characterizing the Poisson transforms $\mathcal{P}_\lambda(\mathcal{A}_o(B))$ where $\mathcal{A}_o(B)$ is some naturally defined subspace of $\mathcal{A}'(B)$. The most natural example for $\mathcal{A}_o(B)$ would be the space $\mathcal{D}'(B)$ of all distributions on B. This problem was taken up by Lewis [1978] and completed by Oshima and Sekiguchi [1980]. To state the result we introduce the subspace $\mathcal{E}^*(X)$ of $\mathcal{E}(X)$ given by

$$\mathcal{E}^*(X) = \{f \in \mathcal{E}(X) : \ f(x) = 0(e^{A\,d(o,x)}) \text{ for some } A > 0\}. \tag{22}$$

Here d denotes the distance function on X. The members of $\mathcal{E}^*(X)$ are labelled smooth functions of *slow growth.* For $\lambda \in \mathfrak{a}_{\mathbf{c}}^*$ we put

$$\mathcal{E}_\lambda^*(X) = \mathcal{E}^*(X) \cap \mathcal{E}_\lambda(X).$$

Theorem 6.7. *Let $\lambda \in \mathfrak{a}_{\mathbf{c}}^*$ be such that $\mathbf{e}_{s^*}(\lambda) \neq 0$. Then*

$$\mathcal{P}_\lambda(\mathcal{D}'(B)) = \mathcal{E}_\lambda^*(X). \tag{23}$$

For X of rank one this was proved by Lewis [1978] (assuming also $\langle i\lambda, \alpha\rangle / \langle \alpha, \alpha\rangle \notin \boldsymbol{Z}$ for $\alpha \in \Sigma_o^+$) who also proved the inclusion $\subset$ in (23) for general X. Theorem 6.7 in general was proved by Oshima and Sekiguchi [1980]. Another independent proof was given by Wallach [1983] and yet another by Ban and Schlichtkrull [1987]. This latter paper also gives the following characterization of $\mathcal{P}_\lambda(\mathcal{E}(B))$. Let $\nu : \boldsymbol{D}(G) \longrightarrow \boldsymbol{E}(X)$ denote the homomorphism induced by the action $f \longrightarrow f^{\tau(g)}$ of G on $\mathcal{E}(X)$ (Ch. I, §3) and let $\mathcal{E}_\lambda^\infty(X)$ denote the set of $f \in \mathcal{E}_\lambda(X)$ satisfying:

$$\text{For some } A > 0, (\nu(D)f)(x) = O(e^{Ad(o,x)}) \text{ for all } D \in \boldsymbol{D}(G).$$

Theorem 6.8. *Let $\lambda \in \mathfrak{a}_c^*$ be such that $\mathbf{e}_{s^*}(\lambda) \neq 0$. Then*

$$\mathcal{P}_\lambda(\mathcal{E}(B)) = \mathcal{E}_\lambda^\infty(X).$$

EXERCISES AND FURTHER RESULTS.

1. ($X = G/K$ of rank one.) Let L denote the Laplace-Beltrami operator and $P(x, b)$ the Poisson kernel. If $f \in \mathcal{D}(X)$ the equation

$$Lu = f$$

has a solution $u \in \mathcal{D}(X)$ if and only if

$$\int_X f(x)dx = \int_X f(x)P(x,b)dx = 0 \qquad \text{for } b \in B.$$

Extend to $\mathcal{E}'(X)$.

2.* ($X = G/K$ of rank one.) Let $a, b \in A$ be such that the equations

$$\varphi_\lambda(a \cdot o) = 0, \qquad \varphi_\lambda(b \cdot o) = 0$$

have no common solution $\lambda \in \mathfrak{a}_c^*$. Then if

$$(M^a u)(x) = 0, \quad (M^b u)(x) = 0, \qquad \text{for all } x \in X,$$

where M^a and M^b are the mean value operators, we have necessarily $u \equiv 0$. (For this extension of Delsarte's two radius theorem, see Berenstein and Zalcman [1980]).

3.* ($X = G/K$ of any rank.) Let $\mathcal{D}'_\infty(X)$ denote the space of K-finite distributions on X. Then

$$D\,\mathcal{D}'_\infty(X) = \mathcal{D}'_\infty(X)$$

for each $D \neq 0$ in $\boldsymbol{D}(X)$. (Use Theorem 5.11 in Ch. III.)

4.* ($X = G/K$ with G complex.) Combining the restriction isomorphism $I(\mathfrak{p}^*) \longrightarrow I(\mathfrak{a}^*)$ with the isomorphism $\Gamma : \boldsymbol{D}(G/K) \longleftrightarrow \boldsymbol{D}(A)$ we have an isomorphism

$$\sigma : \boldsymbol{D}(G/K) \longrightarrow \partial(I(\mathfrak{p}))$$

(Ch. III, §8). Let J denote the ratio of the volume elements in X and in $\mathfrak{p}$ (via Exp.). Let $*$ denote adjoint.

Let $D \in \boldsymbol{D}(G/K)$. The K invariant fundamental solutions T of D are the distributions

$$T : f \longrightarrow T_o(J^{\frac{1}{2}}(f \circ \mathrm{Exp})), \qquad f \in \mathcal{D}(X),$$

where T_o is an arbitrary $\mathrm{Ad}_G(K)$-invariant fundamental solution of $(\sigma(D^*))^*$.

5.* Let S be a compact Riemannian manifold of dimension $n \geq 2$. Consider the subspace

$$\mathcal{E}_o(S) = \{f \in \mathcal{E}(S) : \int_S f(s)ds = 0\}.$$

The Laplace-Beltrami operator L gives bijection of $\mathcal{E}_o(S)$ onto itself, and its inverse is given by the integral operator

$$\Gamma(f) = \int_S g(x,y)f(y)dy,$$

where the "Green's kernel" $g(x,y)$ is characterized by the conditions (deRham [1955], p. 157):

(i) $g(x,y) \equiv g(y,x)$

(ii) $d(x,y)^{n-2}g(x,y)$ is bounded, d being the distance function on S. (If $= 2, d(x,y)^{n-2}$ should be replaced by $(\log d(x,y))^{-1}$).

(iii) $\int_S g(x,y)dy \equiv 0$.

6. Let S in **5.** be a compact Riemannian globally symmetric space U/K and define the function γ on $S - (o)$ $(o = eK)$ by

$$\gamma(x) = g(x,o).$$

Considered as a distribution on S this "Green's function" γ is characterized by

$$(*) \qquad L\gamma = \delta - \frac{1}{|S|}, \qquad \gamma(1) = 0$$

where $|S|$ is the volume of S. Here $\gamma(1) = \int \gamma$. For U/K of rank one an explicit formula for γ is given in [GGA], Ch. II, §5, No. 5.

7. Let U be a compact simply connected semisimple Lie group with Riemannian structure induced by the negative of the Killing form. Let $T \subset U$ be a maximal torus and $J \in \mathcal{D}'(T)$ given by

$$J(F) = \frac{1}{|U|}(-1)^{\frac{1}{2}(\dim U - \dim T)}\left(\delta^{-\frac{1}{2}} \sum_{s \in W} \det s\, F^s\right)(e).$$

Here $|U|$ is the volume of U, W the Weyl group and $\delta^{\frac{1}{2}}$ given by (34) §5. Then the Green's function γ on U has restriction $\overline{\gamma}$ to T given by

$$\overline{\gamma} = \delta^{-\frac{1}{2}}(\gamma_\rho * J),$$

where $\gamma_\rho \in \mathcal{D}'(T)$ is given by a Fourier series over the weight lattice:

$$\gamma_\rho = \sum_{|\lambda| \neq |\rho|} (|\lambda|^2 - |\rho|^2)^{-1} e^\lambda.$$

8. Show that the Green's function on the unit circle determined by $(*)$ in **6.** is given by

$$\gamma(\theta) = \frac{1}{4\pi}\left[\frac{\pi^2}{3} - (\pi - |\theta|)^2\right], \quad -\pi < \theta \leq \pi.$$

9.* Let (G, H) be a symmetric pair, G semisimple noncompact and let L be the Laplace-Beltrami operator on G/H for the pseudo-Riemannian structure given by the Killing form B of the Lie algebra of G. The L is globally solvable, i.e.

$$L\mathcal{E}(G/H) = \mathcal{E}(G/H)$$

(Chang [1979], Kowata and Tanaka [1980]).

10. Let Ω be the Casimir operator on the noncompact connected semisimple Lie group G. Then as a special case of **9** (Rauch and Wigner [1976] Cerèzo and Rouvière [1973] for G complex),

$$\Omega\mathcal{E}(G) = \mathcal{E}(G).$$

Suppose G is complex, $\mathfrak{u}$ a compact real form of the Lie algebra $\mathfrak{g}$ of G such that $\mathfrak{g} = \mathfrak{u} + J\mathfrak{u}$ ($J =$ the complex structure). Let $U_1, \cdots, U_n$ be a basis of $\mathfrak{u}$, orthonormal with respect to $-B$. Prove

(i) The operator

$$\omega = \sum_1^n (JU_i)U_i$$

is a bi-invariant differential operator on G.

(ii) ω is not globally solvable, i.e.

$$\omega\,\mathcal{E}(G) \neq \mathcal{E}(G).$$

(Cerèzo and Rouvière [1973]).

11. Let G/K be a symmetric space of the noncompact type. Suppose $T : \mathcal{E}(G/K) \longrightarrow \mathcal{E}(G/K)$ is a continuous linear operator commuting with the action of G. Prove that T commutes with each $D \in \boldsymbol{D}(G/K)$.

12. Let $D_1, \cdots, D_\ell$ be generators of $\boldsymbol{D}(G/K)$ and according to Theorem 10.1 in Ch. III let $Z_i, W_i \in \boldsymbol{Z}(G/K)$ be such that $D_i Z_i = W_i$ $(1 \leq i \leq \ell)$. Suppose ψ is a joint eigenfunction of $\boldsymbol{Z}(G/K)$,

$$Z\psi = c(Z)\psi, \qquad Z \in \boldsymbol{Z}(G/K),$$

with $c(Z) \in \boldsymbol{C}$. Assuming

$$c(Z_i) \neq 0 \qquad (1 \leq i \leq \ell)$$

ψ is a joint eigenfunction of $\boldsymbol{D}(G/K)$.

For a compact symmetric space U/K on the other hand, each joint eigenfunction of $\boldsymbol{Z}(U/K)$ is a joint eigenfunction of $\boldsymbol{D}(U/K)$.

13. In the notation of §3 put $Q = MAN$ and let db be the normalized K -invariant measure on $K/M = G/Q$. Show that

$$\int_G f(g)e^{-2\rho(H(g))}dg = \int_{G/Q} db(gQ) \int_Q f(gq)dq,$$

where $dq = dm\,da\,dn$ is a left invariant measure on Q.

14. Let D be a bounded symmetric domain in $\boldsymbol{C}^n$ and f a holomorphic function on an open subset of D. Then f is harmonic (Korányi [1971]).

15. (Korányi). Let $\Gamma_1 \subset \Gamma$ where Γ is the set of strongly orthogonal roots in §4 and put

$$b_{\Gamma_1} = \sum_{\gamma \in \Gamma_1} X_{-\gamma}.$$

Then

(i) $A_o \cdot b_{\Gamma_1}$ is a face of the cube

$$\square = \left\{ \sum_{\gamma \in \Gamma} x_\gamma X_{-\gamma} : \ |x_\gamma| < 1 \right\}$$

with center b_{Γ_1}.

(ii) If ℓ_1 is the cardinality of Γ_1 the orbit $M'A_o \cdot b_{\Gamma_1}$ is the union of the $(\ell - \ell_1)$-dimensional faces of $\square$.

16.* The singular support of a distribution T (sing-supp T) is the complement of the set on which the distribution can be represented as a C^∞ function. In $\boldsymbol{R}^n$ one has the following result (Hörmander [1963]). A distribution $S \in \mathcal{E}'(\boldsymbol{R}^n)$ satisfies sing supp$(S) \subset B_A(o)$ if and only if there is a constant N and for each $m \in \boldsymbol{Z}^+$ a constant C_m such that for $\zeta \in \boldsymbol{C}^n$

$$|\widetilde{S}(\zeta)| \leq C_m(1 + |\zeta|)^N e^{A|\mathrm{Im}\,\zeta|} \qquad \text{if} \qquad |\mathrm{Im}\,\zeta| \leq m \log(1 + |\zeta|).$$

For the symmetric space $X = G/K$ the following analog holds (Dadok [1976], [1979]):

Let $T \in \mathcal{E}'(X)$. If sing supp$(T) \subset B_R(o)$ then there is an integer N and for each $m \in \boldsymbol{Z}^+$ a constant C_m such that for $\lambda \in \mathfrak{a}^*_{\boldsymbol{c}}$, $b \in B$,

$$|\widetilde{T}(\lambda, b)| \leq C_m(1 + |\lambda|)^N e^{R|\,\mathrm{Im}\,\lambda|} \qquad \text{if} \qquad |\,\mathrm{Im}\,\lambda| \leq m \log(1 + |\lambda|).$$

If T is K-finite of type δ ($\delta \in \widehat{K}_M$) the converse holds. This K-finiteness assumption cannot be dropped.

17. (Converse of Huygens' principle). For the shifted wave equation (52) §3 on $\boldsymbol{H}^{2n+1}$ assume the Cauchy data v_o and v_1 are exponentially decreasing ([GGA], I, §4). Fix $x \in \boldsymbol{H}^{2n+1}$, $s > 0$. Assume the solution v vanishes in the two half cones

$$\{(z,t) : \ d(x,z) < |t| - s\}$$

with vertex x, s. Then

$$v_o(y) = v_1(y) = 0 \qquad \text{for} \qquad d(x,y) > s.$$

Similar result holds for the wave equation (1) §5 in $\boldsymbol{R}^{2n+1}$.

(Hint: Use the solution formulas (1), §5, [GGA] (5), p. 577, together with the spherical support theorems ([GGA], Ch. I, Lemma 2.7 and Lemma 4.4)).

NOTES

§1. The existence of the fundamental solution (Theorem 1.1) and the solvability results (Theorems 1.4 and 1.8) are from the author's papers [1964a], [1973a]. Alternative proof of Theorem 1.4 were given by Chang [1982] and Dadok [1980]. See also Ban and Schlichtkrull [1993]. The proof of Lemma 1.7 is from Hörmander [1963], Ch. III.

Let G be a noncompact connected Lie group, Z a bi-invariant differential operator on G. Local solvability of Z on G was proved by Raïs [1971] for G nilpotent, for G semisimple by Helgason [1973a], by Duflo and Rais [1976] and Rouvière [1976a] for G solvable and by Duflo [1977] for G arbitrary. For the global solvability question see for example Rouvière [1978]; also Exercise 10.

§2. Corollary 2.2 is due to Godement [1952a]. The other mean value results in §2, in particular, the mean value characterization of joint eigenfunctions, are from the author's paper [1959] and book [1962a]. Theorem 2.11 is proved in a different way in [1959]; the method in the text is motivated by the proof in Hörmander [1983] for the case $\boldsymbol{R}^n$. For $\boldsymbol{H}^n$ the theorem had been proved by Olevsky [1944].

§3. Theorem 3.1, Theorem 3.6 and its proof in the text are from Furstenberg [1963]. These results were also proved in Karpelevič [1965] by different methods. The proof of Theorem 3.1 in the text is from Guivarch [1984], and is based in part on Brunel's Lemma 3.3, the proof of which appears in Revuz [1975]. The Fatou Theorem 3.8 was proved by Helgason and Korányi [1968] (the rank-one case by Knapp [1968]). The modified proof in the text is in part based on Korányi [1972] and Knapp and Williamson [1971], giving respectively Corollary 3.13 and Corollary 3.14. Many variations of the above Fatou theorem have been studied, relaxing the boundedness assumption, allowing more general convergence to the boundary B, replacing the points of the boundary B by other boundary components, and replacing harmonic functions by joint eigenfunctions, generalizing our Theorem 3.16, Ch. II, to non-continuous boundary values. Some representative papers in these fields are Korányi [1970], [1976], Lindahl [1972], Michelson [1973], Schlichtkrull [1984b], Ch. VI, Sjögren [1984] and Stein [1983]. Theorem 3.15 (with the topology of $\mathfrak{M}(B)$) is from Furstenberg [1963], Ch. IV, §2, but Part (i) relating this to the topology of X is proved in Moore [1964], p. 216. The bidisk example was suggested by Korányi.

§4. Theorem 4.10, giving the restricted root structure for bounded symmetric domains G/K was proved by Moore [1964 II] in a more complicated fashion. The main ingrediences in the proof are from Harish-Chandra [1955], [1956a]. Lemmas 4.2–4.7 and Prop. 4.8 are from the two last references, Lemma 4.1 from the first. Its proof and Lemma 4.9, which shortens the proof of Theorem 4.10 substantially, were communicated to me by Korányi. The ball model of $\mathcal{D}$ (Cor. 4.11) was found by Hermann (unpublished; cf. Moore [1964 II]); see also Langlands [1963]. Proposition 4.12 on the polydisk is from Hermann [1964]. The description of the Shilov boundary (Theorem 4.13) is due to Bott and Korányi; its proof as well as that of (47) and Proposition 4.14 (proof by Stein) were shown to me by Korányi. The Poisson representation in Theorem 4.15 is from Furstenberg [1963]; p. 371. The proof in the text uses Korányi's method from Theorem 4.13. Corollary 4.20 is also stated there and justified in Moore [1964 II], Theorem 5. Proposition 4.18 is from Korányi and Wolf [1965] §5; in Part (iii) I have used some helpful remarks by Vogan.

§5. The solution to the wave equation (1) goes back to Poisson [1820] for $n = 3$ and Tedone [1898] in general. For the general Huygens' principle an introduction and a distinguished modern treatment is given by Günther [1991] and [1988]. Proposition 5.1, the solution formulas in Theorem 5.3 and 5.4, and the Huygens' principle are from the author's paper [1984b] (also [1964a]). The extension of Huygens' principle as in Theorem 5.7 was given by Solomatina [1986] (without proof), Ólafsson and Schlichtkrull [1992], and Helgason [1992d]. The first two references rely on the support theorem for the Radon transform (Ch. IV, Theorem 1.1) whereas our proof relies more on the Fourier transform. This is the method we use here in the text because it enters also in the proof of the energy equipartition theorem (Theorem 5.13) of Branson and Ólafsson [1991] which inspired our proof of Theorem 5.7. This paper also contains (42); for further developments in this direction see also Branson, Ólafsson and Schlichtkrull [1993a,b]. The orbital integral theorem (5.10) is from Helgason [1959]; relation (63), Corollary 5.11 with a different proof and the subsequent interesting remark were communicated to me by Schlichtkrull (cf. Schimming and Schlichtkrull [1992]). For the flat case see also Kolk and Varadarajan [1992].

CHAPTER VI

EIGENSPACE REPRESENTATIONS

Already in Chapter II, §2, No. 3 we defined the eigenspace representations for an arbitrary coset space L/H of Lie groups. In §1 below we discuss the role of commutativity of $\boldsymbol{D}(L/H)$ and extend eigenspace representations to spaces of sections of vector bundles. Theorems 6.2 and 8.8 in Chapter III gave irreducibility criteria for the eigenspace representations for the symmetric space G/K and the tangent space G_0/K. The results are here recalled and compared in §2, where the compact case is also settled. In §3 the principal series for G is identified with a specific family of (vector bundle) eigenspace representations; if G is complex it turns out (cf. Corollary 4.4) that the principal series representations are just the eigenspace representations for G/N, the algebra $\boldsymbol{D}(G/N)$ now being commutative. The (complex) finite-dimensional representations of G are then identified with the holomorphic eigenspace representations for G/N.

The spherical principal series is discussed in §3, No. 2 and identified with the family of eigenspace representations for G/MN. We establish irreducibility criteria and construct the intertwining operators by means of the conical distributions.

§1. GENERALITIES

1. A Motivating Example

Let L denote the Laplacian on $\boldsymbol{R}^n$. The group $\boldsymbol{M}(n)$ of motions of $\boldsymbol{R}^n$ leaves L invariant so if f is an eigenfunction of L then the composite $f \circ g$ ($g \in \boldsymbol{M}(n)$) is again an eigenfunction with the same eigenvalue. Thus if for $\lambda \in \boldsymbol{C}$ we consider the eigenspace

$$(1) \qquad \mathcal{E}_\lambda(\boldsymbol{R}^n) = \{f \in \mathcal{E}(\boldsymbol{R}^n) : \; Lf = -\lambda^2 f\}$$

then $\boldsymbol{M}(n)$ leaves $\mathcal{E}_\lambda(\boldsymbol{R}^n)$ invariant and the mapping sending $g \in \boldsymbol{M}(n)$ into the linear transformation $f \longrightarrow f^{\tau(g)}$ of $\mathcal{E}_\lambda(\boldsymbol{R}^n)$ is a representation T_λ of $\boldsymbol{M}(n)$ on $\mathcal{E}_\lambda(\boldsymbol{R}^n)$. With the topology of $\mathcal{E}_\lambda(\boldsymbol{R}^n)$ induced by that of $\mathcal{E}(\boldsymbol{R}^n)$ it is natural to inquire about irreducibility of T_λ in the sense of non-existence of closed invariant subspaces besides 0 and $\mathcal{E}_\lambda(\boldsymbol{R}^n)$. The answer is given by the following result.

Theorem 1.1. *The representation T_λ of $\boldsymbol{M}(n)$ on the eigenspace $\mathcal{E}_\lambda(\boldsymbol{R}^n)$ is irreducible if and only if $\lambda \neq 0$.*

This result is actually a special case of Theorem 8.8 in Chapter III because if G/K is the hyperbolic space $\boldsymbol{H}^n$ then G_0 is just the identity component of $\boldsymbol{M}(n)$. A more elementary proof goes along the following lines. The mapping

$$F \in \mathcal{E}(\boldsymbol{S}^{n-1}) \longrightarrow f \in \mathcal{E}_\lambda(\boldsymbol{R}^n) \tag{2}$$

given by

$$f(x) = \int_{\boldsymbol{S}^{n-1}} e^{i\lambda(x,w)} F(w)\, dw$$

commutes with the action of $\boldsymbol{O}(n)$ and is injective if $\lambda \neq 0$. Also, by [GGA], Chapter II, Theorem 2.7, there exists a sphere $S = S_r(0)$ such that the restriction map

$$f \in \mathcal{E}_\lambda(\boldsymbol{R}^n) \longrightarrow f|_S \in \mathcal{E}(S) \tag{3}$$

not only commutes with the action of $\boldsymbol{O}(n)$ but is injective. Let $\delta \in \boldsymbol{O}(n)^\wedge$ and let the subscript δ denote the subspace of elements of type δ. Then we have from (2) and (3)

$$\dim \mathcal{E}(\boldsymbol{S}^{n-1})_\delta \leq \dim \mathcal{E}_\lambda(\boldsymbol{R}^n)_\delta \leq \dim \mathcal{E}(S)_\delta. \tag{4}$$

It follows that $\dim \mathcal{E}_\lambda(\boldsymbol{R}^n)_\delta$ is finite and equal to $\dim \mathcal{E}(\boldsymbol{S}^{n-1})_\delta$. Now the irreducibility follows just as in the case $n = 2$ (cf. [GGA], Theorem 2.6, and Exercise A2 in Introduction).

2. Eigenspace Representations on Function- and Distribution-spaces.

Since $\boldsymbol{R}^n = \boldsymbol{M}(n)/\boldsymbol{O}(n)$ and since L generates the algebra $\boldsymbol{D}(\boldsymbol{M}(n)/\boldsymbol{O}(n))$ of $\boldsymbol{M}(n)$-invariant differential operators on $\boldsymbol{R}^n$ we are led to the concept of eigenspace representations for an arbitrary coset space L/H of Lie groups L and a closed subgroup H (cf. Chapter II, §2, No. 3).

Given a homomorphism $\chi : \boldsymbol{D}(L/H) \longrightarrow \boldsymbol{C}$ we have a representation T_χ of L on the joint eigenspace

$$E_\chi = \{f \in \mathcal{E}(L/H) \,:\, Df = \chi(D)f \quad \text{for} \quad D \in \boldsymbol{D}(L/H)\}. \tag{5}$$

If L/H has an L-invariant measure dl_H and the adjoint $D \longrightarrow D^*$ of a differential operator is taken with respect to this measure then $D \longrightarrow D^*$ maps $\boldsymbol{D}(L/H)$ into itself and we get a representation $\widetilde{T}_\chi$ of L on the joint distribution eigenspace

$$\mathcal{D}'_\chi = \{T \in \mathcal{D}'(L/H) \,:\, DT = \chi(D)T \text{ for } D \in \boldsymbol{D}(L/H)\}. \tag{6}$$

The topologies of E_χ and $\mathcal{D}'_\chi$ are those induced by $\mathcal{E}(L/H)$ and $\mathcal{D}'(L/H)$.

The question whether the algebra $\boldsymbol{D}(L/H)$ is commutative or not is clearly an important one for a rich supply of eigenspace representations. In case the algebra is not commutative it would be natural to consider some commutative subalgebra $\mathcal{A} \subset \boldsymbol{D}(L/H)$ instead; then a homomorphism $\mu : \mathcal{A} \longrightarrow \boldsymbol{C}$ and the corresponding joint eigenspaces,

$$E_{\mathcal{A},\mu} = \{f \in \mathcal{E}(L/H) : Df = \mu(D)f \quad \text{for} \quad D \in \mathcal{A}\} \tag{7}$$

$$\mathcal{D}'_{\mathcal{A},\mu} = \{T \in \mathcal{D}'(L/H) : DT = \mu(D)T \quad \text{for} \quad D \in \mathcal{A}\}. \tag{8}$$

This gives corresponding *eigenspace representations on* $E_{\mathcal{A},\mu}$ *and* $\mathcal{D}'_{\mathcal{A},\mu}$, *respectively,*

$$T_{\mathcal{A},\mu} \quad \text{and} \quad \widetilde{T}_{\mathcal{A},\mu}. \tag{9}$$

Another point to be mentioned in this connection is that the space L/H may have another coset space representation L'/H' for which $\boldsymbol{D}(L'/H')$ might be commutative even if $\boldsymbol{D}(L/H)$ is not. An example of this is the representation $L = (L \times L)/\Delta L$ where ΔL is the diagonal in $L \times L$.

3. Eigenspace Representations for Vector Bundles

A differential operator D on a manifold X is invariant under a diffeomorphism $\psi : X \longrightarrow X$ if

$$D(f \circ \psi) = Df \circ \psi \quad \text{for} \quad f \in \mathcal{D}(X). \tag{10}$$

We shall now extend this notion to smooth vector bundles. A smooth real *vector bundle* over a manifold X is a manifold E and a C^∞ map $p : E \longrightarrow X$ such that

(i) For $x \in X$, $p^{-1}(x)$ is a vector space over $\boldsymbol{R}$ of dimension n.

(ii) For each $x \in X$ there is a neighborhood U of x and a diffeomorphism τ of $p^{-1}(U)$ onto $U \times \boldsymbol{R}^n$ such that $\tau(e) = (p(e), F(e))$, where F is a linear map of the vector space $p^{-1}(p(e))$ onto $\boldsymbol{R}^n$.

The map p is called the *projection map*, X the *base* and the set $E_x = p^{-1}(x)$ the *fiber* over x. A map τ with the property (ii) is called a *local trivialization.* The standard example of a vector bundle is the *tangent bundle* $\cup_{x \in X} X_x$.

Replacing $\boldsymbol{R}$ by $\boldsymbol{C}$ we have a *complex vector bundle* over X. If X is a complex manifold and one adds the condition in (ii) that τ be holomorphic we have a *holomorphic vector bundle* over X. If $n = 1$ above one speaks of a *line bundle.*

Let E be a complex vector bundle over X, $p : E \longrightarrow X$ the projection map. A *smooth section* is a C^∞ map $s : X \longrightarrow E$ such that $p(s(x)) = x$. The image $s(X)$ then intersects each fiber $p^{-1}(x)$ in a single point $s(x) : s(X) \cap p^{-1}(x) = s(x)$. The smooth sections form a vector space $\Gamma(E)$. A *differential operator* D on E is a linear mapping of $\Gamma(E)$ to $\Gamma(E)$ which via an arbitrary local trivialization of E is expressed by means of ordinary differential operators with linear transformations of the fibers E_x as coefficients. More explicitly, with U as in (ii) let $\boldsymbol{e}_j$ be a basis of $\Gamma(E)|U$ over $\mathcal{E}(U)$. Then if $s \in \Gamma(E)|U$, $s = \sum_j s_j \boldsymbol{e}_j$ $(s_j \in \mathcal{E}(U))$ we have

$$Ds = \sum_{j,k} (D_{jk} s_k) \boldsymbol{e}_j, \tag{11}$$

where D_{jk} is a differential operator on U.

Now suppose $\varphi : E \longrightarrow E$ is a diffeomorphism *compatible* with p in the sense that there is a diffeomorphism ψ of X such that $p \circ \varphi = \psi \circ p$. Suppose further that the restriction map $E_x \longrightarrow E_{\psi(x)}$ is a vector space isomorphism. Then φ operates on $\Gamma(E)$: if $s \in \Gamma(E)$ the map s^φ given by

$$s^\varphi(x) = \varphi(s(\psi^{-1}x)) \qquad (x \in X)$$

belongs to $\Gamma(E)$ and we can define the differential operator

$$D^\varphi : s \longrightarrow \left(Ds^{\varphi^{-1}}\right)^\varphi, \qquad s \in \Gamma(E). \tag{12}$$

The operator D is said to be *invariant* under φ if $D^\varphi = D$. For the trivial bundle $E = X \times \boldsymbol{C}$, sections s become functions, $s(x) = (x, f(x))$, and $\varphi(x, c) = (\psi(x), c_0 c)$ with $c_0 \neq 0$ fixed. Then $s(\psi^{-1}x) = (\psi^{-1}x, f(\psi^{-1}x))$ so

$$s^\varphi(x) = (x, c_0 f(\psi^{-1}x)), \quad s^{\varphi^{-1}}(x) = (x, c_0^{-1} f(\psi(x))).$$

Let $\boldsymbol{e}$ denote the section $x \longrightarrow (x, 1)$ so

$$s = f\boldsymbol{e}, \quad s^\varphi = c_0 f^\psi \boldsymbol{e}, \quad s^{\varphi^{-1}} = c_0 f^{\varphi^{-1}} \boldsymbol{e}.$$

Then by (11),

$$D^\varphi s = c_0 c_0^{-1} \left(D_0 f^{\psi^{-1}}\right)^\psi \boldsymbol{e} = (D_0^\psi f)\boldsymbol{e},$$

where D_0 is a differential operator on X. Thus the invariance condition $D^\varphi = D$ generalizes (10).

Now let X be a coset space G/K of Lie groups and let δ be a finite-dimensional representation of K on a finite-dimensional vector space V over $\boldsymbol{C}$. Let E denote the product $G \times V$ modulo the equivalence relation $(g, v) \sim (gk, \delta(k^{-1})v)$. If $[g, v]$ denotes the equivalence class of the element $(g, v) \in G \times V$ then the mapping $p : [g, v] \longrightarrow gK$ turns E into a vector bundle over G/K, denoted $G \times_K V$. The group G acts on $G \times_K V$ via $g_0 \cdot [g, v] = [g_0 g, v]$ and this action is compatible with p relative to the action $gK \longrightarrow g_0 gK$ of the base. If s is a section, i.e., $s \in \Gamma(G \times_K V)$ we define

$$\widetilde{s}(g) = g^{-1} \cdot s(gK). \tag{13}$$

Identifying the equivalence class $[e, v]$ with the element $v \in V$ we see that $s(gK) = [g, \widetilde{s}(g)]$ so the mapping $s \longrightarrow \widetilde{s}$ is a bijection of $\Gamma(G \times_K V)$ onto the space of smooth maps

$$f : G \longrightarrow V \quad \text{satisfying} \quad f(gk) \equiv \delta(k^{-1}) f(g). \tag{14}$$

Let $\boldsymbol{D}(G \times_K V)$ denote the set of differential operators on $G \times_K V$ which are invariant under the action of G. By (12), $(Ds)^{\varphi} = D^{\varphi} s^{\varphi}$, so $\boldsymbol{D}(G \times_K V)$ is quickly seen to be an algebra. Given a subalgebra $\mathcal{A} \subset \boldsymbol{D}(G \times_K V)$ and a homomorphism $\chi : \mathcal{A} \longrightarrow \boldsymbol{C}$ we can define the *joint eigenspace*

$$\Gamma_\delta(\mathcal{A}, \chi) = \{s \in \Gamma(G \times_K V) : Ds = \chi(D)s \quad \text{for} \quad D \in \mathcal{A}\}. \tag{15}$$

Then the map $T_\chi(g) : s \longrightarrow s^g$ leaves the space $\Gamma_\delta(\mathcal{A}, \chi)$ invariant and $g \longrightarrow T_\chi(g)$ is a representation of G on $\Gamma_\delta(\mathcal{A}, \chi)$, the *eigenspace representation* of G defined by δ, $\mathcal{A}$, and χ.

§2. IRREDUCIBILITY CRITERIA FOR A SYMMETRIC SPACE

1. The Compact Case

Let K be a compact connected Lie group, $M \subset K$ a closed subgroup. We shall now investigate the eigenspace representations for K/M (we do not assume K/M symmetric).

For each $\delta \in \widehat{K}$ let V_δ (with inner product $\langle\, ,\, \rangle$) be a vector space on which δ is realized, put

$$V_\delta^M = \{v \in V_\delta : \delta(m)v = v \quad \text{for} \quad m \in M\}$$

and let

$$\widehat{K}_M = \left\{\delta \in \widehat{K} : V_\delta^M \neq 0\right\}.$$

Put $d(\delta) = \dim V_\delta$, $\ell(\delta) = \dim V_\delta^M$ and let $v_1, \dots, v_{d(\delta)}$ be a basis of V_δ such that $v_1, \dots, v_{\ell(\delta)}$ span V_δ^M. We have the following Hilbert space decompositions ([GGA], Chapter IV, §1, and Chapter V, §3),

$$L^2(K) = \bigoplus_{\delta \in \widehat{K}} H_\delta, \tag{1}$$

where

$$H_\delta = \{F \in C(K) : F(k) = Tr(\delta(k)C),\ C \in \operatorname{Hom}(V_\delta, V_\delta)\} \tag{2}$$

and

$$L^2\,(K/M) = \bigoplus_{\delta \in \widehat{K}_M} C_\delta\,(K/M)\,, \tag{3}$$

where (for $\overset{\vee}{\delta}$ = contragredient to δ and $\delta \in \widehat{K}_M$)

$$C_{\overset{\vee}{\delta}}(K/M) = \{F \in C(K/M) : F(kM) = Tr(C\delta(k)), C \in \operatorname{Hom}(V_\delta, V_\delta^M)\}\,. \tag{4}$$

Viewing K as the symmetric space $(K \times K)/\Delta K$ (ΔK = diagonal) the algebra $\boldsymbol{D}(K \times K)/\Delta K)$ is identified with the center $\boldsymbol{Z}(K)$ of $\boldsymbol{D}(K)$ ([GGA], II, §5) and the normalized characters $d(\delta)^{-1}\chi_\delta$ become the spherical functions. Since each H_δ is contained in an eigenspace of $\boldsymbol{Z}(K)$ (loc. cit., Chapter V, Lemma 1.6) and since different spherical functions correspond to different eigenvalue homomorphisms ([GGA], IV, Cor. 2.3) we see that the H_δ in (1) are the joint eigenspaces of the algebra $\boldsymbol{Z}(K)$.

Next observe that

$$C_\delta\,(K/M) \subset H_{\overset{\vee}{\delta}}, \qquad \delta \in \widehat{K}_M \tag{5}$$

if we view the functions on K/M as functions on K right invariant under M. In particular, the different $C_\delta\,(K/M)$ belong to different eigenspaces for the subalgebra $\boldsymbol{Z}\,(K/M) \subset \boldsymbol{D}\,(K/M)$ induced by the operators in $\boldsymbol{Z}(K)$.

We put $\mathcal{A} = \boldsymbol{Z}\,(K/M)$ and if $D \in \mathcal{A}$ let $\mu_\delta(D)$ denote the eigenvalue of D on $C_\delta\,(K/M)$. Given a homomorphism $\mu : \mathcal{A} \longrightarrow \boldsymbol{C}$ let $T_{\mathcal{A},\mu}$ be the corresponding eigenspace representation of K (cf. (9) §1).

Theorem 2.1. *The eigenspace representations $T_{\mathcal{A},\mu}$ of K on $\mathcal{E}(K/M)$ have the form $T_{\mathcal{A},\mu_\delta}$ ($\delta \in \widehat{K}_M$) acting on $C_\delta(K/M)$ and $T_{\mathcal{A},\mu_\delta}$ decomposes into $\ell(\delta)$ copies of δ.*

Proof. Let us take the basis (v_i) above orthonormal. We know from [GGA], Chapter IV, Theorem 1.6 and Chapter V, §3, No. 2 that if

$$\delta_{ij}(k) = \langle \delta(k)v_j, v_i \rangle \qquad 1 \le i, j \le d(\delta)$$

the functions δ_{ij} form a basis of H_δ and the functions δ_{ij} $(1 \le i \le d(\delta)$, $1 \le j \le \ell(\delta))$ (i.e., the $\ell(\delta)$ first columns) form a basis of $C_{\check{\delta}}(K/M)$ ($\check{\delta}$ = contragredient to δ). By the quoted theorem, $T_{\mathcal{A},\mu_{\check{\delta}}}$ decomposes into $\ell(\delta) = \ell(\check{\delta})$ copies of $\check{\delta}$. In the decomposition (3) we have an absolutely and uniformly convergent decomposition of each $f \in \mathcal{E}(K/M)$ so the $T_{\mathcal{A},\mu_\delta}$ do indeed exhaust the set of eigenspace representations.

We now recall the following result ([GGA], Chapter V, Theorem 3.5 and Chapter IV, Exercise B13). Let $C^\natural(K/M)$ denote the subspace of $C(K)$ consisting of the M-bi-invariant functions.

Theorem 2.2. *The following properties of K/M are equivalent:*

(i) $\ell(\delta) = 1$ for each $\delta = \widehat{K}_M$.

(ii) The convolution algebra $C^\natural(K/M)$ is commutative.

(iii) The algebra $\boldsymbol{D}(K/M)$ is commutative.

We have seen that these properties are present if K/M is symmetric.

Corollary 2.3. *Assume $\boldsymbol{D}(K/M)$ commutative. Then the spaces $C_\delta(K/M)$ $(\delta \in \widehat{K}_M)$ are the joint eigenspaces of $\boldsymbol{D}(K/M)$ and K acts irreducibly on each of them.*

Since $\ell(\delta) = 1$ for each $\delta \in \widehat{K}_M$ we know from Theorem 2.1 that if $\mathcal{A} = \boldsymbol{Z}(K/M)$ then each $T_{\mathcal{A},\mu_\delta}(K)$ acts irreducibly on $C_\delta(K/M)$. Because of this irreducibility it suffices to prove that each $C_\delta(K/M)$ contains a nonzero joint eigenfunction of $\boldsymbol{D}(K/M)$. Fix a unit vector $v_0 \in \boldsymbol{V}_\delta^M$ and put

$$\varphi(kM) = \langle \delta(k)v_0, v_0 \rangle. \tag{6}$$

For $f \in C^\natural(K/M)$ we define $\delta(f)$ by

$$\delta(f) = \int_K f(k)\delta(k)v\,dk \qquad v \in V_\delta.$$

Then the vector $\delta(f)v_0 \in V_\delta^M$ so $\delta(f)v_0 = c_f v_0$ where $c_f \in \boldsymbol{C}$. Hence

$$\int_K f(k)\varphi(kM)\,dk = c_f. \tag{7}$$

Since $\delta(f_1 * f_2) = \delta(f_1)\delta(f_2)$, (7) implies that the mapping $f \longrightarrow \int f\varphi\, dk$ is a homomorphism of $C^\natural(K/M)$ into $\boldsymbol{C}$ so φ is a spherical function ([GGA], Chapter IV, Lemma 3.2). This establishes the corollary.

Remark. For the general case see Exercise 1.

2. The Euclidean Type

Let $\mathfrak{g}$ be a semisimple Lie algebra over $\boldsymbol{R}$, $\mathrm{Int}(\mathfrak{g})$ its adjoint group. Let $\mathfrak{g} = \mathfrak{k} + \mathfrak{p}$ be a Cartan decomposition and $K \subset \mathrm{Int}(\mathfrak{g})$ the analytic subgroup with Lie algebra $\mathrm{ad}_{\mathfrak{g}}(\mathfrak{k}) \sim \mathfrak{k}$. Let G_0 be the group of affine transformations of $\mathfrak{p}$ generated by the translations and the restrictions $k|\mathfrak{p}$ $(k \in K)$. Then $\mathfrak{p} = G_0/(K|\mathfrak{p})$ which we write as G_0/K for simplicity. The algebra $\boldsymbol{D}(G_0/K)$ consists of differential operators with constant coefficients so is commutative. The corresponding eigenspace representations of G_0 are the representations T_λ on the eigenspaces $\mathcal{E}_\lambda(\mathfrak{p})$ in Chapter III, §8, (1). There we proved the following result.

Theorem 2.4. *Let $\lambda \in \mathfrak{a}_{\boldsymbol{c}}^*$. Then the eigenspace representation T_λ is irreducible if and only if λ is regular.*

This result has a simple interpretation in terms of the Fourier transform

$$\widetilde{f}(Y) = \int_{\mathfrak{p}} f(X) e^{-iB(X,Y)}\, dX, \qquad Y \in \mathfrak{p}$$

on $\mathfrak{p}$. Using the notation from Chapter III, §7, No. 4, we can write the inversion formula

$$(8) \qquad f(X) = c \int_K \int_{\mathfrak{a}*} \widetilde{f}(k \cdot A_\lambda) e^{iB(X, kA_\lambda)} |\omega(\lambda)|\, d\lambda\, dk, \qquad f \in \mathcal{S}(\mathfrak{p}),$$

where c is a constant independent of f. Thus we can restate Theorem 2.4 as follows (with ω extended to $\mathfrak{a}_{\boldsymbol{c}}^*$ in the obvious fashion).

Theorem 2.4′. *Let $\lambda = \mathfrak{a}_{\boldsymbol{c}}^*$. Then the eigenspace representation T_λ is reducible if and only if the Plancherel density ω vanishes at λ.*

3. The Noncompact Type

As in Chapter III let $X = G/K$ be a symmetric space of the noncompact type; we adopt the notation used there:

(i) The direct decompositions

$$\mathfrak{g} = \mathfrak{k} + \mathfrak{p}, \quad \mathfrak{g} = \mathfrak{k} + \mathfrak{a} + \mathfrak{n}.$$

(ii) The isomorphism

$$\Gamma : \boldsymbol{D}(G/K) \longrightarrow I(\mathfrak{a}).$$

(iii) The joint eigenspace of $\boldsymbol{D}(G/K)$

$$\mathcal{E}_\lambda(X) = \{f \in \mathcal{E}(X) : Df = \Gamma(D)(i\lambda)f \quad \text{for} \quad D \in \boldsymbol{D}(G/K)\} \tag{9}$$

and the natural representation of T_λ of G on $\mathcal{E}_\lambda(X)$. In Chapter III, §6 we concluded the proof of the following result.

Theorem 2.5. *Let $\lambda \in \mathfrak{a}_{\mathbf{c}}^*$. Then the eigenspace representation T_λ is irreducible if and only if*

$$\frac{1}{\Gamma_X(\lambda)} \neq 0. \tag{10}$$

Here the Gamma function Γ_X of X is defined in Chapter III, §6, (9)–(10). We now recall the inversion formula for the Fourier transform on X (Chapter III, §1)

$$f(x) = w^{-1} \int_B \int_{\mathfrak{a}^*} \tilde{f}(\lambda, b) e^{(i\lambda+\rho)(A(x,b))} |\boldsymbol{c}(\lambda)|^{-2} \, d\lambda \, db, \tag{11}$$

and note that by Chapter II, §3, (18)′, if $\lambda \in \mathfrak{a}^*$,

$$|\boldsymbol{c}(\lambda)|^{-2} = (\boldsymbol{c}(\lambda)\boldsymbol{c}(-\lambda))^{-1} = d_0 \frac{\Gamma_X(\lambda)}{\prod\limits_{\alpha \in \Sigma_0} 2^{-\langle i\lambda, \alpha_0 \rangle} \Gamma(\langle i\lambda, \alpha_0 \rangle)}. \tag{12}$$

Here d_0 is a constant (given by $\mathbf{c}(-i\rho) = 1$) and $\alpha_0 = \alpha/\langle \alpha, \alpha \rangle$. Thus Theorem 2.5 becomes an "opposite" of Theorem 2.4′.

Theorem 2.5′. *Let $\lambda \in \mathfrak{a}_{\mathbf{c}}^*$. Then the eigenspace representation T_λ is reducible if and only if the Plancherel density is singular at λ.*

Here we must quickly add that the singularities which the theorem refers to include those which are removable, that is disappear after cancellation in the fraction (12).

The contrast between Theorems 2.4′ and 2.5′ becomes particularly clear if we specialize to the case when G is complex. Then

$$(\boldsymbol{c}(\lambda)\boldsymbol{c}(-\lambda))^{-1} = c_0 \prod_{\alpha \in \Sigma} \langle i\lambda, \alpha_0 \rangle \qquad (c_0 = \text{const})$$

and using the duplication formula for the Gamma function

$$\Gamma_X(\lambda) = \pi^{\frac{1}{2}} \prod_{\alpha \in \Sigma} 2^{-\langle i\lambda, \alpha_0 \rangle} \Gamma(1 + \langle i\lambda, \alpha_0 \rangle).$$

Here the Plancherel density vanishes exactly if λ is singular whereas T_λ remains irreducible in the tube

$$|\mathrm{Re}\langle i\lambda, \alpha_0 \rangle| < 1 \qquad \text{for} \quad \alpha \in \Sigma.$$

Since the symmetric space G/K is determined by triple $(\mathfrak{a}, \Sigma, m)$ where m is the multiplicity function it is clear *a priori* that the irreducibility question for T_λ should be expressible in terms of these three data. And indeed they do enter explicitly in the formula for Γ_X. But beyond this it could have been expected that the $\mathbf{c}$-function would enter for the following reason. Let $\mathcal{E}_\lambda^0$ denote the closed subspace of $\mathcal{E}_\lambda(X)$ generated by all the G-translates of φ_λ. Then G acts irreducibly on $\mathcal{E}_\lambda^0$ because each closed invariant subspace V will contain a nonzero K-invariant, so $\varphi_\lambda \in V$. Thus the irreducibility question for T_λ amounts to the problem of finding for which $\lambda \in \mathfrak{a}_{\mathbf{c}}^*$ the spaces $\mathcal{E}_\lambda^0$ and $\mathcal{E}_\lambda(X)$ coincide and it is natural that the $\mathbf{c}$-function should enter into this problem.

§3. EIGENSPACE REPRESENTATIONS FOR THE HOROCYCLE SPACE G/MN

1. The Principal Series

Consider again our customary setup as in §2, No. 3 with M as usual the centralizer of A in K. Let δ be a unitary irreducible representation of M on a complex vector space V_δ and let $\lambda \in \mathfrak{a}_{\mathbf{c}}^*$. Let $\Gamma_{\delta,\lambda}$ denote the space of C^∞ functions $f : G \longrightarrow V_\delta$ satisfying

$$f(gman) = \delta(m^{-1}) e^{(i\lambda - \rho)(\log a)} f(g) \tag{1}$$

for $g \in G$, $m \in M$, $a \in A$, $n \in N$ and let $\tau_{\delta,\lambda}$ denote the natural representation of G on $\Gamma_{\delta,\lambda}$. The representations $\tau_{\delta,\lambda}$ ($\delta \in \widehat{M}$, $\lambda \in \mathfrak{a}_{\mathbf{c}}^*$) form the *principal series of* G.

We shall now see how the representations from the principal series can be realized as eigenspace representations. For this we extend δ to a representation of MN on V_δ by putting $\delta(mn) = \delta(m)$ ($m \in M$, $n \in N$). As explained in §1, No. 3 this defines the vector bundle $G \times_{MN} V_\delta$ over G/MN.

The corresponding space $\Gamma(G \times_{MN} V_\delta)$ of sections is (by §1, (14)) identified with the space of smooth functions $F : G \longrightarrow V_\delta$ satisfying

$$F(gmn) \equiv \delta(m^{-1})F(g). \tag{2}$$

The action of G on $\Gamma(G \times_{MN} V_\delta)$ corresponds to left translations on the space of functions F. Therefore the algebra $\boldsymbol{D}(G/MN)$ of left invariant differential operators on G/MN can be viewed as a (commutative) subalgebra of $\boldsymbol{D}(G \times_{MN} V_\delta)$. Letting χ_λ denote the homomorphism of $\boldsymbol{D}(G/MN)$ into $\boldsymbol{C}$ given by

$$\chi_\lambda(D) = \widehat{\Gamma}(D)(i\lambda - \rho)$$

(Chapter II, §2) we see that $\Gamma_{\delta,\lambda}$ above is just the joint eigenspace

$$\Gamma_\delta(\boldsymbol{D}(G/MN), \chi_\lambda) \tag{3}$$

in terms of (15) §1. Thus we have the following result.

Proposition 3.1. *The principal series representation $\tau_{\delta,\lambda}$ is the eigenspace representation of G on the joint eigenspace* (3).

2. The Spherical Principal Series. Irreducibility

We consider now the representations from the principal series which correspond to $\delta = 1$. Then the functions (1) can be viewed as scalar-valued functions on $\Xi = G/MN$, the space $\Gamma_{\delta,\lambda}$ becomes the space

$$\mathcal{E}_\lambda(\Xi) = \left\{\psi \in \mathcal{E}(\Xi) : \psi(ga \cdot \xi_0) \equiv e^{(i\lambda-\rho)(\log a)}\psi(g \cdot \xi_0)\right\} \tag{4}$$

from Chapter II, Prop. 2.6, and the representation $\tau_{\delta,\lambda}$ of G on $\Gamma_{\delta,\lambda}$ becomes the natural representation of G on $\mathcal{E}_\lambda(\Xi)$. Viewing functions ψ on Ξ as distributions $\psi\, d\xi$ ($d\xi$ being the G-invariant measure dg_{MN}) the space $\mathcal{E}_\lambda(\Xi)$ is contained in the distribution space

$$\mathcal{D}'_\lambda(\Xi) = \left\{\Psi \in \mathcal{D}'(\Xi) : \Psi^{\sigma(a)} = e^{-(i\lambda+\rho)(\log a)}\Psi, \qquad a \in A\right\} \tag{5}$$

(cf. Chapter II, Prop. 4.3). As in Chapter II, §§2–3, the natural representation of G on $\mathcal{D}'_\lambda(\Xi)$ is denoted $\widetilde{\tau}_\lambda$. In addition to the spaces $\mathcal{E}_\lambda(\Xi) \subset \mathcal{D}'_\lambda(\Xi)$ we consider for each $\lambda \in \mathfrak{a}^*_c$ the space $\mathcal{K}_\lambda = \mathcal{K}_\lambda(\Xi)$ of equivalence classes of complex-valued measurable functions ψ on $\Xi = G/MN = (K/M) \times A$ defined by

$$\psi(kaMN) = e^{(i\lambda-\rho)(\log a)}F(kM), \tag{6}$$

where F runs through $L^2(K/M)$. Thus the elements $\psi \in \mathcal{K}_\lambda$ are characterized by

(i) $\psi(gaMN) = e^{(i\lambda-\rho)(\log a)}\psi(gMN)$ for $a \in A$, almost all $g \in G$.

(ii) $\int_{K/M} |\psi(kMN)|^2 \, dk_M < \infty$.

Then $\mathcal{K}_\lambda$ is a Hilbert space, the norm $\|\psi\|_\lambda$ being the square root of the expression in (ii). Since $d\xi = e^{2\rho(\log a)}da\, dk_M$ it is clear that each $\psi \in \mathcal{K}_\lambda$ is a locally integrable function on Ξ and thus defines a distribution $\varphi \longrightarrow \int \varphi(\xi)\psi(\xi)\, d\xi$ on Ξ. Since

$$\psi^{\sigma(a)} = e^{(-i\lambda+\rho)(\log a)}\psi, \quad (d\xi)^{\sigma(a)} = e^{-2\rho(\log a)}d\xi$$

we have by (5),

$$\mathcal{E}_\lambda(\Xi) \subset \mathcal{K}_\lambda \subset \mathcal{D}'_\lambda(\Xi). \tag{7}$$

Let τ_λ denote the representation of G on $\mathcal{K}_\lambda$. The vector $\eta_\lambda \in \mathcal{K}_\lambda$ given by

$$\eta_\lambda(kaMN) = e^{(i\lambda-\rho)(\log a)} \tag{8}$$

(cf. Chapter III, §3, (1)) is fixed under $\tau_\lambda(K)$ so for each $\lambda \in \mathfrak{a}_c^*$, τ_λ (as well as $\widetilde{\tau}_\lambda$) is a spherical representation.

Definition. The family $\{\widetilde{\tau}_\lambda : \lambda \in \mathfrak{a}_c^*\}$ (as well as the family $\{\tau_\lambda : \lambda \in \mathfrak{a}_c^*\}$) is called the *spherical principal series.*

Proposition 3.2. *For each* $\lambda \in \mathfrak{a}^*$, τ_λ *is unitary.*

In fact,

$$\psi(gkMN) = e^{(i\lambda-\rho)(H(gk))}\psi(k(gk)MN) \tag{9}$$

so the proposition follows from the often-used formula

$$\int_K F(k)\, dk = \int_K F(k(gk))e^{-2\rho(H(gk))}dk. \tag{10}$$

Consider now the bijection

$$S \longrightarrow \Psi = S \otimes e^{i\lambda-\rho} \tag{11}$$

of $\mathcal{D}'(K/M)$ onto $\mathcal{D}'_\lambda(\Xi)$ given by Prop. 4.4 in Chapter II. Note that since $d\xi = e^{2\rho(\log a)}dk_M\, da$, $S \otimes \tau$ (for $\tau \in \mathcal{E}(A)$) denotes the distribution

$$S \otimes \tau : \varphi \longrightarrow \int \left(\int \varphi(ka \cdot \xi_o)\tau(a)e^{2\rho(\log a)}da \right) dS(k_M).$$

This map is a homeomorphism because a bounded set in $\mathcal{D}(\Xi)$ consists of the functions φ all supported in a fixed compact set on which each derivative $D\varphi$ has a fixed bound C_D (Schwartz [1966], Chapter III, §2). In particular, $\mathcal{K}_\lambda$ is dense in $\mathcal{D}'_\lambda(\Xi)$. It is also known *loc. cit.*, §3) that if (Ψ_i) is a sequence in $\mathcal{D}'(\Xi)$ such that for each $\varphi \in \mathcal{D}(\Xi)$, $\Psi_i(\varphi)$ has a limit $\Psi(\varphi)$ then $\Psi \in \mathcal{D}'(\Xi)$ and $\Psi_i \longrightarrow \Psi$ in the strong topology of $\mathcal{D}'(\Xi)$. In particular, the embedding of $\mathcal{K}_\lambda$ into $\mathcal{D}'_\lambda$ is continuous.

The Compact Models for τ_λ and $\widetilde{\tau}_\lambda$. We have defined above the representation τ_λ as a representation of G on the Hilbert space $\mathcal{K}_\lambda(\Xi)$. However, because of (6) we can realize τ_λ as a representation of G on $L^2(K/M)$. In fact,

$$(\tau_\lambda(g)\psi)(ka\cdot\xi_0) = \psi^{\tau(g)}(ka\cdot\xi_0) = F(k(g^{-1}k)M)e^{(i\lambda-\rho)(H(g^{-1}k)+\log a)}$$

so the action of G on $L^2(K/M)$ is given by

$$(\tau_\lambda(g)F)(kM) = F(k(g^{-1}k)M)e^{(i\lambda-\rho)(H(g^{-1}k))}. \tag{12}$$

Similarly, $\widetilde{\tau}_\lambda$ can be realized as a representation of G on $\mathcal{D}'(K/M)$ by means of the map (11). If T_g denotes the diffeomorphism $kM \longrightarrow k(gk)M$ of K/M we have

$$(\widetilde{\tau}_\lambda(g)S)(kM) = S^{T_g}(kM)e^{(i\lambda-\rho)(H(g^{-1}k))}, \qquad S \in \mathcal{D}'(K/M). \tag{13}$$

The representations (12) and (13) will be referred to as the compact models of τ_λ and $\widetilde{\tau}_\lambda$, respectively.

Proposition 3.3. *Let $\lambda \in \mathfrak{a}_\mathfrak{c}^*$. Then the representation τ_λ is irreducible if and only if $\widetilde{\tau}_\lambda$ is irreducible.*

Proof. Suppose first τ_λ irreducible and $E \subset \mathcal{D}'_\lambda$ a proper closed invariant subspace. Since the identity map $\mathcal{K}_\lambda \longrightarrow \mathcal{D}'_\lambda$ is continuous, $E \cap \mathcal{K}_\lambda$ is closed in $\mathcal{K}_\lambda$ and invariant so either $E \cap \mathcal{K}_\lambda = \mathcal{K}_\lambda$ or $E \cap \mathcal{K}_\lambda = \{0\}$. The first possibility implies the contradiction $E = \mathcal{D}'_\lambda$. Hence $E \cap \mathcal{K}_\lambda = \{0\}$. In order to conclude $E = 0$ consider for $f \in \mathcal{D}(G)$, $\Psi \in \mathcal{D}'(\Xi)$ the distribution

$$\int_G \Psi^{\tau(g)} f(g)\,dg \;:\; \varphi \longrightarrow \int_G \Psi^{\tau(g)}(\varphi)f(g)\,dg. \tag{14}$$

If we approximate f by step functions the property of sequences in $\mathcal{D}'(\Xi)$ mentioned above implies that E is invariant under the map

$$\Psi \longrightarrow \int_G \Psi^{\tau(g)} f(g)\,dg.$$

Now if $\widetilde{\Psi}$ is the "lift" of Ψ to G as defined in Chapter IV, §2, we have

$$\left(\int_G \Psi^{\tau(g)} f(g)\, dg\right)^{\sim} = f * \widetilde{\Psi}. \tag{15}$$

Hence, if $\Psi \in E$, the distribution (14) is a smooth function in E and belongs to $\mathcal{K}_\lambda$. Thus by (15) $f * \widetilde{\Psi} = 0$ for each $f \in \mathcal{D}(G)$ whence $\Psi = 0$. Thus $\widetilde{\tau}_\lambda$ is irreducible.

On the other hand suppose $\widetilde{\tau}_\lambda$ is irreducible and let $E \subset \mathcal{K}_\lambda$ be a proper closed invariant subspace. Then there exists an $\alpha \neq 0$ in $L^2(K/M)$ such that

$$\int_{K/M} \varphi(kMN)\alpha(kM)dk_M = 0 \qquad \varphi \in E. \tag{16}$$

Since E is invariant under convolution by functions in $\mathcal{D}(K/M)$ we may take $\alpha \in \mathcal{D}(K/M)$. The continuous linear functional L_α on $\mathcal{D}'_\lambda$ given by

$$L_\alpha(\Psi) = S(\alpha) \qquad \text{if} \quad \Psi = S \otimes e^{i\lambda - \rho}$$

vanishes on E, hence also on its closure $\overline{E}$ in $\mathcal{D}'_\lambda$. By the irreducibility, $\overline{E} = \mathcal{D}'_\lambda$ or $\overline{E} = 0$. The first possibility is ruled out since $\alpha \neq 0$ so $E = 0$ and τ_λ irreducible.

Theorem 3.4. *Let $\lambda \in \mathfrak{a}^*_c$. Then τ_λ is irreducible $\Longleftrightarrow$ λ and $-\lambda$ are simple.*

Proof. If $\lambda = \xi + i\eta$ ($\xi, \eta \in \mathfrak{a}^*$) put $\overline{\lambda} = \xi - i\eta$ and consider the sesquilinear form $\{\ ,\ \}$ on $\mathcal{K}_\lambda \times \mathcal{K}_{\overline{\lambda}}$ given by

$$\{\psi_1, \psi_2\} = \int_K \psi_1(k \cdot \xi_0)\, \mathrm{conj}(\psi_2(k \cdot \xi_0)\, dk.$$

From (10) and (12) we see that G leaves $\{\ ,\ \}$ invariant. Hence, if $V \subset \mathcal{K}_\lambda$ is a closed invariant subspace, so is the annihilator

$$V^0 = \left\{\psi \in \mathcal{K}_{\overline{\lambda}} : \{v, \psi\} = 0 \quad \text{for} \quad v \in V\right\}.$$

Since $\mathcal{K}_\lambda$ and $\mathcal{K}_{\overline{\lambda}}$ are both isomorphic to $L^2(K/M)$ it is clear from Hilbert space theory that $(V^0)^0 = V$. In particular

$$0 \neq V \neq \mathcal{K}_\lambda \Longleftrightarrow 0 \neq V^0 \neq \mathcal{K}_{\overline{\lambda}}.$$

Hence

$$\tau_\lambda \text{ is irreducible} \Longleftrightarrow \tau_{\overline{\lambda}} \text{ is irreducible.} \tag{17}$$

For each $\lambda \in \mathfrak{a}_c^*$ let $\mathcal{N}_\lambda$ denote the kernel of the map $\psi \longrightarrow \check{\psi}$ of $\mathcal{K}_\lambda$ into $\mathcal{E}_\lambda(X)$. Then by the definition (and Lemma 5.13 in Chapter II) $\mathcal{N}_\lambda \neq 0$ if and only if λ is not simple. For all λ, $\mathcal{N}_\lambda$ is a closed, G-invariant proper subspace of $\mathcal{K}_\lambda$. Thus if λ is not simple, τ_λ is not irreducible. But by the definition λ is simple if and only if $-\overline{\lambda}$ is simple. Combining this with (17) we see that

$$\tau_\lambda \text{ irreducible} \Longrightarrow \lambda \text{ and } -\lambda \text{ are simple.}$$

For the converse we define for each $\lambda \in \mathfrak{a}_c^*$ simple the Hilbert space $\mathcal{H}_\lambda$ by

$$(18)\quad \mathcal{H}_\lambda = \left\{ h \in \mathcal{E}_\lambda(X) : h(x) = \int_B e^{(i\lambda+\rho)(A(x,b))} F(b)\, db, \quad F \in L^2(B) \right\}$$

the norm $|h|_\lambda$ being defined as the L^2-norm of F. Then the mapping $\psi \longrightarrow \check{\psi}$ is an isometry of $\mathcal{K}_\lambda$ onto $\mathcal{H}_\lambda$ commuting with the action of G. Suppose $V \subset \mathcal{H}_\lambda$ is a nonzero G-invariant closed subspace and W the corresponding (closed) subspace of $\mathcal{K}_\lambda$. Select $h \in V$ such that $h(o) \neq 0$ and consider the function

$$h^\natural(x) = \int_K h(k \cdot x)\, dk = \int_B e^{(i\lambda+\rho)(A(x,b))} F^\natural(b)\, db,$$

where $F^\natural$ is the constant $\int F(b)\, db$. Identifying $\mathcal{K}_\lambda$ and $L^2(B)$ let W^0 be the annihilator of W in $\mathcal{K}_\lambda$. If $\varphi \in W^0$ then F and hence $F^\natural$ is orthogonal to φ so $F^\natural \in (W^0)^0 = W$. Thus $h^\natural \in V$ and $h^\natural(o) = h(o) \neq 0$ so $F^\natural \neq 0$.

Suppose now that $-\lambda$ is also simple. If $\varphi \in W^0$ then using (12) for $F = F^\natural$ we deduce (since $H(g^{-1}k) = -A(gK, kM)$)

$$\int_K e^{(-i\lambda+\rho)(A(gK,kM))} \varphi(k \cdot \xi_0)\, dk = 0 \qquad g \in G$$

so $\varphi = 0$. Thus $W = \mathcal{K}_\lambda$ and τ_λ irreducible as desired.

Now we combine this theorem with our simplicity criterion in Chapter III, Theorem 4.4. Using again the notation (Chapter III, §6, (9))

$$\Gamma_X(\lambda) = \prod_{\alpha \in \Sigma_0} \Gamma\left(\tfrac{1}{2}\left(\tfrac{1}{2}m_\alpha + 1 + \langle i\lambda, \alpha_0\rangle\right)\right) \Gamma\left(\tfrac{1}{2}\left(\tfrac{1}{2}m_\alpha + m_{2\alpha} + \langle i\lambda, \alpha_0\rangle\right)\right)$$

we can state the combined result as follows.

Corollary 3.5. *Let $\lambda \in \mathfrak{a}_c^*$. Then*

$$\tau_\lambda \textit{ is irreducible} \Longleftrightarrow \frac{1}{\Gamma_X(\lambda)} \neq 0.$$

Remarks. Let $\mathcal{K}_{\lambda,K}$ denote the space of K-finite vectors in $\mathcal{K}_\lambda$. (This coincides with the space $\mathcal{E}_\lambda(\Xi)_K$ in Chapter III, §3.) The action of G on $\mathcal{K}_\lambda$ induces a representation $d\tau_\lambda$ of $\mathfrak{g}$ and of the universal enveloping algebra $U(\mathfrak{g})$ on $\mathcal{K}_{\lambda,K}$. For G complex, Parthasarathy, Ranga-Rao and Varadarajan [1967], Theorem 4.3, determined the $\lambda \in \mathfrak{a}_{\mathbf{c}}^*$ for which $d\tau_\lambda$ is *algebraically* irreducible (no proper invariant subspace). Their method and irreducibility result were extended by Kostant [1975a] who proved the following criterion for real G.

For each $\alpha \in \Sigma_o^+$ let $I_\alpha \subset \boldsymbol{R}$ denote the interval $|t| < \frac{m_\alpha}{2}$ if $2\alpha \notin \Sigma^+$ and the interval $|t| < \frac{m_\alpha}{2} + 1$ if $2\alpha \in \Sigma^+$. Also put

$$\begin{aligned} \ell_\alpha &= \tfrac{1}{2}m_\alpha, & n_\alpha &= 1 \quad \text{if} \quad 2\alpha \notin \Sigma^+ \\ \ell_\alpha &= \tfrac{1}{2}m_\alpha + 1, & n_\alpha &= 2 \quad \text{if} \quad 2\alpha \in \Sigma^+. \end{aligned}$$

The criterion then reads as follows.

Theorem 3.6. *Let $\lambda \in \mathfrak{a}_{\mathbf{c}}^*$. Then $d\tau_\lambda$ is not (algebraically) irreducible if and only if there exists a root $\alpha \in \Sigma_0^+$ such that*

(i) $\langle i\lambda, \alpha_0\rangle \notin I_\alpha$

(ii) $\langle i\lambda, \alpha_0\rangle - \ell_\alpha \in n_\alpha \boldsymbol{Z}$.

Using the location of the poles of the Gamma function, its duplication formula, and the fact that if $m_{2\alpha} \neq 0$ then it is odd one can verify that the λ satisfying (i) and (ii) coincide with the solution to the equation $1/\Gamma_X(\lambda) = 0$. In order to reconcile this (algebraic) irreducibility criterion with the (topological) irreducibility criterion in Cor. 3.5 one uses a basic result of Harish-Chandra [1953], p. 228 which guarantees *a priori* that $d\tau_\lambda$ is algebraically irreducible if and only if τ_λ is irreducible. Our method of proof of Cor. 3.5 did not require this result, the method being analytic from the start.

Remark. Comparing Cor. 3.5 with Theorem 6.2 in Chapter III we note the remarkable fact that the eigenspace representation T_λ for G/K is irreducible if and only if the eigenspace representation τ_λ for G/MN is irreducible. One might ask whether this could be proved directly by proving *a priori* that T_λ and τ_λ are equivalent in some weak sense. This however is out of the question because while $T_{s\lambda} = T_\lambda$ for each $s \in W$, τ_λ is not in general equivalent to $\tau_{s\lambda}$. For example the condition from Prop. 4.15 in Chapter II

$$\frac{\langle i\lambda - \rho, \alpha\rangle}{\langle \alpha, \alpha\rangle} \in \boldsymbol{Z}^+ \qquad \alpha \in \Sigma^+ \tag{19}$$

is equivalent to $\mathcal{K}_\lambda(\Xi)$ containing a G-finite function and then its G-finite translates span an invariant irreducible finite-dimensional subspace. Since (19) is not invariant under $\lambda \longrightarrow s\lambda$ it is clear that τ_λ is not in general equivalent to $\tau_{s\lambda}$. Simplest example: $\tau_{-i\rho}$ is not equivalent to $\tau_{i\rho}$.

3. Conical Distributions and the Construction of the Intertwining Operators

We shall now prove certain equivalences in the spherical principal series $\{\widetilde{\tau}_\lambda : \lambda \in \mathfrak{a}_c^*\}$ and shall realize these equivalences by intertwining operators constructed by means of the conical distributions $\Psi_{\lambda,s}$. We recall that if (π_1, V_1) and (π_2, V_2) are two equivalent representations of G an *intertwining operator* between π_1 and π_2 is a continuous linear mapping $A : V_1 \longrightarrow V_2$ satisfying

$$A\pi_1(x) = \pi_2(x)A, \qquad x \in G. \tag{20}$$

We also recall that the mapping $S \longrightarrow \Psi$ where

$$\Psi(\varphi) = \int_{K/M} \left(\int_A \varphi(ka \cdot \xi_0) e^{(i\lambda+\rho)(\log a)} da \right) dS(kM) \tag{20}$$

is a bijection of

$$\mathcal{D}'(B) \text{ onto } \mathcal{D}'_\lambda(\Xi), \quad L^2(B) \text{ onto } \mathcal{K}_\lambda. \tag{21}$$

Hereby $\mathcal{D}'(B)$ and $L^2(B)$ become representation spaces for $\widetilde{\tau}_\lambda$ and τ_λ, respectively (cf. (12), (13)). Under the map $S \longrightarrow \Psi$ the distribution $\Psi_{\lambda,s}$ corresponds to the distribution $S_{\lambda,s}$ on B given by

(22)

$$S_{\lambda,s} : F \longrightarrow d_s(\lambda)^{-1} \int_{\overline{N}_s} F(m_s k(\overline{n}_s)M) e^{-(i\lambda+\rho)(H(\overline{n}_s))} d\overline{n}_s, \qquad F \in \mathcal{E}(B)$$

(cf. Chapter II, Theorem 5.4) defined for all $\lambda \in \mathfrak{a}_c^*$ by analytic continuation, $s \in W$ being arbitrary. Let $S_1 \times S_2$ denote the convolution on $\mathcal{D}'(B)$ as defined in Chapter II, §5, No. 2.

Theorem 3.7. *Let $s \in W$. Suppose $\lambda \in \mathfrak{a}_c^*$ satisfies the condition*

$$e_s(\lambda) e_s(-\lambda) \neq 0. \tag{23}$$

Then the convolution operator

$$C_{\lambda,s} : S \longrightarrow S \times S_{s\lambda,s^{-1}}, \qquad S \in \mathcal{D}'(B) \tag{24}$$

is a bijection of $\mathcal{D}'(B)$ onto itself intertwining $\widetilde{\tau}_\lambda$ and $\widetilde{\tau}_{s\lambda}$, i.e.,

$$C_{\lambda,s}\widetilde{\tau}_\lambda(g) = \widetilde{\tau}_{s\lambda}(g)C_{\lambda,s}, \qquad g \in G. \tag{25}$$

In particular, the representations $\widetilde{\tau}_\lambda$ and $\widetilde{\tau}_{s\lambda}$ are equivalent.

Proof. For $\Psi \in \mathcal{D}'_\lambda(\Xi)$ define the distribution $A_{\lambda,s}\Psi$ on Ξ by

$$(A_{\lambda,s}\Psi)(\varphi) = \int_{K/M}\left(\int_A \varphi(ka\cdot\xi_0)e^{(is\lambda+\rho)(\log a)}da\right)d(S\times S_{s\lambda,s^{-1}})(kM). \tag{26}$$

Because of Chapter II, Lemma 5.2 and Theorem 5.9 we obtain

$$(A_{\lambda,s}\Psi)^\vee(x) = \iint e^{(is\lambda+\rho)(A(x,k_1k_2M))}dS(k_1M)\,dS_{s\lambda,s^{-1}}(k_2M).$$

Here we write $A(x,k_1k_2M) = A(k_1^{-1}\cdot x, k_2M)$ and use Chapter II, Theorem 5.10. Then

$$(A_{\lambda,s}\Psi)^\vee(x) = a_{s^{-1}}\boldsymbol{e}_{s^{-1}}(s\lambda)\int_{K/M} e^{(i\lambda+\rho)(A(k^{-1}x,eM))}dS(kM)$$

so, since $\boldsymbol{e}_{s^{-1}}(s\lambda) = \boldsymbol{e}_s(-\lambda)$, and $a_s = a_{s^{-1}}$

$$(A_{\lambda,s}\Psi)^\vee = a_s\boldsymbol{e}_s(-\lambda)\overset{\vee}{\Psi}, \qquad \Psi \in \mathcal{D}'_\lambda(\Xi). \tag{27}$$

The dual transform $\Psi \longrightarrow \overset{\vee}{\Psi}$ commutes with the action of G and $\widetilde{\tau}_\lambda(g)\Psi = \Psi^{\tau(g)}$. Since $\widetilde{\tau}_\lambda(g)\Psi \in \mathcal{D}'_\lambda(\Xi)$, (27) implies

$$\begin{aligned}(A_{\lambda,s}\widetilde{\tau}_\lambda(g)\Psi)^\vee &= (a_s\boldsymbol{e}_s(-\lambda)\widetilde{\tau}_\lambda(g)\Psi)^\vee \\ &= a_s\boldsymbol{e}_s(-\lambda)\left(\overset{\vee}{\Psi}\right)^{\tau(g)} = ((A_{\lambda,s}\Psi)^\vee)^{\tau(g)} \\ &= \left((A_{\lambda,s}(\Psi))^{\tau(g)}\right)^\vee = (\widetilde{\tau}_{s\lambda}(g)A_{\lambda,s}\Psi)^\vee.\end{aligned}$$

Since $s\lambda$ is simple by (23) this implies

$$A_{\lambda,s}\widetilde{\tau}_\lambda(g) = \widetilde{\tau}_{s\lambda}(g)A_{\lambda,s}. \tag{28}$$

This proves (25) and in addition the fact that the operator

$$A_{\lambda,s} : \mathcal{D}'_\lambda(\Xi) \longrightarrow \mathcal{D}'_{s\lambda}(\Xi) \tag{29}$$

intertwines the representations $\widetilde{\tau}_\lambda$ and $\widetilde{\tau}_{s\lambda}$.

On the other hand by Chapter II, Theorem 5.17,

$$S_{\lambda,s} \times S_{s\lambda,s^{-1}} = a_s^2 \boldsymbol{e}_s(\lambda)\boldsymbol{e}_s(-\lambda)\delta_0.$$

Taking convolution with S on the left and noting that $S \times \delta_o = S$ we deduce that $C_{\lambda,s}$ is surjective. Being surjective and preserving K-types it is also injective.

Remark. Because of Theorem 5.9 and Proposition 3.17 in Chapter II, formula (27) can be written in terms of the Poisson transform:

$$\mathcal{P}_{s\lambda}C_{\lambda,s} = a_s\boldsymbol{e}_s(-\lambda)\mathcal{P}_\lambda \qquad \lambda \in \mathfrak{a}_c^* .$$

The proof also shows that

$$C_{\lambda,s}C_{s\lambda,s^{-1}} = a_s^2\boldsymbol{e}_s(\lambda)\boldsymbol{e}_s(-\lambda)I,$$

and the example in our previous remark shows that condition (23) cannot be omitted.

The space $\mathcal{D}'_\lambda(\Xi)$ contains the subspace $\mathcal{E}_\lambda(\Xi)$ which consists of the functions

$$F \otimes e^{i\lambda-\rho} : (kM, a) \longrightarrow F(kM)e^{(i\lambda-\rho)(\log a)}, \qquad F \in \mathcal{E}(B).$$

Then (26) specializes to

$$A_{\lambda,s}(F \otimes e^{i\lambda-\rho}) = (F \times S_{s\lambda,s^{-1}}) \otimes e^{is\lambda-\rho},$$

which by Theorem 5.18 in Ch. II and Lemma 3.2 and (15) in Ch. III, §3 can be written

$$\boldsymbol{d}_s(-\lambda)^{-1}\boldsymbol{c}_s(-\lambda)I_{\lambda,s}(F \otimes e^{i\lambda-\rho}).$$

We thus obtain

(30) $$A_{\lambda,s} = a_s\boldsymbol{e}_s(-\lambda)I_{\lambda,s} \qquad \text{on} \qquad \mathcal{E}_\lambda(\Xi).$$

Because of (27) and (30) we now normalize the intertwining operator $A_{\lambda,s}$ by

(31) $$I_{\lambda,s} = \frac{1}{a_s\boldsymbol{e}_s(-\lambda)}A_{\lambda,s}.$$

Theorem 3.8. *Suppose $\lambda \in \mathfrak{a}^*$ (real dual) and $s \in W$. Then $I_{\lambda,s}$ is an isometry of $\mathcal{K}_\lambda$ onto $\mathcal{K}_{s\lambda}$ and*

$$I_{\lambda,s}\tau_\lambda(g) = \tau_{s\lambda}(g)I_{\lambda,s} \qquad g \in G.$$

Proof. First remark that λ is simple. Next, by (27),

$$(I_{\lambda,s}\Psi)^\vee = \check{\Psi},$$

so the isometry statement above amounts to the fact that if $F \in L^2(B)$ and $S_1 \in \mathcal{D}'(B)$ is such that

$$\int_B e^{(i\lambda+\rho)(A(x,b))} F(b)\, db = \int_B e^{(is\lambda+\rho)(A(x,b))} dS_1(b) \tag{32}$$

for all x then $S_1 \in L^2(B)$ and F and S_1 have the same L^2 norm. Now, if $f \in \mathcal{D}(X)$ the Fourier transform $\widetilde{f}(\lambda, b)$ satisfies

$$\int_B e^{(i\lambda+\rho)(A(x,b))} \widetilde{f}(\lambda, b)\, db = \int_B e^{(is\lambda+\rho)(A(x,b))} \widetilde{f}(s\lambda, b)\, db. \tag{33}$$

Thus if F in (32) is the function $b \longrightarrow \widetilde{f}(\lambda, b)$ the simplicity of $s\lambda$ implies that S_1 is the function $b \longrightarrow \widetilde{f}(s\lambda, b)$. The equality of the L^2 norms is then clear from Chapter III, §1, (12). Now the identity of the norms of F and S_1 in (32) and the surjectivity of $I_{\lambda,s}$ follow from Chapter III, Prop. 1.6.

Corollary 3.9. *Let $\lambda \in \mathfrak{a}^*$, $s \in W$. If $f \in \mathcal{D}(X)$, $I_{\lambda,s}$ maps the function*

$$(kM, a) \longrightarrow \widetilde{f}(\lambda, kM) e^{(i\lambda-\rho)(\log a)}$$

into the function

$$(kM, a) \longrightarrow \widetilde{f}(s\lambda, kM) e^{(is\lambda-\rho)(\log a)}.$$

4. Convolution on G/MN

We have seen in Chapter II, §4 that if

$$\frac{\langle i\lambda - \rho, \alpha\rangle}{\langle \alpha, \alpha\rangle} \in \boldsymbol{Z}^+ \quad \text{for each } \alpha \in \Sigma^+ \tag{34}$$

then $\mathcal{D}'_\lambda(\Xi)$ contains a conical function ψ whose G-translates span the space $\mathcal{D}'_{\lambda,G}$ of G-finite elements in $\mathcal{D}'_\lambda(\Xi)$. Also the natural representation π_ψ of G on $\mathcal{D}'_{\lambda,G}$ is finite-dimensional and irreducible and

$$\psi(g \cdot \xi_0) = \langle \pi_\psi(g^{-1})\boldsymbol{e}, \boldsymbol{e}'\rangle, \tag{35}$$

where e and e', respectively, are highest weight vectors for π_ψ and its contragredient representation.

We shall now prove analogs of (35) for the infinite-dimensional representation $\widetilde{\tau}_\lambda$ in terms of the conical representations $\Psi_{\lambda,s}$ which (cf. Exercise 3) are the *extreme weight vectors* of $\widetilde{\tau}_\lambda$. For this we define a convolution on G/MN.

We have seen how the convolution product $*$ on G induces (by the compactness of K) a convolution product $S \times T$ for distributions on G/K at least one of which has compact support.

Since MN is noncompact one cannot use $*$ to define a convolution product on $\mathcal{E}'(G/MN)$. However, if one of the factors is MN-invariant this can be done as follows. The mapping $f \longrightarrow \overline{f}$, where

$$\overline{f}(gMN) = \int_{MN} f(gmn)\, dm\, dn \tag{36}$$

is a continuous surjection of $\mathcal{D}(G)$ onto $\mathcal{D}(\Xi)$ so the adjoint map $\Psi \longrightarrow \widetilde{\Psi}$ is an injection of $\mathcal{D}'(\Xi)$ into $\mathcal{D}'(G)$. Each $\widetilde{\Psi}$ is then right invariant under MN. If $\varphi \in \mathcal{D}(\Xi)$ and if $\eta \in \mathcal{D}'(\Xi)$ is MN-invariant we put

$$(\varphi \times \eta)(gMN) = \int_G f(gh^{-1})\, d\widetilde{\eta}(h) \tag{37}$$

if $f \in \mathcal{D}(G)$ is any function satisfying $\overline{f} = \varphi$. In order for this to be a valid definition we must show that

$$\int_G f(gh)\, d\widetilde{\eta}'(h) = 0 \qquad \text{if } \overline{f} \equiv 0, \tag{38}$$

$\widetilde{\eta}'$ being the image of $\widetilde{\eta}$ under the diffeomorphism $g \longrightarrow g^{-1}$ of G. Now $\left(\Psi^{\tau(mn)}\right)^{\sim} = (\widetilde{\Psi})^{L(mn)}$ so $\widetilde{\eta}'$ is right invariant under MN. If $\widetilde{\eta}'$ is a continuous function then (38) would be immediate from the integral formula $\int_G f = \int_\Xi \overline{f}$. But in any case $\widetilde{\eta}'$ can be approximated by functions $\alpha * \widetilde{\eta}'$ ($\alpha \in \mathcal{D}(G)$) which are right invariant under MN so (38) (and thus (37)) follows. Relation (38) also justifies defining $\eta^* \in \mathcal{D}'(\Xi)$ by

$$\eta^*(\overline{f}) = \widetilde{\eta}'(f) = \widetilde{\eta}(f'), \qquad f \in \mathcal{D}(G). \tag{39}$$

Then η^* is MN-invariant and we have the following result.

Lemma 3.10. *Let $\varphi \in \mathcal{D}(\Xi)$ and suppose $\eta \in \mathcal{D}'(\Xi)$ is MN-invariant. Then*

$$(\varphi \times \eta)(gMN) = \eta^*(\varphi \circ \tau(g)).$$

Proof. By (37) and (39) we have

$$(\varphi \times \eta)(gMN) = \widetilde{\eta}\left((f^{L(g^{-1})})'\right) = \eta^*\left((f^{L(g^{-1})})^-\right) \\ = \eta^*(\varphi^{\tau(g^{-1})})$$

as claimed.

We consider now the pairing $\langle\,,\,\rangle$ between the spaces $\mathcal{E}_\lambda(\Xi) = \mathcal{E}(\Xi) \cap \mathcal{K}_\lambda$ and $\mathcal{D}'_{-\lambda}$ given by

$$\langle \psi, \Psi \rangle = \int_{K/M} \psi(k \cdot \xi_0)\, dS(kM) \tag{40}$$

if

$$\Psi(\varphi) = \int_{K/M} \left(\int_A e^{(-i\lambda+\rho)(\log a)} \varphi(ka \cdot \xi_0)\, da \right) dS(kM). \tag{41}$$

Theorem 3.11. *With the pairing* (40) *the representations* τ_λ *on* $\mathcal{E}_\lambda$, $\widetilde{\tau}_{-\lambda}$ *on* $\mathcal{D}'_{-\lambda}$, *are contragredient. Moreover, for* $s \in W$, $\lambda \in \mathfrak{a}^*_c$ *we have*

$$\mathcal{D}(\Xi) \times \Psi_{\lambda,s} \in \mathcal{E}_\lambda(\Xi). \tag{42}$$

Finally, we have the following analog of (35): *if* $1/\Gamma_X(\lambda) \neq 0$ *then there is a constant* $a(\lambda, s) \in \boldsymbol{C}$ *such that*

$$a(\lambda, s)\Psi_{-s\lambda,e}(\varphi) = \langle \varphi \times \Psi_{\lambda,s}, \Psi_{-\lambda,s} \rangle, \qquad \varphi \in \mathcal{D}(\Xi). \tag{43}$$

Proof. The first statement amounts to

$$\langle \tau_\lambda(g^{-1})\psi, \widetilde{\tau}_{-\lambda}(g^{-1})\Psi \rangle = \langle \psi, \Psi \rangle, \qquad g \in G. \tag{44}$$

If S in (41) is a function, say F, then by (12) the left hand side of (44) equals

$$\int_{K/M} \psi(k(gk)MN)F(k(gk)M)e^{-2\rho(H(gk))}\, dk_M \\ = \int_{K/M} \psi(kMN)F(kM)\, dk_M = \langle \psi, \Psi \rangle.$$

For a general $\Psi \in \mathcal{D}'_{-\lambda}$ formula (44) then follows by approximation. For (42) we must verify

$$(\varphi \times \Psi_{\lambda,s})^{\sigma(a)} = e^{(-i\lambda+\rho)(\log a)}\varphi \times \Psi_{\lambda,s};$$

however this is obvious from the easy relation

$$\left(\widetilde{\Psi}_{\lambda,s}\right)^{R(a)} = e^{(-i\lambda+\rho)(\log a)}\widetilde{\Psi}_{\lambda,s} \; .$$

Finally for (43) we have, by Lemma 3.10,

$$(\varphi \times \Psi_{\lambda,s})(gMN) = (\Psi_{\lambda,s})^*(\varphi \circ \tau(g)) \tag{45}$$

so we must compute $\Psi_{\lambda,s}{}^*$.

Lemma 3.12. *Under the assumption* $\frac{1}{\Gamma_X(\lambda)} \neq 0$ *we have*

$$\Psi_{\lambda,s}{}^* = c(\lambda, s)\Psi_{-s\lambda,s^{-1}}$$

where $c(\lambda, s) \in \boldsymbol{C}$.

Proof. For simplicity we put $\eta = \Psi_{\lambda,s}$. Then we claim that η^* satisfies

$$(\eta^*)^{\sigma(a)} = e^{(is\lambda-\rho)(\log a)}\eta^* \tag{46}$$

$$(\eta^*)^{\tau(a)} = e^{(i\lambda-\rho)(\log a)}\eta^*. \tag{47}$$

The proofs being similar we just prove the first formula. For this note that

$$(\eta^*)^{\sigma(a)}(\varphi \circ \tau(g)) = \eta^*(\varphi^{\sigma(a)^{-1}} \circ \tau(g))$$

and if $\overline{f} = \varphi$, $f \in \mathcal{D}(G)$

$$\varphi^{\sigma(a)} = e^{2\rho(\log a)}\left(f^{R(a)}\right)^{-} .$$

Thus by (39)

$$(\eta^*)^{\sigma(a)}(\varphi \circ \tau(g)) = e^{-2\rho(\log a)} \int_{\boldsymbol{G}} f(gh^{-1}a)\, d\widetilde{\eta}(h)$$

whereas

$$(\widetilde{\eta}')^{R(a)} = \left(\widetilde{\eta}^{L(a^{-1})}\right)', \quad \widetilde{\eta}^{L(a)} = (\eta^{\tau(a)})^{\sim}$$

and

$$\eta^{\tau(a)} = e^{-(is\lambda+\rho)(\log a)}\eta \,.$$

Thus

$$\int_G f(gh^{-1}a)\, d\widetilde{\eta}(h) = \int_G f(gha)\, d\widetilde{\eta}'(h) = \int_G f(gh)\, d(\widetilde{\eta}')^{R(a)}(h)$$

so

$$(\eta^*)^{\sigma(a)}(\varphi \circ \tau(g)) = e^{-2\rho(\log a)} \int_G f(gh^{-1})\, d(\widetilde{\eta})^{L(a^{-1})}(h)$$

$$= e^{(is\lambda-\rho)(\log a)} \int_G f(gh^{-1})\, d\widetilde{\eta}(h) = e^{(is\lambda-\rho)(\log a)}(\varphi \times \eta)(gMN)\,,$$

which by Lemma 3.10 proves (46). Since η^* is MN-invariant we have proved that $\Psi_{\lambda,s}{}^*$ is a conical distribution in $\mathcal{D}'_{-s\lambda}$.

Now we assume temporarily that λ is regular. Then by Chapter II, Theorem 5.15, $\Psi_{\lambda,s}{}^*$ is a linear combination

$$\Psi_{\lambda,s}{}^* = \sum_{\sigma\in W} c_\sigma \Psi_{-s\lambda,\sigma} \qquad c_\sigma \in \boldsymbol{C}. \tag{48}$$

Invoking now Chapter II, Theorem 5.10 and (47) above we see that in (48) only the term with $\sigma = s^{-1}$ can appear. This proves the lemma if λ is regular. If λ is singular we approximate it with regular λ_n (all λ_n and λ satisfying the assumption of the lemma). Then $\Psi_{-s\mu,s^{-1}} \neq 0$ for $\mu = \lambda_n$ or λ. Select $\varphi \in \mathcal{D}(\Xi)$ such that $\Psi_{-s\lambda,s^{-1}}(\varphi) \neq 0$. Then for n sufficiently large

$$c(\lambda_n, s) = \frac{(\Psi_{\lambda_n,s})^*(\varphi)}{\Psi_{-s\lambda_n,s^{-1}}(\varphi)}. \tag{49}$$

The right hand side tends to a limit as $\lambda_n \longrightarrow \lambda$ and this limit is independent of the choice of the sequence $\lambda_n \longrightarrow \lambda$. The left hand side of (49) shows that the limit is also independent of φ. Hence we denote the limit by $c(\lambda, s)$. From the lemma for regular λ we have

$$\Psi^*_{\lambda_n,s}(\varphi) = c(\lambda_n, s)\Psi_{-s\lambda_n,s^{-1}}(\varphi)$$

for *all* $\varphi \in \mathcal{D}(\Xi)$. Letting $\lambda_n \longrightarrow \lambda$ we get the lemma in general.

We can now conclude the proof of Theorem 3.11, that is formula (43). We know $\varphi \times \Psi_{\lambda,s} \in \mathcal{E}_\lambda$ so by Lemmas 3.10 and 3.12 we just have to compute

$$\int_{K/M} \Psi_{-s\lambda,s^{-1}}(\varphi \circ \tau(k))\, dS_{-\lambda,s}(kM)\,.$$

This expression equals

$$\int_A e^{(-is\lambda+\rho)(\log a)} \left(\int_{K/M} \int_{K/M} \varphi(kk_1a \cdot \xi_0)\, dS_{-s\lambda,s^{-1}}(k_1M)\, dS_{-\lambda,s}(kM) \right),$$

which by Chapter II, Lemma 5.2 equals

$$\int_A e^{(-is\lambda+\rho)(\log a)} \int_{K/M} \varphi(ka \cdot \xi_0)\, d\left(S_{-\lambda,s} \times S_{-s\lambda,s^{-1}}\right)(kM)$$

and now the theorem follows from Chapter II, Theorem 5.17. Q.e.d.

§4. EIGENSPACE REPRESENTATIONS FOR THE COMPLEX SPACE G/N.

1. The algebra $D(G/N)$.

In this section we take the semisimple group G to be complex and simply connected. We can then state Proposition 3.1 and Chapter II, Prop. 4.15 in a sharper form. In particular we shall realize the principal series as well as *all* holomorphic finite-dimensional representations of G as eigenspace representations. Here a finite-dimensional representation π of G is said to be *holomorphic* if its differential $d\pi$ is a complex-linear representation of the Lie algebra $\mathfrak{g}$ of G, that is $d\pi(JX) = i\, d\pi(X)$ where J is the complex structure of $\mathfrak{g}$.

In the present case the Cartan decomposition $\mathfrak{g} = \mathfrak{k} + \mathfrak{p}$ is $\mathfrak{g} = \mathfrak{k} + J\mathfrak{k}$ and $\mathfrak{m}$, the centralizer of $\mathfrak{a}$ in $\mathfrak{k}$ can be chosen as $\mathfrak{m} = J\mathfrak{a}$. Thus M is the centralizer of the torus $\exp(J\mathfrak{a})$ in the compact group K, hence M is connected. The subalgebra $\mathfrak{h} = \mathfrak{a} + J\mathfrak{a}$ is a Cartan subalgebra of $\mathfrak{g}$. The corresponding analytic subgroup of G is denoted H. Then H, N, and $\overline{N}$ are complex analytic subgroups of G. Considering H and G/N as real manifolds let $D(H)$ and $D(G/N)$ denote the corresponding algebras of invariant differential operators. Now given $U \in D(H)$ define D_U by

(1) $$(D_U f)(gN) = \{U_h\,(f(ghN))\}_{h=e}, \qquad f \in \mathcal{E}(G/N).$$

Here U_h denotes the action of the differential operator U in the variable h. Since $hNh^{-1} \subset N$ $(h \in H)$ D_U is well defined. In terms of the fibration $G/N \longrightarrow G/HN$ with general fiber $\{ghN : h \in H\}$, D_U is a differential operator on the bundle G/N given by the constant coefficient operator U on each fiber. It is clear that $D_U \in D(G/N)$.

Theorem 4.1. *The mapping $U \longrightarrow D_U$ is an isomorphism of $\mathbf{D}(H)$ onto $\mathbf{D}(G/N)$. In particular, $\mathbf{D}(G/N)$ is commutative.*

For the proof of this result, which resembles Theorem 2.2 in Chapter II, consider the direct decomposition $\mathfrak{g} = \mathfrak{n} + (\mathfrak{h} + \bar{\mathfrak{n}})$ and the corresponding projection $\sigma : \mathfrak{g} \longrightarrow \bar{\mathfrak{n}} + \mathfrak{h}$. If o denotes the origin $\{N\}$ in G/N and $\pi : G \longrightarrow G/N$ the natural map we have the bijection

$$d\pi_e : \bar{\mathfrak{n}} + \mathfrak{h} \longrightarrow (G/N)_o \tag{2}$$

and it satisfies

$$d\pi_e(\sigma \circ Ad_G(n)X) = d\tau(n)(d\pi_e(X)), \qquad X \in \bar{\mathfrak{n}} + \mathfrak{h}. \tag{3}$$

Thus, in order to determine the polynomials on the tangent space $(G/N)_o$ invariant under the action of N, we must determine the invariants of the group $\sigma \circ Ad_G(N)$ in the symmetric algebra $S(\bar{\mathfrak{n}} + \mathfrak{h})$.

Lemma 4.2. *Let $I(\bar{\mathfrak{n}} + \mathfrak{h})$ denote the set of elements in $S(\bar{\mathfrak{n}} + \mathfrak{h})$ which are invariant under the group $\sigma \circ Ad_G(N)$. Then*

$$I(\bar{\mathfrak{n}} + \mathfrak{h}) = S(\mathfrak{h}).$$

Proof. The inclusion $S(\mathfrak{h}) \subset I(\bar{\mathfrak{n}} + \mathfrak{h})$ is trivial since $\sigma \circ Ad_G(n)$ is the identity on $\mathfrak{h}$. For the converse let $\alpha_1 > \alpha_2 > \cdots > \alpha_p > 0$ be the positive roots of $\mathfrak{g}$ with respect to $\mathfrak{h}$ and put $\beta_i = -\alpha_i$. Select a basis $E_{\beta_i}, F_{\beta_i} = JE_{\beta_i}$ of $\mathfrak{g}^{\beta_i}$ $(1 \leq i \leq p)$. For $X \in \mathfrak{n}$ let $d(X)$ denote the derivation of the symmetric algebra $S(\bar{\mathfrak{n}} + \mathfrak{h})$ extending the endomorphism $\sigma \circ ad_{\mathfrak{n}}(X)$ of $\bar{\mathfrak{n}} + \mathfrak{h}$. Then $P \in I(\bar{\mathfrak{n}} + \mathfrak{h})$ if and only if $d(X)P = 0$ for $X \in \mathfrak{n}$. Let $P \in I(\bar{\mathfrak{n}} + \mathfrak{h})$. Then

$$P = \sum_{(m)(n)} a_{(m,n)} E_{\beta_1}^{m_1} F_{\beta_1}^{n_1} \cdots E_{\beta_p}^{m_p} F_{\beta_p}^{n_p}, \qquad a_{(m,n)} \in S(\mathfrak{h}). \tag{4}$$

Let $X_\alpha \neq 0$ in $\mathfrak{g}^\alpha$, $X_\beta \neq 0$ in $\mathfrak{g}^\beta$ $(\alpha > 0, \beta < 0)$.

$$\begin{aligned} d(X_\alpha)(X_\beta) &= 0 && \text{if } \alpha + \beta > 0, \\ d(X_\alpha)(X_\beta) &\in \mathfrak{h} - \{0\} && \text{if } \alpha + \beta = 0. \end{aligned}$$

Hence $d(X_{\alpha_1})$ annihilates $\mathfrak{h}$ and all $\mathfrak{g}^{\beta_i}$ except for $i = 1$. We now write P in the form

$$P = \sum_{(m_1)(n_1)} b(m_1, n_1) E_{\beta_1}^{m_1} F_{\beta_1}^{n_1}, \tag{5}$$

where $b(m_1, n_1)$ is a polynomial in the vectors E_{β_i} F_{β_i} $(i > 1)$ and coefficients in $S(\mathfrak{h})$.

Since

$$d(X_{\alpha_1})\left(E_{\beta_1}^{m_1}F_{\beta_1}^{n_1}\right) = m_1E_{\beta_1}^{m_1-1}F_{\beta_1}^{n_1}A_1 + n_1E_{\beta_1}^{m_1}F_{\beta_1}^{n_1-1}JA_1\,,$$

where $A_1 = [X_{\alpha_1}, E_{\beta_1}]$, the relation $d(X_{\alpha_1})P = 0$ implies

$$(m+1)b(m+1,n)A_1 + (n+1)b(m,n+1)JA_1 = 0 \qquad (m \geq 0, n \geq 0).$$

We conclude that $b(m+1,n) = b(m,n+1) = 0$ so the expression (5) contains neither E_{β_1} nor F_{β_1}. Applying $d(X_{\alpha_2}),\cdots,d(X_{\alpha_p})$ successively in the similar fashion we get the desired result $P \in S(\mathfrak{h})$.

In Theorem 4.1 the only statement which is not immediate is the surjectivity and for this we only sketch the proof because of the similarity with Theorem 2.2 in Chapter II. Let $D \in \mathbf{D}(G/N)$. Let $X_1,\ldots,X_r,\ldots,X_q$ be a basis of $\mathfrak{h}+\bar{\mathfrak{n}}$ with the r first elements in $\mathfrak{h}$, the rest in $\bar{\mathfrak{n}}$. We express D at o in the coordinate system

$$\exp(\sum_1^q t_iX_i)N \longrightarrow (t_1,\ldots,t_q)\,.$$

Then

$$(Df)(o) = \{P(\partial_1,\ldots,\partial_q)f\left(\exp(\sum_1^q t_iX_i)N\right)\}_{t_i=0} \tag{6}$$

where $\partial_i = \partial/\partial t_i$ and P is a polynomial. We decompose P into monomials and let P^* denote the sum of the terms of highest total degree. For $n \in N$ we write the mapping $\tau(n) : gN \longrightarrow ngN$ in coordinates

$$\tau(n)\exp\left(\sum t_iX_i\right)N = \exp\left(\sum s_iX_i\right)N\,.$$

Putting

$$F(t_1,\ldots,t_q) = f\left(\exp\left(\sum t_iX_i\right)N\right)$$

the invariance $D(f\circ\tau(n))(o) = (Df)(o)$ implies

$$\left\{P\left(\frac{\partial}{\partial t_1},\cdots,\frac{\partial}{\partial t_q}\right)[F(t_1,\ldots,t_q) - F(s_1,\ldots,s_q)]\right\}_{t_i=0} = 0\,. \tag{7}$$

Although the coordinate change $(t_1,\ldots,t_q) \longrightarrow (s_1,\ldots,s_q)$ is not linear the sum $P^*(\partial_1,\ldots,\partial_q)$ of the highest degree terms in $P(\partial_1,\ldots,\partial_q)$ is transformed by the *linear* transformation $d\tau(n)_o$. Since F is arbitrary (7) implies that P^* is invariant under the action of N so by Lemma 4.2, $P^* \in S(\mathfrak{h})$. We can then define D_{P^*} by (1). Then $D - D_{P^*}$ has lower order than D at o, hence everywhere by the invariance. Using induction we deduce $D = D_U$ for some $U \in S(\mathfrak{h})$ proving Theorem 4.1.

2. The Principal Series.

Because of (1) and Theorem 4.1 it is easy to describe the joint eigenfunctions f of the operators in $\boldsymbol{D}(G/N)$. In fact such functions are by (1) characterized by

$$\{U_h\,(f(ghN))\}_{h=e} = \chi(U)f(gN), \qquad \chi(U) \in \boldsymbol{C}. \tag{8}$$

However, U is invariant under all translations on the commutative group H so (8) can be written

$$U_h\,(f(ghN)) = \chi(U)f(ghN)\,, \tag{9}$$

which means that $h \longrightarrow f(ghN)$ is a multiple of a homomorphism ω of H into the multiplicative group $\boldsymbol{C}^* = \boldsymbol{C} - (0)$. Hence by evaluating at $h = e$, we get

$$f(ghN) = f(gN)\omega(h)\,. \tag{10}$$

The converse being obvious we thus have the following result.

Corollary 4.3. *Let $\omega : H \longrightarrow \boldsymbol{C}^*$ be a C^∞ homomorphism and let*

$$E_\omega = \{f \in \mathcal{E}(G/N) \,:\, f(ghN) \equiv f(gN)\omega(h)\}\,.$$

Then E_ω is a joint eigenspace for $\boldsymbol{D}(G/N)$ and all joint eigenspaces of $\boldsymbol{D}(G/N)$ are obtained in this way.

Comparing (10) above with the definition of the principal series ((1) in §3) we can restate this corollary as follows.

Corollary 4.4. *The principal series representations for G are precisely the eigenspace representations for G/N.*

3. The Finite-Dimensional Holomorphic Representations

We shall now realize the irreducible finite-dimensional (holomorphic) representations of G as eigenspace representations for G/N. Let $\mathcal{H}(G/N)$ denote the space of holomorphic functions on G/N and π the natural representation of G on $\mathcal{H}(G/N)$.

Theorem 4.5. *For an algebra homomorphism*

$$\chi : \boldsymbol{D}(G/N) \longrightarrow \boldsymbol{C}$$

let $\mathcal{H}_\chi$ denote the joint holomorphic eigenspace

$$\mathcal{H}_\chi = \{f \in \mathcal{H}(G/N) \, : \, Df = \chi(D)f \qquad \text{for } D \in \boldsymbol{D}(G/N)\}$$

and π_χ the natural representation of G on $\mathcal{H}_\chi$. Then

(i) π_χ is finite-dimensional and irreducible.

(ii) Each finite-dimensional irreducible holomorphic representation of G is equivalent to such a π_χ and is contained in π exactly once.

To prove this we note first from Cor. 4.3 that the holomorphic eigenspaces $\mathcal{H}_\chi$ are just the spaces

(11) $$\mathcal{H}_\omega = \{f \in \mathcal{H}(G/N) \, : \, f(ghN) = f(gN)\omega(h)\}$$

where $\omega : H \longrightarrow \boldsymbol{C}^*$ is a holomorphic homomorphism.

We extend ω to a homomorphism of the group $B = HN$ into $\boldsymbol{C}^*$ by putting $\omega(hn) = \omega(h)$. Then $\mathcal{H}_\omega$ can be identified with the space

(12) $$V_\omega = \{F \text{ holomorphic on } G\,, \quad F(gb) \equiv \omega(b)F(g)\}\ .$$

The homomorphism $\omega : B \longrightarrow \boldsymbol{C}^*$ defines a holomorphic line bundle over G/B in which V_ω is the space of holomorphic sections.

Lemma 4.6. $\dim V_\omega < \infty$.

Proof. The space G/B is compact and the vector space V_ω becomes a Banach space when topologized by uniform convergence on G/B. Using the fact that a uniformly bounded sequence of holomorphic functions has a subsequence converging uniformly on compacts, V_ω is easily seen to be locally compact. Thus V_ω is a locally compact Banach space, hence finite-dimensional.

Lemma 4.7. (The Borel-Weil Theorem). *The natural representation π_ω of G on V_ω is irreducible.*

Proof. By the semisimplicity of G, $V_\omega = \bigoplus_i V_i$ where G acts irreducibly on V_i. Let $F \in V_i$ be a lowest weight vector. Then $F(\overline{n}g) \equiv F(g)$. Thus $F(\overline{n}hn) = F(h) = \omega(h)F(e)$. Since $\overline{N}HN$ contains a neighborhood of e in G and since F is holomorphic, this shows that $\boldsymbol{C}F$ is the same for all i. This proves the lemma.

For Theorem 4.5 the two lemmas above show that π_χ has the stated properties. On the other hand suppose τ is an irreducible finite-dimensional holomorphic representation of G. Let v be a lowest weight vector and let $\omega : H \longrightarrow \boldsymbol{C}^*$ be the homomorphism determined by $\tau(h)v = \omega(h)v$. Then ω is holomorphic and the natural representation π_ω of G on V_ω is irreducible and has the lowest weight vector φ satisfying

$$\varphi(\overline{n}g) = \varphi(g) \qquad \varphi(ghn) = \varphi(g)\omega(h).$$

Thus $\varphi(h^{-1}\overline{n}h_1 n) = \varphi(h^{-1}h_1) = \omega(h^{-1})\varphi(\overline{n}h_1 n)$ so $\varphi(h^{-1}g) = \omega(h^{-1})\varphi(g)$. Hence π_ω is equivalent to the contragredient representation of τ. Also different ω give inequivalent π_ω. This proves the theorem.

Remark. By Corollary 4.4 and Theorem 4.5 the irreducible finite-dimensional holomorphic representations of G are certain subrepresentations of the principal series.

§5. TWO MODELS OF THE SPHERICAL REPRESENTATIONS.

For the symmetric space $X = G/K$ and its horocycle space $\Xi = G/MN$ let T_X denote the natural representation of G on $L^2(X)$ and T_Ξ the natural representation of G on $L^2(\Xi)$. Consider the Hilbert space $\mathcal{H}_\lambda$ from (18) §3 with norm $|\quad|_\lambda$ and let T_λ denote the natural representation of G on $\mathcal{H}_\lambda$. Then Theorems 1.3, 1.5 in Chapter III imply that in terms of direct integral theory

$$L^2(X) = \int_{\mathfrak{a}^*/W} \mathcal{H}_\lambda |\boldsymbol{c}(\lambda)|^{-2} d\lambda, \qquad T_X = \int_{\mathfrak{a}^*/W} T_\lambda |\boldsymbol{c}(\lambda)|^{-2} d\lambda \tag{1}$$

the "component" of $f \in \mathcal{D}(X)$ in $\mathcal{H}_\lambda$ being given by

$$f_\lambda(x) = \int_B \widetilde{f}(\lambda, b) e^{(i\lambda+\rho)(A(x,b))} db. \tag{2}$$

A similar reduction of T_Ξ is quite elementary. In analogy with the Fourier transform on X we define for $\psi \in \mathcal{D}(\Xi)$

$$\psi^*(\lambda, kM) = \int_A \psi(ka \cdot \xi_o) e^{(-i\lambda+\rho)(\log a)} da. \tag{3}$$

Then

$$\psi(ka \cdot \xi_o) = \int_{\mathfrak{a}^*} \psi^*(\lambda, kM) e^{(i\lambda-\rho)(\log a)} d\lambda. \tag{4}$$

As ψ runs through $\mathcal{D}(\Xi)$ the functions ψ_λ given by

$$\psi_\lambda(ka \cdot \xi_o) = \psi^*(\lambda, kM)e^{(i\lambda-\rho)(\log a)} \tag{5}$$

form a dense subspace of the Hilbert space $\mathcal{K}_\lambda$ from §3 with norm $\| \quad \|_\lambda$. Then since $d\xi = e^{2\rho(\log a)} da\, dk_M$ we deduce from (3)–(5),

$$\int_\Xi |\psi(\xi)|^2 d\xi = \int_{\mathfrak{a}^*} \|\psi_\lambda\|_\lambda^2 d\lambda. \tag{6}$$

This amounts to the direct integral decomposition

$$L^2(\Xi) = \int_{\mathfrak{a}^*} \mathcal{K}_\lambda d\lambda, \qquad T_\Xi = \int_{\mathfrak{a}^*} \tau_\lambda d\lambda. \tag{7}$$

We shall now verify that the decompositions (1) and (7) correspond via the Radon transform $f \longrightarrow \widehat{f}$ and its dual $\psi \longrightarrow \overset{\vee}{\psi}$. As before, let $\mathcal{H}_W(\mathfrak{a}_{\mathbf{c}}^*)$ denote the space of W-invariant holomorphic functions of exponential type on $\mathfrak{a}_{\mathbf{c}}^*$.

Theorem 5.1. *For $f \in \mathcal{D}(X)$ we have the following relations connecting (1) and (7):*

$$[\frac{1}{w}\int_{\mathfrak{a}^*} A(\lambda) f_\lambda |\mathbf{c}(\lambda)|^{-2} d\lambda]^\wedge = \int_{\mathfrak{a}^*} A(\lambda)(\widehat{f})_\lambda d\lambda, \qquad A \in \mathcal{H}_W(\mathfrak{a}_{\mathbf{c}}^*). \tag{8}$$

$$f_\lambda = ((\widehat{f})_\lambda)^\vee, \tag{9}$$

$$|f_\lambda|_\lambda = \|(\widehat{f})_\lambda\|_\lambda. \tag{10}$$

Remark. Formula (8) amounts to the function $w^{-1}|\mathbf{c}(\lambda)|^{-2} f_\lambda$ having Radon transform $(\widehat{f})_\lambda$ in a weak sense, the Radon transform $(f_\lambda)^\wedge$ itself not being defined for lack of proper decay at ∞.

Proof. From the Paley-Wiener type theorem for the spherical transform we know that the function

$$a(x) = w^{-1} \int_{\mathfrak{a}^*} A(\lambda)\varphi_\lambda(x)|\mathbf{c}(\lambda)|^{-2} d\lambda, \qquad x \in X, \tag{11}$$

belongs to $\mathcal{D}^{\natural}(X)$ and

$$A(\lambda) = \int_X a(x)\varphi_{-\lambda}(x)dx. \tag{12}$$

Because of (12) and Chapter III, Lemma 1.2,

$$(f \times a)(x) = w^{-1} \int_{\mathfrak{a}^*} A(\lambda) f_\lambda(x) |\mathbf{c}(\lambda)|^{-2} d\lambda. \tag{13}$$

Using now the elementary Exercise 1 in Ch. IV,

$$(f \times a)^\wedge(kb \cdot \xi_o) = \int_A \widehat{f}(kbc^{-1} \cdot \xi_o)\widehat{a}(c \cdot \xi_o)dc \tag{14}$$

and

$$A(\lambda) = \int_A \widehat{a}(c \cdot \xi_o)e^{\rho(\log c)}e^{-i\lambda(\log c)}dc. \tag{15}$$

Inverting (15) and substituting in (14) we obtain

$$(f \times \widehat{a})(kb \cdot \xi_o) = \int_{\mathfrak{a}^*} A(\lambda)\Big(\int_A \widehat{f}(kbc \cdot \xi_o)e^{\rho(\log c)}e^{-i\lambda(\log c)}dc\Big)d\lambda$$

$$= \int_{\mathfrak{a}^*} A(\lambda)(\widehat{f})^*(\lambda, kM)e^{(i\lambda-\rho)(\log b)}d\lambda = \int_{\mathfrak{a}^*} A(\lambda)(\widehat{f})_\lambda(kb \cdot \xi_o)d\lambda.$$

Because of (13) this proves the first relation of the theorem.

Next the function ψ_λ in (5) has dual transform

$$(\psi_\lambda)^\vee(x) = \int_B e^{(i\lambda+\rho)(A(x,b))}\psi^*(\lambda, b)db,$$

which for $\psi = \widehat{f}$ reduces to $f_\lambda(x)$ (cf. (25) in Ch. III, §5). This proves (9) and (10) is obvious from the definition of the norms.

EXERCISES AND FURTHER RESULTS.

1. Show that for a Lie group coset space K/M with K compact, M compact each eigenspace representation is irreducible. Find an example of K/M for which the only joint eigenspace consists of the constants.

2. Fix $\lambda \in \mathfrak{a}^*$ and put $P = MAN$ in our customary notation. Without using Corollary 3.5 prove that the following two statements are equivalent.

(i) The unitary representaion τ_λ on $\mathcal{K}_\lambda$ is irreducible.

(ii) All distributions $S \in \mathcal{D}'(G)$ satisfying the homogeneity condition

$$S^{L(p_2^{-1})R(p_1)} \equiv e^{(i\lambda+\rho)(\log a_1)} e^{(-i\lambda+\rho)(\log a_2)} S$$

$(p_i = m_i a_i n_i)$ are proportional and in fact

$$S(f) = c \int\limits_{MAN} e^{(-i\lambda+\rho)(\log a)} f(man) dm\, da\, dn, \qquad (c = \text{ const.}).$$

3. (Extreme weight vectors and extreme weights.) Let $\lambda \in \mathfrak{a}_{\mathfrak{c}}^*$. A vector $\Psi \in \mathcal{D}'_\lambda(\Xi)$ is called an *extreme weight vector* if there exists a $\mu \in \mathfrak{a}_{\mathfrak{c}}^*$ such that

$$\Psi^{\tau(an)} = e^{\mu(\log a)} \Psi.$$

The corresponding μ are called the *extreme weights* of the representation $\widetilde{\tau}_\lambda$ of G on $\mathcal{D}'_\lambda(\Xi)$.

Suppose $\pi(\lambda) e_{s^*}(\lambda) \neq 0$. Then the extreme weight vectors are the nonzero multiples of $\Psi_{\lambda,s} (s \in W)$ and the corresponding extreme weights are $-is\lambda + \rho$.

4. In the notation of §3 show that the space $\mathcal{D}(\Xi) \times \Psi_{\lambda,e}$ is dense in $\mathcal{K}_\lambda$.

5. Let $\Psi \in \mathcal{E}'(\Xi)$ and suppose $\eta \in \mathcal{D}'(\Xi)$ is MN-invariant. Let $\Psi \times \eta$ denote the functional on $\mathcal{D}(\Xi)$ given by

$$(\Psi \times \eta)(\varphi) = \Psi(\varphi \times \eta^*) \qquad \varphi \in \mathcal{D}(\Xi)$$

with η^* defined as in (39) §3. Show that $\Psi \times \eta$ is a distribution and that if $\Psi = \psi \in \mathcal{D}(\Xi)$ then

$$(\Psi \times \eta)(\varphi) = (\psi \times \eta)(\varphi)$$

in the sense of (37) §3.

6.* Let G be a simply connected nilpotent Lie group, α a nonzero element of the dual $\mathfrak{g}^*$ of the Lie algebra $\mathfrak{g}$ of G and $\mathfrak{k} \subset \mathfrak{g}$ a subalgebra of maximal dimension satisfying $\alpha([\mathfrak{k}, \mathfrak{k}]) = 0$. Let $H \subset G$ be the analytic subgroup with Lie algebra $\mathfrak{h} = \mathfrak{k} \cap \text{kernel}(\alpha)$. Then

(i) $\mathfrak{h}$ is an ideal in $\mathfrak{k}$ of codimension 1.

(ii) Let $Z \in \mathfrak{k}$ such that $\mathfrak{k} = \boldsymbol{R}Z + \mathfrak{h}$ and $\gamma(Z)$ the vector field on G/H given by

$$(\gamma(Z)f)(gH) = \left\{ \frac{d}{dt} f(g \exp(tZ)H) \right\}_{t=0}.$$

Then $\boldsymbol{D}(G/H)$ is generated by $\gamma(Z)$.

(iii) If $\chi : \boldsymbol{D}(G/H) \longrightarrow \boldsymbol{C}$ is a nonzero character the corresponding eigenspace representation T_χ of G (Ch. II §2, No. 3) is irreducible.

(iv) Each unitary irreducible representation of G is a subrepresentation of the representation of G on the eigenspace

$$\{T \in \mathcal{D}'(G/H) : DT = \chi(D)T \quad \text{for} \quad D \in \boldsymbol{D}(G/H)\}$$

for a suitable choice of d, χ and H. (Jacobsen and Stetkær [1981]; for an extension to exponential groups and to solvable groups see Stetkær [1983], Jacobsen [1982] and [1985]).

7.* (Jacobsen [1982]). Let G be a connected Lie group with Lie algebra $\mathfrak{g}$, $H \subset G$ a closed subgroup with Lie algebra $\mathfrak{h} \subset \mathfrak{g}$ and $\mathfrak{n}(\mathfrak{h})$ the normalizer of $\mathfrak{h}$ in $\mathfrak{g}$. Assume

(i) $Ad_G(h)X - X \in \mathfrak{h}$ for all $h \in H$, $X \in \mathfrak{n}(\mathfrak{h})$.

(ii) There is an $\alpha \in \mathfrak{g}^*$ such that $\alpha(\mathfrak{h}) = 0$ and

$$\mathfrak{n}(\mathfrak{h}) = \{X \in \mathfrak{g} \ : \ \alpha([X, \mathfrak{h}]) = 0\}.$$

a) Show there is a natural isomorphism of

$$U(\mathfrak{n}(\mathfrak{h})/\mathfrak{h}) \quad \text{onto} \quad \boldsymbol{D}(G/H),$$

U referring to the universal enveloping algebra.

b) Deduce from a) by verifying assumptions (i) and (ii) that

$$\boldsymbol{D}(A) \text{ is isomorphic to } \boldsymbol{D}(G/MN)$$

(Theorem 2.2, Chapter II);

$$\boldsymbol{D}(MA) \text{ is isomorphic to } \boldsymbol{D}(G/N),$$

(generalizing Theorem 4.1 in this chapter).

8.* A representation π of a group G on a topological vector space is *scalar irreducible* if the only operators commuting with all $\pi(g)$ ($g \in G$) are scalars. Similar definition applies to Lie algebras. (For unitary representations this is equivalent to (topological) irreducibility). For the following spaces the eigenspace representations are all scalar irreducible:

(i) Riemannian symmetric space G/K of the noncompact type and any coset space K/M (K and M compact) (Stetkær [1985a]).

(ii) The tangent space $X_o = G_o/H$ to a symmetric space $X = G/H$ (H not necessarily compact) G_o denoting the group of transformations of X_o generated by the translations and the natural action of H. (Stetkær [1985b], Schlichtkrull and Stetkær [1987]).

(iii) The representation $X \longrightarrow \eta(X)$ of the Lie algebra $\mathfrak{c}$ of conformal vector fields X on the space $\mathcal{E}_o(\boldsymbol{R}^n)$ of harmonic functions on $\boldsymbol{R}^n$ given by

$$\eta(X)f = Xf - \frac{n-2}{2n}(\operatorname{div} X)f \qquad f \in \mathcal{E}_o(\boldsymbol{R}^n)$$

is scalar irreducible (Helgason [1977a]).

9.* Let G/K be a symmetric space of the noncompact type. Assume rank $G = \text{rank} K$ so according to Harish-Chandra [1966] G has a discrete series. Consider the joint eigenspace $\Gamma_\delta(\mathcal{A}, \chi)$ from (15) §1 where $\mathcal{A}$ is the subalgebra of $\boldsymbol{D}(G \times_K V)$ generated by the Casimir operator Ω, $\chi(\Omega) = \langle\lambda+2\rho, \lambda\rangle$ where $\lambda+2\rho_k$ is the lowest weight of δ ($2\rho_k$ (resp. 2ρ) being the sum of the positive compact roots (resp. positive roots)). Let $\mathcal{H}_\lambda$ be the Hilbert space of square-integrable sections in $\Gamma_\delta(\mathcal{A}, \chi)$, and assume $\langle\lambda + \rho, \alpha\rangle < 0$ for all positive roots α. Then there exists a constant $a > 0$ such that if $|\langle\lambda + \rho, \beta\rangle| > a$ for all noncompact roots β then the representation of G on $\mathcal{H}_\lambda$ belongs to the discrete series of G. (Takahashi [1963], Hotta [1971], Schmid [1971].)

Thus most of the discrete series for G is realized by (vector bundle) eigenspace representations for G/K.

NOTES

As a natural generalization of spherical harmonics, eigenspace representations for homogeneous spaces were proposed by the author in [1970a,b], [1977b], [1979] and it has turned out that most representations of Lie groups which are familiar (to this author at least) can be realized in this manner. See Prop. 3.1 for the principal series, Theorem 4.5 for the finite-dimensional representations, Exercise 9 (Hotta) for the discrete series, Exercise 6 (Jacobsen, Stetkær) for nilpotent and solvable groups. However this eigenspace viewpoint does not single out unitary representations in any general fashion.

The Euclidean case (Theorem 1.1 and the more general Theorem 2.4) is from the authors' papers [1974], [1980a]. Many of the notions in §1, No. 3 are discussed in Bott [1965] and Wallach [1973].

The irreducibility criterion for the eigenspace representations for a symmetric space (Theorem 2.5) is proved in Helgason [1970a] for rank one spaces and in [1976] in general. The corresponding problem for non-Riemannian symmetric spaces seems still to be open, but for the non-Riemannian hyperbolic spaces the irreducibility criteria and even the composition series are completely determined by Schlichtkrull [1989].

The criterion in Theorem 3.6 is from Parthasarathy, Ranga-Rao and Varadarajan [1967] for complex G and Kostant [1969], [1975a] for real G. A simplification of the proof was given by Schiffmann [1979]. Corollary 3.5 was obtained by the author in [1970a] and [1976].

Intertwining operators enter significantly in the works of Mackey [1952, 1953] and Bruhat [1956]. For complex G intertwining operators for the principal series were constructed by Kunze and Stein [1967]. For real G the intertwining operators for the spherical principal series were constructed in Helgason [1970a] by means of the conical distributions (Theorems 3.7, 3.8 and Cor. 3.9). Different constructions were given by Schiffmann [1971] and Knapp-Stein [1971]. The relationship of our construction to that of Schiffmann is formula (30) in §3 which amounts to the identity $S^*_{-\lambda,s} = S_{s\lambda,s^{-1}}$ in Theorem 5.18, Ch. II. Intertwining operators have been the subject of much further study (see for example Vogan and Wallach [1990] and Wallach's books [1973], [1989], [1993]).

§3, No.4 and §§4–5 are from the author's paper [1970a]. The proof of Lemma 4.6 given there is quite different and does not use the compactness of G/B. The lemma is essentially known from Bochner [1951], but the proof in the text was communicated to me by Serre.

SOLUTIONS TO EXERCISES

CHAPTER I

A. Radon Transform on $\boldsymbol{R}^n$.

A. 1. By §2 (27) each $E_k \otimes p^\ell$ belongs to $\widetilde{\mathcal{N}}$. Conversely let $\psi \in \widetilde{\mathcal{N}}$. Let $G = \boldsymbol{M}(n)$ with Haar measure dg, let ξ_o be the hyperplane $x_n = 0$ in $\boldsymbol{R}^n$, and let $\widetilde{\psi}(g) = \psi(g \cdot \xi_o)$ for $g \in G$. For $F \in \mathcal{D}(G)$, $\xi = h \cdot \xi_o$ put

$$\psi_F(\xi) = \int_G F(g)\psi(g^{-1}\xi)dg = \int_G F(g)\psi(g^{-1}h \cdot \xi_o)dg$$

which lies in $\mathcal{E}(\boldsymbol{P}^n) \cap \widetilde{\mathcal{N}} = \mathcal{N}$. Let F run through a sequence (F_i) with $F_i \geq 0$, $\int F_i = 1$, $\operatorname{supp}(F_i) \longrightarrow e$. Then $\psi_{F_i} \longrightarrow \psi$ in $C(\boldsymbol{P}^n)$ so statement follows from Theorem 2.5.

A. 2. For the Fourier transform $\widetilde{\varphi}(s)$ we have

$$\varphi^{(k)}(p) = \frac{1}{2\pi} \int_{\boldsymbol{R}} \widetilde{\varphi}(s)(is)^k\, e^{isp} ds.$$

By the definition

$$(Y_k \otimes \varphi)^\vee(x) = \frac{1}{\Omega_n} \int \varphi((x,\eta))Y_k(\eta)d\eta$$
$$= \frac{1}{\Omega_n}\frac{1}{2\pi} \int_{\boldsymbol{R}} \widetilde{\varphi}(s) \left(\int_{\boldsymbol{S}^{n-1}} e^{is(x,\eta)}Y_k(\eta)d\eta \right) ds.$$

On the other hand, we have the classical formula (see e.g. [GGA], p. 25)

$$\int_{\boldsymbol{S}^{n-1}} e^{i\lambda(\eta,\omega)}Y_k(\omega)d\omega = c_{n,k}Y_k(\eta)\frac{J_{k+n/2-1}(\lambda)}{\lambda^{(n/2)-1}},$$

where $c_{n,k} = (2\pi)^{n/2}i^k$ and J_r is the Bessel function. Here we replace k by 0 and n by $n+2k$. Then we obtain

$$\int_{\boldsymbol{S}^{n-1}} e^{i\lambda(\omega,\eta)}Y_k(\eta)d\eta = \left(\frac{i\lambda}{2\pi}\right)^k Y_k(\omega) \int_{\boldsymbol{S}^{n+2k-1}} e^{i\lambda\zeta_1}d\zeta$$

Finally, we put $x = r\omega$ and get

$$(Y_k \otimes \varphi)^\vee(r\omega) = \Omega_n^{-1}(r/2\pi)^k Y_k(\omega) \int_{S^{n+2k-1}} \frac{1}{2\pi} d\zeta \int_{\boldsymbol{R}} \widetilde{\varphi}(s)(is)^k e^{isr\zeta_1} ds$$

as desired.

A. 3. We know from §2, No. 3 that $P_k(2, \cos\theta) = H_k(\sin\theta, \cos\theta)$ where $H_k(x_1, x_2)$ is the unique harmonic polynomial on $\boldsymbol{R}^2$ which is homogeneous of degree k, is invariant under $(x_1, x_2 \longrightarrow (-x_1, x_2)$ and satisfies $H_k(0,1) = 1$. Since $(x_1+ix_2)^k$ and $(x_1-ix_2)^k$ span the space of homogeneous k^{th} degree harmonic polynomials we have

$$H_k(x_1, x_2) = Re((x_2 + ix_1)^k),$$

which gives the desired result.

A. 4. If dk is the normalized Haar measure on K we have

$$(\widehat{f})^\vee(x) = \int_K \Big(\int_{\xi_o} f(x + k \cdot y) dm(y)\Big) dk$$

$$= \int_{\xi_o} dm(y) \int_K f(x + k \cdot y) dk = \int_{\xi_o} (M^{|y|} f)(x) dm(y),$$

where $(M^r f)(z)$ is the average of f over $S_r(z)$. Hence

$$(\widehat{f})^\vee(x) = \Omega_d \int_0^\infty (M^r f)(x) r^{d-1} dr,$$

so, using polar coordiantes around x

$$(\widehat{f})^\vee(x) = \frac{\Omega_d}{\Omega_n} \int_{\boldsymbol{R}^n} |x - y|^{d-n} f(y) dy$$

and now the inversion formula follows from the standard inversion of the Riesz potential, ([GGA], Ch. I, Prop. 2.38).

The statement (i) amounts to that if V is a k-dimensional vector subspace of $\boldsymbol{C}^n$ then $V = k \cdot \xi_o$ for some $k \in \boldsymbol{U}(n)$. This is obvious by choosing a basis of V orthonormal with respect to the standard Hermitian inner product $\langle\ ,\ \rangle$ on $\boldsymbol{C}^n$.

Statement (ii) amounts to proving that if W is a Lagrangian vector subspace of $\boldsymbol{R}^{2n}$ then $W = k \cdot \xi_o$ for some $k \in \boldsymbol{U}(n)$. It is well known that $\dim W = n$. Writing $z = x + iy,\ w = u + iv$ with $x, y, u, v \in \boldsymbol{R}^n$ we have

$$\begin{aligned}\langle z, w\rangle &= x \cdot u + y \cdot v - i(x \cdot v - y \cdot u)\\ &= (x, y) \cdot (u, v) - i\{(x, y), (u, v)\}\end{aligned}$$

so the action of $\boldsymbol{U}(n)$ on $\boldsymbol{C}^n \sim \boldsymbol{R}^{2n}$ preserves both the standard inner product on $\boldsymbol{R}^{2n}$ and the skew symmetric form $\{\ ,\ \}$. If $e_1, \ldots, e_n$ is an orthonormal basis of W over $\boldsymbol{R}$ then the formula above shows, W being isotropic, that $\langle e_i, e_j\rangle = \delta_{ij}$ so the e_i form a complex orthonormal basis of $\boldsymbol{C}^n$. Viewing the standard orthonormal basis of $\boldsymbol{R}^n \times 0$ as a complex orthonormal basis of $\boldsymbol{C}^n$ we see that $W = k \cdot \xi_o$ for a suitable $k \in \boldsymbol{U}(n)$.

A. 5. From [GGA], Ch. I, Theorem 2.20 we have $\operatorname{supp}(T) \subset \overline{B_A(0)}$. For $\epsilon > 0$ let $f \in \mathcal{D}(X)$ have $\operatorname{supp}(f) \subset \overline{B_{A-\epsilon}(0)}$. Then $\operatorname{supp}(\widehat{f}) \subset \overline{\beta}_{A-\epsilon}$ where

$$\beta_R = \{\xi :\ d(o, \xi) < R\}.$$

Also by the inversion formula $cf = (\Lambda \widehat{f})^\vee$, since Λ is now a differential operator,

$$T(cf) = T((\Lambda \widehat{f})^\vee) = \widehat{T}(\Lambda \widehat{f}) = 0$$

so $\operatorname{supp}(T) \cap \overline{B_{A-\epsilon}(0)} = \emptyset$.

A. 6. We have with a constant c

$$\begin{aligned}(\overset{\vee}{\varphi} * f)(x) &= c \int_{\boldsymbol{R}^n} \Big(\int_{\boldsymbol{S}^{n-1}} \varphi(w, (w, x) - (w, y)) dw \Big) f(y) dy\\ &= \int_{\boldsymbol{S}^{n-1}} \Big(\int_{\boldsymbol{R}} \varphi(w, (w, x) - p) \widehat{f}(w, p) dp \Big) dw\\ &= c \int_{\boldsymbol{S}^{n-1}} (\varphi * \widehat{f})(w, (w, x)) dw = (\varphi * \widehat{f})^\vee(x)\end{aligned}$$

(Natterer, [1986], p. 14).

B. Homogeneous Spaces. Grassmann Manifolds.

B. 1. For (ii) we may take $x_2 = x_o$ and write $x_1 = g_1 K,\ \xi = \gamma H$. Then

$$\begin{aligned}&x_o, \xi \text{ incident} \iff \gamma h = k \qquad (\text{ some } h \in H,\ k \in K)\\ &x_1, \xi \text{ incident} \iff g_1 k_1 = \gamma h_1 \qquad (\text{ some } h_1 \in H, k_1 \in K).\end{aligned}$$

Thus if x_o, x_1 are incident to ξ we have $g_1 = kh^{-1}h_1k_1^{-1}$. Conversely, if $g_1 = k'h'k''$ we put $\gamma = k'h'$ and then x_o, x_1 are incident to $\xi = \gamma H$.

For (iii) suppose first $KH \cap HK = K \cup H$. Let $x_1 \neq x_2$ in X. Suppose $\xi_1 \neq \xi_2$ in Ξ both incident to x_1 and x_2. Let $x_i = g_iK$, $\xi_j = \gamma_jH$. Since x_i is incident to ξ_j there exist $k_{ij} \in K$, $h_{ij} \in H$ such that

$$g_ik_{ij} = \gamma_jh_{ij} \qquad i = 1, 2; \quad j = 1, 2.$$

By eliminating g_i and γ_j we obtain

$$k_{22}^{-1}k_{21}h_{21}^{-1}h_{11} = h_{22}^{-1}h_{12}k_{12}^{-1}k_{11}.$$

This being in $KH \cap HK$ it lies in $K \cup H$. If the left hand side is in K, then $h_{21}^{-1}h_{11} \in K$ so we get

$$g_2K = \gamma_1h_{21}K = \gamma_1h_{11}K = g_1K$$

which is a contradiction. Similarly, if the mentioned left hand side is in H we have $k_{22}^{-1}k_{21} \in H$ which gives the contradiction $\gamma_2H = \gamma_1H$.

Conversely, suppose $KH \cap HK \neq K \cup H$. Then there exist h_1, h_2, k_1, k_2 such that $h_1k_1 = k_2h_2$ and $h_1k_1 \notin K \cup H$. Put $x_1 = h_1K$, $\xi_2 = k_2H$. Then $x_o \neq x_1$, $\xi_o \neq \xi_2$, yet both ξ_o and ξ_2 are incident to both x_o and x_1.

B. 2–3. For the first statement see [GGA], Cor. 4.10, Ch. II. For the other suppose the generators $D_i = D_{P_i}$ were not algebraically independent. Let

$$P = \Sigma a_{n_1} \dots n_\ell x_1^{n_1} \dots x_\ell^{n_\ell}$$

be a nonzero polynomial such that $P(D_1, \dots, D_\ell) = 0$. Let $d_i =$ degree (P_i) and $N = \max(\Sigma d_in_i)$, the maximum taken over the set of ℓ-tuples $(n_1, \dots, n_\ell)$ for which $a_{n_1} \dots n_\ell \neq 0$. We write the polynomial

$$S = \Sigma a_{n_1} \dots n_\ell P_1^{n_1} \dots P_\ell^{n_\ell}$$

as the sum $S = Q + R$, where

$$Q = \sum_{\Sigma d_in_i = N} a_{n_1} \dots n_\ell P_1^{n_1} \dots P_\ell^{n_\ell}$$

and degree $(R) < N$. Also $Q \neq 0$ by assumption. Consider the operator

$$\Sigma a_{n_1} \dots_{n_\ell} D_1^{n_1} \cdots D_\ell^{n_\ell} - D_S$$

whose order is $< N$ ([GGA], p. 287). This operator equals $0 - D_Q - D_R$ which by the definition in Exercise **B2** has order N. This gives the desired contradiction.

B. 10. Method of Helgason [1958] or [GGA], Ch. V, Lemma 2.6. First show that it suffices to compute

$$\int_{U(n)} |v_{ij}|^2 |v_{k\ell}|^2 dV$$

and that this integral is given by

(i) $(n(n+1))^{-1}$ if (i,j) and (k,ℓ) are either in the same row or the same column (not both).

(ii) $2(n(n+1))^{-1}$ if $(i,j) = (k,\ell)$

(iii) $(n^2-1)^{-1}$ if (i,j) and (k,ℓ) are neither in the same row nor the same column.

B. 11. The proof is obtained by expanding in a Fourier series on $\boldsymbol{T}^2$ (also observed by Gindikin).

B. 12. If U/K has rank one see [GGA], Ch. I, Cor. 4.19. If U/K has higher rank the result is immediate from Exercise **11** as pointed out by Grinberg.

B. 13. $\check{d}$ is a K-orbit containing $(1,0)$ so equals B. Also $H \cdot o$ is two-dimensional so equals D.

CHAPTER II

A. The Spaces $X = G/K$ and $\Xi = G/MN$.

A. 1. If $kN \subset NK$ then $k \cdot \xi_o \subset \xi_o$ so $k \in M$ by text. If $nK \subset KN$ then $n \cdot o$ belongs to each horocycle through o. If $n \neq e$, $n \cdot o = ka \cdot o$ $(a \neq e)$. But $k \cdot \xi_o$ does not contain $ka \cdot o = n \cdot o$.

Let $\mathfrak{g} = \mathfrak{k} + \mathfrak{a} + \mathfrak{n}$ be the usual Iwasawa decomposition of $\mathfrak{g} = \mathfrak{sl}(2, \boldsymbol{R})$ (as before Lemma 4.9). Let $\mathfrak{g} = \mathfrak{m} + \mathfrak{n} + \mathfrak{q}$ where $\mathfrak{q}$ is MN-invariant. Let $H \in \mathfrak{a}$ have the component H_1 in $\mathfrak{q}$. Then $[H_1, \mathfrak{n}] \subset \mathfrak{n}$ is a contradiction.

A. 2. Use (4) §3.

A. 3. Recall proof of Lemma 4.9 (ii).

A. 4. Consider $V = \boldsymbol{C}^{n+1}$ with the Hermitian form

$$\langle y, w\rangle = y_o\overline{w}_o - y_1\overline{w}_1 - \cdots - y_n\overline{w}_n$$

and put $V^+ = \{y \in \boldsymbol{C}^{n+1} : \langle y, y\rangle > 0\}$. The Hermitian hyperbolic space can be taken as $V^+/\boldsymbol{C}^*$. With non-homogeneous coordinates $z_i = y_i/y_o$, $V^+/\boldsymbol{C}^*$ is identified with the ball

$$B^+ = \{z \in \boldsymbol{C}^n : |z_1|^2 + \cdots + |z_n|^2 < 1\}$$

and the unitary action $\boldsymbol{U}(1,n) = \boldsymbol{U}(V)$ on V induces the action of the projective group $\boldsymbol{PU}(V)$ on B^+ ($\boldsymbol{SU}(1,n)$ mod its center, cf. [DS], X, Exercise D1). Let $\pi : V \longrightarrow V/C^*$ be the natural map. Choose $\ell^* \in \partial B^+$ and choose $y^* \neq 0$ on ℓ^*. The Iwasawa subgroup N (the unipotent radical of the isotropy group $\boldsymbol{PU}(V)_{\ell^*}$ viewed as a subgroup of $\boldsymbol{SU}(1,n)$ fixes y^* and hence also the function

$$d_{y^*}(\ell) = \frac{|\langle y^*, y\rangle|}{|\langle y, y\rangle|^{\frac{1}{2}}}, \quad y \in \pi^{-1}(\ell),\ \ell \in B^+.$$

Thus the equation $d_{y^*} = c$, that is,

$$|\langle y^*, y\rangle|^2 = |\langle y, y\rangle| c^2$$

is a horocycle. In non-homogeneous coordinates this is

$$|1 - z_1^*\overline{z}_1 - \cdots - z_n^*\overline{z}_n|^2 = (1 - |z_1|^2 - \cdots - |z_n|^2)\frac{c^2}{|y_o^*|^2}$$

which is an ellipsoid in the Euclidean metric. A $\boldsymbol{PU}(V)$-invariant metric on B^+ is given by (cf. Mostow [1973], p. 136)

$$d(w, y) = \cosh^{-1}\left(\frac{|\langle w, y\rangle|}{\langle w, w\rangle^{\frac{1}{2}}\langle y, y\rangle^{\frac{1}{2}}}\right)$$

so the sphere $S_r(\pi(w))$ is

$$\frac{|\langle w, y\rangle|}{|\langle y, y\rangle|^{\frac{1}{2}}} = |\langle w, w\rangle|^{\frac{1}{2}} \operatorname{ch}\ r.$$

Let $w \longrightarrow y^*, r \longrightarrow \infty$ with $\langle w, w\rangle^{\frac{1}{2}} \operatorname{ch}\ r = c$ (where $\langle y^*, y^*\rangle = 0$). Then the sphere converges to the horocycle above.

Another verification in terms of the notation of [DS], IX, (§3 and Exercise B4). The horocycle $\overline{N} \cdot o$ is given by

$$(w_1, w_2) = \left(\frac{2it - |z|^2}{2(1-it) + |z|^2}, \frac{-2\overline{z}}{2(1-it) + |z|^2} \right)$$

and therefore equals the ellipsoid

$$2|w_1 + \tfrac{1}{2}|^2 + |w_2|^2 = \tfrac{1}{2}.$$

Similarly the horocycle $N \cdot o$ equals

$$(*) \qquad 2|w_1 - \tfrac{1}{2}|^2 + |w_2|^2 = \tfrac{1}{2}.$$

Let

$$a_r = \begin{pmatrix} \operatorname{ch} r & 0 & \operatorname{sh} r \\ o & 1 & 0 \\ \operatorname{sh} r & 0 & \operatorname{ch} r \end{pmatrix}.$$

Then the sphere $S_r(o)$ equals $Ka_r \cdot o$ which is given by

$$|z_1|^2 + |z_2|^2 = \operatorname{th}^2 r.$$

The image $a_r \cdot S_r(0)$ is by [DS], IX, Exercise B4 given by

$$a_r \begin{pmatrix} z_1 \\ z_2 \end{pmatrix} = (z_1 \operatorname{sh}\ r + \operatorname{ch}\ r)^{-1} \begin{pmatrix} z_1 \operatorname{ch} r + \operatorname{sh} r \\ z_2 \end{pmatrix} = \begin{pmatrix} w_1 \\ w_2 \end{pmatrix}$$

so the equation for $a_r \cdot S_r(o)$ is

$$(1 + \operatorname{th}^2 r)|w_1|^2 - \operatorname{thr}(w_1 + \overline{w}_1) + |w_2|^2 = 0.$$

Thus as $r \longrightarrow \infty$ the sphere $a_r S_r(0)$ converges to the horocycle $(*)$.

A. 5. First reduce the problem to the case $X = \boldsymbol{H}^2$ as follows. Let X_α be a root vector in the Lie algebra of N and let G_α denote the analytic subgroup of G with Lie algebra $\boldsymbol{R}X_\alpha + \boldsymbol{R}\theta X_\alpha + \boldsymbol{R}[X_\alpha, \theta X_\alpha]$. Then $G_\alpha \cdot o$ is a totally geodesic submanifold of X isometric to $\boldsymbol{H}^2$ and the horocycle $\exp tX_\alpha \cdot o$ in $G_\alpha \cdot o$ equals $(G_\alpha \cdot o) \cap (N \cdot o)$. This reduces the problem to $\boldsymbol{H}^2$ with metric

$$ds^2 = \frac{dx^2 + dy^2}{y^2}, \qquad y > 0,$$

where the geodesics are the semicircles

$$\gamma_{u,r} : \ x = u + r\cos\theta, \quad y = r\sin\theta, \qquad 0 < \theta < \pi.$$

We have

$$\widehat{f}(\gamma_{u,r}) = \int_0^\pi f(u + r\cos\theta, r\sin\theta)(\sin\theta)^{-1}d\theta$$

so taking ξ as the line $y = 1$ our assumption amounts to

$$\int_{\gamma_{u,r}} \frac{f(x,y)}{y}dw = 0, \qquad r < 1,$$

where dw is the Euclidean arc element. The rapid decrease of f implies that $f(x,y)/y$ extends to a smooth function F on $\boldsymbol{R}^2$ by $F(x,y) = f(x,|y|)/|y|$. Then

$$(*) \qquad \int_{S_r(x)} F(s)dw(s) = 0 \qquad x \in \boldsymbol{R},\ r < 1.$$

This implies for the corresponding disk $B_r(x)$

$$\int_{B_r(x)} F(u,v)du dv = 0,$$

whence

$$\int_{B_r(o)} (\partial_1 F)(x+u,v)du\ dv = 0$$

with $\partial_1 = \partial/\partial u$. Using the divergence theorem on the vector field $F(x + u,v)\partial/\partial u$ we get

$$\int_{S_r(o)} F(x+u,v)u\ dw(u,v) = 0.$$

Combining this with $(*)$ we deduce

$$(**) \qquad \int_{S_r(x)} F(s)s_1 dw(s) \qquad s = (s_1, s_2).$$

Iterating the implication $(*) \Longrightarrow (**)$ we obtain

$$\int_{S_r(x)} F(s)P(s_1)dw(s) = 0$$

where P is any polynomial so we get the desired conclusion $f \equiv 0$ on the strip $0 < y < 1$.

A. 7. Because of Theorem 2.9 it suffices to prove that the convolution algebra $C_c^\natural(MN)$ of M-bi-invariant functions in $C_c(MN)$ is commutative. This result from Korányi [1980] follows (for $m_{2\alpha} \neq 1$) from Kostant's theorem (Exercise **D3** below) which implies that for each $n \in N$ there exists an $m \in M$ such that $mnm^{-1} = n^{-1}$. Thus $f(n) = f(n^{-1})$ for $f \in C_c^\natural(MN)$ which implies the commutativity. For the case $m_{2\alpha} = 1$ see [GGA], Ch. IV, Exercise B10.

A. 8. With the customary notation we have (as $m^*k(\overline{n})M = k(\overline{n}(m^*\overline{n}))M$),

$$\int_{\overline{N}} F(k(\overline{n})M)e^{-2\rho(H(\overline{n}))}d\overline{n} = \int_{K/M} F(kM)dk_M$$

$$= \int_{\overline{N}} F(k(\overline{n}(m^*\overline{n}))M)e^{-2\rho(H(\overline{n}))}d\overline{n},$$

and since by §6, $H(\overline{n}) = H(\overline{n}(m^*\overline{n})) + B(m^*\overline{n})$, this integral equals

$$\int_{\overline{N}} F(k(J\overline{n})M)e^{-2\rho(H(J\overline{n}))}e^{-2\rho(B(m^*\overline{n}))}\frac{d\overline{n}}{d(J\overline{n})}d(J\overline{n}),$$

proving the result.

A. 9. The vector $\boldsymbol{v}$ is in the center of $\mathfrak{k}_o$ so is fixed under $Ad_{G_o}(K_o)$; also $\boldsymbol{e}$ is in the highest root space so, Ad_{G_o} being spherical, $\boldsymbol{e}$ is M_o-fixed. By computation

$$Ad(a_t)\boldsymbol{v} = \boldsymbol{v} + 3\ \mathrm{sh}t \begin{pmatrix} \mathrm{sh}t\ i & 0 & -\mathrm{ch}t\ i \\ 0 & 0 & 0 \\ \mathrm{ch}t\ i & 0 & -\mathrm{sh}t\ i \end{pmatrix}$$

$$Ad(a_t)\boldsymbol{e} = e^{2t} \begin{pmatrix} i & 0 & -i \\ 0 & 0 & 0 \\ i & 0 & -i \end{pmatrix}.$$

Put

$$v_o = \begin{pmatrix} -\frac{i}{2} & 0 & 0 \\ 0 & i & 0 \\ 0 & 0 & -\frac{i}{2} \end{pmatrix}, \qquad v_1 = \begin{pmatrix} i & 0 & 0 \\ 0 & 0 & 0 \\ 0 & 0 & -i \end{pmatrix}, \qquad v_2 = \begin{pmatrix} 0 & 0 & -i \\ 0 & 0 & 0 \\ i & 0 & 0 \end{pmatrix}.$$

Then the curve

$$t \longrightarrow \mathrm{Ad}(a_t)\boldsymbol{v} = v_o + \tfrac{3}{2}\,\mathrm{ch}\,2t \quad v_1 + \tfrac{3}{2}\,\mathrm{sh}\;\, 2t\; v_2$$

lies in the intersection of X_o with the plane $(s_1, s_2) \longrightarrow v_o + s_1 v_1 + s_2 v_2$.

A. 10. Consider $\mathfrak{a}_o$ as in [DS], Cor. 7.6, Ch. VIII. The geodesic $Ad(\exp\, t(X_\gamma + X_{-\gamma}))v$ is easily computed and lies in the plane

$$(s_1, s_2) \longrightarrow v + s_1(X_\gamma - X_{-\gamma}) + s_2 H_\gamma.$$

B. Conical Functions.

Part (i) is immediate from Theorem 4.8. For (ii) recall that by Corollary 4.13, $-s^*\mu$ is the highest weight of the contragredient π'_ψ. For m^* we choose

$$m^* = \begin{pmatrix} 0 & \dots & 0 & \epsilon \\ 0 & \dots & 1 & 0 \\ \vdots & & & \\ 1 & & 0 & 0 \end{pmatrix}$$

where $\epsilon = \pm 1$, the sign determined by $\det(m^*) = 1$. Also M consists of the diagonal matrices m with diagonal elements ± 1 satisfying $\det(m) = 1$. If $g \in \overline{N}MAN$, $g = \overline{n}(g)m(g)\exp B(g)n_B(g)$ then by [DS], IX, Exercise A2, the diagonal matrix $\exp B(g)$ has entries

$$\exp B(g)_{ii} = \frac{|\Delta_i(g)|}{|\Delta_{i-1}(g)|},$$

where $\Delta_i(g) = \det((g_{\ell m})_{1 \le \ell, m \le i})$ with $g = (g_{\ell m})$. By Theorem 4.7

$$\begin{aligned} \psi(m^* g \cdot \xi_o) &= \psi(m^* \overline{n}(g) \exp B(g) \cdot \xi_o) \\ &= \langle \pi_\psi(\exp(-B(g))\overline{n}(g)^{-1}(m^*)^{-1})e, e' \rangle \\ &= \langle \pi_\psi((m^*)^{-1})e, {}^t\pi_\psi(\exp(B(g))e' \rangle \\ &= \psi(\xi^*) e^{(-s^*\mu)(B(g))}. \end{aligned}$$

Now if $h = m^* g$ so $g = (m^*)^{-1} h$ then $|\Delta_i(g)| = |D_i(h)|$ so the desired formula for $\psi(h \cdot \xi_o)$ follows.

The conical functions in this case are related to "conical polynomials" studied in a forthcoming book by Faraut and Korányi.

C. Hyperbolic Space; Inversion and Support Theorems.

C. 1. (i) By orthogonality with the geodesics, the horocycles are the $(n-1)$-spheres tangential to the boundary $|x|=1$. The induced metric on the horocycle is flat. This is obvious for example in the upper half-space model where $N \cdot o$ is a horizontal plane.

(ii) We see that if $\xi_o = N \cdot o$ and dq the volume element on ξ_o then

$$(\widehat{f})^{\vee}(g \cdot o) = \int_K dk \int_{\xi_o} f(gk \cdot q)dq = \int_{\xi_o} \left[M^{d(o,q)} f\right](p)\, dq,$$

where $(M^r f)(p)$ is the average of f over $S_r(p)$. Thus

$$(\widehat{f})^{\vee}(p) = \Omega_{n-1} \int_o^{\infty} (M^r f)(p) \rho^{n-2} d\rho$$

where $r = d(o,q)$ (d = distance in $\boldsymbol{H}^n$) and $\rho = d'(o,q)$ (d' = distance on horocycle).

It suffices to prove $\rho = \sinh\ r$ when q is in the $x_1 x_n$-plane so we are in the two-dimensional case. From [GGA] p. 36 ($\boldsymbol{R}$ and $N \cdot o$ are isometric under $x \longrightarrow \frac{x}{x+i}$) we see that

$$r = \frac{1}{2} \log \left(\frac{1+|\frac{x}{x+i}|}{1-|\frac{x}{x+i}|} \right) \qquad \rho = |x|.$$

The first formula means

$$\frac{|x|}{|x+i|} = \tanh r \qquad \text{or} \qquad \rho(1+\rho^2)^{-\frac{1}{2}} = \tanh r$$

so $\rho =$ sinh r. Hence

$$(\widehat{f})^{\vee}(p) = \Omega_{n-1} \int_o^{\infty} (M^r f)(p) \operatorname{sh}^{n-2} r \operatorname{ch} r\, dr.$$

(v) (vi) Since the area of $S_r(p)$ is proportional to $\operatorname{sh}^{n-1}(2r)$ the formula in (v) follows from [GGA], Ch. II, Prop. 5.26. For (vi) we can write

$$(\widehat{f})^{\vee}(p) = \tfrac{1}{2}\Omega_{n-1} \int_0^{\infty} (M^r f)(p) \operatorname{sh}(2r) \operatorname{sh}^{n-3}(r) dr.$$

(vi)–(vii) *Let $F(r) = (M^r f)(p)$, let $\Delta_r = \Delta(L)$ and assume k even > 0. Then*

$$\int \operatorname{sh}^k r \operatorname{sh}(2r) \Delta_r F(r) dr = (k+2)(k-2n+4) \int_0^\infty F(r) \operatorname{sh}^k r \operatorname{sh}(2r) dr$$

$$+k(k-n+2) \int_0^\infty F(r) \operatorname{sh}^{k-2} r \operatorname{sh}(2r) dr.$$

If $k = 0$ this should be

$$-2(n-2)\left(\int_0^\infty 2F(r)\operatorname{sh}(2r)dr + F(0)\right).$$

Proof. By the Darboux equation, L applied to $(\widehat{f})^\vee(p)$ amounts to the application of $\Delta(L) = \Delta_r$ to $F(r)$. Now

$$\int_0^\infty \operatorname{sh}^k r \operatorname{sh}(2r)\left(\frac{d^2F}{dr^2} + 2(n-1)\coth(2r)\frac{dF}{dr}\right) dr = \Big[\operatorname{sh}^k r \ \operatorname{sh}(2r) F'\Big]_o^\infty$$

$$-\int_0^\infty F'[\operatorname{sh}^k r \ \operatorname{ch}(2r)2 + k \operatorname{sh}^{k-1} r \operatorname{ch} \ r \operatorname{sh}(2r) - 2(n-1)\operatorname{sh}^k r \ \operatorname{ch}(2r)]dr$$

$$= 2(n-2)\int_0^\infty \operatorname{sh}^k r \ \operatorname{ch}(2r) F' dr - \frac{k}{2}\int_0^\infty \operatorname{sh}^{k-2} r \ \operatorname{sh}^2(2r) F' dr$$

$$= 2(n-2)\Big\{\Big[\operatorname{sh}^k r \operatorname{ch}(2r) F\Big]_o^\infty$$

$$-\int_0^\infty F[2\operatorname{sh}(2r)\operatorname{sh}^k r + k \ \operatorname{sh}^{k-1} r \operatorname{ch} r \operatorname{ch}(2r)]dr\Big\}$$

$$-\frac{k}{2}\Big\{\Big[\operatorname{sh}^{k-2} r \ \operatorname{sh}^2(2r) F\Big]_o^\infty$$

$$-\int_0^\infty F[\operatorname{sh}^{k-2} r 4 \operatorname{sh}(2r)\operatorname{ch}(2r) + (k-2)\operatorname{sh}^{k-3} r \ \operatorname{ch} \ r \ \operatorname{sh}^2(2r)]dr\Big\}$$

$$= -2(n-2)\int_0^\infty F[2\,\mathrm{sh}(2r)\,\mathrm{sh}^k r + \frac{k}{2}\,\mathrm{sh}^{k-2} r\,\mathrm{sh}(2r) + k\,\mathrm{sh}^k\, r\,\mathrm{sh}(2r)]dr$$

$$+\frac{k}{2}\int_0^\infty F[4\,\mathrm{sh}^{k-2} r\,\mathrm{sh}\,2r + 8\,\mathrm{sh}^k\, r\,\mathrm{sh}(2r) + (k-2)(2\,\mathrm{sh}^{k-2}\, r + 2\,\mathrm{sh}^k\, r)\,\mathrm{sh}(2r)]dr$$

$$= \int_0^\infty F\ \mathrm{sh}^k\, r\,\mathrm{sh}(2r)dr\,\{(k+2)(k-2n+4)\}\,dr$$

$$+\int_0^\infty F(r)\ \mathrm{sh}^{k-2}\, r\,\mathrm{sh}(2r)\,\{k(k-n+2)\}\,dr$$

This means

$$(L + (k+2)(2n-k-4))\int_0^\infty F(r)\ \mathrm{sh}^k\, r\ \mathrm{sh}(2r)dr$$

$$= -(n-k-2)k\int_0^\infty F(r)\ \mathrm{sh}^{k-2}\, r\,\mathrm{sh}(2r)dr.$$

By iteration, $k = n-3, n-5, \cdots$, we obtain
$$\big(L + (n-1)(n-1)\big)\ldots\big(L + 2(2n-4)\big)(\widehat{f})^\vee = (-1)^{\frac{n-1}{2}}\Omega_{n-1}(n-2)!f.$$

For a different inversion method see Gelfand, Graev and Vilenkin [1966], Ch. V, §2.

C. 2. Use [GGA], Ch. IV, Exercise C3 (for the case of a hyperbolic space) and combine with [GGA], Ch. I, Lemma 4.4. (For full details see Helgason [1980b]).

D. Conical Distributions.

D. 1. (Sketch) To see first that the theorem is local let $\{V_\alpha\}_{\alpha\in A}$ be a locally finite covering of V by coordinate neighborhoods and $1 = \sum_\alpha \varphi_\alpha$ a partition of 1 subordinate to this covering. Then $T = \Sigma\varphi_\alpha(T|V_\alpha)$ where each restriction $T|V_\alpha$ is assumed to have the indicated representation with distributions $\widetilde{T}_{n_1,\ldots n_p,\alpha}$ on V_α. In order to move the φ_α past the X_i over on $\widetilde{T}_{n_1\ldots n_p,\alpha}$ we repeatedly use the formula $\varphi(Xf) = X(f\varphi) - fX\varphi$. For the local version of the theorem let $\exp tX_i$ be a local 1-parameter group of local diffeomorphisms of a neighborhood of $w \in W$ in V. Then

$$(X_1X_2\varphi)(v) = \left\{\frac{d}{dt_1}(X_2\varphi)(\exp(-tX_1)\cdot v)\right\}_{t_1=0}$$
$$= \left\{\frac{d}{dt_1}\frac{d}{dt_2}\varphi(\exp(-t_2X_2)\exp(-t_1X_1)\cdot v)\right\}_{t_1=t_2=0}$$

and if $\partial_i = \partial/\partial t_i$,

$$((X_1^{n_1}\dots X_p^{n_p})(\varphi))(v) = \{\partial_1^{n_1}\dots\partial_p^{n_p}\varphi(\exp(-t_pX_p)\dots\exp(-t_1X_1)\cdot v)\}_{t=0}.$$

Schwartz' theorem representing T in terms of the ∂_i (Schwartz [1966], Th. XXXV) therefore gives the result of the exercise.

D. 2. Let $\mathfrak{g}^\alpha$ be the subalgebra generated by $\mathfrak{g}_\alpha$ and $\mathfrak{g}_{-\alpha}$. Then $\mathfrak{g}^\alpha$ is semisimple of real rank one and

$$\mathfrak{g}^\alpha = \mathfrak{g}_{-2\alpha} + \mathfrak{g}_{-\alpha} + \mathfrak{g}_\alpha + \mathfrak{g}_{2\alpha} + (\mathfrak{g}^\alpha)_o$$

([DS], IX, §2). Let $e_j \in \mathfrak{g}_{j\alpha}$. Then $e_j, \theta e_j$ and $w = [e_j, \theta e_j]$ span an $\mathfrak{sl}(2, \boldsymbol{C})$ which operates on $(\mathfrak{g}^\alpha)^{\boldsymbol{C}}$. By [GGA], Appendix, Cor. 1.5, $\mathfrak{g}_{j\alpha} \subset [(\mathfrak{g}^\alpha)^{\boldsymbol{C}}, e_j]$ so

$$[(\mathfrak{g}^\alpha)_o, e_j] = \mathfrak{g}_{j\alpha}.$$

A fortiori $[\mathfrak{m} + \mathfrak{a}, e_j] = \mathfrak{g}_{j\alpha}$ so the orbit $M \cdot e_j$ has codimension 1 so if the sphere is connected it must be $M \cdot e_j$ (cf. Kostant [1975], Ch. II).

D. 3. (Sketch following Wallach [1973] and Lepowsky, [1975].

(a) $\mathfrak{g} = \mathfrak{g}_{2\alpha} + \mathfrak{g}_{-\alpha} + \mathfrak{g}_o + \mathfrak{g}_\alpha + \mathfrak{g}_{2\alpha} \qquad \mathfrak{g}_o = \mathfrak{m} + \mathfrak{a}$. Select $X \in \mathfrak{g}_\alpha, Y = -\theta X \in \mathfrak{g}_{-\alpha}$, such that the vector $H = [X, Y] \in \mathfrak{a}$ satisfies

$$[H, X] = 2X, \qquad [H, Y] = -2Y.$$

The algebra $\mathfrak{s} = \boldsymbol{R}X + \boldsymbol{R}Y + \boldsymbol{R}H$ is isomorphic to $\mathfrak{sl}(2, \boldsymbol{R})$ and $\pi = ad_{\mathfrak{g}}|\mathfrak{s}$ is a representation of $\mathfrak{s}$ on $\mathfrak{g}$. Deduce from [GGA], Appendix, Lemma 1.2 (ii) that since the eigenvalues of $ad\ H$ on $\mathfrak{g}$ are $0, \pm 2, \pm 4$ the dimensions of the irreducible components of π can only be $1, 3$ or 5.

(b) Let $\mathfrak{g}^i$ denote the sum of all the $(2i+1)$-dimensional irreducible components of $\mathfrak{g}$ and put

$$\mathfrak{g}_j^i = \mathfrak{g}^i \cap \mathfrak{g}_{j\alpha} \quad (0 \le i \le 2, -2 \le j \le 2).$$

Then

$$\mathfrak{g}^i = \oplus_j \mathfrak{g}^i_j, \mathfrak{g}_{\pm 2\alpha} = \mathfrak{g}^2_{\pm 2}, \mathfrak{g}_{\pm\alpha} = \mathfrak{g}^1_{\pm 1} \oplus \mathfrak{g}^2_{\pm 2}, \mathfrak{g}_o = \mathfrak{g}^o_o \oplus \mathfrak{g}^1_o \oplus \mathfrak{g}^2_o,$$

and the decomposition

$$\mathfrak{g} = \mathfrak{g}^0 \oplus \mathfrak{g}^1 \oplus \mathfrak{g}^2$$

is both B- and B_θ-orthogonal.

(c) Using

$$[X_\alpha, X_{-\alpha}] - B(X_\alpha, X_{-\alpha})A_\alpha \in \mathfrak{m},$$

show that

$$\mathfrak{g}^2_o \subset \mathfrak{m}, \qquad \mathfrak{g}^1_o = \boldsymbol{R}A_\alpha \oplus (\mathfrak{g}^1_o \cap \mathfrak{m}), \qquad \mathfrak{g}^o_o \subset \mathfrak{m}.$$

Let $\mathfrak{m}_i = \mathfrak{g}^i \cap \mathfrak{m}$ $(i = 0, 1, 2)$. The $\mathfrak{m}_o$ is the Lie algebra of M_o.

(d) For $Z \in \mathfrak{g}$ put $Z^* = [X, Z]$, $Z^{**} = (Z^*)^*$, $Z_* = [Y, Z]$, $Z_{**} = (Z_*)_*$. Prove that if $Z \in \mathfrak{m}_2$,

$$(Z^{**})_* = 4Z^*, \qquad (Z_{**})^* = 4Z_*, \qquad (Z^*)_* = 6Z, \qquad (Z_*)^* = 6Z$$

and deduce for $Y, Z \in \mathfrak{m}_2$

$$[Y, Z^{**}] = [Y^{**}, Z] = \tfrac{2}{3}[Z^*, Y^*], \qquad [Y, Z]^{**} = \tfrac{2}{3}[Y^*, Z^*].$$

(e) Given $Z \in \mathfrak{g}$, let Z_i be the component in $\mathfrak{g}^i$ in the decomposition $\mathfrak{g} = \mathfrak{g}^0 \oplus \mathfrak{g}^1 \oplus \mathfrak{g}^2$. Then if $Y, Z \in \mathfrak{m}_2$,

$$[Y, Z]_1 = 0, \qquad [[Y, Z]_0 + 2[Y, Z]_2, Z^{**}] = 0.$$

(f) Suppose $Y, Z \in \mathfrak{m}_2$ and $B_\theta(Y^{**}, Z^{**}) = 0$. Then

$$[Y^{**}, Z_{**}] = -[Z^{**}, Y_{**}] \in \mathfrak{m}, \qquad [Y^{**}, Z_{**}]_1 = 0$$

and

$$[Y^*, Z_{**}] = -6[Y, Z]_*, \qquad [[Y, Z]_o Z^{**}] = \frac{\langle\alpha, \alpha\rangle}{9} B(Y, \theta Y) Y^{**}.$$

(g) Let $U \in \mathfrak{g}_{2\alpha}$ and select $Z \in \mathfrak{m}_2$ such that $U = Z^{**}$. Let V be in the orthocomplement (for B_θ) of U in $\mathfrak{g}_{2\alpha}$ and select $Y \in \mathfrak{m}_2$ such that $Y^{**} = V$. Deduce from (f) that $[W, U] = V$ for some $W \in \mathfrak{m}_o$ and consequently $M_o \cdot U$ fills up a sphere in $\mathfrak{g}_{2\alpha}$.

D. 4. For the existence of S_Ψ one can just repeat the proof of Prop. 4.4. Part (a) is obvious. For Part (b) we have by the definition of Ψ_o, Lemma 3.1 and Cor. 6.2,

$$\Psi_o(\varphi) = \int_{\Xi} (\varphi - \varphi_o)(\xi) e^{\rho(\log a(\xi))} d\xi$$

$$= \int_{\overline{N}A} (\varphi - \varphi_o)(\overline{n}a \cdot \xi_o) e^{-\rho(\log a + B(m^*\overline{n}))} e^{2\rho(\log a)} da\, d\overline{n}.$$

Now take φ of the form $\varphi(\overline{n}a \cdot \xi_o) = f(\overline{n})g(a)$ where $\int g(a)e^{\rho(\log a)} da = 1$. Then (b) follows.

D. 5. (i) Use Theorem 4.1 and Corollary 6.2. (ii) Use Cor. 6.2. (iii) Use the M-invariance of S and S_Ψ. (iv) Prove

$$(u^2 + v^2) D\delta \otimes T_0 \in \operatorname{Con}(\mathcal{D}_0')$$

as an intermediary result. (vi) With the particular g chosen one finds (with $f^{n^{-1}}(\overline{n}) = f(\overline{n}(n\overline{n})))$

$$\Psi((f \otimes g)^{n^{-1}}) = (S + c\Delta\delta) f^{n^{-1}}$$

and for the particular choice of f, $(\Delta\delta)(f^{n^{-1}}) = 0$. Thus $h(s) = S(f^{n^{-1}}) - S(f)$ and the contradiction $h'(0) \neq 0$ is obtained by an elementary computation.

D. 6. Solution is similar to that of Exercise **D5**. For (iv) it is useful to remark the following. Let

$$g = \begin{pmatrix} g_{11} & g_{12} & g_{13} \\ g_{21} & g_{22} & g_{23} \\ g_{31} & g_{32} & g_{33} \end{pmatrix} \in \boldsymbol{SU}(2,1)$$

and

$$\sigma = \begin{pmatrix} p & -\overline{q} & 0 \\ q & \overline{p} & 0 \\ 0 & 0 & 1 \end{pmatrix} \in K$$

such that $k(g)M = \sigma M$. Then

$$p = (g_{11} + g_{13})/(g_{31} + g_{33}), \quad q = (g_{21} + g_{23})/(g_{21} + g_{33})$$

and $k(\overline{n}(n\overline{n}))M = k(n\overline{n})M$.

E. The Heisenberg Group.

E. 1.–E. 2. See Faraut and Harzallah [1987].

E. 3. The homogeneity and the left invariance are obvious. Since $d(g,e) = \|g\|$ only the inequality $\|g_1g_2\| \leq \|g_1\| + \|g_2\|$ remains to be proved and this just involves the Schwarz inequality (Cygan [1981], Korányi [1983] or Faraut and Harzallah [1987]).

For **E. 4, E. 5.** and **E. 6.** see Cowling [1982], Folland [1973] and Korányi [1982b]. For an exposition of these results see Faraut and Harzallah [1987]. Much of the theory is generalized to $\overline{N}$ for G/K of rank one in Cowling, Dooley, Korányi and Ricci [1992].

CHAPTER III

A. Differential Operators.

A. 1. We have for $k \in K$, $g \in G$, $n \in N$, $a \in A$

$$\eta_\lambda(kgn) = \eta_\lambda(g), \qquad \eta_\lambda(ga) = e^{(i\lambda-\rho)(\log a)}\eta_\lambda(g).$$

In the decomposition

$$\boldsymbol{D}(G) = (\mathfrak{k}\boldsymbol{D}(G) + \boldsymbol{D}(G)\mathfrak{n}) \oplus \boldsymbol{D}(A)$$

let $D \longrightarrow D_A$ denote the projection of $\boldsymbol{D}(G)$ onto $\boldsymbol{D}(A)$. If $T \in \mathfrak{k}$, $X \in \mathfrak{n}$ and $D_1, D_2 \in \boldsymbol{D}(G)$ we have

$$D_1X\eta_\lambda = 0, \qquad (TD_2\eta_\lambda)(e) = 0, \qquad (D\eta_\lambda)(e) = (D_A\eta_\lambda)(e),$$

and if $f \in \mathcal{E}(G)$ is right invariant under K,

$$\int_G (TD_2\eta_\lambda)(g)f(g)dg = \int_G (D_2\eta_\lambda)(g)((-T)f)(g)dg = 0.$$

Hence

$$\int_G (D\eta_\lambda)(g)f(g)dg = \int_G (D_A\eta_\lambda)(g)f(g)dg$$
$$= (D_A\eta_\lambda)(e)\int_G \eta_\lambda(g)f(g)dg = (D\eta_\lambda)(e)\int_G \eta_\lambda(g)f(g)dg.$$

A. 2. See Helgason [1992a].

B. Rank One Results.

B. 1. By the Fourier expansion for a $F \in \mathcal{E}(K/M)$ (see e.g. [GGA], Ch. V, §3, (13)) we have

$$F(e) = \sum_{\delta \in \widehat{K}_M} d(\delta) \int_K \widetilde{F}(k) \sum_{i=1}^{d(\delta)} \langle \delta(k) v_i, v_i \rangle dk$$

where $\widetilde{F}(k) = F(kM)$, (v_i) is an orthonormal basis of V_δ such that $v = v_1$ span V_δ^M. Replacing k by km and integrating over M the sum over i can be restricted to $i = 1$.

D. The Compact Case.

D. 1. (i) By calculation $(xt_\theta x^{-1})_1 = \cos\theta$. Alternatively, note that $u \longrightarrow xux^{-1}$ is a rotation fixing t_0 and t_π. (ii) The area of a shpere in S^3 of radius θ is a constant multiple of $\sin^2\theta$. (iii) Calculate $\lim_{\theta \longrightarrow 0} F_f(\theta)/\theta$. (iv) The basis $z^p w^q (p + q = \ell)$ diagonalizes $\pi_\ell(t_\theta)$ giving the formula for $\chi_\ell(t_\theta)$. Then note that by (ii) and the fact that F_f is odd,

$$\chi_\ell(f) = \int_U f(u)\chi_\ell(u)du = \frac{1}{4\pi}\int_0^{2\pi} (e^{-i\theta} - e^{i\theta}) F_f(\theta)\chi_\ell(t_\theta)d\theta$$

$$= \frac{1}{2\pi}\int_0^{2\pi} F_f(\theta) e^{-i(\ell+1)\theta} d\theta.$$

Part (v) follows from the fact that χ_ℓ has L^2 norm on U equal to 1 as a result of (ii) and (iii). Part (vi) follows from (iv). For (vii) suppose $\pi \in \widehat{U}$ is not of the form π_ℓ; using (vi) on $f = \mathrm{Trace}(\pi)$ we get a contradiction.

D. 2. If $k \in K$ we have

$$(\widetilde{\varphi} * \widetilde{f})(u) = \int_U \widetilde{\varphi}(uv^{-1})\widetilde{f}(v)dv = \int_U \widetilde{\varphi}(uv^{-1})\widetilde{f}(vk)dv$$

$$= \int_U \widetilde{\varphi}(uk^{-1}v^{-1})\widetilde{f}(v)dv$$

which by averaging over K becomes

$$\widetilde{\varphi}(u) \int_U \widetilde{\varphi}(v^{-1})\widetilde{f}(v)dv.$$

The generalization follows from [GGA], Proposition 2.4 in Ch. IV.

D. 3. The dual of the symmetric space G/K is now $(U \times U)/U^*$ where the diagonal U^* is isomorphic to K. Formula (24) in §9 gives

$$d(\mu) = \left\{ \prod_{\alpha \in \Sigma^+} \frac{\langle \mu + \rho, \alpha \rangle}{\langle \rho, \alpha \rangle} \right\}^2 .$$

Here $d(\mu)$ is the degree of the irreducible representation τ_μ of $U \times U$ which has a fixed vector under the diagonal group U^* and highest weight μ. The irreducible representations τ of $U \times U$ are of the form

$$\tau(u_1, u_2) = \pi_1(u_1) \otimes \pi_2(u_2)$$

where $\pi_1, \pi_2 \in \widehat{U}$ (cf. Weil [1940], §17). Here τ has a fixed vector under U^* if and only if there is a nonzero vector $A \in V_1 \otimes V_2$ such that

$$\pi_1(u) \otimes \pi_2(u)A = A, \qquad u \in U,$$

V_i being the representation space of π_i. This means for the tensor product $\pi_1 \otimes \pi_2$

$$(\pi_1 \otimes \pi_2)(u)A = A.$$

Because of the identification $V_1 \otimes V_2 = \operatorname{Hom}(V_2', V_1)$ A is a linear transformation of V_2' into V_1 so this equation amounts to $\pi_1(u)A\check{\pi}_2(u^{-1}) = A$ which means π_1 and π_2 contragredient, i.e., $\pi_1 \sim \pi$, $\pi_2 \sim \check{\pi}$. Thus $\mu = (\nu, -s\nu)$ where ν is the highest weight of π (relative to a maximal abelian subalgebra $\mathfrak{t} \subset \mathfrak{u}$) and s is the "maximal" Weyl group element. Considering the relationship between the root system $\Delta(\mathfrak{u}^c, \mathfrak{t}^c)$ and the restricted root system of $\mathfrak{u} \times \mathfrak{u}$ with respect to $\mathfrak{t}^* = \{(H, -H) : \ H \in \mathfrak{t}\}$ ([DS], Ch. VII, §4), where each restricted root has multiplicity 2. Note also for the Killing forms

$$B_{\mathfrak{u}\times\mathfrak{u}}((H, -H), (H', H')) = 2B_{\mathfrak{u}}(H, H').$$

Thus

$$\prod_{\alpha \in \Sigma^+} \frac{\langle \mu + \rho, \alpha \rangle}{\langle \rho, \alpha \rangle} = \prod_{\beta > 0} \frac{\langle \nu + \rho_o, \beta \rangle}{\langle \rho_o, \beta \rangle}$$

where on the left $\langle , \rangle$ refers to $B_{\mathfrak{u}\times\mathfrak{u}}$, on the right to $B_{\mathfrak{u}}$, β runs over the positive roots in $\Delta(\mathfrak{u}^c, \mathfrak{t}^c)$ and ρ_o half their sum. Since $d(\mu) = d(\nu)^2$ the formula above gives the formula for $d(\nu)$ the degree of π.

E. The Flat Case.

E. 2. See Helgason [1980a], §6.

E. 3. We have

$$(M^y M^x f)(z) = \int_K \int_K f(z + \ell \cdot x + k \cdot y) dk\, d\ell$$
$$= \int_K \int_K f(z + \ell \cdot x + \ell k \cdot y) dk\, d\ell = \int_K (M^{x+k\cdot y} f)(z) dk.$$

Hre we take $x = r\boldsymbol{e}_n$, $y = s\boldsymbol{e}_n$ where $\boldsymbol{e}_n = (0, \cdots, 1)$. Then the last integral is constant for k in the subgroup fixing $\boldsymbol{e}_n$ so the integral equals

$$\int_{\boldsymbol{S}^{n-1}(0)} (M^{x+sw} f)(z) dw.$$

Letting θ denote the angle between $\boldsymbol{e}_n$ and w we integrate this last integral with w first varying in the section of $\boldsymbol{S}^{n-1}(0)$ with the plane $(\boldsymbol{e}_n, y) = \cos\theta$. Since

$$|x + sw|^2 = r^2 + s^2 - 2rs\cos\theta$$

this gives the second expression for $(M^y M^x f)(z)$. The last is obtained by the substitution $t = (r^2 + s^2 - 2rs\cos\theta)^{\frac{1}{2}}$. (For a different proof see John [1955], p. 80; see also Àsgeirsson [1937]).

F. The Noncompact Case.

F. 1. If $\lambda \in \mathfrak{a}^*$ then $|\boldsymbol{c}(\lambda)|^2 = \boldsymbol{c}(\lambda)\boldsymbol{c}(-\lambda) = \boldsymbol{c}(s\lambda)\boldsymbol{c}(-s\lambda)$.

F. 2. The formula

$$\int_G f(g)\varphi_{-\lambda}(g) dg = \int_A F_f(a) e^{-i\lambda)\log a)} da$$

converts the statement into an analogous one for the exponentials $e^{i\lambda}$ for which it is obvious.

F. 4. Clearly $\varphi \times f \in L^2(X)^\natural$. If $F \in L^2(X)^\natural$ is orthonal to all $\varphi \times f$ then

$$\int_{\mathfrak{a}^*} \widetilde{F}(\lambda)\widetilde{\varphi}(\lambda)\widetilde{f}(\lambda)|\mathbf{c}(\lambda)|^{-2}d\lambda = 0.$$

Since the functions $\widetilde{\varphi}$ form a uniformly dense subalgebra of $C_o(\mathfrak{a}^*/W)$ and since $\widetilde{f}$ is analytic on $\mathfrak{a}^*$, $F = 0$ a.e.

For the general case let $F \in L^2(X)$ be orthogonal to $f^{\tau(g)}$ for all $g \in G$. Then

$$\int_X F^{\tau(g)}(x) f^{\tau(h)}(x)dx = 0 \qquad g, h \in G.$$

Here we can replace $F^{\tau(g)}$ and $f^{\tau(h)}$ by their K-averages $(F^{\tau(g)})^\natural$ and $(f^{\tau(h)})^\natural$. Integrating against $\varphi(h)$ $(\varphi \in \mathcal{D}^\natural(G))$ then gives $(F^{\tau(g)})^\natural = 0$ by the first part. Hence $F = 0$ a.e.

CHAPTER IV

1. Writing h in G as $h = kan$ according to the Iwasawa decomposition and using the K-invariance of f_2 we have

$$\begin{aligned}(f_1 \times f_2)(g \cdot o) &= \int_G f_1(gh^{-1} \cdot o) f_2(h \cdot o)dh \\ &= \int_{AN} f_1(gn^{-1}a^{-1} \cdot o) f_2(an \cdot o)e^{2\rho \log a)} da\, dn.\end{aligned}$$

Hence

$$(f_1 \times f_2)^\wedge(k_1 a_1 \cdot \xi_o) = \int_N (f_1 \times f_2)(k_1 a_1 n_1 \cdot o)dn,$$

$$\int_{AN} \left(\int_N f_1(k_1 a_1 n_1 a^{-1} \cdot o)dn_1\right) f_2(an \cdot o)e^{2\rho(\log a)} da\, dn\,.$$

Interchanging n_1 and a^{-1} in the inner integral cancels out the factor $e^{2\rho(\log a)}$ so the expression reduces to

$$\int_A \widehat{f_1}(k_1 a_1 a^{-1} \cdot \xi_o)\widehat{f_2}(a \cdot \xi_o)da$$

as desired. Since $*$ is commutative whereas $\times$ is not the K-invariance condition cannot be dropped.

2. Because of the K-invariance of φ we write $\varphi(H)$ instead of $\varphi(k \exp H \cdot \xi_o)$. Then by Ch. II, §3, (56),

$$(f \times \check{\varphi})(x) = \int_G f(g \cdot o) \int_B \varphi(A(g^{-1} \cdot x, b)) e^{2\rho(A(g^{-1} \cdot x, b))} \, db \, dg.$$

Using loc. cit. (47) and (51) this becomes

$$\int_G f(g \cdot o) \int_B \varphi(A(x, g(b)) - A(g \cdot o, g(b))) e^{2\rho(A(x, g(b)))} \, dg(b) dg$$

$$= \int_{K/M} e^{2\rho(A(x,kM))} dk_M \int_{G/K} f(g \cdot o) \varphi(A(x, kM) - A(g \cdot o, kM)) dg_K.$$

Now use the formula

$$\int_{AN} F(kan \cdot o) da \, dn = \int_{G/K} F(kg \cdot o) dg_K = \int_{G/K} F(g \cdot o) dg_K$$

on the function $F(y) = f(y)\varphi(A(x, kM) - A(y, kM))$ whereby our integral over G/K becomes

$$\int_{AN} f(kan \cdot o) \varphi(A(x, kM) - \log a) da \, dn = (\widehat{f} \times \varphi)(k \exp A(x, kM))$$

Substituting and using (56) again this gives

$$(f \times \check{\varphi})(x) = (\widehat{f} * \varphi)^{\vee}$$

as stated.

3. By definition

$$(\mathcal{A}\mu_h)(F) = \mu_h(\mathcal{A}^* F) = \int_K F(\exp H(hk)) e^{-\rho(H(hk))} dk$$

and $\{H(hk) : \ k \in K\} = C(h)$ ([GGA], Ch. IX, Theorem 10.5).

5. (i) The Fourier series (20) §3 converges in the topology of $\mathcal{E}(\boldsymbol{R} \times S^1)$ so

$$\sigma(\psi) = \sum_n \sigma(\psi_{-n}(t) \otimes e^{-in\theta}) = \sum_n (e^{2t}\sigma_n)(\psi_{-n}).$$

(ii) By Theorems 2.4 and 3.4, if $\sigma \in \mathcal{E}'(\Xi)$ then the following conditions are equivalent:

(a) $\sigma \in \mathcal{E}'(\boldsymbol{H}^2)^\wedge$.

(b) $\sigma(\psi) = 0$ for each $\psi \in \mathcal{E}(\Xi)$ satisfying

$$D_n(e^t\psi_n) \qquad \text{is odd} \qquad (n \in \boldsymbol{Z}) \tag{1}$$

where D_n denotes $(D+1)\cdots(D+2|n|-1)$, $D = d/dt$.

(c) $\sigma(\psi) = 0$ for each $\psi \in \mathcal{E}(\Xi)$ satisfying

$$e^t\psi_n \in (D_n^*\mathcal{E}_e'(\boldsymbol{R}))^\perp \qquad (n \in \boldsymbol{Z}) \tag{2}$$

$*$ denoting adjoint and subscript e indicating "even", and $\perp$ denoting annihilator.

If $\sigma \in \mathcal{E}'(\Xi)$ is such that σ_n has the form in (ii) then $e^t\sigma_n = D_n^*\tau_n$ where $\tau_n \in \mathcal{E}_e'(\boldsymbol{R})$. If $\psi \in \mathcal{E}(\Xi)$ satisfies (1) then

$$(e^{2t}\sigma_n)(\psi_{-n}) = (D_n^*\tau_n)(e^t\psi_n) = 0$$

so $\sigma(\psi) = 0$ by (i). Thus by (b) we have $\sigma \in \mathcal{E}'(\boldsymbol{H}^2)^\wedge$.

On the other hand, suppose $\sigma \in \mathcal{E}'(\Xi)$ satisfies (c), that is

$$\sigma(\psi) = 0 \quad \text{whenever} \quad e^t\psi_n \in (D_n^*\mathcal{E}_e'(\boldsymbol{R}))^\perp \qquad (n \in \boldsymbol{Z}).$$

Fix $k \in \boldsymbol{Z}$ and use this on the function $\psi(\xi_{t,\theta}) = \psi_{-k}(t)e^{-ik\theta}$. Then $\sigma(\psi) = 0$ implies $(e^{2t}\sigma_k)(\psi_{-k}) = 0$, that is $(e^t\sigma_k)(e^t\psi_{-k}) = 0$. This means that $e^t\sigma_k$ belongs to the double annihilator $(D_k^*(\mathcal{E}_e'(\boldsymbol{R})))^{\perp\perp}$, which equals $D_k^*(\mathcal{E}_e'(\boldsymbol{R}))$, this latter space being closed in $\mathcal{E}_e'(\boldsymbol{R})$ (cf. Theorem 2.16 in Ch. I). Since $k \in \boldsymbol{Z}$ was arbitrary this shows property (ii) for σ.

CHAPTER V

1. By the symmetry of L

$$\int_X (Lu)(x)e^{2\rho(A(x,b))}\,dx = 0$$

so the conditions are necessary. For the sufficiency, consider the Fourier transform

$$\widetilde{f}(\lambda, b) = \int_X f(x)e^{(-i\lambda+\rho)(A(x,b))}\,dx.$$

The conditions amount to $\widetilde{f}(\pm i\rho, b) = 0$ so $\widetilde{f}(\lambda, b)$ is divisible by $\langle\lambda,\lambda\rangle+\langle\rho,\rho\rangle$ and the quotient is holomorphic of uniform exponential type and satisfies (3) in Ch. III, §5. By the Paley-Wiener theorem, u exists.

3. See Helgason [1976], (Theorem 8.1); another proof is in Dadok [1979].

5. See deRham [1955], Ch. V.

4.,6.–7. See Theorems 5.3, 6.1–6.3 in Helgason [1964a].

8. One has to verify

$$\frac{d^2}{d\theta^2}(\gamma(\theta)) = \delta - \frac{1}{2\pi}, \qquad \int_{-\pi}^{\pi} \gamma(\theta)d\theta = 0$$

and using $(d^2/d\theta^2)(|\theta|) = 2\delta$ this is a simple matter.

10. (i) If $T \in \mathfrak{u}$, $[T, U_i] = \sum_j c_{ij}U_j$ where (c_{ij}) is skew symmetric. Hence

$$\begin{aligned}
[T,\omega] &= \sum_i [T, JU_i]U_i + JU_i[T,U_i] \\
&= \sum_{i,j} c_{ij}(JU_j)U_i + \sum_{i,j} c_{ij}(JU_i)U_j \\
&= \sum_{i,j} c_{ij}(JU_j)U_i - \sum_{i,j} c_{ij}(JU_j)U_i = 0.
\end{aligned}$$

Similarly,

$$\begin{aligned}
[JT,\omega]] &= \Sigma[JT, JU_i]U_i + \sum_i JU_i[JT,U_i] \\
&= -\sum_{i,j} c_{ij}U_jU_i + \sum_{i,j} c_{ij}(JU_i)(JU_j) \\
&= \frac{1}{2}\sum_{i,j} c_{ij}(U_iU_j - U_jU_i) + \frac{1}{2}\sum_{i,j} c_{ij}((JU_i)(JU_j) - (JU_j)(JU_i)) \\
&= \frac{1}{2}\sum_{i,j} c_{ij}[U_i,U_j] + \frac{1}{2}\sum_{i,j} c_{ij}[JU_i, JU_j] = 0.
\end{aligned}$$

This proves (i). For (ii) observe that ω annihilates all C^∞ functions f on G which are right invariant under K. Thus if $\omega u = f$ we find a contradiction by averaging over right translations by K.

11. (From a discussion with Schlichtkrull). Let $\nu : \boldsymbol{D}(G) \longrightarrow \boldsymbol{E}(X)$ be the homomorphism (from Ch. III, §10) given by the action of G on X. Then T commutes with each $\nu(D)$ so by (1) *loc. cit.* $TZ = ZT$ for each $Z \in \boldsymbol{Z}(G/K)$. Let $D \in \boldsymbol{D}(G/K)$. By Theorem 10.1 in Ch. III, $DZ_1 = Z_2$ for some $Z_1 \neq 0$, $Z_2 \in \boldsymbol{Z}(G/K)$. Then $TDZ_1 = TZ_2$, $DTZ_1 = Z_2T$ so $(TD - DT)(Z_1 f) = 0$ for $f \in \mathcal{E}(X)$. By the surjectivity of Z_1 (Theorem 1.4) we conclude $TD = DT$.

12. The first statement is immediate from the theorem quoted. For the necessity of the condition and for the compact case see Helgason [1992a].

13. The equation holds for all f of the form $f(kan) = f_1(k)f_2(a)f_3(n)$, hence for all f.

14. Suppose first f holomorphic on all of D. Since the rotations $z \longrightarrow e^{i\theta}z$ belong to the center of K we have (replacing f by $f^{\tau(k)}$)

$$f(0) = \frac{1}{2\pi}\int_0^{2\pi} f(e^{i\theta}z)d\theta = \frac{1}{2\pi}\int_0^{2\pi} f(e^{i\theta}k \cdot z)d\theta = \int_0^{2\pi} f(k \cdot z)dk.$$

Applying this to the composite function $f \circ g$ $(g \in G_0)$ we see that f satisfies the mean value theorem (Cor. 2.2) so is harmonic. This argument can be localized since Cor. 2.2 can.

15. We have by (38) in §4,

$$A_o \cdot b_{\Gamma_1} = \left\{ \sum_{\gamma \in \Gamma - \Gamma_1} \tanh y_\gamma X_{-\gamma} + b_{\Gamma_1} \; : \; y_\gamma \in \boldsymbol{R} \right\}$$

proving (i). Part (ii) follows from the fact that the Weyl group consists of all signed permutations.

17. See Proposition 5.2 in Helgason [1987]. The flat case is proved in Menzala and Schonbeck [1984] on the basis of the spherical support theorem [GGA], Ch. I, Lemma 2.7.

CHAPTER VI

1. Using a K-invariant Laplace-Beltrami operator on K/M we see that each joint eigenspace E is finite-dimensional. Let $E = \bigoplus_i E_i$ be the direct decomposition into irreducible subspaces. Pick $f_i \in E_i$ such that $f_i(eM) = 1$ and f_i is M-invariant. Then each f_i is a spherical function and $Df_i = \chi(D)f_i$, where the homomorphism $\chi : \boldsymbol{D}(K/M) \longrightarrow \boldsymbol{C}$ is the same for all i. Using [GGA], Ch. IV, Cor. 2.3 we find that all f_i coincide so E is irreducible.

Taking $K = \boldsymbol{SU}(2)$, $M = e$, each joint eigenspace has to contain a character χ of K. If T is a maximal torus with Lie algebra spanned by a vector H it is easily seen that $H\chi$ is not a constant multiple of χ (cf. e.g. [GGA], Ch. V, Ex. A7).

2. This is a basic step in Bruhat's analysis [1956] §6) of the principal series for G. By Schur's lemma (for unitary rpresentations) (i) is equivalent to the statement that all bounded linear operators $A : \mathcal{K}_\lambda \longrightarrow \mathcal{K}_\lambda$ commuting with all $\tau_\lambda(g)$ $(g \in G)$ are scalars. Let A be one such operator, consider the sesquilinear form

$$B(\varphi, \psi) = \int_{K/M} \varphi(kMN)\,\mathrm{conj}((A\psi)(kMN))dk_M$$

and the form

$$\widetilde{B}(f,g) = B(f^\natural, g^\natural) \qquad f,g \in \mathcal{D}(G),$$

where

$$f^\natural(xMAN) = \int_{MAN} f(xam\,n)e^{(-i\lambda+\rho)(\log a)}dm\,da\,dn.$$

Then

$$\widetilde{B}(f^{L(x)R(p_1)}, g^{L(x)R(p_2)}) = e^{-(i\lambda+\rho)(\log a_1)}e^{(i\lambda-\rho)(\log a_2)}\widetilde{B}(f,g)$$

and by the Schwartz kernel theorem (Hörmander [1983], Ch. V)

$$\widetilde{B}(f,g) = \int_{G\times G} f(x)\,\mathrm{conj}(g(y))d\widetilde{T}(x,y),$$

where $\widetilde{T} \in \mathcal{D}'(G \times G)$. Then

$$\widetilde{T}^{L(x,x)R(p_1,p_2)} = e^{(i\lambda+\rho)(\log a_1)}e^{(-i\lambda+\rho)(\log a_2)}\widetilde{T},$$

where $L(x,x)R(p_1,p_2)$ denotes the diffeomorphism $(u,v) \longrightarrow (xup_1, xvp_2)$ of $G \times G$. Consider the diffeomorphism $\varphi : (x,y) \longrightarrow (y^{-1}x, y)$ of $G \times G$. Then, if $h \in \mathcal{D}(G \times G)$, we have by the left invariance of $\widetilde{T}$,

$$\widetilde{T}^\varphi(h) = \widetilde{T}(h^{\varphi^{-1}}) = \widetilde{T}((h^{\varphi^{-1}})^{L(z,z)}).$$

However

$$(h^{\varphi^{-1}})^{L(z,z)}(x,y) = h^{\varphi^{-1}}(z^{-1}x, z^{-1}y) = h(y^{-1}x, z^{-1}y)$$

so,

$$(h^{\varphi^{-1}})^{L(z,z)} = (h^{L(e,z)})^{\varphi^{-1}}.$$

Thus

$$\widetilde{T}^{\varphi}(h) = \widetilde{T}^{\varphi}(h^{L(e,z)})$$

so

$$\widetilde{T}^{\varphi}(h) = \int_G \int_G h(x,y)dS(x)dy,$$

where $S \in \mathcal{D}'(G)$. Since $\varphi^{-1}(x,y) = (yx, y)$ this implies

$$\widetilde{T}(f \otimes g) = \int_G \int_G f(yx)g(y)dS(x)dy.$$

Using the homogeneity of $\widetilde{T}$ under $R(p_1, p_2)$ we obtain the homogeneity condition for S in (ii). The converse follows by reversing the steps. All the commuting operators A are proportional if and only if the corresponding S are proportional and then they must be the example stated.

3. See Helgason [1970a], Ch. III, §6.

4. Using Lemma 3.10 and (39) we see quickly that $\Psi^*_{\lambda,e} = \Psi_{-\lambda,e}$. Thus if $\varphi \in \mathcal{D}(\Xi)$

$$\begin{aligned}(\varphi \times \Psi_{\lambda,e})(kaMN) &= \Psi_{-\lambda,e}(\varphi \circ \tau(ka)) \\ &= e^{(i\lambda-\rho)(\log a)} \int_A \varphi(kcMN)e^{(-i\lambda+\rho)(\log c)}dc.\end{aligned}$$

Taking $\varphi(kaMN) = \beta(kM)\gamma(a)$ the result follows.

5. In the solution below C_i and C_i' denote compact sets and $\overset{\circ}{A}$ denotes the interior of a set A. Let $C_1 \subset \overset{\circ}{C}_2 \subset \Xi$, let $\mathcal{D}_{C_1}(\Xi)$ denote the set of $\varphi \in \mathcal{D}(\Xi)$ with support in C_1, and let $C_i' \subset G$ satisfy $\pi(C_i') = C_i$, $C_1' \subset (C_2')^\circ \subset G$, $\pi: \ G \longrightarrow G/MN$ being the natural mapping. Let C_o be a compact neighborhood of e in MN and put $\widetilde{C}_i = C_i'C_o$ $(i = 1, 2)$. Let $f_1 \in \mathcal{D}(G)$ be ≥ 0 on $G, > 0$ on $\widetilde{C}_1$, and $\operatorname{supp}(f_1) \subset \widetilde{C}_2$. Then the function

$$f(g) = \begin{cases} f_1(g)\frac{\varphi(\pi(g))}{\bar{f}_1(\pi(g))} & \text{if } \pi(g) \in C_1 \\ 0 & \text{if } \pi(g) \notin C_1 \end{cases}$$

satisfies $\overline{f} = \varphi$ (cf. (36) §3). Also $\varphi \longrightarrow f$ is a continuous mapping of $\mathcal{D}_{C_1}(\Xi)$ into $\mathcal{D}_{\tilde{C}_2}(G)$. Thus by (37) in §3, $\Psi \times \eta$ is a distribution. For the last part one must show

$$\int \psi(\xi)(\varphi \times \eta^*)(\xi)d\xi = \int (\psi \times \eta)(\xi)\varphi(\xi)d\xi.$$

Let $f_1 \in \mathcal{D}(G)$ satisfy $\overline{f}_1 = \psi$. Then this last equation amounts to

$$\int_G f_1(g) \int_G f(gh^{-1})d(\eta^*)^{\sim}(h)dg = \int_G f(g) \int_G f_1(gh^{-1})d\tilde{\eta}(h)dg.$$

However, $(\eta^*)^{\sim} = \tilde{\eta}'$ so this last equation is obvious.

BIBLIOGRAPHY

ADIMURTI, KUMARESAN, S.

1979 On the singular support of distributions and Fourier transforms on symmetric spaces. .*Ann. Scuola Norm. Sup. Pisa Cl. Sci.* **6** (1979), 143-150.

AGRANOVSKI, M. L.

1985 Invariant function algebras in symmetric spaces. *Trans. Moscow. Math. Soc.* **47** (1985), 175-197.

ANDERSON, A., and CAMPORESI, R.

1989 Intertwining operators for solving differential equations with applications to symmetric spaces. (Preprint 1989).

ANKER, J-PH.

1990 L_p - Fourier multipliers on Riemannian symmetric spaces of the noncompact type. *Ann. of Math.* **132** (1990), 597-628.

1991a Handling the inverse spherical Fourier transform. In "Harmonic Analysis on Reductive Groups." (W. Barker and P.Sally eds.) pp. 51-56. Birkhäuser, Basel and Boston, 1991.

1991b The spherical Fourier transform of rapidly decreasing functions -- a simple proof of a characterization due to Harish-Chandra, Helgason, Trombi and Varadarajan. *J. Funct. Anal.* **96** (1991), 331-349.

1991c A basic inequality for scattering theory on Riemannian symmetric spaces of the noncompact type. *Amer. J. Math.* **113** (1991), 391-398.

1992 Sharp estimates for some functions of the Laplacian on noncompact symmetric spaces. *Duke Math. J.* **65** (1992), 257-297.

ARTHUR, J.

1983 Paley-Wiener theorems for real reductive groups. *Acta Math.* **150** (1983), 1-89.

ÁSGEIRSSON, L.

1937 Über eine Mittelwertseigenschaft von Lösungen homogener linearer partieller Differentialgleichungen 2.Ordnung mit konstanten Koefficienten. *Math. Ann.* **113** (1937), 321-346.

BADERTSCHER, E.

1990 Pompeiu transforms and Radon transforms on Riemannian symmetric spaces. Habilitationsschrift, Bern, 1990.

BADERTSCHER, E., and KOORNWINDER, T. H.

1992 Continuous Hahn polynomials of differential operator argument and analysis on Riemannian symmetric spaces of constant curvature. *Can. Math. J.* **44** (1992), 750-773.

BADERTSCHER, E. and REIMANN, H.M.

1989 Harmonic analysis for vector fields on hyperbolic spaces. *Math. Zeitschr.* **202** (1989), 431-456.

BAGCI, S., and SITARAM, A.
1979 Spherical mean periodic functions on semisimple Lie groups. *Pacific. J. Math.* **84** (1979), 241-250.
BAN, VAN DEN, E.P.
1982 "Asymptotic Expansions and Integral Formulas for Eigenfunctions on a Semisimple Lie Group." Proefschrift, Utrecht, 1982.
BAN, VAN DEN, E.P., and SCHLICHTKRULL, H.
1987 Asymptotic expansions and boundary values of eigenfunctions on a Riemannian symmetric space. *J. Reine Angew. Math.* **380** (1987), 108-165.
1989 Local boundary data of eigenfunctions on a Riemannian symmetric space. *Invent. Math.* **98** (1989), 639-657.
1993 Convexity for invariant differential operators on semisimple symmetric spaces. *Composittio Math.* (1993).
BARKER, W. H.
1975 The spherical Bochner theorem on semisimple Lie groups. *J. Funct. Anal.* 20 (1975), 179-207.
BARLET, D., and CLERC, J. L.
1986 Le comportement à l'infini des fonctions de Bessel généralisées. I *Advan. Math.* **61** (1986), 165-183.
BARUT , A. D. and RACZKA, R.
1977 "Theory of Group Representations and Applications." Polish Scientific Publishers, Warsaw, 1977.
BEERENDS, R.J.
1987 The Fourier transform of Harish-Chandra's c-function and inversion of the Abel transform. *Math. Ann.* **277** (1987), 1-23.
1988 The Abel transform and shift operators. *Compositio Math.* **66** (1988), 145-197.
BENABDALLAH, A-I., and ROUVIÈRE, F.
1984 Résolubilité des opérateurs bi-invariants sur un groupe de Lie semisimple. *C.R. Acad. Sci. Paris.* **298** (1984), 405-408.
BERENSTEIN, C. and ZALCMAN, L.
1976 Pompeiu's problem on spaces of constant curvature. *J. Analyse Math.* **30** (1976), 113-130.
1980 Pompeiu's problem on symmetric spaces. *Comment. Math. Helv.* **55** (1980), 593-621.
BERENSTEIN, C., and SHAHSHAHANI, M.
1983 Harmonic analysis and the Pompeiu problem. *Amer. J. Math.* **105** (1983), 1217-1229.
BERENSTEIN, C., and TARABUSI, E.C.
1991 Inversion formulas for the k-dimensional Radon transform in real hyperbolic spaces. *Duke Math. J.* **62** (1991), 613-631.
1992 Radon- and Riesz transform in real hyperbolic spaces. *Contemp. Math.* 140 (1992), 1-18.
BERLINE, N., and VERGNE, M.
1981 Équations de Hua et intégrales de Poisson. *C.R. Acad. Sci. Paris Ser. A* **290** (1980), 123-125. In "Non-Commutative Harmonic Analysis and Lie Groups," Lecture Notes in Math.No. 880, pp.1-51, Springer Verlag, New York , 1981 .
BETORI,W., FARAUT,J., and PAGLIACCI, M.
The horocycles of a tree and the Radon transform (preprint).
BIEN, F.V. "$\mathcal{D}$ -Modules and Spherical Represenations. " Math. Notes, Princeton Univ. Press 1990.
BOCHNER, S.
1932 "Vorlesungen über Fouriersche Integrale." Akad. Verlag, Leipzig, 1932.
1951 Tensor fields with finite basis. *Ann. of Math.* **53** (1951), 400-411.

BOMAN, J.
1991 Helgason's support theorem for Radon transforms -- a new proof and a generalization. In "Mathematical Methods in Tomography." Lecture Notes in Math. No. 1497 , 1-5. Springer-Verlag, Berlin and New York, 1991.

BOMAN, J. and QUINTO, E. T.
1987 Support theorems for real-analytic Radon transforms. *Duke Math. J.* **55** (1987), 943-948.
1993 Support theorems for Radon transforms on real-analytic line complexes in $\boldsymbol{R}^3$ · *Trans. Amer. Math. Soc.* **335** (1993), 877-890.

BOTT, R.
1965 Homogeneous differential operators. In "Differential and Combinatorial Topology" (S.S. Cairns, ed.) Princeton Univ,. Press, Princeton, N.J. 1965, 167-186.

BOURBAKI, N.
"Élements de Mathématique," Vol.VI, *Intégration*, Ch. 1-8. Hermann Paris, 1952-1963.

BOURBAKI, N.
"Élements de Mathématique," Vol. V. *Espaces Vectoriels Topologiques* , Chapters I-V. Hermann, Paris, 1953-1955.

BOURBAKI, N.
"Élements de Mathématique. "*Groupes et Algébres de Lie,*" Ch. I-VIII. Hermann, Paris, 1960-1975.

BOUSSEJRA, A. and INTISSAR, A.
1992 Caractérisation des intégrales de Poisson-Szegö de L^2 (∂B) dans la boule de Bergman B^n (n>1). *C.R. Acad. Sci. Paris* **315** (1992), 1353-1357.

BRANSON, T., and ÓLAFSSON, G.
1991 Equipartition of energy for waves in symmetric spaces. *J. Funct. Anal.* **97** (1991), 403-416.

BRANSON, T.P., ÓLAFSSON, G., and SCHLICHTKRULL, H.
1993a Huygens' principle in Riemannian symmetric spaces. Preprint, 1992.
1993b A bundle-valued Radon transform with applications to invariant wave equations. Preprint, 1992.

BRAY, W. O. and SOLMON, D.C.
1990 Paley-Wiener theorems on rank-one symmetric spaces of noncompact type. *Contemp. Math.* **113** (1990), 17-30.

BRUHAT, F.
1956 Sur les représentations induites des groupes de Lie. *Bull. Soc. Math. France* **84** (1956), 97-205.

CAMPOLI, O.
1977 The complex Fourier transform on rank one semisimple Lie groups. Thesis, Rutgers University. 1977.

CAMPORESI, R.
1993 The spherical transform for homogeneous vector bundles over hyperbolic spaces. Preprint 1993.

CARLEMAN, T.
1944 L'integrale de Fourier et les questions qui s'y rattachent. Publ. Sci. Inst. Mittag-Leffler, Uppsala, 1944.

CARTAN, É.
1929 Sur la detérmination d'un système orthogonal complet dans un espace de Riemann symétrique clos. *Rend. Circ. Mat. Palermo* **53** (1929), 217-252.

CARTAN, H., and GODEMENT, R.
1947 Théorie de la dualité et analyse harmonique des groupes abéliens localement compacts. *Ann. Sci. Éc. Norm. Sup.* **64** (1947), 79-99.

CASSELMAN, W., and MILIC'C, D.
1982 Asymptotic behavior of matrix coefficients of admissible representations. *Duke Math. J.* **49** (1982), 869-930.

CERÈZO, A., and ROUVIÈRE, F.
1973 Opérateurs differentiels invariants sur un groupe de Lie. *Séminaire Goulaouic-Schwartz* 1972-1973. École Polytech., Paris., 1973.
CHAMPETIER, C., and DELORME, P.
1981 Sur les représentations des groupes de déplacements de Cartan. *J. Funct. Anal.* **43** (1981), 258-279.
CHANG, W.
1979 Global solvability of the Laplacian on pseudo-Riemannian symmetric spaces. *J. Funct. Anal.* **34** (1979), 481-492.
1982 Invariant differential operators and P-convexity of solvable Lie groups. *Advan. Math.* **46** (1982), 284-304.
CHERN, S. S.
1942 On integral geometry in Klein spaces. *Ann. of Math.* **43** (1942), 178-189.
CHEVALLEY, C.
1946 "Theory of Lie Groups." Vol. I. Princeton Univ. Press, Princeton, N.J. 1946.
CLERC, J. L.
1976 Une formule de type Mehler-Heine pour les zonal es d'un espace riemannien symétrique. *Studia Math.* **57** (1976), 27-32.
1980 Transformation de Fourier sphérique des espaces de Schwartz. *J. Funct. Anal.* **37** (1980), 182-202.
1987 Le comportment à l'infini des fonctions de Bessel généralisées, II. *Advan. Math.* **66** (1987), 31-61.
CLERC, J. L., EYMARD,P., FARAUT, J., RAÏS, M., and TAKAHASHI, R.
1982 "Analyse Harmonique." C.I.M.P.A. Nice 1982.
CLOZEL, L., and DELORME, P.
1984 Le théorème de Paley-Wiener invariant pour les groupes de Lie réductifs. *Invent. Math.* **77** (1984), 427-433.
COHN, L.
1974 Analytic theory of Harish-Chandra's c-function. Lecture Notes in Math. No. 428. Springer-Verlag, Berlin and New York, 1974.
CORMACK, A.M., and QUINTO, T.
1980 A Radon transform on spheres through the origin in $\boldsymbol{R}^n$ and applications to the Darboux equation. *Trans. Amer. Math. Soc.* **260** (1980), 575-581.
COWLING, M.
1982 Unitary and uniformly bounded representations of some simple Lie groups. In "Harmonic Analysis and Group Representations" CIME (1980), Liguori Editore, 1982.
COWLING, M., and KORÁNYI, A.
1984 Harmonic analysis on Heisenberg type groups from a geometric viewpoint. In "Lie Group Representations III." pp. 60-100. Lecture Notes in Math. No. 1077, Springer-Verlag, Berlin and New York, 1984.
COWLING, M., DOOLEY, A. H., KORÁNYI, A., and RICCI, F.
1992 H-type groups and Iwasawa decompositions. *Advan. Math.* **87** (1992), 1-41.
CYGAN, J.
1981 Subadditivity of homogeneous norms on certain nilpotent Lie groups. *Proc. Amer. Math. Soc.* **83** (1981), 69-70.
DADOK, J.
1976 Fourier transforms of distributions on symmetric spaces. Thesis MIT, 1976.
1979 Paley-Wiener theorem for singular support of K-finite distributions on symmetric spaces. *J. Funct. Anal.* **31** (1979), 341-354.
1980 Solvability of invariant differential operators of principal type on certain Lie groups and symmetric spaces. *J. d'Analyse* **37** (1980),118-127.
DAVIDSON, M.G., ENRIGHT, T. J., and STANKE, R. J.
1991 Differential operators and highest weight representations. *Mem. Amer. Math. Soc.* **94** (1991).

DEBIARD A., and GAVEAU, B.
1983 Formule d'inversion en geométrie intégrale Lagrangienne. *C.R. Acad. Sci. Paris,* **296** (1983), Serie I, 423-425.
DE RHAM, G.
1955 "Variétés Différentiables. " Hermann, Paris, 1955.
DELORME, P.
1982 Théorème de type Paley-Wiener pour les groupes de Lie semisimples réels avec une seule classe de conjugaision de sons-groupes de Cartan. *J. Funct. Anal.* **47** (1982). 26-63.
DELORME, P., and FLENSTED-JENSEN, M.
1991 Towards a Paley-Wiener theorem for semisimple symmetric spaces. *Acta Math.* **167** (1991), 127-151.
DIEUDONNE, J.
1978 "Treatise on Analysis," Vol VI, Academic Press, New York, 1978.
DIJK, VAN, G.
1969 Spherical functions on the p-adic group PGL(2). *Indag. Math.* **31** (1969), 213-241.
DIXMIER, J.
1964 "Les C* algèbres et Leurs Représentations," Gauthier-Villars, Paris, 1964.
DUFLO, M.
1977 Opérateurs differentiels bi-invariants sur un groupe de Lie. *Ann . Sci. École Norm. Sup.* **10** (1977), 265-288.
1979 Opérateurs différentiels invariants sur un espace symétrique. *C. R. Acad. Sci. Paris Ser. A* **289** (1979), 135-137.
DUFLO, M., and RAÏS, M.
1976 Sur l'analyse harmonique sur les groupes de Lie résolubles. *Ann. Sci. Éc. Norm. Sup.* **9** (1976), 107-144.
DUFLO, M. ,and WIGNER, D.
1979 Convexité pour les opérateurs différentiels invariants sur les groupes de Lie. *Math. Zeitschr.* **167** (1979), 61-80.
DUISTERMAAT, J. J.
1984 On the similarity between the Iwasawa projection and the diagonal part. *Mém. Soc. Math. France,* **15** (1984), 129-138.
DUISTERMAAT, J. J., KOLK, J. A. C., and VARADARAJAN, V. S.
1983 Functions, flows and oscillatory integrals on flag manifolds and conjugacy classes in real semisimple Lie groups. *Compositio Math.* **49** (1983), 309-398.
EGUCHI, M.
1971 On the Radon transform of the rapidly decreasing functions on symmetric spaces. II. *Hiroshima Math. J.* **1**(1971), 161-169.
1979 Asymptotic expansions of Eisenstein integrals and Fourier transform on symmetric spaces. *J. Funct. Anal.* **34** (1979).
EGUCHI, M., HASHIZUME, M., and OKAMOTO, K.
1973 The Paley-Wiener theorem for distributions on symmetric spaces. *Hiroshima Math. J.* **3** (1973), 109-120.
EGUCHI, M., and KUMAHARA, K.
1982 An L^p Fourier analysis on symmetric spaces. *J. Funct. Anal.* **47** 1982), 230-246.
EGUCHI, M. and OKAMOTO, K.
1977 The Fourier transform of the Schwartz space on a symmetric space. *Proc. Japan Acad.* **53** (1977), 237-241.
EHRENPREIS, L.
1956 Solutions of some problems of division. Part III. *Amer. J. Math.* **78** (1956), 685-715.
1973 The use of partial differential equations for the study of group representations. *Proc. Symp. Pure Math.* Vol. XXVI, Amer. Math. Soc.1973, 317-320.

EHRENPREIS, L., and MAUTNER, F.

1955 -1959 Some properties of the Fourier transform on semisimple Lie groups, I *Ann. of Math.* **61** (1955), 406-443; II, III. *Trans. Amer. Math. Soc.* **84** (1957), 1-55; **90** (1959), 431-484.

EYMARD. P.

1983 Le noyau de Poisson et l'analyse harmonique non-Euclidienne. In "Topics in Modern Harmonic Analysis." Istituto Nat. Alta Mat. Roma, 1983.

EYMARD, P. and LOHOUÉ, N.

1975 Sur la carré du noyau de Poisson dans les espaces symétriques et une conjecture de Stein. *Ann. Sci. Éc. Norm. Sup.* **8** (1975), 179-188.

FARAH, S. B., and KAMOUN, L.

1990 Distributions coniques sur le cone des matrices de rang un et de trace nulle. *Bull. Soc. Math. France* **118** (1990), 251-272.

FARAUT, J.

1979 Distributions sphériques sur les espaces hyperboliques. *J. Math. Pures Appl.* **58** (1979), 369-444.

1982a Analyse harmonique sur les pairs de Guelfand et les espaces hyperboliques. In J.-L Clerc, et. al.. "Analyse Harmonique" C.I.M.P.A. Nice 1982, Ch. IV.

1982b Un théoreme de Paley-Wiener pour la transformation de Fourier sur un espace Riemannien symétrique de rang un. *J. Funct. Anal.* **49** (1982), 230-268.

FARAUT, J., and HARZALLAH, K.

1984 Distributions coniques associées au groupe orthogonal O(p,q). *J. Math. Pure et Appl.* **63** (1984), 81-109.

1987 "Deux Cours d'Analyse Harmonique. " Birkhäuser, Basel and Boston, 1987.

FARAUT, J., and KORÁNYI, A.

1993 "Analysis on Symmetric Cones." Oxford University Press, New York, 1993.

FELIX, R.

1993 Radon Transformation auf nilpotenten Lie gruppen. *Invent. Math.* **112** (1992), 413-443.

FLENSTED-JENSEN, M.

1972 Paley-Wiener theorems for a differential operator connected with symmetric spaces. *Ark. Mat.* **10** (1972), 143-162.

1977a Spherical functions on a simply connected semisimple Lie group. *Amer. J. Math.* **99** (1977), 341-361.

1977b Spherical functions on a simply connected semisimple Lie group., II. *Math. Ann.* **228** (1977), 65-92.

1978 Spherical functions on a real semisimple Lie group. A method of reduction to the complex case. *J. Func. Anal.* **30** (1978), 106-146.

1981 K-finite joint eigenfunctions of $U(g)^K$ on a non-Riemannian semisimple symmetric space G/H In "Non-Commutative Harmonic Analysis and Lie Groups," Lecture Notes in Math. No. 880. Springer-Verlag, Berlin and New York, 1981.

1986 "Analysis on Non-Riemannian Symmetric Spaces." Conf. Board Math. Sci. No. 61. Amer. Math. Soc. Providence, RI 1986.

FLENSTED-JENSEN, M. and KOORNWINDER, T.

1973 The convolution structure for Jacobi function expansions. *Ark. Mat.* **11** (1973), 245-262.

1979a Positive-definite spherical functions on a noncompact rank one symmetric space. In "Analyse Harmonique sur les Groupes de Lie II," Springer Lecture Notes No. 739. Springer-Verlag, Berlin and New York, 1979.

1979b Jacobi functions: the addition formula and the positivity of the dual convolution structure. *Ark. Mat.* **17** (1979), 139-151.

FOLLAND, G. B.

1973 A fundamental solution for a subelliptic operator. *Bull. Amer. Math. Soc.* **79** (1973), 373-376.

FUGLEDE, B.

1958 An integral formula. *Math. Scand.* **6** (1958), 207-212.

FURSTENBERG, H.
1963 A Poisson formula for semisimple Lie groups. *Ann. of Math.* **77** (1963), 335-386.
1965 Translation-invariant cones of functions on semisimple Lie groups. *Bull. Amer. Math. Soc.* **71** (1965), 271-326.
GANGOLLI, R.
1971 On the Plancherel formula and the Paley Wiener theorem for spherical functions on semisimple Lie groups. *Ann. of Math.* **93** (1971), 150-165.
GANGOLLI, R AND VARADARAJAN, V.S.
1988 "Harmonic Analysis of Spherical Functions on Real Reductive Groups. " Springer-Verlag, Berlin and New York, 1988.
GÅRDING, L.
1960 Vecteurs analytiques dans les représentations des groupes de Lie. *Bull. Soc. Math. France* **88** (1960), 73-93.
GELFAND, I. M.
1950 The center of an infinitesimal group algebra,. *Mat. Sb.* **26** (1950), 103-112.
1960 Integral geometry and its relation to group representations, *Russian Math. Surveys* 15 (1960),
143-151.
GELFAND, I. M., GINDIKIN, S. G., and GRAEV, M. I.
1982 Integral geometry in affine and projective spaces. *J. Soviet Math.* **18** (1982), 39-164.
GELFAND, I. M. , and GRAEV, M. I.
1959, 1964 The geometry of homogeneous spaces, group representations in homogeneous spaces and questions in integral geometry related to them. *Amer. Math,. Soc. Transl.* **37** (1964)
GELFAND, I. M., GRAEV, M. I., AND SHAPIRO, S.J.
1969 Differential forms and integral geometry. *Funct. Anal. Appl.* 3 (1969), 24-40.
GELFAND, I. M. , GRAEV, M. I., and VILENKIN, N.
1966 "Generalized Functions." Vol. 5, Integral Geometry and Representation Theory. Academic Press, New York, 1966.
GELFAND, I. M. and NAIMARK, M. A.
1948 An analog of Plancherel's formula for the complex unimodular group. *Dokl. Akad. Nauk USSR* **63** (1948), 609-612.
1952 Unitary representation of the unimodular group containing the identity representation of the unitary subgroup. *Trudy Moscov. Mat. Obsc.* **1** (1952), 423-475.
GELFAND, I. M. , and RAIKOV, D.A.
1943 Irreducible unitary representations of locally compact groups. *Mat. Sb.* **13** (1943), 301-316.
GELLER, D. and STEIN, E. M.
1984 Estimates for singular convolution operators on the Heisenberg group. *Math. Ann.* **267** (1984), 1-15.
GILBERT, J.E., and MURRAY, M.A.M.
1991 "Clifford Algebras and Dirac Operators in Harmonic Analysis." Cambr. Univ. Press, 1991.
GINDIKIN, S. G.
1967 Unitary representations of groups of automorphisms of Riemannian symmetric spaces of null curvature. *Funct. Anal. Appl.* **1** (1967), 28-32.
GINDIKIN, S. G., and KARPELEVICH, F. I.
1962 Plancherel measure of Riemannian symmetric spaces of non-positive curvature. *Dokl. Akad. Nauk. USSR.* **145** (1962), 252-255.
1964 On a problem in integral geometry. Chebotarev Mem. Vol. Kazan Univ. 1964. *Selecta Math. Sovietica* **1** (1981), 169-184.
GLOBEVNIK, J.
1992 A support theorem for the X-ray transform. *J. Math. Anal. Appl.* **165** (1992), 284-287.

GODEMENT, R.
1948 Les fonctions de type positif et la théorie des groupes. *Trans. Amer. Math. Soc.* **63** (1948), 1-84.
1952a Une généralisation du théorème de la moyenne pour les fonctions harmoniques. *C.R. Acad. Sci. Paris* **234** (1952).
1952b A theory of spherical functions I. *Trans. Amer. Math. Soc.* **73** (1952), 496-556.
1957 Introduction aux travaux de A. Selberg. *Séminaire Bourbaki* , **144**, Paris, 1957.
GODIN, P.
1982 Hypoelliptic and Gevrey hypoelliptic invariant differential operators on certain symmetric spaces. *Ann. Scuola Norm. Pisa* **IX** (1982), 175-209.
GONZALEZ, F.
1984 Radon transforms on Grassmann manifolds. Thesis MIT , 1984.
1987 Radon transforms on Grassmann manifolds. *J. Funct. Anal.* **71** (1987), 339-362.
1988 Bi-invariant differential operators on the Euclidean motion group and applications to generalized Radon transforms. *Ark. Mat.* **26** (1988), 191-204.
1990a Bi-invariant differential operators on the complex motion group and the range of the d-plane transform on C^n. *Contemp. Math.* **113** (1990), 97-110.
1990b Invariant differential operators and the range of the Radon d-plane transform. *Math. Ann.* **287** (1990), 627-635.
1991 On the range of the Radon transform and its dual. *Trans. Amer. Math. Soc.* **321** (1991), 601-619.
1993 Range of Radon transform on Grassmann manifolds. In *Proc. Conf. "75 Years of Radon Transform."* , Vienna , Austria, 1992. International Press, Hong Kong, 1993.
GONZALEZ, F., and HELGASON, S.
1986 Invariant differential operators on Grassmann manifolds. *Advan. Math.* **60** (1986), 81-91.
GONZALEZ, F., and QUINTO, E. T.
1994 Support theorems for Radon transforms on higher rank symmetric spaces. *Proc. Amer. Math. Soc.* (1994), to appear.
GOODEY, P. and WEIL, W.
1991 Centrally symmetric convex bodies and the spherical Radon transform. Preprint, 1991.
GOODMAN, R. and WALLACH, N.
1980 Whittaker vectors and conical vectors. *J. Funct. Anal.* **39** (1980), 199-279.
GRINBERG, E.
1985 On the images of Radon transforms. *Duke Math. J.* **52** (1985), 939-972.
1987 Euclidean Radon transforms; ranges and restrictions. *Contemp. Math.* **63** (1987), 109-134.
1992 Aspects of flat Radon transform. *Contemp. Math.* **140** (1992), 73-85
1993a Integration over minimal spheres in Lie groups and symmetric spaces of compact type. Preprint (1993).
1993b Radon transform for maximally curved spheres. In *Proc. Conf. "75 Years of Radon Transform."* , Vienna, Austria, 1992. International Press, Hong Kong., 1993.
GROSS, K., and KUNZE, R.
1976 Bessel functions and representation theory. J. Funct. Anal. **22** (1976), 73-105.
GUARIE, D.
1992 "Symmetries and Laplacians: Introduction to Harmonic Analysis and Applications." North Holland, Amsterdam , 1992.
GUILLEMIN, V.
1976 Radon transform on Zoll surfaces. *Advan. Math.* **22** (1976), 85-99.
1985 The integral geometry of line complexes and a theorem of Gelfand-Graev. *Astérisque*, (1985), 135-149.
1987 Perspectives in integral geometry. *Contemp. Math.* **63** (1987), 135-150.
GUILLEMIN, V. and STERNBERG, S.
1977 "Geometric Asymptotics." *Mathematical Surveys,* Amer. Math. Soc. 1977.

1979 Some problems in integral geometry and some related problems in microlocal analysis, *Amer. J. Math.* **101** (1979), 915-955.

GUIVARCH, Y.

1984 Sur la représentation intégrale des fonctions harmoniques et des fonctions propres positives dans un espace Riemannien symétrique. *Bull. Sci. Math.* **108** (1984), 373-392.

GÜNTHER, P.

1988 "Huygens' Principle and Hyperbolic Equations." Academic Press, Boston, 1988.

1991 Huygens' Principle and Hadamard's conjecture. *Math. Intelligencer* **13** (1991), 56-63.

HARISH-CHANDRA

1951 On some applications of the universal enveloping algebra of a semisimple Lie algebra. *Trans. Amer. Math. Soc.* **70** (1951), 28-96.

1953 Representations of semisimple Lie groups, I. *Trans. Amer. Math. Soc.* **75** (1953), 185-243.

1954 The Plancherel formula for complex semisimple Lie groups. *Trans. Amer. Math. Soc.* **76** (1954), 485-528.

1955 Representations of semisimple Lie groups IV, *Amer. J. Math.* **77** (1955), 743-777.

1956a Representations of semisimple Lie groups, VI. *Amer. J. Math.* **78** (1956), 564-628.

1956b The characters of semisimple Lie groups. *Trans. Amer. Math. Soc.* **83** (1956), 98-163.

1958a Spherical functions on a semisimple Lie group, I. *Amer. J. Math.* **80** (1958), 241-310.

1958b Spherical functions on a semisimple Lie group, II. *Amer. J. Math.* **80** (1958), 553-613.

1959 Some results on differential equations and their applications. *Proc. Nat. Acad. Sci. USA* **45** (1959), 1763-1764.

1960 Differential equations and semisimple Lie groups. Collected Papers,Vol. III, pp. 6-56. Springer -Verlag, New York, 1984.

1966 Discrete series for semisimple Lie groups, II. *Acta Math.* **116** (1966), 1-111.

HARZALLAH, K.

1975 Distributions coniques et représentations associées à $SO_0(1,q)$. In "Analyse Harmonique sur les Groupes de Lie." Lecture Notes in Math. No. 497, pp. 211-229, Springer-Verlag, Berlin and New York, 1975.

HASHIZUME, M., MINEMURA, K., and OKAMOTO, K.

1973 Harmonic functions on hermitian hyperbolic spaces. *Hiroshima Math. J.* **3** (1973), 81-108.

HECKMAN, G., and SCHLICHTKRULL, H.

1994 "Harmonic Analysis and Special Functions on Symmetric Spaces." Academic Press, Orlando, 1994 (to appear).

HELGASON, S.

1959 Differential operators on homogeneous spaces. *Acta Math.* **102** (1959), 239-299.

1962a "Differential Geometry and Symmetric Spaces," Academic Press, New York, 1962

1962b Some results in invariant theory. *Bull. Amer. Math. Soc.* **68** (1962), 367-371.

1963 Duality and Radon transforms for symmetric spaces. *Amer. J. Math.* **85** (1963), 667-692.

1964a Fundamental solutions of invariant differential operators on symmetric spaces. *Amer. J. Math.* **86** (1964), 565-601.

1964b A duality in integral geometry; some generalizations of the Radon transform. *Bull. Amer. Math. Soc.* **70** (1964), 435-446.

1965a The Radon transform on Euclidean spaces, compact two-point homogeneous spaces and Grassmann manifolds. *Acta Math.* **113** (1965),153-180.

1965b Radon-Fourier transforms on symmetric spaces and related group representations. *Bull Amer. Math. Soc.* **71** (1965), 757-763.

1966a A duality in integral geometry on symmetric spaces. Proc. U.S. - Japan Seminar in Differential Geometry, Kyoto, 1965. *Nippon Hyronsha,* Tokyo, 1966.

1966b An analogue of the Paley-Wiener theorem for the Fourier transform on certain symmetric spaces. *Math. Ann.* **165** (1966), 297-308.

1970a A duality for symmetric spaces with applications to group representations. *Advan. Math.* **5** (1970), 1-154.

1970b Group representations and symmetric spaces. *Actes Congr. Internat. Math.* **2** (1970), 313-319.

1973a The surjectivity of invariant differential operators on symmetric spaces. *Ann. of Math.* **98** (1973), 451-480.

1973b Paley-Wiener theorems and surjectivity of invariant differential operators on symmetric spaces and Lie groups. *Bull. Amer. Math. Soc.* **79** (1973), 129-132.

1974 Eigenspaces of the Laplacian; integral representations and irreducibility. *J. Funct. Anal.* **17** (1974), 328-353.

1976 A duality for symmetric spaces with applications to group representations, II. Differential equations and eigenspace representations. *Advan. Math.* **22** (1976), 187-219.

1977a Some results on eigenfunctions on symmetric spaces and eigenspace representations. *Math. Scand.* **41** (1977), 79-89.

1977b Invariant differential equations on homogeneous manifolds. *Bull. Amer. Math. Soc.* **83** (1977), 751-774.

1977c Solvability questions for invariant differential operators. In "Colloquium on Group Theoretical Methods in Physics." Academic Press, New York, 1977.

1978 [DS] "Differential Geometry, Lie groups and Symmetric Spaces," Academic Press, New York, 1978.

1979 Invariant differential operators and eigenspace representations. Pp. 236-286 in "Representation Theory of Lie Groups" (M. Atiyah, Ed.) London Math. Soc. Lecture Notes No. 34. Cambridge Univ. Press, London and New York, 1979.

1980a A duality for symmetric spaces with applications to group representations, III. Tangent space analysis. *Advan. Math.* **30** (1980) 297-323.

1980b Support of Radon transforms. *Advan. Math.* **38** (1980), 91-100.

1983a Ranges of Radon transforms. AMS Short Course on Computerized Tomography, Jan. 1982. *Proc. Symp.Appl. Math.* Amer Math. Soc. Providence , R.I. 1983.

1983b The range of the Radon transform on symmetric spaces. in *Proc. Conf. Representation Theory of Reductive Lie Groups, Utah, 1982* (P. Trombi , Ed.), pp.145-151. Birkhäuser, Basel and Boston, Mass., 1983.

1983c Operational properties of the Radon transform with applications. In *Proc. Conf. Differential Geometry with Applications.* Nové Mesto, 1983, 59-75.

1984a [GGA] "Groups and Geometric Analysis; Integral Geometry, Invariant Differential Operators and Spherical Functions." Academic Press, New York, 1984.

1984b Wave equations on homogeneous spaces. In "Lie Group Representations III." Lecture Notes in Math. No. 1077, pp. 254- 287. Springer -Verlag, New York, 1984.

1987 Some results on Radon transforms, Huygens' principle and X-ray transforms. *Contemp. Math.* **63** (1987), 151-177.

1990 The totally geodesic Radon transform on constant curvature spaces. *Contemp. Math.***113** (1990), 141-149.

1991 Invariant differential operators and Weyl group invariants. In "Harmonic Analysis on Reductive Groups" (W. Barker and P. Sally, eds.) 193-200. Birkhäuser, Boston, 1991.

1992a Some results on invariant differential operators on symmetric spaces. *Amer. J. Math.* **114** (1992), 789-811.

1992b The flat horocycle transform for a symmetric space. *Advan. Math.* **91** (1992), 232-251.

1992c Radon transforms for double fibrations. Examples and viewpoints. In *Proc. Conf. "75 Years of Radon Transform"*, Vienna, Austria 1992 (S.G. Gindikin , P. Michor, eds.) , International Press , Hong Kong, 1993, 163-179.

1992d Huygens' principle for wave equations on symmetric spaces. *J. Funct. Anal.* **107** (1992), 279-288.

HELGASON, S., and JOHNSON, K.
1969 The bounded spherical functions on symmetric spaces. *Advan. Math.* **3** (1969), 586-593.
HELGASON, S., and KORÁNYI, A.
1968 A Fatou-type theorem for harmonic functions on symmetric spaces. *Bull. Amer. Math. Soc.* **74** (1968), 258-263.
HERGLOTZ, G.
1911 Über Potenzreihen mit positivem reellem Teil im Einheitskreis. *Sitz. Ber. Sächs. Akad. Wiss.* **63** (1911), 501-511.
HERMANN, R.
1964 Geometric aspects of potential theory in bounded symmetric domains, III. *Math. Ann.* **153** (1964), 384-394.
HERTLE, A.
1983 Continuity of the Radon transform and its inverse in Euclidean space. *Math. Zeitschr.* **184** (1983), 165-192.
1984 On the range of the Radon transform and its dual. *Math. Ann* . **267** (1984), 91-99.
HILGERT, J.
1993 Radon transform on half planes via group theory. Preprint 1993.
1993 Radon transform on Lie groups. Preprint 1993.
HOLE, A.
1975 Representations of the Heisenberg group of dimension 2n + 1 on eigenspaces. *Math. Scand.* **37** (1975), 129-141.
HÖRMANDER, L.
1963 "Linear Partial Differential Operators." Springer-Verlag, Berlin and New York, 1963.
1983 "The Analysis of Linear Partial Differential Operators I, II. Springer-Verlag, Berlin and New York, 1983.
HOTTA, R.
1971 On realization of the discrete series for semisimple Lie groups. *J. Math. Soc. Japan* **23** (1971), 384-407.
HOWE, R.
1980 On the role of the Heisenberg group in harmonic analysis. *Bull. Amer. Math. Soc.* **3** (1980), 821-843.
HOWE, R., and TAN, E.C.
1992 "Non-Abelian Harmonic Analysis; Applications of $SL(2,\boldsymbol{R})$," Springer-Verlag, Berlin and New York, 1992.
HU, M-C.
1973 Determination of the conical distributions for rank one symmetric spaces. Thesis MIT 1973.
1975 Conical distributions for rank one symmetric spaces. *Bull. Amer. Math. Soc.* **81** (1975), 98-100.
HUA, L. K.
1963 "Harmonic Analysis of Functions of Several Complex Variables in Classical Domains." *Trans. Math. Monographs,* Vol. 6, Amer. Math. Soc. 1963.
JACOBSEN, J.
1982 Invariant differential operators on some homogeneous spaces for solvable Lie groups. Preprint No. 34, Aarhus Univ., 1982.
1983 Eigenspace representations of nilpotent Lie groups, II. *Math. Scand.* **52** (1983), 321-333.
1985 Eigenspace representations of exponential groups. Preprint, Aarhus Univ., 1985.
JACOBSEN, J., and STETKÆR, H.
1981 Eigenspace representations of nilpotent Lie groups. *Math. Scand.* **48** (1981), 41-55.
JOHN, F.
1934 Bestimmung einer Funktion aus ihren Integralen über gewisse Mannigfaltigkeiten. *Math. Ann.* **109** (1934), 488-520.

1938 The ultrahyperbolic differential equation with 4 independent variables. *Duke Math. J.* **4** (1938), 300-322.

1955 "Plane Waves and Spherical Means." Wiley (Interscience), New York, 1955.

JOHNSON, K.

1976a Composition series and intertwining operators for the spherical principal series II. *Trans. Amer. Math. Soc.* **215** (1976), 269-283.

1976b Differential equations and an analog of the Paley-Wiener theorem for semisimple Lie groups. *Nagoya Math, J.* **64** (1976), 17-29.

1977 Remarks on a theorem of Korányi and Malliavin on the Siegel upper half plane of rank two. *Proc. Amer. Math. Soc.* **67** (1977).

1978 Differential equations and the Bergman-Shilov boundary on the Siegel upper half plane. *Ark. för Mat.* **16** (1978), 95-108.

1979a Paley-Wiener theorem for groups of split rank one. *J. Funct. Anal.* **34** (1979), 54-71.

1979b Partial differential equations on semisimple Lie groups. *Proc. Amer. Math. Soc.* **60** (1979), 289-295.

1980 On a ring of invariant polynomials on a Hermitian symmetric space. *J. of Algebra* **67** (1980), 72-81.

1984 Generalized Hua operators and parabolic subgroups. *Ann. of Math.* **120** (1984), 477-495.

1987a A strong generalization of Helgason's theorem. *Trans. Amer. Math. Soc.* **304** (1987), 171-192.

1987b Differential operators and Cartan motion groups. *Contemp. Math.* **63** (1987), 205-219.

JOHNSON, K. and KORÁNYI, A.

1980 The Hua operators on bounded symmetric domains of tube type. *Ann. of Math.* **111** (1980), 589-608.

JOHNSON, K. D., and WALLACH, N.

1972 Composition series and intertwining operators for the spherical principal series. *Bull. Amer. Math. Soc.* **78** (1972), 1053-1059.

1977 Composition series and intertwining operators for the spherical principal series I. *Trans. Amer. Math. Soc.* **229** (1977), 131-173.

KAKEHI, T.

1992 Range characterization of Radon Transforms on complex projective spaces. J. Math. Kyoto Univ. 32 (1992), 387-399.

1993 Range characterization of Radon transforms on S^n and P^nR, ibid. 33 (1993), 315-228.

KARPELEVICH, F. I.

1965 The geometry of geodesics and the eigenfunctions of the Beltrami-Laplace operator on symmetric spaces. *Trans. Moscow Math. Soc.* **14** (1965), 51-199.

KASHIWARA, M., KOWATA, A., MINEMURA, K., OKAMOTO, K., OSHIMA, T., and TANAKA, M.

1978 Eigenfunctions of invariant differential operators on a symmetric space. *Ann. of Math.* **107** (1978), 1-39.

KASHIWARA, M. and OSHIMA, T.

1977 Systems of differential equations with regular singularities and their boundary value problems. *Ann. of Math.* **106** (1977), 145-200.

KAWAZOE, T.

1979 An analog of Paley-Wiener theorem on rank one semisimple Lie groups I, II. *Tokyo J. Math.* **2** (1979), 397-407, 409-421.

1980 An analog of Paley-Wiener theorem on semisimple Lie groups and functional equations for Eisenstein integrals. *Tokyo J. Math.* **3** (1980), 219-248.

KELLEY, J. L.

1963 "Linear Topological Spaces." Van Nostrand Co. Princeton, N.J. 1963.

KNAPP, A.W.

1968 Fatou's theorem for symmetric spaces I. *Ann. of Math.* **88** (1968), 106-127.

1986 "Representation Theory of Semisimple Lie Groups. An Overview Based on Examples." Princeton Univ. Press., Princeton, NJ.,1986.

KNAPP, A.W. and STEIN, E. M.

1971 Intertwining operators on semisimple groups. *Ann. of Math.* **93** (1971), 489-578.

KNAPP, A.W., and WILLIAMSON, R. E.

1971 Poisson integrals and semisimple Lie groups. *J. Anal. Math.* **24** (1971), 53-76.

KOLK, J. and VARADARAJAN, V. S.

1992 Lorentz invariant distributions supported on the forward light cone. *Compositio Math.* **81** (1992), 61-106.

KOORNWINDER, T.H.

1973 The addition formula for Jacobi polynomials and spherical harmonics. *SIAM J. Appl. Math.* **25** (1973), 236-246.

1974 Jacobi polynomials II. An analytic proof of the product formula. *SIAM J. Math. Anal.* **5** (1974),125-137.

1975 A new proof of a Paley-Wiener theorem for the Jacobi transform. *Ark. Mat.* **13** (1975),145-159.

1984 Jacobi functions and analysis on noncompact semisimple Lie groups. In "Special Functions : Group Theoretical Aspects and Applications ." Reidel Press, (1984).

KORÁNYI, A.

1970 Generalizations of Fatou's theorem to symmetric spaces. *Rice Univ. Stud.* **56** (1970), 127-136.

1971 A remark on boundary values in several complex variables. Lecture Notes in Math. No. 185. pp. 1-6. Springer-Verlag, Berlin and New York, 1971.

1972 Harmonic functions on symmetric spaces. In "Symmetric Spaces." (W.M. Boothby and G. L. Weiss, eds.) Marcel Dekker, New York, 1972.

1976 Poisson Integrals and boundary components of symmetric spaces. *Invent. Math.* **34** (1976), 19-35.

1980 Some applications of Gelfand pairs in classical analysis. In "Harmonic Analysis and Group Representations. " CIME, Cortona, 1980.

1982a On the injectivity of the Poisson transform. *J. Funct. Anal.* **45** (1982), 293-296.

1982b Kelvin transform and harmonic polynomials on the Heisenberg group. *J. Funct. Anal.* **49** (1982), 177-185.

1983 Geometric aspects of analysis on the Heisenberg group. In "Topics in Modern Harmonic Analysis" Istituto Nat. Alta. Mat. Roma 1983.

1985 Geometric properties of Heisenberg-type groups. *Advan. Math.* **56** (1985), 28-38.

KORÁNYI, A. and MALLIAVIN, P.

1975 Posson formula and compound diffusion associated to an overdetermined elliptic system on the Siegel halfplane of rank two. *Acta Math.* **134** (1975), 185-209.

KORÁNYI, A., and WOLF, J. A.

1965 Realization of Hermitian symmetric spaces as generalized half planes. *Ann. of Math.* **81** (1965), 265-288.

KOSTANT, B.

1963 Lie group representations on polynomial rings. *Amer. J. Math.* **85** (1963), 327-404.

1969 On the existence and irreducibility of certain series of representations. *Bull. Amer. Math. Soc.* **75** (1969), 627-642.

1975a On the existence and irreducibility of certain series of representations. In. "Lie Groups and Their Representations" (I. M. Gelfand, ed.), pp. 231-329. Halsted, New York, 1975.

1975b Verma modules and the existence of quasiinvariant partial differential operators. In ""Non-Commutative Harmonic Analysis." Lecture Notes in Math. No. 466. Springer-Verlag, New York, 1975.

1991 A formula of Gauss-Kummer and the trace of certain intertwining operators. In "Operator Algebras, Unitary Representations, Enveloping Algebras and Invariant Theory. " Birkhäuser, Basel and Boston, 1991.

KOSTANT, B., and RALLIS, S.
1971 Orbits and Lie group representations associated to symmetric spaces. *Amer. J. Math.* **93** (1971), 753-809.
KÖTHE, G.
1952 Die Randverteilungen analytischer Funktionen. *Math. Zeitschr.* **57** (1952), 13-33.
KOWATA, A., and OKAMOTO, K.
1974 Harmonic functions and the Borel-Weil theorem. *Hiroshima Math. J.* **4** (1974), 89-97.
KOWATA, A., and TANAKA, M.
1980 Global solvability of the Laplace operator on a non-compact affine symmetric space. *Hiroshima Math. J.* **10** (1980), 409-417.
KUCHMENT, P. A.
1981 Representations of solutions of invariant differential equations on certain symmetric spaces. *Sov. Math. Dokl.* **24** (1981), 104-106.
1986 On the spectral synthesis in the spaces of solutions of invariant differential equations. Lecture Notes in Math. No. 1214, pp. 85-100. Springer-Verlag, New York, 1986.
KUNZE, R., and STEIN, E.
1967 Uniformly bounded representations, III. *Amer. J. Math.* **89** (1967), 385-442.
KURUSA, A.
1991a A characterization of the Radon transform's range by a system of PDE's. *J. Math. Anal. Appl.* **161** (1991), 218-226.
1991b The Radon transform on hyperbolic spaces. *Geom. Dedicata* **40** (1991), 325-339.
1992 Support theorems for the totally geodesic Radon transform on constant curvature spaces. Preprint, 1992; Proc. Amer. Math. Soc. (to appear).
LANGLANDS, R.
1963 The dimension of the space of automorphic forms. *Amer. J. Math.* **85** (1963), 99-125.
LASALLE, M.
1982 Séries de Laurent des fonctions holomorphes dans la complexification d'un espace symétrique compact. *Ann. Sci. École Norm. Sup.* **11** (1978), 167-210.
1984 Les équations de Hua d'un domaine borné symétrique de type tube. *Invent. Math.* **77** (1984), 129-161.
LAX, P., and PHILLIPS, R. S.
1978 An example of Huygens' principle. *Comm. Pure Appl. Math.* **31** (1978), 415-423.
1979 Translation representations for the solution of the non-Euclidean wave equation. *Comm. Pure Appl. Math.* **32** (1979), 617-667.
1982 A local Paley-Wiener theorem for the Radon transform of L^2 functions in a non-Euclidean setting. *Comm. Pure Appl. Math.* **35** (1982), 531-554.
LEPOWSKY, J.
1975 Conical vectors in induced modules. *Trans. Amer. Math. Soc.* **208** (1975), 219-272.
1976 Linear factorization of conical polynomials over certain nonassociative algebras. *Trans. Amer. Math. Soc.* **216** (1976), 237-248.
LEWIS, J. B.
1978 Eigenfunctions on symmetric spaces with distribution-valued boundary forms. *J. Func. Anal.* **29,** (1978), 287-307.
LÉVY-BRUHL, P.
1990 Résolubilité d'opérateurs différentiels sur des espaces symétriques nilpotents. *J. Funct. Anal.* **89** (1990), 303-317.
LIMIC, N., NIDERLE, J., and RACZKA, R.
1967 Eigenfunction expansions associated with the second-order invariant operator of hyperboloids and cones, III. *J. Math. Phys.* **8** (1967), 1079-1093.
LINDAHL, L.-Å.
1972 Fatou's theorem for symmetric spaces. *Ark. Mat.* **10** (1972), 33-47.
LIONS, J. L.
1952 Supports dans la transformation de Laplace. *J. Analyse Math.* **2** (1952-53), 123-151.

LIONS, J. L., and MAGENES, E.
1963 Problèmes aux limites non homogènes (VII) *Ann. Mat. Pura Appl.* **4 63** (1963), 201-224.
LOOMIS, L. H.
1953 "Abstact Harmonic Analysis." Van Nostrand Reinhold, New York, 1953.
LOHOUÉ. N. and RYCHENER, T.
1982 Die Resolvente von Δ auf symmetrischen Räumen von nichtkompakten typ. *Comment. Math. Helv.* **57** (1982), 445-468.
LOWDENSLAGER, D.
1958 Potential theory in bounded symmetric homogeneous complex domains. *Ann. of Math.* **67** (1958), 467-484.
LUDWIG, D.
1966 The Radon transform on Euclidean space. *Comm. Pure Appl. Math.* **23** (1966), 49-81.
MACKEY, G.W.
1952 Induced representations of locally compact groups, I. *Ann. of Math.* **55** (1952), 101-139.
1953 Induced representations of locally compact groups, II. *Ann. of Math.* **58** (1953), 193-221.
MADYCH, W. R., and SOLMON, D.C.
1988 A range theorem for the Radon transform. *Proc. Amer. Math. Soc.* **104** (1988), 79-85.
MALGRANGE, B.
1955 Existence et approximation des solutions des équations aux dérivées partielles et des équations de convolution. *Ann. Inst. Fourier Grenoble* **6** (1955-56), 271-355.
MAUTNER, F. I.
1951 Fourier analysis and symmetric spaces. *Proc. Nat. Acad. Sci. USA* **37** (1951), 529-533.
MAYER-LINDENBERG, F.
1981 Zur Dualitätstheorie symmetrischer paare. *J. Reine Angew. Math.* **321** (1981), 36-52.
MEANEY, C.
1986 The inverse Abel transform for $SU(p,q)$. *Ark. Mat.* **24** (1986), 131-140.
MENZALA, G.P. and SCHONBECK, T.
1984 Scattering frequencies for the wave equation with a potential term. *J. Funct. Anal.* **55** (1984), 297-322.
MICHELSON, H. L.
1973 Fatou theorems for eigenfunctions of the invariant differential operators on symmetric spaces. *Trans. Amer. Math. Soc.* **177** (1973), 257-274.
MIZONY, M.
1976 Algébres et noyaux de convolution sur le dual sphérique d'un groupe de Lie semisimple, non-compact et de rang 1. *Publ. Dep. Math. Lyon.* **13** (1976), 1-14.
MOORE, C. C.
1964 Compactifications of symmetric spaces I, *Amer. J. Math.* **86** (1964), 201-218; II, ibid.358-378.
MOSTOW, G.D.
1973 "Strong Rigidity of Locally Symmetric Spaces." *Ann. of Math. Studies,* Princeton Univ. Press, 1973.
NATTERER, F.
1986 "The Mathematics of Computerized Tomography." Wiley, New York, 1986.
NELSON, E.
1959 Analytic vectors. *Ann. of Math.* **70** (1959), 572-615.
ØRSTED, B.
1981 The conformal invariance of Huygens' principle. *J. Differential Geom.* **16** (1981), 1-9.

ÓLAFSSON, G., and SCHLICHTKRULL, H.
1992 Wave propagation on Riemannian symmetric spaces. *J. Funct. Anal.* **107** (1992), 270-278.
OLEVSKY, M.
1944 Some mean value theorems on spaces of constant curvature. *Dokl. Akad. Nauk, USSR* **45** (1944), 95-98.
ORLOFF, J.
1985 Limit formulas and Riesz potentials for orbital integrals on symmetric spaces. Thesis, MIT, 1985.
1990a Invariant Radon transforms on a symmetric space. *Contemp. Math.* **113** (1990), 233 - 242.
1990b Invariant Radon transforms on a symmetric space. *Trans. Amer. Math. Soc.* **318** (1990). 581-600.
OSHIMA, T., SABURI, Y. and WAKAYAMA, M. A note on Ehrenpreis' fundamental principle on a symmetric space. Algebraic Analysis, vol.II. 681-697, Academic Press, Boston, 1988.
OSHIMA, T., and SEKIGUCHI, J.
1980 Eigenspaces of invariant differential operators on an affine symmetric space. *Invent. Math.* **57** (1980), 1-81.
PALAMODOV, V., and DENISJUK, A.
1988 Inversion de la transformation de Radon d'après des données incomplètes. *C.R. Acad. Sci. Paris, Math.* **307** (1988), 181-183.
PALEY, R., and WIENER, N.
1934 "Fourier Transforms in the Complex Domain." Amer. Math,. Soc., Providence , R.I. 1934.
PARTHASARATHY, K. R., RANGA RAO, R., and VARADARAJAN, V.S.
1967 Representations of complex semisimple Lie groups and Lie algebras. *Ann. of Math.* **85** (1967), 383-429.
PHILLIPS, R. S., and SHAHSHAHANI, M.
1993 Scattering theory for symmetric spaces of noncompact type . *Duke Math. J.* **72** (1993), 1-29
POISSON, S.D.
1820 Nouveaux Mémoires de l'Acad. des Sci. Vol III, 1820.
QUINTO, E. T.
1981 Topological restrictions on double fibrations and Radon transforms. *Proc. Amer. Math. Soc.* **81** (1981), 570-574.
1982 Null spaces and ranges for the classical and spherical Radon transforms . *J. Math. Anal. Appl.* **90** (1982), 408-420.
1983 The invertibility of rotation invariant Radon transforms. *J. Math. Anal. Appl.* **91** (1983), 510-521; Erratum, *ibid* . **94** (1983), 602-603.
1987 Injectivity of rotation invariant Radon transforms on complex hyperplanes in C^n. *Contemp. Math.* **63** (1987), 245-260.
1993 Real analytic Radon transforms on rank one symmetric spaces. *Proc. Amer. Math. Soc.* **117** (1993), 179-186.
RADER, C.
1988 Spherical functions on Cartan motion groups. *Trans. Amer. Math. Soc.* **310** (1988), 1-45.
RADON, J.
1917 Über die Bestimmung von Funktionen durch ihre Integralwerte längs gewisser Mannigfaltigkeiten. *Ber. Verh. Sächs. Akad. Wiss. Leipzig, Math. Nat. Kl.* **69** (1917), 262-277.
RAÏS, M.
1971 Solutions élémentaires des opérateurs différentiels bi-invariants sur un groupe de Lie nilpotent. *C.R. Acad. Sci. Paris ,* **273** (1971), 495-498.

1975 Actions de certains groupes dans des espaces de fonctions C^∞. In " Non-commutative Harmonic Analysis ." Lecture Notes in Math. No. 466, pp. 147-150. Springer-Verlag, Berlin and New York, 1975.

1983 Groupes linéaires compacts et fonctions C^∞ covariantes. *Bull. Sc. Math.* **107** (1983), 93-111.

RAUCH, J. and WIGNER, D.

1976 Global solvability of the Casimir operator. *Ann. of Math.* **103** (1976), 229-236.

REVUZ, A.

1975 "Markov Chains." North Holl. Publ. Co. New York, 1975.

RICHTER, F.

1986 "Differentialoperatoren auf Euclidischen k-Ebenräumen und Radon Transformationen. " Dissertation, Humboldt Universität, Berlin, 1986.

1989 On fundamental differential operators and the p-plane Radon transform. Preprint (1989).

ROSSMANN, W.

1978 Analysis on real hyperbolic spaces. *J. Funct. Anal.* **30** (1978), 448-477.

ROUVIÈRE, F.

1976a Sur la résolubilité locale des opérateurs bi-invariants. *Ann. Scuola Norm. Sup. Pisa* **3** (1976), 231-244

1976b Solutions distributions de l'opérateur de Casimir. *C.R. Acad. Sci. Paris Ser. A-B* **282** (1976), 853-856.

1978 Invariant differential equations on certain semisimple Lie groups. *Trans. Amer. Math. Soc.* **243** (1978), 97-114.

1983 Sur la transformation d'Abel de groupes des Lie semisimples de rang un. *Ann. Scuola Norm. Sup. Pisa* **10** (1983), 263-290.

1986 Espaces symétriques et méthode de Kashiwara-Vergne. *Ann. Sc. Éc. Norm. Sup.* **19** (1986), 553-581.

1990- 1991a Invariant analysis and contractions of symmetric spaces. I, II. *Compositio Math.* **73** (1990), 241-270; *ibid.* **80** (1991), 111-136.

1991b Une propriété de symétrie des espaces symétriques. *C.R. Acad. Sci. Paris,* **313** Serie I, (1991), 5-8.

1993 Fibrés en droites sur un espace symétrique et analyse invariante. Preprint 1993. *J. Funct. Anal.* (to appear).

RUDIN, W.

1984 Eigenfunctions of the invariant Laplacian in B. *J. D'Anal. Math.* **43** (1984), 136-148.

SCHIFFMANN, G.

1971 Integrales d'entrelacement et fonctions de Whittaker. *Bull. Soc. Math. France* **99** (1971), 3-72.

1979 Travaux de Kostant sur la série principale. In "Analyse Harmonique sur les Groupes de Lie, II," Lecture Notes in Math.No. 739, pp.460-510. Springer-Verlag, Berlin and New York, 1979.

SCHIMMING, R. and SCHLICHTKRULL, H.

1992 Helmholtz operators on harmonic manifolds. Preprint 1992.

SCHLICHTKRULL,.H.

1984a One-dimensional K-types in finite-dimensional representations of semisimple Lie groups. A generalization of Helgason's theorem. *Math. Scand.* **54** (1984), 279-294.

1984b "Hyperfunctions and Harmonic Analysis on Symmetric Spaces," Birkhäuser, Boston, 1984.

1987 Eigenspaces of the Laplacian on hyperbolic spaces; composition series and integratransforms. *J. Funct. Anal.* **70** (1987), 194-219.

SCHLICHTKRULL, H., and STETKÆR, H.

1987 Scalar irreducibility of eigenspaces on the tangent space of a reductive symmetric space. *J. Funct. Anal.* **74** (1987), 292-299.

SCHMID, W.

1969 Die Randwerte holomorpher Funktionen auf Hermitesch symmetrischen Räumen. *Invent. Math.* **9** (1969), 61-80.

1971 On a conjecture of Langlands. *Ann. of Math.* **93** (1971), 1-42.
1992 Analytic and geometric realization of representations. In "New Developments in Lie Theory and their Applications." (J. Tirao and N. Wallach, eds.). Birkhäuser, Boston, 1992.

SCHWARTZ, L.
1966 "Théorie des Distributions." 2nd ed. Hermann, Paris, 1966.

SEMENOV-TJAN-SHANSKI, M. A.
1976 Harmonic analysis on Riemannian symmetric spaces of negative curvature and scattering theory. *Math. USSR, Izvestija* **10** (1976), 535-563.

SEMYANISTY, V. I.
1961 Homogeneous functions and some problems of integral geometry in spaces of constant curvature. *Sov. Math. Dokl.* **2** (1961), 59-62.

SHAHSHAHANI, M.
1983 Invariant hyperbolic systems on symmetric spaces. In "Differential Geometry." (R. Brooks et al. eds.) Birkhäuser, Basel and Boston, pp. 203-233, 1983.
1989 Poincaré inequality, uncertainty principle and scattering theory on symmetric spaces. *Amer. J. Math.* **111** (1989), 197-224.

SHAHSHAHANI, M., and SITARAM, A.
1987 The Pompeiu problem in exterior domains in symmetric spaces. *Contemp. Math.* **63** (1987), 267-277.

SHERMAN, T.
1975 Fourier analysis on the sphere. *Trans. Amer. Math.* Soc. **209** (1975), 1-31.
1977 Fourier analysis on compact symmetric spaces. Bull. Amer. Math. Soc. **83** (1977), 378-380.
1990 The Helgason Fourier transform for compact Riemannian symmetric spaces of rank one. *Acta Math.* **164** (1990), 73 -144.

SITARAM, A.
1980 An analog of the Wiener Tauberian theorem for the spherical transform on semisimple Lie groups. *Pac. J. Math.* **89** (1980), 439-445.
1988 On an analog of the Wiener Tauberian theorem for symmetric spaces of the noncompact type. *Pac. J. Math.* **133** (1988), 197-208.

SJÖGREN, P.
1981 Characterizations of Poisson integrals on symmetric spaces. *Math. Scand.* **49** (1981), 229-249.
1984 A Fatou theorem for eigenfunctions of the Laplace-Beltrami operator in a symmetric space. *Duke Math. J.* **51** (1984), 47-56.
1988 Asymtotic behaviour of generalized Poisson integrals in rank one symmetric spaces and trees. *Ann. Scuola Norm. Sup. Pisa Cl. Sci.* **15** (1988), 98-113.

SOLMON, D. C.
1976 The X-ray transform. *J. Math. Anal. Appl.* **56** (1976), 61-83.
1987 Asymtotic formulas for the dual Radon transform. *Math. Zeitschr.* **195** (1987), 321-343.

SOLOMATINA, L. E.
1986 Translation representation and Huygens' principle for an invariant wave equation in a Riemannian symmetric space. *Soviet Math. Izv.* **30** (1986), 108-111.
1988 Fundamental solutions of invariant differential equations on symmetric spaces. Function Spaces and Eqs. of Mathematical Physics. 51-61 87 Voronezh, Gos. Univ. Voronezh , 1988.

SPEH, B., and VOGAN, D.
1980 Reducibility of generalized principal series representations. *Acta Math.* **145** (1980), 227-299.

STANTON, R. J.
1976 On mean convergence of Fourier series on compact Lie groups. *Trans. Amer. Math. Soc.* **218** (1976), 61-87.

STANTON, R. J. and TOMAS, P. A.
1978 Expansions for spherical functions on noncompact symmetric spaces. *Acta Math.* **140** (1978), 251-276.
1979 Pointwise inversion of the spherical transform on L^p ($1 \leq p < 2$). *Proc. Amer. Math. Soc.* **73** (1979), 398-404.
STEIN, E.M.
1970 "Singular Integrals and Differentiability Properties of Functions." Princeton Univ. Press, 1970.
1983 Boundary behavior of harmonic functions on symmetric spaces. *Invent. Math.* **74** (1983), 63-83.
STEIN, E.M., and WEISS, G.
1968 Generalization of the Cauchy-Riemann equations and representations of the rotation group. *Amer. J. Math.* **90** (1968), 163-196.
STETKÆR, H.
1983 Remarks on irreducibility of eigenspace representations. *Ann. Glob. Anal. and Geometry.* **1** (1983), 35-48.
1985a Scalar irreducibility of eigenspace representations associated to a symmetric space. *Math. Scand.* **57** (1985), 289-292.
1985b Scalar irreducibility of certain eigenspace representations. *J. Funct. Anal.* **61** (1985), 295-306.
1986 Representations of groups on eigenspaces of invariant differential operators. *Banach Center Publ.* **19** (1986).
1991 Ultra-irreducibility of induced representations of semi-direct products. *Trans. Amer. Math. Soc.* **324** (1991), 543-554.
1993 Complete irreducibility and χ-spherical representations. *J. Funct. Anal.* **113** (1993), 413-425.
STRASBURGER, A.
1984 On a differential equation for conical distributions, Case $\mathbf{SO}_0(n, 1)$. In "Operator Algebras and Group Representations." (G. Arsene et al. ed.) Pitman Publ. Ltd. London 1984.
STRICHARTZ, R. S.
1973 Harmonic analysis on hyperboloids. *J. Funct. Anal.* **12** (1973), 341-383.
1981 L^p-estimates for Radon transforms in Euclidean and non-Euclidean spaces. *Duke Math. J.* **48** (1981), 699-727.
1982a Radon inversion-variations on a theme. *Amer. Math. Monthly* **89** (1982), 377-384 & 420-425.
1982b Explicit solutions of Maxwell's equations on a space of constant curvature. *J. Funct. Anal.* **46** (1982), 58-87.
1984 Mean value properties of the Laplacian via spectral theory. Trans. Amer. Math. Soc. **284** (1984), 219-228.
1986 Harmonic analysis on Grassmann bundles. *Trans. Amer. Math. Soc.* **296** (1986), 387-409.
1989 Harmonic analysis as spectral theory of Laplacians. *J. Funct. Anal.* **87** (1989), 51-148.
1991 L^p harmonic analysis and Radon transform on the Heisenberg group. *J. Funct. Anal.* **96** (1991), 350-406.
SUGIURA, M.
1990 "Unitary Representations and Harmonic Analysis." 2nd ed. North Holland and Kodansha, Amsterdam and Tokyo, 1990.
SULANKE, R.
1966 Croftonsche Formeln in Kleinschen Räumen. *Math. Nachr.* **32** (1966), 217-241.
TAKAHASHI, R.
1963 Sur les représentations untaires des groupes de Lorentz généralisés. *Bull. Soc. Math. France* **91** (1963), 289-433.

TAKEUCHI, M.

1973 Polynomial representations associated with symmetric bounded domains. *Osaka J. Math.* **10** (1973), 441-473.

TAYLOR, M.E.

1986 "Noncommutative Harmonic Analysis" Math. Surveys and Monogr. Amer. Math. Soc. Providence RI, 1986.

TEDONE, O.

1898 Sull' integrazione dell'equazione $\partial^2 I/\partial t^2 - \Sigma\, \partial^2 I/\partial x_i^2 = 0$. *Ann. Mat.* **1** (1898), 1-24.

TERRAS, A.

1985,1988 "Harmonic Analysis on Symmetric Spaces and Applications, I, II. Springer-Verlag, Berlin and New York, 1985, 1988.

TITCHMARSH, E. C.

1939 "The Theory of Functions," 2nd ed. Oxford Univ. Press, London and New York, 1939.

TORASSO, P.

1977 Le théorème de Paley-Wiener pour l'espace des fonctions indéfinement differentiables et a support compact sur un espace symétrique de type non compact. *J. Funct. Anal.* **26** (1977), 201-213.

TRÈVES, F.

1966 "Linear Partial Differential Equations with Constant Coefficients." Gordon and Breach, New York, 1966.

TRIMÈCHE, K.

1991 Opérateurs de permutations et analyse harmonique associés à des opérateurs aux derivées partielles. *J. Math. Pure Appl.* **9** (1991), 1-73.

TROMBI, P., and VARADARAJAN, V. S.

1971 Spherical transforms on semisimple Lie groups. *Ann. of Math.* **94** (1971), 246-303.

VARADARAJAN, V.S

1977 "Harmonic Analysis on Real Reductive Groups." Lecture Notes in Math. No. 576. Springer-Verlag, Berlin and New York, 1977.

VILENKIN, N.

1968 "Special Functions and the Theory of Group Representations, " Translation of Math. Monogr. Vol 22. Amer. Math. Soc. Providence, R.I. 1968.

VILENKIN, N., and KLIMYK, A.U.

1991-'93 "Representations of Lie Groups and Special Functions. "Vols. I, II, III. Kluwer, Dordrecht, 1991, 1993.

VOGAN, D.

1981 "Representations of Real Reductive Groups." Birkhäuser, Basel and Boston, 1981.

VOGAN, D. and WALLACH, N.

1990 Intertwining operators for real reductive groups. *Advan. Math.* **82** (1990), 203-243.

VRETARE,.L.

1976 Elementary spherical functions on symmetric spaces. *Math. Scand.* **39** (1976), 343-358.

1977 On a recurrence formula for elementary spherical functions on symmetric space and its applications. *Math. Scand.* **41** (1977), 99-112.

WALLACH, N.

1973 "Harmonic Analysis in Homogeneous Spaces." Dekker, New York, 1973.

1975 On Harish Chandra's generalized c-functions. *Amer. J. Math.* **97** (1975), 386-403.

1983 Asymtotic expansions of generalized matrix entries of representations of real reductive groups. Lecture Notes in Math. No. 1024, pp. 287-369. Springer-Verlag, Berlin and New York, 1983.

1988, 1992 "Real Reductive Groups I, II . " Academic Press, New York, 1988, 1992.

1990 The powers of the resolvent on a locally symmetric space. *Bull. Soc. Math. Belg.* **62** (1990), 777-790.

WARNER, G.
1972 "Harmonic Analysis on Semisimple Lie Groups," Vols I, II. Springer-Verlag, Berlin and New York, 1972.
WAWRZYNCZYK, A.
1985 Spectral analysis and mean periodic functions on rank-one symmetric spaces. *Bol. Soc. Mat. Mex.* 30 (1985), 15-29.
1984 "Group Representations and Special Functions." Reidel, Dordrecht, 1984.
WEIL, A.
1940 "L' intégration dans les Groupes Topologiques et ses Applications." Hermann, Paris, 1940.
WHITTAKER, E.T. and WATSON, G.N. "A Course of Modern Analysis", Cambr. Univ. Press, 1927.
WIEGERINCK, J. J. O. O.
1985 A support theorem for the Radon transform on $\boldsymbol{R}^3$. *Nederl. Akad. Wetensch. Proc. A* **88** 1985.
WIGNER, D.
1977 Bi-invariant operators on nilpotent Lie groups. *Invent. Math.* **41** (1977), 259-264.
WILLIAMS, F.L.
1985 Formula for the Casimir operator in Iwasawa coordinates. *Tokyo J. Math.* **8** (1985), 99-105.
WILLIAMS, G.D.
1978 The principal series of a p-adic group. *Quarterly J. Math.* **29** (1978), 31-56.
WOLF, J. A.
1964 Self-adjoint function spaces on Riemannian symmetric manifolds. *Trans. Amer. Math. Soc.* **113** (1964), 299-315
ZHELOBENKO, D. P.
1974 Harmonic analysis on complex semisimple Lie groups. *Proc. Int. Congr. Math.* Vancouver, 1974, Vol. II.
ZHU, CHEN-BO.
1993 Invariant differential operators on symplectic Grassmann manifolds. Preprint, Nat. Univ. of Singapore, 1993.
1993 On a technique in invariant theory. Preprint, Nat. Univ. of Singapore, 1993.
ZORICH, A.V.
1991 Inversion of horospherical integral transform on Lorentz group and on some other real semisimple Lie groups. *RIMS, Kyoto,* 1991, 1-37.

Symbols Frequently Used

∂_i: partial derivative, 4
$\mathcal{D}(X)$: $C_c^\infty(X)$, 5
$\mathcal{D}'(X)$: set of distributions on X, 5
$\mathcal{D}'_\lambda$: eigenspace, 94
$\mathcal{D}_K(X)$: set of $f \in \mathcal{D}$ with support in K, 5
$\mathcal{D}_H(\boldsymbol{P}^n)$: subspace of $\mathcal{D}(\boldsymbol{P}^n)$, 13
$\mathcal{D}_H(\boldsymbol{G}(d,n))$: subspace of $\mathcal{D}(\boldsymbol{G}(d,n))$, 60
$\mathcal{D}^\delta(X)$: space of K-commuting functions, 283
$\mathcal{D}_\delta(X)$: K-finite functions of type δ, 283
$\mathcal{D}^\natural(X), \mathcal{D}'_\natural(X)$: space of K-invariant elements in $\mathcal{D}(X), \mathcal{D}'(X)$, 232, 396
$\mathcal{D}^\natural(G), \mathcal{D}'_\natural(G)$: space of K-bi-invariant members of $\mathcal{D}(G), \mathcal{D}'(G)$, 108
$\boldsymbol{D}(G)$: set of left-invariant differential operators on G, 87
$\boldsymbol{D}_H(G)$: subalgebra of $\boldsymbol{D}(G)$, 87
$\boldsymbol{D}(G/H)$: set of G-invariant differential operators on G/H, 88, 93
$\boldsymbol{D}_W(A)$: W-invariants in $\boldsymbol{D}(A)$, 87
$\boldsymbol{D}(X), \boldsymbol{D}(\Xi)$: invariant operators on X, Ξ, 87
$d(\delta)$ or d_δ: dimension (= degree) of a representation, 16
$\Delta(D)$: radial part of D, 87
$\Delta_{MN}(D), \Delta_K(D), \Delta_N(D)$: radial parts of D, 92, 87
$\Delta(\mathfrak{g}^{\boldsymbol{c}}, \mathfrak{h}^{\boldsymbol{c}})$: set of roots, 149
$\boldsymbol{d}_s(\lambda), \boldsymbol{e}_s(\lambda)$: factors in $\boldsymbol{c}_s(\lambda)$, 163
$\boldsymbol{E}(M)$: set of all differential operators on M, 40
$\mathcal{E}(X)$: $C^\infty(X)$, 4
$\mathcal{E}'(X)$: space of distributions of compact support, 5
$\mathcal{E}^\natural(X), \mathcal{E}'_\natural(X)$: space of K-invariant elements in $\mathcal{E}(X), \mathcal{E}'(X)$, 232, 396
$\mathcal{E}_\lambda, \mathcal{E}_{(\lambda)}, \mathcal{E}^\infty_{(\lambda)}, \mathcal{E}^*, \mathcal{E}^\infty_\lambda, \mathcal{E}_{\lambda,\delta}$: eigenspaces, 94, 234, 235, 236, 294, 508
$\mathcal{E}^\natural(G)$: space of K-bi-invariant members of $\mathcal{E}(G)$, 241
E_k: eigenspace of Laplacian, 13
$e_{\lambda,b}$: plane wave eigenfunction, 118
$F(a,b;c;z)$: hypergeometric function, 344
$f \to \dot{f}$: map from $C_c(G)$ to $C_c(G/H)$, 29, 177
f^δ: K-commuting function, 276
$\mathcal{F}$: spherical transform, 398
g^ϕ, T^ϕ, D^ϕ: images of $g \in \mathcal{E}(M), T \in \mathcal{D}'(M)$, operator D under φ, 5
φ_λ: spherical function, 105
$\Phi_{\lambda,\delta}$: generalized spherical function, 233
G_0: a group of linear transformations of X_0, 297
$\boldsymbol{G}(d,n), \boldsymbol{G}_{d,n}$: manifolds of d-planes, 43, 46
$\mathcal{H}^A(\boldsymbol{C}^n), \mathcal{H}_W(\mathfrak{a}^*_c), \mathcal{H}(\mathfrak{a}^* \times B)_W$: exponential type, 269, 285, 547
$\mathrm{Hom}(V,W)$: space of linear transformations of V into W, 283
$\boldsymbol{H}^n$: hyperbolic space, 50
$H_l, H(\mathfrak{p})$: space of harmonic polynomials, 18, 235
$\mathcal{H}_\lambda$: Hilbert space inside $\mathcal{E}_\lambda(X), \mathcal{E}_\lambda(\mathfrak{p})$, 296, 324, 531
$\mathcal{H}^\delta(\mathfrak{a}^*)$: special holomorphic functions on $\mathfrak{a}^*_c$, 285
$\mathcal{H}$: Hilbert transform, 8, 403
η_λ: K-fixed vector in $\mathcal{E}_\lambda(\Xi)$, 250
$H(g)$: component in $g = k \exp H(g) n$, 118
Im: imaginary part, 269
$I(E)$: space of invariant polynomials on E, 235
I^γ: Riesz potential, 7

sgn(x): signum function, 8
$\mathcal{S}_H(\boldsymbol{P}^n)$: subspace of $\mathcal{S}(\boldsymbol{P}^n)$, 13
sh x: sinh x, 2
$\sigma(F, G)$: weak topology, 32
$\sigma(a)$: diffeomorphism of Ξ, 126
$\Sigma(\mathfrak{g}, \mathfrak{a})$: set of restricted roots, 149
σ_R: sphere in Ξ, 382
$\Sigma, \Sigma^+, \Sigma_0^+, \Sigma_s^+, \Sigma_*$: sets of restricted roots, 77, 109, 150, 158
${}^tA$: transpose of A, 33
Tr(A): trace of A, 16
$T_\lambda, \tau_\lambda, \tilde{\tau}_\lambda$: eigenspace representations, 94, 296
τ: homomorphism of $\boldsymbol{D}(G)$, 238
$\tau(x)$: translation on G/H, 6
θ: Cartan involution, 77
$\boldsymbol{U}(n)$: unitary group, 67
V_δ: representation space of δ, 233
V_δ^M: space of fixed vectors under $\delta(M)$, 233
$V(\delta)$: space of K-finite vectors of type δ, 16
W: Weyl group, 78, 338
Ξ: dual space, 78
Ξ^*: open orbit in Ξ, 80
Ξ_0: space of horocycle planes, 400
ξ^*: origin in Ξ^*, 80
$\xi(x, b)$: horocycle determined by x and b, 118
$\Psi_{\lambda,s}, \Psi'_{\lambda,s}$: conical distributions, 163, 156
$\Psi_{\lambda,\delta}$: generalized Bessel function, 301
$\boldsymbol{Z}, \boldsymbol{Z}^+$: the integers, the nonnegative integers, 4
$\boldsymbol{Z}(G)$: center of $\boldsymbol{D}(G)$, 338
$\boldsymbol{Z}(G/K)$: image of $\boldsymbol{Z}(G)$ in $\boldsymbol{D}(G/K)$, 338
$\sim$: Fourier transform, spherical transform, lift of functions, distributions, 5, 95, 177, 223
$\wedge$: Radon transform, incidence, 1, 35
$\vee$: Dual Radon transform, incidence, 1, 35
$*, \times$: convolutions, adjoint operation, pullback, star operator, Fourier transform, 7, 8, 16, 30, 98, 100, 115, 158, 225, 537
$\oplus$: direct sum, 504
$\otimes$: tensor product, 13, 127, 131
$\langle\, ,\, \rangle$: inner product, 32
$\natural, E^\natural$: space of K-invariants in E, 104, 108
$\square$: operator, 10, 116
$\square_p$: operator on $\boldsymbol{G}(p, n)$, 46
—: closure, 4; restriction, 136
$^\circ$: interior, 4
$\perp$: annihilator, 18

INDEX

J

K

L

M

N

O

P

Q

R

MATHEMATICAL SURVEYS AND MONOGRAPHS SERIES LIST

Volume

8 **Discontinuous groups and automorphic functions,** J. Lehner

7.2 **The algebraic theory of semigroups, Volume II,** A. H. Clifford and G. B. Preston

7.1 **The algebraic theory of semigroups, Volume I,** A. H. Clifford and G. B. Preston

6 **Introduction to the theory of algebraic functions of one variable,** C. C. Chevalley

5 **The kernel function and conformal mapping,** S. Bergman

4 **The theory of valuations,** O. F. G. Schilling

3 **Geometry of polynomials,** M. Marden

2 **The theory of rings,** N. Jacobson

1 **The problem of moments,** J. A. Shohat and J. D. Tamarkin

ISBN 0-8218-1538-5

9 780821 815380